http://chemistry.brookscole.com/seag... P9-ECO-472

What do you need to learn now?

GOB
Chemistry◆Now™

Take charge of your learning with *GOB ChemistryNow*, a powerful and interactive study tool that will help you manage and maximize your study time. This collection of dynamic technology resources will help you determine your unique study needs and then gives you a personalized learning plan that will enhance your chemical problem-solving skills and conceptual understanding. And best of all, access to *GOB ChemistryNow* is **FREE** with every new copy of this text!

Totally integrated with this text!

Look for these media flags in the text. ———————→
They direct you to the corresponding media-enhanced activities on *GOB ChemistryNow*.

> **GOB**
> **Chemistry◆Now™**
> Go to GOB Now and click to learn how to test the identification of organic functional groups.

With a click of the mouse, *GOB ChemistryNow's* unique interactive activities allow you to:

✓ Review for the Nursing Boards with the **Nursing Board Concept Review**, featuring questions modeled after questions from the Nursing School and Allied Health Entrance Exam (16e) and the NCLEX-LPN Certification Exam.

✓ Learn about careers in the health professions at the **Career Corner.**

✓ Master the essentials with a chapter-specific **Math Review.**

✓ Create your *Personalized Learning Plan* or review for an exam using the **Pre-Test Web Quizzes.**

✓ Explore chemical principles with simulations, animations, and movies. by selecting **Chemistry Interactive.**

✓ Solve **Coached Problems** with interactive media-based exercises, *Guided Simulations* and *Intelligent Tutors.*

✓ Complete the **Chapter Quiz** to assess your mastery of core chapter concepts and skills.

Get Started!

Log on to *GOB ChemistryNow* at **http://chemistry.brookscole.com/seager5e** with the passcode access card packaged with your text. The *GOB ChemistryNow* system is made up of three interrelated parts:

✓ **What Do I Know?**   ✓ **What Do I Need to Learn?**   ✓ **What Have I Learned?**

These three interrelated elements work together, but are distinct enough to allow you the freedom to explore only those lessons and exercises that meet your personal needs. To maximize the effectiveness of the system, start with the chapter-specific Pre-Test Web quiz to create a *Personalized Learning Plan* that outlines the elements you need to review to master the chapter's most essential concepts. After you've worked through all of the problems and tutorials highlighted in your Plan, you'll move on to a Chapter Quiz. Your results can be e-mailed to your instructor, helping both you and your instructor assess your progress. By providing you with a better understanding of exactly what you need to focus on to succeed in the course, *GOB ChemistryNow* puts you in the driver's seat!

Log on today!
http://chemistry.brookscole.com/seager5e

Welcome to your *GOB ChemistryNow*™ Media Integration Guide!

The *Media Integration Guide* on the next several pages provides grids that link each chapter to the wealth of interactive media resources you will find at *GOB ChemistryNow*, the first Web-based, assessment–centered personalized learning system for Allied Health Chemistry students.

Get ready for your next chemistry exam!

GOB ChemistryNow™

http://chemistry.brookscole.com/seager5e

Chapter	Selected Nursing Board Concept Review	Coached Problems	Chemistry Interactive
1	(1.1–1.6) Matter; Mass vs. Weight; Mass vs. Volume; Property vs. Change; Chemical vs. Physical: Pure Substance vs. Mixture; Solutions; Prefixes *hetero*- and *homo*-; calorie vs. Calorie (1.7–1.11) Factors; Percentages; Mass vs. Volume; Density Is Constant	◆ Pure Substances and Mixtures ◆ Unit Conversions Using the Metric System ◆ Volume-Displacement Method ◆ Measurements That Involve Calculating Density	◆ Reaction of Phosphorus with Chlorine ◆ Water, Sand, and Copper Sulfate Separation ◆ Paper Chromatography of Ink ◆ Transfer of Thermal Energy ◆ Crystal Structures: Ag and Au
2	(2.1–2.4) Electrons; Protons; Neutrons; Location of Subatomic Particles; Neutral Atoms; Atomic Number; Mass Number; Isotope Symbols and Names; Calculating Ratios; Atomic Mass Units (2.5–2.7) Calculating a Weighted Average; Conversion Factors for Atoms and Mass, Moles and Mass, Atoms and Moles	◆ Atomic Symbols, Numbers, and Mass Numbers ◆ Atomic Weight of an Element Based on Masses and Abundance of Isotopes ◆ Calculation of Percent Composition of an Element	◆ Electron Cloud; Neutron; Proton; Nucleus
3	(3.1–3.3) Periods vs. Groups/Families; Chemical Properties of Elements in a Group or Family; Valence Shells; Electrons in Valence Shells (3.4–3.6) Shells and Subshells; Orbitals; Noble Gas Configurations; Electron Configurations; Classifying the Elements; Properties of Metals, Metalloids, and Nonmetals	◆ Different Parts of the Periodic Table ◆ Filling Electrons into Electron Orbital Diagrams ◆ Reporting Abbreviated Electron Configurations	◆ Periods of the Periodic Table ◆ Rows of the Periodic Table ◆ Electron Promotion and Decay in a Hydrogen Atom ◆ Orbital Shapes ◆ Electron Configurations and Periodic Blocks
4	(4.1–4.6) Octet Rule; Formation of Simple Ions; Binary Compounds; Sharing Electrons; Orbital Overlap; Drawing Lewis Structures (4.7–4.11) Electron Pairs; Bond Polarization; Polar/Nonpolar Molecule; Compound Nomenclature; Dispersion Forces; Dipolar Forces	◆ Recognition of the Charges of Ions Formed from Atoms ◆ Naming Ionic Compounds ◆ Interpreting and Drawing Lewis Structures ◆ Determination of Molecular Shape	◆ Formation of NaCL on the Molecular Scale ◆ Formation of a Crystal Lattice ◆ Interatomic Electrostatic Interactions ◆ VSEPR Electron Shapes ◆ Ion-Dipole and Dipole-Dipole Forces
5	(5.1–5.6) Law of Conservation of Matter/Balancing; States of Matter; Oxidation/Reduction; Oxidation Number; Single and Double Replacement Reactions (5.7–5.11) Molecular Equation vs. Net Ionic Equation; Prefixes *endo-/exo-*; Atomic and Molar Ratios; Limiting vs. Excess Reactant; Theoretical Yield; Percent Yield	◆ Balancing Chemical Equations ◆ Calculating Oxidation Numbers ◆ Writing Ionic Equations for Reactions in an Aqueous Solution ◆ Calculating Percent Yield for a Reaction	◆ Reaction of Al and Br_2 ◆ Burning of White Phosphorus ◆ Heated Iron and Gaseous Chlorine ◆ Catalyzed Decomposition of Hydrogen Peroxide ◆ Precipitation of AgCl and $Fe(OH)_3$
6	(6.1–6.8) Molecules in Constant Motion; Kinetic Energy and T Relationship; Pressure; Kelvin; P vs. V; V vs. T; P vs. T; Combined Gas Law; STP; Molar Volume (6.9–6.15) Partial Pressure; Evaporation/Condensation Equilibrium; Vapor Pressure; Specific Heat	◆ Relationship of Molecular Speed of Gas Molecules, T, V, and P ◆ Calculating Gas Pressure ◆ Graham's Law	◆ Motion of Molecules of a Solid, Liquid, and Gas ◆ Graph of Volume/Pressure Relationship ◆ Temperature/Volume Relationship for a Gas

Media Integration Guide

Media Integration Guide

GOB Chemistry❂Now™

http://chemistry.brookscole.com/seager5e

Chapter	Selected Nursing Board Concept Review	Coached Problems	Chemistry Interactive
7	◆ (7.1–7.6) Like Dissolves Like; Interparticle Forces; Rate of Dissolving; Ratio Solute; Density vs. Concentration; Volumetric Glassware; Dilution Formula; Moles ⇌ Liters ◆ (7.7–7.10) Freezing Point Depression; Vapor Pressure Depression; Boiling Point Elevation; Dissociation; Osmotic Pressure; Osmolarity; Dialyzing Membranes; Dialysis and Kidney Malfunction	◆ Using Solubility Guidelines to Predict Ionic Compound Solubilities ◆ Calculating Molarity ◆ Calculating the Theoretical Yield of a Reaction ◆ Calculating Boiling Point, Freezing Point, and Osmotic Pressure	◆ Precipitation of Lead Iodide: Saturated Solution ◆ Potassium Permanganate Dissolving ◆ Preparation of an Aqueous Solution by Direct Addition and Dilution ◆ Stoichiometric Ratio ◆ Vapor and Osmotic Pressure
8	◆ (8.1–8.4) Change in Concentration; Observation of Concentration; Reaction Mechanisms; Activation Energy; Exothermic vs. Endothermic ◆ (8.5–8.8) Relating Concentration of Reactants and Rate; Relationship Between Energy Diagrams, Reaction Rate, and Progress; Equilibrium Constants; Concentration; Temperature; Catalysts	◆ Energy Diagrams, Reaction Rate, and Progress ◆ Relating Concentration of Reactants and Rate ◆ Constructing Equilibrium Constant Expressions ◆ The Position of an Equilibrium	◆ Reaction of Bleach and Food Dye ◆ Collision Theory ◆ Lycopodium Demonstration ◆ Water Tank Equilibria ◆ Effect of the NO_2/N_2O_4 System of Changing Temperature and Volume
9	◆ (9.1–9.6) Calculation; Interpretation; Reactions; Activity Series ◆ (9.7–9.9) Cation/Anion; Hydrates; Equivalents; Dilute vs. Weak; Dissociation ◆ (9.10–9.13) Endpoint; Equivalence Point; Indicators; Equilibrium; Buffer Capacity; Calculating pH	◆ Relationship and Converting between $[H_3O^+]$, $[OH^-]$, and pH ◆ pH Solutions of Different Acids ◆ Use of Titration Data to Determine the Concentration of an Unknown Acid	◆ Reaction Between Ammonia and Water ◆ Autoionization Between H_2O Molecules ◆ Action of a Strong Acid and Weak Acid ◆ Action of a Strong Base and Weak Base ◆ Demonstration of Titration
10	◆ (10.1–10.5) Nuclear Radiation; Composition of Radiation; Stability; Calculations; Low Level vs. Intense Exposure; Ionizing Radiation; Radiation Sickness; Physical vs. Biological Units ◆ (10.6–10.9) Tracers; Therapeutic Uses of Radioisotopes; Radioactive Dating; Cyclotron; Matter–Energy Conversion; Fission vs. Fusion	◆ Equations for Nuclear Reactions ◆ Calculating the Degree to Which a Radioactive Isotope Will Decay in a Given Amount of Time	◆ Radioactive Emissions ◆ Alpha Particles Deflected by a High Positive Charge ◆ Rutherford Experiment ◆ Half-life Demonstration
11	◆ (11.1–11.3) Importance/Abundance of Carbon Containing Compounds; Carbon Can Make 4 Covalent Bonds; Isomerization is Possible for Carbon Atoms ◆ (11.4–11.9) Structures of Key Functional Groups; Prefixes; Stem Names; -ane Ending; IUPAC Nomenclature Rules; Alkyl Groups; Free Rotation ◆ (11.10–11.11) Combustion Production; Exothermic	◆ Identification of Organic Functional Groups ◆ Naming Alkanes ◆ Identifying Axial and Equatorial Positions in Cycloalkanes ◆ Identifying cis- and trans- Isomers in Cycloalkanes	◆ Free Rotation of Alkanes
12	◆ (12.1–12.6) General Formula of Alkenes; -ene Ending; 120° Bond Angle; Geometric Isomers; Alkene Properties; Addition Reactions; Monomer vs. Polymer; Addition Polymerization ◆ (12.7–12.8) Common Names of Benzene Derivatives; Phenyl vs. Phenol; Substitution Reactions; Carcinogens; Vitamins	◆ Naming System Used for Alkenes ◆ Naming Alkenes ◆ Reactivity of Alkenes ◆ Markovnikov's Rule ◆ Naming Aromatic Compounds	◆ Molecular Structure of Alkenes ◆ Sigma and Pi Orbitals of Alkenes ◆ Propene Hydrochlorination Reaction ◆ Molecular Structure of Benzene ◆ Sigma and Pi Orbitals of Benzene

Introductory Chemistry for Today

Fifth Edition

Spencer L. Seager
Weber State University

Michael R. Slabaugh
Weber State University

THOMSON

BROOKS/COLE

Australia • Canada • Mexico • Singapore • Spain
United Kingdom • United States

THOMSON

BROOKS/COLE

Chemistry Editor: *John Holdcroft*
Assistant Editor: *Lauren Raike*
Editorial Assistant: *Annie Mac*
Technology Project Manager: *Ericka Yeoman-Saler*
Executive Marketing Manager: *Julie Conover*
Marketing Assistant: *Melanie Banfield*
Advertising Project Manager: *Stacey Purviance*
Project Manager, Editorial Production: *Lisa Weber*
Print Buyer: *Karen Hunt*
Permissions Editor: *Joohee Lee*
Production Service: *Lachina Publishing Services*

Text Designer: *Delgado Design, Inc.*
Photo Researcher: *Lynne-Marie Sanders*
Copy Editor: *Becky Strehlow*
Illustrators: *John and Judy Waller, Scientific Illustrators, and Lachina Publishing Services*
Cover Designer: *Laurie Anderson*
Cover Image: © *Terje Rakke/Getty Images*
Cover Printer: *Von Hoffmann Corporation*
Compositor: *Lachina Publishing Services*
Printer: *Quebecor World Versailles*

For more information about our products, contact us at:
Thomson Learning Academic Resource Center
1-800-423-0563

For permission to use material from this text, contact us by:
Phone: 1-800-730-2214 **Fax:** 1-800-730-2215
Web: http://www.thomsonrights.com

Library of Congress Control Number: 2003115241

Student Edition: ISBN 0-534-39591-0

Brooks/Cole—Thomson Learning
10 Davis Drive
Belmont, CA 94002
USA

Asia
Thomson Learning
5 Shenton Way #01-01
UIC Building
Singapore 068808

Australia/New Zealand
Thomson Learning
102 Dodds Street
Southbank, Victoria 3006
Australia

Canada
Nelson
1120 Birchmount Road
Toronto, Ontario M1K 5G4
Canada

Europe/Middle East/Africa
Thomson Learning
High Holborn House
50/51 Bedford Row
London WC1R 4LR
United Kingdom

To our grandchildren:
Nate and Braden Barlow, and Megan and Bradley Seager
Alexander, Elyse, Megan, and Mia Slabaugh

H work   1, 68, 70, 72, 74 82 84

2 page sheet   1, 2, 3, 5, 6, 7

ABOUT THE AUTHORS

SPENCER L. SEAGER

Spencer L. Seager is a professor of chemistry at Weber State University, where he served as chemistry department chairman from 1969 until 1993. He teaches general chemistry at the university and is also active in projects to help improve chemistry and other science education in local elementary schools. He received his B.S. degree in chemistry and Ph.D. degree in physical chemistry from the University of Utah. Other interests include making minor home repairs, reading history of science and technology, listening to classical music, and walking for exercise.

MICHAEL R. SLABAUGH

Michael R. Slabaugh is a professor of chemistry at Weber State University, where he teaches the yearlong sequence of general, organic, and biochemistry. He received his B.S. degree in chemistry from Purdue University and his Ph.D. degree in organic chemistry from Iowa State University. His interest in plant alkaloids led to a year of postdoctoral study in biochemistry at Texas A&M University. His current professional interests are chemistry education and community involvement in science activities, particularly the State Science and Engineering Fair in Utah. He also enjoys the company of family, hiking in the mountains, and fishing the local streams.

BRIEF CONTENTS

CONTENTS

➤ **Chapter 7**

Solutions and Colloids *203*

➤ **Chapter 8**

Reaction Rates and Equilibrium *243*

➤ **Chapter 9**

Acids, Bases, and Salts *267*

> # Chapter 10

Radioactivity and Nuclear Processes 311

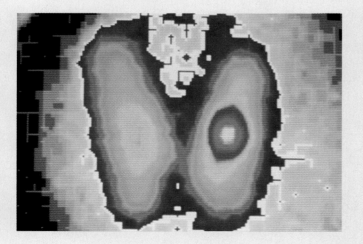

> # Chapter 11

Organic Compounds: Alkanes 341

> # Chapter 12

Unsaturated Hydrocarbons 379

PREFACE

➤ THE IMAGE OF CHEMISTRY

We, as authors, are pleased that the acceptance of the previous four editions of this textbook by students and their teachers has made it possible to publish this fifth edition. In the earlier editions, we expressed our concern about the negative image of chemistry held by many of our students, and their genuine fear of working with chemicals in the laboratory. Unfortunately, little seems to have changed in the last few years. Reports in the media related to chemicals or to chemistry continue to be primarily negative, and in many cases seem to be designed to increase the fear and concern of the general public. With this edition, we continue to hope that those who use this book will gain a more positive understanding and appreciation of the important contributions that chemistry makes in their lives.

➤ THEME AND ORGANIZATION

This edition continues the theme of the positive and useful contributions made by chemistry in our world. Consistent with that theme, we continue to use the chapter opening focus on health care professionals introduced in the second edition. The photos and accompanying brief descriptions of the role of chemistry in each profession continue to emphasize positive contributions of chemistry in our lives.

This text is designed to be used in either a two-semester or three-quarter course of study that provides an introduction to general chemistry, organic chemistry, and biochemistry. Most students who take such courses are majoring in nursing, other health professions, or the life sciences, and consider biochemistry to be the most relevant part of the course of study. However, an understanding of biochemistry depends upon a sound background in organic chemistry, which in turn depends upon a good foundation in general chemistry. We have attempted to present the general and organic chemistry in sufficient depth and breadth to make the biochemistry understandable.

As with previous editions, this textbook is published in a complete hardcover form and a two-volume paperback edition. One volume of the paperback edition contains all the general chemistry and the first two chapters of organic chemistry from the hardcover text. The second volume of the paperback edition contains all the organic and biochemistry of the hardcover edition. The availability of the textbook in these various forms has been a very popular feature among those who use the text because of the flexibility it affords them.

The decisions about what to include and what to omit from the text were based on our combined 60-plus years of teaching, input from numerous reviewers and adopters, and our philosophy that a textbook functions as a personal tutor to each student. In the role of a personal tutor, a text must be more than just a collection of facts, data, and exercises. It should also help students relate to the material they are studying, carefully guide them through more difficult material, provide them with interesting and relevant examples of chemistry in their lives, and become a reference and a resource that they can use in other courses or their professions.

➤ NEW TO THIS EDITION

In this fifth edition of the text, we have retained features that received a positive reception by our own students, the students of other adopters, other teachers, and reviewers. The retained features are 12 *Study Skills* boxes with 1 reaction map; 1 *How Reactions Occur* boxes; 21 *Chemistry Around Us* boxes with 6 new to this edition; 12 *Over the Counter* boxes with 2 new to this edition; and 11 *For Future Reference* boxes with 1 new to this edition. A new feature of this fifth edition is 12 boxes entitled *Chemistry and Your Health*. Various chemically related health topics are discussed. A majority deal with health issues and questions that are quite current, such as the dangers of obesity, safety questions surrounding genetically modified foods and the relationship between C-Reactive protein and heart disease. In addition to this new box feature, approximately 25% of the end-of-chapter exercises have been changed.

At the end of each chapter, we have linked a *Learning Objectives Assessment* section to appropriate end-of-chapter exercises to test a student's knowledge of the initial *Learning Objectives* at the beginning of the chapter. We have also added color-coding to the end-of-chapter exercises to rank the questions by difficulty level. This is just one more step to allow students to gauge their progress in mastering chapter material.

CHEMISTRY AND YOUR HEALTH • 3.1

PROTECTING CHILDREN FROM IRON POISONING

The element iron plays a vital role in a number of body processes. Perhaps the most well-known of these functions is the role of iron as a component of hemoglobin, the oxygen-transporting protein of blood. When blood is iron poor, body tissues do not receive enough oxygen, and *anemia*, a general weakening of the body, results. Recommended Dietary Allowances (RDA) have been established for iron because of this and other important contributions to good health. The general RDA is 10 to 15 mg per day depending on age and sex, and 30 mg per day for pregnant women.

A well-balanced diet that includes meat, whole grains, and dark green vegetables will generally provide enough iron to satisfy the RDA for most individuals other than pregnant women. In an attempt to enhance the effectiveness of the diet in providing iron, many foods are enriched or fortified with iron—primarily breads, other flour products, and cereals. About 25% of all dietary iron consumed in the United States comes from such foods.

The emphasis on the health benefits of iron and the attempts to get enough iron into the diet make it somewhat surprising to be told that iron is also a serious poisoning threat for children. In fact, iron is the leading cause of poisoning deaths in children under the age of six. Since 1986, 110 thousand iron poisoning incidents in children have been reported, including 33 deaths. The iron-containing products involved range from innocent-appearing nonprescription daily multivitamin/mineral supplements for children to high-potency prescription iron supplements for pregnant women. Children showed symptoms of poisoning from consuming as few as 5 to as many as 98 iron-containing tablets. Immediate symptoms include nausea, vomiting, and diarrhea. Deaths occurred from ingesting as little as 200 mg to as much as 5850 mg of iron.

In an attempt to address this problem, the U.S. Food and Drug Administration (FDA) published regulations in 1997 that require warnings on all iron-containing drugs and dietary supplements about the risk of iron poisoning in children and the need to keep such products out of children's reach. In addition, the regulations require that most products containing 30 mg or more of iron per dosage unit have to be packaged as individual doses (such as in blister packages). This rule is designed to limit the number of pills or capsules a child might consume because of the difficulty encountered by a child in opening small individual packets. These 1997 FDA regulations added to rules already in place, including a U.S. Consumer Product Safety Commission regulation given in 1987. According to this regulation, most drugs and food supplements containing more than 250 mg of iron per container have to be packaged in child-resistant containers.

The primary responsibility for protecting young children from iron poisoning still rests with parents, older siblings, and other caregivers. These individuals must first recognize the hazards presented by iron-containing products and must then be very diligent in keeping such products out of the reach of young children.

LEARNING OBJECTIVES ASSESSMENT

You can get an approximate but quick idea of how well you have met the learning objectives given at the beginning of this chapter by working the selected end-of-chapter exercises given below. The answer to each exercise is given in Appendix B of the book.

Objective 1 (Section 3.1): Exercise 3.4

Objective 2 (Section 3.2): Exercise 3.12

Objective 3 (Section 3.3): Exercise 3.18

Objective 4 (Section 3.3): Exercise 3.22

Objectives 5 and 6 (Section 3.4): Exercise 3.24

Objective 7 (Section 3.4): Exercise 3.28

Objective 8 (Section 3.5): Exercise 3.34

Objective 9 (Section 3.5): Exercise 3.36

Objectives 10 and 11 (Section 3.6): Exercises 3.40 and 3.42

➤ Features

Each chapter has features especially designed to help students organize, study effectively, understand, and enjoy the material in the course.

Chapter Opening Photos. Each chapter opens with a photo of one of the many health care professionals that provide us with needed services. These professionals represent some of the numerous professions that require an understanding of chemistry.

Chapter Outlines/Learning Objectives. At the beginning of each chapter, a list of learning objectives provides students with a convenient overview of what they should gain by studying the chapter. At the end of the chapter, they encounter a learning objectives assessment consisting of a list of end-of-chapter exercises keyed to each learning objective. By working the suggested exercises, students get a quick indication of how well they have met the stated learning objectives. Thus, students begin each chapter with a set of objectives and end with an indication of how well they satisfied the objectives.

Key Terms. Identified within the text by the use of bold type, key terms are defined in the margin near the place where they are introduced. Students reviewing a chapter can quickly identify the important concepts on each page with this marginal glossary. A full glossary of key terms and concepts appears at the end of the text.

Over the Counter. These boxed features contain useful information about health-related products that are readily available to consumers without a prescription. The information in each box provides a connection between the chemical behavior of the product and its effect on the body.

Chemistry Around Us. These boxed features present everyday applications of chemistry that emphasize in a real way the important role of chemistry in our lives. Twenty percent of these are new to this edition and emphasize health-related applications of chemistry.

Chemistry and Your Health. These new boxed features contain current chemistry-related health issues and questions such as the dangers of obesity, safety questions surrounding genetically modified foods, and the relationship between C-reactive protein and heart disease.

CHAPTER 10 Radioactivity and Nuclear Processes

Scientists are continually developing new technology for use in the health sciences; the goal is to provide new information for improved diagnosis and better treatment of patients. Here, a medical imaging technologist uses magnetic resonance imaging (MRI) equipment to visualize soft tissues of the body). MRI is one of the special topics included in this chapter.

Examples. To reinforce students in their problem-solving skill development, carefully worked out solutions in numerous examples are included in each chapter.

Learning Checks. Short self-check exercises follow examples and discussions of key or difficult concepts. A complete set of solutions is included in Appendix C. These allow students to immediately measure their understanding and progress.

Study Skills. Most chapters contain a *Study Skills* feature in which a challenging topic, skill, or concept of the chapter is addressed. Study suggestions, analogies, and approaches are provided to help students master these ideas.

How Reactions Occur. The mechanisms of representative organic reactions are presented in one boxed insert to help students dispel the mystery of how these reactions take place.

OVER THE COUNTER • 2.1

CALCIUM SUPPLEMENTS

Some meanings of the word *supplement* are "to add to," "to fill up," and "to complete." When used in a nutritional context, a supplement provides an amount of a substance that is in addition to the amount obtained from the diet. An important question, then, is: Who, if anyone, should take dietary supplements? The obvious answer is that anyone who does not get enough of a particular nutrient from the diet to satisfy the needs of the body should take a supplement. How do we apply this obvious answer to the question of whether or not to take a calcium supplement?

Calcium in various forms performs numerous functions in the body. However, about 99% is used to build bones and teeth. During the body's lifetime, all bones undergo a continuous natural process of buildup and breakdown. The rate of buildup exceeds the rate of breakdown for the first 25–30 years in women and the first 30–35 years in men. After these ages, the rate of breakdown catches and exceeds the rate of buildup, resulting in a gradual decrease in bone density. As bone density decreases, the bones become increasingly weakened, brittle, and susceptible to breaking—a condition called *osteoporosis*. About 50% of women and 13% of men over age 50 suffer a broken bone as a result of osteoporosis.

groups are the tendency to skip meals and the substitution of soft drinks and other nondairy drinks in place of milk.

If a calcium supplement is needed, a number of factors should be considered. Vitamin D is essential for optimal calcium absorption by the body. For this reason, many calcium supplements include vitamin D in their formulation, and clearly indicate this on their labels. Calcium supplements are most efficiently absorbed when taken in individual doses of 500 mg or less. The dose per tablet or capsule is generally indicated on the label. The dosages found in the calcium supplements carried by a typical pharmacy range from a low of 333 mg to a high of 630 mg. The calcium comes in various compound forms, including calcium carbonate (often from oyster shells), calcium citrate, and calcium phosphate. Different forms can be advantageous for different body conditions. For example, a person with high gastric acid production should take calcium carbonate with food to improve absorption.

Osteoporosis might seem too far away in the future to concern a teenager or young adult. However, simple changes in lifestyle now—such as improving the diet or taking a calcium supplement—might provide substantial and much-appreciated health benefits in that (not so far) distant future.

For Future Reference. Each chapter concludes with a box containing interesting information that should be especially useful to students as they continue their studies, practice their professions, or just live their lives. For example, the boxes provide information about good food sources of various vitamins and minerals, deal with the effects of drugs, and discuss the meaning of terms used on food labels.

Concept Summary. Located at the end of each chapter, this feature provides a concise review to reinforce the major ideas.

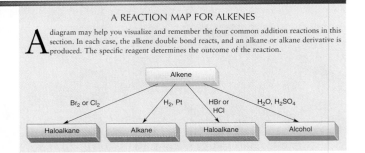

> STUDY SKILLS 12.2

Key Terms and Concepts. These are listed at the end of the chapter for easy review, with a reference to the chapter section in which they are presented.

Key Equations. This boxed feature provides a useful summary of general equations and reactions from the chapter. This feature is particularly helpful to students in the organic chemistry chapters.

Exercises. Nearly 800 end-of-chapter exercises are arranged by section. They are color-coded according to their level of difficulty. Approximately half of the exercises are answered in the back of the text. Completely worked out solutions to these answered exercises are included in the Student Study Guide. Solutions and answers to the remaining exercises are provided in the Instructor's Manual. We have included a significant number of clinical and other familiar applications of chemistry in the exercises.

Chemistry for Thought. Included at the end of each chapter are special questions designed to encourage students to expand their reasoning skills. Some of these exercises are based on photographs found in the chapter, while others emphasize clinical or other useful applications of chemistry.

➤ POSSIBLE COURSE OUTLINES

This text may be used effectively in either a two-semester or three-quarter course of study:

First semester: Chapters 1–13 (general chemistry and three chapters of organic chemistry)
Second semester: Chapters 14–25 (organic chemistry and biochemistry)

First semester: Chapters 1–10 (general chemistry)
Second semester: Chapters 11–21 (organic chemistry and some biochemistry)

First quarter: Chapters 1–10 (general chemistry)
Second quarter: Chapters 11–18 (organic chemistry)
Third quarter: Chapters 19–25 (biochemistry)

➤ ANCILLARIES

The following ancillaries are available to qualified adopters. Please consult your local Brooks/Cole • Thomson sales representative for details.

Print Resources

Safety-Scale Laboratory Experiments for General, Organic, and Biochemistry, 5th Edition. Prepared by Spencer L. Seager and Michael R. Slabaugh, this well-tested collection of experiments has been developed during more than 30 years of laboratory instruction with students at Weber State University. This manual provides a blend of training in laboratory skills and experiences illustrating concepts from the authors' textbook. The experiments are designed to use small quantities of chemicals, and emphasize safety and proper disposal of used materials. ISBN 0-534-39970-3

Instructor's Guide for Safety-Scale Laboratory Experiments. Prepared by the authors of the laboratory manual, this useful resource gives complete directions for preparing the reagents and other materials used in each experiment. It also contains useful comments concerning the experiments, answers to questions included in the experiments, and suggestions for the proper disposal of used materials. This product is available online. ISBN 0-534-39975-4

Study Guide and Solutions Manual. Prepared by Jennifer P. Harris of Portland Community College, each chapter contains a chapter outline, learning objectives, a programmed review of important topics and concepts, detailed solutions to the even-numbered exercises answered in the text, and self-test questions. ISBN 0-534-39973-8

Instructor's Manual and Testbank. Prepared by James K. Hardy of the University of Akron, each chapter contains a summary chapter outline, learning objectives, instructor resource materials, solutions to *Chemistry for Thought* questions, answers and solutions to odd-numbered exercises not answered in the text, and more than 1,300 exam questions. ISBN 0-534-39974-6

Transparency Acetates. The publisher provides 200 full-color transparencies illustrating key figures from the text for use in class. ISBN 0-534-39976-2

Media Resources

GOB ChemistryNow at http://www.chemistry.brookscole.com/seager5e Completely integrated with the fifth edition of the text, *GOB ChemistryNow* is an assessment-based program. In every chapter of the text, icons with descriptive captions direct students to a corresponding module within the *GOB ChemistryNow* Web site. Students can review for the Nursing Board Exams, create a *Personalized Learning Plan*, practice for an exam, and much more. Please see the inside front cover of the book for a complete description of *GOB ChemistryNow*.

InfoTrac® College Edition is an online library available FREE with each new copy of *Chemistry for Today*. It gives students full-length articles—not simply abstracts—from more than 700 popular and scholarly periodicals, updated daily and dating back as far as four years. Student subscribers receive a personalized account ID that gives them four months of unlimited access—at any hour of the day. New to this edition, InfoTrac citations are available at the end of each chapter to provide students with relevant articles for the material being learned.

CNN Chemistry Videos, produced by Turner Learning, can stimulate and engage classes by launching a lecture, sparking a discussion, or demonstrating an application. Each chemistry-related segment from recent CNN broadcasts clearly demonstrates the relevancy of chemistry to everyday life. ISBN 0-534-36924-3

Multimedia Manager is a digital library and presentation tool that is available on one convenient multi-platform CD-ROM. With its easy-to-use interface, a professor can take advantage of Brooks/Cole's already created text-specific presentations, which consist of text art, photos, tables, and more, in a variety of e-formats that are easily exported into presentation software or used on Web-based course support materials. It allows professors to customize their own presentations by importing personal lecture slides or other material of their choosing. The result is an interactive and fluid lecture that truly engages your students. ISBN 0-534-39986-X

ExamView allows professors to create, deliver, and customize tests and study guides (both print and online) in minutes with this easy-to-use assessment and tutorial system. ExamView offers both a Quick Test Wizard and an Online Test Wizard that guide professors step-by-step through the process of creating tests, and the unique "WYSI-WYG" capability allows professors to see the test being created on the screen exactly as it will print or display online. Tests of up to 250 questions can be built using up to 12 question types. Using ExamView's complete word processing capabilities, an unlimited number of new questions can be entered or existing questions can be edited. ISBN 0-534-39977-0

WebTutor Toolbox for WebCT or Blackboard is preloaded with content and available free via a pincode when packaged with this text, WebTutor ToolBox for WebCT or Blackboard pairs all the content of this text's rich book companion Web site with all the sophisticated course management functionality of a WebCT or Blackboard product. Professors can assign materials (including online quizzes) and have the results flow AUTOMATICALLY to their gradebook. ToolBox is ready to use as soon as the user logs on—or, it can be customized with its preloaded content by uploading images and other resources, adding Web links, or creating new practice materials. Students only have access to student resources on the Web site. Instructors can enter a pincode for access to password-protected Instructor Resources. Contact your Thomson representative for information on packaging WebTutor ToolBox with this text. ISBN 0-534-27488-9

Brooks/Cole Chemistry Resource Center. http://www.brookscole.com/chemistry. At Brooks/Cole's chemistry Web site, instructors and students can access a homepage for *Chemistry for Today*, Fifth Edition. All information is arranged according to the *Chemistry for Today* table of contents. Students can access flash cards for all glossary terms, practice quizzes for every chapter, and hyperlinks that relate to each chapter's contents. In addition, students can research the accomplishments of past and present contributors to the field of chemistry in timeline format.

➤ ACKNOWLEDGMENTS

We express our sincere appreciation to the following reviewers, who read and commented on the third edition and offered helpful advice and suggestions for improving this edition:

Bruce Banks
University of North Carolina–Greensboro

Jean Gade
Northern Kentucky University

Galen George
Santa Rosa Junior College

Mary Herrmann
University of Cincinnati

Jack Hefley
Blinn College

Jim Johnson
Sinclair Community College

Carol Larocque
Cambrian College

Richard Lavallee
Santa Monica College

James Petrich
San Antonio College

William Scovell
Bowling Green State University

We also express appreciation to the following reviewers, who helped us revise the first four editions:

Hugh Akers
Lamar University–Beaumont

Johanne I. Artman
Del Mar College

Gabriele Backes
Portland Community College

Bruce Banks
University of North Carolina–Greensboro

David Boykin
Georgia State University

Deb Breiter
Rockford College

Lorraine C. Brewer
University of Arkansas

Martin Brock
Eastern Kentucky University

Christine Brzezowski
University of Utah

Sybil K. Burgess
University of North Carolina–Wilmington

Sharmaine S. Cady
East Stroudsburg University

Linda J. Chandler
Salt Lake Community College

Sharon Cruse
Northern Louisiana University

Thomas D. Crute
Augusta College

Jack L. Dalton
Boise State University

Lorraine Deck
University of New Mexico

Kathleen A. Donnelly
Russell Sage College

Jan Fausset
Front Range Community College

Patricia Fish
The College of St. Catherine

Harold Fisher
University of Rhode Island

John W. Francis
Columbus State Community

Wes Fritz
College of DuPage

Galen George
Santa Rosa Junior College

Linda Thomas-Glover
Guilford Technical Community College

Jane D. Grant
Florida Community College

James K. Hardy
University of Akron

Leland Harris
University of Arizona

Robert H. Harris
University of Nebraska–Lincoln

Claudia Hein
Diablo Valley College

John Henderson
Jackson Community College

Laura Kibler-Herzog
Georgia State University

Arthur R. Hubscher
Ricks College

Kenneth Hughes
University of Wisconsin–Oshkosh

Jeffrey A. Hurlbut
Metropolitan State College of Denver

Jim Johnson
Sinclair Community College

Richard. F. Jones
Sinclair Community College

Frederick Jury
Collin County Community College

Howard K. Ono
California State University–Fresno

Lidija Kampa
Kean College of New Jersey

James A. Petrich
San Antonio College

James F. Kirby
Quinnipiac College

Thomas G. Richmond
University of Utah

Peter J. Krieger
Palm Beach Community College

James Schreck
University of Northern Colorado

Terrie L. Lampe
De Kalb College–Central Campus

William M. Scovall
Bowling Green State University

Leslie J. Lovett
Fairmont State College

Jean M. Shankweiler
El Camino Community College

Armin Mayr
El Paso Community College

Francis X. Smith
King's College

Evan McHugh
Pikes Peak Community College

J. Donald Smith
University of Massachusetts–Dartmouth

Trudy McKee
Thomas Jefferson University

Malcolm P. Stevens
University of Hartford

Melvin Merken
Worcester State College

Eric R. Taylor
University of Southwestern Louisiana

W. Robert Midden
Bowling Green State University

James A. Thomson
University of Waterloo

Pamela S. Mork
Concordia College

Mary Lee Trawick
Baylor University

Phillip E. Morris, Jr.
University of Alabama–Birmingham

Katherin Vafeades
University of Texas–San Antonio

Robert N. Nelson
Georgia Southern University

Cary Willard
Grossmont College

Elva Mae Nicholson
Eastern Michigan University

Don Williams
Hope College

H. Clyde Odom
Charleston Southern University

Les Wynston
California State University–Long Beach

We also give special thanks to Lauren Raike, assistant editor for Brooks/Cole who guided and encouraged us in the preparation of this fifth edition. Her suggestions on special features, content changes, and organization of topics were invaluable. John Holdcroft, chemistry editor, Lisa Weber, production project manager, and Ericka Yeoman-Saler, technology project manager, were also essential to the team and contributed greatly to the success of the project. We are very grateful for the superb work of Lachina Publishing Services, especially to Jeff Lachina, for outstanding work in coordinating the production. We appreciate the significant help of three associates: Mary Ann Francis, Wayne April, and Kristanne Willden who did excellent work in researching special topics, typing, working exercises, and proofreading.

Finally, we extend our love and heartfelt thanks to our families for their patience, support, encouragement, and understanding during a project that occupied much of our time and energy.

Spencer L. Seager

Michael R. Slabaugh

CHAPTER 1

Matter, Measurements, and Calculations

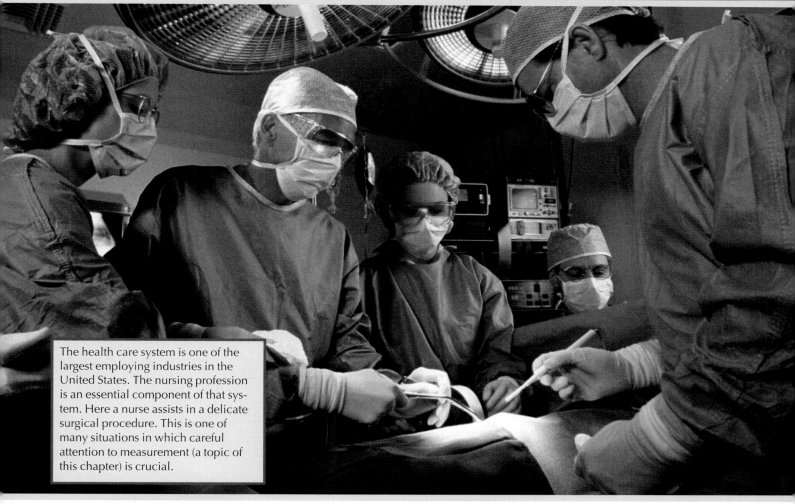

The health care system is one of the largest employing industries in the United States. The nursing profession is an essential component of that system. Here a nurse assists in a delicate surgical procedure. This is one of many situations in which careful attention to measurement (a topic of this chapter) is crucial.

LEARNING OBJECTIVES

When you have completed your study of this chapter, you should be able to:

1. Explain what matter is. (Section 1.1)

2. Explain differences between the terms *physical* and *chemical* as applied to:
   a. Properties of matter (Section 1.2)
   b. Changes in matter (Section 1.2)

3. Describe matter in terms of the accepted scientific model. (Section 1.3)

4. On the basis of observation or information given to you, classify matter into the correct category based on the following pairs:
   a. Heterogeneous or homogeneous (Section 1.4)
   b. Solution or pure substance (Section 1.4)
   c. Element or compound (Section 1.4)

5. Recognize the units of the metric system. (Section 1.6)

6. Convert metric-system measurement units into related units. (Section 1.6)

7. Express numbers using scientific notation. (Section 1.7)

8. Do calculations involving numbers expressed in scientific notation. (Section 1.7)

9. Express the results of measurements and calculations using the correct number of significant figures. (Section 1.8)

10. Use the factor-unit method to solve numerical problems. (Section 1.9)

11. Do calculations involving percentages. (Section 1.10)

12. Do calculations involving densities. (Section 1.11)

GOB
Chemistry ⚛ Now ™
This icon, appearing throughout the book, indicates an opportunity to explore interactive tutorials, animations, and practice problems; available on the GOB Now Web site at **http://chemistry. brookscole.com/seager5e**

Chemistry is often described as the scientific study of matter. In a way, almost any study is a study of matter, because matter is the substance of everything. Chemists, however, are especially interested in matter; they study it and attempt to understand it from nearly every possible point of view.

The chemical nature of all matter makes an understanding of chemistry useful and necessary for individuals who are studying in a wide variety of areas, including the health sciences, the natural sciences, home economics, education, environmental science, and law enforcement.

Matter comes in many shapes, sizes, and colors that are interesting to look at and describe. Early chemists did little more than describe what they observed, and their chemistry was a descriptive science that was severely limited in scope. It became a much more useful science when chemists began to make quantitative measurements, do calculations, and incorporate the results into their descriptions. Some fundamental ideas about matter are presented in this chapter, along with some ideas about quantitative measurement, the scientific measurement system, and calculations.

➤ 1.1 What Is Matter?

Definitions are useful in all areas of knowledge; they provide a common vocabulary for both presentations to students and discussions between professionals. You will be expected to learn a number of definitions as you study chemistry, and the first one is a definition of *matter*. Earlier, we said that matter is the substance of everything. That isn't very scientific, even though we think we know what it means. If you stop reading for a moment and look around, you will see a number of objects that might include people, potted plants, walls, furniture, books, windows, and a TV set or radio. The objects you see have at least two things in common: Each one has mass, and each one occupies space. These two common characteristics provide the basis for the scientific definition of matter. **Matter** is anything that has mass and occupies space. You probably understand what is meant by an object occupying space, especially if you have tried to occupy the same space as some other object. The resulting physical bruises leave a lasting mental impression.

You might not understand the meaning of the term *mass* quite as well, but it can also be illustrated painfully. Imagine walking into a very dimly lit room and being able to just barely see two large objects of equal size on the floor. You know that one is a bowling ball and the other is an inflated plastic ball, but you can't visually identify which is which. However, a hard kick delivered to either object easily allows you to identify each one. The bowling ball resists being moved much more strongly than does the inflated ball. Resistance to movement depends on the amount of matter in an object, and **mass** is an actual measurement of the amount of matter present.

The term *weight* is probably more familiar to you than *mass*, but the two are related. All objects are attracted to each other by gravity, and the greater their mass, the stronger the attraction between them. The **weight** of an object on Earth is a measurement of the gravitational force pulling the object toward Earth. An object with twice the mass of a second object is attracted with twice the force, and therefore has twice the weight of the second object. The mass of an object is constant no matter where it is located (even if it is in a weightless condition in outer space). However, the weight of an object depends on the strength of the gravitational attraction to which it is subjected. For exam-

Matter
Anything that has mass and occupies space.

Mass
A measurement of the amount of matter in an object.

Weight
A measurement of the gravitational force acting on an object.

CHEMISTRY AROUND US • 1.1

A CENTRAL SCIENCE

Chemistry is often referred to as the "central science" because it serves as a necessary foundation for many other scientific disciplines. Regardless of which scientific field you are interested in, every single substance you will discuss or work with is made up of chemicals. Also, many processes important to those fields will be based on an understanding of chemistry.

boxes focus on specific substances that play essential roles in meeting the needs of society.

Chemistry is the foundation for many other scientific disciplines.

We also consider chemistry a central science because of its crucial role in responding to the needs of society. We use chemistry to discover new processes, develop new sources of energy, produce new products and materials, provide more food, and ensure better health.

As you read this text, you will encounter chapter opening photos dealing with applications of chemistry in the health care professions. Within the chapters, other Chemistry Around Us

Chemicals are present in everything we can touch, smell, or see. Chemistry is all around us.

ple, a rock that weighs 16 pounds on Earth would weigh about 2.7 pounds on the moon because the gravitational attraction is only about one-sixth that of Earth. However, the rock contains the same amount of matter and thus has the same mass whether it is located on Earth or on the moon.

Despite the difference in meaning between mass and weight, the determination of mass is commonly called "weighing." We will follow that practice in this book, but we will use the correct term *mass* when referring to an amount of matter.

➤ 1.2 Properties and Changes

When you looked at your surroundings earlier, you didn't have much trouble identifying the various things you saw. For example, unless the decorator of your room had unusual tastes, you could easily tell the difference between a TV set and a potted plant by observing such characteristics as shape, color, and size. Our ability to identify objects or materials and discriminate between them depends on such characteristics. Scientists prefer to use the term *property* instead of characteristic, and they classify properties into two categories, physical and chemical.

CHEMISTRY AROUND US • 1.2

COSMETICS: COMPLEX MIXTURES AND COMPLEX REGULATIONS

The federal Food, Drug, and Cosmetic (FD&C) Act defines a cosmetic as anything applied directly to the human body for cleansing, beautifying, promoting attractiveness, or altering the appearance without affecting the body's structure or functions. According to this definition, mixtures as diverse as a modern roll-on deodorant and henna, a colored plant extract used in ancient times as well as today to dye hair, are classified as cosmetics. However, it is interesting to note that according to the FD&C Act, soap is not legally considered to be a cosmetic.

The sale of cosmetics in the United States is regulated by the federal Food and Drug Administration (FDA), but the regulatory requirements applied to the sale of cosmetics are not nearly as stringent as those applied to other FDA-regulated products. With the exception of color additives and a few prohibited substances, cosmetics manufacturers may use any ingredient or raw material in their products and market the products without obtaining FDA approval. The regulation that provides consumers with the greatest amount of information about the chemical composition of cosmetics comes not from the FDA, but from the Fair Packaging and Labeling Act. This act requires that every cosmetic product must be labeled with a list of all ingredients in order of decreasing quantity. For example, many skin-care products contain more water than any other ingredient, so water is listed first.

Any cosmetic product that is also designed to treat or prevent disease, or otherwise affect the structure or functions of the human body, is regulated as both a drug and a cosmetic, and must meet the labeling requirements for both. Some well-known examples of this type of product are dandruff shampoos, fluoride toothpastes, and antiperspirants/deodorants. A good way to tell if you are buying a cosmetic that is also regulated as a drug is to see if the first item on the ingredient label is listed as an "active ingredient." Regulations require that the active ingredient be identified and listed first, followed by the cosmetic ingredients in order of decreasing amounts.

Many different types of products are classified as cosmetics. Each one must have a list of ingredients on the label.

Physical properties
Properties of matter that can be observed or measured without trying to change the composition of the matter being studied.

Chemical properties
Properties matter demonstrates when attempts are made to change it into new substances.

Physical changes
Changes matter undergoes without changing composition.

Chemical changes
Changes matter undergoes that involve changes in composition.

Physical properties are those that can be observed or measured without changing or trying to change the composition of the matter in question—no original substances are destroyed, and no new substances appear. For example, you can observe the color or measure the size of a sheet of paper without attempting to change the paper into anything else. Color and size are physical properties of the paper. **Chemical properties** are the properties matter demonstrates when attempts are made to change it into other kinds of matter. For example, a sheet of paper can be burned; in the process, the paper is changed into new substances. On the other hand, attempts to burn a piece of glass under similar conditions fail. The ability of paper to burn is a chemical property, as is the inability of glass to burn.

You can easily change the size of a sheet of paper by cutting off a piece. The paper sheet is not converted into any new substance by this change, but it is simply made smaller. **Physical changes** can be carried out without changing the composition of a substance. However, there is no way you can burn a sheet of paper without changing it into new substances. Thus, the change that occurs when paper burns is called a **chemical change.** ■ Figure 1.1 shows an example of a chemical change, the burning of magnesium metal. The bright light produced by this chemical change led to the use of magnesium in the flash powder used in early photography. Magnesium is still used in fireworks to produce a brilliant white light.

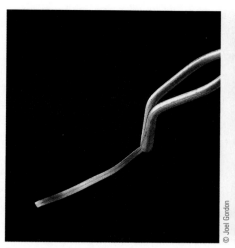

A strip of magnesium metal held with tongs.

After being ignited with a flame, the magnesium burns with a blinding white light.

The white ash of magnesium oxide from the burning of several magnesium strips.

➤ **FIGURE 1.1** A chemical change occurs when magnesium metal burns.

Example 1.1

Classify each of the following changes as physical or chemical: (a) a match is burned; (b) iron is melted; (c) limestone is crushed; (d) limestone is heated, producing lime and carbon dioxide; (e) an antacid seltzer tablet is dissolved in water; and (f) a rubber band is stretched.

Solution

Changes b, c, and f are physical changes because no composition changes occurred and no new substances were formed.

The others are chemical changes because new substances were formed. A match is burned—combustion gases are given off, and matchstick wood is converted to ashes. Limestone is heated—lime and carbon dioxide are the new substances. A seltzer tablet is dissolved in water—the fizzing that results is evidence that at least one new material (a gas) is produced.

Learning Check 1.1

Classify each of the following changes as physical or chemical, and, in the cases of chemical change, describe one observation or test that indicates new substances have been formed: (a) milk sours, (b) a wet handkerchief dries, (c) fruit ripens, (d) a stick of dynamite explodes, (e) air is compressed into a steel container, and (f) water boils.

Among the most common physical changes are changes in state, such as the melting of solids to form liquids, the sublimation of solids to form gases, or the evaporation of liquids to form gases. These changes take place when heat is added to or removed from matter, as represented in ■ Figure 1.2. We will discuss changes in state in more detail in Chapter 6.

➤ 1.3 A Model of Matter

Model building is a common activity of scientists, but the results in many cases would not look appropriate on a fireplace mantle. **Scientific models** are

Scientific models
Explanations for observed behavior in nature.

(a) Pure iodine is a dark-colored solid at room temperature.

(b) When heated, solid iodine is converted to a purple-colored gas.

(c) Pure benzene is a colorless liquid at room temperature.

(d) When cooled in an ice/salt bath, benzene freezes to a solid.

➤ **FIGURE 1.2** Examples of physical change: Solid iodine (a) becomes gaseous iodine (b); liquid benzene (c) becomes solid benzene (d).

➤ **FIGURE 1.3** A hang glider soars far above the ground. How does this feat confirm that air is matter?

explanations for observed behavior. Some, such as the well-known representation of the solar system, can easily be depicted in a physical way. Others are so abstract that they can be represented only by mathematical equations.

Our present understanding of the nature of matter is a model that has been developed and refined over many years. Based on careful observations and measurements of the properties of matter, the model is still being modified as more is learned. In this book, we will concern ourselves with only some very basic concepts of this model, but even these basic ideas will provide a powerful tool for understanding the behavior of matter.

The study of the behavior of gases such as air, oxygen, and carbon dioxide by some of the earliest scientists led to a number of important ideas about matter. The volume of a gas kept at a constant temperature was found to change with pressure. An increase in pressure caused the gas volume to decrease, whereas a decrease in pressure permitted the gas volume to increase. It was also discovered that the volume of a gas maintained at constant pressure increased as the gas temperature was increased. Gases were also found to have mass and to mix rapidly with one another when brought together.

A simple model for matter was developed that explained these gaseous properties, as well as many properties of solids and liquids. Some details of the model are discussed in Chapter 6, but one conclusion is important to us now. All matter is made up of particles that are too small to see (see ■ Figure 1.3). The early framers of this model called the small particles *molecules*. It is now known that molecules are the constituent particles of many, but not all, substances. In this chapter, we will limit our discussion to substances made up of molecules. Substances that are not made of molecules are discussed in Sections 4.3 and 4.11.

The results of some simple experiments will help us formally define the term *molecule*. Suppose you have a container filled with oxygen gas and you perform a number of experiments with it. You find that a glowing splinter of wood bursts into flames when placed in the gas. A piece of moist iron rusts much faster in the oxygen than it does in air. A mouse or other animal can safely breathe the gas.

Now suppose you divide another sample of oxygen the same size as the first into two smaller samples. The results of similar experiments done with these samples would be the same as before. Continued subdivision of an oxygen sample into smaller and smaller samples does not change the ability of the oxygen in the samples to behave just like the oxygen in the original sample. We conclude that the physical division of a sample of oxygen gas into smaller and smaller samples does not change the oxygen into anything else—it is still oxygen. Is there a limit to such divisions? What is the smallest sample of oxygen that will behave like the larger sample? We hope you have concluded that the smallest sample must be a single molecule. Although its very small size would make a one-molecule sample difficult to handle, it would nevertheless behave just as a larger sample would—it could be stored in a container, it would make wood burn rapidly, it would rust iron, and it could be breathed safely by a mouse.

We are now ready to formally define the term *molecule*. A **molecule** is the smallest particle of a pure substance that has the properties of that substance and is capable of a stable independent existence. Alternatively, a molecule is defined as the limit of physical subdivision for a pure substance.

In less formal terms, these definitions indicate that a sample of pure substance—such as oxygen, carbon monoxide, or carbon dioxide—can be physically separated into smaller subsamples only until there is a single molecule. Any further separation cannot be done physically, but if it were done (chemically), the resulting sample would no longer have the properties of the larger samples.

The idea that it might be possible to chemically separate a molecule into smaller units grew out of continued study and experimentation by early scientists. In modern terminology, the smaller particles that make up molecules are called *atoms*. Early scientists used graphic symbols such as circles and squares to represent the few different atoms that were known at the time. Instead of different shapes, we will use representations such as those in ■ Figure 1.4 for oxygen, carbon monoxide, and carbon dioxide molecules.

The three pure substances just mentioned illustrate three types of molecules found in matter. Oxygen molecules consist of two oxygen atoms, and are called **diatomic molecules** to indicate that fact. Molecules such as oxygen that contain only one kind of atom are also called **homoatomic molecules** to indicate that the atoms are all of the same kind. Carbon monoxide molecules also contain two atoms and therefore are diatomic molecules. However, in this case the atoms are not identical, a fact indicated by the term **heteroatomic molecule.** Carbon dioxide molecules consist of three atoms that are not all identical, so carbon dioxide molecules are described by the terms **triatomic** and heteroatomic. The words *diatomic* and *triatomic* are commonly used to indicate two- or three-atom molecules, but the word **polyatomic** is usually used to describe molecules that contain more than three atoms.

Molecule
The smallest particle of a pure substance that has the properties of that substance and is capable of a stable independent existence. Alternatively, a molecule is the limit of physical subdivision for a pure substance.

Diatomic molecules
Molecules that contain two atoms.

Homoatomic molecules
Molecules that contain only one kind of atom.

Heteroatomic molecules
Molecules that contain two or more kinds of atoms.

Triatomic molecules
Molecules that contain three atoms.

Polyatomic molecules
Molecules that contain more than three atoms.

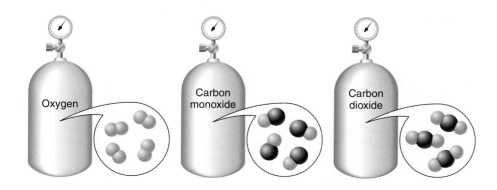

➤ **FIGURE 1.4** Symbolic representations of molecules.

Example 1.2

Use the terms *diatomic, triatomic, polyatomic, homoatomic,* or *heteroatomic* to classify the following molecules correctly:

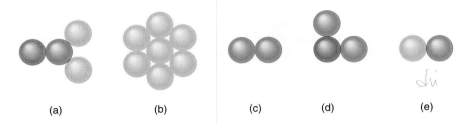

(a)   (b)   (c)   (d)   (e)

Solution

a. Polyatomic and heteroatomic (more than three atoms, and the atoms are not all identical)
b. Polyatomic and homoatomic (more than three atoms, and the atoms are identical)
c. Diatomic and homoatomic (two atoms, and the atoms are identical)
d. Triatomic and heteroatomic (three atoms, and the atoms are not identical)
e. Diatomic and heteroatomic (two atoms, and the atoms are not identical)

Learning Check 1.2

Use the terms *diatomic, triatomic, polyatomic, homoatomic,* or *heteroatomic* to classify the following molecules correctly:

a. Water molecules have been found to contain two hydrogen atoms and one oxygen atom.
b. Molecules of ozone contain three oxygen atoms.
c. Paper molecules (cellulose) are very large. When molecules of paper are burned in pure oxygen, two substances are formed. One contains carbon atoms and oxygen atoms, and the other contains hydrogen atoms and oxygen atoms. Classify only the paper molecules.

Atom
The limit of chemical subdivision for matter.

How far can the chemical subdivision of molecules go? You are probably a step ahead of us and have guessed that the answer is atoms. In fact, this provides us with a definition of atoms. An **atom** is the limit of chemical subdivision. In less formal terms, atoms are the smallest particles of matter that can be produced as a result of chemical changes. However, all chemical changes do not necessarily break molecules into atoms. In some cases, chemical changes might just divide a large molecule into two or more smaller molecules. Also, as we will see later, some chemical changes form larger molecules from smaller ones. The important point is that only chemical changes will work if we want to divide molecules, and the smallest particles of matter that can possibly be produced by such a division are called atoms.

➤ 1.4 Classifying Matter

Unknown substances are often analyzed to determine their compositions. An analyst, upon receiving a sample to analyze, will always ask an important question: Is the sample a pure substance or a mixture? Any sample of matter must be one or the other. Pure water and sugar are both pure substances, but you can create a mixture by stirring together some sugar and pure water.

```
                    ┌─────────────┐
                    │   Matter    │
                    └─────────────┘
                           │
            ┌──────────────┴──────────────┐
     ┌─────────────┐              ┌─────────────────┐
     │   Mixture   │              │ Pure substance  │
     └─────────────┘              └─────────────────┘
```

Proportions of components may vary

Properties vary with composition

Can be physically separated
into two or more pure substances

Constant composition

Fixed set of properties

Cannot be physically
separated into simpler
substances

➤ **FIGURE 1.5** Mixtures and pure substances.

What is the difference between a pure substance and a mixture? Two differences are that a **pure substance** has a constant composition and a fixed set of physical and chemical properties. Pure water, for example, always contains the same proportions of hydrogen and oxygen and freezes at a specific temperature. A **mixture** of sugar and water, however, can vary in composition, and the properties will be different for the different compositions. For example, a glass of sugar water could contain a few crystals of sugar or several spoonfuls. Properties such as the sweetness and freezing point would vary depending on the amount of sugar present in the mixture.

Another difference is that a pure substance cannot be physically separated into simpler substances, whereas a mixture can theoretically be separated into its components. For example, if we heat a sugar-and-water mixture, the water evaporates, and the sugar remains. We say mixtures can theoretically be separated because some separations are very difficult to achieve. ■ Figure 1.5 summarizes these ideas.

Pure substances, and mixtures such as sugar water, are examples of **homogeneous matter**—matter that has a uniform appearance and the same properties throughout. Homogeneous mixtures such as sugar water are called **solutions** (see ■ Figure 1.6). Mixtures in which the properties and appearance are not uniform throughout the sample are examples of **heterogeneous matter.** The mixture of rock salt and sand that is spread on snowy roads during the winter is an example.

Commonly, the word *solution* is used to describe homogeneous liquid mixtures such as sugar water, but solutions of gases and solids also exist. The air around us is a gaseous solution, containing primarily nitrogen and oxygen. The alloys of some metals are solid solutions. For example, small amounts of copper are often added to the gold used in making jewelry. The resulting solid solution is harder than gold and has greater resistance to wear.

Pure substance
Matter that has a constant composition and fixed properties.

Mixture
A physical blend of matter that can be physically separated into two or more components.

Homogeneous matter
Matter that has the same properties throughout the sample.

Solutions
Homogeneous mixtures of two or more pure substances.

Heterogeneous matter
Matter with properties that are not the same throughout the sample.

(a)   (b)

➤ **FIGURE 1.6** Sugar and water (a) form a solution when mixed (b).

GOB
Chemistry ⚛ Now™
Go to GOB Now and click to discover the differences among pure substances, homogeneous mixtures, and heterogeneous mixtures.

© Spencer L. Seager

➤ **FIGURE 1.7** Common household mixtures. The milk and gelatin dessert are homogeneous to the unaided eye, whereas the fruit salad and pizza are heterogeneous.

Element
A pure substance consisting of only one kind of atom in the form of homoatomic molecules or individual atoms.

Compound
A pure substance consisting of two or more kinds of atoms in the form of heteroatomic molecules or individual atoms.

Most matter is found in nature in the form of heterogeneous mixtures. The properties of such mixtures depend on the location from which samples are taken. In some cases, the heterogeneity is obvious (see ■ Figure 1.7). In a slice of tomato, for example, the parts representing the skin, juice, seeds, and pulp can be easily seen and identified because they look different. Thus, at least one property (e.g., color or texture) is different for the different parts. However, a sample of clean sand from a seashore must be inspected very closely before slight differences in appearance can be seen for different grains. At this point, you might be thinking that even the solutions described earlier as homogeneous mixtures would appear to be heterogeneous if they were looked at closely enough. We could differentiate between sugar and water molecules if sugar solutions were observed under sufficient magnification. We will generally limit ourselves to differences normally visible when we classify matter as heterogeneous on the basis of appearance.

Earlier, we looked at three examples of pure substances—oxygen, carbon monoxide, and carbon dioxide—and found that the molecules of these substances are of different types. Oxygen molecules are diatomic and homoatomic, carbon monoxide molecules are diatomic and heteroatomic, and carbon dioxide molecules are triatomic and heteroatomic. Many pure substances have been found to consist of either homoatomic or heteroatomic molecules—a characteristic that permits them to be classified into one of two categories. Pure substances made up of homoatomic molecules are called **elements,** and those made up of heteroatomic molecules are called **compounds.** Thus, oxygen is an element, whereas carbon monoxide and carbon dioxide are compounds.

It is useful to note a fact here that is discussed in more detail later in Section 4.11. The smallest particles of some elements and compounds are individual atoms rather than molecules. However, in elements of this type, the individual atoms are all of the same kind, whereas in compounds, two or more kinds of atoms are involved. Thus, the classification of a pure substance as an element or a compound is based on the fact that only one kind of atom is found in elements and two or more kinds are found in compounds. In both cases, the atoms may be present individually or in the form of homoatomic molecules (elements) or heteroatomic molecules (compounds). Some common household materials are pure substances (elements or compounds), as shown in ■ Figure 1.8.

| **Learning Check 1.3** | Classify the molecules represented below as those of an element or a compound: |

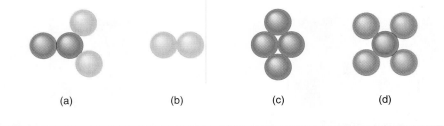

(a) (b) (c) (d)

➤ **FIGURE 1.8** Pure substances found around the home include the elements copper (wire) and aluminum (foil), and the compounds salt and baking soda.

The characteristics of the molecules of elements and compounds lead us to some conclusions about their chemical behavior. Elements cannot be chemically subdivided into simpler pure substances, but compounds can. Because elements contain only one kind of atom and the atom is the limit of chemical subdivision, there is no chemical way to break an element into any simpler pure substance—the simplest pure substance is an element. On the other hand,

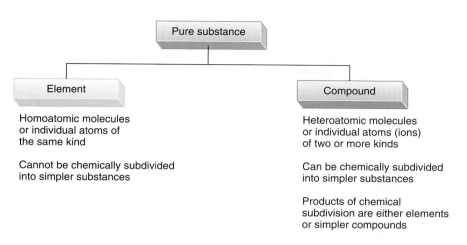

➤ **FIGURE 1.9** Elements and compounds.

because the molecules of compounds contain more than one kind of atom, breaking such molecules into simpler pure substances is possible. For example, a molecule of table sugar can be chemically changed into two simpler molecules (which are also sugars) or into atoms or molecules of the elements carbon, hydrogen, and oxygen. Thus, compounds can be chemically subdivided into simpler compounds or elements. ■ Figure 1.9 summarizes these ideas, and ■ Figure 1.10 illustrates a classification scheme for matter based on the ideas we have discussed.

Example 1.3

When sulfur, an element, is heated in air, it combines with oxygen to form sulfur dioxide. Classify sulfur dioxide as an element or a compound.

Solution

Because sulfur and oxygen are both elements and they combine to form sulfur dioxide, the molecules of sulfur dioxide must contain atoms of both sulfur and oxygen. Thus, sulfur dioxide is a compound because its molecules are heteroatomic.

Suppose an element and a compound combine to form only one product. Classify the product as an element or a compound.

Learning Check 1.4

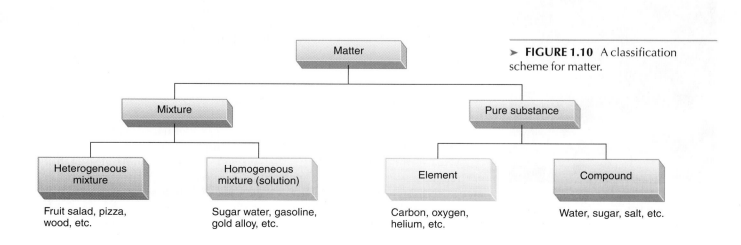

➤ **FIGURE 1.10** A classification scheme for matter.

➤ 1.5 Measurement Units

Matter can be classified and some physical or chemical properties can be observed without making any measurements. However, the use of quantitative measurements and calculations greatly expands our ability to understand the chemical nature of the world around us. A measurement consists of two parts, a number and a unit. A number expressed without a unit is generally useless, especially in scientific work. We constantly make and express measurements in our daily lives. We measure the gallons of gasoline put into our cars, the time it takes to drive a certain distance, and the temperature on a hot or cold day. In some of our daily measurements, the units might be implied or understood. For example, if someone said the temperature outside was 39, you would probably assume this was 39° Fahrenheit if you lived in the United States, but in most other parts of the world, it would be 39° Celsius. Such confusion is avoided by expressing both the number and the unit of a measurement.

All measurements are based on units agreed on by those making and using the measurements. When a measurement is made in terms of an agreed-on unit, the result is expressed as some multiple of that unit. For example, when you purchase 10 pounds of potatoes, you are buying a quantity of potatoes equal to 10 times the standard quantity called 1 pound. Similarly, 3 feet of string is a length of string 3 times as long as the standard length that has been agreed on and called 1 foot.

The earliest units used for measurements were based on the dimensions of the human body. For example, the foot was the length of some important person's foot, and the biblical cubit was the length along the forearm from the elbow to the tip of the middle finger. One problem with such units is obvious; the size of the units changed when the person on whom they were based changed because of death, change in political power, and so on.

As science became more quantitative, scientists found that the lack of standard units became more and more of a problem. A standard system of units was developed in France about the time of the French Revolution and was soon adopted by scientists throughout the world. This system, called the *metric system*, has since been adopted and is used by almost all nations of the world. The United States adopted the system but has not yet put it into widespread use.

In an attempt to further standardize scientific measurements, an international agreement in 1960 established certain basic metric units, and units derived from them, as preferred units to be used in scientific measurements. Measurement units in this system are known as SI units after the French Système International d'Unités. SI units have not yet been totally put into widespread use. Many scientists continue to express certain quantities, such as volume, in non-SI units. The metric system in this book is generally based on accepted SI units but also includes a few of the commonly used non-SI units.

➤ 1.6 The Metric System

The metric system has a number of advantages compared with other measurement systems. One of the most useful of these advantages is that the metric system is a decimal system in which larger and smaller units of a quantity are related by factors of 10. See ■ Table 1.1 for a comparison between the metric and English units of length—a meter is slightly longer than a yard (see ■ Figure 1.11). Notice in Table 1.1 that the units of length in the metric system are related by multiplying a specific number of times by 10—remember,

➤ **FIGURE 1.11** The length of a golf club is about 1 m.

© Jim Pickerell/Stock Boston Inc.

TABLE 1.1   Metric and English units of length

	Base unit	Larger unit	Smaller unit
Metric	1 meter	1 kilometer = 1000 meters	10 decimeters = 1 meter 100 centimeters = 1 meter 1000 millimeters = 1 meter
English	1 yard	1 mile = 1760 yards	3 feet = 1 yard 36 inches = 1 yard

$100 = 10 \times 10$ and $1000 = 10 \times 10 \times 10$. On the other hand, the relationships between the units of the English system show no such pattern.

The relationships between units of the metric system that are larger or smaller than a **basic** (defined) **unit** are indicated by prefixes attached to the name of the basic unit. Thus, 1 kilometer (km) is a unit of length that is 1000 times longer than the basic unit of 1 meter (m), and a millimeter (mm) is only $\frac{1}{1000}$ the length of 1 m. Some commonly used prefixes are given in ■ Table 1.2.

Area and volume are examples of **derived units** of measurement; they are obtained or derived from the basic unit of length:

$$\text{area} = (\text{length})(\text{length}) = (\text{length})^2$$

$$\text{volume} = (\text{length})(\text{length})(\text{length}) = (\text{length})^3$$

The unit used to express an area depends on the unit of length used.

Basic unit of measurement
A specific unit from which other units for the same quantity are obtained by multiplication or division.

Derived unit of measurement
A unit obtained by multiplication or division of one or more basic units.

Example 1.4

Calculate the area of a rectangle that has sides of 1.5 and 2.0 m. Express the answer in units of square meters and square centimeters.

TABLE 1.2   Common prefixes of the metric system

Prefix[a]	Abbreviation	Relationship to basic unit	Exponential relationship to basic unit[b]
mega-	M	$1{,}000{,}000 \times$ basic unit	$10^6 \times$ basic unit
kilo-	k	$1000 \times$ basic unit	$10^3 \times$ basic unit
deci-	d	$1/10 \times$ basic unit	$10^{-1} \times$ basic unit
centi-	c	$1/100 \times$ basic unit	$10^{-2} \times$ basic unit
milli-	m	$1/1000 \times$ basic unit	$10^{-3} \times$ basic unit
micro-	μ	$1/1{,}000{,}000 \times$ basic unit	$10^{-6} \times$ basic unit
nano-	n	$1/1{,}000{,}000{,}000 \times$ basic unit	$10^{-9} \times$ basic unit
pico-	p	$1/1{,}000{,}000{,}000{,}000 \times$ basic unit	$10^{-12} \times$ basic unit

[a]The prefixes in boldface (heavy) type are the most common ones. [b]The use of exponents to express large and small numbers is discussed in Section 1.7.

Solution

$$\text{area} = (\text{length})(\text{length})$$

In terms of meters, area = (1.5 m)(2.0 m) = 3.0 m^2. Note that m^2 represents meter squared, or square meters. In terms of centimeters, area = (150 cm)(200 cm) = 30,000 cm^2. The lengths expressed in centimeters were obtained by remembering that 1 m = 100 cm.

Learning Check 1.5

The area of a circle is given by the formula $A = \pi r^2$, where r is the radius and $\pi = 3.14$. Calculate the area of a circle that has a radius of 3.5 cm.

The unit used to express volume also depends on the unit of length used in the calculation. Thus, a volume could have such units as cubic meters (m^3), cubic decimeters (dm^3), or cubic centimeters (cm^3). The abbreviation cc is also used to represent cubic centimeters, especially in medical work. The liter (L), a non-SI unit of volume, has been used as a basic unit of volume by chemists for many years (see ■ Figure 1.12). For all practical purposes, 1 L and 1 dm^3 are equal volumes. This also means that 1 milliliter (mL) is equal to 1 cm^3. Most laboratory glassware is calibrated in liters or milliliters.

Example 1.5

A circular Petri dish with vertical sides has a radius of 7.50 cm. You want to fill the dish with a liquid medium to a depth of 2.50 cm. What volume of medium in milliliters and liters will be required?

GOB Chemistry•Now™

Go to GOB Now and click to learn how to perform unit conversions using the metric system.

Solution

The volume of medium required will equal the area of the circular dish (in square centimeters, cm^2) multiplied by the liquid depth (in centimeters, cm). Note that the unit of this product will be cubic centimeters (cm^3). According to Learning Check 1.5, the area of a circle is equal to πr^2, where $\pi = 3.14$. Thus, the liquid volume will be

$$V = (3.14)(7.50\ \text{cm})^2(2.50\ \text{cm}) = 442\ \text{cm}^3$$

Because 1 cm^3 = 1 mL, the volume equals 442 mL. Also, because 1 L = 1000 mL, the volume can be converted to liters:

$$442\ \cancel{\text{mL}} \times \frac{1\ \text{L}}{1000\ \cancel{\text{mL}}} = 0.442\ \text{L}$$

Notice that the milliliter units canceled in the calculation. This conversion to liters is an example of the factor-unit method of problem solving, which is discussed in Section 1.9.

Learning Check 1.6

A rectangular aquarium has sides with lengths of 30.0 cm and 20.0 cm, and a height of 15.0 cm. Calculate the volume of the aquarium, and express the answer in milliliters and liters.

The basic unit of mass in the metric system is 1 kilogram (kg), which is equal to about 2.2 pounds in the English system. A kilogram is too large to be conveniently used in some applications, so it is subdivided into smaller units. Two of these smaller units that are often used in chemistry are the gram (g)

➤ **FIGURE 1.12** A liter is slightly larger than a quart.

➤ **FIGURE 1.13** Metric masses of some common items as found in a 0.4-g paper clip, 3.0-g razor blade, 3.1-g penny, and 4.7-g nickel.

and milligram (mg) (see ■ Figure 1.13). The prefixes *kilo-* (k) and *milli-* (m) indicate the following relationships between these units:

$$1 \text{ kg} = 1000 \text{ g}$$

$$1 \text{ g} = 1000 \text{ mg}$$

$$1 \text{ kg} = 1,000,000 \text{ mg}$$

Example 1.6

All measurements in international track and field events are made using the metric system. Javelins thrown by female competitors must have a mass of no less than 600 g. Express this mass in kilograms and milligrams.

Solution

Because 1 kg = 1000 g, 600 g can be converted to kilograms as follows:

$$600 \text{ g} \times \frac{1 \text{ kg}}{1000 \text{ g}} = 0.600 \text{ kg}$$

Also, because 1 g = 1000 mg,

$$600 \text{ g} \times \frac{1000 \text{ mg}}{1 \text{ g}} = 600,000 \text{ mg}$$

Once again, the units of the original quantity (600 g) were canceled, and the desired units were generated by this application of the factor-unit method (see Section 1.9).

Learning Check 1.7

The javelin thrown by male competitors in track and field meets must have a minimum mass of 0.800 kg. A javelin is weighed and found to have a mass of 0.819 kg. Express the mass of the weighed javelin in grams.

Temperature is difficult to define but easy for most of us to measure—we just read a thermometer. However, thermometers can have temperature scales that represent different units. For example, a temperature of 293 would probably be considered quite high until it was pointed out that it is just room

➤ **FIGURE 1.14** Fahrenheit, Celsius, and Kelvin temperature scales. The lowest temperature possible is absolute zero, 0 K.

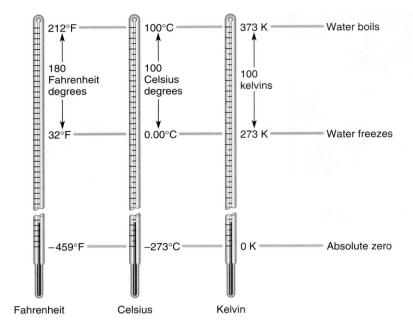

temperature as measured using the Kelvin temperature scale. Temperatures on this scale are given in kelvins, K. (Notice that the abbreviation is K, not °K.)

The Celsius scale (formerly known as the centigrade scale) is the temperature scale used in most scientific work. On this scale, water freezes at 0°C and boils at 100°C under normal atmospheric pressure. A Celsius degree (division) is the same size as a kelvin of the Kelvin scale, but the two scales have different zero points. ■ Figure 1.14 compares the two scientific temperature scales and the familiar Fahrenheit scale. There are 100 Celsius degrees (divisions) between the freezing point (0°C) and the boiling point (100°C) of water. On the Fahrenheit scale, these same two temperatures are 180 degrees (divisions) apart (the freezing point is 32°F and the boiling point is 212°F). Readings on these two scales are related by the following equations:

$$°C = \frac{5}{9}(°F - 32) \tag{1.1}$$

$$°F = \frac{9}{5}(°C) + 32 \tag{1.2}$$

As mentioned, the difference between the Kelvin and Celsius scales is simply the zero point; consequently, readings on the two scales are related as follows:

$$°C = K - 273 \tag{1.3}$$

$$K = °C + 273 \tag{1.4}$$

Notice that Equation 1.2 can be obtained by solving Equation 1.1 for Fahrenheit degrees, and Equation 1.4 can be obtained by solving Equation 1.3 for kelvins. Thus, you need to remember only Equations 1.1 and 1.3, rather than all four.

Example 1.7

A temperature reading of 77°F is measured with a Fahrenheit thermometer. What reading would this temperature give if a Celsius thermometer were used?

Solution

The change is from a Fahrenheit reading to a Celsius reading, so Equation 1.1 is used:

$$°C = \frac{5}{9}(°F - 32) = \frac{5}{9}(77° - 32) = \frac{5}{9}(45°) = 25°$$

Thus, the reading on a Celsius thermometer would be 25°C.

What Kelvin thermometer reading would correspond to the 77°F reading described in Example 1.7?

Learning Check 1.8

The last units discussed at this point are derived units of energy. Other units will be introduced later in the book as they are needed. The metric system unit of energy is a joule (J), pronounced "jewl." A joule is quite small, as shown by the fact that a 50-watt light bulb uses 50 J of energy every second. A typical household in the United States uses several billion joules of electrical energy in a month.

The calorie (cal), a slightly larger unit of energy, is sometimes used by chemists. One calorie is the amount of heat energy required to increase the temperature of 1 g of water by 1°C. The calorie and joule are related as follows:

$$1 \text{ cal} = 4.184 \text{ J}$$

The nutritional calorie of the weight watcher is actually 1000 scientific calories, or 1 kcal. It is represented by writing *calorie* with a capital C (Calorie, abbreviated Cal). ■ Table 1.3 contains a list of the commonly used metric units, their relationship to basic units, and their relationship to English units.

TABLE 1.3 Commonly used metric units

Quantity	Metric unit	Relationship to metric basic unit	Relationship to English unit
Length	meter (m)	Basic unit	1 m = 1.094 yd
	centimeter (cm)	100 cm = 1 m	1 cm = 0.394 in.
	millimeter (mm)	1000 mm = 1 m	1 mm = 0.0394 in.
	kilometer (km)	1 km = 1000 m	1 km = 0.621 mi
Volume	cubic decimeter (dm^3)	Basic unit	1 dm^3 = 1.057 qt
	cubic centimeter (cm^3 or cc)	1000 cm^3 = 1 dm^3	1 cm^3 = 0.0338 fl oz
	liter (L)	1 L = 1 dm^3	1 L = 1.057 qt
	milliliter (mL)[a]	1000 mL = 1 dm^3	1 mL = 0.0338 fl oz
Mass	gram (g)	1000 g = 1 kg	1 g = 0.035 oz
	milligram (mg)	1,000,000 mg = 1 kg	1 mg = 0.015 grain
	kilogram (kg)	Basic unit	1 kg = 2.20 lb
Temperature	degree Celsius (°C)	1°C = 1 K	1°C = 1.80°F
	kelvin (K)	Basic unit	1 K = 1.80°F
Energy	calorie (cal)	1 cal = 4.184 J	1 cal = 0.00397 BTU[b]
	kilocalorie (kcal)	1 kcal = 4184 J	1 kcal = 3.97 BTU
	joule (J)	Basic unit	1 J = 0.000949 BTU
Time	second (s)	Basic unit	Same unit used

[a]NOTE: 1 mL = 1 cm^3. [b]A BTU (British thermal unit) is the amount of heat required to increase the temperature of 1 pound of water 1°F.

CHEMISTRY AROUND US • 1.3

EFFECTS OF TEMPERATURE ON BODY FUNCTION

We humans are classified as warm-blooded animals; our body temperature remains relatively constant even when the temperature of our surroundings increases or decreases. Just how constant is our body temperature? Normal body temperature is considered to be 37.0°C when measured orally. However, a normal individual can have an oral temperature as low as 36.1°C upon awakening in the morning and as high as 37.2°C before bedtime in the late evening.

Besides this regular variation, body temperature also fluctuates in response to extremes in the temperature of the surroundings. In extremely hot environments, the capacity of the body's cooling mechanism can be overtaxed, and the body temperature will increase. Body temperatures more than 3.5°C above normal begin to interfere with bodily functions. Temperatures higher than 41.1°C can result in convulsions and can cause permanent brain damage, especially in children.

Hypothermia develops when the body's internal heat generation is not sufficient to balance heat lost to very cold surroundings. The body's temperature drops, and at 28.5°C, the person appears pale and cold and may have an irregular heartbeat. Unconsciousness usually results if the body temperature drops below 26.7°C. The respiration also slows down and becomes shallow, and the oxygenation of body tissues decreases.

Thus, the body temperatures of warm-blooded humans are not really so constant after all. Also, even though we have built-

in cooling and heating systems, their capacity and ability to maintain a constant normal body temperature are limited.

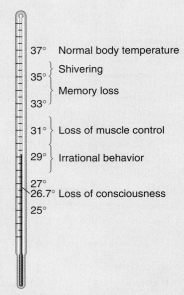

The effects of low body temperature on body function using the Celsius scale.

➤ 1.7 Large and Small Numbers

Numbers are used in all measurements and calculations. Many numbers are readily understood and represented because of common experience with them. A price of 10 dollars, a height of 7 feet, a weight of 165 pounds, and a time of 40 seconds are examples of such numbers. But how do we handle numbers like the diameter of a hydrogen atom (about one hundred-millionth of a centimeter) or the distance light travels in 1 year—a light-year (about 6 trillion miles)? These numbers are so small and large, respectively, that they defy understanding in terms of relationships to familiar distances. Even if we can't totally relate to them, it is important in scientific work to be able to conveniently represent and work with such numbers.

Scientific notation provides a method for conveniently representing any number including those that are very large or very small. In **scientific notation,** numbers are represented as the product of a nonexponential term and an exponential term in the general form $M \times 10^n$. The nonexponential term M is a number between 1 and 10 (but not equal to 10) written with a decimal to the right of the first nonzero digit in the number. This position of the decimal is called the **standard position.** The exponential term is a 10 raised to a whole number exponent n that may be positive or negative. The value of n is the number of places the decimal must be moved from the standard position in M to be at the original position in the number when the number is written normally without using scientific notation. If n is positive, the origi-

Scientific notation
A way of representing numbers consisting of a product between a nonexponential number and 10 raised to a whole-number exponent that may be positive or negative.

Standard position for a decimal
In scientific notation, the position to the right of the first nonzero digit in the nonexponential number.

nal decimal position is to the right of the standard position. If *n* is negative, the original decimal position is to the left of the standard position.

Example 1.8

The following numbers are written using scientific notation. Write them without using scientific notation.

a. 3.72×10^5 **b.** 8.513×10^{-7}

Solution

a. The exponent 5 indicates that the original position of the decimal is located 5 places to the right of the standard position. Zeros are added to accommodate this change:

$$3.72 \times 10^5 = 372,000. = 372,000$$

standard position   original position

b. The exponent -7 indicates that the original position of the decimal is 7 places to the left of the standard position. Again, zeros are added as needed:

$$8.513 \times 10^{-7} = .0000008513$$

original position   standard position

Example 1.9

Write the following numbers using scientific notation:

a. 8725.6 **b.** 0.000729

Solution

a. The standard decimal position is between the 8 and 7: 8.7256. However, the original position of the decimal is 3 places to the right of the standard position. Therefore, the exponent must be $+3$:

$$8725.6 = 8.7256 \times 10^3$$

b. The standard decimal position is between the 7 and 2: 7.29. However, the original position is 4 places to the left of standard. Therefore, the exponent must be -4:

$$0.000729 = 7.29 \times 10^{-4}$$

The zeros to the left of the 7 are dropped because they are not significant figures (see Section 1.8); they only locate the decimal in the nonscientific notation and are not needed in the scientific notation.

Some of the following numbers are written using scientific notation, and some are not. In each case, rewrite the number using the notation in which it is not written.

Learning Check 1.9

a. 5.88×10^2 **c.** 3.915×10^{-4} **e.** 36.77
b. 0.000439 **d.** 9870 **f.** 0.102

Example 1.10

Determine which of the following numbers are written correctly using scientific notation. For those that are not, rewrite them correctly.

a. 001.5×10^{-3} **b.** 28.0×10^2 **c.** 0.35×10^4

Solution

a. Incorrect; the zeros to the left are not needed. The correct answer is 1.5×10^{-3}.

b. Incorrect; the decimal is not in the standard position. Move the decimal 1 position to the left and increase the exponent by 1 to give the correct answer of 2.80×10^3.

c. Incorrect; the decimal is not in the standard position. Move the decimal 1 position to the right and decrease the exponent by 1 to give the correct answer of 3.5×10^3.

Learning Check 1.10

Determine which of the following numbers are written correctly using scientific notation. For those that are not, rewrite them correctly.

a. 62.5×10^4 **b.** 0.0098 **c.** 0.0041×10^{-3} **d.** 7.85×10^2

The multiplication and division of numbers written in scientific notation can be done quite simply by using some characteristics of exponentials. Consider the following multiplication:

$$(a \times 10^y)(b \times 10^z)$$

The multiplication is done in two steps. First, the nonexponential terms a and b are multiplied in the usual way. The exponential terms 10^y and 10^z are multiplied by adding the exponents y and z and using the resulting sum as a new exponent of 10. Thus, we can write

$$(a \times 10^y)(b \times 10^z) = (a \times b)(10^{y+z})$$

Division is done similarly. The nonexponential terms are divided in the usual way, and the exponential terms are divided by subtracting the exponent of the bottom term from that of the top term. The final answer is then written as a product of the resulting nonexponential and exponential terms:

$$\frac{a \times 10^y}{b \times 10^z} = \left(\frac{a}{b}\right)(10^{y-z})$$

Multiplication and division calculations involving scientific notation are easily done using a hand calculator. ■ Table 1.4 gives the steps, the typical calculator procedures (buttons to press), and typical calculator readout or display for the division of 7.2×10^{-3} by 1.2×10^4.

Example 1.11

Do the following operations:

a. $(3.5 \times 10^4)(2.0 \times 10^2)$ **c.** $(4.6 \times 10^{-7})(5.0 \times 10^3)$

b. $\dfrac{3.8 \times 10^5}{1.9 \times 10^2}$ **d.** $\dfrac{1.2 \times 10^3}{3.0 \times 10^{-2}}$

TABLE 1.4	Using a calculator for scientific notation calculations	
Step	Procedure	Calculator display
1. Enter 7.2	Press buttons 7, ., 2	7.2
2. Enter 10^{-3}	Press button that activates exponential mode (EE, Exp, etc.)	7.2 00
	Press 3	7.2 03
	Press change-sign button (±, etc.)	7.2 −03
3. Divide	Press divide button (÷)	7.2 −03
4. Enter 1.2	Press buttons 1, ., 2	1.2
5. Enter 10^4	Press button that activates exponential mode (EE, Exp, etc.)	1.2 00
	Press 4	1.2 04
6. Obtain answer	Press equals button (=)	6. −07

Solution

a. $(3.5 \times 10^4)(2.0 \times 10^2) = (3.5 \times 2.0)(10^4 \times 10^2)$
$$= (7.0)(10^{4+2}) = 7.0 \times 10^6$$

b. $\dfrac{3.8 \times 10^5}{1.9 \times 10^2} = \dfrac{3.8}{1.9} \times \dfrac{10^5}{10^2} = (2.0)(10^{5-2}) = 2.0 \times 10^3$

c. $(4.6 \times 10^{-7})(5.0 \times 10^3) = (4.6 \times 5.0)(10^{-7} \times 10^3)$
$$= (23)(10^{-7+3}) = 23 \times 10^{-4}$$

To get the decimal into standard position, move it 1 place to the left. This changes the exponent from −4 to −3, so the final result is 2.3×10^{-3}. (This number in decimal form, 0.0023, can be written correctly as either 23×10^{-4} or 2.3×10^{-3}, but scientific notation requires that the decimal point be to the right of the first nonzero number.)

d. $\dfrac{1.2 \times 10^3}{3.0 \times 10^{-2}} = \dfrac{1.2}{3.0} \times \dfrac{10^3}{10^{-2}} = (0.40)(10^{3-(-2)}) = 0.40 \times 10^5$

Adjust the decimal to standard position and get 4.0×10^4. If these examples were done using a calculator, the displayed answers would normally be given in correct scientific notation.

Perform the following operations, and express the result in correct scientific notation:

Learning Check 1.11

a. $(2.4 \times 10^3)(1.5 \times 10^4)$ c. $\dfrac{6.3 \times 10^5}{2.1 \times 10^3}$

b. $(3.5 \times 10^2)(2.0 \times 10^{-3})$ d. $\dfrac{4.4 \times 10^{-2}}{8.8 \times 10^{-3}}$

The diameter of a hydrogen atom, mentioned earlier as one hundred-millionth of a centimeter, is written in scientific notation as 1.0×10^{-8} cm. Similarly, 1 light-year of 6 trillion miles is 6.0×10^{12} mi.

➤ 1.8 Significant Figures

Every measurement contains an uncertainty that is characteristic of the device used to make the measurement. These uncertainties are represented by the numbers used to record the measurement. Consider the following small square:

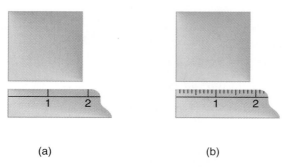

(a)   (b)

In (a), the length of one side of the square is measured with a ruler divided into centimeters. It is easy to see that the length is greater than 1 cm, but not quite 2 cm. The length is recorded by writing the number that is known with certainty to be correct (the 1) and writing an estimate for the uncertain number. The result is 1.9 cm, where the .9 is the estimate. In (b), the ruler is divided into tenths of centimeters. It is easy to see that the length is at least 1.8 cm, but not quite 1.9 cm. Once again the certain numbers (1.8) are written, and an estimate is made for the uncertain part. The result is 1.86 cm. When measurements are recorded this way, the numbers representing the certain measurement plus one number representing the estimate are called **significant figures**. Thus, the first measurement of 1.9 cm contains two significant figures, and the second measurement of 1.86 cm contains three significant figures.

The maximum number of significant figures possible in a measurement is determined by the design of the measuring device and cannot be changed by expressing the measurement in different units. The 1.8-cm length determined earlier can also be represented in terms of meters and millimeters as follows:

$$1.8 \text{ cm} = 0.018 \text{ m} = 18 \text{ mm}$$

In this form, it appears that the length expressed as 0.018 m contains four significant figures, but this is impossible; a measurement made with a device doesn't become more certain simply by changing the unit used to express the number. Thus, the zeros are not significant figures; their only function is to locate the correct position for the decimal. Zeros located to the left of nonzero numbers, such as the two zeros in 0.018 cm, are never considered to be significant. Thus, 12.5 mg, 0.0125 g, and 0.0000125 kg all represent the same measured mass, and all contain three significant figures.

Zeros located between nonzero numbers or trailing zeros located at the end of numbers will be considered significant. Thus, 2050 μL, 2.050 mL, and 0.002050 L all represent the same volume measurement, and all contain four significant figures.

The rule that specifies counting trailing zeros as significant is generally followed by scientists, but some quantities are expressed with trailing zeros that are not significant. For example, suppose you read in a newspaper that the population of a city is 1,250,000 people. Should the four trailing zeros be considered significant? If they are, it means that the population is known with certainty to the nearest 10 people and that the measurement has an uncertainty of only plus or minus 1 person. A more reasonable conclusion is that the census is correct to the nearest 1000 people. This could be represented as 1250 thousand or, using scientific notation, 1.250×10^6. In either of these representations, only four significant figures are used. In scientific notation, the correct number of significant figures is used in the nonexponential term, and the location of the decimal is determined by the exponent. In this book, large numbers will always be represented by exponential notation instead of using nonsignificant trailing zeros. However, you are likely to encounter nonsignificant trailing zeros in other reading materials. The rules for determining the significance of zeros are summarized as follows:

1. Zeros not preceded by nonzero numbers are not significant figures. These zeros are sometimes called *leading zeros*.
2. Zeros located between nonzero numbers are significant figures. These zeros are sometimes called *buried* or *confined zeros*.
3. Zeros located at the end of a number are significant figures. These zeros are sometimes called *trailing zeros*.

Example 1.12

Determine the number of significant figures in each of the following measurements, and use scientific notation to express each measurement using the correct number of significant figures:

a. 24.6°C   b. 0.036 g   c. 15.0 mL   d. 0.0020 m

Solution

a. All the numbers are significant: three significant figures, 2.46×10^1 °C.
b. The leading zeros are not significant: two significant figures, 3.6×10^{-2} g.
c. The trailing zero is significant: three significant figures, 1.50×10^1 mL.
d. The leading zeros are not significant, but the trailing zero is: two significant figures, 2.0×10^{-3} m.

Determine the number of significant figures in each of the following measurements:

Learning Check 1.12

a. 250 mg   c. 0.0108 kg   e. 0.001 mm
b. 18.05 mL   d. 37°C   f. 101.0 K

Use scientific notation to express each of the following measurements using the correct number of significant figures:

Learning Check 1.13

a. 101 m   c. 0.00230 kg   e. 21.65 mL
b. 1200 g   d. 1296°C   f. 0.015 km

Most measurements that are made do not stand as final answers. Instead, they are usually used to make calculations involving multiplication, division, addition, or subtraction. The answer obtained from such a calculation cannot have more certainty than the least certain measurement used in the calculation. It should be written to reflect an uncertainty equal to that of the most uncertain measurement. This is accomplished by the following rules:

1. The answer obtained by multiplication or division must contain the same number of significant figures as the quantity with the fewest significant figures used in the calculation.
2. The answer obtained by addition or subtraction must contain the same number of places to the right of the decimal as the quantity in the calculation with the fewest number of places to the right of the decimal.

To follow these rules, it is often necessary to reduce the number of significant figures by rounding answers. The following are rules for rounding:

1. If the first of the nonsignificant figures to be dropped from an answer is 5 or greater, all the nonsignificant figures are dropped, and the last significant figure is increased by 1.
2. If the first of the nonsignificant figures to be dropped from an answer is less than 5, all nonsignificant figures are dropped, and the last significant figure is left unchanged.

Remember, if you use a calculator, it will often express answers with too few or too many figures (see ■ Figure 1.15). It will be up to you to determine the proper number of significant figures to use and to round the calculator answer correctly.

Example 1.13

Do the following calculations, and round the answers to the correct number of significant figures:

a. $(4.95)(12.10)$ **b.** $\dfrac{3.0954}{0.0085}$ **c.** $\dfrac{(9.15)(0.9100)}{3.117}$

Solution

All calculations are done with a hand calculator, and the calculator answer is written first. Appropriate rounding is done to get the final answer.

a. Calculator answer: 59.895
The number 4.95 has three significant figures, and 12.10 has four. Thus, the answer must have three significant figures:

$$59.895$$

significant figures — nonsignificant figures

The first of the nonsignificant figures to be dropped is 9, so after both are dropped, the last significant figure is increased by 1. The final answer containing three significant figures is 59.9.

b. Calculator answer: 364.16471
The number 3.0954 has five significant figures, and 0.0085 has two. Thus, the answer must have only two:

$$364.16471$$

significant figures — nonsignificant figures

The first of the nonsignificant figures to be dropped is 4, so the last significant figure remains unchanged after the nonsignificant figures are dropped. The correct answer then is 360. If this is written 360, it will contain three significant figures. However, the answer can be written with the proper number of significant figures by using scientific notation. The final correct answer containing two significant figures is 3.6×10^2.

➤ **FIGURE 1.15** Calculators usually display the sum of 4.362 and 2.638 as 7 (too few digits), and the product of 0.67 and 10.14 as 6.7938 (too many digits).

c. Calculator answer: 2.6713186

The number 9.15 has three significant figures, 0.9100 has four, and 3.117 has four. Thus, the answer must have only three:

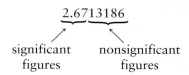

2.6713186

significant   nonsignificant
figures   figures

Appropriate rounding gives 2.67 as the correct answer.

Learning Check 1.14

Do the following calculations, and round the answers to the correct number of significant figures:

a. (0.0019)(21.39)　　**b.** $\dfrac{8.321}{4.1}$　　**c.** $\dfrac{(0.0911)(3.22)}{1.379}$

Example 1.14

Do the following additions and subtractions, and write the answers with the correct number of significant figures:

a. 1.9 + 18.65　　**c.** 1500 + 10.9 + 0.005
b. 15.00 − 8.1　　**d.** 5.1196 − 5.02

Solution

In each case, the numbers are arranged vertically with the decimals in a vertical line. The answer is then rounded so it contains the same number of places to the right of the decimal as the smallest number of places in the quantities added or subtracted.

a.　1.9　　The answer must be expressed with one place to the right of
　　18.65　　the decimal to match the one place in 1.9.
　　20.55

Correct answer: 20.6 (Why was the final 5 increased to 6?)

b.　15.00　　The answer must be expressed with one place to the right of
　−8.1　　the decimal to match the 8.1.
　　6.90

Correct answer: 6.9 (What rounding rule was followed?)

c. 1500　　The answer must be expressed with no places to the right of
　10.9　　the decimal to match the 1500.
　　0.005
　1510.905

Correct answer: 1511 (Why was the final 0 of the answer increased to 1?)

d.　5.1196　　The answer must be expressed with two places to the right
　−5.02　　of the decimal to match the 5.02.
　　0.0996

Correct answer: 0.10 (What rounding rule was followed?)

Notice that the answer to part a has more significant figures than the least significant number used in the calculation. The answer to part d, on the other hand, has fewer significant figures than either number used in the calculation. This happened because the rule for dealing with addition and subtraction focuses on the number of figures located to the right of the decimal and is not concerned with the figures to the left of the decimal. Thus, the number of significant figures in the answer is sometimes increased as a result of addition and decreased as a result of subtraction. You should be aware of this and not be confused when it happens.

| Learning Check 1.15 | Do the following additions and subtractions, and round the answers to the correct number of significant figures: |

a. $8.01 + 3.2$ **c.** $4.33 - 3.12$
b. $3000 + 20.3 + 0.009$ **d.** $6.023 - 2.42$

Exact numbers
Numbers that have no uncertainty; numbers from defined relationships, counting numbers, and numbers that are part of reduced simple fractions.

Some numbers used in calculations are **exact numbers** that have no uncertainty associated with them and are considered to contain an unlimited number of significant trailing zeros. Such numbers are not used when the appropriate number of significant figures is determined for calculated answers. In other words, exact numbers do not limit the number of significant figures in calculated answers. One kind of exact number is a number used as part of a defined relationship between quantities. For example, 1 m contains exactly 100 cm:

$$1 \text{ m} = 100 \text{ cm}$$

Thus, the numbers 1 and 100 are exact. A second kind of exact number is a counting number obtained by counting individual objects. A dozen eggs contains exactly 12 eggs, not 11.8 and not 12.3. The 12 is an exact counting number. A third kind of exact number is one that is part of a reduced simple fraction such as $\frac{1}{4}$, $\frac{2}{3}$, or the $\frac{5}{9}$ used in Equation 1.1 to convert Fahrenheit temperature readings into Celsius readings.

➤ 1.9 Using Units in Calculations

Some beginning chemistry students are concerned about not being able to solve numerical chemistry problems. They may say, "I can work the problems, I just can't set them up." What they are really saying is "I can do the arithmetic once the numbers are properly arranged, but I can't do the arranging."

This section presents a method for arranging numbers that will work for most of the numerical problems you will encounter in this course. This method has a number of names, including the factor-unit method, the factor-label method, and dimensional analysis. We will call it the factor-unit method. It is a systematic approach to solving numerical problems and consists of the following steps:

Step 1. Write down the known or given quantity. Include both the numerical value and units of the quantity.

Step 2. Leave some working space and set the known quantity equal to the unit of the unknown quantity.

Step 3. Multiply the known quantity by one or more factors, such that the units of the factor cancel the units of the known quantity and generate the units of the unknown quantity. These **factors** are fractions derived from fixed relationships between quantities. These relationships can be definitions or experimentally measured quantities. For example, the defined relationship 1 m = 100 cm provides the following two factors:

$$\frac{1 \text{ m}}{100 \text{ cm}} \quad \text{and} \quad \frac{100 \text{ cm}}{1 \text{ m}}$$

Factors used in the factor-unit method
Fractions obtained from fixed relationships between quantities.

Step 4. After you get the desired units, do the necessary arithmetic to produce the final answer.

Example 1.15

Use the factor-unit method and fixed relationships from Table 1.3 to calculate the number of yards in 100 m.

Solution

The known quantity is 100 m, and the unit of the unknown quantity is yard (yd).

Step 1.   100 m

Step 2.   100 m   = yd

Step 3. $100 \text{ m} \times \dfrac{1.094 \text{ yd}}{1 \text{ m}} = \text{yd}$

The factor $\dfrac{1.094 \text{ yd}}{1 \text{ m}}$ came from the fixed relationship 1 m = 1.094 yd found in Table 1.3.

Step 4. $\dfrac{(100)(1.094) \text{ yd}}{1} = 109.4 \text{ yd}$

This answer should be rounded to 109 yd, an answer that contains three significant figures, just as 100 m does. The 1 m in the factor is an exact number used as part of a defined relationship, so it doesn't influence the number of significant figures in the answer.

Example 1.16

A laboratory technician uses a micropipet to measure a 50-μL (50-microliter) sample of blood serum for analysis. Express the sample volume in liters (L).

Solution

The known quantity is 50 μL, and the unit of the unknown quantity is liters.

Step 1.   50 μL

Step 2.   50 μL   = L

Step 3. $50 \text{ μL} \times \dfrac{1 \times 10^{-6} \text{ L}}{1 \text{ μL}} = \text{L}$

The factor $\dfrac{1 \times 10^{-6} \text{ L}}{1 \text{ μL}}$ came from the fixed relationship 1 μL = 1×10^{-6} L described in Table 1.2.

Step 4. $\dfrac{(50)(1 \times 10^{-6} \text{ L})}{1} = 5.0 \times 10^{-5} \text{ L}$

The answer is expressed using the same number of significant figures as the 50 μL because 1 and 1×10^{-6} are exact numbers by definition.

Creatinine is a substance found in the blood. An analysis of a blood serum sample detected 1.1 mg of creatinine. Express this amount in grams by using the factor-unit method.

Learning Check 1.16

Example 1.17

Chemistry·Now™

Go to GOB Now and click to review the factor-unit method by viewing a tutorial showing multistep problem solving.

One of the fastest-moving nerve impulses in the body travels at a speed of 400 feet per second (ft/s). What is the speed in miles per hour?

Solution

The known quantity is 400 ft/s, and the unit of the unknown quantity is miles per hour (mi/h).

Step 1. $\dfrac{400 \text{ ft}}{\text{s}}$

Step 2. $\dfrac{400 \text{ ft}}{\text{s}} = \dfrac{\text{mi}}{\text{h}}$

Step 3. $\left(\dfrac{400 \text{ ft}}{\text{s}}\right)\left(\dfrac{1 \text{ mi}}{5280 \text{ ft}}\right)\left(\dfrac{60 \text{ s}}{1 \text{ min}}\right)\left(\dfrac{60 \text{ min}}{1 \text{ h}}\right) = \dfrac{\text{mi}}{\text{h}}$

The factors came from the following well-known fixed relationships: 1 mi = 5280 ft, 1 min = 60 s, 1 h = 60 min. All numbers in these factors are exact numbers based on definitions.

Step 4. $\dfrac{(400)(1)(60)(60) \text{ mi}}{(5280)(1)(1) \text{ h}} = 272.7 \dfrac{\text{mi}}{\text{h}}$

Rounding to three significant figures, the same as in 400 ft/s, gives 273 mi/h.

Learning Check 1.17

A world-class sprinter can run 100 m in 10.0 s. This corresponds to a speed of 10.0 m/s. Convert this speed to miles per hour. Use information from Table 1.3.

► 1.10 Calculating Percentages

The word *percent* literally means per one hundred. It is the number of specific items in a group of 100 such items. Since items are seldom found in groups of exactly 100, we usually have to calculate the number of specific items that would be in the group if it did contain exactly 100 items. This number is the percentage, and the calculation follows:

$$\text{percent} = \frac{\text{number of specific items}}{\text{total items in the group}} \times 100 \tag{1.5}$$

$$\% = \frac{\text{part}}{\text{total}} \times 100 \tag{1.6}$$

In Equation 1.6, the word *part* is used to represent the number of specific items included in the total.

Example 1.18

A college has 4517 female and 3227 male students enrolled. What percentage of the student body is female?

Solution

The total student body consists of 7744 people, of which 4517 are female.

$$\% \text{ female} = \frac{\text{number of females}}{\text{total number of students}} \times 100$$

$$\% \text{ female} = \frac{4517}{7744} \times 100 = 58.33$$

Rearrangement of Equation 1.6 gives another useful percent relationship. Because % = part/total × 100,

$$\text{part} = \frac{(\%)(\text{total})}{100} \tag{1.7}$$

According to Equation 1.7, the number of specific items corresponding to a percentage can be calculated by multiplying the percentage and total, then dividing the product by 100.

Example 1.19

The human body is approximately 70% water by mass. What is the mass of water in a 170-pound (lb) person?

Solution

The part in this problem is the mass of water in a person who weighs a total of 170 lb. Substitution into Equation 1.7 gives

$$\text{mass of water} = \frac{(70)(170 \text{ lb})}{100} = 119 \text{ lb}$$

We should round our answer to only two significant figures to match the two in the 70%. Do this by using scientific notation; the answer is 1.2×10^2 lb.

➤ STUDY SKILLS 1.1

HELP WITH CALCULATIONS

Many students feel uneasy about working chemistry problems that involve the use of mathematics. The uneasiness is often increased if the problem to be solved is a story problem. One tip that will help you solve such problems in this textbook is to remember that almost all of these problems are one of two types: those for which a specific formula applies and those where the factor-unit method is used. When you do homework or take quizzes or examinations and encounter a math-type problem, your first task should be to decide which type of problem it is, formula or factor-unit.

In this chapter, formula-type problems were those that dealt with percentage calculations (see Examples 1.18 and 1.19) and the conversion from one temperature scale to another (see Example 1.7). If you decide a problem is a formula type, Step 1 in solving it is to write down the formula that applies. This important step makes it easier to do the next step because a formula is like a road map that tells you how to proceed from one point to another. Step 2 is to identify the given quantity and put its number and units into the formula. This will leave only the answer missing in most formula-type problems. You can then obtain the answer by doing appropriate calculations.

In this textbook, the factor-unit method discussed in Section 1.9 is used for most problems that require mathematical calculations. This method simplifies problem solving and should be mastered so it can be used where it applies. The beauty of this method is that it mimics your natural, everyday way of solving problems. This real-life method usually involves identifying where you are, where you want to go, and how to get there. The factor-unit method follows the same pattern: Step 1, identify the given number and its units; Step 2, write down the unit of the desired answer; Step 3, put in factors that will convert the units of the given quantity into the units of the desired answer.

Thus, we see that working story problems is not as difficult a task as it might first appear. A key is to see through all the words and find what is given (number and units). Then look for what is wanted by focusing on key words or phrases like "how much," "what is," and "calculate." Finally, use one of the two methods, formula or factor-unit, to solve the problem.

a. A student is saving money to buy a computer that will cost $1200. The student has saved $988. What percentage of the purchase price has been saved?

b. In a chemistry class of 83 students, 90.4% voted not to have the final exam. How many students wanted to take the exam?

➤ 1.11 Density

A discussion of the density of matter will provide a final review of some of the important topics of this chapter. Density is a physical property of matter, so it can be measured without changing the composition of the sample of matter under investigation. **Density** is the ratio of the mass of a sample of matter to the volume of the same sample.

Density
The number given when the mass of a sample of a substance is divided by the volume of the same sample.

$$\text{density} = \frac{\text{mass}}{\text{volume}} \tag{1.8}$$

$$d = \frac{m}{V} \tag{1.9}$$

We see from these equations that once a numerical value has been obtained for the density, two factors are available that relate mass and volume, and these may be used to solve problems.

Example 1.20

The density of iron metal has been determined to be 7.2 g/cm^3.

a. Use the density value to calculate the mass of an iron sample that has a volume of 35.0 cm^3.
b. Use the density value to calculate the volume occupied by 138 g of iron.

Solution

The value of the density tells us that one cubic centimeter (cm^3) of iron has a mass of 7.2 g. This may be written as 7.2 g = 1.0 cm^3, using two significant figures for the volume. This relationship gives two factors that can be used to solve our problems:

$$\frac{7.2 \text{ g}}{1.0 \text{ cm}^3} \quad \text{and} \quad \frac{1.0 \text{ cm}^3}{7.2 \text{ g}}$$

a. The sample volume is 35.0 cm^3, and we wish to use a factor to convert this to grams. The first factor given above will work.

$$35.0 \text{ cm}^3 \times \frac{7.2 \text{ g}}{1.0 \text{ cm}^3} = 252 \text{ g (calculator answer)}$$

$$= 2.5 \times 10^2 \text{ g (properly rounded answer)}$$

b. The sample mass is 138 g, and we wish to convert this to cubic centimeters (cm^3). The second factor given above will work.

$$138 \text{ g} \times \frac{1.0 \text{ cm}^3}{7.2 \text{ g}} = 19.17 \text{ cm}^3 \text{ (calculator answer)}$$

$$= 19 \text{ cm}^3 \text{ (properly rounded answer)}$$

CHEMISTRY AND YOUR HEALTH • 1.1

HOW HEALTHY ARE YOU?

It is almost impossible to read a newspaper or news magazine, watch a television news program, or listen to a radio news show without encountering a story about a health issue. In some ways, such stories provide a valuable service. For example, most people are aware that a significant percentage of the population in the United States is overweight, a condition with many health implications. In other ways, such stories sometimes create unnecessary anxiety and fear in the general population. For instance, during the middle 1990s a great deal of attention was focused by the media on possible adverse health effects associated with the magnetic fields generated by high-voltage electric power lines. The results of numerous scientific studies generally disproved such effects, but a great deal of concern was generated among the general population for several years. When such news coverage generates anxiety and fear, it is often the result of the public not understanding the science involved.

In this book you will find numerous box features with the title *Chemistry and Your Health*. Included in these boxes will be brief discussions of health issues that are related to the chemistry topics you will study and, we hope, understand. The human body is a very complex machine that requires numerous parts and systems to function properly if we are to experience normal good health. The normal function of most of these parts and systems is related in some way to the chemical processes you will study in this book.

According to some health experts, the values and behavior of five key body characteristics provide a good assessment of a person's overall health. These characteristics are blood pressure, blood sugar and insulin levels, blood cholesterol and triglyceride levels, overall physical fitness, and body fat. Information about these and numerous other health-related topics will be found in this book, some in the form of special feature boxes; watch for them.

A healthful diet is one of the keys to maintaining good health.

Aluminum metal has a density of 2.7 g/cm^3.

Learning Check 1.19

a. Calculate the mass of an aluminum sample with a volume of 60.0 cm^3.
b. Calculate the volume of an aluminum sample that has a mass of 98.5 g.

For some substances density is rather easily determined experimentally by direct measurement. The mass of a sample is obtained by weighing the sample. The sample volume can be calculated if the sample is a regular solid such as a cube. If the sample is a liquid or an irregular solid, the volume can be measured by using volumetric apparatus such as those shown in ■ Figure 1.16. Densities of solids are often given in units of g/cm^3, and those of liquids in units of g/mL because of the different ways the volumes are determined. However, according to Table 1.3, 1 cm^3 = 1 mL, so the numerical value is the same regardless of which of the two volume units is used.

Example 1.21

a. A hypodermic syringe was used to deliver 5.0 cc (cm^3) of alcohol into an empty container that had a mass of 25.12 g when empty. The container with the alcohol sample weighed 29.08 g. Calculate the density of the alcohol.

> **FIGURE 1.16** Glassware for measuring volumes of liquids. Left to right: graduated cylinders, buret, volumetric pipet, graduated pipet, and syringe.

b. A cube of copper metal measures 2.00 cm on each edge and weighs 71.36 g. What is the density of the copper sample?

c. According to its owner, a chain necklace is made of pure gold. In order to check this, the chain was weighed and found to have a mass of 19.21 g. The chain was then put into a graduated cylinder that contained 20.8 mL of water. After the chain was put into the cylinder, the water level rose to 21.9 mL. The density of pure gold was looked up in a handbook and found to be 19.2 g/cm³. Is the chain made of pure gold?

Solution

a. According to Table 1.3, 1 cc (or cm³) = 1 mL, so the volume is 5.0 mL. The sample mass is equal to the difference between the mass of the container with the sample inside and the mass of the empty container:

$$m = 29.08 \text{ g} - 25.12 \text{ g} = 3.96 \text{ g}$$

The density of the sample is equal to the sample mass divided by the sample volume:

$$d = \frac{m}{V} = \frac{3.96 \text{ g}}{5.0 \text{ mL}} = 0.79 \text{ g/mL (rounded value)}$$

OVER THE COUNTER • 1.1

NONPRESCRIPTION MEDICINES

The important role of pharmacy in the modern practice of medicine is well known. In a simplified but familiar scenario, a sick patient consults a physician, who diagnoses the ailment and writes a prescription for medicine. The patient takes the prescription to a pharmacist, who prepares and packages the medication, which the patient takes according to the directions on the container. However, it is estimated that nearly 40% of the common, everyday health problems of people in the United States are treated without the aid of either a physician or a pharmacist. This is possible because of the availability of an estimated 100 thousand medicinal products that may be purchased by consumers without a prescription. These nonprescription medications are often called over-the-counter (OTC) drugs or medicines. They are available not only in pharmacies, but also in such places as supermarkets and convenience stores.

The range of health problems that can be treated with OTC medicines has grown significantly in recent years as a result of action by the FDA that transferred more than 600 effective prescription-only medicines to nonprescription (OTC) status. However, the widespread availability and use of these products raised significant questions about how best to provide consumers with the adequate directions for proper use that are required by law. In an attempt to improve upon the nonstandardized labeling practices used for OTC medicines, the FDA in 1997 proposed new OTC medicine labeling rules:

1. All wording must be in "plain English."
2. A large, easy-to-read type must be used.
3. Labels must follow a consistent design style.
4. Information must be given in a standardized order.
5. Standardized headings and subheadings must be used.

As is often true, progress of this type does not come without a price tag. The FDA estimated it would cost about $14 million to comply with the new labeling rules. However, the Nonprescription Drug Manufacturers Association estimated implementation costs at $155 million for compliance within the required two years and more than $400 million if packaging changes were required to accommodate the new labeling format. The actual costs have not yet been determined, but consumers will certainly pay a significant part of these costs through increased prices in OTC products.

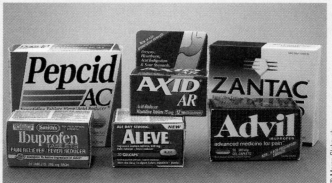

The active ingredients in all these products were originally available only by prescription.

(a) (b)

> **FIGURE 1.17** Measuring the volume of irregular metal pieces.

GOB
Chemistry·∴·Now™

Go to GOB Now and click to learn how to use the volume-displacement method to calculate the density of an irregularly shaped solid.

GOB
Chemistry·∴·Now™

Go to GOB Now and click to engage in a tutorial to examine how to calculate the length of a cube's edge using measurements that involve calculating density.

b. The volume of a cube is equal to the product of the three sides:

$$V = (2.00 \text{ cm})(2.00 \text{ cm})(2.00 \text{ cm}) = 8.00 \text{ cm}^3$$

The density is equal to the mass divided by the volume:

$$d = \frac{m}{V} = \frac{71.36 \text{ g}}{8.00 \text{ cm}^3} = 8.92 \text{ g/cm}^3 \text{ (rounded value)}$$

c. The volume of the chain is equal to the difference in the water level in the cylinder with and without the chain present:

$$V = 21.9 \text{ mL} - 20.8 \text{ mL} = 1.1 \text{ ml} = 1.1 \text{ cm}^3$$

The density is equal to the mass divided by the volume:

$$d = \frac{m}{V} = \frac{19.21 \text{ g}}{1.1 \text{ cm}^3} = 17 \text{ g/cm}^3 \text{ (rounded value)}$$

The experimentally determined density is less than the density of pure gold, so the chain is not made of pure gold.

---

a. A pipet was used to put a 10.00-mL sample of a liquid into an empty container. The empty container weighed 51.22 g, and the container with the liquid sample weighed 64.93 g. Calculate the density of the liquid in g/mL.

b. A box of small irregular pieces of metal was found in a storage room. It was known that the metal was either nickel or chromium. A 35.66-g sample of the metal was weighed and put into a graduated cylinder that contained 21.2 mL of water (■ Figure 1.17a). The water level after the metal was added was 25.2 mL (Figure 1.17b). Was the metal nickel (density = 8.9 g/cm³) or chromium (density = 7.2 g/cm³)?

Learning Check 1.20

Most of the chapters in this text will conclude with a section titled the same as this one. Each of these sections will contain information that may prove useful to you in the future as you continue your studies, practice the profession for which you are training, or just live your life. Most of the information is drawn from health or nutrition topics, areas of interest to most of us regardless of our professions.

In this chapter you were introduced to units and their symbols used in metric measurements. Doctors communicate with pharmacists by using a number of symbols when they write prescriptions. The symbols in the table below are the ones most often used for this purpose, so you can now break the prescription code.

Symbol	Meaning	Symbol	Meaning	Symbol	Meaning
mg	milligram	tid	three times daily	tab	tablet
ml	milliliter	qid	four times daily	cap	capsule
gr	grain = 65 mg	ud	as directed	c̄	with meals
dr	one dram = 5 mL	Ṫ	one	ac	before meals
	= 1 teaspoon	ṪṪ	two	pc	after meals
q̄h	every hour	ṪṪṪ	three	hs	at bedtime
q̄d	every day	sig:	directions to patient	prn	as needed
q̄od	every other day	DAW	dispense as written	stat	at once
bid	twice daily				

······· **CONCEPT SUMMARY** ·······

What Is Matter? Matter, the substance of everything, is defined as anything that has mass and occupies space. Mass is a measurement of the amount of matter present in an object. Weight is a measure of the gravitational force pulling on an object.

Properties and Changes. Chemical properties cannot be determined without attempting to change one kind of matter into another. Physical properties can be determined without attempting such composition changes. Any change in matter that is accompanied by a composition change is a chemical change. Physical changes take place without the occurrence of any composition changes.

A Model of Matter. Scientific models are explanations for observed behavior. The results of many observations led scientists to a model for matter in which all matter is composed of tiny particles. In many substances, these particles are called molecules, and they represent the smallest piece of such substances that is capable of a stable existence. Molecules, in turn, are made up of atoms, which represent the limit of chemical subdivision for matter. The terms *diatomic, triatomic, polyatomic, homoatomic,* and *heteroatomic* are commonly used to describe the atomic composition of molecules.

Classifying Matter. It often simplifies things to classify the items being studied. Some useful categories into which matter can be classified are heterogeneous, homogeneous, solution, pure substance, element, and compound. All matter is either heterogeneous or homogeneous. Heterogeneous matter is a mixture in which the properties and appearance are not uniform. Homogeneous matter is either a mixture of two or more pure substances (and the mixture is called a solution), or it is a pure substance. If it is a pure substance, it is either an element (containing atoms of only one kind) or a compound (containing two or more kinds of atoms).

Measurement Units. All measurements are based on standard units that have been agreed on and adopted. The earliest measurements were based on human body dimensions, but the changeable nature of such basic units made the adoption of a worldwide standard desirable.

The Metric System. The metric system of measurement is used by most scientists worldwide and all major nations except the United States. It is a decimal system in which larger and smaller units of a quantity are related by factors of 10. Prefixes are used to designate relationships between the basic unit and larger or smaller units of a quantity.

Large and Small Numbers. Because of difficulties in working with very large or very small numbers in calculations, a system of scientific notation has been devised to represent such numbers. In scientific notation, numbers are represented as products of a nonexponential number and 10 raised to some power. The nonexponential number is always written with the decimal in the standard position (to the right of the first nonzero digit in the number). Numbers written in scientific notation can be manipulated in calculations by following a few rules.

Significant Figures. In measured quantities, the significant figures are the numbers representing the part of the measurement that is certain, plus one number representing an estimate. The maximum number of significant figures possible in a measurement is determined by the design of the measuring device. The results of calculations made using numbers from measurements can be expressed with the proper number of significant figures by following simple rules.

Using Units in Calculations. The factor-unit method for doing calculations is based on a specific set of steps. One crucial step involves the use of factors that are obtained from fixed numerical relationships between quantities. The units of the factor must always cancel the units of the known quantity and generate the units of the unknown or desired quantity.

Calculating Percentages. The word *percent* means per one hundred, and a percentage is literally the number of specific items contained in a group of 100 items. Few items always occur in groups of exactly 100, so a calculation can be done that determines how many specific items would be in a group if the group actually did contain exactly 100 items.

Density. The density of a substance is the ratio of the mass of a sample to the volume of the same sample. Measured values of density provide two factors that can be used with the factor-unit method to calculate the mass of a substance if the volume is known, or the volume if the mass is known.

LEARNING OBJECTIVES ASSESSMENT

You can get an approximate but quick idea of how well you have met the learning objectives given at the beginning of this chapter by working the selected end-of-chapter exercises given below. The answer to each exercise is given in Appendix B of the book.

Objective 1 (Section 1.1): Exercise 1.2

Objective 2 (Section 1.2): Exercises 1.8 a & b and 1.10 b & c

Objective 3 (Section 1.3): Exercise 1.12

Objective 4 (Section 1.4): Exercises 1.18, 1.22, and 1.24

Objective 5 (Section 1.6): Exercise 1.30

Objective 6 (Section 1.6): Exercise 1.40

Objective 7 (Section 1.7): Exercise 1.48

Objective 8 (Section 1.7): Exercise 1.60

Objective 9 (Section 1.8): Exercises 1.64 and 1.66

Objective 10 (Section 1.9): Exercise 1.82

Objective 11 (Section 1.10): Exercise 1.92

Objective 12 (Section 1.11): Exercise 1.98

KEY TERMS AND CONCEPTS

Atom (1.3)
Basic unit of measurement (1.6)
Chemical changes (1.2)
Chemical properties (1.2)
Compound (1.4)
Density (1.11)
Derived unit of measurement (1.6)
Diatomic molecules (1.3)
Element (1.4)
Exact numbers (1.8)

Factors used in the factor-unit method (1.9)
Heteroatomic molecules (1.3)
Heterogeneous matter (1.4)
Homoatomic molecules (1.3)
Homogeneous matter (1.4)
Mass (1.1)
Matter (1.1)
Mixture (1.4)
Molecule (1.3)
Physical changes (1.2)

Physical properties (1.2)
Polyatomic molecules (1.3)
Pure substance (1.4)
Scientific models (1.3)
Scientific notation (1.7)
Significant figures (1.8)
Solutions (1.4)
Standard position for a decimal (1.7)
Triatomic molecules (1.3)
Weight (1.1)

KEY EQUATIONS

1. Conversion of temperature readings from one scale to another (Section 1.6)

$$°C = \frac{5}{9}(°F - 32)$$

Equation 1.1

$$°F = \frac{9}{5}(°C) + 32$$

Equation 1.2

$$°C = K - 273 \qquad \text{Equation 1.3}$$

$$K = °C + 273 \qquad \text{Equation 1.4}$$

2. Calculation of percentage (Section 1.10)

$$\text{percent} = \frac{\text{number of specific items}}{\text{total items in the group}} \times 100 \qquad \text{Equation 1.5}$$

$$\% = \frac{\text{part}}{\text{total}} \times 100 \qquad \text{Equation 1.6}$$

3. Calculation of number of items representing a specific percentage of a total (Section 1.10)

$$\text{part} = \frac{(\%)(\text{total})}{100} \qquad \text{Equation 1.7}$$

4. Calculation of density from mass and volume data (Section 1.11)

$$d = \frac{m}{V} \qquad \text{Equation 1.9}$$

EXERCISES

LEGEND: 1 = straightforward, **2** = intermediate, **3** = challenging. All even-numbered exercises are answered in Appendix B.

WHAT IS MATTER? (SECTION 1.1)

1.1 A heavy steel ball is suspended by a thin wire. The ball is hit from the side with a hammer but hardly moves. Describe what you think would happen if this identical experiment were carried out on the moon.

1.2 Explain how the following are related to each other: matter, mass, and weight.

1.3 Tell how you would try to prove to a doubter that air is matter.

1.4 Which of the following do you think is likely to change the most when done on Earth and then on the moon? Carefully explain your reasoning.
   a. The distance you can throw a bowling ball through the air
   b. The distance you can roll a bowling ball on a flat, smooth surface

1.5 The attractive force of gravity for objects near Earth's surface increases as you move toward Earth's center. Suppose you are transported from a deep mine to the top of a tall mountain.
   a. How would your mass be changed by the move?
   b. How would your weight be changed by the move?

1.6 Earth's rotation causes it to bulge at the equator. How would the weights of people of equal mass differ when one was determined at the equator and one at the North Pole? (See Exercise 1.5.)

PROPERTIES AND CHANGES (SECTION 1.2)

1.7 Classify each of the following as a physical or chemical change, and give at least one observation, fact, or reason to support your answer.
   a. A plum ripens.
   b. Water boils.
   c. A glass window breaks.
   d. Food is digested.

1.8 Classify each of the following as a physical or chemical change, and give at least one observation, fact, or reason to support your answer.
   a. A stick is broken into two pieces.
   b. A candle burns.
   c. Rock salt is crushed by a hammer.
   d. Tree leaves change color in autumn.

1.9 Classify each of the following properties as physical or chemical. Explain your reasoning in each case.
   a. Iron melts at 1535°C.
   b. Alcohol is very flammable.
   c. The metal used in artificial hip-joint implants is not corroded by body fluids.
   d. A 1-in. cube of aluminum weighs less than a 1-in. cube of lead.
   e. An antacid tablet neutralizes stomach acid.

1.10 Classify each of the following properties as physical or chemical. Explain your reasoning in each case.
   a. Mercury metal is a liquid at room temperature.
   b. Sodium metal reacts vigorously with water.
   c. Water freezes at 0°C.
   d. Gold does not rust.
   e. Chlorophyll molecules are green in color.

A MODEL OF MATTER (SECTION 1.3)

1.11 A sample of methane gas is compressed to a small volume and then allowed to expand back to a larger volume. Have the methane molecules been changed by the process? Explain your answer.

1.12 Succinic acid, a white solid that melts at 182°C, is heated gently, and a gas is given off. After the gas evolution stops, a white solid remains that melts at a temperature different from 182°C.
   a. Have the succinic acid molecules been changed by the process? Explain your answer.

b. Is the white solid that remains after heating still succinic acid? Explain your answer.

c. In terms of the number of atoms contained, how do you think the size of succinic acid molecules compares with the size of the molecules of white solid produced by this process? Explain your answer.

d. Classify molecules of succinic acid by using the term *homoatomic* or *heteroatomic*. Explain your reasoning.

1.13 A sample of solid elemental phosphorus that is deep red in color is burned. While the phosphorus is burning, a white smoke is produced that is actually a finely divided solid that is collected.

a. Have the molecules of phosphorus been changed by the process of burning? Explain your answer.

b. Is the collected white solid a different substance from the phosphorus? Explain your answer.

c. In terms of the number of atoms contained, how do you think the size of the molecules of the white solid compares with the size of the molecules of phosphorus? Explain your answer.

d. Classify molecules of the collected white solid using the term *homoatomic* or *heteroatomic*. Explain your reasoning.

1.14 Oxygen gas and solid carbon are both made up of homoatomic molecules. The two react to form a single substance, carbon dioxide. Use the term *homoatomic* or *heteroatomic* to classify molecules of carbon dioxide. Explain your reasoning.

1.15 Under appropriate conditions, hydrogen peroxide can be changed to water and oxygen gas. Use the term *homoatomic* or *heteroatomic* to classify molecules of hydrogen peroxide. Explain your reasoning.

1.16 Water can be decomposed to hydrogen gas and oxygen gas by passing electricity through it. Use the term *homoatomic* or *heteroatomic* to classify molecules of water. Explain your reasoning.

1.17 Methane gas, a component of natural gas, is burned in pure oxygen. The only products of the process are water and carbon dioxide. Use the term *homoatomic* or *heteroatomic* to classify molecules of methane. Explain your reasoning.

CLASSIFYING MATTER (SECTION 1.4)

1.18 Classify each pure substance represented below by a capital letter as an element or a compound. Indicate when such a classification cannot be made, and explain why.

a. Substance A is composed of heteroatomic molecules.

b. Substance D is composed of homoatomic molecules.

c. Substance E is changed into substances G and J when it is heated.

1.19 Classify each pure substance represented below by a capital letter as an element or a compound. Indicate when such a classification cannot be made, and explain why.

a. Two elements when mixed combine to form only substance L.

b. An element and a compound when mixed form substances M and Q.

c. Substance X is not changed by heating it.

1.20 Consider the following experiments, and answer the questions pertaining to classification:

a. A pure substance R is heated, cooled, put under pressure, and exposed to light but does not change into anything else. What can be said about classifying R as an element or a compound? Explain your reasoning.

b. Upon heating, solid pure substance T gives off a gas and leaves another solid behind. What can be said about classifying T as an element or compound? Explain your reasoning.

c. What can be said about classifying the solid left in part b as an element or compound? Explain your reasoning.

1.21 Early scientists incorrectly classified calcium oxide (lime) as an element for a number of years. Discuss one or more reasons why you think they might have done this.

1.22 Classify each of the following as homogeneous or heterogeneous:

a. A gold chain

b. Liquid eyedrops

c. Chunky peanut butter

d. A slice of watermelon

e. Cooking oil

f. Italian salad dressing

g. Window glass

1.23 Classify each of the following as homogeneous or heterogeneous:

a. Muddy flood water

b. Gelatin dessert

c. Normal urine

d. Smog-filled air

e. An apple

f. Mouthwash

g. Petroleum jelly

1.24 Classify as pure substance or solution each of the materials of Exercise 1.22 that you classified as homogeneous.

1.25 Classify as pure substance or solution each of the materials of Exercise 1.23 that you classified as homogeneous.

MEASUREMENT UNITS (SECTION 1.5)

1.26 Briefly discuss why a system of measurement units is an important part of our modern society.

1.27 In the distant past, 1 in. was defined as the length resulting from laying a specific number of grain kernels such as corn in a row. Discuss the disadvantages of such a system.

1.28 An old British unit used to express weight is a stone. It is equal to 14 lb. What sort of weighings might be expressed in stones? Suggest some standard that might have been used to establish the unit.

THE METRIC SYSTEM (SECTION 1.6)

1.29 Which of the following quantities are expressed in metric units?

a. The amount of aspirin in a tablet: 5 grains

b. The distance between two cities: 55 km

c. The internal displacement of an auto engine: 5 L

d. The time for a race: 4 min, 5.2 s

e. The area of a field: 3.6 acres

f. The temperature on a hot day: 104°F

1.30 Which of the following are expressed in metric units?

a. Normal body temperature: 37°C

b. The amount of soft drink in a bottle: 2 L

   c. The height of a ceiling in a room: 8.0 ft
   d. The amount of aspirin in a tablet: 81 mg
   e. The volume of a cooking pot: 4 qt
   f. The time for a short race to be won: 10.2 s

1.31 Referring to Table 1.3, suggest an appropriate metric system unit for each nonmetric unit in Exercise 1.29.

1.32 Referring to Table 1.3, suggest an appropriate metric system unit for each nonmetric unit in Exercise 1.30.

1.33 Referring only to Table 1.2, answer the following questions:
   a. A computer has 12 megabytes of memory storage. How many bytes of storage is this?
   b. A 10-km race is 6.2 mi long. How many meters long is it?
   c. A chemical balance can detect masses as small as 0.1 mg. What is this detection limit in grams?
   d. A micrometer is a device used to measure small lengths. If it lives up to its name, what is the smallest metric length that could be measured using a micrometer?

1.34 Referring only to Table 1.2, answer the following questions:
   a. Devices are available that allow liquid volumes as small as one microliter (μL) to be measured. How many microliters would be contained in 1.00 liter?
   b. Electrical power is often measured in kilowatts. How many watts would equal 75 kilowatts?
   c. Ultrasound is sound of such high frequency that it cannot be heard. The frequency is measured in hertz (vibrations per second). How many hertz corresponds to 15 megahertz?
   d. A chlorine atom has a diameter of 200 picometers. How many meters is this diameter?

1.35 One foot is approximately equal to 30.5 cm. Express this length in millimeters and meters.

1.36 Cookbooks are going metric. In such books, 1 cup is equal to 240 mL. Express 1 cup in terms of liters and cubic centimeters.

1.37 One pound of butter weighs about 0.454 kg. How many grams is this? How many milligrams?

1.38 The shotput used by female track and field athletes has a mass of 4.0 kg. What would be the weight of such a shotput in pounds?

1.39 Referring to Table 1.3, answer the following questions:
   a. Which is larger, a liter or a quart?
   b. How many milliliters are in a 12.0-fl-oz soft drink?
   c. Which is larger, a BTU or a kilocalorie?

1.40 Referring to Table 1.3, answer the following questions:
   a. Approximately how many inches longer is a meter stick than a yardstick?
   b. A temperature increases by 65°C. How many kelvins would this increase be?
   c. You have a 5-lb bag of sugar. Approximately how many kilograms of sugar do you have?

1.41 Do the following, using appropriate values from Table 1.3:
   a. Calculate the area in square meters of a circular skating rink that has a 12.5-m radius. For a circle, the area (A) is related to the radius (r) by $A = \pi r^2$, where $\pi = 3.14$.
   b. Calculate the floor area and volume of a rectangular room that is 5.0 m long, 2.8 m wide, and 2.1 m high. Express your answers in square meters and cubic meters (meters cubed).
   c. A model sailboat has a triangular sail that is 25 cm high (h) and has a base (b) of 15 cm. Calculate the area (A) of the sail in square centimeters. $A = \dfrac{(b)(h)}{2}$ for a triangle.

1.42 Using appropriate values from Table 1.3, answer the following questions:
   a. One kilogram of water has a volume of 1.0 dm^3. What is the mass of 1.0 cm^3 of water?
   b. One quart is 32 fl oz. How many fluid ounces are contained in a 2.0-L bottle of soft drink?
   c. Approximately how many milligrams of aspirin are contained in a 5-grain tablet?

1.43 The weather report says the temperature is 23°F. What is this temperature on the Celsius scale? On the Kelvin scale?

1.44 Recall from Chemistry Around Us 1.3 that a normal body temperature might be as low as 36.1°C in the morning and as high as 37.2°C at bedtime. What are these temperatures on the Fahrenheit scale?

1.45 One pound of body fat releases approximately 4500 kcal of energy when it is metabolized. How many joules of energy is this? How many BTUs?

LARGE AND SMALL NUMBERS (SECTION 1.7)

1.46 Which of the following numbers are written using scientific notation correctly? For those that are not, explain what is wrong.
   a. 02.7×10^{-3}
   b. 4.1×10^2
   c. 71.9×10^{-6}
   d. 10^3
   e. $.0405 \times 10^{-2}$
   f. 0.119

1.47 Which of the following numbers are written using scientific notation correctly? For those that are not, explain what is wrong.
   a. 3.6×10^{25}
   b. 3.9^{-2}
   c. 295×10^3
   d. 0.05×10^{-3}
   e. 10^{-4}
   f. 13.1×10^6

1.48 Write each of the following numbers using scientific notation:
   a. 14 thousand
   b. 365
   c. 0.00204
   d. 461.8
   e. 0.00100
   f. 9.11 hundred

1.49 Write each of the following numbers using scientific notation:
   a. 1.02 thousand
   b. 0.07102
   c. 3050
   d. 1.51 million
   e. three thousand
   f. 31.05

1.50 The speed of light is about 186 thousand mi/s, or 1100 million km/h. Write both numbers using scientific notation.

1.51 A sheet of paper is 0.0106 cm, or 0.0042 in., thick. Write both numbers using scientific notation.

1.52 A single water molecule has a mass of 2.99×10^{-23} g. Write this number in a decimal form without using scientific notation.

1.53 In 2.0 g of hydrogen gas, there are approximately 6.02×10^{23} hydrogen molecules. Write this number without using scientific notation.

1.54 Do the following multiplications, and express each answer using scientific notation:
  a. $(8.2 \times 10^{-3})(1.1 \times 10^{-2})$
  b. $(2.7 \times 10^2)(5.1 \times 10^4)$
  c. $(3.3 \times 10^{-4})(2.3 \times 10^2)$
  d. $(9.2 \times 10^{-4})(2.1 \times 10^4)$
  e. $(4.3 \times 10^6)(6.1 \times 10^5)$

1.55 Do the following multiplications, and express each answer using scientific notation:
  a. $(5.0 \times 10^{-5})(7.1 \times 10^{-2})$
  b. $(6.3 \times 10^{-9})(3.7 \times 10^7)$
  c. $(3.2 \times 10^{-4})(1.0 \times 10^4)$
  d. $(2.7 \times 10^2)(3.8 \times 10^4)$
  e. $(7.1 \times 10^4)(6.9 \times 10^7)$

1.56 Express each of the following numbers using scientific notation, then carry out the multiplication. Express each answer using scientific notation.
  a. $(144)(0.0876)$
  b. $(751)(106)$
  c. $(0.0422)(0.00119)$
  d. $(128,000)(0.0000316)$

1.57 Express each of the following numbers using scientific notation, then carry out the multiplication. Express each answer using scientific notation.
  a. (one million)(0.00022)
  b. $(5280)(12)$
  c. $(1200)(463)$
  d. $(52.0)(0.0184)$

1.58 Do the following divisions, and express each answer using scientific notation:
  a. $\dfrac{3.1 \times 10^{-3}}{1.2 \times 10^2}$

  b. $\dfrac{7.9 \times 10^4}{3.6 \times 10^2}$

  c. $\dfrac{4.7 \times 10^{-1}}{7.4 \times 10^2}$

  d. $\dfrac{0.00229}{3.16}$

  e. $\dfrac{119}{3.8 \times 10^3}$

1.59 Do the following divisions, and express each answer using scientific notation:
  a. $\dfrac{9.1 \times 10^3}{3.7 \times 10^1}$

  b. $\dfrac{7.7 \times 10^4}{4.2 \times 10^{-6}}$

  c. $\dfrac{4.4 \times 10^{-5}}{2.2 \times 10^6}$

  d. $\dfrac{3.9 \times 10^{-5}}{5.7 \times 10^{-2}}$

  e. $\dfrac{172}{2.15}$

1.60 Do the following calculations, and express each answer using scientific notation:
  a. $\dfrac{(5.3)(0.22)}{(6.1)(1.1)}$

  b. $\dfrac{(3.8 \times 10^{-4})(1.7 \times 10^{-2})}{6.3 \times 10^3}$

  c. $\dfrac{4.8 \times 10^6}{(7.4 \times 10^3)(2.5 \times 10^{-4})}$

  d. $\dfrac{5.6}{(0.022)(109)}$

  e. $\dfrac{(4.6 \times 10^{-3})(2.3 \times 10^2)}{(7.4 \times 10^{-4})(9.4 \times 10^{-5})}$

1.61 Do the following calculations, and express each answer using scientific notation:
  a. $\dfrac{(7.4 \times 10^{-3})(1.3 \times 10^4)}{5.5 \times 10^{-2}}$

  b. $\dfrac{6.4 \times 10^5}{(8.8 \times 10^3)(1.9 \times 10^{-4})}$

  c. $\dfrac{(6.4 \times 10^{-2})(1.1 \times 10^{-8})}{(2.7 \times 10^{-4})(3.4 \times 10^{-4})}$

  d. $\dfrac{(963)(1.03)}{(0.555)(412)}$

  e. $\dfrac{1.15}{(0.12)(0.73)}$

SIGNIFICANT FIGURES (SECTION 1.8)

1.62 Indicate to what decimal position readings should be estimated and recorded (nearest 0.1, .01, etc.) for measurements made with the following devices:
  a. A ruler with smallest scale marking of 0.1 cm
  b. A measuring telescope with smallest scale marking of 0.1 mm
  c. A protractor with smallest scale marking of 1°
  d. A tire pressure gauge with smallest scale marking of 1 lb/in.^{2}

1.63 Indicate to what decimal position readings should be estimated and recorded (nearest 0.1, .01, etc.) for measurements made with the following devices:
  a. A buret with smallest scale marking of 0.1 mL
  b. A graduated cylinder with smallest scale marking of 1 mL
  c. A thermometer with smallest scale marking of 0.1°C
  d. A barometer with smallest scale marking of 1 torr

1.64 Write the following measured quantities as you would record them, using the correct number of significant figures based on the device used to make the measurement:
  a. Exactly 6 mL of water measured with a graduated cylinder that has smallest markings of 0.1 mL

b. A temperature that appears to be exactly 37 degrees using a thermometer with smallest markings of 1°C

c. A time of exactly nine seconds measured with a stopwatch that has smallest markings of 0.1 second

d. Fifteen and one-half degrees measured with a protractor that has 1-degree scale markings

1.65 Write the following measured quantities as you would record them, using the correct number of significant figures based on the device used to make the measurements.

a. A length of two and one-half centimeters measured with a measuring telescope with smallest scale markings of 0.1 mm

b. An initial reading of exactly 0 for a buret with smallest scale markings of 0.1 mL

c. A length of four and one-half centimeters measured with a ruler that has a smallest marking of 0.1 cm

d. An atmospheric pressure of exactly 690 torr measured with a barometer that has smallest markings of 1 torr

1.66 In each of the following, identify the measured numbers and exact numbers. Do the indicated calculation, and write your answer using the correct number of significant figures.

a. A bag of potatoes is found to weigh 5.06 lb. The bag contains 16 potatoes. Calculate the weight of an average potato.

b. The foul-shooting percentages for the five starting players of a women's basketball team are 71.2%, 66.9%, 74.1%, 80.9% and 63.6%. What is the average shooting percentage of the five players?

1.67 In each of the following, identify the measured numbers and exact numbers. Do the indicated calculations, and write your answer using the correct number of significant figures.

a. An individual has a job of counting the number of people who enter a store between 1 P.M. and 2 P.M. each day for 5 days. The counts were 19, 24, 17, 31, and 40. What was the average number of people entering the store per day for the 5-day period?

b. The starting five members of a women's basketball team have the following heights: 6′1″, 5′8″, 5′6″, 5′1″, and 4′11″. What is the average height of the starting five?

1.68 Determine the number of significant figures in each of the following:

a. 0.0400
b. 309
c. 4.006
d. 4.4×10^{-3}
e. 1.002
f. 255.02

1.69 Determine the number of significant figures in each of the following:

a. 132.0
b. 2.00×10^3
c. 0.0004
d. 4796
e. 0.00200
f. 1769.0

1.70 Do the following calculations and use the correct number of significant figures in your answers. Assume all numbers are the results of measurements.

a. (3.71)(1.4)
b. (0.0851)(1.2262)
c. $\dfrac{(0.1432)(2.81)}{(0.7762)}$
d. $(3.3 \times 10^4)(3.09 \times 10^{-3})$
e. $\dfrac{(760)(2.00)}{6.02 \times 10^{20}}$

1.71 Do the following calculations and use the correct number of significant figures in your answers. Assume all numbers are the results of measurements.

a. (4.09)(3.0)
b. $\dfrac{(3.192 \times 10^6)(0.0041)}{105}$
c. $\dfrac{(19.3)(100)}{1000}$
d. $(1.02 \times 10^{-21})(1.1 \times 10^9)^2$
e. $\dfrac{(251)(3.1 \times 10^{-1})}{(24)(3.0)}$

1.72 Do the following calculations and use the correct number of significant figures in your answers. Assume all numbers are the results of measurements.

a. 0.208 + 4.9 + 1.11
b. 228 + 0.999 + 1.02
c. 8.543 − 7.954
d. $(3.2 \times 10^{-2}) + (5.5 \times 10^{-1})$
(HINT: Write in decimal form first, then add.)
e. 336.86 − 309.11
f. 21.66 − 0.02387

1.73 Do the following calculations and use the correct number of significant figures in your answers. Assume all numbers are the results of measurements.

a. 2.1 + 5.07 + 0.119
b. 0.051 + 8.11 + 0.02
c. 4.337 − 3.211
d. $(2.93 \times 10^{-1}) + (6.2 \times 10^{-2})$
(HINT: Write in decimal form first, then add.)
e. 471.19 − 365.09
f. 17.76 − 0.0479

1.74 Do the following calculations and use the correct number of significant figures in your answers. Assume all numbers are the results of measurements. In calculations involving both addition/subtraction and multiplication/division, it is usually better to do additions/subtractions first.

a. $\dfrac{(0.0267 + 0.00119)(4.626)}{28.7794}$
b. $\dfrac{212.6 - 21.88}{86.37}$
c. $\dfrac{27.99 - 18.07}{4.63 - 0.88}$
d. $\dfrac{18.87}{2.46} - \dfrac{18.07}{0.88}$
(HINT: Do divisions first, then subtract.)
e. $\dfrac{(8.46 - 2.09)(0.51 + 0.22)}{(3.74 + 0.07)(0.16 + 0.2)}$
f. $\dfrac{12.06 - 11.84}{0.271}$

1.75 Do the following calculations and use the correct number of significant figures in your answers. Assume all numbers are the results of measurements. In calculations involving both addition/subtraction and multiplication/division, it is usually better to do additions/subtractions first.

a. $\dfrac{132.15 - 32.16}{87.55}$

b. $\dfrac{(0.0844 + 0.1021)(7.174)}{19.1101}$

c. $\dfrac{(2.78 - 0.68)(0.42 + 0.4)}{(1.058 + 0.06)(0.22 + 0.2)}$

d. $\dfrac{27.65 - 21.71}{4.97 - 0.36}$

e. $\dfrac{12.47}{6.97} - \dfrac{203.4}{201.8}$

(HINT: Do divisions first, then subtract.)

f. $\dfrac{19.37 - 18.49}{0.822}$

1.76 The following measurements were obtained for the length and width of a series of rectangles. Each measurement was made using a ruler with a smallest scale marking of 0.1 cm.
Black rectangle: $l = 12.00$ cm, $w = 10.40$ cm
Red rectangle: $l = 20.20$ cm, $w = 2.42$ cm
Green rectangle: $l = 3.18$ cm, $w = 2.55$ cm
Orange rectangle: $l = 13.22$ cm, $w = 0.68$ cm

a. Calculate the area (length × width) and perimeter (sum of all four sides) for each rectangle and express your results in square centimeters and centimeters, respectively, and give the correct number of significant figures in the result.
b. Change all measured values to meters and then calculate the area and perimeter of each rectangle. Express your answers in square meters and meters, respectively, and give the correct number of significant figures.
c. Does changing the units used change the number of significant figures in the answers?

USING UNITS IN CALCULATIONS (SECTION 1.9)

1.77 Determine a single factor derived from Table 1.3 that could be used as a multiplier to make each of the following conversions:
a. 4 yd to meters
b. 125,000 BTU to kilocalories
c. 400 mm to inches
d. 200 cm to inches

1.78 Determine a single factor from Table 1.3 that could be used as a multiplier to make each of the following conversions:
a. 20 mg to grains
b. 350 mL to fl oz
c. 4 qt to liters
d. 5 yd to meters

1.79 Obtain a factor from Table 1.3 and calculate the number of liters in 1.00 gal (4 qt) by using the factor-unit method of calculation.

1.80 A marathon race is about 26 miles. Obtain a factor from Table 1.3 and use the factor-unit method to calculate the distance of a marathon in kilometers.

1.81 A metric cookbook calls for a baking temperature of 200°C. Your oven settings are in degrees Fahrenheit. What Fahrenheit setting should you use?

1.82 A metric cookbook calls for 250 mL of milk. Your measuring cup is in English units. About how many cups of milk should you use? (NOTE: You will need two factors, one from Table 1.3 and one from the fact that 1 cup = 8 fl oz.)

1.83 An Olympic competitor threw the javelin 96.33 m. What is this distance in feet?

1.84 You have a 40-lb baggage limit for a transatlantic flight. When your baggage is put on the scale, you think you are within the limits because it reads 18.0. But then you realize that weight is in kilograms. Do a calculation to determine whether your baggage is overweight.

1.85 You need 2.50 lb of meat that sells for $2.70/lb (i.e., 1 lb = $2.70). Use this price to determine a factor to calculate the cost of the meat you need using the factor-unit method.

1.86 During a glucose tolerance test, the serum glucose concentration of a patient was found to be 131 mg/dL. Convert the concentration to grams per liter.

CALCULATING PERCENTAGES (SECTION 1.10)

1.87 Retirement age is 65 years in many companies. What percentage of the way from birth to retirement is a 45-year-old person?

1.88 A salesperson made a sale of $467.80 and received a commission of $25.73. What percent commission was paid?

1.89 After drying, 140 lb of grapes yields 28 lb of raisins. What percentage of the grapes' mass was lost during the drying process?

1.90 The recommended daily intake of thiamin is 1.4 mg for a male adult. Suppose such a person takes in only 1.0 mg/day. What percentage of the recommended intake is he receiving?

1.91 The recommended daily caloric intake for a 20-year-old woman is 2000. How many calories should her breakfast contain if she wants it to be 45% of her recommended daily total?

1.92 Immunoglobulin antibodies occur in five forms. A sample of serum is analyzed with the following results. Calculate the percentage of total immunoglobulin represented by each type.

Type:	IgG	IgA	IgM	IgD	IgE
Amount (mg):	987.1	213.3	99.7	14.4	0.1

DENSITY (SECTION 1.11)

1.93 Calculate the density of the following materials for which the mass and volume of samples have been measured. Express the density of liquids in g/mL, the density of solids in g/cm^3, and the density of gases in g/L.
a. 250 mL of liquid mercury metal (Hg) has a mass of 3400 g.
b. 500 mL of concentrated liquid sulfuric acid (H_2SO_4) has a mass of 925 g.
c. 5.00 L of oxygen gas has a mass of 7.15 g.
d. A 200-cm^3 block of magnesium metal (Mg) has a mass of 350 g.

1.94 Calculate the density of the following materials for which the mass and volume of samples have been measured. Express the density of liquids in g/mL, the density of solids in g/cm^3, and the density of gases in g/L.

   a. A 50.0-mL sample of liquid acetone has a mass of 39.6 g.

   b. A 1.00-cup (236-mL) sample of homogenized milk has a mass of 243 g.

   c. 20.0 L of dry carbon dioxide gas (CO_2) has a mass of 39.54 g.

   d. A 25.0-cm^3 block of nickel metal (Ni) has a mass of 222.5 g.

1.95 Calculate the volume and density of a rectangular block of platinum metal (Pt) with edges of 7.50 cm, 10.9 cm, and 3.00 cm. The block weighs 5273 g.

1.96 Calculate the volume and density of a cube of lead metal (Pb) that has a mass of 718.3 g and has edges that measure 3.98 cm.

1.97 The volume of an irregularly shaped solid can be determined by immersing the solid in a liquid and measuring the volume of liquid displaced. Find the volume and density of the following:

   a. An irregular piece of the mineral quartz is found to weigh 12.4 g. It is then placed into a graduated cylinder that contains some water. The quartz does not float. The water in the cylinder was at a level of 25.2 mL before the quartz was added and at 29.9 mL afterward.

   b. The volume of a sample of lead shot is determined using a graduated cylinder, as in part a. The cylinder readings are 16.3 mL before the shot is added and 21.7 mL after. The sample of shot weighs 61.0 g.

   c. A sample of coarse rock salt is found to have a mass of 11.7 g. The volume of the sample is determined by the graduated-cylinder method described in part a, but kerosene is substituted for water because the salt will not dissolve in kerosene. The cylinder readings are 20.7 mL before adding the salt and 26.1 mL after.

1.98 The density of ether is 0.736 g/mL. What is the volume in mL of 280 g of ether?

1.99 Calculate the mass in grams of 125 mL of chloroform ($d = 1.49$ g/mL).

> CHEMISTRY FOR THOUGHT

1.1 The following pairs of substances represent heterogeneous mixtures. For each pair, describe the steps you would follow to separate the components and collect them.

   a. Wood sawdust and sand

   b. Sugar and sand

   c. Iron filings and sand

   d. Sand soaked with oil

1.2 Explain why a bathroom mirror becomes foggy when someone takes a hot shower. Classify any changes that occur as physical or chemical.

1.3 A 175-lb patient is to undergo surgery and will be given an anesthetic intravenously. The safe dosage of anesthetic is 12 mg/kg of body weight. Determine the maximum dose of anesthetic that should be used.

1.4 A 20-year-old student was weighed and found to have a mass of 44.5 kg. She converted this to pounds and got an answer of 20.2 lb. Describe the mistake she probably made in doing the calculation.

1.5 Liquid mercury metal freezes to a solid at a temperature of $-38.9°C$. Suppose you want to measure a temperature that is at least as low as $-45°C$. Can you use a mercury thermometer? If not, propose a way to make the measurement.

1.6 Answer the question contained in Figure 1.3. How does hang gliding confirm that air is an example of matter?

1.7 Show how the factor-unit method can be used to prepare an oatmeal breakfast for 27 guests at a family reunion. The directions on the oatmeal box say that 1 cup of dry oatmeal makes 3 servings.

1.8 A chemist is brought a small solid figurine. The owner wants to know if it is made of silver but doesn't want it damaged during the analysis. The chemist decides to determine the density, knowing that silver has a density of 10.5 g/mL. The figurine is put into a graduated cylinder that contains 32.6 mL of water. The reading while the figurine is in the water is 60.1 mL. The mass of the figurine is 240.8 g. Is the figurine made of silver? Explain your reasoning.

1.9 Refer to Chemistry Around Us 1.2, then check the labels on your toiletries and see if you can identify one product that is regulated as both a drug and a cosmetic.

1.10 Refer to Figure 1.6, then use the model of matter described in Section 1.3 to propose an explanation for the following observation. When two teaspoons of sugar are dissolved in a small glass of water, the volume of the resulting solution is not significantly larger than the original volume of the water.

InfoTrac College Edition Readings

"Dark-matter heretic," *American Scientist*, Jan–Feb 2003, 91(1): 23(3). Record number A95688261.

"Switching drugs from prescription to OTC raises concerns," *Medical Letter on the CDC & FDA*, Nov 3, 2002:13. Record number A93300224.

"The origin of the universe," *National Forum*, Wntr 1996, 76(1): 9(7). Record number A19746565.

CHAPTER 2

Atoms and Molecules

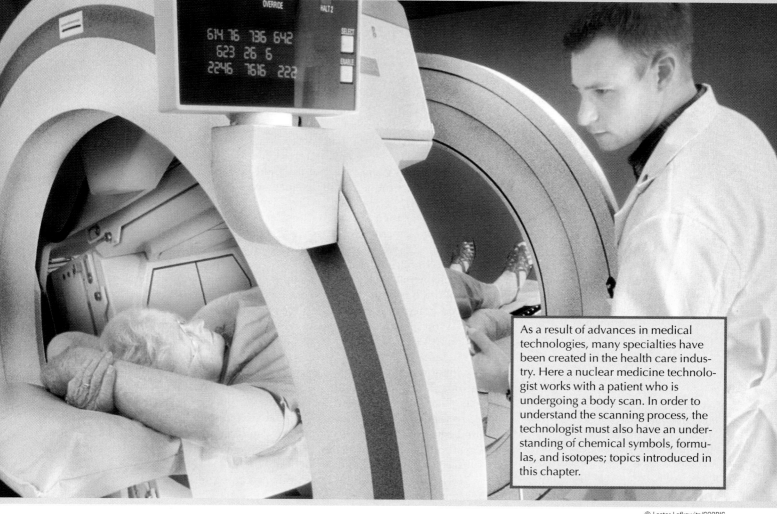

As a result of advances in medical technologies, many specialties have been created in the health care industry. Here a nuclear medicine technologist works with a patient who is undergoing a body scan. In order to understand the scanning process, the technologist must also have an understanding of chemical symbols, formulas, and isotopes; topics introduced in this chapter.

LEARNING OBJECTIVES

When you have completed your study of this chapter, you should be able to:

1. Use chemical element symbols to write formulas for compounds. (Section 2.1)

2. Identify the characteristics of protons, neutrons, and electrons. (Section 2.2)

3. Use the concepts of atomic number and mass number to determine the number of subatomic particles in isotopes and to write correct symbols for isotopes. (Section 2.3)

4. Use atomic weights of the elements to calculate molecular weights of compounds. (Section 2.4)

5. Use isotope percent abundances and masses to calculate atomic weights of elements. (Section 2.5)

6. Use the mole concept to obtain relationships between number of moles, number of grams, and number of atoms for elements, and use those relationships to obtain factors for use in factor-unit calculations. (Section 2.6)

7. Use the mole concept and molecular formulas to obtain relationships between number of moles, number of grams, and number of atoms or molecules for compounds, and use those relationships to obtain factors for use in factor-unit calculations. (Section 2.7)

e introduced some fundamental ideas about matter, atoms, molecules, measurements, and calculations in Chapter 1. In this chapter, these ideas are applied, the mole is defined, and the quantitative nature of chemistry becomes more apparent. A system of symbols is introduced that simplifies the way atoms and molecules are represented.

➤ 2.1 Symbols and Formulas

In Chapter 1, we defined elements as homogeneous pure substances made up of identical atoms. At the last count, 113 different elements were known to exist. This leads to the conclusion that a minimum of 113 different kinds of atoms exist. Eighty-eight of the elements are naturally occurring and therefore are found in Earth's crust, oceans, or atmosphere. The others are synthetic elements produced in the laboratory.

Each element can be characterized and identified by its unique set of physical and chemical properties, but it would be very cumbersome to list all these properties each time a specific element was discussed. For this reason, each element has been assigned a unique name and symbol. Many elements have been named by their discoverer, and as a result the names are varied. Some are based on elemental properties, others come from the names of famous scientists or places, while others are derived from the names of astronomical bodies or mythological characters.

An **elemental symbol** is based on the element's name and consists of a single capital letter or a capital letter followed by a lowercase letter. The symbols for 11 elements are based on the element's name in Latin or German. Elemental symbols are sometimes used to represent an element in a general way or to represent a single atom of an element. ■ Table 2.1 lists the elements whose names and symbols have been agreed on.

Compounds are pure substances made up of two or more different kinds of atoms. The atoms found in compounds are the same ones found in elements. Thus, the symbols used to represent elements can be combined and used to represent compounds. A molecular compound is symbolized by a **compound formula,** in which each atom is represented by an appropriate elemental symbol. When more than one atom of an element is present in a molecule, a subscript is used to indicate the number.

Notice the similarity between this practice and the molecular representations in Figure 1.4. The carbon dioxide molecules in Figure 1.4 are represented by the formula CO_2, where C represents an atom of carbon and O an atom of oxygen. The subscript 2 on the oxygen indicates that the molecule contains two atoms of oxygen. Notice that the single carbon atom in the molecule is not indicated by a subscript 1. Subscript 1 is never used in molecular formulas; it is understood. Formulas for molecular compounds are sometimes used to represent a compound in a general way, or to represent a single molecule of a compound. See ■ Table 2.2 and ■ Figure 2.1 for other examples of compound formulas.

Elemental symbol
A symbol assigned to an element based on the name of the element, consisting of one capital letter or a capital letter followed by a lowercase letter.

Compound formula
A symbol for the molecule of a compound, consisting of the symbols of the atoms found in the molecule. Atoms present in numbers higher than 1 have the number indicated by a subscript.

Example 2.1

Write formulas for the following compounds:

a. Nitrogen dioxide: one nitrogen (N) atom and two oxygen (O) atoms
b. Sulfuric acid: two hydrogen (H) atoms, one sulfur (S) atom, and four oxygen (O) atoms

TABLE 2.1	The chemical elements and their symbols

Ac	actinium	Er	erbium	Mn	manganese	Ru	ruthenium
Ag	silver (*argentum*)[a]	Es	einsteinium	Mo	molybdenum	S	sulfur
Al	aluminum	Eu	europium	Mt	meitnerium	Sb	antimony (*stibium*)[a]
Am	americium	F	flourine	N	nitrogen	Sc	scandium
Ar	argon	Fe	iron (*ferrum*)[a]	Na	sodium (*natrium*)[a]	Se	selenium
As	arsenic	Fm	fermium	Nb	niobium	Sg	seaborgium
At	astatine	Fr	francium	Nd	neodymium	Si	silicon
Au	gold (*aurum*)[a]	Ga	gallium	Ne	neon	Sm	samarium
B	boron	Gd	gadolinium	Ni	nickel	Sn	tin (*stannum*)[a]
Ba	barium	Ge	germanium	No	nobelium	Sr	strontium
Be	beryllium	H	hydrogen	Np	neptunium	Ta	tantalum
Bh	bohrium	He	helium	O	oxygen	Tb	terbium
Bi	bismuth	Hf	hafnium	Os	osmium	Tc	technetium
Bk	berkelium	Hg	mercury (*hydrargyrum*)[a]	P	phosphorus	Te	tellurium
Br	bromine			Pa	protactinium	Th	thorium
C	carbon	Ho	holmium	Pb	lead (*plumbum*)[a]	Ti	titanium
Ca	calcium	Hs	hassium	Pd	palladium	Tl	thallium
Cd	cadmium	I	iodine	Pm	promethium	Tm	thulium
Ce	cerium	In	indium	Po	polonium	U	uranium
Cf	californium	Ir	iridium	Pr	praseodymium	V	vanadium
Cl	chlorine	K	potassium (*kalium*)[a]	Pt	platinum	W	tungsten (*wolfram*)[a]
Cm	curium	Kr	krypton	Pu	plutonium	Xe	xenon
Co	cobalt	La	lanthanum	Ra	radium	Y	yttrium
Cr	chromium	Li	lithium	Rb	rubidium	Yb	ytterbium
Cs	cesium	Lu	lutetium	Re	rhenium	Zn	zinc
Cu	copper (*cuprum*)[a]	Lr	lawrencium	Rf	rutherfordium	Zr	zirconium
Db	dubnium	Md	mendelevium	Rh	rhodium		
Dy	dysprosium	Mg	magnesium	Rn	radon		

[a]Elements with symbols not derived from their English names.

TABLE 2.2	Examples of compound formulas

Compound name	Molecular representation	Molecular formula
Methane		CH_4
Water		H_2O
Carbon monoxide		CO
Hydrogen peroxide		H_2O_2

Solution

a. The symbols for the atoms are obtained from Table 2.1. The single N atom will not have a subscript because ones are understood and never written. The two O atoms will be represented by writing a subscript 2. The molecular formula is NO_2.

➤ **FIGURE 2.1** Iron pyrite is a mineral that contains iron and sulfur atoms in a 1:2 ratio, respectively. The golden crystals, called "fool's gold" by experienced miners, have caused (temporary) excitement for many novice prospectors. What is the formula for iron pyrite?

© Joel Gordon

b. Using similar reasoning, the H atom will have a subscript 2, the S atom will have no subscript, and the O atom will have a subscript 4. Therefore, the molecular formula is H_2SO_4.

| Learning Check 2.1 | Write molecular formulas for the following compounds: |

a. Phosphoric acid: three hydrogen (H) atoms, one phosphorus (P) atom, and four oxygen (O) atoms
b. Sulfur trioxide: one sulfur (S) atom and three oxygen (O) atoms
c. Glucose: six carbon (C) atoms, twelve hydrogen (H) atoms, and six oxygen (O) atoms

➤ 2.2 Inside the Atom

An atom has been defined as the limit of chemical subdivision for matter. On the basis of the characteristics of atoms that have been discussed, you probably have a general (and correct) idea that atoms can be considered to be the units from which matter is made. However, the question of how atoms interact to form matter has not yet been addressed. This interesting topic is discussed in Chapters 3 and 4, but a bit more must be learned about atoms first.

Extensive experimental evidence collected since the middle of the 19th century indicates that atoms are made up of many smaller particles. More than 100 of these subatomic particles have been discovered, and the search for more continues. As yet there is no single theory that can explain all observations involving subatomic particles, but three fundamental particles are included in all current theories: the proton, the neutron, and the electron. Most chemical behavior of matter can be explained in terms of a few of the well-known characteristics of these particles. These important characteristics are mass, electrical charge, and location in atoms. They are summarized in ■ Table 2.3. The atomic mass unit (u) shown in the table is discussed in Section 2.4.

In Chapter 1, atoms were described as being very tiny particles. The masses of single atoms of the elements are now known to fall within the range of 1.67 $\times 10^{-24}$ g for the least massive to 5.00 $\times 10^{-22}$ g for the most massive. Thus, it is not surprising to find that the particles that make up atoms have very small masses as well. Even though atomic masses are very small, the mass information given in Table 2.3 indicates that most of the mass of an atom comes from the

CHEMISTRY AROUND US • 2.1

"SEEING" ATOMS

Prior to 1970, scientists' belief in the existence of atoms was based on the successful use of atomic theory to explain the properties of matter. Direct visual evidence of atoms, however, eluded scientists until 1970, when an electron microscope, which uses electron beams instead of light, obtained pictures of large, heavy atoms of thorium and uranium.

In 1981, the scanning tunneling microscope (STM) was introduced. A different type of microscope, the STM does not have lenses and a light source like the microscopes with which you are familiar. Instead, an STM uses a very thin, needlelike probe that moves horizontally over the surface of a sample. A high-voltage current passes between the tip of the probe and the sample. The magnitude of the current depends on the distance between the probe tip and the sample surface. As the probe passes over the atom-sized bumps on the sample surface, the STM keeps the current constant by moving the probe up and down to maintain a very tiny but constant separation distance between the atoms and the probe tip. A computer detects and processes these tiny up-and-down motions, which are then converted to a greatly enlarged, three-dimensional image. Because the motion resulted from the atomic bumps on the surface, the image shows the location of atoms on the surface. The inventors of the STM, Gerd Benning, Heinrich Rohrer, and Ernst Ruska, shared the 1986 Nobel Prize in physics for their accomplishment.

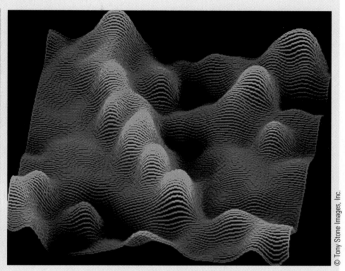

A view from a scanning tunneling microscope (STM) shows the locations of atoms protruding from a short strand of a DNA double helix.

© Tony Stone Images, Inc.

protons and neutrons it contains. The protons and neutrons are tightly bound together to form the central portion of an atom called the **nucleus.** Because protons have a +1 charge each, and neutrons have no charge, the nucleus of an atom has a positive charge that is equal to the number of protons it contains.

Electrons are negatively charged particles located outside the nucleus of an atom. Protons and electrons carry equal but opposite charges, so a neutral atom that has no electrical charge must have the same number of protons in its nucleus as it has electrons outside the nucleus. The electrons of an atom are thought to move very rapidly throughout a relatively large volume of space surrounding the small but very heavy nucleus (see ■ Figure 2.2).

Even though subatomic particles exist, the atom itself is the particle of primary interest in chemistry because subatomic particles do not lead an independent existence for any appreciable length of time. The only way they gain long-term stability is by combining with other particles to form an atom.

Nucleus

The central core of atoms that contains protons, neutrons, and most of the mass of atoms.

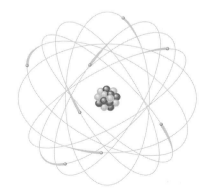

➤ **FIGURE 2.2** Electrons move rapidly around a massive nucleus. This figure is not drawn to scale. For a nucleus of the size shown, the closest electrons would be at least 80 m away.

| TABLE 2.3 | Characteristics of the fundamental subatomic particles |

Particle	Common symbols	Characteristics			
		Charge (±)	Mass (g)	Mass (u)	Location
Electron	e^-	1−	9.07×10^{-28}	1/1836	Outside nucleus
Proton	p, p^+, H^+	1+	1.67×10^{-24}	1	Inside nucleus
Neutron	n	0	1.67×10^{-24}	1	Inside nucleus

➤ 2.3 Isotopes

Atomic number of an atom

A number equal to the number of protons in the nucleus of an atom. Symbolically it is represented by Z.

Neutral atoms have no electrical charge and must contain identical numbers of protons and electrons. However, because neutrons have no electrical charge, their numbers in an atom do not have to be the same as the numbers of protons or electrons. The number of protons in the nucleus of an atom is given by the **atomic number** for the atom. Atomic numbers are represented by the symbol Z. All atoms of a specific element must have the same atomic number. The atomic numbers for each element are the numbers above the elemental symbols of the periodic table inside the front cover of this book. Remember, this is also the number of electrons in the neutral atoms of each element.

The possibility of having different numbers of neutrons combined with one given number of protons to form nuclei leads to some interesting results. For example, three different kinds of hydrogen atoms are known to exist. Each kind of atom contains one proton and one electron, so all have an atomic number of 1. However, the nuclei of the different kinds of atoms contain different numbers of neutrons. The most common kind has no neutrons in the nucleus, the next most common has one, and the least common kind has two. The sum of the number of protons and the number of neutrons in a nucleus is called the **mass number** and is represented by the symbol A. Thus, the three kinds of hydrogen atoms all have atomic numbers of 1 and mass numbers of 1, 2, and 3, respectively. Atoms that have the same atomic number but different mass numbers are called **isotopes**. Most elements are made up of mixtures of two or more isotopes. When it is important to distinguish between them, the following notation is used for each isotope: $^A_Z E$, where E is the symbol for the element.

Mass number of an atom

A number equal to the sum of the number of protons and neutrons in the nucleus of an atom. Symbolically it is represented by A.

Isotopes

Atoms that have the same atomic number but different mass numbers. That is, they are atoms of the same element that contain different numbers of neutrons in their nuclei.

The three isotopes of hydrogen are represented as follows using this notation: $^1_1 H$, $^2_1 H$, and $^3_1 H$. When these symbols are not convenient to use, as in written or spoken references to the isotopes, the elemental name followed by the mass number is used. Thus, the three hydrogen isotopes are hydrogen-1, hydrogen-2, and hydrogen-3. These three isotopes have specific names: protium ($A = 1$), deuterium ($A = 2$), and tritium ($A = 3$).

Example 2.2

GOB
Chemistry‑⚛‑Now™

Go to GOB Now and click to learn how to examine the relationships among atomic symbols, atomic numbers, and atomic mass numbers.

Use the periodic table inside the front cover to answer the following questions about isotopes:

a. What are the mass number, atomic number, and isotope symbol ($^A_Z E$) for an atom that contains 7 protons and 8 neutrons?
b. How many neutrons are contained in an atom of nickel-60?
c. How many protons and how many neutrons are contained in an atom with a mass number of 26 and the symbol Mg?

Solution

a. The mass number, A, equals the sum of the number of protons and the number of neutrons: $A = 7 + 8 = 15$. The atomic number, Z, equals the number of protons: $Z = 7$. According to the periodic table, the element with an atomic number of 7 is nitrogen, with the symbol N. The isotope symbol is $^{15}_7 N$.
b. According to the periodic table, nickel has the symbol Ni, and an atomic number, Z, of 28. The mass number, 60, is equal to the sum of the number of protons and the number of neutrons: $A = \#p + \#n$, or $\#p + \#n = 60$. The number of protons is equal to the atomic number, 28. Therefore, $\#n = 60 - \#p$. $\#n = 60 - 28 = 32$. The atom contains 32 neutrons.

OVER THE COUNTER • 2.1

CALCIUM SUPPLEMENTS

Some meanings of the word *supplement* are "to add to," "to fill up," and "to complete." When used in a nutritional context, a supplement provides an amount of a substance that is in addition to the amount obtained from the diet. An important question, then, is: Who, if anyone, should take dietary supplements? The obvious answer is that anyone who does not get enough of a particular nutrient from the diet to satisfy the needs of the body should take a supplement. How do we apply this obvious answer to the question of whether or not to take a calcium supplement?

Calcium in various forms performs numerous functions in the body. However, about 99% is used to build bones and teeth. During the body's lifetime, all bones undergo a continuous natural process of buildup and breakdown. The rate of buildup exceeds the rate of breakdown for the first 25–30 years in women and the first 30–35 years in men. After these ages, the rate of breakdown catches and exceeds the rate of buildup, resulting in a gradual decrease in bone density. As bone density decreases, the bones become increasingly weakened, brittle, and susceptible to breaking—a condition called *osteoporosis*. About 50% of women and 13% of men over age 50 suffer a broken bone as a result of osteoporosis.

One of the best ways to reduce the risk of osteoporosis in later life is to build as much bone as possible during early life, when the rate of buildup exceeds the rate of breakdown. In order to build as much bone as possible during the most active bone-building years, the following daily calcium intakes have been suggested:

- Birth to 6 months: 400 mg
- Age 6 months to 1 year: 600 mg
- Age 1–10: 800 mg
- Age 11–24: 1200–1500 mg
- Age 25–50: 1500 mg
- Age 51–65: 100–1500 mg (depending on hormone replacement therapy)
- Over age 65: 1500 mg

Sufficient calcium for building bones is provided by a balanced diet that includes calcium-rich foods such as dairy products, certain vegetables (broccoli, kale, and turnip and collard greens), tofu, some canned fish, legumes (beans, peas, etc.), and seeds and nuts. Unfortunately, many people in the prime of their bone-building years (preadolescents, adolescents, and young adults) follow diets that fall far short of the recommended levels of calcium for optimal bone-building. Two very significant reasons for this nutritional shortfall among people in these age groups are the tendency to skip meals and the substitution of soft drinks and other nondairy drinks in place of milk.

If a calcium supplement is needed, a number of factors should be considered. Vitamin D is essential for optimal calcium absorption by the body. For this reason, many calcium supplements include vitamin D in their formulation, and clearly indicate this on their labels. Calcium supplements are most efficiently absorbed when taken in individual doses of 500 mg or less. The dose per tablet or capsule is generally indicated on the label. The dosages found in the calcium supplements carried by a typical pharmacy range from a low of 333 mg to a high of 630 mg. The calcium comes in various compound forms, including calcium carbonate (often from oyster shells), calcium citrate, and calcium phosphate. Different forms can be advantageous for different body conditions. For example, a person with high gastric acid production should take calcium carbonate with food to improve absorption.

Osteoporosis might seem too far away in the future to concern a teenager or young adult. However, simple changes in lifestyle now—such as improving the diet or taking a calcium supplement—might provide substantial and much-appreciated health benefits in that (not so far) distant future.

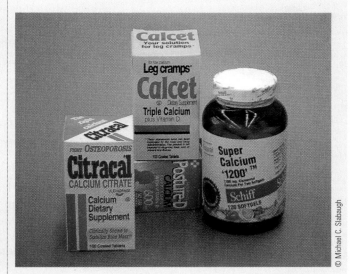

The source of calcium in over-the-counter calcium supplements varies from product to product. Some common sources are bone meal, crushed oyster shells, and calcium compounds such as calcium carbonate, calcium phosphate, and calcium citrate.

c. According to the periodic table, the element with the symbol Mg is magnesium, which has an atomic number of 12. Therefore, the atom contains 12 protons. Since $A = \#p + \#n$, we see that $\#p + \#n = 26$, and $\#n = 26 - \#p = 26 - 12 = 14$. The atom contains 14 neutrons.

Learning Check 2.2

Use the periodic table inside the front cover to answer the following questions about isotopes:

a. What are the atomic number, mass number, and isotope symbol for an atom that contains 4 protons and 5 neutrons?
b. How many neutrons are contained in an atom of chlorine-37?
c. How many protons and how many neutrons are contained in an atom with a mass number of 28 and the symbol Si?

➤ 2.4 Relative Masses of Atoms and Molecules

Because of their extremely small size, it is very inconvenient to use the actual masses of atoms when the atoms are being characterized or when quantitative calculations are done. In fact, the earliest chemists had no way of determining the actual masses of atoms. For this reason, a system was devised that utilized relative or comparative masses for the atoms. These relative masses are the numbers that are given beneath the symbol and name for each element in the periodic table inside the front cover.

Relative masses provide a simple way of comparing the masses of atoms. For example, the mass of neon atoms, Ne, from the periodic table is given as 20.18. Similarly, the mass of calcium atoms, Ca, is given as 40.08. These numbers simply indicate that calcium atoms have a mass that is about twice the mass of neon atoms. The exact relationship between the two masses is calculated as follows, using the correct number of significant figures:

$$\frac{\text{Ca atom mass}}{\text{Ne atom mass}} = \frac{40.08}{20.18} = 1.986$$

In a similar way, we arrive at the conclusion that helium atoms are about four times as massive as hydrogen atoms:

$$\frac{\text{He atom mass}}{\text{H atom mass}} = \frac{4.003}{1.008} = 3.971$$

In each case, we have been able to determine the relationship between the masses of the atoms without using the actual masses.

Modern instruments called mass spectrometers allow the actual masses of individual atoms to be measured. These measured masses show the same relationships to each other as do the relative masses:

$$\frac{\text{Ca atom mass}}{\text{Ne atom mass}} = \frac{6.655 \times 10^{-23}\,\text{g}}{3.351 \times 10^{-23}\,\text{g}} = 1.986$$

$$\frac{\text{He atom mass}}{\text{H atom mass}} = \frac{6.647 \times 10^{-24}\,\text{g}}{1.674 \times 10^{-24}\,\text{g}} = 3.971$$

Thus, we see that the relative masses used in the periodic table give the same results as the actual masses when the masses of atoms are compared with one another.

Actual atomic masses are given in mass units such as grams, but the relative values used in the periodic table are given in units referred to as **atomic mass units.** Until recently, the abbreviation for an atomic mass unit was amu. However, the accepted abbreviation is now u. The actual mass represented by a single atomic mass unit is $\frac{1}{12}$ the mass of a single carbon-12 atom.

Atomic mass unit (u)
A unit used to express the relative masses of atoms. One u is equal to $\frac{1}{12}$ the mass of an atom of carbon-12.

CHEMISTRY AND YOUR HEALTH • 2.1

ARE YOU AT RISK FOR OSTEOPOROSIS?

Osteoporosis, the abnormal thinning of bones that often accompanies aging, may lead to bone fractures, disability, and even death. While women are most susceptible, this serious condition also affects men but usually at a more advanced age than women. A number of significant risk factors have been identified, including the following:

1. A poor diet, especially one low in calcium
2. Advanced age
3. The onset of menopause (or having had ovaries removed)
4. A sedentary life style
5. Smoking
6. A family history of osteoporosis or hip fracture
7. Heavy drinking
8. The long-term use of certain steroid drugs
9. Vitamin D deficiency

Ninety-nine percent of the calcium found in the body is located in the skeleton and teeth, so it is not surprising that the behavior of this metallic element in the body plays a central role in a number of these risk factors for osteoporosis. For example, it is known that in later life the body's ability to absorb calcium from food in the small intestine decreases (factor 2) and such absorption requires the presence of vitamin D (factor 9).

It is well recognized that the best insurance against developing osteoporosis in later life is to build and strengthen as much bone as possible during the first 25 to 35 years of life. Two essential components of this process are eating a healthful diet containing adequate amounts of calcium and vitamin D, and following a healthful lifestyle that includes regular weight-bearing exercise such as walking, jogging, weight lifting, stair climbing, or physical labor. Good dietary sources of calcium are given in the For Future Reference feature at the end of this chapter. Dietary calcium supplements provide an additional way to enhance calcium intake, especially for individuals who are at risk to develop osteoporosis (See Over the Counter 2.1).

A bone density test provides an effective way to diagnose the presence or extent of osteoporosis in an individual. Such tests measure the absorption of X rays by bones, and are not invasive or uncomfortable. Bone density tests were not routinely recommended in the past, but some health and wellness organizations now suggest that women who are at high risk should have such a test by the age of 50, and all women over age 65 should be routinely tested.

The relative masses of the elements as given in the periodic table are referred to as atomic masses or atomic weights. We will use the term **atomic weights** in this book. In those cases where the naturally occurring element exists in the form of a mixture of isotopes, the recorded atomic weight is the average value for the naturally occurring isotope mixture. This idea is discussed in Section 2.5.

Atomic weight
The mass of an average atom of an element expressed in atomic mass units.

Example 2.3

Use the periodic table inside the front cover to answer the following questions:

a. Which element has atoms that are nearest to being twice as massive as atoms of calcium, Ca?
b. Which element has atoms that are nearest in mass to atoms of calcium, Ca?
c. To the nearest whole number, how many hydrogen atoms, H, would be required to have a mass equal to the mass of a single carbon-12 atom?

Solution

a. Calcium atoms have a mass of 40.08 u. Twice this value is 80.16 u. The element with atoms nearest this value is bromine, Br, with atoms that have a mass of 79.90 u.
b. Calcium atoms have a mass of 40.08 u, and the element with atoms nearest this value is argon, Ar, with a value of 39.95 u.
c. By definition, one atomic mass unit is equal to $\frac{1}{12}$ the mass of a single carbon-12 atom, so a single carbon-12 atom has a mass of 12 u. Each hydrogen atom, H, has a mass of 1.008 u, so it would take 12 hydrogen atoms to equal the mass of a single carbon-12 atom.

Learning Check 2.3

Use the periodic table inside the front cover to answer the following questions:

a. Which element has atoms that are nearest to being twice as massive as atoms of copper, Cu?

b. Which element has atoms that are nearest to being $\frac{1}{4}$ the mass of atoms of iron, Fe?

c. To the nearest whole number, how many helium atoms, He, would be required to have a mass equal to the mass of a single carbon-12 atom?

Molecular weight
The relative mass of a molecule expressed in atomic mass units and calculated by adding the atomic weights of the atoms in the molecule.

Molecules are made up of atoms, so the relative mass of a molecule can be calculated by adding the atomic weights of the atoms that make up the molecule. Relative molecular masses calculated in this way are called **molecular weights** and are also given in atomic mass units.

Example 2.4

GOB
Chemistry ⚛ Now™
Go to GOB Now and click to review how to calculate the molecular weight for a compound.

Use atomic weights from the periodic table inside the front cover to determine the molecular weight of urea, CH_4N_2O, the chemical form in which much nitrogenous body waste is excreted in the urine.

Solution

According to the formula given, a urea molecule contains one carbon atom, C, two nitrogen atoms, N, four hydrogen atoms, H, and one oxygen atom, O. The molecular weight is calculated as follows:

$$MW = 1(at.\ wt.\ C) + 2(at.\ wt.\ N) + 4(at.\ wt.\ H) + 1(at.\ wt.\ O)$$

$$MW = 1(12.01\ u) + 2(14.01\ u) + 4(1.008\ u) + 1(16.00\ u) = 60.062\ u$$

Rounded to four significant figures, the correct answer is 60.06 u.

Learning Check 2.4

a. Use atomic weights from the periodic table to determine the molecular weight of sulfuric acid. Each molecule contains two hydrogen (H) atoms, one sulfur (S) atom, and four oxygen (O) atoms.

b. Determine the molecular weight of isopropyl alcohol (C_3H_8O), the active ingredient in most rubbing alcohol sold commercially.

➤ 2.5 Isotopes and Atomic Weights

The atomic weights discussed in Section 2.4 were defined as the relative masses of atoms of the elements. It would have been more correct to define them as the relative masses of average atoms of the elements. Why include the idea of an average atom? Remember, the mass number of an isotope is the sum of the number of protons and neutrons in the nucleus of atoms of the isotope. Also, protons and neutrons both have masses of 1 u (see Table 2.3). The masses of electrons are quite small, so the atomic weights of isotopes are very nearly equal to their mass numbers. For example, all naturally occurring phosphorus is made up of atoms containing 15 protons and 16 neutrons. The mass number of this isotope is 31, and its symbol is $^{31}_{15}P$. The relative mass (atomic weight) of atoms of this isotope is 30.97 u, the same as the atomic weight of the element given in the periodic table. The atomic weight of the isotope and the listed atomic weight of the element are the same because all atoms contained in the element are identical and have the same relative mass.

Naturally occurring chlorine is different. It is a mixture of two isotopes, $^{35}_{17}Cl$ and $^{37}_{17}Cl$. Chlorine-35 has a mass number of 35 and a relative mass of 34.97 u. Chlorine-37 has a mass number of 37 and a relative mass of 36.97 u. Note that the mass numbers and relative masses (atomic weights) of the atoms of an isotope are essentially identical. However, naturally occurring chlorine is a mixture containing both isotopes. Thus, the determined relative mass of chlorine atoms will be the average relative mass of the atoms found in the mixture (see ■ Figure 2.3).

The average mass of each particle in a group of particles is simply the total mass of the group divided by the number of particles in the group. This calculation requires that the total number of particles be known. However, the percentage of each isotope in a mixture of atoms is easier to determine than the actual number of atoms of each isotope present. The percentages can be used to calculate average masses. Remember that percent means per 100, so we use an imaginary sample of an element containing 100 atoms in the calculation. On this basis, the number of atoms of each isotope in the 100-atom sample will be the percentage of that isotope in the sample. The mass contributed to the sample by each isotope will be the product of the number of atoms of the isotope (the percentage of the isotope) and the mass of the isotope. The total mass of the sample will be the sum of the masses contributed by each isotope. This total mass divided by 100, the number of atoms in our imaginary sample, gives the mass of an average atom, which is the atomic weight of the element.

© West

▶ **FIGURE 2.3** The pieces of fruit in a bowl are somewhat like atoms of the isotopes of an element. Each piece of fruit may have the same color, taste, and texture, but it is unlikely that any two have *exactly* the same mass. The 12 oranges in the bowl weigh a total of 2.36×10^3 g. What is the average mass of each orange in the bowl?

Example 2.5

Calculate the atomic weight of chlorine, given that the naturally occurring element consists of 75.53% chlorine-35 (mass = 34.97 u) and 24.47% chlorine-37 (mass = 36.97 u).

Solution

$$\text{Atomic weight} = \frac{(\% \text{ chlorine-35})(\text{mass chlorine-35}) + (\% \text{ chlorine-37})(\text{mass chlorine-37})}{100}$$

$$= \frac{(75.53)(34.97 \text{ u}) + (24.47)(36.97 \text{ u})}{100}$$

$$= \frac{2641.28 \text{ u} + 904.66 \text{ u}}{100} = \frac{3545.94 \text{ u}}{100} = 35.4594 \text{ u}$$

$$= 35.46 \text{ u (rounded value)}$$

This result is slightly different from the periodic table atomic weight value of 35.45 because of slight errors introduced in rounding the isotope masses to four significant figures.

a. Naturally occurring fluorine consists of a single isotope, fluorine-19, with a mass of 19.00 u (using four significant figures). Determine the atomic weight of fluorine, and compare your answer with the value given in the periodic table.
b. Naturally occurring magnesium has three isotopes: magnesium-24, magnesium-25, and magnesium-26. Their relative masses and percent abundances are, respectively, 23.99 u (78.70%), 24.99 (10.13%), and 25.98 (11.17%). Determine the atomic weight of magnesium, and compare it with the value given in the periodic table.

Learning Check 2.5

GOB
Chemistry ⚛ Now™

Go to GOB Now and click to learn how to calculate the atomic weight of an element based on masses and abundances of isotopes.

➤ **FIGURE 2.4** Samples of magnesium (top) and carbon with masses of 24.31 g and 12.01 g, respectively.

➤ 2.6 Avogadro's Number: The Mole

The atomic weights of the elements given in the periodic table have far more uses than simply comparing the masses of the atoms of various elements. However, these uses are not apparent until some additional ideas are developed. According to the periodic table, the atomic weight of magnesium, Mg, is 24.31 u, and the atomic weight of carbon, C, is 12.01 u. As we have seen, this means that an average magnesium atom has about twice the mass of an average carbon atom. Modern instruments allow us to determine that the actual mass of an average magnesium atom is 4.037×10^{-23} g, and the actual mass of an average carbon atom is 1.994×10^{-23} g.

Even though the actual masses of magnesium and carbon atoms are extremely small, it should still be possible to collect together enough atoms of either element to give a sample of any desired mass. Suppose we wish to collect enough atoms of each element to give a sample with a mass in grams equal to the atomic weight of each element. That is, we want to collect enough magnesium atoms to give a sample with a total mass of 24.31 g, and enough carbon atoms to give a sample with a total mass of 12.01 g (see ■ Figure 2.4). How many atoms of each element will be required?

We know the mass of one atom of each element, and this fact will provide us with a factor that will allow us to convert the desired sample mass into the number of atoms required. The given mass of one atom of magnesium can be written as

$$1 \text{ Mg atom} = 4.037 \times 10^{-23} \text{ g Mg}$$

This known relationship provides two factors that can be used to solve factor-unit problems:

$$\frac{1 \text{ Mg atom}}{4.037 \times 10^{-23} \text{ g Mg}} \quad \text{and} \quad \frac{4.037 \times 10^{-23} \text{ g Mg}}{1 \text{ Mg atom}}$$

Our task is to convert the desired sample mass of 24.31 g Mg into Mg atoms. The first factor will cancel the units "g Mg" and will generate the units "Mg atoms."

$$(24.31 \text{ g Mg}) \times \frac{1 \text{ Mg atom}}{4.037 \times 10^{-23} \text{ g Mg}} = 6.022 \times 10^{23} \text{ Mg atoms}$$

In a similar way, the number of C atoms needed to produce a sample with a mass of 12.01 g is calculated:

$$(12.01 \text{ g C}) \times \frac{1 \text{ C atom}}{1.994 \times 10^{-23} \text{ g C}} = 6.022 \times 10^{23} \text{ C atoms}$$

The result that the same number of atoms is required for each sample might seem surprising, but it is a consequence of the sample sizes we wanted to produce. If we collected just one atom of each element, the ratio of the mass of Mg to the mass of C would be equal to the atomic weight of Mg divided by the atomic weight of C:

$$\frac{\text{Mg mass}}{\text{C mass}} = \frac{4.037 \times 10^{-23} \text{ g}}{1.994 \times 10^{-23} \text{ g}} = 2.024 = \frac{\text{At. wt. Mg}}{\text{At. wt. C}} = \frac{24.31 \text{ u}}{12.01 \text{ u}}$$

This result reflects the fact we discussed in Section 2.4, that magnesium atoms have about twice the mass of carbon atoms. If we collected samples of 100 atoms of each element, the ratio of the mass of the Mg sample to the mass of the C sample would still be 2.024 because each sample would have a mass 100 times greater than the mass of a single atom:

$$\frac{\text{Mg mass}}{\text{C mass}} = \frac{(100)(4.037 \times 10^{-23} \text{ g})}{(100)(1.994 \times 10^{-23} \text{ g})} = 2.024$$

© Mark Slabaugh

It follows that if we collected samples containing 6.022×10^{23} atoms of each element, the ratio of the mass of the Mg sample to the mass of the C sample would still be 2.024 because each sample would have a mass that would be 6.022×10^{23} times greater than the mass of a single atom:

$$\frac{\text{Mg mass}}{\text{C mass}} = \frac{(6.022 \times 10^{23})(4.037 \times 10^{-23} \ \cancel{g})}{(6.022 \times 10^{23})(1.994 \times 10^{-23} \ \cancel{g})} = 2.024$$

These results lead to the following conclusion: Any samples of magnesium and carbon that have mass ratios equal to 2.024 will contain the same number of atoms.

Example 2.6

Show that samples of magnesium and carbon with masses of 9.663 g and 4.774 g, respectively, have a mass ratio of 2.024 and contain the same number of atoms.

Solution

The mass ratio is obtained by dividing the mass of the magnesium sample by the mass of the carbon sample:

$$\frac{\text{Mg mass}}{\text{C mass}} = \frac{9.663 \ \text{g}}{4.774 \ \text{g}} = 2.024$$

The number of atoms in each is calculated using the factors obtained earlier from the mass of one atom of each element:

$$(9.663 \ \cancel{g} \ \cancel{Mg}) \times \frac{1 \ \text{Mg atom}}{4.037 \times 10^{-23} \ \cancel{g} \ \cancel{Mg}} = 2.394 \times 10^{23} \ \text{Mg atoms}$$

$$(4.744 \ \cancel{g} \ \cancel{C}) \times \frac{1 \ \text{C atom}}{1.994 \times 10^{-23} \ \cancel{g} \ \cancel{C}} = 2.394 \times 10^{23} \ \text{C atoms}$$

Show that samples of magnesium and carbon with masses of 13.66 g and 6.749 g, respectively, have a mass ratio of 2.024 and contain the same number of atoms.

Learning Check 2.6

The preceding results for magnesium and carbon may be generalized for all the elements of the periodic table as follows. Any samples of two elements that have a mass ratio equal to the ratio of their atomic weights will contain identical numbers of atoms. In addition, we have seen that if the number of grams of sample is equal numerically to the atomic weight of an element, the number of atoms in the sample is equal to 6.022×10^{23}. ■ Figure 2.5 illustrates these ideas for particles that are familiar to most of us.

➤ **FIGURE 2.5** An average jelly bean has a mass that is 1.60 times the mass of an average dry bean. Each jar contains the same number of beans. The total mass of jelly beans is 472 g. What is the total mass of the dry beans?

© Spencer L. Seager

With modern equipment, it is possible to determine the number of atoms in any size sample of an element. However, before this was possible, the practice of focusing on samples with a mass in grams equal to the atomic weights of the elements became well established. It continues today. We have learned that the number of atoms in such samples is 6.022×10^{23}.

We saw in Section 2.4 that molecular weights, the relative masses of molecules, are calculated by adding the atomic weights of the atoms contained in the molecules. The resulting molecular weights are expressed in atomic mass units, just as are the atomic weights of the atoms. The same ideas we used to discuss the actual and relative masses of atoms can be applied to molecules. We will not go through the details but simply state the primary conclusion: A sample of compound with a mass in grams equal to the molecular weight of the compound contains 6.022×10^{23} molecules of the compound.

The number 6.022×10^{23} is called *Avogadro's number* in honor of Amadeo Avogadro (1776–1856), an Italian scientist who made important contributions to the concept of atomic weights. As we have seen, this number represents the number of atoms or molecules in a specific sample of an element or compound. Because of its importance in calculations, the number of particles represented by Avogadro's number is given a specific name; it is called a **mole**, abbreviated mol.

It is sometimes helpful to remember that the word *mole* represents a specific number, just as the word *dozen* represents 12, regardless of the objects being counted. Thus, 6.022×10^{23} people would be called 1 mol of people, just as 12 people would be called 1 dozen people. The immensity of Avogadro's number is illustrated by the results of a few calculations based on 1 mol of people. One mol of people would be enough to populate about 1×10^{14} Earths at today's level. That is 100 trillion Earths. Or, put another way, the present population of Earth is $1 \times 10^{-12}\%$ (0.000000000001%) of 1 mol.

In the development of these ideas to this point, we have used four significant figures for atomic weights, molecular weights, and Avogadro's number to minimize the introduction of rounding errors. However, in calculations throughout the remainder of the book, three significant figures will generally be sufficient and will be used.

In a strict sense, 1 mol is a specific number of particles. However, in chemistry it is customary to follow the useful practice of also letting 1 mol stand for the mass of a sample of element or compound that contains Avogadro's number of particles. Thus, the application of the mole concept to sulfur (at. wt. = 32.1 u) gives the following relationships:

$$1 \text{ mol S atoms} = 6.02 \times 10^{23} \text{ S atoms} = 32.1 \text{ g S}$$

When written individually, these three relationships can be used to generate six factors for use in factor-unit calculations involving sulfur:

$$1 \text{ mol S atoms} = 6.02 \times 10^{23} \text{ S atoms}$$

$$6.02 \times 10^{23} \text{ S atoms} = 32.1 \text{ g S}$$

$$1 \text{ mol S atoms} = 32.1 \text{ g S}$$

Mole
The number of particles (atoms or molecules) contained in a sample of element or compound with a mass in grams equal to the atomic or molecular weight, respectively. Numerically, 1 mol is equal to 6.022×10^{23} particles.

Example 2.7

Determine the following using the factor-unit method of calculation and factors obtained from the preceding three relationships given for sulfur (S):

a. The mass in grams of 1.35 mol of S
b. The number of moles of S atoms in 98.6 g of S
c. The number of S atoms in 98.6 g of S
d. The mass in grams of one atom of S

Solution

a. The known quantity is 1.35 mol of S, and the unit of the unknown quantity is grams of S. The factor comes from the relationship 1 mol S atoms = 32.1 g S.

$$(1.35 \ \text{mol S atoms}) \left(\frac{32.1 \ \text{g S}}{1 \ \text{mol S atoms}} \right) = 43.3 \ \text{g S}$$

b. The known quantity is 98.6 g of S, and the unit of the unknown quantity is moles of S atoms. The factor comes from the same relationship used in (a).

$$(98.6 \ \text{g S}) \left(\frac{1 \ \text{mol S atoms}}{32.1 \ \text{g S}} \right) = 3.07 \ \text{mol S atoms}$$

c. The known quantity is, again, 98.6 g of S, and the unit of the unknown is the number of S atoms. The factor comes from the relationship 6.02×10^{23} S atoms = 32.1 g S.

$$(98.6 \ \text{g S}) \left(\frac{6.02 \times 10^{23} \ \text{S atoms}}{32.1 \ \text{g S}} \right) = 1.85 \times 10^{24} \ \text{S atoms}$$

d. The known quantity is one S atom, and the unit of the unknown is grams of S. The factor comes from the same relationship used in (c), 6.02×10^{23} S atoms = 32.1 g S. Note that the factor is the inverse of the one used in (c) even though both came from the same relationship. Thus, we see that each relationship provides two factors.

$$(1 \ \text{S atom}) \left(\frac{32.1 \ \text{g S}}{6.02 \times 10^{23} \ \text{S atoms}} \right) = 5.33 \times 10^{-23} \ \text{g S}$$

Learning Check 2.7

Calculate the mass of a single oxygen atom in grams. How does the ratio of this mass and the mass of a carbon atom given earlier (1.994×10^{-23} g) compare with the ratio of the atomic weights of oxygen and carbon given in the periodic table?

➤ 2.7 The Mole and Chemical Formulas

According to Section 2.1, the formula for a compound is made up of the symbols for each element present. Subscripts following the elemental symbols indicate the number of each type of atom in the molecule represented. Thus, chemical formulas represent the numerical relationships that exist among the atoms in a compound. Application of the mole concept to formulas provides additional useful information.

Consider water as an example. The formula H_2O represents a 2:1 ratio of hydrogen atoms to oxygen atoms in a water molecule. Since this ratio is fixed, the following statements can be written:

1. 2 H_2O molecules contain 4 H atoms and 2 O atoms.
2. 10 H_2O molecules contain 20 H atoms and 10 O atoms.
3. 100 H_2O molecules contain 200 H atoms and 100 O atoms.
4. 6.02×10^{23} H_2O molecules contain 12.04×10^{23} H atoms and 6.02×10^{23} O atoms.

Statement 4 is significant because 6.02×10^{23} particles is 1 mol. Thus, Statement 4 can be changed to Statement 5:

5. 1 mol of H_2O molecules contains 2 mol of H atoms and 1 mol of O atoms.

■ Figure 2.6 contains another example of this concept.

➤ **FIGURE 2.6** Liquid carbon disulfide (CS_2) is composed of carbon (left) and sulfur (right), elements that are solids. How many moles of sulfur atoms would be contained in 1.5 mol of CS_2 molecules?

© Mark Slabaugh

HELP WITH MOLE CALCULATIONS

Problems involving the use of the mole often strike fear into the hearts of beginning chemistry students. The good news is that problems involving the use of moles, atoms, molecules, and grams are made easier by using the factor-unit method. The method focuses your attention on the goal of eliminating the unit of the known, or given, quantity and of generating the unit of the answer, or unknown, quantity. Remember, Step 1 is to write down the number and unit of the given quantity. In Step 2, write down the unit of the answer. In Step 3, multiply the known quantity by one or more factors whose units will cancel that of the known quantity and will generate the unit of the answer. In Step 4, obtain the answer by doing the required arithmetic using the numbers that were introduced in Steps 1–3.

The ability to write the necessary factors for use in Step 3 is essential if you are to become proficient in solving mole problems using the factor-unit method. The factors come from fixed relationships between quantities that are obtained from definitions, experimental measurements, or combinations of the two. The definition of the mole, coupled with experimentally determined atomic and molecular weights, gives the following fixed relationships:

Atom Y: $1 \text{ mol Y atoms} = 6.02 \times 10^{23} \text{ Y atoms} = y \text{ g Y}$

where y is the atomic weight of element Y.

Molecule Z: $1 \text{ mol Z molecules} = 6.02 \times 10^{23} \text{ Z molecules} = z \text{ g Z}$

where z is the molecular weight of compound Z. Each of these sets of three related quantities will give six factors that can be used in factor-unit problems. Each factor is simply a ratio between any two of the three related quantities such as

$$\frac{1 \text{ mol Y atoms}}{6.02 \times 10^{23} \text{ Y atoms}}$$

Because each ratio can be inverted, six different factors result from each set of three quantities. See if you can write the five other factors for atom Y. If you can, you are on your way to mastering the factor-unit method for problems involving the mole. If you can't, go back and review Section 2.6.

Example 2.8

How many moles of ears, tails, and legs are contained in 1 mol of normal rabbits?

Solution

This example is nonchemical, but it might help you grasp the relationships that exist between the individual parts of a formula and the formula as a whole. The parts of a rabbit are related to a rabbit just as the parts of a formula are related to the entire formula.

Ears: Each rabbit has two ears and a 2:1 ratio exists between the number of ears and the number of rabbits. Therefore, 1 mol of rabbits contains 2 mol of ears.

Tails: The 1:1 ratio of tails to rabbits leads to the result that 1 mol of rabbits contains 1 mol of tails.

Legs: The 4:1 ratio of legs to rabbits leads to the result that 1 mol of rabbits contains 4 mol of legs.

Example 2.9

How many moles of each type of atom are contained in 1 mol of chloroform ($CHCl_3$)?

Solution

Just as each rabbit has two ears, each chloroform molecule contains one C atom, one H atom, and three Cl atoms. Therefore, 1 mol of $CHCl_3$ contains 1 mol of C atoms, 1 mol of H atoms, and 3 mol of Cl atoms.

How many moles of each type of atom would be contained in 0.50 mol of glucose ($C_6H_{12}O_6$)?

The usefulness of this approach can be increased by remembering and using the mass relationships of the mole concept. Thus, Statement 5 written earlier for water can be changed to Statement 6:

6. 18.0 g of water contains 2.0 g of H and 16.0 g of O.

Or, in a more concise form:

$$18.0 \text{ g } H_2O = 2.0 \text{ g H} + 16.0 \text{ g O}$$

Mass relationships such as these allow percent compositions to be calculated easily.

Example 2.10

Ammonia (NH_3) and ammonium nitrate (NH_4NO_3) are commonly used agricultural fertilizers. Which one of the two contains the higher mass percentage of nitrogen (N)?

GOB
Chemistry··Now™
Go to GOB Now and click to calculate the percent composition of an element in a chemical compound.

Solution

In each case, the mass percentage of N is given by

$$\% \text{ N} = \frac{\text{part}}{\text{total}} \times 100 = \frac{\text{mass of N}}{\text{mass of compound}} \times 100$$

We will use 1 mol of each compound as a sample because the mass in grams of 1 mol of compound and the mass in grams of N in the 1 mol of compound are readily determined. One mol of NH_3 weighs 17.0 g and contains 1 mol of N atoms, which weighs 14.0 g.

$$\% \text{ N} = \frac{14.0 \text{ g}}{17.0 \text{ g}} \times 100 = 82.4\%$$

Similarly, 1 mol of NH_4NO_3 weighs 80.0 g and contains 2 mol of N atoms, which weigh 28.0 g.

$$\% \text{ N} = \frac{28.0 \text{ g}}{80.0 \text{ g}} \times 100 = 35.0\%$$

Determine the mass percentage of carbon in carbon dioxide (CO_2) and carbon monoxide (CO).

> **FOR FUTURE REFERENCE** GOOD DIETARY SOURCES OF CALCIUM

Calcium is the most abundant mineral in the body. About 99% is in the form of compounds in bones and teeth, and much of the remaining 1% circulates as Ca^{2+} ions in blood and other body fluids. Despite their appearance as unchanging, inert structures, bones are constantly involved in processes of being built up or broken down. The net result of these processes generally depends on age. Growing children build up bone faster than it breaks down, so they gain more bone than they lose. Healthy adults are usually in a state of balance. In some older individuals, the breakdown rate exceeds the buildup rate, and conditions such as osteoporosis develop.

Even though it represents only a small part of the body's calcium, the amount circulating in the blood is vital to life, and is maintained at an essentially constant level. If the blood calcium level is not maintained by the absorption of calcium from dietary sources, the difference is made up by drawing on the calcium reserves

(continued)

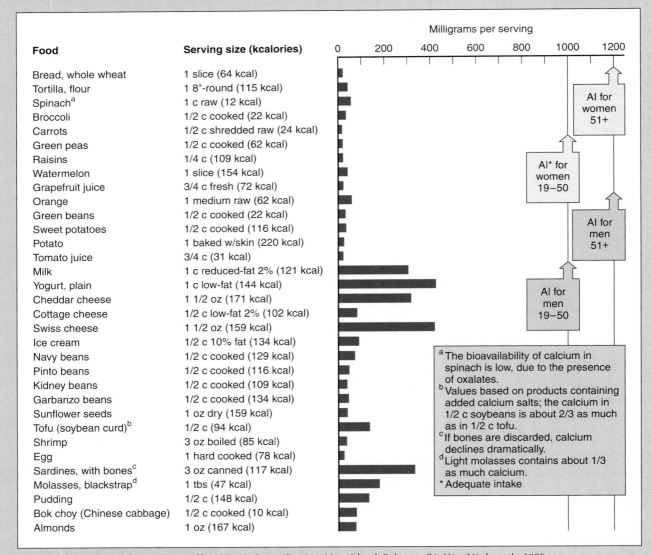

Food	Serving size (kcalories)
Bread, whole wheat	1 slice (64 kcal)
Tortilla, flour	1 8"-round (115 kcal)
Spinach[a]	1 c raw (12 kcal)
Broccoli	1/2 c cooked (22 kcal)
Carrots	1/2 c shredded raw (24 kcal)
Green peas	1/2 c cooked (62 kcal)
Raisins	1/4 c (109 kcal)
Watermelon	1 slice (154 kcal)
Grapefruit juice	3/4 c fresh (72 kcal)
Orange	1 medium raw (62 kcal)
Green beans	1/2 c cooked (22 kcal)
Sweet potatoes	1/2 c cooked (116 kcal)
Potato	1 baked w/skin (220 kcal)
Tomato juice	3/4 c (31 kcal)
Milk	1 c reduced-fat 2% (121 kcal)
Yogurt, plain	1 c low-fat (144 kcal)
Cheddar cheese	1 1/2 oz (171 kcal)
Cottage cheese	1/2 c low-fat 2% (102 kcal)
Swiss cheese	1 1/2 oz (159 kcal)
Ice cream	1/2 c 10% fat (134 kcal)
Navy beans	1/2 c cooked (129 kcal)
Pinto beans	1/2 c cooked (116 kcal)
Kidney beans	1/2 c cooked (109 kcal)
Garbanzo beans	1/2 c cooked (134 kcal)
Sunflower seeds	1 oz dry (159 kcal)
Tofu (soybean curd)[b]	1/2 c (94 kcal)
Shrimp	3 oz boiled (85 kcal)
Egg	1 hard cooked (78 kcal)
Sardines, with bones[c]	3 oz canned (117 kcal)
Molasses, blackstrap[d]	1 tbs (47 kcal)
Pudding	1/2 c (148 kcal)
Bok choy (Chinese cabbage)	1/2 c cooked (10 kcal)
Almonds	1 oz (167 kcal)

[a] The bioavailability of calcium in spinach is low, due to the presence of oxalates.
[b] Values based on products containing added calcium salts; the calcium in 1/2 c soybeans is about 2/3 as much as in 1/2 c tofu.
[c] If bones are discarded, calcium declines dramatically.
[d] Light molasses contains about 1/3 as much calcium.
* Adequate intake

SOURCE: Adapted from Whitney, E.N.; Rolfes, S.R. *Understanding Nutrition*, 8th ed. Belmont, CA: West/Wadsworth, 1999.

of the bones, thereby speeding up the breakdown processes. Because of this, it is essential that people of all ages include adequate sources of calcium in their diets, or use supplements (see Over the Counter 2.1).

Dietary calcium is found most abundantly in milk and milk products. However, some people avoid milk for a variety of reasons. Some perceive milk products as too high in fat and calories, and omit them from their diet in an attempt to lose or control weight. Some cultural groups do not use milk in their food, some vegetarians exclude milk as well as meat from their diets, some people are allergic to milk protein, and other people suffer from lactose intolerance. Such individuals need to include nonmilk sources of calcium in their diets in order to meet their daily calcium requirements. The chart on the previous page lists some good sources.

CONCEPT SUMMARY

Symbols and Formulas. Symbols based on names have been assigned to every element. Most consist of a single capital letter followed by a lowercase letter. A few consist of a single capital letter. Compounds are represented by formulas made up of elemental symbols. The number of atoms of each element in a molecule is shown by subscripts.

Inside the Atom. Atoms are made up of numerous smaller particles of which the most important to chemical studies are the proton, neutron, and electron. Positively charged protons and neutral neutrons have a relative mass of 1 u each and are located in the nuclei of atoms. Negatively charged electrons with a mass of 1/1836 u are located outside the nuclei of atoms.

Isotopes. Most elements in their natural state are made up of more than one kind of atom. These different kinds of atoms of a specific element are called isotopes and differ from one another only in the number of neutrons in their nuclei. A symbol incorporating atomic number, mass number, and elemental symbol is used to represent specific isotopes.

Relative Masses of Atoms and Molecules. Relative masses called atomic weights have been assigned to each element and are tabulated in the periodic table. The units used are atomic mass units, abbreviated u. Relative masses for molecules, called molecular weights, are determined by adding the atomic weights of the atoms making up the molecules.

Isotopes and Atomic Weights. The atomic weights measured for elements are average weights that depend on the percentages and masses of the isotopes in the naturally occurring element. If the isotope percent abundances and isotope masses are known for an element, its atomic weight can be calculated.

Avogadro's Number: The Mole. Avogadro's number of the atoms of an element has a mass in grams equal to the atomic weight of the element. Avogadro's number of molecules has a mass in grams equal to the molecular weight. Avogadro's number of particles is called a mole, abbreviated mol.

The Mole and Chemical Formulas. The mole concept when applied to molecular formulas gives numerous relationships that yield useful factors for factor-unit calculations.

LEARNING OBJECTIVES ASSESSMENT

You can get an approximate but quick idea of how well you have met the learning objectives given at the beginning of this chapter by working the selected end-of-chapter exercises given below. The answer to each exercise is given in Appendix B of the book.

Objective 1 (Section 2.1): Exercise 2.4

Objective 2 (Section 2.2): Exercises 2.10 and 2.12

Objective 3 (Section 2.3): Exercises 2.16 and 2.22

Objective 4 (Section 2.4): Exercise 2.32

Objective 5 (Section 2.5): Exercise 2.38

Objective 6 (Section 2.6): Exercises 2.44 a & b, and 2.46 a & b

Objective 7 (Section 2.7): Exercises 2.50 b and 2.52 b

KEY TERMS AND CONCEPTS

Atomic mass unit (u) (2.4)
Atomic number of an atom (2.3)
Atomic weight (2.4)
Compound formula (2.1)

Elemental symbol (2.1)
Isotopes (2.3)
Mass number of an atom (2.3)
Mole (2.6)

Molecular weight (2.4)
Nucleus (2.2)

LEGEND: 1 = straightforward, 2 = intermediate, 3 = challenging. All even-numbered exercises are answered in Appendix B.

You will find it useful to refer to Table 2.1 and the periodic table inside the front cover as you work these exercises.

SYMBOLS AND FORMULAS (SECTION 2.1)

2.1 Draw a "formula" for each of the following molecules using circular symbols of your choice to represent atoms:
  a. A diatomic molecule of an element
  b. A diatomic molecule of a compound
  c. A triatomic molecule of an element
  d. A molecule of a compound containing one atom of one element and four atoms of another element

2.2 Draw a "formula" for each of the following molecules using circular symbols of your choice to represent atoms:
  a. A triatomic molecule of a compound
  b. A molecule of a compound containing two atoms of one element and two atoms of a second element
  c. A molecule of a compound containing two atoms of one element, one atom of a second element, and four atoms of a third element
  d. A molecule containing two atoms of one element, six atoms of a second element, and one atom of a third element

2.3 Write formulas for the following molecules using elemental symbols from Table 2.1 and subscripts. Compare these formulas with those of Exercise 2.1.
  a. A diatomic molecule of fluorine gas
  b. A diatomic molecule of hydrogen chloride (one hydrogen atom and one chlorine atom)
  c. A triatomic molecule of ozone (a molecular form of the element oxygen)
  d. A molecule of methane (one carbon atom and four hydrogen atoms)

2.4 Write formulas for the following molecules using elemental symbols from Table 2.1 and subscripts. Compare these formulas with those of Exercise 2.2.
  a. A molecule of water (two hydrogen atoms and one oxygen atom)
  b. A molecule of hydrogen peroxide (two hydrogen atoms and two oxygen atoms)
  c. A molecule of sulfuric acid (two hydrogen atoms, one sulfur atom, and four oxygen atoms)
  d. A molecule of ethyl alcohol (two carbon atoms, six hydrogen atoms, and one oxygen atom)

2.5 Determine the number of each type of atom in molecules represented by the following formulas:
  a. nitrous acid (HNO_2)
  b. chlorine dioxide (ClO_2)
  c. ethyl alcohol (C_2H_6O)
  d. chloroform ($CHCl_3$)

2.6 Determine the number of each type of atom in molecules represented by the following formulas:
  a. methane (CH_4)
  b. perchloric acid ($HClO_4$)
  c. aluminum chloride ($AlCl_3$)
  d. propane (C_3H_8)

2.7 Tell what is wrong with each of the following molecular formulas and write a correct formula:
  a. H3PO3 (phosphorous acid)
  b. SICl$_4$ (silicon tetrachloride)
  c. SOO (sulfur dioxide)
  d. 2HO (hydrogen peroxide—two hydrogen atoms and two oxygen atoms)

2.8 Tell what is wrong with each of the following formulas and write a correct formula:
  a. HSH (hydrogen sulfide)
  b. HCLO$_2$ (chlorous acid)
  c. 2HN$_2$ (hydrazine—two hydrogen atoms and four nitrogen atoms)
  d. C2H6 (ethane)

INSIDE THE ATOM (SECTION 2.2)

2.9 Determine the charge and mass (in u) of nuclei made up of the following particles:
  a. 3 protons and 4 neutrons
  b. 10 protons and 12 neutrons
  c. 35 protons and 46 neutrons
  d. 56 protons and 81 neutrons

2.10 Determine the charge and mass (in u) of nuclei made up of the following particles:
  a. 4 protons and 5 neutrons
  b. 9 protons and 10 neutrons
  c. 20 protons and 23 neutrons
  d. 47 protons and 60 neutrons

2.11 Determine the number of electrons that would have to be associated with each nucleus described in Exercise 2.9 to produce a neutral atom.

2.12 Determine the number of electrons that would have to be associated with each nucleus described in Exercise 2.10 to produce a neutral atom.

ISOTOPES (SECTION 2.3)

2.13 Determine the number of electrons and protons contained in an atom of the following elements:
  a. sulfur
  b. As
  c. element number 24

2.14 Determine the number of electrons and protons contained in an atom of the following elements:
  a. potassium
  b. Cd
  c. element number 51

2.15 Determine the number of protons, number of neutrons, and number of electrons in atoms of the following isotopes:
  a. $^{7}_{3}Li$
  b. $^{22}_{10}Ne$
  c. $^{44}_{20}Ca$

2.16 Determine the number of protons, number of neutrons, and number of electrons in atoms of the following isotopes:
   a. $^{34}_{16}S$
   b. $^{91}_{40}Zr$
   c. $^{131}_{54}Xe$

2.17 Write symbols like those given in Exercises 2.15 and 2.16 for the following isotopes:
   a. cadmium-110
   b. cobalt-60
   c. uranium-235

2.18 Write symbols like those given in Exercises 2.15 and 2.16 for the following isotopes:
   a. silicon-28
   b. argon-40
   c. strontium-88

2.19 Determine the mass number and atomic number for atoms containing the nuclei described in Exercise 2.9. Write symbols for each atom like those given in Exercises 2.15 and 2.16.

2.20 Determine the mass number and atomic number for atoms containing the nuclei described in Exercise 2.10. Write symbols for each atom like those given in Exercises 2.15 and 2.16.

2.21 Write isotope symbols for atoms with the following characteristics:
   a. Contains 15 electrons and 16 neutrons
   b. A radon atom with a mass number of 211
   c. An oxygen atom that contains 10 neutrons

2.22 Write isotope symbols for atoms with the following characteristics:
   a. Contains 17 electrons and 20 neutrons
   b. A copper atom with a mass number of 65
   c. A zinc atom that contains 36 neutrons

RELATIVE MASSES OF ATOMS AND MOLECULES (SECTION 2.4)

2.23 Write the symbols and names for two elements whose average atoms have masses that are within 0.3 u of each other. Don't look beyond element number 83.

2.24 How many average helium atoms would be needed to balance the mass of a single average carbon atom?

2.25 What are the symbol and name for an element whose average atoms have a mass very close to three times the mass of an average beryllium atom?

2.26 What are the symbol and name for an element whose average atoms have a mass that is 77.1% of the mass of an average chromium atom?

2.27 In the first 36 elements, 6 elements have atoms whose average mass is within 0.2 u of being twice the atomic number of the element. Write the symbols and names for these 6 elements.

2.28 What are the symbol and name of the element whose average atoms have a mass very nearly half the mass of an average silicon atom?

2.29 Determine the molecular weights of the following in u:
   a. oxygen (O_2)
   b. carbon monoxide (CO)
   c. chloric acid ($HClO_3$)
   d. glycerine ($C_3H_8O_3$)
   e. sulfur dioxide (SO_2)

2.30 Determine the molecular weights of the following in u:
   a. sulfur trioxide (SO_3)
   b. glycerin ($C_3H_8O_3$)
   c. sulfuric acid (H_2SO_4)
   d. nitrogen (N_2)
   e. propane (C_3H_8)

2.31 The molecular weight was determined for a gas that is known to be an oxide of nitrogen. The value obtained experimentally was 43.98 u. Which of the following is most likely to be the formula of the gas? NO, N_2O, NO_2.

2.32 A flammable gas is known to contain only carbon and hydrogen. Its molecular weight is determined and found to be 28.05 u. Which of the following is the likely identity of the gas? acetylene (C_2H_2), ethylene (C_2H_4), ethane (C_2H_6).

2.33 Glycine, an amino acid found in proteins, has a molecular weight of 75.07 u and is represented by the formula $C_2H_xNO_2$. What number does x stand for in the formula?

2.34 Serine, an amino acid found in proteins, has a molecular weight of 105.10 u and is represented by the formula $C_yH_7NO_3$. What number does y stand for in the formula?

ISOTOPES AND ATOMIC WEIGHTS (SECTION 2.5)

2.35 Naturally occurring beryllium has a single isotope. Determine the following for the naturally occurring atoms of beryllium:
   a. The number of neutrons in the nucleus
   b. The mass (in u) of the nucleus (to three significant figures)

2.36 Naturally occurring aluminum has a single isotope. Determine the following for the naturally occurring atoms of aluminum:
   a. The number of neutrons in the nucleus
   b. The mass (in u) of the nucleus (to three significant figures)

2.37 Calculate the atomic weight of lithium on the basis of the following percent composition and atomic weights of the naturally occurring isotopes. Compare the calculated value with the atomic weight listed for lithium in the periodic table.

$$\text{lithium-6} = 7.42\% \ (6.0151 \text{ u})$$
$$\text{lithium-7} = 92.58\% \ (7.0160 \text{ u})$$

2.38 Calculate the atomic weight of boron on the basis of the following percent composition and atomic weights of the naturally occurring isotopes. Compare the calculated value with the atomic weight listed for boron in the periodic table.

$$\text{boron-10} = 19.78\% \ (10.0129 \text{ u})$$
$$\text{boron-11} = 80.22\% \ (11.0093 \text{ u})$$

2.39 Calculate the atomic weight of silicon on the basis of the following percent composition and atomic weights of the naturally occurring isotopes. Compare the calculated value with the atomic weight listed for silicon in the periodic table.

$$\text{silicon-28} = 92.21\% \; (27.9769 \; u)$$

$$\text{silicon-29} = 4.70 \; (28.9765 \; u)$$

$$\text{silicon-30} = 3.09\% \; (29.9738 \; u)$$

2.40 Calculate the atomic weight of copper on the basis of the following percent composition and atomic weights of the naturally occurring isotopes. Compare the calculated value with the atomic weight listed for copper in the periodic table.

$$\text{copper-63} = 69.09\% \; (62.9298 \; u)$$

$$\text{copper-65} = 30.91\% \; (64.9278 \; u)$$

AVOGADRO'S NUMBER: THE MOLE (SECTION 2.6)

2.41 Refer to the periodic table and determine how many grams of phosphorus contain the same number of atoms as 0.12 g of carbon.

2.42 Refer to the periodic table and determine how many grams of fluorine contain the same number of atoms as 1.60 g of oxygen.

2.43 Write three relationships (equalities) based on the mole concept for each of the following elements:
   a. potassium
   b. magnesium
   c. tin

2.44 Write three relationships (equalities) based on the mole concept for each of the following elements:
   a. silicon
   b. calcium
   c. argon

2.45 Use a factor derived from the relationships written in Exercise 2.43 and the factor-unit method to determine the following:
   a. The number of moles of potassium atoms in a 50.0-g sample of potassium
   b. The number of magnesium atoms in a 1.82-mol sample of magnesium
   c. The number of tin atoms in a 200-g sample of tin

2.46 Use a factor derived from the relationships written in Exercise 2.44 and the factor-unit method to determine the following:
   a. The number of grams of silicon in 1.25 mol of silicon
   b. The mass in grams of one calcium atom
   c. The number of argon atoms in a 20.5-g sample of argon

THE MOLE AND CHEMICAL FORMULAS (SECTION 2.7)

2.47 Refer to the periodic table and calculate the molecular weights for the compounds PH_3 and SO_2. Then, determine how many grams of PH_3 contain the same number of molecules as 6.41 g of SO_2.

2.48 Refer to the periodic table and calculate the molecular weights for the compounds BF_3 and H_2S. Then, determine how many grams of BF_3 contain the same number of molecules as 0.34 g of H_2S.

2.49 For each formula given below, write statements equivalent to Statements 1–6 (see Section 2.7):
   a. methane (CH_4)
   b. ammonia (NH_3)
   c. chloroform ($CHCl_3$)

2.50 For each formula given below, write statements equivalent to Statements 1–6 (see Section 2.7):
   a. benzene (C_6H_6)
   b. nitrogen dioxide (NO_2)
   c. hydrogen chloride (HCl)

2.51 Answer the following questions based on information contained in the statements you wrote for Exercise 2.49.
   a. How many moles of hydrogen atoms are contained in 1 mol of CH_4 molecules?
   b. How many grams of nitrogen are contained in 1.00 mol of NH_3?
   c. What is the mass percentage of chlorine in $CHCl_3$?

2.52 Answer the following questions based on information contained in the statements you wrote for Exercise 2.50.
   a. How many moles of hydrogen atoms are contained in 0.75 mol of benzene?
   b. How many oxygen atoms are contained in 0.50 mol of nitrogen dioxide?
   c. What is the mass percentage of chlorine in HCl?

2.53 How many moles of $C_4H_{10}O$ contain the same number of carbon atoms as 1 mol of $C_2H_3O_2F$?

2.54 How many grams of C_2H_6O contain the same number of oxygen atoms as 0.75 mol of H_2O?

2.55 Determine the mass percentage of nitrogen in N_2O and NO_2.

2.56 Determine the mass percentage of hydrogen in CH_4 and C_2H_6.

2.57 Any of the statements based on a mole of substance (Statements 4–6) can be used to obtain factors for problem solving by the factor-unit method. Write statements equivalent to 4, 5, and 6 for nitrophenol ($C_6H_5NO_3$). Use a single factor obtained from the statements to solve each of the following. A different factor will be needed in each case.
   a. How many grams of nitrogen are contained in 70.0 g of $C_6H_5NO_3$?
   b. How many moles of oxygen atoms are contained in 1.50 mol of $C_6H_5NO_3$?
   c. How many atoms of carbon are contained in 9.00×10^{22} molecules of $C_6H_5NO_3$?

2.58 Any of the statements based on a mole of substance (Statements 4–6) can be used to obtain factors for problem solving by the factor-unit method. Write statements equivalent to 4, 5, and 6 for phosphoric acid (H_3PO_4). Use a single factor obtained from the statements to solve each of the following. A different factor will be needed in each case.
   a. How many grams of hydrogen are contained in 46.8 g of H_3PO_4?
   b. How many moles of oxygen atoms are contained in 1.25 mol of H_3PO_4?
   c. How many atoms of phosphorus are contained in 8.42×10^{21} molecules of H_3PO_4?

2.59 Urea (CH$_4$N$_2$O) and ammonium sulfate (N$_2$H$_8$SO$_4$) are both used as agricultural fertilizers. Which one contains the higher mass percentage of nitrogen?

2.60 Two iron ores that have been used as sources of iron are magnetite (Fe$_3$O$_4$) and hematite (Fe$_2$O$_3$). Which one contains the higher mass percentage of iron?

2.61 Both calcite (CaCO$_3$) and dolomite (CaMgC$_2$O$_6$) are used as dietary calcium supplements. Calculate the mass percentage of calcium in each mineral.

> CHEMISTRY FOR THOUGHT

2.1 a. Explain how atoms of different elements differ from one another.

  b. Explain how atoms of different isotopes of the same element differ from one another.

2.2 The atomic weight of aluminum is 26.98 u, and the atomic weight of nickel is 58.69 u. All aluminum atoms have a mass of 26.98 u, but not a single atom of nickel has a mass of 58.69 u. Explain.

2.3 Answer the question in the caption of Figure 2.3. Would you expect any orange in the bowl to have the exact mass you calculated as an average? Explain.

2.4 Answer the question in the caption of Figure 2.5. Use your answer and the fact that an average jelly bean has a mass of 1.18 g to calculate the number of beans in each jar.

2.5 Answer the question in the caption of Figure 2.6. How many CS$_2$ molecules would be required to contain 0.25 mol of sulfur atoms?

2.6 In Section 2.4 it was pointed out that an atomic mass unit, u, is equal to $\frac{1}{12}$ the mass of an atom of carbon-12. Suppose one atomic mass unit was redefined as being equal to $\frac{1}{24}$ the mass of a carbon-12 atom. How would this change influence the value of the atomic weight of magnesium?

2.7 How would the change of question 2.6 influence the ratio of the atomic weight of magnesium divided by the atomic weight of hydrogen?

2.8 How would the change of question 2.6 influence the value of Avogadro's number?

InfoTrac College Edition Readings

"The elemental man: an interview with Glenn T. Seaborg," *Skeptical Inquirer,* Nov–Dec 1997, 21(6):43(5). Record number A20379232.

"Newly identified molecules contribute to normal silencing of most human genes," *Genomics & Genetics Weekly,* April 4, 2003:39. Record number A99267058.

"What you need to know about calcium," *Harvard Health Letter,* April 2003, 28(6):0. Record number A98575559.

GOB
Chemistry ·:· Now ™

Assess your understanding of this chapter's topics with additional quizzing and conceptual-based problems at http://chemistry.brookscole.com/seager5e

CHAPTER 3

Electronic Structure and the Periodic Law

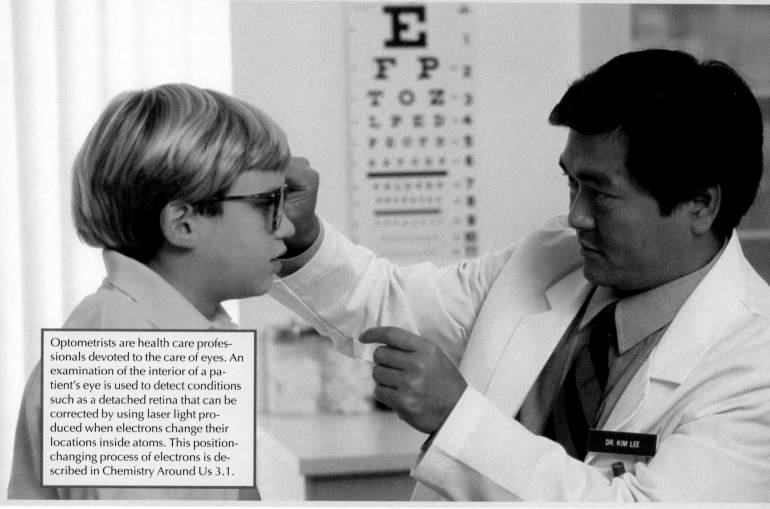

Optometrists are health care professionals devoted to the care of eyes. An examination of the interior of a patient's eye is used to detect conditions such as a detached retina that can be corrected by using laser light produced when electrons change their locations inside atoms. This position-changing process of electrons is described in Chemistry Around Us 3.1.

© Michael Keller/CORBIS

LEARNING OBJECTIVES

When you have completed your study of this chapter, you should be able to:

1. Locate elements in the periodic table on the basis of group and period designations. (Section 3.1)

2. Determine numerical relationships such as the number of electrons in designated orbitals, shells, or subshells. (Section 3.2)

3. Determine the number of electrons in the valence shell. (Section 3.3)

4. Relate electronic structure to the location of elements in the periodic table. (Section 3.3)

5. Write electronic configurations for elements. (Section 3.4)

6. Determine the number of unpaired electrons for elements. (Section 3.4)

7. Identify elements from their electronic configurations. (Section 3.4)

8. Determine the shell and subshell location of the distinguishing electron in elements. (Section 3.5)

9. Based on their location in the periodic table, correctly classify elements into the following categories: representative element, transition element, inner-transition element, noble gas, metal, nonmetal, and metalloid. (Section 3.5)

10. Recognize property trends within the periodic table. (Section 3.6)

11. Use property trends in the periodic table to predict selected properties of elements. (Section 3.6)

n Chapter 1, we defined atoms as particles that represent the limit of chemical subdivision. According to this idea, atoms of a specific element cannot be divided into smaller particles or converted into atoms of another element by any physical or chemical change. Then in Chapter 2, we introduced the idea that atoms are, in fact, made up of particles that are smaller than the atoms themselves. Two of these particles, protons and neutrons, form the nuclei of atoms, whereas electrons are located outside the nuclei.

The idea that atoms are made up of subatomic particles implies that it should be possible to obtain even smaller particles from atoms. Scientists have found that it is possible. During chemical changes, electrons are transferred from one atom to another or are shared between atoms. Some details of these processes are given in Chapter 4, but it is known that they depend on the arrangements of the electrons around the nuclei of atoms. These electronic arrangements are one of the major topics of this chapter.

➤ 3.1 The Periodic Law and Table

By the early 19th century, detailed studies of the elements known at that time had produced an abundance of chemical information. Scientists looked for order in these facts, with the hope of providing a systematic approach to the study of chemistry. Two scientists independently, and almost simultaneously, made the same important contribution to this end. Julius Lothar Meyer, a German, and Dmitri Mendeleev, a Russian, each produced classification schemes for the elements in 1869. Both schemes were based on the **periodic law,** which in its present form is stated as follows: When all the elements are arranged in order of increasing atomic numbers, elements with similar chemical properties will occur at regular (periodic) intervals.

A convenient way to compactly represent such behavior is to use tables. The arrangement of the elements in a table based on the periodic law is called a *periodic table.* In a modern periodic table, such as the one inside the front cover of this book, elements with similar chemical properties are found in vertical columns called **groups** or **families.**

The groups are designated by a roman numeral and a letter at the top of each column. These group designations have not been universally accepted by chemists throughout the world. An effort has been under way since 1979 to establish a universally acceptable group designation. The simple numerical designation given in parentheses over the traditional designation appears to be the one that will be adopted. In this book, references to groups will be given using both designations, with the new one in parentheses. The horizontal rows in the table are called **periods** and are numbered from top to bottom. Thus, each element belongs to both a period and a group of the periodic table.

Periodic law
A statement about the behavior of the elements when they are arranged in a specific order. In its present form, it is stated as follows: Elements with similar chemical properties occur at regular (periodic) intervals when the elements are arranged in order of increasing atomic numbers.

Group or family of the periodic table
A vertical column of elements that have similar chemical properties.

Period of the periodic table
A horizontal row of elements.

Example 3.1

Identify the group and period to which each of the following belongs:

a. P   **b.** Cr   **c.** element number 30   **d.** element number 53

Solution

a. Phosphorus (P) is in group VA(15) and period 3.
b. Chromium (Cr) is in group VIB(6) and period 4.

c. The element with atomic number 30 is zinc (Zn), which is found in group IIB(12) and period 4.
d. Element number 53 is iodine (I) found in group VIIA(17) and period 5.

Learning Check 3.1

Write the symbol for the element found in the following places of the periodic table:

a. Group IVA(14) and period 4 b. Group VIIB(7) and period 6

It should be noted that the periodic table as given inside the front cover appears to violate the practice of arranging the elements according to increasing atomic number. Element 72 follows element 57, and 104 follows 89, whereas elements 58–71 and 90–103 are arranged in two rows at the bottom of the table. Technically, these two rows should be included in the body of the table as shown in ■ Figure 3.1a. To save horizontal space, they are placed in the position shown in Figure 3.1b. This exception presents no problem, as long as it is understood.

Example 3.2

a. How many elements are found in period 6 of the periodic table?
b. How many elements are found in group VA(15) of the periodic table?

1																															2
3	4																									5	6	7	8	9	10
11	12																									13	14	15	16	17	18
19	20	21															22	23	24	25	26	27	28	29	30	31	32	33	34	35	36
37	38	39															40	41	42	43	44	45	46	47	48	49	50	51	52	53	54
55	56	57	58	59	60	61	62	63	64	65	66	67	68	69	70	71	72	73	74	75	76	77	78	79	80	81	82	83	84	85	86
87	88	89	90	91	92	93	94	95	96	97	98	99	100	101	102	103	104	105	106	107	108	109	110	111	112						

(a)

1																	2
3	4											5	6	7	8	9	10
11	12											13	14	15	16	17	18
19	20	21	22	23	24	25	26	27	28	29	30	31	32	33	34	35	36
37	38	39	40	41	42	43	44	45	46	47	48	49	50	51	52	53	54
55	56	57	72	73	74	75	76	77	78	79	80	81	82	83	84	85	86
87	88	89	104	105	106	107	108	109	110	111	112						

58	59	60	61	62	63	64	65	66	67	68	69	70	71
90	91	92	93	94	95	96	97	98	99	100	101	102	103

(b)

➤ **FIGURE 3.1** Forms of the periodic table. (a) The periodic table with elements 58–71 and 90–103 (area in color) in their proper positions. (b) The periodic table modified to conserve space, with elements 58–71 and 90–103 (in color) placed at the bottom.

OVER THE COUNTER • 3.1

ZINC FOR COLDS? THE JURY IS STILL OUT

Zinc, element number 30 of the periodic table, might be a key to relieving the cold symptoms suffered by millions every year. In preliminary studies, it has been shown that compounds of zinc have the ability to inhibit the reproduction of viruses and possibly to promote the body's production of interferon, a virus-fighter. In an attempt to determine the effectiveness of zinc compounds against the viruses that cause the common cold, 100 adult patients were studied at the Cleveland Clinic. The patients were given lozenges within 24 hours of contracting a cold. During their waking hours, the patients dissolved a lozenge in their mouths every 2 hours. This treatment was continued until the patients no longer showed any cold symptoms. Some of the patients were given lozenges that contained zinc gluconate, while others were given placebo (non-zinc-containing) lozenges.

The results of the study showed that cold symptoms lasted an average of 4.4 days in the patients who received the zinc gluconate, compared to an average duration of 7.6 days in the patients who did not receive the zinc compound. The sales of over-the-counter zinc products, especially zinc lozenges, skyrocketed following the publication of these results in 1996. However, this enthusiasm was tempered a bit by the results of another study published in 1998. In this study, the effects of zinc lozenges on the cold symptoms of 249 students in grades 1 through 12 were investigated. Researchers concluded that the zinc lozenges were not effective against cold symptoms in children and teenagers.

Various explanations have been proposed for the discrepancy in the results of the two studies, but it has been generally concluded that further studies are needed to clarify what role, if any, zinc compounds may play in treating cold symptoms.

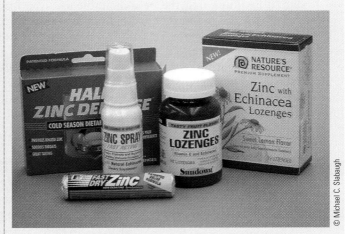

Despite conflicting research results about their effectiveness, many brands of zinc lozenges are available for use in treating the common cold.

Solution

a. Period 6 includes elements 55–86, even though elements 58–71 are shown below the main table. Therefore, period 6 contains 32 elements.

b. A count shows group VA(15) to contain 5 elements: N, P, As, Sb, and Bi.

Learning Check 3.2

How many elements are found in the following?

a. Period 1 of the periodic table

b. Group IIB(12) of the periodic table

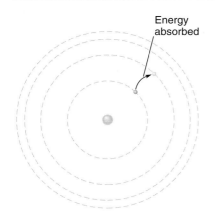

> **FIGURE 3.2** A diagram of the Bohr hydrogen atom (not drawn to scale; the orbits are actually much larger than the nucleus). The electron is elevated to a higher-energy orbit when energy is absorbed.

➤ 3.2 Electronic Arrangements in Atoms

In 1913 Niels Bohr, a young Danish physicist, made an important contribution to our understanding of atomic structure. He was working under the direction of Ernest Rutherford, a British scientist, who had proposed a solar system model for atoms in which negative electrons moved in circular orbits around the positive nucleus, much like the planets move around the sun. Bohr built on this model by proposing that the single electron of a hydrogen atom could occupy orbits only at specific distances from the nucleus, and thus the electron could have only specific energies (see ■ Figure 3.2). Bohr further proposed that the electron changed orbits only by absorbing or releasing energy. The addition of energy to a hydrogen atom elevated the electron to a higher-energy orbit located farther from the nucleus. The energy released when an electron dropped from a higher- to a lower-energy orbit appeared as emitted light.

Research into the nature of atoms continued after Bohr's proposal, and in 1926 a revised model of atomic structure was proposed by Erwin Schrödinger, an Austrian physicist, who received the Nobel Prize in physics in 1933 in recognition of this achievement. According to Schrödinger's quantum mechanical model, the precise paths of electrons cannot be determined accurately, as Bohr's model required. It was found that the location and energy of electrons around a nucleus can be specified using three terms: shell, subshell, and orbital. This is somewhat like locating an individual in a city by specifying a street, building, and apartment.

The location of electrons in a **shell** is indicated by assigning a number, n, to the shell and to all the electrons within the shell. The n value of the lowest-energy shell is 1, that of the next higher energy is 2, the next is 3, and so on. Higher n values for a shell correspond to higher energies and greater distances from the nucleus for the electrons of the shell. Electrons in the third shell all have an n value of 3, all have an energy higher than the energies of electrons in shells 1 and 2, and also are located farther from the nucleus than those of shells 1 and 2.

Each shell is made up of **subshells** that are designated by a letter from the group s, p, d, and f. Because all subshells are designated by one of these letters regardless of the shell in which the subshell is found, a combination of both shell number and subshell letter is used to identify subshells clearly. Thus, a p subshell in shell number 2 is referred to as a $2p$ subshell. The number of subshells found in a shell is the same as the value of n for the shell. Thus, shell number 2 ($n = 2$) contains two subshells. The subshells are the $2p$ mentioned earlier and a $2s$. Electrons located in specific subshells are often referred to in terms of the same number and letter as the subshell. For example, we might refer to an atom as having three $2p$ electrons. All electrons within a specific subshell have the same energy.

The description of the location and energy of electrons moving around a nucleus is completed in the quantum mechanical model by specifying an orbital. Each subshell consists of one or more **atomic orbitals**, which are specific volumes of space around nuclei in which electrons move. These atomic orbitals must not be confused with the fixed electron orbits of the original Bohr theory; they are *not* the same. These volumes of space around nuclei have different shapes, depending on the energy of the electrons they contain (see ■ Figure 3.3). All s subshells consist of a single orbital that is also designated by the letter s and further identified by the n value of the shell to which the subshell belongs. Thus, the $2s$ subshell mentioned earlier consists of a single $2s$ orbital. All p subshells consist of three p orbitals that also carry the n value of the shell. Thus, the $2p$ subshell of shell number 2 consists of three $2p$ orbitals. Since all the electrons in a subshell have the same energy, an electron in any one of the three $2p$ orbitals has the same energy, regardless of which orbital of the three it occupies. All d subshells contain five orbitals, and all f subshells contain seven orbitals. According to the quantum mechanical model, each orbital within a subshell can contain a maximum of two electrons.

Shell
A location and energy of electrons around a nucleus that is designated by a value for n, where $n = 1, 2, 3$, etc.

Subshell
A component of a shell that is designated by a letter from the group s, p, d, and f.

Atomic orbital
A volume of space around atomic nuclei in which electrons of the same energy move. Groups of orbitals with the same n value form subshells.

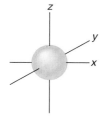
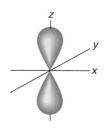
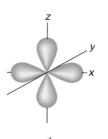

s p d

➤ **FIGURE 3.3** Shapes of typical s, p, and d orbitals.

The shapes of orbitals such as those given in Figure 3.3 must not be interpreted incorrectly. The fact that *s* orbitals are spherical in shape does not mean that the electrons move around on the spherical surface. According to the quantum mechanical model, electrons in *s* orbitals move around inside the spherical volume of the orbital in paths that cannot be determined. All that can be determined about their behavior within the orbital is the probability of finding them in a specific location. Thus, if it were determined that an electron had a 2% probability of being at a specific location, it would simply mean that the electron could be found at that location within an orbital 2 times out of every 100 times we looked for it there. Similarly, electrons in *p* or *d* orbitals do not move on the surface of the dumbbell- or cloverleaf-shaped orbitals; they move *within* the three-dimensional dumbbell- or cloverleaf-shaped volumes.

The energy of electrons located in an orbital within a subshell and shell is determined by two factors. As mentioned earlier, the higher the value of *n*, the higher the energy. In addition, if *n* is the same, but the subshell is different, the energy of the contained electrons increases in the order *s*, *p*, *d*, *f*. Thus, a 3*p* electron (an electron in a 3*p* subshell) has a higher energy than a 3*s* electron (an electron in a 3*s* subshell). ■ Figure 3.4 is a diagrammatic representation of the fourth shell of an atom completely filled with electrons.

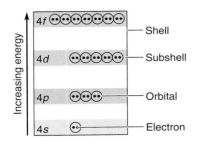

Increasing energy

4*f* — Shell

4*d* — Subshell

4*p* — Orbital

4*s* — Electron

➤ **FIGURE 3.4** The fourth shell of an atom, completely filled with electrons.

Example 3.3

Determine the following for the third shell of an atom:

a. The number of subshells
b. The designation for each subshell
c. The number of orbitals in each subshell
d. The maximum number of electrons that can be contained in each subshell
e. The maximum number of electrons that can be contained in the shell

Solution

a. The number of subshells is the same as the number used to designate the shell. Therefore, the third shell contains three subshells.
b. Subshells increase in energy according to the order *s*, *p*, *d*, *f*. The subshells in the shell are therefore designated 3*s*, 3*p*, and 3*d*.
c. The number of orbitals in the subshells is 1, 3, and 5 because *s* subshells always contain a single orbital. *p* subshells always contain three orbitals, and *d* subshells always contain five orbitals.
d. Each atomic orbital can contain a maximum of two electrons, independent of the type of orbital under discussion. Therefore, the 3*s* subshell (one orbital) can hold a maximum of two electrons, the 3*p* subshell (three orbitals) a maximum of 6 electrons, and the 3*d* subshell (five orbitals) a maximum of 10 electrons.
e. The maximum number of electrons that can be contained in the shell is simply the sum of the maximum number for the subshells, or 18.

Learning Check 3.3

In the following, what is the maximum number of electrons that can be found?

a. In a 4*p* orbital   b. In a 5*d* subshell   c. In shell number 1

It might seem to you at this point that the modifications to the Bohr theory have created a number of hard-to-remember relationships between shells, subshells, orbitals, and electrons. However, some patterns help make the relationships easy to remember. ■ Table 3.1 summarizes them for the first four shells of an atom.

TABLE 3.1 Relationships between shells, subshells, orbitals, and electrons

Shell number (n)	Number of subshells in shell	Subshell designation	Number of orbitals in subshell	Orbital designation	Maximum number of electrons in subshell	Maximum number of electrons in shell
1	1	1s	1	1s	2	2
2	2	2s	1	2s	2	
		2p	3	2p	6	8
3	3	3s	1	3s	2	
		3p	3	3p	6	
		3d	5	3d	10	18
4	4	4s	1	4s	2	
		4p	3	4p	6	
		4d	5	4d	10	
		4f	7	4f	14	32

➤ 3.3 The Shell Model and Chemical Properties

The arrangement of electrons into orbitals, subshells, and shells provides an explanation for the similarities in chemical properties of various elements. ■ Table 3.2 gives the number of electrons in each shell for the first 20 elements of the periodic table.

In Table 3.2, notice that the third shell stops filling when 8 electrons are present, even though the shell can hold a maximum of 18 electrons. The reasons for this are discussed in Section 3.4. Also, note that all the elements in a specific group of the periodic table have the same number of electrons in the outermost occupied shell. This outermost occupied shell (the one of highest energy) is called the **valence shell.** Similarities in elemental chemical properties result from identical numbers of electrons in the valence shells of the atoms (see ■ Figure 3.5).

Valence shell
The outermost (highest-energy) shell of an element that contains electrons.

➤ **FIGURE 3.5** Left to right: Magnesium, calcium, and strontium, members of group IIA(2) of the periodic table, have similar chemical properties and appearances.

© Spencer L. Seager

TABLE 3.2 Electron occupancy of shells

Element	Belongs to group	Symbol	Atomic number	Electrons in shell number			
				1	2	3	4
Hydrogen	IA(1)	H	1	1			
Helium	Noble gas(18)	He	2	2			
Lithium	IA(1)	Li	3	2	1		
Beryllium	IIA(2)	Be	4	2	2		
Boron	IIIA(13)	B	5	2	3		
Carbon	IVA(14)	C	6	2	4		
Nitrogen	VA(15)	N	7	2	5		
Oxygen	VIA(16)	O	8	2	6		
Fluorine	VIIA(17)	F	9	2	7		
Neon	Noble gas(18)	Ne	10	2	8		
Sodium	IA(1)	Na	11	2	8	1	
Magnesium	IIA(2)	Mg	12	2	8	2	
Aluminum	IIIA(13)	Al	13	2	8	3	
Silicon	IVA(14)	Si	14	2	8	4	
Phosphorus	VA(15)	P	15	2	8	5	
Sulfur	VIA(16)	S	16	2	8	6	
Chlorine	VIIA(17)	Cl	17	2	8	7	
Argon	Noble gas(18)	Ar	18	2	8	8	
Potassium	IA(1)	K	19	2	8	8	1
Calcium	IIA(2)	Ca	20	2	8	8	2

Example 3.4

Referring to Table 3.2, indicate the number of electrons in the valence shell of elements in groups IA(1), IIA(2), IIIA(13), and IVA(14).

Solution

According to Table 3.2, the elements in group IA(1) are hydrogen, lithium, sodium, and potassium. Each element has one electron in the valence shell. Hydrogen belongs in group IA(1) on the basis of its electronic structure, but its properties differ significantly from other group members.

The group IIA(2) elements are beryllium, magnesium, and calcium. Each has two electrons in the valence shell. The group IIIA(13) elements are boron and aluminum. Both have three electrons in the valence shell. The group IVA(14) elements are carbon and silicon; each has four valence-shell electrons.

Learning Check 3.4

Referring to Table 3.2, indicate the number of electrons in the valence shell of elements in groups VA(15), VIA(16), and VIIA(17), and the noble gases(18).

Example 3.4 and Learning Check 3.4 emphasize the fact that elements belonging to the same periodic table group have the same number of electrons in the valence shell (helium, the first element in the noble gases, is an exception). Notice also that the number of electrons in the valence shell is identical to the roman numeral that designates the group number. Elements of group IIIA(13), for example, have three electrons in the valence shell. It is also apparent that the n value for the valence shell increases by 1 with each heavier member of a group, that is, $n = 2$ for Li, $n = 3$ for Na, and $n = 4$ for K.

Example 3.5

Using the periodic table and Example 3.4 and Learning Check 3.4, determine the n value for the valence shell and the number of electrons in the valence shell for the following elements:

a. Ba   **b.** Br   **c.** element number 50
d. The third element of group VIA(16)

Solution

a. Ba is the fifth element of group IIA(2), and because $n = 4$ for Ca (the third element of the group), $n = 6$ for the fifth element. The number of valence-shell electrons is two, the same as the group number.
b. In a similar way, Br, the third element of group VIIA(17), has $n = 4$ and seven valence-shell electrons.
c. Element number 50 is tin (Sn), which is the fourth element of group IVA(14). Therefore, $n = 5$ and the number of valence-shell electrons is four.
d. The third element of group VIA(16) is selenium (Se), for which $n = 4$ and the number of valence-shell electrons is six.

How many electrons will be found in the following?

a. The valence shell of Sr
b. The valence shell of the third element in group IVA(14)
c. The valence shell of the fifteenth element in period 4

Learning Check 3.5

➤ 3.4 Electronic Configurations

According to Section 3.3, similarities in chemical properties between elements are related to the number of electrons that occupy the valence shells of their atoms. We now look at the electronic arrangements of atoms in more detail. These detailed arrangements, called **electronic configurations**, indicate the number of electrons that occupy each subshell and orbital of an atom.

Imagine that electrons are added one by one to the orbitals that belong to subshells and shells associated with a nucleus. The first electron will go to as low an energy state as possible, which is represented by the $1s$ orbital of the $1s$ subshell of the first shell. The second electron will join the first and completely fill the $1s$ orbital, the $1s$ subshell, and the first shell (remember, an orbital is filled when it contains two electrons). The third electron will have to occupy the lowest-energy subshell ($2s$) of the second shell. The fourth electron will also occupy (and fill) the $2s$ subshell. The fifth electron must seek out the next–highest-energy subshell, which is the $2p$. The $2p$ subshell contains three $2p$ orbitals, so the sixth electron can either join the fifth in one of

Electronic configurations
The detailed arrangement of electrons indicated by a specific notation, $1s^2 2s^2 2p^4$, etc.

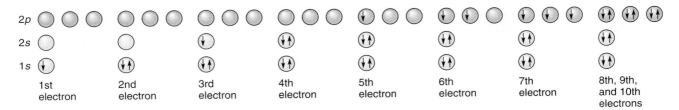

| 1st electron | 2nd electron | 3rd electron | 4th electron | 5th electron | 6th electron | 7th electron | 8th, 9th, and 10th electrons |

> **FIGURE 3.6** The filling order for the first 10 electrons.

Hund's rule
A statement of the behavior of electrons when they occupy orbitals: Electrons will not join other electrons in an orbital if an empty orbital of the same energy is available for occupancy.

Pauli exclusion principle
A statement of the behavior of electrons when they occupy orbitals: Only electrons spinning in opposite directions can simultaneously occupy the same orbital.

the $2p$ orbitals or go into an empty $2p$ orbital. It will go into an empty orbital in compliance with **Hund's rule,** which states: Electrons will not join other electrons in an orbital if an empty orbital of the same energy is available.

It has been found that electrons behave as if they spin on an axis, and only electrons spinning in opposite directions (indicated by ↑ and ↓) can occupy the same orbital. This principle, known as the **Pauli exclusion principle,** explains why orbitals can contain a maximum of two electrons. Hund's rule and the Pauli exclusion principle can be combined: Electrons will pair with other electrons in an orbital only if there is no empty orbital of the same energy available and if there is one electron with opposite spin already in the orbital.

The seventh added electron will occupy the last empty $2p$ orbital, and the eighth, ninth, and tenth electrons will pair up with electrons already in the $2p$ orbitals. The tenth electron fills the $2p$ subshell, thus completing the second shell. This filling order is illustrated in ■ Figure 3.6.

The eleventh electron, with no empty orbital available that has the same energy as a $2p$ and no partially filled orbitals, will occupy the empty lowest-energy subshell of the third shell. The filling order for electrons beyond the tenth follows the pattern given in ■ Figure 3.7, where shells are indicated by

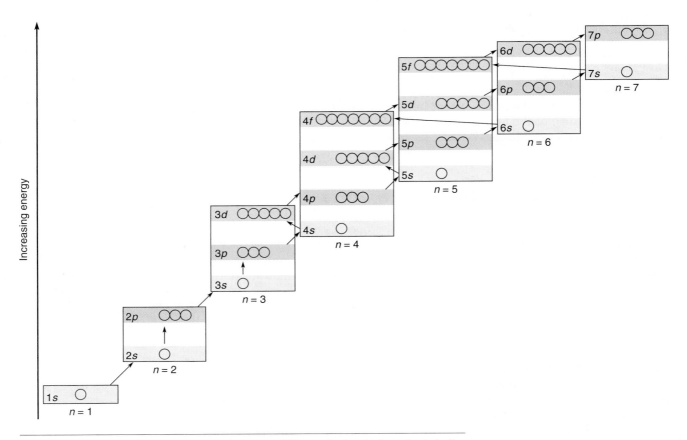

> **FIGURE 3.7** The relative energies and electron-filling order for shells and subshells.

large rectangles, subshells by colored rectangles, and orbitals by circles. The filling order is obtained by following the arrows.

As shown in Figure 3.7, some low-energy subshells of a specific shell have energies lower than the upper subshell of a preceding shell. For example, the 4s subshell has a lower energy than, and fills before, the 3d subshell. Figure 3.7 indicates that the third shell will not accept more than 8 electrons until the 4s subshell is complete. Thus, electrons 21 through 30 go into the 3d subshell and complete the filling of the third shell.

It is often convenient to represent the electronic configuration of an atom in a concise way. This is done by writing the subshells in the correct filling order and then indicating the number of electrons in each subshell by a superscript.

GOB
Chemistry⋅⚛⋅Now™
Go to GOB Now and click to view a simulation exploring the electronic configurations of the elements, emphasizing the relationships between electronic configuration and position on the periodic table.

Example 3.6

Write the electronic configurations for the following, and indicate the number of unpaired electrons in each case:

a. An atom that contains 7 electrons
b. An atom that contains 17 electrons
c. An atom of element number 22
d. An atom of arsenic (As)

Solution

The correct filling order of subshells from Figure 3.7 is

1s, 2s, 2p, 3s, 3p, 4s, 3d, 4p, 5s, 4d, 5p, 6s, 4f, 5d, 6p, 7s, 5f, 6d, 7p

---

CHEMISTRY AROUND US • 3.1

EXCITED ELECTRONS AND NEON LIGHTS

The electronic configurations discussed in this chapter represent the lowest-energy arrangements of electrons in atoms. We can add energy to the atoms in forms such as heat, light, electron bombardment, or chemical reactions. When we add energy to atoms, part or all of it can elevate electrons to higher-energy levels. When the excited electrons fall back to lower-energy levels, the atoms will often emit visible light.

When sodium-containing substances are heated in a flame, the flame appears yellow-orange as the electrons of sodium are excited by the heat of the flame and then fall back to lower-energy levels. Sodium-vapor streetlights also show the characteristic yellow-orange color as metallic sodium is vaporized, and the electrons of the vaporized atoms are excited as a result of electron bombardment.

A similar process produces a bluish white light in mercury-vapor streetlights and fluorescent lights. In these useful light sources, a small amount of liquid mercury is vaporized and excited by the passage of an electrical current.

The exciting of electrons of the atoms of some gases followed by their return to lower-energy levels produces bright, vivid colors such as the various colors from so-called neon signs. Actually, neon is only one of several gases used in such signs; helium and argon are examples of other gases that are used. Sometimes, the gases are put into colored glass tubes to give the desired colors.

Brightly colored visible light is emitted by a neon sign as excited electrons of the gas fall back to their most stable low-energy states.

a. Even though s subshells can hold 2 electrons, p subshells 6, d subshells 10, and f subshells 14, only enough subshells to hold 7 electrons will be used. Therefore, the configuration is written as shown below, starting on the left, with circles representing orbitals and arrows representing electrons. It is apparent that no subshells beyond $2p$ are needed because that subshell contains only 3 electrons. Note that both the $1s$ and $2s$ subshells are full; the $2p$ subshell is half-full, with one electron in each of the three orbitals (Hund's rule). These three electrons are unpaired.

$$1s^2 \quad\quad 2s^2 \quad\quad 2p^3$$

b. Similarly, the configuration for 17 electrons is shown below. Here the $1s$, $2s$, $2p$, and $3s$ subshells are full, as are the first and second shells. The $3p$ subshell is not full and can accept 1 more electron. One electron in the $3p$ subshell is unpaired.

$$1s^2 \quad\quad 2s^2 \quad\quad 2p^6 \quad\quad 3s^2 \quad\quad 3p^5$$

c. An atom of element number 22 contains 22 protons in the nucleus and must therefore contain 22 electrons. The electronic configuration is

$$1s^2 \quad 2s^2 \quad\quad 2p^6 \quad\quad 3s^2 \quad\quad 3p^6 \quad\quad 4s^2 \quad\quad 3d^2$$

Note here that only two of the $3d$ orbitals of the $3d$ subshell are occupied, and each of these orbitals contains a single electron. Thus, there are two unpaired electrons.

d. Arsenic (As) is element number 33 and therefore contains 33 electrons. The electronic configuration is

$$1s^2 2s^2 2p^6 3s^2 3p^6 4s^2 3d^{10} 4p^3$$

> **STUDY SKILLS 3.1**

THE CONVENTION HOTELS ANALOGY

A new concept is often made easier to understand by relating it to something familiar. The concept of electronic configurations is very likely new to you, but you are probably familiar with hotels. The way electrons fill up orbitals, subshells, and shells around a nucleus can be compared to the way rooms, floors, and hotels located near a convention center will fill with convention delegates. To make our analogy work, imagine that the hotels are located on a street that runs uphill from the convention center, as shown at right. Further imagine that none of the hotels has elevators, so the only way to get to upper floors is to climb the stairs.

In this analogy, the convention center is equivalent to the nucleus of an atom, and each hotel represents a shell, each floor represents a subshell, and each room represents an orbital. If you were a delegate who wanted to describe to a friend where you were staying, you would indicate the hotel (shell), floor (subshell), and room (orbital) assigned to you. Electronic configurations such as $1s^2 2s^2 2p^1$ give similar information for each electron: The numbers preceding each letter indicate the shells, the letters indicate the subshells, and the superscripts coupled with Hund's rule indicate which orbitals are occupied.

Three more assumptions allow us to extend the analogy: (1) No more than two delegates can be assigned to a room, (2) delegates prefer not to have roommates if an empty room on the same floor is available, and (3) all delegates want to use as little energy as possible when they walk from the convention center to their rooms (remember, there are no elevators).

With these assumptions in mind, it is obvious that the first delegate to check in will choose to stay in the single room of the s floor of Hotel One (a small but very exclusive hotel). The

1s

2s 2p

3s 3p 3d

4s 4p 4d 4f

5s 5p 5d 5f

6s 6p 6d

7s

(a)

1s
2s 2p
3s 3p 3d
4s 4p 4d 4f
5s 5p 5d 5f
6s 6p 6d
7s

(b)

> **FIGURE 3.8** An aid to remembering subshell-filling order.

This time, the circles and arrows have not been used to indicate orbitals and electrons. You should satisfy yourself that the following facts are clear: The first, second, and third shells are full. The fourth shell is partially full, with the 4s subshell being full and the 4p subshell being half-full. The 4p subshell contains 3 unpaired electrons.

Write the electronic configurations for the following, and indicate the number of unpaired electrons in each case:

a. Element number 9

b. Mg

c. The element found in group VIA(16) and period 3

d. An atom that contains 23 protons

Learning Check 3.6

Although Figure 3.7 gives the details of subshell-filling order, a more concise diagram is available and easy to remember. It is shown in ■ Figure 3.8, where the subshells are first arranged as in (a) and then diagonal arrows are drawn as in (b). To get the correct subshell-filling order, follow the arrows from top to bottom, going from the head of one arrow to the tail of the next.

second delegate will also choose this same floor and room of Hotel One and will fill the hotel to capacity. The third delegate will have to go to Hotel Two and will choose the one room on the s floor. The fourth delegate will also choose the one room of the s floor, and fill that floor. The fifth delegate will choose a room on the p floor of Hotel Two because the s floor is full. The sixth delegate will also choose a room on the p floor of Hotel Two but will choose one that is not occupied. The seventh delegate will also choose an empty room on the p floor of Hotel Two. Delegates eight, nine, and ten can either pair up with the delegates already in the rooms of the p floor of Hotel Two or go uphill to Hotel Three. They choose to save energy by walking up the stairs and staying in rooms with roommates on the p floor of Hotel Two. Additional arriving delegates will occupy the floors of Hotels Three and Four in the order dictated by these same energy and pairing considerations. Thus, we see that the delegates in their rooms are analogous to the electrons in their orbitals.

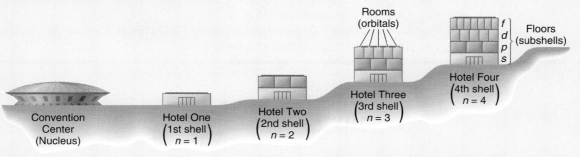

The electronic configurations described to this point provide details of the shells, subshells, and orbitals involved but are somewhat cumbersome. In some applications, these details are not needed, and simplified representations are used that emphasize the electrons in the valence shell.

We see from Table 3.2 that the noble gases neon and argon both have electronic configurations that end with a completely filled p subshell. This is true for all noble gases except helium, which ends with a filled $1s$ subshell. In Chapter 4, we will see that these **noble gas configurations** are important in understanding the bonding that occurs between atoms.

Noble gas configuration
An electronic configuration in which the last eight electrons occupy and fill the s and p subshells of the highest-occupied shell.

Noble gas configurations can be used to write abbreviated electronic configurations. Instead of writing the configurations in their entirety, the symbols for the noble gases are used in brackets to represent the electrons found in their configurations. Electrons that are present in addition to those of the noble gases are written following the symbol. For example, the electronic configuration for sodium can be represented as $1s^2 2s^2 2p^6 3s^1$ or $[Ne]3s^1$.

Example 3.7

Write abbreviated electronic configurations for the following:

a. An atom that contains 7 electrons
b. An atom that contains 17 electrons
c. An atom of element number 22
d. An atom of arsenic (As)

GOB
Chemistry·⚛·Now™
Go to GOB Now and click to learn how to fill electrons into electron orbital diagrams.

Solution
All these configurations were written in conventional form in Example 3.6.

a. From Example 3.6, the configuration is $1s^2 2s^2 2p^3$. We see that the two electrons in the $1s$ subshell represent the noble gas configuration of helium (He), so we can write the configuration as $[He]2s^2 2p^3$.

GOB
Chemistry·⚛·Now™
Go to GOB Now and click to engage in a tutorial showing how to report abbreviated electronic configurations.

b. From Example 3.6, the configuration is $1s^2 2s^2 2p^6 3s^2 3p^5$. We see that the $1s^2 2s^2 2p^6$ portion is the configuration of neon (Ne), so we can write the configuration as $[Ne]3s^2 3p^5$.

c. From Example 3.6, the configuration is $1s^2 2s^2 2p^6 3s^2 3p^6 4s^2 3d^2$. We see that the first 18 electrons represented by $1s^2 2s^2 2p^6 3s^2 3p^6$ correspond to the electrons of argon (Ar). Thus, we can write the configuration as $[Ar]4s^2 3d^2$.

d. From Example 3.6, the configuration of arsenic is $1s^2 2s^2 2p^6 3s^2 3p^6 4s^2 3d^{10} 4p^3$. Once again, the first 18 electrons can be represented by the symbol for argon. The configuration is $[Ar]4s^2 3d^{10} 4p^3$.

Learning Check 3.7

Write abbreviated electronic configurations for the following. These are the same elements used in Learning Check 3.6.

a. An atom of element number 9
b. An atom of magnesium (Mg)
c. An atom of the element found in group VIA(16) and period 3
d. An atom that contains 23 protons

➤ 3.5 Another Look at the Periodic Table

Now that you know more about electronic configurations, you can better understand the periodic law and table. First, consider the electronic configurations of the elements belonging to the same group of the periodic table. Group IA(1), for example, contains Li, Na, K, Rb, Cs, and Fr, with electronic configurations for the first four elements as shown below:

Element symbol	Conventional form	Abbreviated form
Li	$1s^2 2s^1$	$[\text{He}]2s^1$
Na	$1s^2 2s^2 2p^6 3s^1$	$[\text{Ne}]3s^1$
K	$1s^2 2s^2 2p^6 3s^2 3p^6 4s^1$	$[\text{Ar}]4s^1$
Rb	$1s^2 2s^2 2p^6 3s^2 3p^6 4s^2 3d^{10} 4p^6 5s^1$	$[\text{Kr}]5s^1$

Notice that each of these elements has a single electron in the valence shell. Further, each of these valence-shell electrons is located in an s subshell. Elements belonging to other groups also have valence-shell electronic configurations that are the same for all members of the group, and all members have similar chemical properties. It has been determined that the similar chemical properties of elements in the same group result from similar valence-shell electronic configurations.

The periodic table becomes more useful when we interpret it in terms of the electronic configurations of the elements in various areas. One relationship is shown in ■ Figure 3.9, where the periodic table is divided into four areas on the basis of the type of subshell occupied by the highest-energy electron in the atom. This last electron added to an atom is called the **distinguishing electron**. Note that the s area is 2 columns (elements) wide, the p area is 6 columns wide, the d area is 10 columns wide, and the f area is 14 columns wide—exactly the number of electrons required to fill the s, p, d, and f subshells, respectively.

Distinguishing electron
The last and highest-energy electron found in an element.

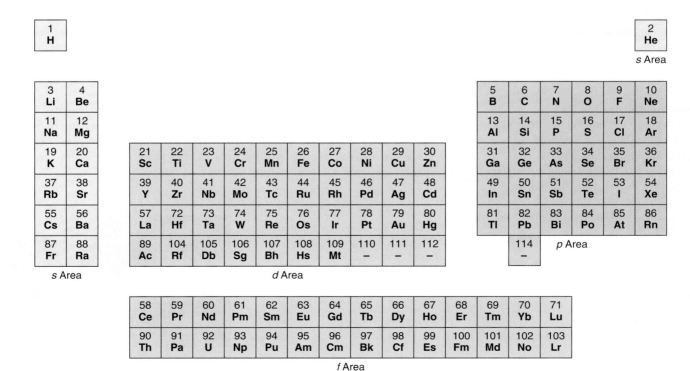

➤ **FIGURE 3.9** The periodic table divided by distinguishing electrons of the elements.

Representative elements

																	Noble gases
(1) I A	(2) II A	(3) III B	(4) IV B	(5) V B	(6) VI B	(7) VII B	(8)	(9) VIII B	(10)	(11) I B	(12) II B	(13) III A	(14) IV A	(15) V A	(16) VI A	(17) VII A	(18) VIII A
1 **H**					Transition elements												2 **He**
3 **Li**	4 **Be**											5 **B**	6 **C**	7 **N**	8 **O**	9 **F**	10 **Ne**
11 **Na**	12 **Mg**	21 **Sc**?										13 **Al**	14 **Si**	15 **P**	16 **S**	17 **Cl**	18 **Ar**
19 **K**	20 **Ca**	21 **Sc**	22 **Ti**	23 **V**	24 **Cr**	25 **Mn**	26 **Fe**	27 **Co**	28 **Ni**	29 **Cu**	30 **Zn**	31 **Ga**	32 **Ge**	33 **As**	34 **Se**	35 **Br**	36 **Kr**
37 **Rb**	38 **Sr**	39 **Y**	40 **Zr**	41 **Nb**	42 **Mo**	43 **Tc**	44 **Ru**	45 **Rh**	46 **Pd**	47 **Ag**	48 **Cd**	49 **In**	50 **Sn**	51 **Sb**	52 **Te**	53 **I**	54 **Xe**
55 **Cs**	56 **Ba**	57 **La**	72 **Hf**	73 **Ta**	74 **W**	75 **Re**	76 **Os**	77 **Ir**	78 **Pt**	79 **Au**	80 **Hg**	81 **Tl**	82 **Pb**	83 **Bi**	84 **Po**	85 **At**	86 **Rn**
87 **Fr**	88 **Ra**	89 **Ac**	104 **Rf**	105 **Db**	106 **Sg**	107 **Bh**	108 **Hs**	109 **Mt**	110 **–**	111 **–**	112 **–**	114 **–**					

Period 1–7 (left axis label: **Period**)

Inner-transition elements

58 **Ce**	59 **Pr**	60 **Nd**	61 **Pm**	62 **Sm**	63 **Eu**	64 **Gd**	65 **Tb**	66 **Dy**	67 **Ho**	68 **Er**	69 **Tm**	70 **Yb**	71 **Lu**
90 **Th**	91 **Pa**	92 **U**	93 **Np**	94 **Pu**	95 **Am**	96 **Cm**	97 **Bk**	98 **Cf**	99 **Es**	100 **Fm**	101 **Md**	102 **No**	103 **Lr**

➤ **FIGURE 3.10** Elemental classification on the basis of electronic configurations.

Representative element
An element in which the distinguishing electron is found in an *s* or a *p* subshell.

Transition element
An element in which the distinguishing electron is found in a *d* subshell.

Inner-transition element
An element in which the distinguishing electron is found in an *f* subshell.

Electronic configurations are also used to classify elements of the periodic table, as shown in ■ Figure 3.10.

The *noble gases* make up the group of elements found on the extreme right of the periodic table. They are all gases at room temperature and are unreactive with other substances (hence, the group name). With the exception of helium, the first member of the group, noble gases are characterized by filled *s* and *p* subshells of the highest occupied shell.

Representative elements are those found in the *s* and *p* areas of the periodic table, not including the noble gases. The distinguishing electrons of representative elements partially or completely fill an *s* subshell—groups IA(1) and IIA(2)—or partially fill a *p* subshell—groups IIIA(13), IVA(14), VA(15), VIA(16), and VIIA(17). Most of the common elements are representative elements.

The *d* area of the periodic table contains the **transition elements** (Figure 3.10) in which the distinguishing electron is found in a *d* subshell. Some transition elements are used for everyday applications (■ Figure 3.11). **Inner-transition elements** are those in the *f* area of the periodic table, and the distinguishing electron is found in an *f* subshell.

➤ **FIGURE 3.11** Transition elements, such as gold, silver, copper, nickel, platinum, and zinc, are often used in coins and medals. List some chemical and physical properties that would be desirable in metals used for such purposes.

Example 3.8

Use the periodic table and Figures 3.9 and 3.10 to determine the following for Ca, Fe, S, and Kr:

a. The type of distinguishing electron
b. The classification based on Figure 3.10

Solution

a. On the basis of the location of each element in Figure 3.9, the distinguishing electrons are of the following types:

Ca: *s*   Fe: *d*   S: *p*   Kr: *p*

b. The classifications based on Figure 3.10 are: Ca, representative element; Fe, transition element; S, representative element; Kr, noble gas.

Determine the following for element numbers 38, 47, 50, and 86:

a. The type of distinguishing electron
b. The classification according to Figure 3.10

The elements can also be classified into the categories of metals, non-metals, and metalloids. This approach, used in ■ Figure 3.12, shows that most elements are classified as metals. It is also apparent from Figure 3.10 that all nonmetals and all metalloids are representative elements.

Most **metals** have the following properties. (Are these properties physical or chemical?)

GOB
Chemistry⚛Now™
Go to GOB Now and click to examine the different parts of the periodic table.

Metals
Elements found in the left two-thirds of the periodic table. Most have the following properties: high thermal and electrical conductivities, high malleability and ductility, and a metallic luster.

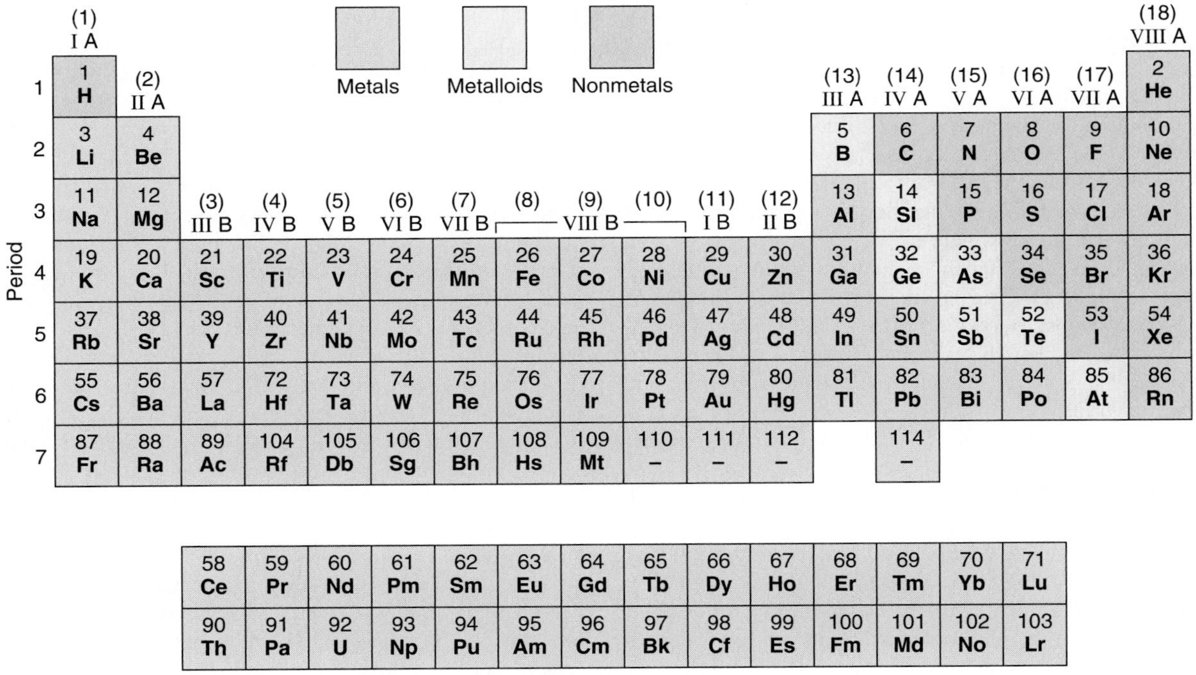

➤ **FIGURE 3.12** Locations of metals, nonmetals, and metalloids in the periodic table of the elements.

Nonmetals
Elements found in the right one-third of the periodic table. They often occur as brittle, powdery solids or gases and have properties generally opposite those of metals.

Metalloids
Elements that form a narrow diagonal band in the periodic table between metals and nonmetals. They have properties somewhat between those of metals and nonmetals.

High thermal conductivity—they transmit heat readily.
High electrical conductivity—they transmit electricity readily.
Ductility—they can be drawn into wires.
Malleability—they can be hammered into thin sheets.
Metallic luster—they have a characteristic "metallic" appearance.

Nonmetals, the elements found in the right one-third of the periodic table, generally have chemical and physical properties opposite those of metals. Under normal conditions, they often occur as brittle, powdery solids or as gases.

Metalloids, such as boron (B) and silicon (Si) are the elements that form a diagonal separation zone between metals and nonmetals in the periodic table. Metalloids have properties somewhat between those of metals and nonmetals, and they often exhibit some of the characteristic properties of each type.

Learning Check 3.9	Classify each of the following elements as metal, nonmetal, or metalloid:

a. Xe   **b.** As   **c.** Hg   **d.** Ba   **e.** Th

➤ 3.6 Property Trends Within the Periodic Table

In Figure 3.12, the elements are classified into the categories of metal, metalloid, or nonmetal according to their positions in the table and properties such

CHEMISTRY AND YOUR HEALTH • 3.1

PROTECTING CHILDREN FROM IRON POISONING

The element iron plays a vital role in a number of body processes. Perhaps the most well-known of these functions is the role of iron as a component of hemoglobin, the oxygen-transporting protein of blood. When blood is iron poor, body tissues do not receive enough oxygen, and *anemia*, a general weakening of the body, results. Recommended Dietary Allowances (RDA) have been established for iron because of this and other important contributions to good health. The general RDA is 10 to 15 mg per day depending on age and sex, and 30 mg per day for pregnant women.

A well-balanced diet that includes meat, whole grains, and dark green vegetables will generally provide enough iron to satisfy the RDA for most individuals other than pregnant women. In an attempt to enhance the effectiveness of the diet in providing iron, many foods are enriched or fortified with iron—primarily breads, other flour products, and cereals. About 25% of all dietary iron consumed in the United States comes from such foods.

The emphasis on the health benefits of iron and the attempts to get enough iron into the diet make it somewhat surprising to be told that iron is also a serious poisoning threat for children. In fact, iron is the leading cause of poisoning deaths in children under the age of six. Since 1986, 110 thousand iron poisoning incidents in children have been reported, including 33 deaths. The iron-containing products involved range from innocent-appearing nonprescription daily multivitamin/mineral supplements for children to high-potency prescription iron supplements for pregnant women. Children showed symptoms of poisoning from consuming as few as 5 to as many as 98 iron-containing tablets. Immediate symptoms include nausea, vomiting, and diarrhea. Deaths occurred from ingesting as little as 200 mg to as much as 5850 mg of iron.

In an attempt to address this problem, the U.S. Food and Drug Administration (FDA) published regulations in 1997 that require warnings on all iron-containing drugs and dietary supplements about the risk of iron poisoning in children and the need to keep such products out of children's reach. In addition, the regulations require that most products containing 30 mg or more of iron per dosage unit have to be packaged as individual doses (such as in blister packages). This rule is designed to limit the number of pills or capsules a child might consume because of the difficulty encountered by a child in opening small individual packets. These 1997 FDA regulations added to rules already in place, including a U.S. Consumer Product Safety Commission regulation given in 1987. According to this regulation, most drugs and food supplements containing more than 250 mg of iron per container have to be packaged in child-resistant containers.

The primary responsibility for protecting young children from iron poisoning still rests with parents, older siblings, and other caregivers. These individuals must first recognize the hazards presented by iron-containing products and must then be very diligent in keeping such products out of the reach of young children.

➤ **FIGURE 3.13** The elements of group VA(15) of the periodic table. Phosphorus must be stored under water because it will ignite when exposed to the oxygen in air. It has a slight color because of reactions with air.

as thermal conductivity and metallic luster. It is generally true that within a period of the periodic table, the elements become less metallic as we move from left to right. Within a group, the metallic character increases from top to bottom.

Consider group VA(15) as an example. We see that the elements of the group consist of the two nonmetals nitrogen (N) and phosphorus (P), the two metalloids arsenic (As) and antimony (Sb), and the metal bismuth (Bi). We see the trend toward more metallic character from top to bottom that was mentioned above. According to what has been discussed concerning the periodic law and periodic table, these elements should have some similarities in chemical properties because they belong to the same group of the periodic table. Studies of their chemical properties show that they are similar but not identical. Instead, they follow trends according to the location of the elements within the group.

Certain physical properties also follow such trends, and some of these are easily observed, as shown in ■ Figure 3.13. At room temperature and ordinary atmospheric pressure, nitrogen is a colorless gas, phosphorus is a white nonmetallic solid, arsenic is a brittle gray solid with a slight metallic luster, antimony is a brittle silver-white solid with a metallic luster, and bismuth is a lustrous silver-white metal.

We see that these physical properties are certainly not identical, but we do see that they change in a somewhat regular way (they follow a trend) as we move from element to element down the group. Nitrogen is a gas, but the others are solids. Nitrogen is colorless, phosphorus is white, then the other three become more and more metallic-looking. Also, as mentioned above, we note a change from nonmetal to metalloid to metal as we come down the group.

The trends in these properties occur in a regular way that would allow us to predict some of them for one element if they were known for the other elements in the group. For example, if the properties of bismuth were unknown, we would have predicted that it would have a silvery white color and a metallic luster based on the appearance of arsenic and antimony. The three elements from group VIIA(17) shown in ■ Figure 3.14 also demonstrate some obvious predictable trends in physical properties.

Trends in properties occur in elements that form periods across the periodic table as well as among those that form vertical groups. We will discuss two of these properties and their trends for representative elements. Our focus will be on the general trends, recognizing that some elements show deviations from these general behaviors. We will also propose explanations for the trends based on the electronic structure of atoms discussed in the chapter.

The first property we will consider is the size of the atoms of the representative elements. The size of an atom is considered to be the radius of a sphere extending from the center of the nucleus of the atom to the location of the outermost electrons around the nucleus. The behavior of this property

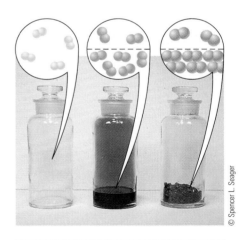

➤ **FIGURE 3.14** Chlorine, bromine, and iodine (left to right) all belong to group VIIA(17) of the periodic table. Their atoms all have the same number of electrons in the valence shell and therefore similar chemical properties. However, they do not have similar appearances. At room temperature and under normal atmospheric pressure, chlorine is a pale yellow gas, bromine is a dark red liquid that readily changes into a gas, and iodine is a gray-black solid that changes into a purple gas when heated slightly. Astatine is the next member of the group after iodine. Try to think like Mendeleev and predict its color and whether it will be a liquid, solid, or gas under normal conditions.

➤ **FIGURE 3.15** Scale drawings of the atoms of some representative elements enlarged about 60 million times. The numbers are the atomic radii in picometers (10^{-12} m).

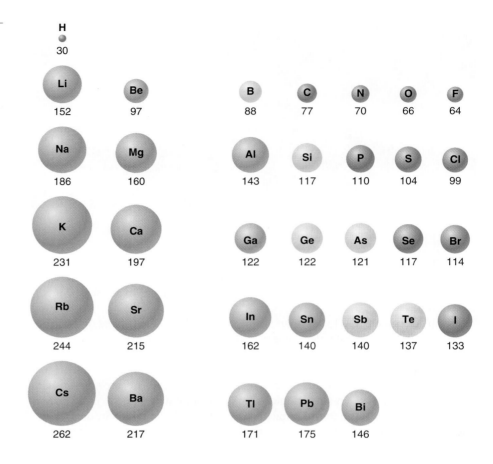

across a period and down a group is shown in ■ Figure 3.15. We see that the size increases from the top to the bottom of each group, and decreases from left to right across a period.

Remember that when we go from element to element down a group, a new electronic shell is being filled. Consider group IA(1). In lithium atoms (Li), the second shell ($n = 2$) is beginning to fill, in sodium atoms (Na), the third shell ($n = 3$) is beginning to fill, and so on for each element in the group until the sixth shell ($n = 6$) begins to fill in cesium atoms (Cs). As n increases, the distance of the electrons from the nucleus in the shell designated by n increases, and the atomic radius as defined above also increases.

The decrease in atomic radius across a period can also be understood in terms of the outermost electrons around the nucleus. Consider period 3, for example. The abbreviated electronic structure for sodium atoms (Na) is $[Ne]3s^1$, and the outermost electron is in the third shell ($n = 3$). In magnesium atoms (Mg), the electronic structure is $[Ne]3s^2$, and we see that the outermost electron is still in the third shell. In aluminum atoms (Al), the electronic structure is $[Ne]3s^23p^1$, and we see that, once again, the outermost electron is in the third shell. In fact, the outermost electrons in all the atoms across period 3 are in the third shell.

Because all these electrons are in the same third shell, they should all be the same distance from the nucleus, and the atoms should all be the same size. However, each time another electron is added to the third shell of these elements, another positively charged proton is added to the nucleus. Thus, in sodium atoms there are 11 positive nuclear charges attracting the electrons, but in aluminum atoms there are 13. The effect of the increasing nuclear charge attracting electrons in the same shell is to pull all the electrons of the shell closer to the nucleus and cause the atomic radii to decrease as the nuclear charge increases.

H 1311							**He** 2370
Li 521	**Be** 899	**B** 799	**C** 1087	**N** 1404	**O** 1314	**F** 1682	**Ne** 2080
Na 496	**Mg** 737	**Al** 576	**Si** 786	**P** 1052	**S** 1000	**Cl** 1245	**Ar** 1521
K 419	**Ca** 590	**Ga** 576	**Ge** 784	**As** 1013	**Se** 939	**Br** 1135	**Kr** 1351
Rb 402	**Sr** 549	**In** 559	**Sn** 704	**Sb** 834	**Te** 865	**I** 1007	**Xe** 1170
Cs 375	**Ba** 503	**Tl** 590	**Pb** 716	**Bi** 849	**Po** 791	**At** 926	**Rn** 1037

> **FIGURE 3.16** First ionization energies for selected representative elements. The values are given in kJ/mole. (SOURCE: Data from Bard, A. J.; Parsons, R.; Jordan, J. *Standard Potentials in Aqueous Solution*. New York: Dekker, 1985, pp. 24–27.)

The chemical reactivity of elements is dependent on the behavior of the electrons of the atoms of the elements, especially the valence electrons. One property that is related to the behavior of the electrons of atoms is the ionization energy.

The *ionization energy* of an element is the energy required to remove an electron from an atom of the element in the gaseous state. The removal of one electron from an atom leaves the atom with a net 1+ charge because the nucleus of the resulting atom contains one more proton than the number of remaining electrons. The resulting charged atom is called an *ion*. We will discuss ions and the ionization process in more detail in Sections 4.2 and 4.3. A reaction for the removal of one electron from an atom of sodium is

$$Na(g) \rightarrow Na^+(g) + e^-$$

Because this process represents the removal of the first electron from a neutral sodium atom, the energy necessary to accomplish the process is called the **first ionization energy.** If a second electron were removed, the energy required would be called the second ionization energy, and so forth. We will focus on only the first ionization energy for representative elements. ■ Figure 3.16 contains values for the first ionization energy of a number of representative elements.

The general trend is seen to be a decrease from the top to the bottom of a group, and an increase from left to right across a period. The higher the value of the ionization energy, the more difficult it is to remove an electron from the atoms of an element. Thus, we see that in general, the electrons of metals are more easily removed than are the electrons of nonmetals. Also, the farther down a group a metal is located, the easier it is to remove an electron.

The metals of group IA(1) all react with ethyl alcohol, C_2H_5OH, to produce hydrogen gas, H_2, as follows, where M is a general representation of the metals of the group:

$$2M + 2C_2H_5OH \rightarrow 2C_2H_5OM + H_2$$

In this reaction, electrons are removed from the metal and transferred to the hydrogen. The reaction is shown in progress for the first three members of the group in ■ Figure 3.17. The rate (speed) of the reaction is indicated by the amount of hydrogen gas being released.

First ionization energy
The energy required to remove the first electron from a neutral atom.

Refer to Figure 3.17, and do the following:

Learning Check 3.10

a. Arrange the three metals of the group vertically in order of the rate of the reaction with ethyl alcohol. Put the slowest reaction at the top and the fastest at the bottom.

(a)

(b)

(c)

© Spencer L. Seager

➤ **FIGURE 3.17** The reaction of group IA(1) elements with ethyl alcohol. (a) Lithium (Li), (b) sodium (Na), (c) potassium (K). Each metal sample was wrapped in a wire screen to keep it from floating.

b. Compare the trend in reaction rate coming down the group with the trend in ionization energy coming down the group.

c. Compare the trend in reaction rate coming down the group with the trend in ease of removing an electron from an atom of the metal coming down the group.

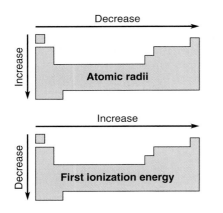

➤ **FIGURE 3.18** General trends for atomic size and ionization energy of representative elements.

The trends in ionization energy and ease of removing an electron can be explained using arguments similar to those for explaining the sizes of atoms. As we come down a group, the valence electrons are located farther and farther away from the nucleus because they are located in higher-energy shells. The farther the electrons are away from the nucleus, the weaker is the attraction of the positively charged nucleus for the negatively charged electrons, and the easier it is to pull the electrons away.

Similarly, as we go from left to right in a period, the valence electrons are going into the same shell and therefore should be about the same distance from the nucleus. But as we saw before, the nuclear charge increases as well as the number of electrons. As a result, there is a greater nuclear charge attracting the electrons the farther we move to the right. Thus, a valence electron of an atom farther to the right is more difficult to remove than a valence electron of an atom farther to the left.

The general trends we have discussed are summarized in ■ Figure 3.18. It is easiest to remember the trends by always going in the same direction in the table and noting how the properties change in that direction. We have chosen to summarize by always going from top to bottom for groups, and left to right for periods.

➤ **FOR FUTURE REFERENCE** GOOD DIETARY SOURCES OF ZINC

Whthe effectiveness of zinc in treating the common cold has not been clearly established or disproved (see Over the Counter 3.1), it is well-known that it functions in many important ways in the body. Zinc plays an essential role in the action of numerous proteins in the body, including a variety of enzymes. It is also involved in processes such as some immune responses, blood clotting, proper thyroid function, wound healing, and fetal development.

Unlike the situation with calcium, the body does not have any readily available storage forms of zinc that can be used to maintain blood zinc levels. For this reason, a person must eat zinc-rich foods frequently to maintain good health. The chart below lists good dietary sources of zinc.

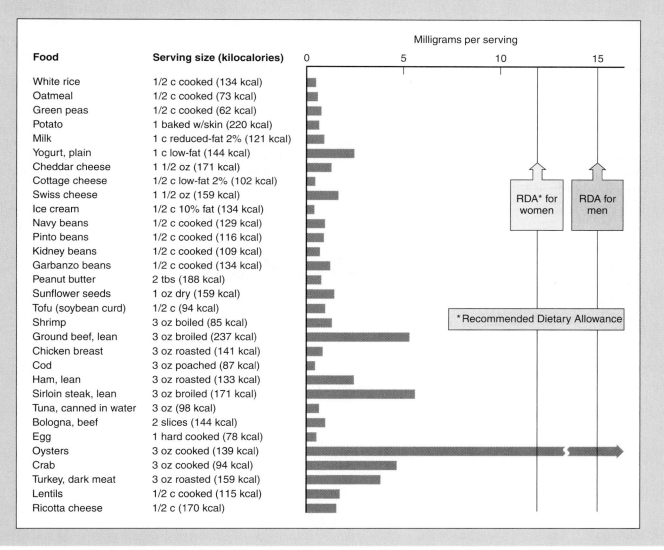

Food	Serving size (kilocalories)
White rice	1/2 c cooked (134 kcal)
Oatmeal	1/2 c cooked (73 kcal)
Green peas	1/2 c cooked (62 kcal)
Potato	1 baked w/skin (220 kcal)
Milk	1 c reduced-fat 2% (121 kcal)
Yogurt, plain	1 c low-fat (144 kcal)
Cheddar cheese	1 1/2 oz (171 kcal)
Cottage cheese	1/2 c low-fat 2% (102 kcal)
Swiss cheese	1 1/2 oz (159 kcal)
Ice cream	1/2 c 10% fat (134 kcal)
Navy beans	1/2 c cooked (129 kcal)
Pinto beans	1/2 c cooked (116 kcal)
Kidney beans	1/2 c cooked (109 kcal)
Garbanzo beans	1/2 c cooked (134 kcal)
Peanut butter	2 tbs (188 kcal)
Sunflower seeds	1 oz dry (159 kcal)
Tofu (soybean curd)	1/2 c (94 kcal)
Shrimp	3 oz boiled (85 kcal)
Ground beef, lean	3 oz broiled (237 kcal)
Chicken breast	3 oz roasted (141 kcal)
Cod	3 oz poached (87 kcal)
Ham, lean	3 oz roasted (133 kcal)
Sirloin steak, lean	3 oz broiled (171 kcal)
Tuna, canned in water	3 oz (98 kcal)
Bologna, beef	2 slices (144 kcal)
Egg	1 hard cooked (78 kcal)
Oysters	3 oz cooked (139 kcal)
Crab	3 oz cooked (94 kcal)
Turkey, dark meat	3 oz roasted (159 kcal)
Lentils	1/2 c cooked (115 kcal)
Ricotta cheese	1/2 c (170 kcal)

Milligrams per serving

RDA* for women

RDA for men

*Recommended Dietary Allowance

SOURCE: Adapted from Whitney, E. N.; Rolfes, S. R. *Understanding Nutrition*, 8th ed. Belmont, CA: West/Wadsworth, 1999.

CONCEPT SUMMARY

The Periodic Law and Table. The chemical properties of the elements tend to repeat in a regular (periodic) way when the elements are arranged in order of increasing atomic numbers. This periodic law is the basis for the arrangement of the elements called the periodic table. In this table, each element belongs to a vertical grouping, called a group or family, and a horizontal grouping, called a period. All elements in a group or family have similar chemical properties.

Electronic Arrangements in Atoms. Niels Bohr proposed a theory for the electronic structure of hydrogen based on the idea that the electrons of atoms move around atomic nuclei in fixed circular orbits. Electrons change orbits only when they absorb or release energy. The Bohr model was modified as a result of continued research. It was found that precise Bohr orbits for electrons could not be determined. Instead, the energy and location of electrons could be specified in terms of shells, subshells, and orbitals, which are indicated by a notation system of numbers and letters.

The Shell Model and Chemical Properties. The modified Bohr model, or shell model, of electronic structure provides an explanation for the periodic law. The rules governing electron occupancy in shells, subshells, and orbitals result in a repeating pattern of valence-shell electron arrangements. Elements with similar chemical properties turn out to be elements with identical numbers and types of electrons in their valence shells.

Electronic Configurations. The arrangements of electrons in orbitals, subshells, and shells are called electronic configurations. Rules and patterns have been found that allow these configurations to be represented in a concise way. Written electronic configurations allow details of individual orbital, subshell, and shell electron occupancies to be seen readily. Also, the number of unpaired electrons in elements is easily determined. Electronic configurations can be represented in an abbreviated form by using noble gas symbols to represent some of the inner electrons.

Another Look at the Periodic Table. Correlations between electronic configurations for the elements and the periodic table arrangement of elements make it possible to determine a number of details of electronic structure for an element simply on the basis of the location of the element in the periodic table. Special attention is paid to the last or distinguishing electron in an element. Elements are classified according to the type of subshell (*s*, *p*, *d*, *f*) occupied by this electron. The elements are also classified on the basis of other properties as metals, nonmetals, or metalloids.

Property Trends Within the Periodic Table. Chemical and physical properties of elements follow trends within the periodic table. These trends are described in terms of changes in properties of elements from the top to the bottom of groups, and from the left to the right of periods. The sizes of atoms and first ionization energies are two properties that show distinct trends.

LEARNING OBJECTIVES ASSESSMENT

You can get an approximate but quick idea of how well you have met the learning objectives given at the beginning of this chapter by working the selected end-of-chapter exercises given below. The answer to each exercise is given in Appendix B of the book.

Objective 1 (Section 3.1): Exercise 3.4

Objective 2 (Section 3.2): Exercise 3.12

Objective 3 (Section 3.3): Exercise 3.18

Objective 4 (Section 3.3): Exercise 3.22

Objectives 5 and 6 (Section 3.4): Exercise 3.24

Objective 7 (Section 3.4): Exercise 3.28

Objective 8 (Section 3.5): Exercise 3.34

Objective 9 (Section 3.5): Exercise 3.36

Objectives 10 and 11 (Section 3.6): Exercises 3.40 and 3.42

KEY TERMS AND CONCEPTS

Atomic orbital (3.2)
Distinguishing electron (3.5)
Electronic configuration (3.4)
First ionization energy (3.6)
Group or family of the periodic table (3.1)
Hund's rule (3.4)
Inner-transition element (3.5)

Metalloid (3.5)
Metal (3.5)
Noble gas configuration (3.4)
Nonmetal (3.5)
Pauli exclusion principle (3.4)
Periodic law (3.1)
Period of the periodic table (3.1)

Representative element (3.5)
Shell (3.2)
Subshell (3.2)
Transition element (3.5)
Valence shell (3.3)

LEGEND: 1 = straightforward, 2 = intermediate, 3 = challenging. All even-numbered exercises are answered in Appendix B.

THE PERIODIC LAW AND TABLE (SECTION 3.1)

3.1 Identify the group and period to which each of the following elements belongs:
   a. Ca
   b. element number 22
   c. nickel
   d. tin

3.2 Identify the group and period to which each of the following elements belongs:
   a. Sb
   b. cadmium
   c. element number 31
   d. Zr

3.3 Write the symbol and name for the elements located in the periodic table as follows:
   a. Belongs to group VIA(16) and period 3
   b. The first element (reading down) in group VIB(6)
   c. The fourth element (reading left to right) in period 3
   d. Belongs to group IB(11) and period 5

3.4 Write the symbol and name for the elements located in the periodic table as follows:
   a. The noble gas belonging to period 4
   b. The fourth element (reading down) in group IVA(14)
   c. Belongs to group VIB(6) and period 5
   d. The sixth element (reading left to right) in period 6

3.5 **a.** How many elements are located in group VIIIB(8, 9, 10) of the periodic table?
   b. How many elements are found in period 2 of the periodic table?
   c. How many total elements are in groups IIA(2) and VIA(16) of the periodic table?

3.6 **a.** How many elements are located in group VIIB(7) of the periodic table?
   b. How many total elements are found in periods 1 and 2 of the periodic table?
   c. How many elements are found in period 5 of the periodic table?

3.7 The following statements either define or are closely related to the terms *periodic law, period,* and *group.* Match the terms to the appropriate statements.
   a. This is a vertical arrangement of elements in the periodic table.
   b. The chemical properties of the elements repeat in a regular way as the atomic numbers increase.
   c. The chemical properties of elements 11, 19, and 37 demonstrate this principle.
   d. Elements 4 and 12 belong to this arrangement.

3.8 The following statements either define or are closely related to the terms *periodic law, period,* and *group.* Match the terms to the appropriate statements.
   a. This is a horizontal arrangement of elements in the periodic table.
   b. Element 11 begins this arrangement in the periodic table.

   c. The element nitrogen is the first member of this arrangement.
   d. Elements 9, 17, 35, and 53 belong to this arrangement.

ELECTRONIC ARRANGEMENTS IN ATOMS (SECTION 3.2)

3.9 According to the Bohr theory, which of the following would have the higher energy?
   a. An electron in an orbit close to the nucleus
   b. An electron in an orbit located farther from the nucleus

3.10 What particles in the nucleus cause the nucleus to have a positive charge?

3.11 What is the maximum number of electrons that can be contained in each of the following?
   a. A 2*s* orbital
   b. A 2*s* subshell
   c. The first shell

3.12 What is the maximum number of electrons that can be contained in each of the following?
   a. A 2*p* orbital
   b. A 2*p* subshell
   c. The second shell

3.13 How many orbitals are found in the third shell? Write designations for the orbitals.

3.14 How many orbitals are found in the second shell? Write designations for the orbitals.

3.15 How many orbitals are found in a 3*d* subshell? What is the maximum number of electrons that can be located in this subshell?

3.16 How many orbitals are found in a 4*f* subshell? What is the maximum number of electrons that can be located in this subshell?

3.17 Identify the subshells found in the fourth shell; indicate the maximum number of electrons that can occupy each subshell and the total number of electrons that can occupy the shell.

THE SHELL MODEL AND CHEMICAL PROPERTIES (SECTION 3.3)

3.18 Look at the periodic table and tell how many electrons are in the valence shell of the following elements:
   a. element number 54
   b. The first element (reading down) in group VA(15)
   c. Sn
   d. The fourth element (reading left to right) in period 3

3.19 Look at the periodic table and tell how many electrons are in the valence shell of the following elements:
   a. element number 53
   b. Ge
   c. tellurium
   d. The fifth element in group IVA(14)

3.20 What period 6 element has chemical properties most like sodium? How many valence-shell electrons does this element have? How many valence-shell electrons does sodium have?

3.21 What period 2 element has chemical properties most like silicon? How many valence-shell electrons does this element have? How many valence-shell electrons does silicon have?

3.22 If you discovered an ore deposit containing copper, what other two elements might you also expect to find in the ore? Explain your reasoning completely.

3.23 Radioactive isotopes of strontium were produced by the explosion of nuclear weapons. They were considered serious health hazards because they were incorporated into the bones of animals that ingested them. Explain why strontium would be likely to be deposited in bones.

ELECTRONIC CONFIGURATIONS (SECTION 3.4)

3.24 Write an electronic configuration for each of the following elements, using the form $1s^2 2s^2 2p^6$, and so on. Indicate how many electrons are unpaired in each case.
   a. element number 37
   b. Si
   c. titanium
   d. Ar

3.25 Write an electronic configuration for each of the following elements, using the form $1s^2 2s^2 2p^6$, and so on. Indicate how many of the electrons are unpaired in each case.
   a. Br
   b. element number 36
   c. cadmium
   d. Sb

3.26 Write electronic configurations and answer the following:
   a. How many total s electrons are found in an atom of potassium?
   b. How many unpaired electrons are found in an atom of aluminum?
   c. How many subshells are completely filled in an atom of magnesium?

3.27 Write electronic configurations and answer the following:
   a. How many total electrons in Ge have a number designation (before the letters) of 4?
   b. How many unpaired p electrons are found in sulfur? What is the number designation of these unpaired electrons?
   c. How many $3d$ electrons are found in tin?

3.28 Write the symbol and name for each of the elements described. More than one element will fit some descriptions.
   a. Contains only two $2p$ electrons
   b. Contains an unpaired $3s$ electron
   c. Contains two unpaired $3p$ electrons
   d. Contains three $4d$ electrons
   e. Contains three unpaired $3d$ electrons

3.29 Write the symbol and name for each of the elements described. More than one element will fit some descriptions.
   a. Contains one unpaired $5p$ electron
   b. Contains a half-filled $5s$ subshell

   c. Contains a half-filled $6p$ subshell
   d. The last electron completes the $4d$ subshell
   e. The last electron half fills the $4f$ subshell

3.30 Write abbreviated electronic configurations for the following:
   a. arsenic
   b. an element that contains 25 electrons
   c. silicon
   d. element number 53

3.31 Write abbreviated electronic configurations for the following:
   a. lead
   b. element number 53
   c. an element that contains 24 electrons
   d. selenium

3.32 Refer to the periodic table and write abbreviated electronic configurations for all elements in which the noble gas symbol used will be [Ne].

3.33 Refer to the periodic table and determine how many elements have the symbol [Kr] in their abbreviated electronic configurations.

ANOTHER LOOK AT THE PERIODIC TABLE (SECTION 3.5)

3.34 Classify each of the following elements into the s, p, d, or f area of the periodic table on the basis of the distinguishing electron:
   a. lead
   b. element 27
   c. Tb
   d. Rb

3.35 Classify each of the following elements into the s, p, d, or f area of the periodic table on the basis of the distinguishing electron:
   a. Kr
   b. tin
   c. Pu
   d. element 40

3.36 Classify the following elements as representative, transition, inner-transition, or noble gases:
   a. iron
   b. element 15
   c. U
   d. xenon
   e. tin

3.37 Classify the following elements as representative, transition, inner-transition, or noble gases:
   a. W
   b. Cm
   c. element 10
   d. helium
   e. barium

3.38 Classify the following as metals, nonmetals, or metalloids:
   a. argon
   b. element 3
   c. Ge
   d. boron
   e. Pm

3.39 Classify the following as metals, nonmetals, or metalloids:
a. rubidium
b. arsenic
c. element 50
d. S
e. Br

PROPERTY TRENDS WITHIN THE PERIODIC TABLE (SECTION 3.6)

3.40 Use trends within the periodic table to predict which member of each of the following pairs is more metallic:
a. K or Ti
b. As or Bi
c. Mg or Sr
d. Sn or Ge

3.41 Use trends within the periodic table to predict which member of each of the following pairs is more metallic:
a. C or Sn
b. Sb or In
c. Ca or As
d. Al or Mg

3.42 Use trends within the periodic table and indicate which member of each of the following pairs has the larger atomic radius:
a. Ga or Se
b. N or Sb

c. O or C
d. Te or S

3.43 Use trends within the periodic table and indicate which member of each of the following pairs has the larger atomic radius:
a. Mg or Sr
b. Rb or Ca
c. S or Te
d. I or Sn

3.44 Use trends within the periodic table and indicate which member of each of the following pairs gives up one electron more easily:
a. Si or Cl
b. Mg or Ba
c. F or Br
d. Ca or S

3.45 Use trends within the periodic table and indicate which member of each of the following pairs gives up one electron more easily:
a. Mg or Al
b. Ca or Be
c. S or Al
d. Te or O

CHEMISTRY FOR THOUGHT

3.1 Samples of three metals that belong to the same group of the periodic table are shown in Figure 3.5. When magnesium reacts with bromine, a compound with the formula $MgBr_2$ results. What would be the formulas of the compounds formed by reactions of bromine with each of the other metals shown? Explain your reasoning.

3.2 Answer the problem posed in Figure 3.14, then predict the same things for fluorine, the first member of the group. Explain the reasoning that led you to your answers.

3.3 Answer the problem posed in Figure 3.11. What property that makes gold suitable for coins and medals also makes it useful in electrical connectors for critical electronic parts such as computers in spacecraft?

3.4 Calcium metal reacts with cold water as follows:

$$Ca + 2H_2O \rightarrow Ca(OH)_2 + H_2$$

Magnesium metal does not react with cold water. What behavior toward cold water would you predict for strontium and barium? Write equations to represent any predicted reactions.

3.5 Refer to the hotels analogy in Study Skills 3.1 and determine the number of floors and the number of rooms on the top floor of Hotel Five.

3.6 A special sand is used by a company as a raw material. The company produces zirconium metal that is used to contain the fuel in nuclear reactors. What other metals are likely to be produced from the same raw material? Explain your answer.

InfoTrac College Edition Readings

"Electrons grab unexpected energy share," *Science News*, Jan 12, 2002, 161(2):31(1). Record number A82374346.

"Hassium holds its place at the table," *Science News*, June 23, 2001, 159(25):392. Record number A76627432.

"Atomic-size 'sorter' performs vital cellular function," *Heart Disease Weekly*, Dec 2, 2001:6. Record number A80112391.

CHAPTER 4

Forces Between Particles

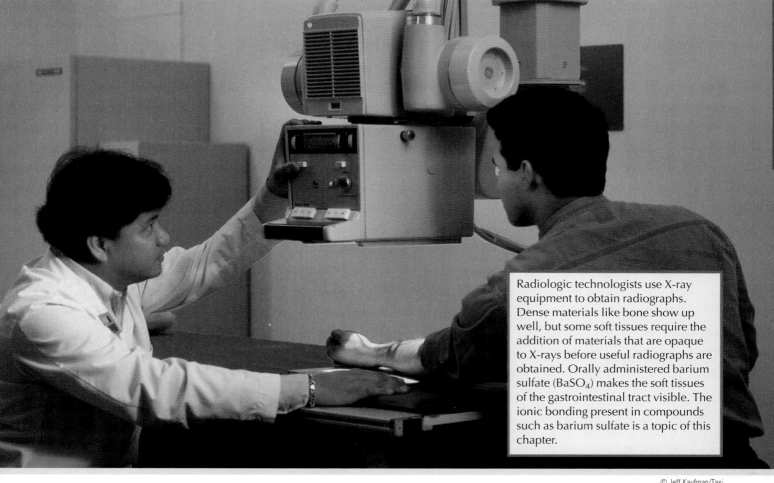

Radiologic technologists use X-ray equipment to obtain radiographs. Dense materials like bone show up well, but some soft tissues require the addition of materials that are opaque to X-rays before useful radiographs are obtained. Orally administered barium sulfate ($BaSO_4$) makes the soft tissues of the gastrointestinal tract visible. The ionic bonding present in compounds such as barium sulfate is a topic of this chapter.

© Jeff Kaufman/Taxi

LEARNING OBJECTIVES

When you have completed your study of this chapter, you should be able to:

1. Draw correct Lewis structures for atoms of representative elements. (Section 4.1)

2. Use electronic configurations to determine the number of electrons gained or lost by atoms as they achieve noble gas configurations. (Section 4.2)

3. Use the octet rule to correctly predict the ions formed during the formation of ionic compounds. (Section 4.3)

4. Write correct formulas for the following types of ionic compounds:

   a. Binary compounds containing a representative metal and a representative nonmetal. (Section 4.3)

   b. Compounds containing a representative metal and a polyatomic ion. (Section 4.10)

5. Correctly name the following:

   a. Binary ionic compounds (Section 4.4)

   b. Binary covalent compounds (Section 4.10)

   c. Ionic compounds containing a representative metal and a polyatomic ion. (Section 4.10)

6. Determine formula weights for ionic compounds. (Section 4.5)

7. Draw Lewis structures for molecules and polyatomic ions. (Section 4.8)

8. Use the VSEPR theory to predict the shapes of molecules and polyatomic ions. (Section 4.8)

9. Use electronegativities to determine the type of bonding that is likely to occur between pairs of representative elements. (Section 4.9)

10. Determine whether a covalent molecule is polar or nonpolar. (Section 4.9)

11. Relate melting points and boiling points of pure substances to the strength and types of interparticle forces present in the substances. (Section 4.11)

I n the discussion to this point, we have emphasized that matter is composed of tiny particles. However, we have not yet discussed the forces that hold these particles together to form the matter familiar to us. It is these forces that produce many of the properties associated with various types of matter—properties such as the electrical conductivity of copper wire, the melting point of butter, and the boiling point of water.

➤ 4.1 Noble Gas Configurations

The octet rule, proposed in 1916 by G. N. Lewis (an American) and Walter Kossel (a German), was the first widely accepted theory to describe bonding between atoms. The basis for the rule was the observed chemical stability of the noble gases. Because the noble gases did not react readily with other substances and because chemical reactivity depends on electronic structure, chemists concluded that the electronic structure of the noble gases represented a stable (low-energy) configuration. This noble gas configuration (see Section 3.4) is characterized by two electrons in the valence shell of helium and eight valence-shell electrons for the other members of the group (Ne, Ar, Kr, Xe, and Rn).

Example 4.1

Use the techniques described in Section 3.4 to write the electronic configurations for the noble gases. Identify the valence-shell electrons, and verify the noble gas configurations just mentioned.

Solution

The subshell-filling order in Figure 3.7 gives the electronic configurations shown below. Valence-shell electrons are in darker type, and the noble gas configurations are verified.

		Number of valence-shell electrons
He:	$1s^2$	2
Ne:	$1s^2 2s^2 2p^6$	8
Ar:	$1s^2 2s^2 2p^6 3s^2 3p^6$	8
Kr:	$1s^2 2s^2 2p^6 3s^2 3p^6 4s^2 3d^{10} 4p^6$	8
Xe:	$1s^2 2s^2 2p^6 3s^2 3p^6 4s^2 3d^{10} 4p^6 5s^2 4d^{10} 5p^6$	8
Rn:	$1s^2 2s^2 2p^6 3s^2 3p^6 4s^2 3d^{10} 4p^6 5s^2 4d^{10} 5p^6 6s^2 4f^{14} 5d^{10} 6p^6$	8

Learning Check 4.1

Write the electronic configurations for F and K. What would have to be done to change these configurations to noble gas configurations (how many electrons would have to be added or removed)?

Lewis structure

A representation of an atom or ion in which the elemental symbol represents the atomic nucleus and all but the valence-shell electrons. The valence-shell electrons are represented by dots arranged around the elemental symbol.

A simplified way to represent the valence-shell electrons of atoms was invented by G. N. Lewis. In these representations, called electron-dot formulas or **Lewis structures,** the symbol for an element represents the nucleus and all electrons around the nucleus except those in the valence shell. Valence-shell electrons are shown as dots around the symbol. Thus, rubidium is represented as Rb·.

When a Lewis structure is to be written for an element, it is necessary to determine the number of valence-shell electrons in the element. This can be done by writing the electronic configuration using the methods described in Chapter 3 and identifying the valence-shell electrons as those with the highest *n* value. A simpler alternative for representative elements is to refer to the periodic table and note that the number of valence-shell electrons in the atoms of an element is the same as the number of the group in the periodic table to which the element belongs. The group number used must be the one that precedes the letter A, not the one in parentheses as given in the periodic table inside the front cover of this book. Using this method, we see that rubidium belongs to group IA(1) and therefore has one valence-shell electron.

Example 4.2

Draw Lewis structures for atoms of the following:

a. element number 4
b. cesium (Cs)
c. aluminum (Al)
d. selenium (Se)

The accepted procedure is to write the elements symbol and put a dot for each valence electron in one of four equally spaced locations around the symbol. Imagine a square around the symbol. Each side of the square represents one of the four locations. An element with four valence electrons would have one dot in each of the four locations. A fifth electron would be represented by one additional dot in one of the locations. Each location can have a maximum of two dots.

Solution

a. Element number 4 is beryllium; it belongs to group IIA(2) and thus has two valence-shell electrons. The Lewis structure is Be· .
b. Cesium is in group IA(1), has one valence-shell electron, and has the Lewis structure Cs· .
c. Aluminum is in group IIIA(13). It has three valence-shell electrons and is represented by the following Lewis structure: Al· .
d. Selenium is in group VIA(16) and thus has six valence-shell electrons. The Lewis structure is :Se· .

Draw Lewis structures for atoms of the following:

a. element number 9   *Fluorine 7 e⁻ VIIA (9)*
b. magnesium (Mg)   *#12 IIA (12) 2 e⁻*
c. sulfur (S)   *VIA(16) 6 e⁻*
d. krypton (Kr)   *XVIIA (36) 8 e⁻*

It is important to clearly understand the relationship between Lewis structures and electronic configurations for atoms. The following example and learning check will help you review this relationship.

Example 4.3

Represent the following using abbreviated electronic configurations and Lewis structures:

a. F   b. K   c. Mg   d. Si

Solution

a. Fluorine (F) contains 9 electrons, with 7 of them classified as valence electrons (those in the 2s and 2p subshells). The configuration and Lewis structure are

$$[He]2s^2 2p^5 \quad \text{and} \quad :\!\ddot{F}\!\cdot$$

b. Potassium (K) contains 19 electrons, with 1 of them classified as a valence electron (the 1 in the 4s subshell). The configuration and Lewis structure are

$$[Ar]4s^1 \quad \text{and} \quad K\cdot$$

c. Magnesium (Mg) contains 12 electrons, with 2 of them classified as valence electrons (the 2 in the 3s subshell). The configuration and Lewis structures are

$$[Ne]3s^2 \quad \text{and} \quad \dot{Mg}\cdot$$

d. Silicon (Si) contains 14 electrons, with 4 of them classified as valence electrons (those in the 3s and 3p subshells). The configuration and Lewis structures are

$$[Ne]3s^2 3p^2 \quad \text{and} \quad \cdot\dot{Si}\cdot$$

| Learning Check 4.3 | Using the two methods illustrated in Example 4.3, write electronic configurations and Lewis structures for the following atoms: |

a. Li   **b.** Br   **c.** Sr   **d.** S

Octet rule
A rule for predicting electron behavior in reacting atoms. It says that atoms will gain or lose sufficient electrons to achieve an outer electron arrangement identical to that of a noble gas. This arrangement usually consists of eight electrons in the valence shell.

Simple ion
An atom that has acquired a net positive or negative charge by losing or gaining electrons.

Ionic bond
The attractive force that holds together ions of opposite charge.

➤ 4.2 Ionic Bonding

According to the **octet rule** of Lewis and Kossel, atoms tend to interact through electronic rearrangements that produce a noble gas electronic configuration for each atom involved in the interaction. Except for the lowest-energy shell, this means that each atom ends up with eight electrons in the valence shell. There are exceptions to this octet rule, but it is still used because of the amount of information it provides. It is especially effective in describing reactions between the representative elements of the periodic table (see Figure 3.10).

During some chemical interactions, the octet rule is satisfied when electrons are transferred from one atom to another. As a result of the transfers, neutral atoms acquire net positive or negative charges and become attracted to one another. These charged atoms are called **simple ions,** and the attractive force between oppositely charged atoms constitutes an **ionic bond.** A second type of interaction that also satisfies the octet rule is discussed in Section 4.6.

Example 4.4

Show how the following atoms can achieve a noble gas configuration and become ions by gaining or losing electrons:

a. Na   **b.** Cl

Solution

a. The electronic structure of sodium (Na) is represented below using an abbreviated configuration and a Lewis structure:

$$[Ne]3s^1 \quad \text{and} \quad Na\cdot$$

The first representation makes it obvious that an Ne configuration with 8 electrons in the valence shell would result if the Na atom lost the single electron located in the 3s subshell. The loss is represented by the following equation:

$$Na \rightarrow Na^+ + 1e^-$$

Notice that the removal of a single negative electron from a neutral Na atom leaves the atom with 11 positive protons in the nucleus and 10 negative electrons. This gives the atom a net positive charge. The atom has become a positive ion.

b. The electronic structure for chlorine (Cl) is shown below using an abbreviated configuration and a Lewis structure:

$$[Ne]3s^2 3p^5 \quad \text{and} \quad :\overset{\cdot\cdot}{\underset{\cdot\cdot}{Cl}}\cdot$$

The Cl atom can achieve an Ne configuration by losing the 7 valence electrons. However, it is energetically much more favorable to achieve the configuration of argon (Ar) by adding 1 electron to the valence shell. This electron would complete an octet and change a Cl atom into a negative chloride ion as represented by the following equation, where Cl^- represents a Cl atom with 17 protons and 18 electrons, giving the atom a net negative charge:

$$Cl + 1e^- \rightarrow Cl^-$$

GOB
Chemistry·⚛·Now™
Go to GOB Now and click to view a simulation that explores the relationship between the electronic configurations of atoms and their commonly formed ions through a simulation.

Learning Check 4.4

In Learning Check 4.1, you described what had to be done to change the electronic configurations of F and K to noble gas configurations. Write equations to illustrate these changes.

As a general rule, metals lose electrons and nonmetals gain electrons during ionic bond formation. The number of electrons lost or gained by a single atom rarely exceeds three, and this number can be predicted accurately for representative elements by using the periodic table. The number of electrons easily lost by a representative metal atom is the same as the group number (represented by roman numerals preceding the A in the periodic table). The number of electrons that tend to be gained by a representative nonmetal atom is equal to eight minus the group number. However, the nonmetal hydrogen is an exception. In most reactions, it loses one electron, consistent with its placement in group IA(1). In other, less common, reactions it gains one electron just as group VIIA(17) elements do.

GOB
Chemistry·⚛·Now™
Go to GOB Now and click to learn how to test the recognition of the charges of ions formed from atoms.

Example 4.5

Use the periodic table to predict the number of electrons lost or gained by atoms of the following elements during ionic bond formation. Write an equation to represent the process in each case.

a. Li
b. Any group IIA(2) element, represented by the symbol M
c. element number 15
d. carbon

Solution

a. Lithium (Li), a metal, is in group IA(1); therefore it will lose 1 electron per atom:

$$Li \rightarrow Li^+ + 1e^-$$

		Number of protons in nucleus	Number of electrons around nucleus	Net charge on particle
TABLE 4.1 A metal atom, a metal ion, and a noble gas atom compared				
Particle	Symbol			
Magnesium atom	Mg	12	12	0
Magnesium ion	Mg^{2+}	12	10	+2
Neon atom	Ne	10	10	0

b. Any group IIA(2) metal will lose 2 electrons; therefore,

$$M \rightarrow M^{2+} + 2e^-$$

c. Element number 15 is phosphorus (P), a nonmetal of group VA(15). It will gain $8 - 5 = 3$ electrons. The equation is

$$P + 3e^- \rightarrow P^{3-}$$

d. Carbon (C) is in group IVA(14) and is a nonmetal. It should, therefore, gain $8 - 4 = 4$ electrons. However, no more than 3 electrons generally become involved in ionic bond formation, so we conclude that carbon will not react readily to form ionic bonds.

Learning Check 4.5

Predict the number of electrons that would be gained or lost during ionic bond formation for each of the following elements. Write an equation to represent each predicted change.

a. element number 34   **b.** Rb   **c.** element number 18   **d.** In

---

While your attention is focused on noble gas configurations, it is appropriate to emphasize the following point: The attainment of a noble gas electronic configuration by an atom does not mean that the atom is converted into a noble gas. Instead, as seen above, the atoms are converted into simple ions (charged atoms). This point is emphasized by the comparison given in ■ Table 4.1. Notice that the electronic configuration of Mg^{2+} is the same as that of Ne, but the number of protons (the atomic number) of Mg^{2+} is still the same as that of Mg. Atoms and simple ions that have identical electronic configurations are said to be **isoelectronic.**

Isoelectronic
A term that literally means "same electronic," used to describe atoms or ions that have identical electronic configurations.

➤ 4.3 Ionic Compounds

So far the discussion has focused on the electron-transfer process as it occurs for isolated atoms. In reality, the electrons lost by a metal are the same ones gained by the nonmetal with which it is reacting. Substances formed from such reactions are called *ionic compounds.* No atom can lose electrons unless another atom is available to accept them, as shown in the formulas used to represent ionic compounds. These formulas represent the combining ratio of the positive and negative ions found in the compounds. This ratio is determined by the charges on the ions, which are determined by the number of electrons transferred.

CHEMISTRY AND YOUR HEALTH • 4.1

IMPORTANT IONS IN BODY PROCESSES

Fluids within the human body contain numerous ions that perform important functions. The major positive ions present are sodium (Na^+), potassium (K^+), calcium (Ca^{2+}), and magnesium (Mg^{2+}). The major negative ions are chloride (Cl^-), sulfate (SO_4^{2-}), hydrogen phosphate (HPO_4^{2-}), and bicarbonate (HCO_3^-). These ions help maintain not only the electrical neutrality of body and cellular fluids but also the proper volume of blood and other liquid systems. The distribution of ions in these fluids is specific; for example, fluids inside cells are rich in potassium and magnesium ions, whereas fluids surrounding cells are rich in sodium and calcium ions.

Sodium and potassium ions are involved in maintaining nerve sensitivity and muscle control. Magnesium ions are necessary for the proper transmission of nerve impulses. Calcium ions are essential for the formation of bones and teeth, for proper blood clotting, and for the digestive precipitation of milk protein in the stomach. Sodium and potassium ions induce the heart to beat, and calcium ions regulate its rhythm. Chloride ions play a special role in maintaining the electrical neutrality of red blood cells and are required for the preparation of hydrochloric acid (HCl) found in gastric juices. The bicarbonate ion is an important part of a chemical system in the blood that transports carbon dioxide from the tissue cells to the lungs for exhalation.

All these major ions, as well as others present in smaller amounts, must be obtained from the diet. If any are not present in sufficient amounts, diseases result. For example, an iron (Fe^{2+}) deficiency results in anemia. Excesses of ions can also present problems—some people retain excess water and suffer from high blood pressure if too much salt (sodium chloride) is ingested.

Example 4.6

Represent the electron-transfer process that takes place when the following pairs of elements react ionically. Determine the formula for each resulting ionic compound.

a. Na and Cl   **b.** Mg and F

Solution

a. Sodium (Na) is in group IA(1), and we can write

$$Na \rightarrow Na^+ + 1e^-$$

Chlorine (Cl), in group VIIA(17), reacts as follows:

$$Cl + 1e^- \rightarrow Cl^-$$

The resulting ions, Na^+ and Cl^-, will combine in a 1:1 ratio because the total positive and total negative charges in the final formula must add up to zero. The formula then is NaCl. Note that the metal is written first in the formula, a generally followed practice. Also, the formula is written using the smallest number of each ion possible—in this case, one of each.

The actual electron-transfer process and the achievement of octets can more easily be visualized as follows:

$$[Ne]3s^1 + [Ne]3s^23p^5 \rightarrow [Ne] + [Ne]3s^23p^6$$
$$\;\; Na \qquad\quad Cl \qquad\quad Na^+ \qquad\quad Cl^-$$

In this representation, it must be remembered that the $[Ne]3s^23p^6$ electronic configuration of Cl^- is the same configuration as argon, the noble gas that follows chlorine in the periodic table.

b. Magnesium (Mg) of group IIA(2) will lose two electrons per atom, whereas fluorine (F) of group VIIA(17) will gain one electron per atom. Therefore, we can write

$$Mg \rightarrow Mg^{2+} + 2e^-$$

$$F + 1e^- \rightarrow F^-$$

➤ **FIGURE 4.1** The reaction of sodium metal and chlorine gas. The vigorous reaction releases energy that heats the flask contents to a high temperature. Would you expect a similar reaction to occur between potassium metal and fluorine gas?

© Joel Gordon, courtesy of West Publishing Company/ CHEMISTRY by Radel & Navidi, 2d ed.

It is apparent that two fluorine atoms will be required to accept the electrons from one magnesium atom. From another point of view, two F^- ions will be needed to balance the charge of a single Mg^{2+} ion. Both observations lead to the formula MgF_2 for the compound. Note the use of subscripts to indicate the number of ions involved in the formula. The subscript 1, on Mg, is understood and never written.

Learning Check 4.6

Write equations to represent the formation of ions for each of the following pairs of elements. Write a formula for the ionic compound that would form in each case.

a. Mg and O   **b.** K and S   **c.** Ca and Br

Binary compound
A compound made up of two different elements.

Ionic compounds of the types used in Example 4.6 and Learning Check 4.6 are called **binary compounds** because each contains only two different kinds of atoms, Na and Cl in one case (see ■ Figure 4.1) and Mg and F in the other.

➤ 4.4 Naming Binary Ionic Compounds

GOB
Chemistry⚛Now™

Go to GOB Now and click to view the relationship between the formulas and names of ionic compounds.

Names for binary ionic compounds are easily assigned when the names of the two elements involved are known. The name of the metallic element is given first, followed by the stem of the nonmetallic elemental name to which the suffix *-ide* has been added.

name = metal + nonmetal stem + *-ide*

The stem of the name of a nonmetal is the name of the nonmetal with the ending dropped. ■ Table 4.2 gives the stem of the names of some common nonmetallic elements.

Example 4.7

GOB
Chemistry⚛Now™

Go to GOB Now and click to learn how to name ionic compounds.

Name the following binary ionic compounds:

a. KCl   **b.** SrO   **c.** Ca_3N_2

Solution

a. The metal is potassium (K), and the nonmetal is chlorine (Cl). Thus, the compound name is potassium <u>chlor</u>ide (the stem of the nonmetallic name is underlined).

TABLE 4.2	Stem names and ion formulas of common nonmetallic elements	
Element	**Stem**	**Formula of ion**
Bromine	brom-	Br^-
Chlorine	chlor-	Cl^-
Fluorine	fluor-	F^-
Iodine	iod-	I^-
Nitrogen	nitr-	N^{3-}
Oxygen	ox-	O^{2-}
Phosphorus	phosph-	P^{3-}
Sulfur	sulf-	S^{2-}

b. Similarly, strontium (Sr) and oxygen (O) give the compound name strontium <u>ox</u>ide.

c. The elements are calcium (Ca) and nitrogen (N), hence the name calcium <u>nitr</u>ide.

Learning Check 4.7

Assign names to the binary compounds whose formulas you wrote in Learning Check 4.6.

Names for the constituent ions of a binary compound are obtained in the same way as the compound name. Thus, K^+ is a potassium ion, whereas Cl^- is a chloride ion.

Some metal atoms, especially those of transition and inner-transition elements, form more than one type of charged ion. Copper, for example, forms both Cu^+ and Cu^{2+}, and iron forms Fe^{2+} and Fe^{3+}. The names of ionic compounds containing such elements must indicate which ion is present in the compound. A nomenclature system that does this well indicates the ionic charge of the metal ion by a roman numeral in parentheses following the name of the metal. Thus, CuCl is copper(I) chloride and $CuCl_2$ is copper(II) chloride. These names are expressed verbally as "copper one chloride" and "copper two chloride."

An older system is still in use but works only for naming compounds of metals that can form only two different charged ions. In this method, the endings *-ous* and *-ic* are attached to the stem of the metal name. For metals with elemental symbols derived from non-English names, the stem of the non-English name is used. The *-ous* ending is always used with the ion of lower charge and the *-ic* ending with the ion of higher charge. Thus, CuCl is cuprous chloride, and $CuCl_2$ is cupric chloride. In the case of iron, $FeCl_2$ is ferrous chloride, and $FeCl_3$ is ferric chloride (see ■ Figure 4.2). Notice that the Cu^{2+} ion was called cupric, whereas the Fe^{2+} was called ferrous. The designations do not tell the actual ionic charge, but only which of the two is higher (Cu^{2+} and Fe^{3+}) or lower (Cu^+ and Fe^{2+}) for the metal in question.

© Mark Slabaugh

➤ **FIGURE 4.2** Chloride compounds of copper and iron. Top: Cuprous and cupric chloride. Bottom: Ferrous and ferric chloride.

Example 4.8

Write formulas for ionic compounds that would form between the following simple ions. Note that the metal forms two different simple ions, and name each compound two ways.

a. Cr^{2+} and S^{2-} **b.** Cr^{3+} and S^{2-}

Solution

In each case, the metal ion is from the metal chromium (Cr).

a. Because the ionic charges are equal in magnitude but opposite in sign, the ions will combine in a 1:1 ratio. The formula is CrS. Chromium forms simple ions with +2 and +3 charges. This one is +2, the lower of the two; thus, the names are chromium(II) sulfide and chromous sulfide.

b. In this case, the charges on the combining ions are +3 and −2. The smallest combining ratio that balances the charges is two Cr^{3+} (a total of +6 charge) and three S^{2-} (a total of −6 charge). The formula is Cr_2S_3. The names are chromium(III) sulfide and chromic sulfide.

Learning Check 4.8   Write formulas for the ionic compounds that would form between the following simple ions. The metal is cobalt, and it forms only the two simple ions shown. Name each compound, using two methods.

a. Co^{2+} and Br^- b. Co^{3+} and Br^-

➤ 4.5 The Smallest Unit of Ionic Compounds

In Section 1.3, a molecule was defined as the smallest unit of a pure substance capable of a stable, independent existence. As you will see later, some compound formulas are used to represent single molecules. True molecular formulas represent the precise numbers of atoms of each element that are found in a molecule. However, as we have seen, formulas for ionic compounds represent only the simplest combining ratio of the ions in the compounds.

The stable form of an ionic compound is not a molecule, but a crystal in which many ions of opposite charge occupy **lattice sites** in a rigid three-dimensional arrangement called a **crystal lattice**. For example, the lattice of sodium chloride (ordinary table salt) represented in ■ Figure 4.3 is the stable form of pure sodium chloride.

Lattice site
The individual location occupied by a particle in a crystal lattice.

Crystal lattice
A rigid three-dimensional arrangement of particles.

➤ **FIGURE 4.3** Crystal lattice for sodium chloride (table salt).

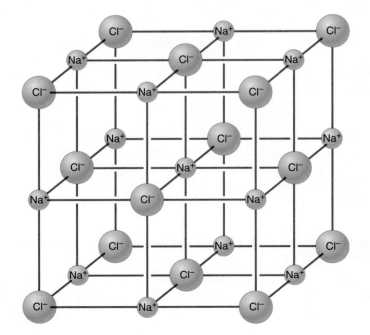

CHEMISTRY AROUND US • 4.1

WATER: ONE OF EARTH'S SPECIAL COMPOUNDS

An adequate supply of clean water is essential to our health and well-being. We can live without food for many days, but life would end in only a few days without water. Our circulatory system is an aqueous stream that distributes an amazing variety of substances throughout the body. Our cells are filled with water solutions in which the chemical reactions of life take place. Without water and its unique properties, life would not be possible on Earth.

Three-fourths of Earth's surface is covered with water, which gives the planet a blue color when it is viewed from space. The oceans represent the world's largest liquid solution. Water participates in most of the chemical reactions that occur in nature.

Most water used for human consumption comes from reservoirs, lakes, rivers, and wells. Much of this water is used and reused by numerous cities as it travels downstream. Such reused water may become seriously polluted with waste from previous users and with pathogenic microorganisms. As a safety precaution, most of the water we use undergoes chemical and physical treatment to purify it. This treatment process includes settling, whereby specific materials are added to bring down suspended solids. This is followed by filtration through sand and gravel to remove still more suspended matter. The filtered water may then be aerated by spraying it into the air. This part of the treatment process removes some odors and improves the taste of the water. Charcoal filtration, a treatment process that is gaining in use, also removes odors along with colored materials. In a final step, chlorine or another disinfectant is added to kill any remaining bacteria.

© Lawrence Migdale/Tony Stone Worldwide

The treatment of water before use is an important safety precaution.

Even though the formulas of ionic compounds do not represent true molecular formulas, they are often used as if they did. This is especially true when equations representing chemical reactions are written, or when the mole concept is applied to chemical formulas (see Section 2.7). When the atomic weights of the atoms making up a true molecular formula are added together, the result is called the molecular weight of the compound (Section 2.4). A similar quantity obtained by adding up the atomic weights of the atoms shown in the formula of an ionic compound is called a **formula weight**. The mole concept is applied to formula weights in a manner similar to the way it is applied to molecular weights.

Formula weight
The sum of the atomic weights of the atoms shown in the formula of an ionic compound.

Example 4.9

Carbon dioxide, CO_2, is a molecular compound, whereas magnesium chloride, $MgCl_2$, is an ionic compound.

a. Determine the molecular weight for CO_2 and the formula weight for $MgCl_2$ in atomic mass units.
b. Determine the mass in grams of 1.00 mol of each compound.
c. Determine the number of CO_2 molecules in 1.00 mol and the number of Mg^{2+} and Cl^- ions in 1.00 mol of $MgCl_2$.

Solution

a. For CO_2, the molecular weight is the sum of the atomic weights of the atoms in the formula:

$$MW = (1)(\text{at. wt. C}) + (2)(\text{at. wt. O}) = (1)(12.0 \text{ u}) + (2)(16.0 \text{ u})$$
$$MW = 44.0 \text{ u}$$

For $MgCl_2$, the formula weight is also equal to the sum of the atomic weights of the atoms in the formula:

$$FW = (1)(\text{at. wt. Mg}) + (2)(\text{at. wt. Cl}) = (1)(24.3 \text{ u}) + (2)(35.5 \text{ u})$$
$$FW = 95.3 \text{ u}$$

b. According to Section 2.6, 1.00 mol of a molecular compound has a mass in grams equal to the molecular weight of the compound. Thus, 1.00 mol $CO_2 = 44.0$ g CO_2.

   For ionic compounds, 1.00 mol of compound has a mass in grams equal to the formula weight of the compound. Thus, 1.00 mol $MgCl_2 = 95.3$ g $MgCl_2$.

c. In Section 2.6 we learned that 1.00 mol of a molecular compound contains Avogadro's number of molecules. Thus, 1.00 mol $CO_2 = 6.02 \times 10^{23}$ molecules of CO_2.

   In the case of ionic compounds, 1.00 mol of compound contains Avogadro's number of formula units. That is, 1.00 mol of magnesium chloride contains Avogadro's number of $MgCl_2$ units, where each unit represents one Mg^{2+} ion and two Cl^- ions. Thus,

$$1.00 \text{ mol } MgCl_2 = 6.02 \times 10^{23} \text{ } MgCl_2 \text{ units}$$

or

$$1.00 \text{ mol } MgCl_2 = 6.02 \times 10^{23} \text{ } Mg^{2+} \text{ ions} + 1.20 \times 10^{24} \text{ } Cl^- \text{ ions}$$

Note that the number of Cl^- ions is simply twice Avogadro's number.

Learning Check 4.9

Determine the same quantities that were determined in Example 4.9, but use the molecular compound hydrogen sulfide, H_2S, and the ionic compound calcium oxide, CaO.

➤ 4.6 Covalent Bonding

In Example 4.5, we found that carbon, a nonmetal, would have to gain four electrons in order to form ionic bonds. We concluded that carbon does not generally form ionic bonds. Yet carbon is known to form many compounds with other elements. Also, electron transfers would not be expected to take place between two atoms of the same element because such an exchange would not change the electronic configuration of the atoms involved. Yet molecules of elements containing two atoms, such as chlorine (Cl_2), oxygen (O_2), and nitrogen (N_2), are known to exist and, in fact, represent the stable form in which these elements occur in nature. What is the nature of the bonding in these molecules?

Again, G. N. Lewis provided an answer by suggesting that the valence-shell electrons of the atoms in such molecules are shared in a way that satisfies the octet rule for each of the atoms. This process is symbolized below for fluorine, (F_2), using Lewis structures:

$$:\ddot{F}\cdot + \cdot\ddot{F}: \longrightarrow :\ddot{F}:\ddot{F}:$$

shared pair (counted in octet of each atom)

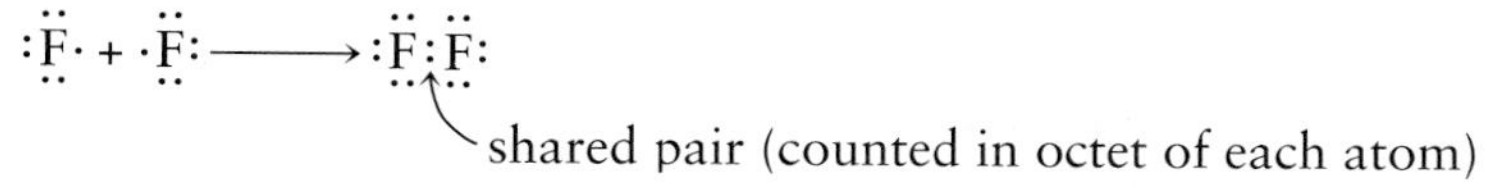

Sometimes the shared electron pairs are shown as straight lines, and sometimes the nonshared pairs are not shown, as illustrated in Example 4.10.

Example 4.10

Represent the following reactions using Lewis structures:

a. $Cl + Cl \rightarrow Cl_2$ **b.** $H + H \rightarrow H_2$

GOB
Chemistry⋅⦿⋅Now™
Go to GOB Now and click to learn how to interpret Lewis structures.

Solution

a. Each chlorine (Cl) atom has seven valence-shell electrons. Because one electron from each atom is shared, the octet rule is satisfied for each atom

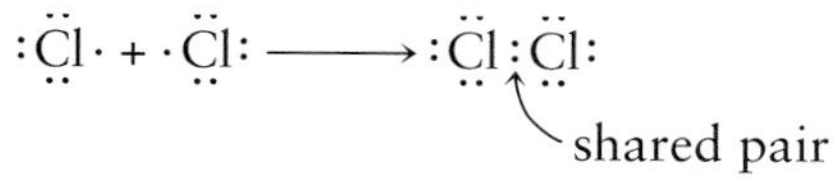

$$:\ddot{Cl}\cdot + \cdot\ddot{Cl}: \longrightarrow :\ddot{Cl}:\ddot{Cl}:$$

shared pair

The structure of Cl_2 can also be represented as

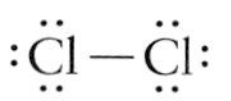

$$:\ddot{Cl}-\ddot{Cl}: \quad \text{or} \quad Cl-Cl$$

b. Each hydrogen (H) atom has one valence-shell electron. By sharing the two electrons, each H atom achieves the helium (He) structure:

$$H\cdot + \cdot H \longrightarrow H:H \quad \text{or} \quad H-H$$

Represent the following reactions using Lewis structures:

a. $N + N \rightarrow N_2$ (more than two electrons will be shared)   **b.** $Br + Br \rightarrow Br_2$

Learning Check 4.10

To understand the origin of the attractive force between atoms that results from electron sharing, consider the H_2 molecule and the concept of atomic orbital overlap. Suppose two H atoms are moving toward each other. Each atom has a single electron in a spherical $1s$ orbital. While the atoms are separated, the orbitals are independent of each other; but as the atoms get closer together, the orbitals overlap and blend to create an orbital common to both atoms called a *molecular orbital*. The two shared electrons then move throughout the overlap region but have a high probability of being found somewhere between the two nuclei. As a result, the positive nuclei are both attracted toward the negative pair of electrons and hence toward each other. This process, represented in ■ Figure 4.4,

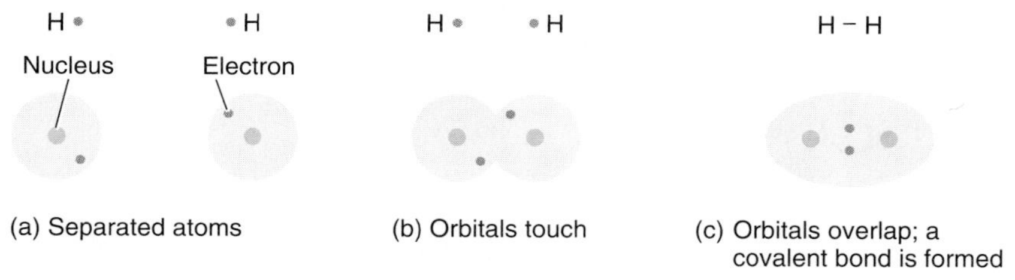

(a) Separated atoms   (b) Orbitals touch   (c) Orbitals overlap; a covalent bond is formed

➤ FIGURE 4.4 Orbital overlap during covalent bond formation.

Covalent bond
The attractive force that results between two atoms that are both attracted to a shared pair of electrons.

produces a net attractive force between the nuclei. This attractive force is the **covalent bond** that holds the atoms together. Because forces cannot be seen, we represent only the presence of the shared electron pair between the atoms (by a pair of dots or a line) and remember that an attractive force is also there.

Covalent bonding between like atoms has been described, but it also occurs between unlike atoms. Examples of both types are given in ■ Table 4.3. Lewis structures, like those in Table 4.3, are most easily drawn by using a systematic approach such as the one represented by the following steps:

1. Use the molecular formula to determine how many atoms of each kind are in the molecule.
2. Use the given connecting pattern of the atoms to draw an initial structure for the molecule, with the atoms arranged properly.
3. Determine the total number of valence-shell electrons contained in the atoms of the molecule.
4. Put one pair of electrons between each bonded pair of atoms in the initial structure drawn in Step 2. Subtract the number of electrons used in this step from the total number determined in Step 3. Use the remaining electrons to complete the octets of all atoms in the structure, beginning with the atoms that are present in greatest number in the molecule. Remember, hydrogen atoms require only one pair to achieve the electronic configuration of helium.
5. If all octets cannot be satisfied with the available electrons, move nonbonding pairs (those that are not between bonded atoms) to positions

TABLE 4.3 Examples of covalent bonding

Molecule	Atomic Lewis structure	Sharing pattern	Molecular Lewis structure
Nitrogen gas (N_2)	$\cdot\ddot{N}\cdot$ $(1s^22s^22p^3)$	:N: ⟶ ⟵ :N:	:N::N:
Carbon dioxide (CO_2) (each O is bonded to the C)	$\cdot\dot{C}\cdot$ $(1s^22s^22p^2)$ $:\ddot{O}\cdot$ $(1s^22s^22p^4)$	:O: C :O:	:O::C::O:
Formaldehyde (H_2CO) (each H and the O are bonded to the C)	$\cdot\dot{C}\cdot$ $(1s^22s^22p^2)$ $:\ddot{O}\cdot$ $(1s^22s^22p^4)$ $H\cdot$ $(1s^1)$	H· C O: H·	H :C::O: H
Methane (CH_4) (each H is bonded to the C)	$\cdot\dot{C}\cdot$ $(1s^22s^22p^2)$ $H\cdot$ $(1s^1)$	H H C H H	H H:C:H H
Ammonia (NH_3) (each H is bonded to the N)	$\cdot\ddot{N}\cdot$ $(1s^22s^22p^3)$ $H\cdot$ $(1s^1)$	H N H H	H:N:H H
Water (H_2O) (each H is bonded to the O)	$:\ddot{O}\cdot$ $(1s^22s^22p^4)$ $H\cdot$ $(1s^1)$	:O: H H	:O:H H
Oxygen diflouride (OF_2) (each F is bonded to the O)	$:\ddot{O}\cdot$ $(1s^22s^22p^4)$ $:\ddot{F}\cdot$ $(1s^22s^22p^5)$	:O: F: :F:	:O:F: :F:

between bonded atoms to complete octets. This will create **double** or **triple bonds** between some atoms.

Double and triple bonds
The bonds resulting from the sharing of two and three pairs of electrons, respectively.

Example 4.11

Draw Lewis structures for the following molecules:

a. NH_3 **b.** SO_3 **c.** C_2H_2

Solution

a. *Step 1.* The formula indicates that the molecule contains 1 nitrogen (N) atom and 3 hydrogen (H) atoms.

Step 2. The connecting pattern will have to be given to you in some form. Table 4.3 shows that each H atom is connected to the N atom; thus, we draw

$$H \ N \ H$$
$$H$$

Step 3. Nitrogen is in group VA(15) of the periodic table and so has 5 valence electrons; hydrogen is in group IA(1) and has 1 valence electron. The total is 8 (5 from one N atom and 3 from the three H atoms).

Step 4. We put one pair of electrons between each H atom and the N atom in the initial structure drawn in Step 2:

$$H{:}N{:}H$$
$$\overset{..}{H}$$

This required 6 of the 8 available electrons. The remaining pair is used to complete the octet of nitrogen. Remember, hydrogen achieves the noble gas configuration of helium with only 2 electrons:

$$H{:}\overset{..}{N}{:}H \qquad \text{or} \qquad H - \overset{..}{\underset{|}{N}} - H$$
$$\overset{..}{H} \qquad\qquad\qquad\qquad H$$

Step 5. All octets are satisfied by Step 4, so nothing more needs to be done.

b. *Step 1.* The formula indicates that 1 sulfur (S) and 3 oxygen (O) atoms are present in a molecule.

Step 2. Each O atom is bonded only to the S atom; thus, we draw

$$O \ S \ O$$
$$O$$

Step 3. Sulfur and oxygen are both in group VIA(16), and so each atom has 6 valence electrons. The total is 24 (6 from the one S atom and 18 from the three O atoms).

Step 4. We put one pair of electrons between each O atom and the S atom:

$$O{:}S{:}O$$
$$\overset{..}{O}$$

This required 6 of the 24 available electrons. The remaining 18 are used to complete the octets, beginning with the O atoms:

$$:\overset{..}{\underset{..}{O}}{:}S{:}\overset{..}{\underset{..}{O}}{:}$$
$$:\overset{..}{\underset{..}{O}}{:}$$

Step 5. We see that the octet of sulfur is not completed, even though all available electrons have been used. If one nonbonding pair of any one

GOB
Chemistry·ᐧ·Now™
Go to GOB Now and click to see the step-by-step process for drawing Lewis structures.

GOB
Chemistry·ᐧ·Now™
Go to GOB Now and click to look at a tutorial examining the drawing of Lewis structures.

of the three O atoms is moved to a bonding position between the oxygen and sulfur, it will help satisfy the octet of both atoms. The final Lewis structure is given below. Note that it contains a double bond between one of the oxygens and the sulfur:

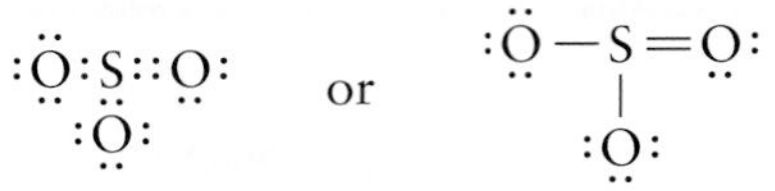

$$:\ddot{O}:S::\ddot{O}: \qquad \text{or} \qquad :\ddot{O}-S=\ddot{O}:$$
$$:\ddot{O}: \qquad\qquad\qquad\qquad\quad |$$
$$\qquad\qquad\qquad\qquad\qquad :\ddot{O}:$$

If a nonbonding pair of electrons from either of the other two oxygens is used, the following structures result. These are just as acceptable as the preceding structures:

$$:\ddot{O}=S-\ddot{O}: \qquad\qquad :\ddot{O}-S-\ddot{O}:$$
$$\qquad\quad | \qquad\qquad\qquad\qquad\quad ||$$
$$\qquad :\ddot{O}: \qquad\qquad\qquad\qquad \ddot{O}:$$

c. *Step 1.* The formula indicates 2 carbon (C) atoms and 2 hydrogen (H) atoms per C_2H_2 molecule.

Step 2. The C atoms are bonded to each other, and 1 H atom is bonded to each C atom.

$$H \; C \; C \; H$$

Step 3. Carbon is in group IVA(14) and has 4 valence electrons; hydrogen is in group IA(1) and has 1 valence electron. The total electrons available is 10 (8 from the two C atoms and 2 from the two H atoms).

Step 4. We put one pair of electrons between the two C atoms and between each C atom and H atom.

$$H : C : C : H$$

This required 6 of the available 10 electrons. The remaining 4 electrons are used to complete the C octets (remember, hydrogen needs only 2 electrons):

$$H : C : \ddot{C} : H$$

Step 5. All electrons are used in Step 4, but the octet of 1 C atom is still incomplete. It needs two more pairs. If both nonbonding pairs on the second C atom were shared with the first C atom, both carbons would have complete octets. The result is

$$H : C ::: C : H \qquad \text{or} \qquad H - C \equiv C - H$$

This molecule contains a triple bond (three shared pairs) between the C atoms.

Learning Check 4.11

Draw Lewis structures for the following molecules:

a. CH_4 (each H atom is bonded to the C atom)
b. H_2CO (the O atom and two H atoms are bonded to the C atom)
c. HNO_3 (the O atoms are each bonded to the N atom, and the H atom is bonded to one of the O atoms)

➤ 4.7 Polyatomic Ions

Polyatomic ions
Covalently bonded groups of atoms that carry a net electrical charge.

An interesting combination of ionic and covalent bonding is found in compounds that contain **polyatomic ions.** These ions are covalently bonded groups of atoms that carry a net electrical charge. With the exception of the ammonium ion, NH_4^+, the common polyatomic ions are negatively charged.

Lewis structures can be drawn for polyatomic ions with just a slight modification in the steps given earlier for covalently bonded molecules. In Step 3, the total number of electrons available is obtained by first determining the total number of valence-shell electrons contained in the atoms of the ion. To this number are added electrons representing the number required to give the negative charge to the ion.

For example, in the SO_4^{2-} ion, the total number of valence-shell electrons is 30, with 6 coming from the sulfur atom and 6 from each of the four oxygen atoms. To this number we add 2 because two additional electrons are required to give the group of neutral atoms the charge of 2− found on the ion. This gives 32 total available electrons.

The only exception to this procedure for the common polyatomic ions is for the positively charged ammonium ion, NH_4^+. In this case, the number of electrons available is the total number of valence electrons minus 1, because one electron would have to be removed from the neutral group of atoms to produce the + charge on the ion.

Example 4.12

Draw Lewis structures for the following polyatomic ions. The connecting patterns of the atoms to each other are indicated.

a. SO_4^{2-} (each O atom is bonded to the S atom)
b. NO_3^- (each O atom is bonded to the N atom)
c. $H_2PO_4^-$ (each O atom is bonded to the P atom, and each H atom is bonded to an O atom)

Solution

a. The only difference between drawing Lewis structures for molecules and polyatomic ions comes in Step 3, where the total number of valence-shell electrons is determined.

   Step 1. The formula indicates the ion contains one sulfur (S) atom and four oxygen (O) atoms.

   Step 2. The given bonding relationships lead to the following initial structure:

$$O$$
$$O \; S \; O$$
$$O$$

   Step 3. Both the S and O atoms are in group VIA(16), so each atom contributes 6 valence electrons. The total number of valence-shell electrons from the atoms is therefore $5 \times 6 = 30$. However, the ion has a 2− charge. This charge comes from the presence of 2 electrons in the ion in addition to the electrons from the constituent atoms. These two additional electrons must be included in the total valence-shell electrons. Thus, the total number of electrons is $30 + 2 = 32$.

   Step 4. The following is obtained when the 32 electrons are distributed among the atoms of the initial structure of Step 2:

Step 5. All octets are satisfied by Step 4. However, it is conventional to enclose Lewis structures of polyatomic ions in brackets and to indicate the net charge on the ion. Thus, the Lewis structures are

$$\left[\begin{array}{c} :\ddot{O}: \\ :\ddot{O}:\overset{..}{S}:\ddot{O}: \\ :\ddot{O}: \end{array}\right]^{2-} \quad \text{or} \quad \left[\begin{array}{c} O \\ | \\ O-S-O \\ | \\ O \end{array}\right]^{2-}$$

where the second structure does not show nonbonding (unshared) electron pairs, and the bonding pairs are represented by a line.

b. *Step 1.* Three oxygen (O) atoms and one nitrogen (N) atom are found in the ion.

Step 2.

$$\begin{array}{c} O \\ O \ N \ O \end{array}$$

Step 3. Electrons from oxygen = $3 \times 6 = 18$.
Electrons from nitrogen = $1 \times 5 = 5$.
One electron comes from the ionic charge = 1.
Total electrons = $18 + 5 + 1 = 24$.

Step 4.

$$\begin{array}{c} :\ddot{O}: \\ :\ddot{O}:\overset{..}{N}:\ddot{O}: \end{array}$$

Step 5. The octet of nitrogen is not satisfied by Step 4, so an unshared pair of electrons on an O atom must be shared with the N atom. This will form a double bond in the ion. The resulting Lewis structures are

$$\left[\begin{array}{c} :\ddot{O}: \\ :\ddot{O}:\overset{..}{N}::\ddot{O}: \end{array}\right]^{-} \quad \text{or} \quad \left[\begin{array}{c} O \\ | \\ O-N=O \end{array}\right]^{-}$$

c. *Step 1.* Two hydrogen (H) atoms, one phosphorus (P) atom, and four oxygen (O) atoms are found in the ion.

Step 2.

$$\begin{array}{c} O \\ O \ P \ O \ H \\ O \\ H \end{array}$$

Step 3. Electrons from hydrogen = $2 \times 1 = 2$.
Electrons from oxygen = $4 \times 6 = 24$.
Electrons from phosphorus = $1 \times 5 = 5$.
Electron from ionic charge = 1.
Total electrons = $2 + 24 + 5 + 1 = 32$.

Step 4.

Step 5. All octets are satisfied by Step 4 (remember, H atoms are satisfied by a single pair of electrons). The resulting Lewis structures are

Draw Lewis structures for the following polyatomic ions:

Learning Check 4.12

a. PO_4^{3-} (each O atom is bonded to the P atom)
b. SO_3^{2-} (each O atom is bonded to the S atom)
c. NH_4^{+} (each H atom is bonded to the N atom)

➤ 4.8 Shapes of Molecules and Polyatomic Ions

Most molecules and polyatomic ions do not have flat, two-dimensional shapes like those implied by the molecular Lewis structures of Table 4.3. In fact, the atoms of most molecules and polyatomic ions form distinct three-dimensional shapes. Being able to predict the shape is important because the shape contributes to the properties of the molecule or ion. This can be done quite readily for molecules composed of representative elements.

To predict molecular or ionic shapes, first draw Lewis structures for the molecules or ions using the methods discussed in Sections 4.6 and 4.7. Once you have drawn the Lewis structure for a molecule or ion, you can predict its shape by applying a simple theory to the structure. The theory is called the *valence-shell electron-pair repulsion theory*, or **VSEPR theory** (sometimes pronounced "vesper" theory). According to VSEPR theory, electron pairs in the valence shell of an atom are repelled and get as far away from one another as possible. Any atom in a molecule or ion that is bonded to two or more other atoms is called a *central atom*. When the VSEPR theory is applied to the valence-shell electrons of central atoms, the shape of the molecule or ion containing the atoms can be predicted. Two rules are followed:

VSEPR theory
A theory based on the mutual repulsion of electron pairs. It is used to predict molecular shapes.

1. All valence-shell electron pairs around the central atom are counted equally, regardless of whether they are bonding or nonbonding pairs.

2. Double or triple bonds between atoms are treated like a single pair of electrons when predicting shapes.

The electron pairs around a central atom will become oriented in space to get as far away from one another as possible. Thus, two pairs will be oriented with one pair on each opposite side of the central atom. Three pairs will form a triangle around the central atom, and four pairs will be located at the corners of a regular tetrahedron with the central atom in the center (see ■ Figures 4.5 and 4.6). The VSEPR theory can be used to predict shapes of molecules and ions with five or more pairs on the central atom, but we will not go beyond four pairs in this book.

➤ **FIGURE 4.5** Arrangements of electron pairs around a central atom *(E).*

GOB
Chemistry ⚛ Now™
Go to GOB Now and click to examine the relationship between electron-pair geometry and Lewis structures.

Two electron pairs   Three electron pairs   Four electron pairs

Example 4.13

Draw Lewis structures for the following molecules, apply the VSEPR theory, and predict the shape of each molecule:

a. CO_2 (each O atom is bonded to the C atom)
b. C_2Cl_4 (the C atoms are bonded together, and two Cl atoms are bonded to each C atom)
c. NH_3 (each H atom is bonded to the N atom)
d. H_2O (each H atom is bonded to the O atom)

Solution

a. The Lewis structure is $\ddot{\text{O}}=\text{C}=\ddot{\text{O}}$: . The central atom is carbon (C) and it has two pairs of electrons in the valence shell surrounding it (remember, each double bond is treated like a single pair of electrons in the VSEPR theory). The two pairs will be located on opposite sides of the C atom, so the molecule has the shape drawn above with the O, C, and O atoms in a line; it is a linear molecule.

b. The Lewis structure is

$$:\ddot{\text{Cl}} \diagdown \diagup \ddot{\text{Cl}}:$$
$$\text{C}=\text{C}$$
$$:\ddot{\text{Cl}} \diagup \diagdown \ddot{\text{Cl}}:$$

Both C atoms are central atoms in this molecule, and each one has three electron pairs in the valence shell (again, the double bond is treated as only

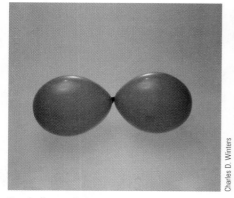

Two balloons, linear

Three balloons, triangular

Four balloons, tetrahedral

➤ **FIGURE 4.6** When balloons of the same size and shape are tied together, they will assume positions like those taken by pairs of valence electrons around a central atom. Where would the central atom be located in these balloon models?

one pair). The three pairs around each C atom will be arranged in the shape of a triangle:

Thus, the molecule has the shape of two flat triangles in the same plane connected together at one point:

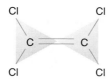

c. The Lewis structure is

$$H - \ddot{N} - H$$
$$|$$
$$H$$

The central atom is nitrogen (N), and it has four electron pairs in the valence shell. The four pairs will be located at the corners of a tetrahedron with the N atom in the middle:

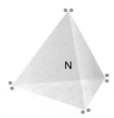

The shape of the molecule is determined only by the positions of the atoms, not by the position of the unshared pair of electrons. Thus, the NH_3 molecule has the shape of a pyramid with a triangular base. The N atom is at the peak of the pyramid, and an H atom is at each corner of the base:

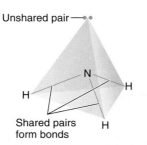

d. The Lewis structure is $H \!:\! \ddot{O} \!:\! H$. The central atom is oxygen (O), and it is seen to have four pairs of electrons in its valence shell. The four pairs will be located at the corners of a tetrahedron with the O atom in the middle:

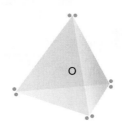

GOB
Chemistry⋅Now™
Go to GOB Now and click to view the step-by-step process of determining molecular shape.

Once again, the shape of the molecule is determined by the locations of the atoms. Thus, H_2O is seen as having a bent or angular shape:

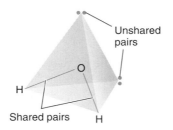

Unshared pairs

O

H

Shared pairs   H

Learning Check 4.13

Chemistry·I·Now™

Go to GOB Now and click to learn the determination of molecular shape.

Predict the shapes of the following molecules by applying the VSEPR theory:

a. BF_3 (the Lewis structure is

$$:\ddot{F}:$$
$$\ddot{B}:\ddot{F}:$$
$$:\ddot{F}:$$

for this molecule, which does not obey the octet rule).

b. CCl_4 (each Cl atom is bonded to the C atom)

c. SO_2 (each O atom is bonded to the S atom)

d. CS_2 (each S atom is bonded to the C atom)

OVER THE COUNTER • 4.1

VERSATILE ZINC OXIDE

While the jury is still out on the usefulness of zinc compounds in treating the symptoms of the common cold (see Over the Counter 3.1), the usefulness of zinc oxide as a topical astringent, antiseptic, and skin protectant is well established. Zinc oxide, with the simple formula ZnO, is a coarse white or grayish powder that is insoluble in water or in alcohol. Because of this insolubility, it is often blended with other ingredients to produce a suspension of the solid in such forms as lotions, creams, or salves. The solid compound in a finely divided form is sometimes added to products such as baby powders.

In these forms, zinc oxide is used topically on the skin as a sunscreen to block harmful ultraviolet rays, and as a shield from various chemical irritants. In addition, products containing the compound are used to relieve a number of common skin conditions, including diaper rash; poison ivy, oak, and sumac rashes; external hemorrhoids; and insect bites. In these applications its antiseptic property helps prevent infections, and it is thought to promote healing by attracting protein to the affected areas of the skin, thereby encouraging new tissue growth. In these topical applications, the insolubility and consistency of zinc oxide prevent it from being absorbed into the skin. As a result, side effects or allergic reactions to the compound are rare, even with long-term use.

In addition to its valuable medical uses, this versatile compound is also found in a wide variety of commercial applica-

tions. It functions as a pigment and reinforcing agent in some rubber products, as a pigment and mold-growth inhibitor in paints; as a pigment in ceramics, floor tile, and glass; as a thickener in cosmetics; as a feed additive for cattle; and as a dietary supplement for humans.

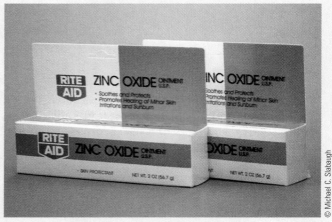

Zinc oxide in the form of an ointment is convenient and easy to use.

Example 4.14

Use the VSEPR theory to predict the shapes of the following polyatomic ions. The Lewis structures were drawn in Example 4.12.

a. SO_4^{2-} **b.** NO_3^-

Solution

a. The Lewis structure from Example 4.12 is

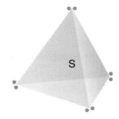

We see that sulfur is the central atom, and it has four electron pairs around it. The four pairs will be located at the corners of a tetrahedron with the S in the middle.

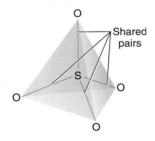

The shape of the ion is determined by the location of the oxygen atoms. Thus, the ion has a tetrahedral shape:

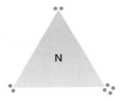

b. The Lewis structure from Example 4.12 is

We see that nitrogen is the central atom, and it has three electron pairs around it. Remember, the double bond formed by the two shared pairs between the nitrogen and one of the oxygens counts as only one pair in the VSEPR theory. The three pairs will form a triangle around the N.

An oxygen is located at each corner of the triangle, so the ion has a flat triangular shape:

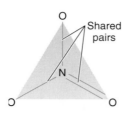

Use the VSEPR theory to predict the shapes of the following polyatomic ions. The Lewis structures were drawn in Learning Check 4.12.

a. PO_4^{3-} **b.** SO_3^{2-} **c.** NH_4^+

➤ 4.9 The Polarity of Covalent Molecules

Nonpolar covalent bond
A covalent bond in which the bonding pair of electrons is shared equally by the bonded atoms.

Electronegativity
The tendency of an atom to attract shared electrons of a covalent bond.

Bond polarization
A result of shared electrons being attracted to the more electronegative atom of a bonded pair of atoms.

Polar covalent bond
A covalent bond that shows bond polarization; that is, the bonding electrons are shared unequally.

Molecules of chlorine (Cl—Cl) and hydrogen chloride (H—Cl) have a number of similarities. For example, they contain two atoms each and are therefore diatomic molecules, and the atoms are held together by a single covalent bond. Some differences also exist, the most obvious being the homoatomic nature of the chlorine molecule (an element) and the heteroatomic nature of the hydrogen chloride molecule (a compound). This difference influences the distribution of the shared electrons in the two molecules.

An electron pair shared by two identical atoms is attracted equally to each of the atoms, and the probability of finding the electrons close to an atom is equal for the two atoms. Thus, on average, the electrons spend exactly the same amount of time associated with each atom. Covalent bonds of this type are called **nonpolar covalent bonds.**

Different atoms generally have different tendencies to attract the shared electrons of a covalent bond. A measurement of this tendency is called **electronegativity.** The electronegativity of the elements is a property that follows trends in the periodic table. It increases from left to right across a period and decreases from top to bottom of a group. These trends are shown in ■ Table 4.4, which contains electronegativity values for the common representative elements. As a result of electronegativity differences, the bonding electrons shared by two different atoms are shared unequally. The electrons spend more of their time near the atom with the higher electronegativity. The resulting shift in average location of bonding electrons is called **bond polarization** and gives a **polar covalent bond.**

As a result of the unequal electron sharing, the more electronegative atom acquires a partial negative charge ($\delta-$), and the less electronegative atom has a partial positive charge ($\delta+$). The molecule as a whole has no net charge, just an uneven charge distribution.

Example 4.15

Use only the periodic table to determine the following for the diatomic covalent molecules listed below: (1) the more electronegative element, (2) the direction of bond polarization, and (3) the charge distribution resulting from the polarization.

a. I—Cl   **b.** Br—Br   **c.** C≡O

| TABLE 4.4 | Electronegativities for the common representative elements |

Increasing electronegativity →

			H			
			2.1			
Li	Be	B	C	N	O	F
1.0	1.5	2.0	2.5	3.0	3.5	4.0
Na	Mg	Al	Si	P	S	Cl
0.9	1.2	1.5	1.8	2.1	2.5	3.0
K	Ca	Ga	Ge	As	Se	Br
0.8	1.0	1.6	1.8	2.0	2.4	2.8
Rb	Sr	In	Sn	Sb	Te	I
0.8	1.0	1.7	1.8	1.9	2.1	2.5
Cs	Ba					
0.7	0.9					

Decreasing electronegativity ↓

Solution

a. Both chlorine (Cl) and iodine (I) belong to group VIIA(17); chlorine is higher in the group and therefore is the more electronegative. The bond polarization (shift in average bonding-electron location) will be toward chlorine, as indicated by the arrow shown below the bond. The result is a partial negative charge on Cl and a partial positive charge on I:

$$\overset{\delta+}{I}\underset{\rightarrow}{-}\overset{\delta-}{Cl}$$

b. Since both bromine (Br) atoms have the same electronegativity, no bond polarization or unequal charge distribution results. The molecule is non-polar covalent and is represented as Br—Br.

c. Oxygen (O) and carbon (C) both belong to the second period. Oxygen, being located farther to the right in the period, is the more electronegative. Thus, the molecule is represented as

$$\overset{\delta+}{C}\underset{\Rightarrow}{\equiv}\overset{\delta-}{O}$$

Use only the periodic table and show the direction of any bond polarization and resulting charge distribution in the following molecules:

Learning Check 4.15

a. N≡N **b.** I—Br **c.** H—Br

The extent of bond polarization depends on electronegativity differences between the bonded atoms and forms the basis for a classification of bonds. When the electronegativity difference (ΔEN) between the bonded atoms is 0.0, the bond is classified as nonpolar covalent (the electrons are shared equally). When ΔEN is 2.1 or greater, the bond is classified as ionic (the electrons are transferred). When ΔEN is between 0.0 and 2.1, the bond is classified as polar covalent (the electrons are unequally shared, and the bond is polarized). Note, however, that the change from purely covalent to completely ionic compounds is gradual and continuous as ΔEN values change. Thus, a bond between hydrogen (H) and boron (B) ($\Delta EN = 0.1$) is classified as polar, but the electron sharing is only slightly unequal and the bond is only slightly polar (see

➤ **FIGURE 4.7** The darker the shading, the larger ΔEN is for the bonded atoms. Larger ΔEN values correspond to greater inequality in the sharing of the bonding electrons.

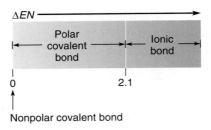

■ Figure 4.7). As the values in Table 4.4 indicate, compounds formed by a reaction of a metal with a nonmetal are generally ionic, while those between nonmetals are either nonpolar or polar covalent.

Example 4.16

Using Table 4.4, classify the bonds in the following compounds as nonpolar covalent, ionic, or polar covalent:

a. ClF   **b.** MgO   **c.** PI_3

Solution

a. The electronegativities of chlorine (Cl) and fluorine (F) are 3.0 and 4.0, respectively. Obtain the ΔEN value by subtracting the smaller from the larger, regardless of the order of the elements in the formula. Thus,

$$\Delta EN = 4.0 - 3.0 = 1$$

and the bond is polar covalent.

b. Similarly,

$$\Delta EN = 3.5 - 1.2 = 2.3$$

and the bond is ionic.

c. The ΔEN is determined for each phosphorus–iodine bond:

$$\Delta EN = 2.5 - 2.1 = 0.4$$

Thus, each bond is classified as polar covalent.

Learning Check 4.16

Using Table 4.4, classify the bonds in the following compounds as nonpolar covalent, ionic, or polar covalent:

a. KF   **b.** NO   **c.** AlN   **d.** K_2O

You can now predict the polarity of bonds within molecules, but it is also important to be able to predict the polar nature of the molecules themselves. The terms *polar* and *nonpolar,* when used to describe molecules, indicate their electrical charge symmetry (see ■ Figure 4.8). In **polar molecules** (often called dipoles), the charge distribution resulting from bond polarizations is nonsymmetric. In **nonpolar molecules,** any charges caused by bond polarization are symmetrically distributed through the molecule. ■ Table 4.5 illustrates polar and nonpolar molecules. The molecular shapes were determined by using the VSEPR theory.

Polar molecule
A molecule that contains polarized bonds and in which the resulting charges are distributed unsymmetrically throughout the molecule.

Nonpolar molecule
A molecule that contains no polarized bonds, or a molecule containing polarized bonds in which the resulting charges are distributed symmetrically throughout the molecule.

➤ 4.10 More about Naming Compounds

In Section 4.4, we discussed the rules for naming binary ionic compounds. In this section, we expand the rules to include the naming of binary covalent compounds and ionic compounds that contain polyatomic ions.

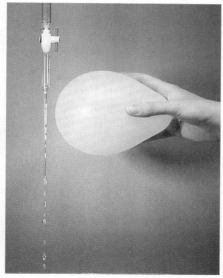

© Spencer L. Seager

© Spencer L. Seager

A stream of polar water molecules is attracted by a charged balloon.

A stream of nonpolar carbon tetrachloride (CCl_4) molecules is not affected by a charged balloon.

TABLE 4.6	Greek numerical prefixes
Number	**Prefix**
1	*mono-*
2	*di-*
3	*tri-*
4	*tetra-*
5	*penta-*
6	*hexa-*
7	*hepta-*
8	*octa-*
9	*nona-*
10	*deca-*

It is apparent from Table 4.4 that covalent bonds (including polar covalent) form most often between representative elements classified as nonmetals. It is not possible to predict formulas for these compounds in a simple way, as was done for ionic compounds in Section 4.4. However, simple rules do exist for naming binary covalent compounds on the basis of their formulas. The rules are similar to those used to name binary ionic compounds: (1) Give the name of the less electronegative element first (the element given first in the formula), (2) give the stem of the name of the more electronegative element next and add the suffix *-ide,* and (3) indicate the number of each type of atom in the molecule by means of the Greek prefixes listed in ■ Table 4.6.

TABLE 4.5	The polarity of molecules				
Molecular formula	Geometric structure	Bond polarization and molecular charge distribution	Geometry of charge distribution	Classification of charge distribution	Classification of molecule
H_2	H — H	H — H	No charges	No charges	Nonpolar
HCl	H — Cl	$\overset{\delta+}{H} \Rightarrow \overset{\delta-}{Cl}$	+ −	Nonsymmetric	Polar
CO_2	O = C = O	$\overset{\delta-}{O} \equiv \overset{\delta2+}{C} \Rrightarrow \overset{\delta-}{O}$	− 2+ −	Symmetric	Nonpolar
BF_3	F — B — F, F	F — B — F, F (with δ−, δ3+, δ−)	− / 3+ / − −	Symmetric	Nonpolar
N_2O	N ≡ N — O	$N \equiv \overset{\delta+}{N} \Rightarrow \overset{\delta-}{O}$	+ −	Nonsymmetric	Polar
H_2O	H — O, H	$\overset{\delta+}{H} \Rightarrow \overset{\delta2-}{O}$, $\overset{\delta+}{H}$	+ 2− / +	Nonsymmetric	Polar

HELP WITH POLAR AND NONPOLAR MOLECULES

When concepts are closely related, it is very easy to become confused. In this chapter, the terms *polar* and *nonpolar* are used to describe both bonds and molecules. What is the correct way to apply the terms?

It may be helpful to remember the numbers 3 and 2: three types of bonds, but only two types of molecules. The bonds are nonpolar covalent ($\Delta EN = 0$), polar covalent ($\Delta EN = 0.1-2.0$), and ionic ($\Delta EN = 2.1$ or greater). Remember that compounds bonded by ionic bonds do not form molecules, so there are only two types of molecules: nonpolar and polar.

The polarity of a molecule depends on two factors: (1) the polarity of the bonds between the atoms of the molecule, and (2) the geometric arrangement of the atoms in space. First, determine whether any of the bonds between atoms are polar. If only nonpolar bonds are present, the molecule will be nonpolar regardless of the geometric arrangement of the atoms. However, if one or more polar bonds are found, the arrangement of the atoms must be considered as a second step.

For diatomic (two-atom) molecules, a polar bond between the atoms results in a polar molecule:

$A-A$	$\Delta EN = 0$ because atoms are identical; nonpolar
$A-B$	If $\Delta EN = 0$; nonpolar
$A-B$	If $\Delta EN = 0.1-2.0$; polar

Thus, we see that for diatomic molecules, the type of bond between atoms and the resulting type of molecule is the same.

For a molecule that contains polar bonds and three or more atoms, the molecular geometry must be known and taken into account before the polar nature of the molecule can be determined. Remember, the unequal sharing of electrons between two atoms bonded by a polar bond will cause one of the bonded atoms to have a partial positive charge and one a partial negative charge. If these atoms with their charges are symmetrically arranged in space to form the molecule, the molecule will be nonpolar. Thus, each of the following planar (flat) molecules will be nonpolar even if the bonds between the atoms are polar:

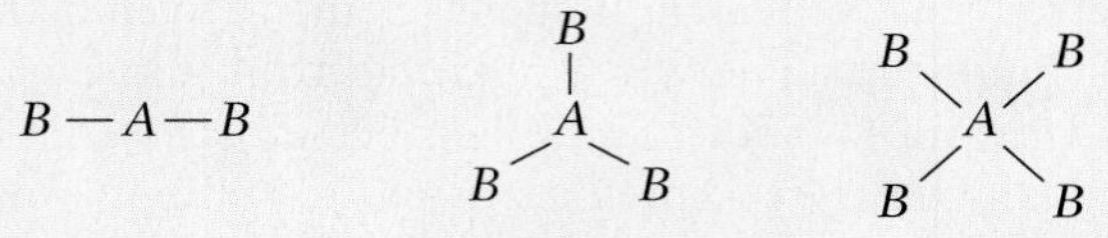

On the other hand, each of the following molecules will be polar if they contain polar bonds because the resulting charges on the atoms are not distributed symmetrically in space:

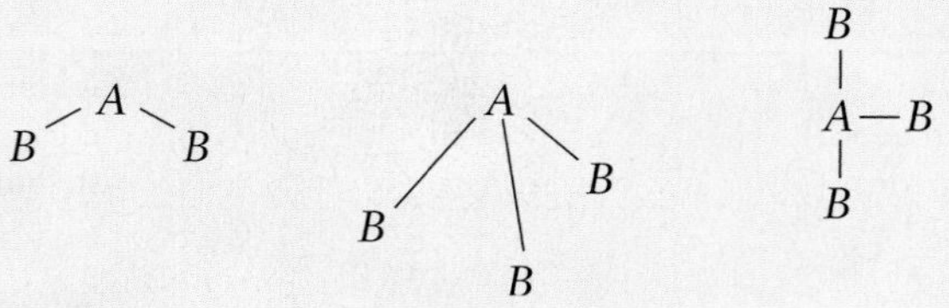

Example 4.17

Name the following binary covalent compounds:

a. CO_2 **b.** CO **c.** NO_2 **d.** N_2O_5 **e.** CS_2

Solution

a. The elements are carbon (C) and oxygen (O), and carbon is the less electronegative. Because the stem of oxygen is *ox-*, the two portions of the name will be carbon and oxide. Only one C atom is found in the molecule, but the prefix *mono-* is dropped when it appears at the beginning of a name. The two O atoms in the molecule are indicated by the prefix *di-*. The name therefore is carbon dioxide.

b. Similarly, we arrive at the name carbon monoxide for CO.

c. NO_2 is assigned the name nitrogen dioxide.

d. N_2O_5 is assigned the name dinitrogen pentoxide.

e. CS_2 is named carbon disulfide.

Name the following binary covalent compounds:

a. SO_3 **b.** BF_3 **c.** S_2O_7 **d.** CCl_4

Learning Check 4.17

The formulas and names of some common polyatomic ions are given in ■ Table 4.7. From this information, the formulas and names for compounds containing polyatomic ions can be written. The rules are essentially the same as those used earlier for binary ionic compounds. In the formulas, the metal (or ammonium ion) is written first, the positive and negative charges must add up to zero, and parentheses are used around the polyatomic ions if more than one is used. In names, the positive metal (or ammonium) ion is given first, followed by the name of the negative polyatomic ion (see ■ Figure 4.9). No numerical prefixes are used except where they are a part of the polyatomic ion name. Names and formulas of acids (compounds in which hydrogen is bound to polyatomic ions) will be given in Chapter 9.

Example 4.18

Write formulas and names for compounds composed of ions of the following metals and the polyatomic ions indicated:

a. Na and NO_3^- **c.** K and HPO_4^{2-}

b. Ca and ClO_3^- **d.** NH_4^+ and NO_3^-

Solution

a. Sodium (Na) is a group IA(1) metal and forms Na^+ ions. Electrical neutrality requires a combining ratio of one Na^+ for one NO_3^-. The formula is $NaNO_3$. The name is given by the metal name plus the polyatomic ion name: sodium nitrate.

b. Calcium (Ca), a group IIA(2) metal, forms Ca^{2+} ions. Electrical neutrality therefore requires a combining ratio of one Ca^{2+} for two ClO_3^- ions. The

GOB
Chemistry⊕Now™

Go to GOB Now and click to engage in a tutorial testing memorization of the names of common polyatomic ions.

TABLE 4.7 Some common polyatomic ions			
Very common		**Common**	
NH_4^+	ammonium	CrO_4^{2-}	chromate
$C_2H_3O_2^-$	acetate	$Cr_2O_7^{2-}$	dichromate
CO_3^{2-}	carbonate	NO_2^-	nitrite
ClO_3^-	chlorate	MnO_4^-	permanganate
CN^-	cyanide	SO_3^{2-}	sulfite
HCO_3^-	hydrogen carbonate (bicarbonate)	ClO^-	hypochlorite
OH^-	hydroxide	HPO_4^{2-}	hydrogen phosphate
NO_3^-	nitrate	$H_2PO_4^-$	dihydrogen phosphate
PO_4^{3-}	phosphate	HSO_4^-	hydrogen sulfate (bisulfate)
SO_4^{2-}	sulfate	HSO_3^-	hydrogen sulfite (bisulfite)

© Mark Slabaugh

➤ **FIGURE 4.9** Examples of compounds that contain polyatomic ions. Referring to Table 4.7, write formulas for these compounds (clockwise from top): potassium carbonate, potassium chromate, potassium phosphate, and potassium permanganate.

formula is $Ca(ClO_3)_2$. (NOTE: The use of parentheses around the poly-atomic ion prevents the confusion resulting from writing $CaClO_{32}$, which implies that there are 32 oxygen atoms in the formula. Parentheses are always used when multiples of a specific polyatomic ion in a formula are indicated by a subscript.) The name is calcium chlorate.

c. Potassium (K), a group IA(1) metal, forms K^+ ions. The required combin-ing ratio of 2:1 gives the formula K_2HPO_4. The name is potassium hydro-gen phosphate.

d. The NH_4^+ is not a metallic ion, but it behaves like one in numerous com-pounds. The 1:1 combining ratio gives the formula NH_4NO_3. (NOTE: The polyatomic ions are written separately, and the nitrogen atoms are not grouped to give a formula such as $N_2H_4O_3$.) The name is ammonium nitrate.

Learning Check 4.18

Write formulas and names for compounds containing ions of the following metals and the polyatomic ions indicated:

a. Ca and HPO_4^{2-} c. K and MnO_4^-
b. Mg and PO_4^{3-} d. NH_4^+ and $Cr_2O_7^{2-}$

➤ 4.11 Other Interparticle Forces

Ionic and covalent bonding can account for certain properties of many sub-stances. However, some experimental observations can be explained only by proposing the existence of other types of forces between particles.

Earlier in this chapter, a crystal lattice was described in connection with ionic bonding (see Figure 4.3). Most pure substances (elements or compounds) in the solid state also exist in the form of a crystal lattice. However, in some of these solids, neutral atoms or molecules occupy the lattice sites instead of ions. When solids are melted, the forces holding the lattice particles in place are overcome, and the particles move about more freely in what is called the liquid state. The addition of more heat overcomes the attractive interparticle forces to a still greater extent, and the liquid is converted into a gas or vapor; the liquid boils. Particles in the vapor state move about very freely and are influenced only slightly by interparticle attractions. Thus, the temperatures at which melting and boiling take place give an indication of the strength of the interparticle forces that are being overcome. (These states of matter—solid, liquid, and gas—are discussed in more detail in Chapter 6.)

Suppose an experiment is carried out with several pure substances. Some are familiar to you, but others you probably have never seen in the solid state. The substances are listed in ■ Table 4.8, together with some pertinent information.

TABLE 4.8   Some characteristics of selected pure substances

Substance	Formula or symbol	Classification	Particles occupying lattice sites
Sodium chloride	NaCl	Compound	Na^+ and Cl^- ions
Water	H_2O	Compound	H_2O molecules
Carbon monoxide	CO	Compound	CO molecules
Quartz (pure sand)	SiO_2	Compound	Si and O atoms
Copper metal	Cu	Element	Cu atoms
Oxygen	O_2	Element	O_2 molecules

CHEMISTRY AROUND US • 4.2

NITRIC OXIDE: A SIMPLE BUT VITAL BIOLOGICAL MOLECULE

Nitric oxide (NO) is a covalently bonded compound that is a toxic gas under ordinary conditions of temperature and pressure. The diatomic molecules of NO are only slightly polar, as indicated by an electronegativity difference (ΔEN) of only 0.5.

Until 1987, NO was regarded only as an environmental pollutant involved in numerous environmental problems, including the production of smog and acid precipitation. In 1987, researchers discovered that nitric oxide was produced by blood vessels. When NO was produced on the inside of blood vessels, it relaxed nearby muscles of the vessels, thereby reducing blood pressure. This discovery explained how a group of drugs, including amyl nitrite and nitroglycerine, worked to stop painful attacks of *angina*. During an angina attack, blood vessels to the heart constrict and reduce the supply of blood and oxygen to this vital organ. Drugs such as amyl nitrite and nitroglycerine produce NO inside the vessels, cause the vessels to relax, and restore the blood supply.

A second function of nitric oxide is protecting the body against unwanted foreign particles such as bacteria. Blood cells called macrophages seek out and destroy foreign particles by injecting them with a fatal dose of toxic NO.

In the early 1990s, researchers discovered that NO functions as a neurotransmitter, a chemical that carries messages from one nerve cell to another. This discovery was surprising because other known neurotransmitters are larger, more complicated molecules, and none are gases. The small size and low polarity of NO molecules allows them to diffuse quickly through cell membranes, a characteristic that enhances the role of a neurotransmitter. In this role, NO is known to be involved in long-term memory functions of the brain, the maintenance of blood pressure, central nervous system functions, and the immune system's response to infections caused by some types of viruses.

The experiment is started at the very low temperature of $-220°C$ to have all substances in the solid state. The temperature is then increased slowly and uniformly to 2600°C, at which point all the substances will be in the gaseous state. See ■ Table 4.9, which gives only those temperatures corresponding to a specific change in one of the substances.

The melting points of the substances used in this experiment show that the weakest interparticle forces are found in solid oxygen. The forces then increase in the order CO, H_2O, NaCl, Cu, and SiO_2. With two exceptions,

TABLE 4.9   The behavior of selected pure substances in response to heating

Temperature (°C)	Behavior or state of substance					
	Oxygen (O₂)	Carbon monoxide (CO)	Water (H₂O)	Salt (NaCl)	Copper (Cu)	Quartz (SiO₂)
−220	Solid	Solid	Solid	Solid	Solid	Solid
−218	**Melts**	Solid	Solid	Solid	Solid	Solid
−199	Liquid	**Melts**	Solid	Solid	Solid	Solid
−192	Liquid	**Boils**	Solid	Solid	Solid	Solid
−183	**Boils**	Gas	Solid	Solid	Solid	Solid
0	Gas	Gas	**Melts**	Solid	Solid	Solid
100	Gas	Gas	**Boils**	Solid	Solid	Solid
801	Gas	Gas	Gas	**Melts**	Solid	Solid
1083	Gas	Gas	Gas	Liquid	**Melts**	Solid
1413	Gas	Gas	Gas	**Boils**	Liquid	Solid
1610	Gas	Gas	Gas	Gas	Liquid	**Melts**
2230	Gas	Gas	Gas	Gas	Liquid	**Boils**
2595	Gas	Gas	Gas	Gas	**Boils**	Gas
2600	Gas	Gas	Gas	Gas	Gas	Gas

Network solid
A solid in which the lattice sites are occupied by atoms that are covalently bonded to each other.

Metallic bond
An attractive force responsible for holding solid metals together. It originates from the attraction between positively charged atomic kernels that occupy the lattice sites and mobile electrons that move freely through the lattice.

Dipolar force
The attractive force that exists between the positive end of one polar molecule and the negative end of another.

Hydrogen bonding
The result of attractive dipolar forces between molecules in which hydrogen atoms are covalently bonded to very electronegative elements (O, N, or F).

GOB
Chemistry⚛Now™
Go to GOB Now and click to learn how to determine the types of forces existing between two molecules.

the boiling points follow the same order. How do these melting and boiling points relate to the lattice particles of these substances?

The lattice particles in solid silicon dioxide are individual atoms of silicon and oxygen. They are held together in the lattice by covalent bonds. Solids of this type are called **network solids,** and when such solids are melted or vaporized, strong covalent bonds must be broken.

The individual atoms that are the lattice particles of copper metal are held together in the lattice by what is called a **metallic bond.** As described in Section 4.2, the atoms of metals lose valence-shell electrons readily. Imagine a large number of metal atoms occupying lattice sites. Now imagine that the valence-shell electrons of each of the atoms move readily throughout the lattice. The attraction of the positive kernels (the metal atom nuclei plus low-level electrons) to the mobile electrons, and hence to one another, constitutes a metallic bond. The mobile electrons of the metallic bond are responsible for a number of the observed properties of metals, including high thermal conductivity, high electrical conductivity, and the characteristic metallic luster.

The bonds broken when sodium chloride (NaCl) melts or boils are ionic bonds resulting from attractions between positive (Na^+) and negative (Cl^-) ions, the lattice particles in Figure 4.3. As shown by Table 4.9, these bonds are generally quite strong.

The lattice particles of solid carbon monoxide (CO) are the nonsymmetric CO molecules. These molecules are polar because of the nonsymmetric distribution of charges between the carbon and oxygen atoms. They are held in the solid lattice by **dipolar forces** resulting from the attraction of the positive end of one polar molecule to the negative end of another polar molecule. These forces are usually weak, and melting and vaporization take place at very low temperatures. Such substances are usually thought of as gases because that is their normal state at room temperature.

Water molecules, the lattice particles of ice, are also held in place by dipolar forces. However, these forces are stronger than those of the dipoles for CO. In water molecules, the hydrogens carry a partial positive charge, and the oxygen has a partial negative charge (see Table 4.5). Thus, the hydrogens of one molecule are attracted to the oxygens of other molecules. This attraction, called **hydrogen bonding,** is stronger than most other dipolar attractions because of the small size of the hydrogen atom and the high electronegativity of oxygen (see ■ Figure 4.10). Hydrogen bonding occurs in gases, liquids, and

(a) Polar water molecule   (b) Hydrogen bonding in liquid water   (c) Hydrogen bonding in solid water

➤ **FIGURE 4.10** Hydrogen bonding in water. Hydrogen bonds are dotted, and covalent bonds are solid.

➤ **FIGURE 4.11** Icebergs move as a result of wind and ocean currents. They can be very dangerous to ships, so their positions are reported and their probable courses estimated by an International Ice Patrol. The patrol was established in 1914 following the sinking of the *Titanic*.

© Gregory Dimigian/Photo Researchers

solids composed of polar molecules in which hydrogen atoms are covalently bonded to highly electronegative elements (generally O, N, or F).

Water is the most well known substance in which hydrogen bonding dramatically influences the properties. These properties make water useful in many processes, including those characteristic of living organisms. Because of this widespread use, water is not often thought of as being a peculiar substance. However, it is often the peculiarities that make it so useful.

As we saw in Example 4.13d, water molecules are angular. Because of this shape and the large difference in electronegativity between hydrogen and oxygen, water molecules are polar, with partial positive charges on the hydrogens and a partial negative charge on the oxygen, as shown in Figure 4.10a. The attractions between these oppositely charged parts of the molecules result in hydrogen bonding. This is represented for the liquid and solid states of water in Figures 4.10b and c.

Let's look at some of the peculiar properties of water. As shown in Table 4.9, water has a normal boiling point of 100°C. This is much higher than would be predicted from the measured boiling points of compounds containing hydrogen and the other members of group VIA(16) of the periodic table. The boiling point of H_2Te is $-2.2°C$, of H_2Se is $-41.3°C$, and of H_2S is $-60.3°C$. Thus, it appears that the boiling point decreases with decreasing compound molecular weight. On the basis of this trend, water should boil at approximately $-64°C$. The 164-degree difference between predicted and measured boiling points is caused by strong hydrogen bonds between water molecules in the liquid. The other three compounds do not have strong hydrogen bonds between molecules because the electronegativites of the elements to which hydrogen is covalently bonded are too low. This deviation from prediction makes water a liquid at normal temperatures and allows it to be used in many of the ways familiar to us.

It is also common knowledge that solid water (ice) floats on liquid water (see ■ Figure 4.11). Nothing peculiar here—or is there? In fact, there is. Water, like most other liquids, increases in density as it is cooled. This means that a specific mass of liquid decreases in volume as its temperature is lowered. Unlike most liquids, water behaves this way only until it reaches a temperature of about 4°C, as shown by the data in ■ Table 4.10. Then, its density decreases as it is cooled further. When its temperature reaches 0°C, water freezes and, in the process, undergoes a dramatic decrease in density. Put another way, a quantity of liquid water at 0°C expands significantly when it freezes to solid at 0°C. Thus, a specific volume of ice has a lower mass than the same volume of liquid water, so the ice floats on the water. Hydrogen bonds between water molecules are, once again, responsible. The strong hydrogen bonds orient water molecules into a very open three-dimensional crystal lattice when it freezes (see Figure 4.10c). This open lattice of the solid occupies more space than is occupied by the molecules in the liquid state.

GOB
Chemistry⚛Now™

Go to GOB Now and click to learn how to examine the types of forces existing between two molecules.

TABLE 4.10 Water density as a function of temperature

Temperature (°C)	Water density (g/mL)
100	0.9586
80	0.9719
60	0.9833
40	0.9923
20	0.9982
10	0.9997
5	0.9999
4 (actually 3.98)	1.0000
2	0.9999
0 (liquid H_2O)	0.9998
0 (solid H_2O)	0.9170

TABLE 4.11 The normal melting and boiling points of group VIIA(17) elements

Substance	Melting point (°C)	Boiling point (°C)
F_2	−223	−188
Cl_2	−103	−34.6
Br_2	−7.2	58.8
I_2	113.9	184.3

Dispersion forces
Very weak attractive forces acting between the particles of all matter. They result from momentary nonsymmetric electron distributions in molecules or atoms.

Even though containers or engines might be ruptured or cracked if water is allowed to freeze in them, the overall effect of this characteristic of water is beneficial. If ice did not float, it would form on natural waters in the winter and sink to the bottom. Gradually, ponds, lakes, and other bodies of water would fill with ice. In the spring, the ice would melt from the top down because the heavier solid ice would stay under the water. As a result, most natural waters in areas that experience freezing winters would contain significant amounts of ice year-round. It would be a strange (and possibly hostile) environment for us and for wildlife.

The amount of heat required to melt a quantity of solid and the amount of heat required to vaporize a quantity of liquid also depend on the attractive forces between molecules. As expected, these are high for water, compared with the hydrogen compounds of the three other elements in group VIA(16). The beneficial results of these high values for water will be discussed in Chapter 6.

The forces between O_2 molecules, the lattice particles of solid oxygen, do not fit into any of the classifications we have discussed to this point. The O_2 molecule is not polar and contains no ionic or metallic bonds, and solid oxygen melts at much too low a temperature to fit into the category of a network solid. The forces between O_2 molecules are called **dispersion forces** and result from momentary nonsymmetric electron distributions in the molecules.

The nonbonding electrons in an O_2 molecule can be visualized as being uniformly distributed. However, there is a small statistical chance that, during their normal movement, more of the electrons will momentarily be on one side of a molecule than on the other. This condition causes the molecule to become dipolar for an instant. The resulting negative side of the dipole will tend to repel electrons of adjoining molecules and induce them also to become dipolar (they become induced dipoles). The original (statistical) dipole and all induced dipoles are then attracted to one another. This happens many times per second throughout the solid or liquid oxygen. The net effect is a weak dispersion force of attraction between the molecules. The net force is weak because it represents the average result of many weak, short-lived attractions per second between molecules.

Dispersion forces exist in all matter, but because they are very weak, their contribution is negligible when other, stronger forces are also present. Thus, solid water is held together by dispersion forces and hydrogen bonds, but the properties are almost entirely the result of the hydrogen bonds. The ease with which a dipole can be induced increases with the size of the particle (atoms or molecules). The larger the particle, the stronger the resulting dipolar attraction (dispersion force), and the harder to separate the particles by melting or boiling. Thus, we expect to find melting and boiling points of the elements increase as we move down a group of the periodic table.

Example 4.19

Illustrate the behavior of dispersion forces by tabulating the melting and boiling temperatures for F_2, Cl_2, Br_2, and I_2.

Solution

The molecules increase in size in the order F_2, Cl_2, Br_2, I_2. The strength of dispersion forces increases in the same order, as shown in ■ Table 4.11.

Learning Check 4.19

Using only the periodic table, predict which member of each of the following pairs of elements would have the higher melting and boiling points:

a. O and Se   **b.** Sb and P   **c.** He and Ne

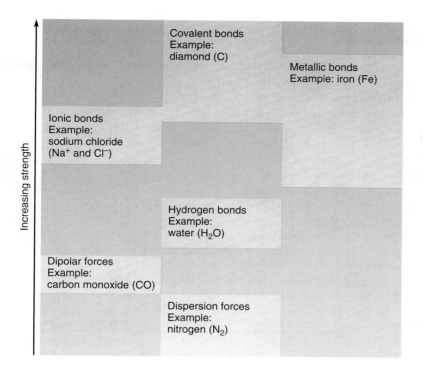

➤ **FIGURE 4.12** The relative strengths of interparticle forces.

The melting and boiling points in Example 4.19 indicate the variation in magnitude of dispersion forces. Similar variations are found for the other forces described in this section; sometimes their strengths overlap. For example, some metallic bonds are weaker than ionic bonds, whereas others are stronger. This is illustrated in ■ Figure 4.12, which summarizes the relative strengths of interparticle forces.

FOR FUTURE REFERENCE   POTASSIUM-RICH FOODS

The element potassium occurs in the body as the simple, positively charged K$^+$ ion. It is the most abundant positive ion found *inside* the cells of the body. Potassium ions are involved in numerous important body functions, including the maintenance of normal fluid and electrolyte balance, the transmission of nerve impulses, and the inducing of muscles, such as the heart, to contract (see Chemistry and Your Health 4.1).

A minimum potassium intake of 2000 mg per day has been established for adults. However, people using diuretics to control high blood pressure should be aware that some diuretics cause significant amounts of potassium to be excreted in the urine. Such individuals might have to take potassium supplements in addition to their normal dietary sources. The best dietary sources of potassium are *fresh* foods of all kinds, because potassium is abundant in all living cells, and the cells of foods remain intact until the foods are processed. The chart below lists some common foods that provide a minimum of 150 mg of potassium per serving.

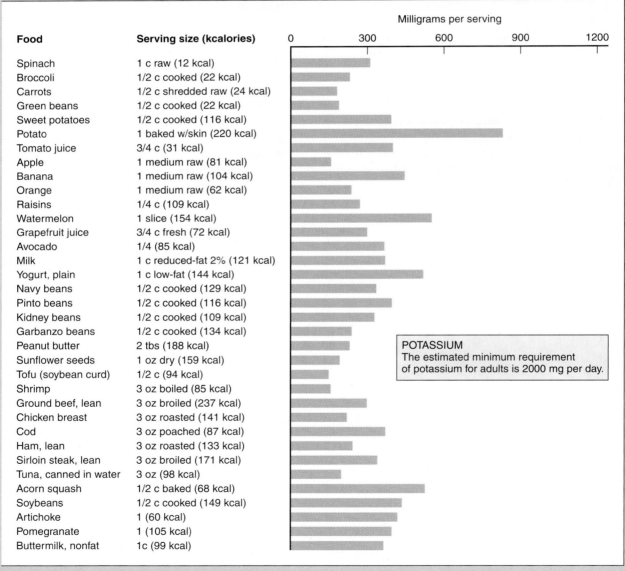

Food	Serving size (kcalories)
Spinach	1 c raw (12 kcal)
Broccoli	1/2 c cooked (22 kcal)
Carrots	1/2 c shredded raw (24 kcal)
Green beans	1/2 c cooked (22 kcal)
Sweet potatoes	1/2 c cooked (116 kcal)
Potato	1 baked w/skin (220 kcal)
Tomato juice	3/4 c (31 kcal)
Apple	1 medium raw (81 kcal)
Banana	1 medium raw (104 kcal)
Orange	1 medium raw (62 kcal)
Raisins	1/4 c (109 kcal)
Watermelon	1 slice (154 kcal)
Grapefruit juice	3/4 c fresh (72 kcal)
Avocado	1/4 (85 kcal)
Milk	1 c reduced-fat 2% (121 kcal)
Yogurt, plain	1 c low-fat (144 kcal)
Navy beans	1/2 c cooked (129 kcal)
Pinto beans	1/2 c cooked (116 kcal)
Kidney beans	1/2 c cooked (109 kcal)
Garbanzo beans	1/2 c cooked (134 kcal)
Peanut butter	2 tbs (188 kcal)
Sunflower seeds	1 oz dry (159 kcal)
Tofu (soybean curd)	1/2 c (94 kcal)
Shrimp	3 oz boiled (85 kcal)
Ground beef, lean	3 oz broiled (237 kcal)
Chicken breast	3 oz roasted (141 kcal)
Cod	3 oz poached (87 kcal)
Ham, lean	3 oz roasted (133 kcal)
Sirloin steak, lean	3 oz broiled (171 kcal)
Tuna, canned in water	3 oz (98 kcal)
Acorn squash	1/2 c baked (68 kcal)
Soybeans	1/2 c cooked (149 kcal)
Artichoke	1 (60 kcal)
Pomegranate	1 (105 kcal)
Buttermilk, nonfat	1c (99 kcal)

Milligrams per serving

POTASSIUM
The estimated minimum requirement of potassium for adults is 2000 mg per day.

SOURCE: Adapted from Whitney, E. N.; Rolfes, S. R. *Understanding Nutrition*, 8th ed. Belmont, CA: West/Wadsworth, 1999.

CONCEPT SUMMARY

Noble Gas Configurations. The lack of reactivity for noble gases led to the proposal that the electronic configurations of the noble gases represented stable configurations. These configurations, usually consisting of eight electrons in the valence shell, can be represented in several useful ways.

Ionic Bonding. Ionic compounds are formed when reacting atoms gain or lose electrons to achieve a noble gas configuration of eight electrons in the valence shell. This octet rule predicts that atoms will be changed into charged particles called simple ions. Ions of opposite charge are attracted to each other; the attractive force is called an ionic bond.

Ionic Compounds. Oppositely charged ions group together to form compounds in ratios determined by the positive and negative charges of the ions. The formulas representing these ratios contain the lowest number of each ion possible in a proportion such that the total positive charges and total negative charges used are equal.

Naming Binary Ionic Compounds. Binary ionic compounds contain a metal and a nonmetal. They are named by naming the metal, then adding the suffix *-ide* to the stem of the nonmetal.

The Smallest Unit of Ionic Compounds. Ionic compounds do not exist in the form of molecules but as three-dimensional arrangements of oppositely charged ions. The sum of the atomic weights of the elements in the formula of ionic compounds is called the formula weight and is used in calculations somewhat like the molecular weight of molecular compounds.

Covalent Bonding. Elements with little or no tendency to gain or lose electrons often react and achieve noble gas electronic configurations by sharing electrons. Lewis structures are useful in representing electron sharing. Shared pairs of electrons exert an attractive force on both atoms that share them. The atoms are held together by this attraction to form a covalent bond.

Polyatomic Ions. Polyatomic ions are groups of two or more covalently bonded atoms that carry a net electrical charge. They are conveniently represented using Lewis structures.

Shapes of Molecules and Polyatomic Ions. The shapes of many molecules and polyatomic ions can be predicted by using the valence-shell electron-pair repulsion theory (VSEPR). According to the VSEPR theory, electron pairs in the valence shell of the central atom of a molecule or ion repel one another and become arranged so as to maximize their separation distances. The resulting arrangement determines the molecular or ionic shape when one or all of the electron pairs involved form bonds between the central atom and other atoms.

The Polarity of Covalent Molecules. Shared electrons may be shared equally, or they may be attracted more strongly to one of the atoms they bond together. The tendency of a covalently bonded atom to attract shared electrons is called the electronegativity of the atom. Unequally shared bonding-electron pairs form polar covalent bonds. The extent of bond polarization can be estimated from the electronegativity differences between the bonded atoms. The higher the electronegativity difference, the more polar (or ionic) is the bond. Polar covalent bonds cause partial positive and negative charges to form within molecules. When these charges are symmetrically distributed in the molecule, it is said to be nonpolar. An unsymmetric distribution gives rise to a polar molecule.

More about Naming Compounds. Binary covalent compounds are named using the name of the less electronegative element first, followed by the stem plus *-ide* of the more electronegative element. Greek prefixes are used to represent the number of each type of atom in molecules of the compounds. Ionic compounds that contain a metal ion (or ammonium ion) plus a polyatomic ion are named by first naming the metal (or ammonium ion) followed by the name of the polyatomic ion.

Other Interparticle Forces. Forces other than ionic and covalent bonds are also known to hold the particles of some pure substances together in the solid and liquid states. These forces include metallic bonds, dipolar attractions, hydrogen bonds, and dispersion forces. The strength of the predominant force acting in a substance is indicated by the melting and boiling points of the substance.

LEARNING OBJECTIVES ASSESSMENT

You can get an approximate but quick idea of how well you have met the learning objectives given at the beginning of this chapter by working the selected end-of-chapter exercises given below. The answer to each exercise is given in Appendix B of the book.

Objective 1 (Section 4.1): Exercise 4.2

Objective 2 (Section 4.2): Exercise 4.12

Objective 3 (Section 4.3): Exercise 4.20

Objective 4 (Sections 4.3 and 4.10): Exercises 4.22 and 4.70

Objective 5 (Sections 4.4 and 4.10): Exercises 4.30, 4.66 and 4.72

Objective 6 (Section 4.5): Exercise 4.38

Objectives 7 and 8 (Section 4.8): Exercise 4.52

Objective 9 (Section 4.9): Exercise 4.58

Objective 10 (Section 4.9): Exercise 4.64

Objective 11 (Section 4.11): Exercise 4.78

Binary compound (4.3)
Bond polarization (4.9)
Covalent bond (4.6)
Crystal lattice (4.5)
Dipolar force (4.11)
Dispersion force (4.11)
Double bond (4.6)
Electronegativity (4.9)
Formula weight (4.5)

Hydrogen bonding (4.11)
Ionic bond (4.2)
Isoelectronic (4.2)
Lattice site (4.5)
Lewis structure (4.1)
Metallic bond (4.11)
Network solid (4.11)
Nonpolar covalent bond (4.9)

Nonpolar molecule (4.9)
Octet rule (4.2)
Polar covalent bond (4.9)
Polar molecule (4.9)
Polyatomic ion (4.7)
Simple ion (4.2)
Triple bond (4.6)
VSEPR theory (4.8)

EXERCISES

LEGEND: 1 = straightforward, 2 = intermediate, 3 = challenging. All even-numbered exercises are answered in Appendix B.

NOBLE GAS CONFIGURATIONS (SECTION 4.1)

4.1 Refer to the group numbers of the periodic table and draw Lewis structures for atoms of the following:
  a. lithium
  b. sodium
  c. chlorine
  d. boron

4.2 Refer to the group numbers of the periodic table and draw Lewis structures for atoms of the following:
  a. iodine
  b. strontium
  c. tin
  d. sulfur

4.3 Write abbreviated electronic configurations for the following:
  a. iodine
  b. element number 38
  c. As
  d. phosphorus

4.4 Write abbreviated electronic configurations for the following:
  a. antimony
  b. Rb
  c. element number 52
  d. cesium

4.5 Draw Lewis structures for the elements given in Exercise 4.3.

4.6 Draw Lewis structures for the elements given in Exercise 4.4.

4.7 Use the symbol E to represent an element in a general way and draw Lewis structures for atoms of the following:
  a. Any group IA(1) element
  b. Any group IVA(14) element

4.8 Use the symbol E to represent an element in a general way and draw Lewis structures for atoms of the following:
  a. Any group IIIA(13) element
  b. Any group VIA(16) element

IONIC BONDING (SECTION 4.2)

4.9 Indicate both the minimum number of electrons that would have to be added and the minimum number that would have to be removed to change the electronic configuration of each element listed in Exercise 4.3 to a noble gas configuration.

4.10 Indicate both the minimum number of electrons that would have to be added and the minimum number that would have to be removed to change the electronic configuration of each element listed in Exercise 4.4 to a noble gas configuration.

4.11 Use the periodic table and predict the number of electrons that will be lost or gained by the following elements as they change into simple ions. Write an equation using elemental symbols, ionic symbols, and electrons to represent each change.
  a. Ca
  b. aluminum
  c. fluorine
  d. element number 34

4.12 Use the periodic table and predict the number of electrons that will be lost or gained by the following elements as they change into simple ions. Write an equation using elemental symbols, ionic symbols, and electrons to represent each change.
  a. Cs
  b. oxygen
  c. element number 7
  d. iodine

4.13 Write a symbol for each of the following ions:
  a. A bromine atom that has gained one electron
  b. A sodium atom that has lost one electron
  c. A sulfur atom that has gained two electrons

4.14 Write a symbol for each of the following ions:
  a. A selenium atom that has gained two electrons
  b. A rubidium atom that has lost one electron
  c. An aluminum atom that has lost three electrons

4.15 Identify the element in period 2 that would form each of the following ions. E is used as a general symbol for an element.
  a. E^-
  b. E^{2+}
  c. E^{3-}
  d. E^+

4.16 Identify the element in period 3 that would form each of the following ions. E is used as a general symbol for an element.
  a. E^{2-}
  b. E^{3+}
  c. E^+
  d. E^-

4.17 Identify the noble gas that is isoelectronic with each of the following ions:
   a. Mg^{2+}
   b. Te^{2-}
   c. N^{3-}
   d. Be^{2+}

4.18 Identify the noble gas that is isoelectronic with each of the following ions:
   a. Rb^+
   b. P^{3-}
   c. Se^{2-}
   d. Mg^{2+}

IONIC COMPOUNDS (SECTION 4.3)

4.19 Write equations to represent positive and negative ion formation for the following pairs of elements. Then write a formula for the ionic compound that results when the ions combine.
   a. Mg and S
   b. strontium and nitrogen
   c. elements number 3 and 34

4.20 Write equations to represent positive and negative ion formation for the following pairs of elements. Then write a formula for the ionic compound that results when the ions combine.
   a. Ca and Cl
   b. lithium and bromine
   c. elements number 12 and 16

4.21 Write the formula for the ionic compound formed from Sr^{2+} and each of the following ions:
   a. S^{2-}
   b. Br^-
   c. N^{3-}
   d. Cl^-

4.22 Write the formula for the ionic compound formed from Ba^{2+} and each of the following ions:
   a. Te^{2-}
   b. N^{3-}
   c. F^-
   d. P^{3-}

4.23 Classify each of the following as a binary compound or not a binary compound:
   a. HF
   b. OF_2
   c. H_2SO_4
   d. H_2S
   e. $MgBr_2$

4.24 Classify each of the following as a binary compound or not a binary compound:
   a. CaS
   b. Na_3P
   c. $MgSO_4$
   d. $NaHCO_3$
   e. P_2O_5

NAMING BINARY IONIC COMPOUNDS (SECTION 4.4)

4.25 Name the following metal ions:
   a. Ca^{2+}
   b. K^+
   c. Al^{3+}
   d. Rb^+

4.26 Name the following metal ions:
   a. Li^+
   b. Mg^{2+}
   c. Ba^{2+}
   d. Cs^+

4.27 Name the following nonmetal ions:
   a. Cl^-
   b. N^{3-}
   c. S^{2-}
   d. Se^{2-}

4.28 Name the following nonmetal ions:
   a. Br^-
   b. O^{2-}
   c. P^{3-}
   d. Te^{2-}

4.29 Name the following binary ionic compounds:
   a. K_2O
   b. $SrCl_2$
   c. Al_2O_3
   d. LiBr
   e. CaS

4.30 Name the following binary ionic compounds:
   a. $CaCl_2$
   b. BaS
   c. $ZnBr_2$
   d. Al_2S_3
   e. SrF_2

4.31 Name the following binary ionic compounds, using a roman numeral to indicate the charge on the metal ion:
   a. $CrCl_2$ and $CrCl_3$
   b. CoS and Co_2S_3
   c. FeO and Fe_2O_3
   d. $PbCl_2$ and $PbCl_4$

4.32 Name the following binary ionic compounds, using a roman numeral to indicate the charge on the metal ion:
   a. SnS and SnS_2
   b. $FeCl_2$ and $FeCl_3$
   c. Cu_2O and CuO
   d. AuCl and $AuCl_3$

4.33 Name the binary compounds of Exercise 4.31 by adding the endings -*ous* and -*ic* to indicate the lower and higher ionic charges of the metal ion in each pair of compounds. The non-English root for lead (Pb) is *plumb-*.

4.34 Name the binary compounds of Exercise 4.32 by adding the endings -*ous* and -*ic* to indicate the lower and higher ionic charges of the metal ion in each pair of compounds. The non-English root for gold (Au) is *aur-*, and that of tin (Sn) is *stann*.

4.35 Write formulas for the following binary ionic compounds:
   a. manganese(II) chloride
   b. iron(III) sulfide
   c. chromium(II) oxide
   d. iron(II) bromide
   e. tin(II) chloride

4.36 Write formulas for the following binary ionic compounds:
   a. lead (IV) oxide
   b. cobalt(II) chloride
   c. copper(II) sulfide
   d. nickel(III) nitride
   e. platinum(II) phosphide

THE SMALLEST UNIT OF IONIC COMPOUNDS (SECTION 4.5)

4.37 Determine the formula weight in atomic mass units for each of the following binary ionic compounds:
   a. Na_2O
   b. FeO
   c. PbS_2
   d. $AlCl_3$

4.38 Determine the formula weight in atomic mass units for each of the following binary ionic compounds:
   a. $NaBr$
   b. CaF_2
   c. Cu_2S
   d. Li_3N

4.39 Identify the ions that would occupy lattice sites in a solid sample of each compound given in Exercise 4.37.

4.40 Identify the ions that would occupy lattice sites in a solid sample of each compound given in Exercise 4.38.

4.41 Calculate the mass in grams of positive ions and negative ions contained in 1 mol of each compound given in Exercise 4.37.

4.42 Calculate the mass in grams of positive ions and negative ions contained in 1 mol of each compound given in Exercise 4.38.

4.43 Calculate the number of positive ions and negative ions contained in 1.00 mol of each compound given in Exercise 4.37.

4.44 Calculate the number of positive ions and negative ions contained in 1.00 mol of each compound given in Exercise 4.38.

COVALENT BONDING (SECTION 4.6)

4.45 Represent the following reaction using Lewis structures:

$$I + I \rightarrow I_2$$

4.46 Represent the following reaction using Lewis structures:

$$8S \rightarrow S_8 \text{ (the atoms form a ring)}$$

4.47 Represent the following molecules by Lewis structures:
   a. HF
   b. IBr
   c. PH_3 (each H atom is bonded to the P atom)
   d. $HClO_2$ (the O atoms are each bonded to the Cl, and the H is bonded to one of the O atoms)

4.48 Represent the following molecules by Lewis structures:
   a. H_2S (each H atom is bonded to the S atom)
   b. ClF
   c. HBr
   d. $HClO$ (the H and Cl are each bonded to O)

POLYATOMIC IONS (SECTION 4.7)

4.49 Draw Lewis structures for the following polyatomic ions:
   a. ClO_3^- (each O atom is bonded to the Cl atom)
   b. CN^-
   c. CO_3^{2-} (each O atom is bonded to the C atom)

4.50 Draw Lewis structures for the following polyatomic ions:
   a. PH_4^+ (each H atom is bonded to the P atom)
   b. HPO_4^{2-} (each O atom is bonded to the P atom, and the H atom is bonded to an O atom)
   c. HSO_4^- (each O atom is bonded to the S atom, and the H atom is bonded to an O atom)

SHAPES OF MOLECULES AND POLYATOMIC IONS (SECTION 4.8)

4.51 Draw Lewis structures for the following molecules:
   a. O_3 (the O atoms are bonded together, like beads on a string)
   b. CS_2 (each S atom is bonded to the C atom)
   c. SeO_2 (each O atom is bonded to the Se atom)
   d. H_2SO_3 (each O atom is bonded to the S atom, and one H atom is bonded to each of two O atoms)

4.52 Predict the shape of each of the following molecules by first drawing a Lewis structure, then applying the VSEPR theory:
   a. H_2S (each H atom is bonded to the S atom)
   b. PCl_3 (each Cl atom is bonded to the P atom)
   c. OF_2 (each F atom is bonded to the O atom)
   d. SnF_4 (each F atom is bonded to the Sn atom)

4.53 Predict the shape of each of the following molecules by first drawing a Lewis structure, then applying the VSEPR theory:
   a. O_3 (see Exercise 4.51 for Lewis structure)
   b. SeO_2 (see Exercise 4.51 for Lewis structure)
   c. PH_3 (each H atom is bonded to the P atom)
   d. SO_3 (each O atom is bonded to the S atom)

4.54 Predict the shape of each of the following polyatomic ions by first drawing a Lewis structure, then applying the VSEPR theory:
   a. NO_2^- (each O is bonded to N)
   b. ClO_3^- (each O is bonded to Cl)
   c. CO_3^{2-} (each O is bonded to C)
   d. H_3O^+ (each H is bonded to O)
   Note the positive charge; compare with NH_4^+.

4.55 Predict the shape of each of the following polyatomic ions by first drawing a Lewis structure, then applying the VSEPR theory:
   a. NH_2^- (each H atom is bonded to N)
   b. PO_3^{3-} (each O is bonded to P)
   c. $BeCl_4^{2-}$ (each Cl is bonded to Be)
   d. ClO_4^- (each O is bonded to Cl)

THE POLARITY OF COVALENT MOLECULES (SECTION 4.9)

4.56 Use the periodic table and Table 4.4 to determine which of the following bonds will be polarized. Show the resulting charge distribution in those molecules that contain polarized bonds.
   a. $H—I$
   b.
   $$\begin{array}{c} \quad\quad S = O \\ O \end{array}$$

c. $O—O$
$\diagdown$
O

b. $H—C≡N$

c. $F—O$
$\diagdown$
F

4.57 Use the periodic table and Table 4.4 to determine which of the following bonds will be polarized. Show the resulting charge distribution in those molecules that contain polarized bonds.
   a. $Cl—F$
   b. $H—Se$
      $\diagdown$
      H
   c. H $\diagdown$ $\diagup H$
      $B—B$
      H $\diagup$ $\diagdown H$

MORE ABOUT NAMING COMPOUNDS (SECTION 4.10)

4.66 Name the following binary covalent compounds:
   a. PCl_3
   b. N_2O_5
   c. CCl_4
   d. BF_3
   e. CS_2

4.67 Name the following binary covalent compounds:
   a. SiO_2
   b. SiF_4
   c. P_2O_5
   d. $AlBr_3$
   e. CBr_4

4.58 Use Table 4.4 and classify the bonds in the following compounds as nonpolar covalent, polar covalent, or ionic:
   a. $LiBr$
   b. HCl
   c. PH_3 (each H is bonded to P)
   d. SO_2 (each O is bonded to S)
   e. CsF

4.59 Use Table 4.4 and classify the bonds in the following compounds as nonpolar covalent, polar covalent, or ionic:
   a. MgI_2 (each I is bonded to Mg)
   b. NCl_3 (each Cl is bonded to N)
   c. H_2S (each H is bonded to S)
   d. RbF
   e. SrO

4.68 Write formulas for the following binary covalent compounds:
   a. dinitrogen tetroxide
   b. sulfur hexachoride
   c. silicon dioxide
   d. oxygen difluoride

4.69 Write formulas for the following binary covalent compounds:
   a. disulfur monoxide
   b. sulfur hexafluoride
   c. silicon tetrachloride
   d. carbon diselenide

4.60 On the basis of the charge distributions you drew for the molecules of Exercise 4.56, classify each of the molecules as polar or nonpolar.

4.61 On the basis of the charge distributions you drew for the molecules of Exercise 4.57, classify each of the molecules as polar or nonpolar.

4.70 Write the formulas and names for compounds composed of ions of the following metals and the indicated polyatomic ions:
   a. calcium and the nitrite ion
   b. magnesium and the hypochlorite ion
   c. Cs and $Cr_2O_7{}^{2-}$
   d. K and $SO_3{}^{2-}$

4.62 Use Table 4.4 and predict the type of bond you would expect to find in compounds formed from the following elements:
   a. nitrogen and oxygen
   b. magnesium and oxygen
   c. N and H

4.63 Use Table 4.4 and predict the type of bond you would expect to find in compounds formed from the following elements:
   a. sulfur and oxygen
   b. aluminum and bromine
   c. C and Cl

4.71 Write the formulas and names for compounds composed of ions of the following metals and the indicated polyatomic ions:
   a. calcium and the phosphate ion
   b. sodium and the dichromate ion
   c. Li and $CO_3{}^{2-}$
   d. Na and $PO_4{}^{3-}$

4.64 Show the charge distribution in the following molecules, and predict which are polar molecules:
   a. $C≡O$
   b. $H—Se$
      $|$
      H
   c. I
      $|$
      Al
      $\diagup$ $\diagdown$
      I I

4.72 Write formulas for the following compounds:
   a. barium hydroxide
   b. magnesium sulfite
   c. calcium carbonate
   d. ammonium sulfate
   e. lithium hydrogen carbonate

4.73 Write formulas for the following compounds:
   a. potassium permanganate
   b. calcium hydroxide
   c. calcium phosphate
   d. ammonium dihydrogenphosphate
   e. calcium hypochlorite

4.65 Show the charge distribution in the following molecules, and predict which are polar molecules:
   a. $S=C=S$

4.74 Write a formula for the following compounds, using M with appropriate charges to represent the metal ion:
   a. Any group IA(1) element and $SO_3{}^{2-}$
   b. Any group IA(1) element and $C_2H_3O_2{}^-$

c. Any metal that forms M^{2+} ions and $Cr_2O_7{}^{2-}$
d. Any metal that forms M^{3+} ions and $PO_4{}^{3-}$
e. Any metal that forms M^{3+} ions and $NO_3{}^-$

4.75 Write a formula for the following compounds, using M with appropriate charges to represent the metal ion:
a. Any group IIA(2) element and $HSO_3{}^-$
b. Any group IIA(2) element and $HPO_4{}^{2-}$
c. Any metal that forms M^+ ions and $NO_2{}^-$
d. Any metal that forms M^{3+} ions and $CO_3{}^{2-}$
e. Any metal that forms M^{2+} ions and $HPO_4{}^{2-}$

OTHER INTERPARTICLE FORCES (SECTION 4.11)

4.76 The covalent compounds ethyl alcohol and dimethyl ether both have the formula C_2H_6O. However, the alcohol melts at $-117.3°C$ and boils at $78.5°C$, whereas the ether melts at $-138.5°C$ and boils at $-23.7°C$. How could differences in forces between molecules be used to explain these observations?

4.77 The following structural formulas represent molecules of ethyl alcohol and dimethyl ether. Assign the correct name to each formula and explain how your choice is consistent with your answer to Exercise 4.76.

$$
\begin{array}{ccc}
& H & \quad\quad H \\
& | & \quad\quad | \\
H- & C-O-C & -H \\
& | & \quad\quad | \\
& H & \quad\quad H
\end{array}
\qquad
\begin{array}{ccc}
H & H & \\
| & | & \\
H-C- & C-O & -H \\
| & | & \\
H & H &
\end{array}
$$

4.78 Describe the predominant forces that exist between molecules of the noble gases. Arrange the noble gases in a predicted order of increasing boiling point (lowest first) and explain the reason for the order.

4.79 Use the concept of interparticle forces to propose an explanation for the fact that CO_2 is a soft, low-melting solid (dry ice), whereas SiO_2 is a hard solid (sand). Focus on the nature of the particles that occupy lattice sites in the solid.

4.80 Table sugar, sucrose, melts at about 185°C. Which interparticle forces do you think are unlikely to be the predominant ones in the lattice of solid sucrose?

4.81 The formula for sucrose is $C_{12}H_{22}O_{11}$, where many of the hydrogens and oxygens are combined to form OH groups that are bonded to carbon atoms. What type of predominant interparticle bonding would you now propose for solid sucrose (see Exercise 4.80)?

CHEMISTRY FOR THOUGHT

4.1 Refer to Figure 4.1, and answer the question in the caption. What other metals and nonmetals would you predict might react in a similar way?

4.2 The colors of some compounds, such as those shown in Figure 4.2, result from the presence of water in the compounds. Propose an experiment you could perform to see if this was true for the compounds shown in Figure 4.2.

4.3 Refer to Figure 4.6, and answer the question in the caption. Propose the shapes that would be assumed by a group of five balloons and by a group of six balloons.

4.4 Refer to Figure 4.9. Two of the compounds are highly colored (other than white). All the compounds consist of potassium and a polyatomic ion. If you have not yet done so, write formulas for the compounds and see if you can find a characteristic of the polyatomic ions of the colored compounds that is not found in the white compounds. Then refer to Table 4.7 and predict which of the other polyatomic anions would form colored compounds of potassium.

4.5 Recall how a metal atom changes to form a positively charged metal ion. How do you think the sizes of a metal atom and a positive ion of the same metal will compare?

4.6 Recall how a nonmetal atom such as chlorine changes to form a negatively charged ion. How do you think the size of a nonmetal atom and a negatively charged ion of the same nonmetal will compare?

4.7 Neon atoms do *not* combine to form Ne_2 molecules. Explain.

4.8 Refer to Figure 4.8, and answer the question in the caption. The balloon carries a negative charge. What would happen if a positively charged object was used in place of the balloon?

4.9 In Chemistry Around Us 4.2, NO was described as a vital biological molecule. Explain how NO forms when a fuel such as natural gas, CH_4, is burned in air at a high temperature.

InfoTrac College Edition Readings

"Seeming sedate, some solid surfaces seethe," *Science News*, Feb 24, 2001, 159(8):118. Record number A72058419.

"Scanning tunneling microscope forms bonds between molecules," *Advanced Materials & Processes*, Feb 2000, 157(2):21. Record number A60121992.

"Longest carbon-carbon bonds discovered," *Science News*, August 4, 2001, 160(5):79. Record number A77557217.

CHAPTER 5

Chemical Reactions

© Photodisk Collection

Medical technologists provide data to help physicians diagnose and treat patients. Much of these data come from analyses of body fluids. These analyses are performed by reacting body fluid samples with reagents that react with specific materials such as glucose. In this chapter, several different types of chemical reactions are presented, some of which are used in body fluid analysis.

LEARNING OBJECTIVES

When you have completed your study of this chapter, you should be able to:

1. Identify the reactants and products in written reaction equations. (Section 5.1)

2. Balance simple reaction equations by inspection. (Section 5.1)

3. Assign oxidation numbers to elements in chemical formulas. (Section 5.3)

4. Identify the oxidizing and reducing agents in redox reactions. (Section 5.3)

5. Classify reactions into the categories of redox or nonredox. (Section 5.4)

6. Classify reactions into the categories of decomposition, combination, single replacement, or double replacement. (Sections 5.4, 5.5, and 5.6)

7. Write molecular equations in total ionic and net ionic forms. (Section 5.7)

8. Classify reactions as exothermic or endothermic. (Section 5.8)

9. Use the mole concept to do calculations based on reaction equations. (Section 5.9)

10. Use the mole concept to do calculations based on the limiting-reactant principle. (Section 5.10)

11. Use the mole concept to do percentage-yield calculations. (Section 5.11)

I n previous chapters, we introduced the terms *molecule, element, compound,* and *chemical change.* In Chapter 1, you learned that chemical changes result in the transformation of one or more substances into one or more new substances. The processes involved in such changes are called *chemical reactions.* In this chapter, you will learn to write and read chemical equations that represent chemical reactions, to classify reactions, and to do calculations based on the application of the mole concept to chemical equations.

➤ 5.1 Chemical Equations

A simple chemical reaction between elemental hydrogen and oxygen has been used to power the engines of a number of spacecraft, including the space shuttle. The products are water and much heat. For the moment, we will focus only on the substances involved; we will deal with the heat later. The reaction can be represented by a word equation:

$$\text{hydrogen} + \text{oxygen} \rightarrow \text{water} \tag{5.1}$$

The word equation gives useful information, including the reactants and products of the reaction. The **reactants** are the substances that undergo the chemical change; by convention, these are written on the left side of the equation. The **products,** or substances produced by the chemical change, are written on the right side of the equation. The reactants and products are separated by an arrow that points to the products. A plus sign ($+$) is used to separate individual reactants and products.

Word equations convey useful information, but the chemical equations used by chemists convey much more:

$$2H_2(g) + O_2(g) \rightarrow 2H_2O(\ell) \tag{5.2}$$

In this equation, the reactants and products are represented by molecular formulas that tell much more than the names used in the word equation. For example, it is apparent that both hydrogen and oxygen molecules are diatomic. The equation is also consistent with a fundamental law of nature called the **law of conservation of matter.** According to this law, atoms are neither created nor destroyed in chemical reactions but are rearranged to form new substances. Thus, atoms are conserved in chemical reactions, but molecules are not. The numbers written as coefficients to the left of the molecular formulas make the equation consistent with this law by making the total number of each kind of atom equal in the reactants and products. Note that coefficients of 1 are never written but are understood. Equations written this way are said to be **balanced.** In addition, the symbol in parentheses to the right of each formula indicates the state or form in which the substance exists. Thus, in Equation 5.2, the reactants hydrogen and oxygen are both in the form of gases (g), and the product water is in the form of a liquid (ℓ). Other common symbols you will encounter are (s) to designate a solid and (aq) to designate a substance dissolved in water. The symbol (aq) comes from the first two letters of *aqua,* the Latin word for water.

Reactants of a reaction
The substances that undergo chemical change during the reaction. They are written on the left side of the equation representing the reaction.

Products of a reaction
The substances produced as a result of the reaction taking place. They are written on the right side of the equation representing the reaction.

Law of conservation of matter
Atoms are neither created nor destroyed in chemical reactions.

Balanced equation
An equation in which the number of atoms of each element in the reactants is the same as the number of atoms of that same element in the products.

Example 5.1

Determine the number of atoms of each type on each side of Equation 5.2.

Solution

The coefficient 2 written to the left of H_2 means that two hydrogen molecules with the formula H_2 are reacted. Since each molecule contains two hydrogen atoms, a total of four hydrogen atoms is represented. The coefficient to the left of O_2 is 1, even though it is not written in the equation. The oxygen molecule contains two atoms, so a total of two oxygen atoms is represented. The coefficient 2 written to the left of H_2O means that two molecules of H_2O are produced. Each molecule contains two hydrogen atoms and one oxygen atom. Therefore, the total number of hydrogen atoms represented is four, and the number of oxygen atoms is two. These are the same as the number of hydrogen and oxygen atoms represented in the reactants.

Learning Check 5.1

Determine the number of atoms of each type on each side of the following balanced equation:

$$N_2(g) + 3H_2(g) \rightarrow 2NH_3(g)$$

When the identity and formulas of the reactants and products of a reaction are known, the reaction can be balanced by applying the law of conservation of matter.

Example 5.2

Nitrogen dioxide (NO_2) is an air pollutant that is produced in part when nitric oxide (NO) reacts with oxygen gas (O_2). Write a balanced equation for the production of NO_2 by this reaction.

GOB
Chemistry·⚛·Now™
Go to GOB Now and click to learn how to balance chemical equations.

Solution
The reactants written to the left of the arrow will be NO and O_2. The product is NO_2:

$$NO(g) + O_2(g) \rightarrow NO_2(g) \tag{5.3}$$

A quick inspection shows that the reactants contain three oxygen atoms, whereas the product has just two. The number of nitrogen atoms is the same in both the reactants and the products. A practice sometimes resorted to by beginning chemistry students is to change the formula of oxygen gas to O by changing the subscript and to write

$$NO(g) + O(g) \rightarrow NO_2(g) \tag{5.4}$$

This is not allowed. The natural molecular formulas of any compounds or elements cannot be adjusted; they are fixed by the principles of chemical bond formation described in Chapter 4. All that can be done to balance a chemical equation is to change the coefficients of the reactants and products. In this case, inspection reveals that the following is a balanced form of the equation:

$$2NO(g) + O_2(g) \rightarrow 2NO_2(g) \tag{5.5}$$

Notice that the following forms of the equation are also balanced:

$$4NO(g) + 2O_2(g) \rightarrow 4NO_2(g) \tag{5.6}$$

and

$$NO(g) + \frac{1}{2}O_2(g) \rightarrow NO_2(g) \tag{5.7}$$

In Equation 5.6, the coefficients are double those used in Equation 5.5; in Equation 5.7, one coefficient is a fraction. The lowest possible whole-number coefficients are used in balanced equations. Thus, Equation 5.5 is the correct form.

Learning Check 5.2	Write and balance an equation that represents the reaction of sulfur dioxide (SO_2) with oxygen gas (O_2) to give sulfur trioxide (SO_3).

➤ 5.2 Types of Reactions

A large number of chemical reactions are known to occur; only a relatively small number of them will be studied in this book. This study is made easier by classifying the reactions according to certain characteristics. Such a classification scheme could be developed in a number of ways, but we have chosen to first classify reactions as being either oxidation–reduction (redox) reactions or nonredox reactions. As you will see, redox reactions are very important in numerous areas of study, including metabolism. Once reactions are classified as redox or nonredox, many can be further classified into one of several other categories, as shown in ■ Figure 5.1. Notice that according to Figure 5.1, single-replacement, or substitution, reactions are redox reactions, whereas double-replacement, or metathesis, reactions are nonredox. Combination and decomposition reactions can be either redox or nonredox.

➤ 5.3 Redox Reactions

Almost all elements react with oxygen to form oxides. The process is so common that the word *oxidation* was coined to describe it. Some examples are the rusting of iron,

$$4Fe(s) + 3O_2(g) \rightarrow 2Fe_2O_3(s) \tag{5.8}$$

and the burning of hydrogen,

$$2H_2(g) + O_2(g) \rightarrow 2H_2O(\ell) \tag{5.9}$$

The reverse process, reduction, originally referred to the technique of removing oxygen from metal oxide ores to produce the free metal. Some examples are

$$CuO(s) + H_2(g) \rightarrow Cu(s) + H_2O(\ell) \tag{5.10}$$

and

$$2Fe_2O_3(s) + 3C(s) \rightarrow 4Fe(s) + 3CO_2(g) \tag{5.11}$$

➤ **FIGURE 5.1** A classification of chemical reactions.

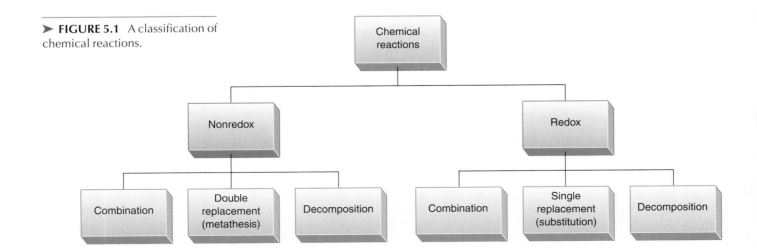

Today, the words **oxidation** and **reduction** are used in a rather broad sense. ■ Table 5.1 contains most of the common meanings. To understand oxidation and reduction in terms of electron transfer and oxidation number (O.N.) change, you must become familiar with the concept of oxidation numbers.

Oxidation numbers, sometimes called **oxidation states,** are positive or negative numbers assigned to the elements in chemical formulas according to a set of rules. The following rules will be used; be sure to note that Rule 1 applies only to uncombined elements—that is, elements in their free state. Rules 2 through 7 apply to elements combined to form compounds or ions.

Rule 1. The oxidation number (O.N.) of any uncombined element is 0.
Examples: $Al(0)$, $O_2(0)$, $Br_2(0)$, and $Na(0)$

Rule 2. The O.N. of a simple ion is equal to the charge on the ion.
Examples: $Na^+(+1)$, $Mg^{2+}(+2)$, $S^{2-}(-2)$, and $Br^-(-1)$

Rule 3. The O.N.s of group IA(1) and IIA(2) elements are $+1$ and $+2$, respectively.
Examples: $Na_2CO_3(Na = +1)$, $Sr(NO_3)_2(Sr = +2)$, and $CaCl_2$ $(Ca = +2)$

Rule 4. The O.N. of hydrogen is $+1$.
Example: HCl $(H = +1)$ and H_3PO_4 $(H = +1)$

Rule 5. The O.N. of oxygen is -2 except in peroxides, where it is -1.
Examples: CaO $(O = -2)$, H_2SO_4 $(O = -2)$, H_2O $(O = -2)$, and H_2O_2 $(O = -1)$

Rule 6. The algebraic sum of the O.N.s of all atoms in a complete compound formula equals zero.
Examples: K_2CO_3: 2(O.N. of K) + (O.N. of C) + 3(O.N. of O) = 0
$$2(+1) \qquad +4 \qquad +3(-2) = 0$$
$$+2 \qquad +4 \qquad +(-6) = 0$$

HNO_2: (O.N. of H) + (O.N. of N) + 2(O.N. of O) = 0
$$+1 \qquad +3 \qquad +2(-2) = 0$$
$$+1 \qquad +3 \qquad +(-4) = 0$$

Rule 7. The algebraic sum of the O.N.s of all atoms in a polyatomic ion is equal to the charge on the ion.
Examples: MnO_4^-: (O.N. of Mn) + 4(O.N. of O) = -1
$$+7 \qquad +4(-2) = -1$$
$$+7 \qquad +(-8) = -1$$

HPO_4^{2-}: (O.N. of H) + (O.N. of P) + 4(O.N. of O) = -2
$$+1 \qquad +5 \qquad +4(-2) = -2$$
$$+1 \qquad +5 \qquad +(-8) = -2$$

Oxidation
Originally, a process involving a reaction with oxygen. Today it means a number of things, including a process in which electrons are given up, hydrogen is lost, or an oxidation number increases.

Reduction
Originally, a process in which oxygen was lost. Today it means a number of things, including a process in which electrons are gained, hydrogen is accepted, or an oxidation number decreases.

Oxidation numbers or oxidation states
Positive or negative numbers assigned to the elements in chemical formulas according to a specific set of rules.

TABLE 5.1	Common uses of the terms oxidation and reduction
Term	**Meaning**
Oxidation	To combine with oxygen
	To lose hydrogen
	To lose electrons
	To increase in oxidation number
Reduction	To lose oxygen
	To combine with hydrogen
	To gain electrons
	To decrease in oxidation number

Example 5.3

a. Assign O.N.s to the blue element in each of the following:

$$CO_2, \quad NO_2^-, \quad H_2O, \quad N_2, \quad K^+$$

b. Assign O.N.s to each element in the following:

$$CO_2, \quad Mg(NO_3)_2, \quad NO_3^-, \quad CH_2O$$

GOB
Chemistry·*·Now™
Go to GOB Now and click to learn how to calculate oxidation numbers.

Solution

a. CO_2, $O = -2$ (Rule 5); NO_2^-, $O = -2$ (Rule 5); H_2O, $H = +1$ (Rule 4); N_2, $N = 0$ (Rule 1); K^+, $K = +1$ (Rule 2).

b. CO_2: The O.N. of O is -2 (Rule 5), and the O.N. of C can be calculated by using Rule 6 as follows:

$$(O.N. \text{ of } C) + 2(O.N. \text{ of } O) = 0$$

$$(O.N. \text{ of } C) + 2(-2) = 0$$

$$(O.N. \text{ of } C) + (-4) = 0$$

Therefore,

$$O.N. \text{ of } C = +4$$

$Mg(NO_3)_2$: The O.N. of Mg is $+2$ (Rule 3), the O.N. of O is -2 (Rule 5), and the O.N. of N can be calculated by using Rule 6 as follows:

$$(O.N. \text{ of } Mg) + 2(O.N. \text{ of } N) + 6(O.N. \text{ of } O) = 0$$

$$(+2) + 2(O.N. \text{ of } N) + 6(-2) = 0$$

$$2(O.N. \text{ of } N) + 2 + (-12) = 0$$

$$2(O.N. \text{ of } N) - 10 = 0$$

Therefore,

$$O.N. \text{ of } N = +5$$

NO_3^-: The O.N. of O is -2 (Rule 5), and the O.N. of N can be calculated by using Rule 7 as follows:

$$(O.N. \text{ of } N) + 3(O.N. \text{ of } O) = -1$$

$$(O.N. \text{ of } N) + 3(-2) = -1$$

$$(O.N. \text{ of } N) + (-6) = -1$$

Therefore,

$$O.N. \text{ of } N = +5$$

Note that N has the same O.N. in NO_3^- and in $Mg(NO_3)_2$. This is expected because the polyatomic NO_3^- ion is present in $Mg(NO_3)_2$.

CH_2O: The O.N. of H is $+1$ (Rule 4), the O.N. of O is -2 (Rule 5), and the O.N. of C can be calculated by using Rule 6 as follows:

$$(O.N. \text{ of } C) + 2(O.N. \text{ of } H) + (O.N. \text{ of } O) = 0$$

$$(O.N. \text{ of } C) + 2(+1) + (-2) = 0$$

$$(O.N. \text{ of } C) + 2 - 2 = 0$$

Therefore,

$$O.N. \text{ of } C = 0$$

This example shows that an O.N. of 0 may be found for some combined elements as well as those in an uncombined state (Rule 1).

Learning Check 5.3　　Assign O.N.s to each element in the following:

a. SO_3　　b. $Ca(ClO_3)_2$　　c. ClO_4^-

Now that you can assign O.N.s to the elements involved in reactions, you are ready to look at redox processes in terms of O.N.s and electron transfers. The reaction when sulfur is burned in oxygen is represented by the equation

$$S(s) + O_2(g) \rightarrow SO_2(g) \qquad (5.12)$$

You can see that this reaction represents an oxidation because sulfur has combined with oxygen (remember Table 5.1). When oxidation numbers are assigned to sulfur, the reactant S has an oxidation number of 0, while the S in the SO_2 product has an oxidation number of +4. Thus, the oxidation that has taken place resulted in an increase in the oxidation number of sulfur. Oxidation always corresponds to an increase in oxidation number.

The same conclusion is arrived at by thinking of oxidation numbers as charges. Thus, a sulfur atom with 0 charge is oxidized as it acquires a +4 charge. This change in charge results when the sulfur atom releases four electrons that have a −1 charge each. This leads to another definition of oxidation: Oxidation takes place when electrons are lost.

The oxygen of Equation 5.12 undergoes an oxidation number change from 0 in O_2 to −2 in SO_2. This decrease in oxidation number corresponds to a reduction of the oxygen. In terms of electrons, this same change is accomplished when each uncharged oxygen atom in the O_2 molecule accepts two electrons. Thus, the four electrons lost by sulfur as it was oxidized is the same total number accepted by the oxygen as it was reduced.

An oxidation process must always be accompanied by a reduction process, and all the electrons released during oxidation must be accepted during reduction. Thus, oxidation and reduction processes always take place simultaneously, hence the term redox. In redox reactions, the substance that is oxidized (and releases electrons) is called the **reducing agent** because it is responsible for reducing another substance. Similarly, the substance that is reduced (accepts electrons) is called the **oxidizing agent** because it is responsible for oxidizing another material (see ■ Figure 5.2). These characteristics are summarized in ■ Table 5.2.

A word of caution is appropriate here. Although electron transfers are a useful concept for understanding redox reactions involving covalent substances, it must be remembered that such transfers actually take place only during the formation of ionic compounds (see Section 4.3). In covalent substances, the electrons are actually shared (see Section 4.6). The oxidation number assignment rules that were given, as well as the electron-transfer idea, are based on the arbitrary practice of assigning shared electrons to the more electronegative element sharing them. However, it must be remembered that none of the atoms in covalent molecules actually acquire a net charge.

➤ **FIGURE 5.2** Combustion, the first reaction known to be carried out by humans, is a rapid redox reaction. Identify the oxidizing and reducing agents of the reaction.

Reducing agent
The substance that contains an element that is oxidized during a chemical reaction.

Oxidizing agent
The substance that contains an element that is reduced during a chemical reaction.

Example 5.4

Determine oxidation numbers for each atom represented in the following equations and identify the oxidizing and reducing agents:

a. $4Al(s) + 3O_2(g) \rightarrow 2Al_2O_3(s)$

TABLE 5.2 Properties of oxidizing and reducing agents

Oxidizing agent	Reducing agent
Gains electrons	Loses electrons
Oxidation number decreases	Oxidation number increases
Becomes reduced	Becomes oxidized

b. $CO(g) + 3H_2(g) \rightarrow H_2O(g) + CH_4(g)$

c. $S_2O_8{}^{2-}(aq) + 2I^-(aq) \rightarrow I_2(aq) + 2SO_4{}^{2-}(aq)$

Solution

The O.N. under each elemental symbol was calculated by using the methods demonstrated in Example 5.3.

a. $4Al + 3O_2 \rightarrow 2Al_2\ O_3$
   $\ 0\ \ \ \ 0\ \ \ \ +3\ \ -2$

The O.N. of Al has changed from 0 to +3. Therefore, Al has been oxidized and is the reducing agent. The O.N. of O has decreased from 0 to −2. The oxygen has been reduced and is the oxidizing agent.

b. $C\ O + 3H_2 \rightarrow H_2\ O + C\ H_4$
   $+2\ -2\ \ \ \ \ \ 0\ \ \ \ \ +1\ -2\ \ \ -4\ +1$

The O.N. of H_2 increased from 0 to +1. H_2 has been oxidized and is the reducing agent. The O.N. of C has decreased from +2 to −4. Carbon has been reduced and could be called the oxidizing agent. However, when one element in a molecule or ion is the oxidizing or reducing agent, the convention is to refer to the entire molecule or ion by the appropriate term. Thus, carbon monoxide (CO) is the oxidizing agent.

OVER THE COUNTER • 5.1

ANTISEPTICS AND DISINFECTANTS

Antiseptics and disinfectants are both used for the same purpose: to kill bacteria. The difference between these two categories of bacteria killers is where they are used. Antiseptics are used to kill bacteria on living tissue, such as wounds. Disinfectants are used to kill bacteria on inanimate objects. Some antiseptics, such as iodine and hydrogen peroxide, operate by oxidizing and thus destroying compounds essential to the normal functioning of the bacteria. A solution containing 3% hydrogen peroxide dissolved in water is an antiseptic found in most pharmacies, and it is often used to treat minor cuts and abrasions. A 2% solution of iodine dissolved in alcohol, called tincture of iodine, is also generally available, and it is used in a way similar to hydrogen peroxide. One disadvantage of the iodine solution is that it stains the skin a yellow-brown color.

Oxidizing antiseptics are often regarded as being too harsh. They may damage skin and other normal tissue, as well as kill the bacteria. For this reason, they have been replaced in many products by antiseptics derived from phenol. Water solutions of phenol, called carbolic acid, were first introduced as hospital antiseptics in 1867 by the English surgeon Joseph Lister. Before that time, antiseptics had not been used, and very few patients survived even minor surgery because of postoperative infections. These phenolic derivatives can often be recognized on ingredient labels by the characteristic *-ol* ending of their names. Some examples are thymol, eucalyptol, and eugenol.

Because disinfectants are used on inanimate objects, there is much less concern about the damage they might do to living tissue, and many of them are oxidizing agents. Sodium hypochlorite is one of the most widely used disinfectant compounds. In 5% solutions, it is marketed as liquid laundry bleach. This solution is an effective disinfectant for sinks, toilets, and similar fixtures. A chemically similar compound called calcium hypochlorite is the active ingredient in bleaching powder, and it is also used in hospitals as a disinfectant for clothing and bedding. Chlorine gas and ozone gas are two widely used, strong, oxidizing disinfectants. Their most well-known use is water treatment; they are added in small quantities to municipal water supplies to kill any harmful bacteria that may be present.

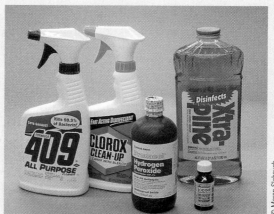

Antiseptics and disinfectants are both used to kill bacteria.

c. $S_2O_8^{2-} + 2I^- \rightarrow I_2 + 2SO_4^{2-}$
　　$+7 \ -2 \qquad -1 \quad 0 \quad +6 \ -2$

The O.N. of I has increased from -1 to 0, and the I^- ion is the reducing agent. The O.N. of S in $S_2O_8^{2-}$ has decreased from $+7$ to $+6$. Thus, S has been reduced, and the $S_2O_8^{2-}$ is the oxidizing agent.

Assign oxidation numbers to each atom represented in the following equations and identify the oxidizing and reducing agents:

a. $Zn(s) + 2H^+(aq) \rightarrow Zn^{2+}(aq) + H_2(g)$
b. $2KI(aq) + Cl_2(aq) \rightarrow 2KCl(aq) + I_2(aq)$
c. $IO_3^-(aq) + 3HSO_3^-(aq) \rightarrow I^-(aq) + 3HSO_4^-(aq)$

➤ 5.4 Decomposition Reactions

In **decomposition reactions,** a single substance is broken down to form two or more simpler substances, as shown in ■ Figure 5.3. In this and other "box" representations of reactions, the number of molecules in the boxes will not match the coefficients of the reaction equation. However, they will be in the correct proportions. In Figure 5.3, for example, the box on the left contains the same number of H_2O_2 molecules as the number of H_2O molecules on the right. Also, the number of H_2O molecules on the right is twice the number of O_2 molecules. The general form of the equation for a decomposition reaction is

$$A \rightarrow B + C \qquad (5.13)$$

Some decomposition reactions are also redox reactions, whereas others are not. Examples of decomposition reactions are given by Equations 5.14 and 5.15:

$$2HgO(s) \rightarrow 2Hg(\ell) + O_2(g) \qquad (5.14)$$

$$CaCO_3(s) \rightarrow CaO(s) + CO_2(g) \qquad (5.15)$$

Equation 5.14 represents the redox reaction that takes place when mercury(II) oxide (HgO) is heated. Mercury metal (Hg) and oxygen gas (O_2) are the products. This reaction was used by Joseph Priestley in 1774 when he discovered oxygen. Equation 5.15 represents a nonredox reaction used commercially to produce lime (CaO) by heating limestone ($CaCO_3$) to a high temperature. The decomposition of H_2O_2 represented by Figure 5.3 is shown in ■ Figure 5.4.

Decomposition reaction
A chemical reaction in which a single substance reacts to form two or more simpler substances.

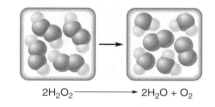
$2H_2O_2 \longrightarrow 2H_2O + O_2$

➤ **FIGURE 5.3** A decomposition reaction.

➤ **FIGURE 5.4** The decomposition of hydrogen peroxide. The catalyst is an enzyme provided by a piece of freshly cut potato. How can you tell that one of the products of the reaction is a gas?

A solution of hydrogen peroxide (H_2O_2) in water does not decompose rapidly at room temperature.

When a catalyst (see Chapter 8) is added, the decomposition takes place very rapidly.

➤ **FIGURE 5.5** A combination reaction.

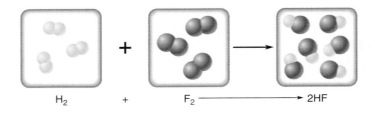

$$H_2 \quad + \quad F_2 \longrightarrow 2HF$$

➤ 5.5 Combination Reactions

Combination reaction
A chemical reaction in which two or more substances react to form a single substance.

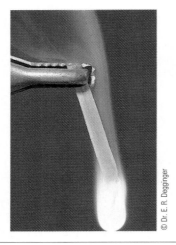

© Dr. E. R. Degginger

➤ **FIGURE 5.6** Magnesium metal burns in air to form magnesium oxide.

Single-replacement reaction
A chemical reaction in which an element reacts with a compound and displaces another element from the compound.

Combination reactions are sometimes called *addition* or *synthesis reactions*. Their characteristic is that two or more substances react to form a single substance (see ■ Figure 5.5). The reactants can be any combination of elements or compounds, but the product is always a compound. The general form of the equation is

$$A + B \rightarrow C \tag{5.16}$$

At high temperatures, a number of metals will burn and give off very bright light. This burning is a redox combination reaction, represented for magnesium metal by Equation 5.17 and shown in ■ Figure 5.6.

$$2Mg(s) + O_2(g) \rightarrow 2MgO(s) \tag{5.17}$$

A nonredox combination reaction that takes place in the atmosphere contributes to the acid rain problem. An air pollutant called sulfur trioxide (SO_3) reacts with water vapor and forms sulfuric acid. The reaction is represented by Equation 5.18:

$$SO_3(g) + H_2O(\ell) \rightarrow H_2SO_4(aq) \tag{5.18}$$

➤ 5.6 Replacement Reactions

Single-replacement reactions, also called *substitution reactions,* are always redox reactions and take place when one element reacts with a compound and displaces another element from the compound. A single-replacement reaction is represented in ■ Figure 5.7 and demonstrated in ■ Figure 5.8. The general equation for the reaction is shown in Equation 5.19.

$$A + BX \rightarrow B + AX \tag{5.19}$$

This type of reaction is useful in a number of processes used to obtain metals from their oxide ores. Iron, for example, can be obtained by reacting iron(III) oxide ore (Fe_2O_3) with carbon. The carbon displaces the iron from the oxide, and carbon dioxide is formed. The equation for the reaction is

$$3C(s) + 2Fe_2O_3(s) \rightarrow 4Fe(s) + 3CO_2(g) \tag{5.20}$$

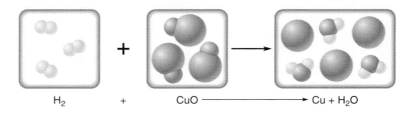

$$H_2 \quad + \quad CuO \longrightarrow Cu + H_2O$$

➤ **FIGURE 5.7** A single-replacement reaction.

> **FIGURE 5.8** When a piece of copper wire (Cu) is placed in a solution of silver nitrate ($AgNO_3$) in water, crystals of silver metal (Ag) form on the wire, and the liquid solution that was originally colorless turns blue as copper nitrate [$Cu(NO_3)_2$] forms in solution. Write a balanced equation for this single-replacement reaction.

Double-replacement reactions, also called *metathesis reactions*, are never redox reactions. These reactions, represented in ■ Figure 5.9, often take place between substances dissolved in water. The general Equation 5.21 shows the partner-swapping characteristic of these reactions:

$$AX + BY \rightarrow BX + AY \qquad (5.21)$$

Double-replacement reaction
A chemical reaction in which two compounds react and exchange partners to form two new compounds.

The reaction that takes place when a base is used to neutralize an acid is a good example of a double-replacement reaction:

$$HCl(aq) + NaOH(aq) \rightarrow NaCl(aq) + H_2O(\ell) \qquad (5.22)$$

Example 5.5

Classify each of the reactions represented by the following equations as redox or nonredox. Further classify them as decomposition, combination, single-replacement, or double-replacement reactions.

a. $SO_2(g) + H_2O(\ell) \rightarrow H_2SO_3(aq)$
b. $2K(s) + 2H_2O(\ell) \rightarrow 2KOH(aq) + H_2(g)$
c. $N_2(g) + 3H_2(g) \rightarrow 2NH_3(g)$
d. $BaCl_2(aq) + Na_2CO_3(aq) \rightarrow BaCO_3(s) + 2NaCl(aq)$

Solution

The O.N. under each elemental symbol was calculated by using the methods demonstrated in Example 5.3.
a. $\underset{+4\ -2}{S\ O_2} + \underset{+1\ -2}{H_2\ O} \rightarrow \underset{+1\ +4\ -2}{H_2\ S\ O_3}$

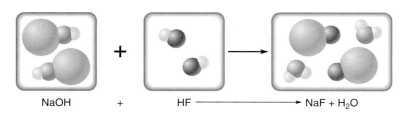

NaOH + HF ⟶ NaF + H_2O

> **FIGURE 5.9** A double-replacement, or metathesis, reaction.

CHEMISTRY AND YOUR HEALTH • 5.1

THE IMPORTANCE OF COLOR IN YOUR DIET

Scientific evidence accumulated during the 1990s suggested that diets rich in fruits and vegetables had a protective effect against a number of different types of cancer. Studies showed that simply increasing the levels of vitamins and minerals in the diet did not provide the increased protection. This led to research into the nature of other substances found in fruits and vegetables that are important for good health. As a result of this research, a number of chemical compounds found in plants and called *phytonutrients* have been shown to be involved in the maintenance of healthy tissues and organs. The mechanism for their beneficial action in the body is not understood for all phytonutrients, but a significant number are known to work as antioxidants that stop harmful oxidation reactions from occurring.

The colors of fruits and vegetables help identify those containing beneficial compounds. The table below contains a list of some of the more well-known phytonutrients together with sources, colors, and beneficial actions. The amount of evidence supporting the existence of benefits from phytonutrients is not the same for all those listed in the table. In some cases, the experimental evidence is extensive (e.g., the cancer-blocking behavior of isothiocyanates), while in other cases the listed benefits are based on a limited amount of research and more studies are being done (e.g., the contribution to eye health by anthocyanins).

Fruit/vegetable color	Fruit/vegetable examples	Phytonutrients	Possible benefits
Red	Tomatoes, watermelon, pink grapefruit	Lycopene (a carotenoid)	Protect against prostate; cervical and pancreatic cancer and heart and lung disease
Red/purple	Red and blue grapes, blueberries, strawberries, beets, eggplant, red cabbage, red peppers, plums, red apples	Anthocyanins (flavonoids)	Antioxidants; block formation of blood clots and help maintain good eye health
Orange	Carrots, mangoes, sweet potatoes, cantaloupe, winter squash	Alpha- and beta-carotenes	Cancer fighters; protect skin against free radicals, promote repair of damaged DNA
Orange/yellow	Oranges, peaches, papaya, nectarines	Beta-cryptoxanthin	May help prevent heart disease
Yellow/green	Spinach, collards, corn, green peas, avocado, honeydew	Lutein and zeaxanthin (both are carotenoids)	Reduce risk of cataracts and age-related macular degeneration
Green	Broccoli, brussels sprouts, cabbage, kale, bok choy	Sulforaphane, isothiocyanates, and indoles	Cancer blocking
White/green	Onions, leeks, garlic, celery, asparagus, pears, green grapes	Allicin (in onions) and flavanoids	Antitumor agent; antioxidants

No O.N. changes take place; therefore, the reaction is nonredox. Because two substances combine to form a third, the reaction is a combination reaction.

b. $2K + 2H_2O \rightarrow 2K\ OH + H_2$

$0 \qquad +1\ -2 \quad +1-2+1 \qquad 0$

The O.N. of K increases from 0 to +1 and that of H decreases from +1 to 0. The reaction is a redox reaction. Because K displaces H, it is a single-replacement reaction.

c. $N_2 + 3H_2 \rightarrow 2N\ H_3$

$0 \qquad 0 \qquad\quad -3\ +1$

The O.N.s of both N and H change; therefore, the reaction is a redox reaction. Two substances combine to form a third, so it is a combination reaction.

d. $Ba\ Cl_2 + Na_2C\ O_3 \rightarrow Ba\ C\ O_3 + 2Na\ Cl$

$+2\ -1 \quad +1\ +4\ -2 \quad\ +2\ +4\ -2 \qquad +1\ -1$

No changes in O.N. occur. The reaction is nonredox and is an example of a double-replacement, or partner-swapping, reaction.

Classify each of the reactions represented by the following equations as redox or nonredox. Further classify them as decomposition, combination, single-replacement, or double-replacement reactions.

a. $2HI(g) \rightarrow H_2(g) + I_2(g)$
b. $2H_2O_2(aq) \rightarrow 2H_2O(\ell) + O_2(g)$ (NOTE: H_2O_2 is a peroxide)
c. $NaCl(aq) + AgNO_3(aq) \rightarrow AgCl(s) + NaNO_3(aq)$
d. $4P(s) + 5O_2(g) \rightarrow 2P_2O_5(s)$
e. $2NaI(aq) + Cl_2(aq) \rightarrow 2NaCl(aq) + I_2(aq)$

➤ 5.7 Ionic Equations

Many of the reactions of interest in the course you are taking occur between compounds or elements dissolved in water. Ionic compounds and some polar covalent compounds break apart (dissociate) into ions when they are dissolved in water. Thus, a water solution of sodium hydroxide (NaOH), an ionic compound, does not contain molecules of NaOH but, rather, contains equal numbers of sodium ions (Na^+) and hydroxide ions (OH^-). Covalently bonded hydrogen chloride, HCl, dissolves readily in water to form H^+ and Cl^- ions. Equations for reactions between substances that form ions in solution can be written in several ways. For example, Equation 5.22 contains three substances that form ions, HCl, NaOH, and NaCl. Equation 5.22 is written in the form of a **molecular equation** in which each compound is represented by its formula. This same reaction, when represented by a **total ionic equation,** becomes

$$H^+(aq) + Cl^-(aq) + Na^+(aq) + OH^-(aq) \rightarrow$$
$$Na^+(aq) + Cl^-(aq) + H_2O(\ell) \quad (5.23)$$

In this equation each ionic compound is shown dissociated into ions, the form it takes when it is dissolved in water. Some of the ions appear as both reactants and products. These so-called **spectator ions** do not actually undergo any changes in the reaction. Because of this, they are dropped from the equation when it is written as a **net ionic equation:**

$$H^+(aq) + OH^-(aq) \rightarrow H_2O(\ell) \quad (5.24)$$

The net ionic equation makes the partner-swapping characteristics of this double-replacement reaction less obvious, but it does emphasize the actual chemical changes that take place. ■ Figure 5.10 shows an experiment in which an ionic reaction occurs.

GOB
Chemistry·ʌ·Now™
Go to GOB Now and click to study how to write ionic equations for reactions in aqueous solutions.

Molecular equation
An equation written with each compound represented by its formula.

Total ionic equation
An equation written with all soluble ionic substances represented by the ions they form in solution.

Spectator ions
The ions in a total ionic reaction that are not changed as the reaction proceeds. They appear in identical forms on the left and right sides of the equation.

Net ionic equation
An equation that contains only un-ionized or insoluble materials and ions that undergo changes as the reaction proceeds. All spectator ions are eliminated.

➤ **FIGURE 5.10** The liquid in the large container is a solution of solid sodium chloride (NaCl) in water. The liquid being added is a solution of solid silver nitrate ($AgNO_3$) in water. When the two liquids are mixed, an insoluble white solid forms. The solid is silver chloride (AgCl). Write the molecular, total ionic, and net ionic equations for the reaction.

© Joel Gordon, courtesy of West Publishing Company/CHEMISTRY by Radel & Navidi, 2d ed

Example 5.6

Write equations for the following double-replacement reaction in total ionic and net ionic forms. Note that barium sulfate ($BaSO_4$) does not dissolve in water and should not be written in dissociated form. All other compounds are ionic and soluble in water.

$$Na_2SO_4(aq) + BaCl_2(aq) \rightarrow BaSO_4(s) + 2NaCl(aq)$$

Solution

Total ionic:

$$2Na^+(aq) + SO_4{}^{2-}(aq) + Ba^{2+}(aq) + 2Cl^-(aq) \rightarrow$$
$$BaSO_4(s) + 2Na^+(aq) + 2Cl^-(aq)$$

Net ionic:

$$SO_4{}^{2-}(aq) + Ba^{2+}(aq) \rightarrow BaSO_4(s)$$

The Na^+ and Cl^- ions appear in equal numbers as reactants and products. Thus, they are spectator ions and are not shown in the net ionic equation.

Learning Check 5.6

Write equations for the following reactions in total ionic and net ionic forms. Consider all ionic compounds to be soluble except $CaCO_3$ and $BaSO_4$, and remember that covalent molecules do not form ions when they dissolve.

a. $2NaI(aq) + Cl_2(aq) \rightarrow 2NaCl(aq) + I_2(aq)$
b. $CaCl_2(aq) + Na_2CO_3(aq) \rightarrow 2NaCl(aq) + CaCO_3(s)$
c. $Ba(OH)_2(aq) + H_2SO_4(aq) \rightarrow 2H_2O(\ell) + BaSO_4(s)$

➤ 5.8 Energy and Reactions

Besides changes in composition, energy changes accompany all chemical reactions. The equation for the reaction between hydrogen and oxygen given at the beginning of this chapter (Equation 5.2) can also be written

$$2H_2(g) + O_2(g) \rightarrow 2H_2O(g) + energy \qquad (5.25)$$

Most of the energy of this reaction appears as heat, but the energy released or absorbed during chemical changes can take many forms including sound, electricity, light, high-energy chemical bonds, and motion. Some of these forms are discussed in detail in later chapters, but here the focus will be on energy in general, and it will be expressed in calories or joules, as if it all took the form of heat.

On the basis of heat, Equation 5.25 can be written

$$2H_2(g) + O_2(g) \rightarrow 2H_2O(g) + 115.6 \text{ kcal (483.7 kJ)} \qquad (5.26)$$

This equation assumes that the hydrogen and oxygen are gases when they react and that the water produced is also a gas (vapor). If the water was produced as a liquid, a total of 136.6 kcal, or 571.5 kJ, would be released. According to Equation 5.26, whenever 2 mol of water vapor is formed from 2 mol of hydrogen gas and 1 mol of oxygen gas, 115.6 kcal, or 483.7 kJ, of energy is also released.

The reaction represented by Equation 5.26 is an example of an **exothermic** (heat out) **reaction,** in which heat is released as the reaction takes place. In **endothermic** (heat in) **reactions,** heat is absorbed. A reaction of this type is utilized in emergency cold packs, which consist of a small plastic pouch of liquid sealed inside a larger pouch that also contains a solid substance. When the large pouch is squeezed firmly, the small pouch of liquid breaks, the

Exothermic reaction
A reaction that liberates heat.

Endothermic reaction
A reaction that absorbs heat.

CHEMISTRY AROUND US • 5.1

REDOX REACTIONS AND ENERGY FOR LIFE

All living organisms require energy to carry out the processes necessary to sustain life. This energy is obtained through redox reactions of various types. The most familiar to most of us are the redox reactions that are part of the complicated process called metabolism. An example of a metabolic redox process that releases energy is the oxidation of a carbohydrate such as glucose, $C_6H_{12}O_6$. In the body, the overall process occurs as the result of a series of several reactions, but it can be represented by the following equation:

$$C_6H_{12}O_6 + 6O_2 \rightarrow 6CO_2 + 6H_2O + \text{energy}$$

Organisms that depend on such a process for energy must obtain the carbohydrate from a food source, and the primary food source of carbohydrates is plants. Plants derive the energy needed to form carbohydrates and carry out their life processes from the sun in the process called photosynthesis. Photosynthesis is another complicated process that consists of many individual reactions, but the overall process can be represented in a simplified way as:

$$\text{sunlight} + CO_2 + H_2O \rightarrow \text{carbohydrates} + O_2$$

Thus, we see that the energy used by animals is actually solar energy stored in the form of carbohydrates by plants. Not only do the plants store solar energy, they regenerate oxygen that can be used in the oxidation process of the carbohydrates by animals. Green plants, algae, and some bacteria use photosynthesis to capture and store solar energy.

Some bacteria that live in soil as well as fresh and marine waters use other redox reactions as sources of energy. For example, nitrifying bacteria that live in the soil convert both ammonia, NH_3, and nitrites, NO_2^-, to nitrates, NO_3^-. In this oxidation process energy is released that can be used by the bacteria, and soil fertility is increased by the nitrate that is also produced.

In 1977 scientists discovered a large number of previously unknown animals living so deep in the ocean that sunlight cannot reach them. Organisms living in more shallow water where sunlight can penetrate depend on photosynthesis for their food and energy supplies, but what provides energy to these bottom dwellers? It was discovered that these organisms lived in the vicinity of hydrothermal vents in the ocean floor. These vents are chimneylike structures that form over fissures or cracks in the ocean floor. Mineral-rich hot water flows upward through the fissures into the cold surrounding ocean water.

The hot water that pours from the vents forms when cold water seeps downward through cracks in the ocean floor and comes in contact with hot lava. The cold water contains sulfate, SO_4^{2-}, which is reduced to hydrogen sulfide, H_2S, under the conditions of high temperature and high pressure created as the water is heated. The heated water containing dissolved H_2S is then forced up through the vents.

The water surrounding the vents was discovered to be inhabited by large numbers of sulfur bacteria that oxidize the H_2S back to sulfate, SO_4^{2-}. During the oxidation process, energy is released and is used by the bacteria to convert dissolved carbon dioxide, CO_2, in the water into organic nutrients. These nutrients are used as food by the animals living in the vicinity of the vents. Thus, the energy to sustain life in these dark regions comes from hot lava in the form of high-energy bonds in H_2S. This energy is released and used to form food when the H_2S is oxidized to SO_4^{2-}.

An ocean thermal vent.

liquid and solid react, heat is absorbed, and the liquid contents of the larger pouch become cold. This reaction is described in more detail in Chapter 7.

➤ 5.9 The Mole and Chemical Equations

The mole concept introduced in Section 2.6 and applied to chemical formulas in Section 2.7 can also be used to calculate mass relationships in chemical reactions. The study of such mass relationships is called **stoichiometry**, a word derived from the Greek *stoicheion* (element) and *metron* (measure).

Stoichiometry calculations require that balanced equations be used for the reactions being studied. Consider the following equation for the redox reaction that takes place when methane (CH_4), a major constituent of natural gas, is burned:

$$CH_4(g) + 2O_2(g) \rightarrow CO_2(g) + 2H_2O(\ell) \qquad (5.27)$$

Stoichiometry
The study of mass relationships in chemical reactions.

Remember, coefficients in balanced equations refer to the formula that follows the coefficient. With this in mind, note that all of the following statements are consistent with this balanced equation:

1. 1 CH_4 molecule + 2 O_2 molecules → 1 CO_2 molecule + 2 H_2O molecules
2. 10 CH_4 molecules + 20 O_2 molecules → 10 CO_2 molecules + 20 H_2O molecules
3. 100 CH_4 molecules + 200 O_2 molecules → 100 CO_2 molecules + 200 H_2O molecules
4. 6.02×10^{23} CH_4 molecules + 12.0×10^{23} O_2 molecules → 6.02×10^{23} CO_2 molecules + 12.0×10^{23} H_2O molecules
5. 1 mol CH_4 + 2 mol O_2 → 1 mol CO_2 + 2 mol H_2O
6. 16.0 g CH_4 + 64.0 g O_2 → 44.0 g CO_2 + 36.0 g H_2O

Statements 1 through 4 are derived directly from the equation coefficients. Statement 4 leads to Statement 5 when it is remembered that 6.02×10^{23} particles is equal to 1 mol of particles. Statement 6 is obtained from Statement 5 when it is remembered that 1 mol of molecules has a mass in grams equal to the molecular weight of the compound in question. Although Statements 1 through 3 are true, they are not very useful. However, Statements 4 through 6 are all based on the definition of the mole and as a result are quite useful.

Suppose you were asked this question: How many moles of CO_2 could be formed by reacting together 2 mol CH_4 and 4 mol O_2? Statement 5 can be used to solve this problem quickly. We see that the amounts reacted correspond to twice the amounts represented by Statement 5. Thus, 2 mol CO_2 would be formed, which is also twice the amount represented by Statement 5. However, many stoichiometric problems cannot be solved quite as readily, so it is helpful to learn a general approach that works well for many problems of this type. This approach is based on the factor-unit method described earlier in Section 1.9. The needed factors are obtained from Statement 5 (or the coefficients of the balanced equation), the molecular or formula weights, and the mole definition.

GOB **Chemistry Now**™

Go to GOB Now and click to examine how to calculate the amount of a product formed given a mass of a reactant.

■ Figure 5.11 summarizes the important quantities and relationships between quantities that are used to solve stoichiometric problems. The boxes contain the quantities that are known or that are to be obtained (the unknown). Arrows between two boxes indicate that a single factor can be used to convert the quantity in one box to the quantity in the second box. The arrows are labeled with the source of the factor needed to carry out the calculations using the factor-unit method.

Example 5.7

The following questions are based on Equation 5.27.

a. How many moles of oxygen gas (O_2) will be required to react with 1.72 mol of methane (CH_4)?
b. How many grams of H_2O will be produced when 1.09 mol of CH_4 reacts with an excess of O_2?
c. How many grams of O_2 must react with excess CH_4 to produce 8.42 g of carbon dioxide (CO_2)?

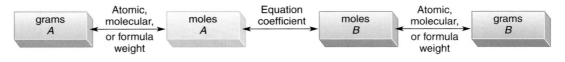

➤ **FIGURE 5.11** Relationships for problem solving based on balanced equations.

Solution

a. According to Figure 5.11, this problem has the pattern

$$\text{mol } A \rightarrow \text{mol } B$$

It is important to remember that the A and B of Figure 5.11 only indicate two different substances A and B, without concern for where they are located in the reaction equation. One might be a reactant and one a product, or both might be reactants, or both might be products. The factor needed to solve the problem comes from the equation coefficients (or Statement 5). Two factors are possible:

$$\frac{1 \text{ mol CH}_4}{2 \text{ mol O}_2} \quad \text{or} \quad \frac{2 \text{ mol O}_2}{1 \text{ mol CH}_4}$$

The steps to follow in the factor-unit method are as follows (see Section 1.9):

1. Write down the known or given quantity. It will include both a number and a unit.
2. Leave working space, and set the known quantity equal to the unit of the unknown quantity.
3. Multiply the known quantity by a factor such that the unit of the known quantity is canceled and the unit of the unknown quantity is generated.
4. After the units on each side of the equation match, do the necessary arithmetic to get the final answer.

In this problem, the known quantity is 1.72 mol CH_4, and the unknown quantity has the unit of mol O_2.

Step 1. 1.72 mol CH_4

Step 2. 1.72 mol CH_4 $\qquad\qquad$ = mol O_2

Step 3. 1.72 $\cancel{\text{mol CH}_4} \times \dfrac{2 \text{ mol O}_2}{1 \cancel{\text{mol CH}_4}}$ = mol O_2

The factor

$$\frac{2 \text{ mol O}_2}{1 \text{ mol CH}_4}$$

was used rather than the other possibility of

$$\frac{1 \text{ mol CH}_4}{2 \text{ mol O}_2}$$

because it properly canceled the unit of the known and generated the unit of the unknown.

Step 4. $(1.72)\left(\dfrac{2 \text{ mol O}_2}{1}\right) = 3.44$ mol O_2

The answer contains the proper number of significant figures because the 2 and 1 of the factor are exact numbers (see Section 1.8).

b. This problem has the pattern

$$\text{mol } A \rightarrow \text{g } B$$

According to Figure 5.11, the required pathway is

$$\text{mol CH}_4 \rightarrow \text{mol H}_2\text{O} \rightarrow \text{g H}_2\text{O}$$

The two parts of the pathway can be done individually or combined into one calculation. Both methods will be shown here. After you gain experience in working such problems, you should use the combined approach to save time.

Individual Parts Method: The pattern for the first part is

$$mol\ CH_4 \rightarrow mol\ H_2O$$

where the necessary factor comes from the equation coefficients and is

$$\frac{1\ mol\ CH_4}{2\ mol\ H_2O} \quad or \quad \frac{2\ mol\ H_2O}{1\ mol\ CH_4}$$

The known quantity is 1.09 mol CH_4, and the unit of the unknown quantity is mol H_2O. The steps in the factor-unit method are as follows:

Step 1. 1.09 mol CH_4

Step 2. 1.09 mol CH_4 $\qquad\qquad\qquad$ = mol H_2O

Step 3. 1.09 mol $CH_4 \times \dfrac{2\ mol\ H_2O}{1\ mol\ CH_4}$ = mol H_2O

Once again, the choice of which of the two factors to use was dictated by the necessity to cancel the unit of the known quantity and generate the unit of the unknown quantity.

Step 4. $(1.09)\left(\dfrac{2\ mol\ H_2O}{1}\right)$ = 2.18 mol H_2O

The pattern for the second part is

$$mol\ H_2O \rightarrow g\ H_2O$$

The answer of the first part (2.18 mol H_2O) becomes the known quantity for the second part. The factor comes from the molecular weight of water (18.01 u) and is

$$\frac{18.0\ g\ H_2O}{1\ mol\ H_2O} \quad or \quad \frac{1\ mol\ H_2O}{18.0\ g\ H_2O}$$

The steps in the factor-unit method are as follows:

Step 1. 2.18 mol H_2O

Step 2. 2.18 mol H_2O $\qquad\qquad\qquad$ = g H_2O

Step 3. 2.18 mol $H_2O \times \dfrac{18.0\ g\ H_2O}{1\ mol\ H_2O}$ = g H_2O

Step 4. $(2.18)\left(\dfrac{18.0\ g\ H_2O}{1}\right)$ = 39.2 g H_2O

Combined Method: The primary difference between the individual and combined approaches is that two factors are used in the combined method. They are the same two used in the individual method, and they are chosen, again, on the basis of the appropriate cancellation and generation of units.

The steps in the factor-unit method are:

Step 1. 1.09 mol CH_4

Step 2. 1.09 mol CH_4 $\qquad\qquad\qquad\qquad\qquad$ = g H_2O

Step 3. 1.09 mol $CH_4 \times \dfrac{2\ mol\ H_2O}{1\ mol\ CH_4} \times \dfrac{18.0\ g\ H_2O}{1\ mol\ H_2O}$ = g H_2O

Notice that the units of the factors cancel unwanted units and generate the unit of the unknown.

Step 4. $(1.09)\left(\dfrac{2}{1}\right)\left(\dfrac{18.0\ g\ H_2O}{1}\right)$ = 39.2 g H_2O

As it should be, the answer is the same as the answer obtained by doing the parts individually.

CHEMISTRY AROUND US • 5.2

AIR BAG CHEMISTRY

Air bags installed in automobiles have proven to be effective, yet somewhat controversial, safety devices. According to the National Transportation Safety Administration, air bags have saved an estimated 1100 lives and prevented many more serious injuries from automobile accidents. This effectiveness is tempered by the fact that air bags have caused the deaths of a number of small children and infants. Short adults have also died as a result of being hit in the face rather than the chest by an expanding air bag. Research continues in an effort to improve this safety record.

An air bag is essentially a nylon fabric bag that fills very rapidly with a gas when a collision occurs. Inflated air bags cushion and protect the driver and front-seat passenger. The gas that inflates an air bag is nitrogen, N_2, that is produced in a gas generator by a redox reaction of a toxic, explosive material called sodium azide, NaN_3. When a collision occurs, an electrical impulse triggers the rapid decomposition of the sodium azide according to the following equation:

$$2NaN_3(s) \rightarrow 2Na(g) + 3N_2(g)$$

The sodium vapor produced by this reaction would be very hazardous to the occupants of the automobile, but potassium nitrate, KNO_3, also included in the gas-generating mixture, reacts with the sodium vapor to produce potassium oxide, K_2O, sodium oxide, Na_2O, and some additional nitrogen gas. The equation for the reaction is

$$10Na(g) + 2KNO_3(s) \rightarrow K_2O(s) + 5Na_2O(s) + N_2(g)$$

The potassium oxide and sodium oxide are also hazardous materials; they are both basic oxides that react with water to form very caustic potassium hydroxide and sodium hydroxide, respectively. However, silicon dioxide, SiO_2, a third substance included in the gas-generating mixture of chemicals, is an acidic oxide that reacts with the basic potassium and sodium oxides, neutralizes their caustic characteristics, and converts them into a safe silicate-glass powder.

Thus, we see that from a chemical standpoint, the products of an air bag inflation are rendered safe by a series of carefully designed reactions. However, most air bags are never inflated, which means that old cars sent to scrap yards still contain a small amount of very toxic and explosive sodium azide. This is a recycling or disposal problem that is yet to be resolved.

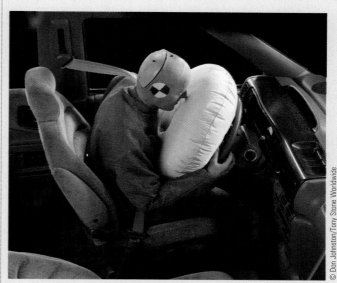

Air bags: Protection provided by chemical reactions.

© Don Johnston/Tony Stone Worldwide

c. According to Figure 5.11, this problem has the pattern

$$g\ A \rightarrow g\ B$$

The required pathway is

$$g\ CO_2 \rightarrow mol\ CO_2 \rightarrow mol\ O_2 \rightarrow g\ O_2$$

Thus, the pathway involves three calculations, which can be done individually or combined. We will show the solution in combined form.

The steps in the factor-unit method are as follows:

Step 1. 8.42 g CO_2

Step 2. 8.42 g CO_2 $\hspace{6cm}$ = g O_2

Step 3. $8.42 \ \cancel{g\ CO_2} \times \dfrac{1\ \cancel{mol\ CO_2}}{44.0\ \cancel{g\ CO_2}} \times \dfrac{2\ \cancel{mol\ O_2}}{1\ \cancel{mol\ CO_2}} \times \dfrac{32.0\ g\ O_2}{1\ \cancel{mol\ O_2}} = g\ O_2$

Step 4. $(8.42)\left(\dfrac{1}{44.0}\right)\left(\dfrac{2}{1}\right)\left(\dfrac{32.0\ g\ O_2}{1}\right) = 12.2\ g\ O_2$

Learning Check 5.7

The following reaction equation is balanced:

$$N_2(g) + 3H_2(g) \rightarrow 2NH_3(g)$$

a. Write two factors that could be used in factor-unit calculations involving the pathway

$$\text{mol } H_2 \rightarrow \text{mol } NH_3$$

b. Calculate the mol NH_3 that would be produced if 2.11 mol H_2 was reacted with excess N_2.

c. Calculate the g N_2 required to react with 9.47 g H_2.

Chemistry·⌬·Now™

Go to GOB Now and click to view a simulation showing the effect of a single reactant controlling the extent of a reaction.

Limiting-reactant principle

The maximum amount of product possible from a reaction is determined by the amount of reactant present in the least amount, based on its reaction coefficient and molecular weight.

Limiting reactant

The reactant present in a reaction in the least amount, based on its reaction coefficients and molecular weight. It is the reactant that determines the maximum amount of product that can be formed.

➤ 5.10 The Limiting Reactant

Nearly everyone is familiar with two characteristics of automobiles. They stop running when they run out of gasoline, and they stop running when the air intake to the engine becomes clogged. Most people also know that gasoline and air are mixed, and the mixture is burned inside the automobile engine. However, the combustion reaction occurs only as long as both reactants, gasoline and air, are present. The reaction (and engine) stops when gasoline is absent, despite the presence of ample amounts of air. On the other hand, the reaction stops when air is absent, even though plenty of gasoline is available.

The behavior of the auto engine illustrates the **limiting-reactant principle** that a chemical reaction will take place only as long as all necessary reactants are present. The reaction will stop when one of the reactants, the **limiting reactant,** is used up. According to this principle, the amount of product formed depends on the amount of limiting reactant present because the reaction will stop (and no more product will form) when the limiting reactant is used up. This is represented in ■ Figure 5.12 for the reaction $F_2(g) + H_2(g) \rightarrow 2HF(g)$.

Example 5.8

A mixture containing 20.0 g CH_4 and 100 g O_2 is ignited and burns according to Equation 5.27. What substances will be found in the mixture after the reaction stops?

Solution

The task is to identify the limiting reactant. It will be completely used up, and the other reactant will be found mixed with the products (CO_2 and H_2O) after the reaction stops. A simple way to identify the limiting reactant is to find out which reactant will give the least amount of products. To do this, two prob-

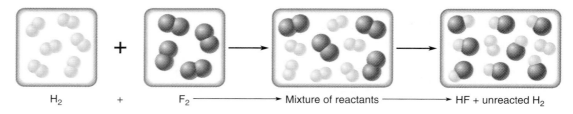

| H_2 | + | F_2 ————————→ Mixture of reactants ————————→ HF + unreacted H_2 |

➤ **FIGURE 5.12** The limiting reactant is used up in a reaction.

lems must be solved, one in which the known quantity is 20.0 g CH_4 and one in which the known quantity is 100.0 g O_2. In each case, the theoretical amount of CO_2 produced is calculated, although the same information would be obtained by solving for the amount of water produced.

$$(20.0 \text{ g } CH_4)\left(\frac{1 \text{ mol } CH_4}{16.0 \text{ g } CH_4}\right)\left(\frac{1 \text{ mol } CO_2}{1 \text{ mol } CH_4}\right)\left(\frac{44.0 \text{ g } CO_2}{1 \text{ mol } CO_2}\right) = 55.0 \text{ g } CO_2$$

$$(100 \text{ g } O_2)\left(\frac{1 \text{ mol } O_2}{32.0 \text{ g } O_2}\right)\left(\frac{1 \text{ mol } CO_2}{2 \text{ mol } O_2}\right)\left(\frac{44.0 \text{ g } CO_2}{1 \text{ mol } CO_2}\right) = 68.8 \text{ g } CO_2$$

Since the 20.0 g CH_4 produces the smaller amount of CO_2, CH_4 is the limiting reactant. Therefore, the final mixture will contain the products CO_2 and H_2O and the leftover O_2.

Example 5.9

The reaction shown in Figure 5.8 was carried out by a student who wanted to produce some silver metal. The equation for the reaction is

$$Cu(s) + 2AgNO_3(aq) \rightarrow 2Ag(s) + Cu(NO_3)_2(aq)$$

The student dissolved 17.0 g of $AgNO_3$ in distilled water, then added 8.50 g of copper metal, Cu, to the resulting solution. What mass of silver metal, Ag, was produced?

Solution

The limiting reactant must be identified because it will determine the amount of silver produced. To find the limiting reactant, we calculate the theoretical amount of silver produced by each reactant. The reactant that will produce the smallest mass of silver is the limiting reactant, and the mass of silver calculated for it is the amount produced by the reaction.

$$17.0 \text{ g } AgNO_3\left(\frac{1 \text{ mol } AgNO_3}{170 \text{ g } AgNO_3}\right)\left(\frac{2 \text{ mol } Ag}{2 \text{ mol } AgNO_3}\right)\left(\frac{108 \text{ g } Ag}{1 \text{ mol } Ag}\right) = 10.8 \text{ g } Ag$$

$$8.50 \text{ g } Cu\left(\frac{1 \text{ mol } Cu}{63.6 \text{ g } Cu}\right)\left(\frac{2 \text{ mol } Ag}{1 \text{ mol } Cu}\right)\left(\frac{108 \text{ Ag}}{1 \text{ mol } Ag}\right) = 28.9 \text{ g } Ag$$

We see that the $AgNO_3$ is the limiting reactant, and it will produce 10.8 g of Ag.

Learning Check 5.8

Refer to the reaction used for Learning Check 5.7. Assume that 2.00 mol H_2 and 15.5 g N_2 are reacted. What is the maximum mass of ammonia (NH_3) that can be produced? Which reactant is the limiting reactant?

➤ 5.11 Reaction Yields

In Sections 5.9 and 5.10, calculations were done to determine how much product could be obtained from the reaction of specified amounts of reactants. In many instances, the amount determined by such calculations would be greater than the amount produced by reacting the specified reactants in a laboratory. Does this mean matter is destroyed when reactions are actually done in the laboratory? No, this would violate the law of conservation of matter, a

Chemistry·⚛·Now™

Go to GOB Now and click to engage in a tutorial showing how to calculate the amount of a product formed when one of two reactants controls the extent of a reaction.

Chemistry·⚛·Now™

Go to GOB Now and click to learn how to calculate percent yield for a reaction.

Side reactions
Reactions that do not give the desired product of a reaction.

fundamental law of nature. Less product might be obtained because some of the reactants form compounds other than the desired product. Such reactions are called **side reactions.** Carbon dioxide gas, CO_2, is produced when carbon-containing fuels burn in an ample supply of oxygen. However, a side reaction produces toxic carbon monoxide gas, CO, when the oxygen supply is limited.

Another cause of less product than expected might be poor laboratory technique resulting in such things as losses when materials are transferred from one container to another. Whatever the cause, the mass of product obtained in an experiment is called the *actual yield,* and the mass calculated according to the methods of Sections 5.9 and 5.10 is called the *theoretical yield.* The **percentage yield** is the actual yield divided by the theoretical yield multiplied by 100 (see Section 1.10):

Percentage yield
The percentage of the theoretical amount of a product actually produced by a reaction.

$$\% \text{ yield} = \frac{\text{actual yield}}{\text{theoretical yield}} \times 100 \qquad (5.28)$$

Example 5.10

A chemist wants to produce urea (N_2CH_4O) by reacting ammonia (NH_3) and carbon dioxide (CO_2). The balanced equation for the reaction is

$$2NH_3(g) + CO_2(g) \rightarrow N_2CH_4O(s) + H_2O(\ell)$$

The chemist reacts 5.11 g NH_3 with excess CO_2 and isolates 3.12 g of solid N_2CH_4O. Calculate the percentage yield of the experiment.

Solution

First, the theoretical yield must be calculated. This problem has the pattern

$$g\ A \rightarrow g\ B$$

and follows this pathway:

$$g\ NH_3 \rightarrow mol\ NH_3 \rightarrow mol\ N_2CH_4O \rightarrow g\ N_2CH_4O$$

The known quantity is 5.11 g NH_3, and the unit of the unknown quantity is g N_2CH_4O.

The steps in the factor-unit method are as follows:

Step 1. 5.11 g NH_3

Step 2. 5.11 g NH_3 $\hspace{7cm}$ = g N_2CH_4O

Step 3. $5.11\ \cancel{g\ NH_3}\left(\dfrac{1\ \cancel{mol\ NH_3}}{17.0\ \cancel{g\ NH_3}}\right)\left(\dfrac{1\ \cancel{mol\ N_2CH_4O}}{2\ \cancel{mol\ NH_3}}\right)\left(\dfrac{60.1\ g\ N_2CH_4O}{1\ \cancel{mol\ N_2CH_4O}}\right)$ = g N_2CH_4O

Step 4. $(5.11)\left(\dfrac{1}{17.0}\right)\left(\dfrac{1}{2}\right)\left(\dfrac{60.1\ g\ N_2CH_4O}{1}\right) = 9.03\ g\ N_2CH_4O$

The actual yield was 3.12 g, so the percentage yield is

$$\% \text{ yield} = \frac{\text{actual yield}}{\text{theoretical yield}} \times 100 = \frac{3.12\ g}{9.03\ g} \times 100 = 34.6\%$$

Learning Check 5.9

a. A chemist isolates 17.43 g of product from a reaction that has a calculated theoretical yield of 21.34 g. What is the percentage yield?

b. Lime (CaO) is produced by heating calcium carbonate. The equation for the reaction is $CaCO_3(s) \rightarrow CaO(s) + CO_2(g)$. Suppose 510 g $CaCO_3$ is heated, and 235 g CaO is isolated after the reaction mixture cools. What is the percentage yield for the reaction?

> **STUDY SKILLS 5.1**

HELP WITH OXIDATION NUMBERS

Assigning oxidation numbers is a task that seems to be straightforward, but mistakes caused by carelessness often show up when it is done on exams. You can minimize such mistakes by using a systematic approach to the task. First, have the seven rules for assigning oxidation numbers well in mind. Then, use an organized method such as the boxlike approach (shown below) to assign an oxidation number to the Mn in Mn_2O_3. In this approach, we begin with an oxidation number that is given by one of the rules (O is -2 according to Rule 5):

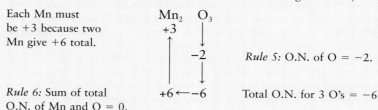

Each Mn must be $+3$ because two Mn give $+6$ total.

$Mn_2 \quad O_3$

Rule 5: O.N. of O $= -2$.

Rule 6: Sum of total O.N. of Mn and O $= 0$.

$+6 \leftarrow -6$

Total O.N. for 3 O's $= -6$

Follow the arrows in a clockwise direction to solve for the O.N. of each atom of Mn ($+3$ is correct). Remember that oxidation numbers are reported for individual atoms. For example, the total oxidation number due to Mn atoms is $+6$, but because there are two atoms of Mn in the formula, the oxidation number for each atom is just $+3$.

> **FOR FUTURE REFERENCE** FOODS HIGH IN ANTIOXIDANTS

Many oxidation reactions normally take place in the body. Some are beneficial, such as reactions that generate heat and other forms of energy, or that produce substances not obtained directly from the diet but needed for good health. However, some normal reactions also produce undesireable byproducts that behave as oxidizing agents (oxidants) in reactions that damage the body or cause diseases such as some types of cancer. Substances called *free radicals* make up one type of oxidant that has been implicated in numerous diseases, including some cancers. Free radicals consist of molecules that contain unpaired electrons. Oxygen is an important component of many such free radicals that are often called *oxygen radicals.*

Research done during the last decade of the 20th century suggested that many beneficial compounds, called *phytonutrients,* are present in plants and should be included in a healthful diet (see Chemistry and Your Health 5.1). The benefits derived from a significant number of these compounds have been attributed to the ability of the compounds to act as antioxidants, which destroy or deactivate oxygen radicals. The fruits and vegetables listed below are especially high in such antioxidants. The numbers following each name represent the oxygen radical absorbance capacity (ORAC) scores corresponding to a 3.5-ounce serving. A higher ORAC score corresponds to a greater ability to deactivate oxygen radicals.

FRUITS
Prunes, 5770
Raisins, 2830
Blueberries, 2400
Strawberries, 1540
Raspberries, 1220
Plums, 949
Oranges, 750
Red grapes, 739
Cherries, 670

VEGETABLES
Kale, 1770
Spinach, 1260
Brussels sprouts, 980
Broccoli florets, 890
Beets, 840
Red bell peppers, 710
Yellow corn, 400
Eggplant, 390
Carrots, 210

Chemical Equations. Chemical reactions are conveniently represented by equations in which reacting substances, called reactants, and the substances produced, called products, are written in terms of formulas. Coefficients are placed before reactant and product formulas to balance the equation. A balanced equation satisfies the law of conservation of matter.

Types of Reactions. To facilitate their study, reactions are classified into the general category redox or nonredox, and further classified as decomposition, combination, single-replacement, or double-replacement reactions.

Redox Reactions. Reactions in which reactants undergo oxidation or reduction are called redox reactions. Oxidation and reduction are indicated conveniently by the oxidation number changes that occur. Oxidation numbers are assigned according to a specific set of rules. A substance is oxidized when the oxidation number of a constituent element increases, and it is reduced when the oxidation number of a constituent element decreases.

Decomposition Reactions. Decomposition reactions are characterized by one substance reacting to give two or more products. Decomposition reactions can be redox or nonredox.

Combination Reactions. Combination reactions, also called addition or synthesis reactions, are characterized by two or more reactants that form a single compound as a product. Combination reactions can be redox or nonredox.

Replacement Reactions. Single-replacement reactions, sometimes called substitution reactions, are always redox reactions.

One element reacts with a compound and displaces another element from the compound. Double-replacement reactions, also called metathesis reactions, are always nonredox reactions. They can be recognized by their partner-swapping characteristics.

Ionic Equations. Many water-soluble compounds separate (dissociate) into ions when dissolved in water. Reactions of such materials can be represented by molecular equations in which no ions are shown, total ionic equations in which all ions are shown, or net ionic equations in which only ions actually undergoing a change are shown.

Energy and Reactions. Energy changes accompany all chemical reactions. The energy can appear in a variety of forms, but heat is a common one. Exothermic reactions liberate heat, and endothermic reactions absorb it.

The Mole and Chemical Equations. The mole concept, when applied to chemical equations, yields relationships that can be used to obtain factors for doing factor-unit calculations.

The Limiting Reactant. The limiting reactant is the reactant present in a reaction in an amount that determines the maximum amount of product that can be made. Factor-unit calculations can be used to determine which reactant is limiting.

Reaction Yields. The mass of product isolated after a reaction is often less than the mass theoretically possible. The ratio of the actual isolated mass (actual yield) to the calculated theoretical yield multiplied by 100 is the percentage yield.

You can get an approximate but quick idea of how well you have met the learning objectives given at the beginning of this chapter by working the selected end-of-chapter exercises given below. The answer to each exercise is given in Appendix B of the book.

Objective 1 (Section 5.1): Exercise 5.2

Objective 2 (Section 5.1): Exercise 5.6

Objective 3 (Section 5.3): Exercise 5.10

Objective 4 (Section 5.3): Exercise 5.18

Objectives 5 and 6 (Sections 5.4, 5.5, and 5.6)): Exercise 5.20

Objective 7 (Section 5.7): Exercise 5.30 a, b, & c

Objective 8 (Section 5.8): Exercise 5.34

Objective 9 (Section 5.9): Exercise 5.42

Objective 10 (Section 5.10): Exercise 5.52

Objective 11 (Section 5.11): Exercise 5.56

Balanced equation (5.1)
Combination reaction (5.5)
Decompostion reaction (5.4)
Double-replacement reaction (5.6)
Endothermic reaction (5.8)
Exothermic reaction (5.8)
Law of conservation of matter (5.1)
Limiting reactant (5.10)

Limiting-reactant principle (5.10)
Molecular equation (5.7)
Net ionic equation (5.7)
Oxidation (5.3)
Oxidation number (oxidation state) (5.3)
Oxidizing agent (5.3)
Percentage yield (5.11)
Product of a reaction (5.1)

Reactant of a reaction (5.1)
Reducing agent (5.3)
Reduction (5.3)
Side reaction (5.11)
Single-replacement reaction (5.6)
Spectator ion (5.7)
Stoichiometry (5.9)
Total ionic equation (5.7)

.... **KEY EQUATIONS** ....

1. Decomposition reaction—one substance changes to two or more new substances (Section 5.4):

$$A \rightarrow B + C$$

Equation 5.13

2. Combination reaction—two or more substances react to produce one new substance (Section 5.5):

$$A + B \rightarrow C$$

Equation 5.16

3. Single-replacement reaction—one element reacts with a compound to produce a new compound and new element (Section 5.6):

$$A + BX \rightarrow B + AX$$

Equation 5.19

4. Double-replacement reaction—partner-swapping reaction (Section 5.6):

$$AX + BY \rightarrow BX + AY$$

Equation 5.21

5. Percentage yield (Section 5.11):

$$\% \text{ yield} = \frac{\text{actual yield}}{\text{theoretical yield}} \times 100$$

Equation 5.28

.... **EXERCISES** ....

LEGEND: 1 = straightforward, 2 = intermediate, 3 = challenging. All even-numbered exercises are answered in Appendix B.

CHEMICAL EQUATIONS (SECTION 5.1)

5.1 Identify the reactants and products in each of the following reaction equations:
  a. $BaO_2(s) + H_2SO_4(\ell) \rightarrow BaSO_4(s) + H_2O_2(\ell)$
  b. $2H_2O_2(aq) \rightarrow 2H_2O(\ell) + O_2(g)$
  c. methane + water $\rightarrow$ carbon monoxide + hydrogen
  d. copper(II) oxide + hydrogen $\rightarrow$ copper + water

5.2 Identify the reactants and products in each of the following reaction equations:
  a. $2H_2(g) + O_2(g) \rightarrow 2H_2O(\ell)$
  b. $CaCO_3(s) \rightarrow CaO(s) + CO_2(g)$
  c. sodium + water $\rightarrow$ hydrogen + sodium hydroxide
  d. copper + silver nitrate $\rightarrow$ copper nitrate + silver

5.3 Identify which of the following are consistent with the law of conservation of matter. For those that are not, explain why they are not.
  a. $4Al(s) + 3O_2(g) \rightarrow 2Al_2O_3(s)$
  b. $P_4(s) + O_2(g) \rightarrow P_4O_{10}(s)$
  c. 3.20 g oxygen + 3.21 g sulfur $\rightarrow$ 6.41 g sulfur dioxide
  d. $CH_4(g) + 2O_2(g) \rightarrow CO_2(g) + 2H_2O(g)$

5.4 Identify which of the following are consistent with the law of conservation of matter. For those that are not, explain why they are not.
  a. $Mg(s) + O_2(g) \rightarrow MgO(s)$
  b. $2SO_2(g) + O_2(g) \rightarrow 2SO_3(g)$
  c. 1.60 g oxygen + 1.00 g hydrogen $\rightarrow$ 2.80 g water
  d. $CuO(s) + H_2(g) \rightarrow H_2O(\ell) + Cu(s)$

5.5 Determine the number of atoms of each element on each side of the following equations and decide which equations are balanced:

  a. $H_2S(aq) + I_2(aq) \rightarrow 2HI(aq) + S(s)$
  b. $KClO_3(s) \rightarrow KCl(s) + O_2(g)$
  c. $SO_2(g) + H_2O(\ell) \rightarrow H_2SO_3(aq)$
  d. $Ba(ClO_3)_2(aq) + H_2SO_4(aq) \rightarrow$
$$2HClO_3(aq) + BaSO_4(s)$$

5.6 Determine the number of atoms of each element on each side of the following equations and decide which equations are balanced:
  a. $Ag(s) + Cu(NO_3)_2(aq) \rightarrow Cu(s) + AgNO_3(aq)$
  b. $2N_2O(g) + 3O_2(g) \rightarrow 4NO_2(g)$
  c. $Mg(s) + O_2(g) \rightarrow 2MgO(s)$
  d. $H_2SO_4(aq) + Ca(OH)_2(aq) \rightarrow CaSO_4(s) + 2H_2O(\ell)$

5.7 Balance the following equations:
  a. $Cl_2(aq) + NaBr(aq) \rightarrow NaCl(aq) + Br_2(aq)$
  b. $CaF_2(s) + H_2SO_4(aq) \rightarrow CaSO_4(s) + HF(g)$
  c. $Cl_2(g) + NaOH(aq) \rightarrow$
$$NaOCl(aq) + NaCl(aq) + H_2O(\ell)$$
  d. $KClO_3(s) \rightarrow KClO_4(s) + KCl(s)$
  e. dinitrogen monoxide $\rightarrow$ nitrogen + oxygen
  f. dinitrogen pentoxide $\rightarrow$ nitrogen dioxide + oxygen
  g. $P_4O_{10}(s) + H_2O(\ell) \rightarrow H_4P_2O_7(aq)$
  h. $CaCO_3(s) + HCl(aq) \rightarrow CaCl_2(aq) + H_2O(\ell) + CO_2(g)$

5.8 Balance the following equations:
  a. $KClO_3(s) \rightarrow KCl(s) + O_2(g)$
  b. $C_2H_6(g) + O_2(g) \rightarrow CO_2(g) + H_2O(\ell)$
  c. nitrogen + oxygen $\rightarrow$ dinitrogen pentoxide
  d. $MgCl_2(s) + H_2O(g) \rightarrow MgO(s) + HCl(g)$
  e. $CaH_2(s) + H_2O(\ell) \rightarrow Ca(OH)_2(s) + H_2(g)$
  f. $Al(s) + Fe_2O_3(s) \rightarrow Al_2O_3(s) + Fe(s)$
  g. aluminum + bromine $\rightarrow$ aluminum bromide
  h. $HgNO_3(aq) + NaCl(aq) \rightarrow Hg_2Cl_2(s) + NaNO_3(aq)$

REDOX REACTIONS (SECTION 5.3)

5.9 Assign oxidation numbers to the blue element in each of the following formulas:
  a. Cl_2O_5
  b. $KClO_4$
  c. Ba^{2+}
  d. F_2
  e. $H_4P_2O_7$
  f. H_2S

5.10 Assign oxidation numbers to the blue element in each of the following formulas:
  a. $HClO$
  b. $NaNO_2$
  c. N_2
  d. $Ca(ClO)_2$
  e. Al^{3+}
  f. N_2O_5

5.11 Find the element with the highest oxidation number in each of the following formulas:
  a. N_2O_5
  b. $KHCO_3$
  c. $NaOCl$
  d. $NaNO_3$
  e. $HClO_4$
  f. $Ca(NO_3)_2$

5.12 Find the element with the highest oxidation number in each of the following formulas:
  a. HNO_2
  b. B_2O_3
  c. $CaCO_3$
  d. $NaHSO_3$
  e. $Mg(ClO_4)_2$
  f. KIO_3

5.13 For each of the following equations, indicate whether the blue element has been oxidized, reduced, or neither oxidized nor reduced.
  a. $2Mg(s) + O_2(g) \rightarrow 2MgO(s)$
  b. $CuO(s) + H_2(g) \rightarrow Cu(s) + H_2O(g)$
  c. $Ag^+(aq) + Cl^-(aq) \rightarrow AgCl(s)$
  d. $BaCl_2(aq) + H_2SO_4(aq) \rightarrow BaSO_4(s) + 2HCl(aq)$
  e. $Zn(s) + 2H^+(aq) \rightarrow Zn^{2+}(aq) + H_2(g)$

5.14 For each of the following equations, indicate whether the blue element has been oxidized, reduced, or neither oxidized nor reduced.
  a. $2Al(s) + 6HCl(aq) \rightarrow 2AlCl_3(aq) + 3H_2(g)$
  b. $Na_2CO_3(s) + 2HCl(aq) \rightarrow$
  $ 2NaCl(aq) + CO_2(g) + H_2O(\ell)$
  c. $Zn(s) + CuCl_2(aq) \rightarrow Cu(s) + ZnCl_2(aq)$
  d. $CuO(s) + H_2(g) \rightarrow Cu(s) + H_2O(g)$
  e. $3MnO_2(s) + 4Al(s) \rightarrow 2Al_2O_3(s) + 3Mn(s)$

5.15 Assign oxidation numbers to each element in the following equations and identify the oxidizing and reducing agents:
  a. $H_2(g) + Cl_2(g) \rightarrow 2HCl(g)$
  b. $H_2O(g) + CH_4(g) \rightarrow CO(g) + 3H_2(g)$
  c. $CuO(s) + H_2(g) \rightarrow Cu(s) + H_2O(g)$
  d. $B_2O_3(s) + 3Mg(s) \rightarrow 2B(s) + 3MgO(s)$
  e. $Fe_2O_3(s) + CO(g) \rightarrow 2FeO(s) + CO_2(g)$
  f. $Cr_2O_7^{2-}(aq) + 2H^+(aq) + 3Mn^{2+}(aq) \rightarrow$
  $ 2Cr^{3+}(aq) + 3MnO_2(s) + H_2O(\ell)$

5.16 Assign oxidation numbers to each element in the following equations and identify the oxidizing and reducing agents:
  a. $4Al(s) + 3O_2(g) \rightarrow 2Al_2O_3(s)$
  b. $2Na(s) + 2H_2O(\ell) \rightarrow 2NaOH(aq) + H_2(g)$
  c. $FeO(s) + CO(g) \rightarrow Fe(s) + CO_2(g)$
  d. $2MnO_4^-(aq) + 10Cl^-(aq) + 16H^+(aq) \rightarrow$
  $ 5Cl_2(g) + 2Mn^{2+}(aq) + 8H_2O(\ell)$
  e. $2N_2O(g) + 3O_2(g) \rightarrow 4NO_2(g)$
  f. $2TiCl_4(s) + Zn(s) \rightarrow 2TiCl_3(s) + ZnCl_2(s)$

5.17 The tarnish of silver objects is a coating of silver sulfide (Ag_2S), which can be removed by putting the silver in contact with aluminum metal in a dilute solution of baking soda or salt. The equation for the cleaning reaction is

$$3Ag_2S(s) + 2Al(s) \rightarrow 6Ag(s) + Al_2S_3(s)$$

The sulfur in these compounds has a -2 oxidation number. What are the oxidizing and reducing agents in the cleaning reaction?

5.18 Aluminum metal reacts rapidly with highly basic solutions to liberate hydrogen gas and a large amount of heat. This reaction is utilized in a popular solid drain cleaner that is composed primarily of lye (sodium hydroxide) and aluminum granules. When wet, the mixture reacts as follows:

$$6NaOH(aq) + 2Al(s) \rightarrow 3H_2(g) + 2Na_3AlO_3(aq) + heat$$

The liberated H_2 provides agitation that, together with the heat, breaks the drain stoppage loose. What are the oxidizing and reducing agents in the reaction?

5.19 Identify the oxidizing and reducing agents in the Haber process for producing ammonia from elemental nitrogen and hydrogen. The equation for the reaction is

$$N_2(g) + 3H_2(g) \rightarrow 2NH_3(g)$$

DECOMPOSITION, COMBINATION, AND REPLACEMENT REACTIONS (SECTIONS 5.4–5.6)

5.20 Classify each of the reactions represented by the following equations, first as a redox or nonredox reaction. Then further classify each redox reaction as a decomposition, single-replacement, or combination reaction, and each nonredox reaction as a decomposition, double-replacement, or combination reaction.
  a. $K_2CO_3(s) \rightarrow K_2O(s) + CO_2(g)$
  b. $Ca(s) + 2H_2O(\ell) \rightarrow Ca(OH)_2(s) + H_2(g)$
  c. $BaCl_2(aq) + H_2SO_4(aq) \rightarrow BaSO_4(s) + 2HCl(aq)$
  d. $SO_2(g) + H_2O(\ell) \rightarrow H_2SO_3(aq)$
  e. $2NO(g) + O_2(g) \rightarrow 2NO_2(g)$
  f. $2Zn(s) + O_2(g) \rightarrow 2ZnO(s)$

5.21 Classify each of the reactions represented by the following equations, first as a redox or nonredox reaction. Then further classify each redox reaction as a decomposition, single-replacement, or combination reaction, and each nonredox reaction as a decomposition, double-replacement, or combination reaction.
  a. $2Na(s) + Cl_2(g) \rightarrow 2NaCl(s)$
  b. $BaCO_3(s) \rightarrow CO_2(g) + BaO(s)$
  c. $3MnO_2(s) + 4Al(s) \rightarrow 2Al_2O_3(s) + 3Mn(s)$
  d. $2HI(g) \rightarrow H_2(g) + I_2(g)$
  e. $3NaOH(aq) + FeCl_3(aq) \rightarrow Fe(OH)_3(s) + 3NaCl(aq)$
  f. $CO_2(g) + H_2O(\ell) \rightarrow H_2CO_3(aq)$

5.22 Baking soda ($NaHCO_3$) can serve as an emergency fire extinguisher for grease fires in the kitchen. When heated, it liberates CO_2, which smothers the fire. The equation for the reaction is

$$2NaHCO_3(s) \xrightarrow{\text{Heat}} Na_2CO_3(s) + H_2O(g) + CO_2(g)$$

Classify the reaction into the categories used in Exercises 5.20 and 5.21.

5.23 Baking soda may serve as a source of CO_2 in bread dough. It causes the dough to rise. The CO_2 is released when $NaHCO_3$ reacts with an acidic substance:

$$NaHCO_3(aq) + H^+(aq) \rightarrow Na^+(aq) + H_2O(\ell) + CO_2(g)$$

Classify the reaction as redox or nonredox.

5.24 Many homes are heated by the energy released when natural gas (represented by CH_4) reacts with oxygen. The equation for the reaction is

$$CH_4(g) + 2O_2(g) \rightarrow CO_2(g) + 2H_2O(g)$$

Classify the reaction as redox or nonredox.

5.25 Hydrogen peroxide will react and liberate oxygen gas. In commercial solutions, the reaction is prevented to a large degree by the addition of an inhibitor. The equation for the oxygen-liberating reaction is

$$2H_2O_2(aq) \rightarrow 2H_2O(\ell) + O_2(g)$$

Classify the reaction as redox or nonredox.

5.26 Chlorine, used to treat drinking water, undergoes the reaction in water represented by the following equation:

$$Cl_2(aq) + H_2O(\ell) \rightarrow HOCl(aq) + HCl(aq)$$

Classify the reaction as redox or nonredox.

5.27 Triple superphosphate, an ingredient of some fertilizers, is prepared by reacting rock phosphate (calcium phosphate) and phosphoric acid. The equation for the reaction is

$$Ca_3(PO_4)_2(s) + 4H_3PO_4(aq) \rightarrow 3Ca(H_2PO_4)_2(s)$$

Classify the reaction into the categories used in Exercises 5.20 and 5.21.

IONIC EQUATIONS (SECTION 5.7)

5.28 Consider all of the following ionic compounds to be water soluble, and write the formulas of the ions that would be formed if the compounds were dissolved in water. Table 4.7 will be helpful.
   a. $CaCl_2$
   b. $Mg(NO_3)_2$
   c. $(NH_4)_3PO_4$
   d. $LiOH$
   e. K_2CrO_4
   f. $Ca(HCO_3)_2$

5.29 Consider all the following ionic compounds to be water soluble, and write the formulas of the ions that would be formed if the compounds were dissolved in water. Table 4.7 will be helpful.
   a. $K_2Cr_2O_7$
   b. H_2SO_4
   c. NaH_2PO_4
   d. Na_3PO_4

   e. NH_4Cl
   f. $KMnO_4$

5.30 Reactions represented by the following equations take place in water solutions. Write each molecular equation in total ionic form, then identify spectator ions and write the equations in net ionic form. Solids that do not dissolve are designated by (s), gases that do not dissolve are designated by (g), and substances that dissolve but do not dissociate appear in blue.
   a. $SO_2(aq) + H_2O(\ell) \rightarrow H_2SO_3(aq)$
   b. $CuSO_4(aq) + Zn(s) \rightarrow Cu(s) + ZnSO_4(aq)$
   c. $2KBr(aq) + 2H_2SO_4(aq) \rightarrow$
   $$Br_2(aq) + SO_2(aq) + K_2SO_4(aq) + 2H_2O(\ell)$$
   d. $AgNO_3(aq) + NaOH(aq) \rightarrow AgOH(s) + NaNO_3(aq)$
   e. $BaCO_3(s) + 2HNO_3(aq) \rightarrow Ba(NO_3)_2(aq) + CO_2(g) + H_2O(\ell)$
   f. $N_2O_5(aq) + H_2O(\ell) \rightarrow 2HNO_3(aq)$

5.31 Reactions represented by the following equations take place in water solutions. Write each molecular equation in total ionic form, then identify spectator ions and write the equations in net ionic form. Solids that do not dissolve are designated by (s), gases that do not dissolve are designated by (g), and substances that dissolve but do not dissociate appear in blue.
   a. $H_2O(\ell) + Na_2SO_3(aq) + SO_2(aq) \rightarrow 2NaHSO_3(aq)$
   b. $3Cu(s) + 8HNO_3(aq) \rightarrow$
   $$3Cu(NO_3)_2(aq) + 2NO(g) + 4H_2O(\ell)$$
   c. $2HCl(aq) + CaO(s) \rightarrow CaCl_2(aq) + H_2O(\ell)$
   d. $CaCO_3(s) + 2HCl(aq) \rightarrow$
   $$CaCl_2(aq) + CO_2(aq) + H_2O(\ell)$$
   e. $MnO_2(s) + 4HCl(aq) \rightarrow$
   $$MnCl_2(aq) + Cl_2(aq) + 2H_2O(\ell)$$
   f. $2AgNO_3(aq) + Cu(s) \rightarrow Cu(NO_3)_2(aq) + 2Ag(s)$

5.32 The following molecular equations all represent neutralization reactions of acids and bases. These reactions will be discussed further in Chapter 9. Write each equation in total ionic form, identify the spectator ions, then write the net ionic equation. Water is the only substance that does not dissociate. What do you notice about all the net ionic equations?
   a. $HI(aq) + KOH(aq) \rightarrow KI(aq) + H_2O(\ell)$
   b. $H_2SO_4(aq) + 2NH_4OH(aq) \rightarrow$
   $$(NH_4)_2SO_4(aq) + 2H_2O(\ell)$$
   c. $HCl(aq) + LiOH(aq) \rightarrow LiCl(aq) + H_2O(\ell)$

5.33 The following molecular equations all represent neutralization reactions of acids and bases. These reactions will be discussed further in Chapter 9. Write each equation in total ionic form, identify the spectator ions, then write the net ionic equation. Water is the only substance that does not dissociate. What do you notice about all the net ionic equations?
   a. $HBr(aq) + RbOH(aq) \rightarrow RbBr(aq) + H_2O(\ell)$
   b. $H_2SO_4(aq) + 2LiOH(aq) \rightarrow Li_2SO_4(aq) + 2H_2O(\ell)$
   c. $HCl(aq) + CsOH(aq) \rightarrow CsCl(aq) + H_2O(\ell)$

ENERGY AND REACTIONS (SECTION 5.8)

5.34 In addition to emergency cold packs, emergency hot packs are available that heat up when water is mixed with a solid. Is the process that takes place in such packs exothermic or endothermic? Explain.

5.35 In refrigeration systems, the area to be cooled has pipes running through it. Inside the pipes, a liquid evaporates and becomes a gas. Is the evaporation process exothermic or endothermic? Explain.

5.36 An individual wants to keep some food cold in a portable picnic cooler. A piece of ice is put into the cooler, but it is wrapped in a thick insulating blanket to slow its melting. Comment on the effectiveness of the cooler in terms of the direction of heat movement inside the cooler.

5.37 The human body cools itself by the evaporation of perspiration. Is the evaporation process endothermic or exothermic? Explain.

THE MOLE AND CHEMICAL EQUATIONS (SECTION 5.9)

5.38 For the reactions represented by the following equations, write statements equivalent to Statements 1, 4, 5, and 6 given in Section 5.9.
  a. $S(s) + O_2(g) \rightarrow SO_2(g)$
  b. $Sr(s) + 2H_2O(\ell) \rightarrow Sr(OH)_2(s) + H_2(g)$
  c. $2H_2S(g) + 3O_2(g) \rightarrow 2H_2O(g) + 2SO_2(g)$
  d. $4NH_3(g) + 5O_2(g) \rightarrow 4NO(g) + 6H_2O(g)$
  e. $CaO(s) + 3C(s) \rightarrow CaC_2(s) + CO(g)$

5.39 For the reactions represented by the following equations, write statements equivalent to Statements 1, 4, 5, and 6 given in Section 5.9.
  a. $2H_2(g) + O_2(g) \rightarrow 2H_2O(g)$
  b. $PCl_3(\ell) + Cl_2(g) \rightarrow PCl_5(s)$
  c. $2PbS(s) + 3O_2(g) \rightarrow 2PbO(s) + 2SO_2(g)$
  d. $BaCO_3(s) \rightarrow BaO(s) + CO_2(g)$
  e. $SO_2(g) + NO_2(g) \rightarrow SO_3(g) + NO(g)$

5.40 For the following equation, write statements equivalent to Statements 1, 4, 5, and 6 given in Section 5.9. Then write at least six factors (including numbers and units) based on Figure 5.11 and the mole definition that could be used to solve problems by the factor-unit method.

$$2SO_2(g) + O_2(g) \rightarrow 2SO_3(g)$$

5.41 Calculate the number of grams of SO_2 that must react to produce 350 g SO_3. Use the statements written in Exercise 5.40 and express your answer using the correct number of significant figures.

5.42 Calculate the mass of limestone ($CaCO_3$) that must be decomposed to produce 500 g lime (CaO). The equation for the reaction is

$$CaCO_3(s) \rightarrow CaO(s) + CO_2(g)$$

5.43 Calculate the number of moles of CO_2 generated by the reaction of Exercise 5.42 when 500 g CaO is produced.

5.44 Calculate the number of grams of bromine (Br_2) needed to react exactly with 50.1 g aluminum (Al). The equation for the reaction is

$$2Al(s) + 3Br_2(\ell) \rightarrow 2AlBr_3(s)$$

5.45 Calculate the moles of $AlBr_3$ produced by the process of Exercise 5.44.

5.46 How many grams of $AlBr_3$ are produced by the process in Exercise 5.44?

5.47 In Exercise 5.17 you were given the following equation for the reaction used to clean tarnish from silver:

$$3Ag_2S(s) + 2Al(s) \rightarrow 6Ag(s) + Al_2S_3(s)$$

  a. How many grams of aluminum would need to react to remove 0.250 g Ag_2S tarnish?
  b. How many moles of Al_2S_3 would be produced by the reaction described in part a?

5.48 Pure titanium metal is produced by reacting titanium(IV) chloride with magnesium metal. The equation for the reaction is

$$TiCl_4(s) + 2Mg(s) \rightarrow Ti(s) + 2MgCl_2(s)$$

How many grams of Mg would be needed to produce 1.00 kg of pure titanium?

5.49 An important metabolic process of the body is the oxidation of glucose to water and carbon dioxide. The equation for the reaction is

$$C_6H_{12}O_6(aq) + 6O_2(aq) \rightarrow 6CO_2(aq) + 6H_2O(\ell)$$

  a. What mass of water in grams is produced when the body oxidizes 1.00 mol of glucose?
  b. How many grams of oxygen are needed to oxidize 1.00 mol of glucose?

5.50 Caproic acid is oxidized in the body as follows:

$$C_6H_{12}O_2(aq) + 8O_2(aq) \rightarrow 6CO_2(aq) + 6H_2O(\ell)$$

How many grams of oxygen are needed to oxidize 1.00 mol of caproic acid?

THE LIMITING REACTANT (SECTION 5.10)

5.51 A sample of 4.00 g of methane (CH_4) is mixed with 15.0 g of chlorine (Cl_2).
  a. Determine which is the limiting reactant according to the following equation:

$$CH_4(g) + 4Cl_2(g) \rightarrow CCl_4(\ell) + 4HCl(g)$$

  b. What is the maximum mass of CCl_4 that can be formed?

5.52 Nitrogen and oxygen react as follows:

$$N_2(g) + 2O_2(g) \rightarrow 2NO_2(g)$$

Suppose 1.25 mol N_2 and 50.0 g O_2 are mixed together.
  a. Which one is the limiting reactant?
  b. What is the maximum mass in grams of NO_2 that can be produced from the mixture?

5.53 Suppose you want to use acetylene (C_2H_2) as a fuel. You have a cylinder that contains 500 g C_2H_2 and a cylinder that contains 2000 g of oxygen (O_2). Do you have enough oxygen to burn all the acetylene? The equation for the reaction is

$$2C_2H_2(g) + 5O_2(g) \rightarrow 4CO_2(g) + 2H_2O(g)$$

5.54 Ammonia, carbon dioxide, and water vapor react to form ammonium bicarbonate as follows:

$$NH_3(aq) + CO_2(aq) + H_2O(\ell) \rightarrow NH_4HCO_3(aq)$$

Suppose 50.0 g NH_3, 80.0 g CO_2, and 2.00 mol H_2O are reacted. What is the maximum number of grams of NH_4HCO_3 that can be produced?

5.55 Chromium metal (Cr) can be prepared by reacting the oxide with aluminum. The equation for the reaction is

$$Cr_2O_3(s) + 2Al(s) \rightarrow 2Cr(s) + Al_2O_3(s)$$

What three substances will be found in the final mixture if 150 g Cr_2O_3 and 150 g Al are reacted?

REACTION YIELDS (SECTION 5.11)

5.56 The actual yield of a reaction was 12.18 g of product, while the calculated theoretical yield was 15.93 g. What was the percentage yield?

5.57 A product weighing 14.37 g was isolated from a reaction. The amount of product possible according to a calculation was 17.55 g. What was the percentage yield?

5.58 For a combination reaction, it was calculated that 7.59 g A would exactly react with 4.88 g B. These amounts were reacted, and 9.04 g of product was isolated. What was the percentage yield of the reaction?

5.59 A sample of calcium metal with a mass of 2.00 g was reacted with excess oxygen. The following equation represents the reaction that took place:

$$2Ca(s) + O_2(g) \rightarrow 2CaO(s)$$

The isolated product (CaO) weighed 2.26 g. What was the percentage yield of the reaction?

5.60 Upon heating, mercury(II) oxide undergoes a decomposition reaction:

$$2HgO(s) \rightarrow 2Hg(\ell) + O_2(g)$$

A sample of HgO weighing 7.22 g was heated. The collected mercury weighed 5.95 g. What was the percentage yield of the reaction?

CHEMISTRY FOR THOUGHT

5.1 In experiments where students prepare compounds by precipitation from water solutions, they often report yields of dry product greater than 100%. Propose an explanation for this.

5.2 Refer to Figure 5.2 and follow the instructions. Then explain how the concept of limiting reactant is used to extinguish a fire.

5.3 Refer to Figure 5.4 and answer the question. Suggest another material that might provide the enzyme catalyst.

5.4 What do the observations of Figure 5.10 indicate to you about the abilities of the following solids to dissolve in water: NaCl, $AgNO_3$, and AgCl? What would you expect to observe if you put a few grams of solid AgCl into a test tube containing 3 mL of water and shook the mixture? What would you expect to observe if you repeated this experiment but used a few grams of $NaNO_3$ instead of AgCl?

5.5 The concentration of alcohol (CH_3CH_2OH) in the breath of an individual who has been drinking is measured by an instrument called a Breathalyzer. The breath sample is passed through a solution that contains the orange-colored $Cr_2O_7^{2-}$ ion. The equation for the net ionic redox reaction that occurs is:

$$2Cr_2O_7^{2-}(aq) + 3CH_3CH_2OH(aq) + 16H^+(aq) \rightarrow$$
$$4Cr^{3+}(aq) + 3CH_3COOH(aq) + 11H_2O(\ell)$$

The Cr^{3+} ion has a pale violet color in solution. Explain how color changes in the solution could be used to indicate the amount of alcohol in the breath sample.

5.6 Certain vegetables and fruits, such as potatoes and apples, darken quickly when sliced. Submerging the slices in water slows this process. Explain.

5.7 Products used to whiten teeth at home are available. They all contain some type of peroxide. Propose a chemical method by which such products work.

5.8 In an ordinary flashlight battery, an oxidation reaction and a reduction reaction take place at different locations to produce an electrical current that consists of electrons. In one of the reactions, the zinc container of the battery slowly dissolves as it is converted into zinc ions. Is this the oxidation or the reduction reaction? Is this reaction the source of electrons, or are electrons used to carry out the reaction? Explain your answers with a reaction equation.

InfoTrac College Edition Readings

"Mechanical reactions," *Scientific American*, August 2002, 287(2):29(1). Record number A88730998.

"Antioxidants," *Chemistry Review*, Nov 2001, 11(2):S16(2). Record number A80604846.

"Is that newfangled cookware safe?" *FDA Consumer*, Oct 1990, 24(8):12(4). Record number A9073336.

CHAPTER 6

The States of Matter

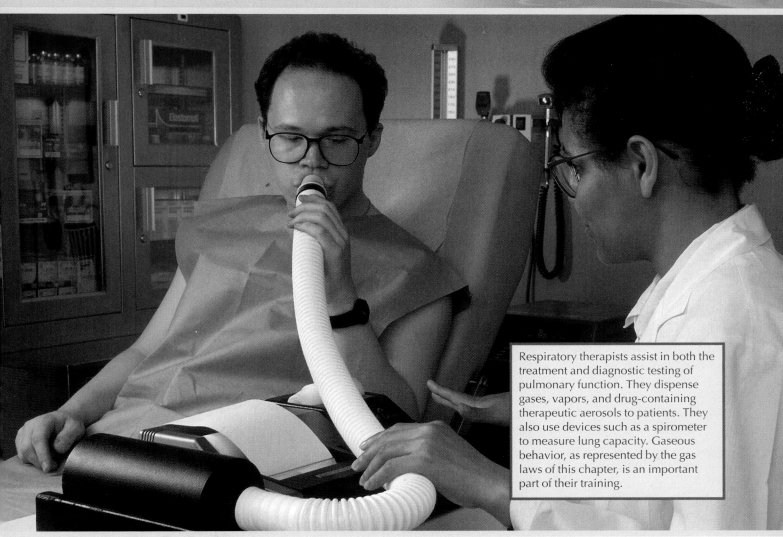

Respiratory therapists assist in both the treatment and diagnostic testing of pulmonary function. They dispense gases, vapors, and drug-containing therapeutic aerosols to patients. They also use devices such as a spirometer to measure lung capacity. Gaseous behavior, as represented by the gas laws of this chapter, is an important part of their training.

© Jeff Kaufman/Taxi

LEARNING OBJECTIVES

When you have completed your study of this chapter, you should be able to:

1. Do calculations based on the property of density. (Section 6.1)

2. Use the kinetic molecular theory to explain the properties of matter in different states. (Sections 6.3, 6.4, and 6.5)

3. Do calculations based on Boyle's law, Charles's law, the combined gas law, Dalton's law, and Graham's law. (Sections 6.6, 6.7, 6.9, and 6.10)

4. Do calculations based on the ideal gas law. (Section 6.8)

5. Do calculations based on the energy changes that accompany the heating, cooling, or changing of state of a substance. (Sections 6.11, 6.12, 6.13, and 6.15)

I f you live in an area that has cold winters, you have probably seen water in the three different forms used to categorize the states in which matter occurs. On a cold day, you can usually find solid water (ice) floating in a pool of cold liquid water, and at the same time you can see a small cloud of tiny water droplets that forms when gaseous water condenses as you exhale into the cold air.

Most matter is not as easily observed in all three states as the water in the preceding example. In fact, most matter is classified as a solid, a liquid, or a gas on the basis of the form in which it is commonly observed. However, according to Section 4.11, the state of a substance depends on temperature. You will see in this chapter that the state also depends on pressure. Therefore, when a substance is classified as a solid, a liquid, or a gas, we are usually simply stating its form under normal atmospheric pressure and at a temperature near 25°C. ■ Figure 6.1 gives the states of the elements under those conditions.

In this chapter, we will study the characteristics of the common states of matter, along with a theory that relates these states to molecular behavior. We will also investigate the energy relationships that accompany changes of state.

➤ 6.1 Observed Properties of Matter

Solids, liquids, and gases can easily be recognized and distinguished by differences in physical properties. Four properties that can be used are density, shape, compressibility, and thermal expansion.

Density was defined in Section 1.11 as the mass of a sample of matter divided by the volume of the same sample. A failure to remember this is the basis for the wrong answers given to the children's riddle, "Which is heavier, a pound of lead or a pound of feathers?" Of course, they weigh the same, but many people say lead is heavier because they think in terms of samples having the same volume. Of course, if you have lead and feather samples of the same volume, the lead sample will be much heavier because its density is much greater.

The **shape** of matter is sometimes independent of a container (for solids), or it may be related to the shape of the container (for liquids and gases). In

Shape
Shape depends on the physical state of matter.

➤ **FIGURE 6.1** The common states of the elements at normal atmospheric pressure and 25°C.

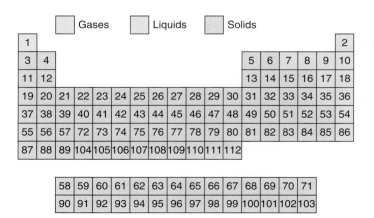

Solids have a shape and volume that does not depend on the container.

Liquids take the shape of the part of the container they fill. Each sample above has the same volume.

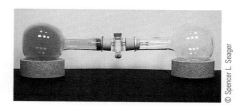

Gases completely fill and take the shape of their container. When the valve separating the two parts of the container is opened, the gas fills the entire container volume (bottom photo).

➤ **FIGURE 6.2** Characteristics of the states of matter.

the case of liquids, the shape depends on the extent to which the container is filled. Gases always fill the container completely (see ■ Figure 6.2).

Compressibility, the change in volume resulting from a pressure change, is quite high for gases. Compressibility allows a lot of gas to be squeezed into a small volume if the gas is put under sufficient pressure—think of automobile tires, or a tank of compressed helium used to fill toy balloons.

Thermal expansion, the change in volume resulting from temperature changes, is a property used in thermometers. As the temperature of the liquid increases, the liquid expands and fills more of the fine capillary tube on which the temperature scale is engraved. You see the liquid move up the tube and know the temperature is increasing.

These four properties are compared for the three states of matter in ■ Table 6.1.

Compressibility
The change in volume of a sample resulting from a pressure change acting on the sample.

Thermal expansion
The change in volume of a sample resulting from a change in temperature of the sample.

TABLE 6.1 Physical properties of solids, liquids, and gases

Property	State		
	Solid	Liquid	Gas
Density	High	High—usually lower than that of corresponding solid	Low
Shape	Definite	Indefinite—takes shape of container to the extent it is filled	Indefinite—takes shape of container it fills
Compressibility	Small	Small—usually greater than that of corresponding solid	Large
Thermal expansion	Very small	Small	Moderate

Example 6.1

Samples of plumber's solder, rubbing alcohol, and air are collected. The volume and mass of each sample are determined at 20°C as follows: solder—volume = 28.6 mL, mass = 268.8 g; alcohol—volume = 100.0 mL, mass = 78.5 g; air—volume = 500.0 mL, mass = 0.602 g. Calculate the density of each substance in grams per milliliter.

Solution

The density (d) is obtained by dividing the sample mass (in grams) by the sample volume (in milliliters):

solder:
$$d = \frac{268.8 \text{ g}}{28.6 \text{ mL}} = 9.40 \text{ g/mL}$$

alcohol:
$$d = \frac{78.5 \text{ g}}{100.0 \text{ mL}} = 0.785 \text{ g/mL}$$

air:
$$d = \frac{0.602 \text{ g}}{500.0 \text{ mL}} = 0.00120 \text{ g/mL}$$

The calculated densities follow the pattern given in Table 6.1.

Learning Check 6.1

Samples of copper, glycerin, and helium are collected and weighed at 20°C. The results are as follows: copper—volume = 12.8 mL, mass = 114.2 g; glycerin—volume = 50.0 mL, mass = 63.0 g; helium—volume = 1500 mL, mass = 0.286 g.

a. Calculate the density of each sample at 20°C.
b. Describe qualitatively (increase, decrease, little change, no change, etc.) what happens to the following for each sample when the pressure on it is doubled: sample volume, sample mass, sample density (see Table 6.1).

The density calculations done in Example 6.1 used Equation 1.9, which is repeated here:

$$d = \frac{m}{V}$$

Remember, in this equation d is density, m is mass, and V is volume. This useful equation can be rearranged so that any one quantity in it can be calculated if the other two are known.

Example 6.2

The sample of rubbing alcohol used in Example 6.1 is heated to 50°C. The density of the sample is found to decrease to a value of 0.762 g/mL. What is the volume of the sample at 50°C?

Solution

It is assumed that no alcohol is lost by evaporation, so the sample mass has the same value of 78.5 g it had in Example 6.1. Equation 1.9 is rearranged and solved for volume:

$$V = \frac{m}{d} \tag{6.1}$$

The new volume is

$$V = \frac{m}{d} = \frac{78.5 \text{ g}}{0.762 \text{ g/mL}} = 103 \text{ mL}$$

Thus, the heating causes the liquid to expand from a volume of 100 mL to 103 mL.

Calculate the mass in grams of a 1200-mL air sample at a temperature where the air has a density of 1.18×10^{-3} g/mL.

Learning Check 6.2

➤ 6.2 The Kinetic Molecular Theory of Matter

The long title of this section is the name scientists have given to a model or theory used to explain the behavior of matter in its various states. Some theories, including this one, are made up of a group of generalizations or postulates. This useful practice makes it possible to study and understand each postulate individually, instead of the entire theory.

The postulates of the kinetic molecular theory are:

1. Matter is composed of tiny particles called molecules.
2. The particles are in constant motion and therefore possess kinetic energy.
3. The particles possess potential energy as a result of attracting or repelling each other.
4. The average particle speed increases as the temperature increases.
5. The particles transfer energy from one to another during collisions in which no net energy is lost from the system.

These postulates contain two new terms, kinetic energy and potential energy. **Kinetic energy** is the energy a particle has as a result of its motion. Mathematically, kinetic energy is calculated as

$$\text{KE} = \frac{1}{2} m v^2 \tag{6.2}$$

where m is the particle mass and v is its velocity. Thus, if two particles of different mass are moving at the same velocity, the heavier particle will possess more kinetic energy than the other particle. Similarly, the faster moving of two particles with equal masses will have more kinetic energy.

GOB
Chemistry·⚛·Now™
Go to GOB Now and click to explore the relationship between molecular speed of gas molecules, temperature, volume, and pressure on the molecular scale.

Kinetic energy
The energy a particle has as a result of its motion. Mathematically, it is $\text{KE} = \frac{1}{2}mv^2$.

Example 6.3

Calculate the kinetic energy of two particles with masses of 2.00 g and 3.00 g if they are both moving with a velocity of 15.0 cm/s.

Solution

The kinetic energy of the 2.00-g particle is

$$\text{KE} = \frac{1}{2}(2.00 \text{ g})\left(15.0 \frac{\text{cm}}{\text{s}}\right)^2 = 225 \frac{\text{g cm}^2}{\text{s}^2}$$

The kinetic energy of the 3.00-g particle is

$$KE = \frac{1}{2}(3.00 \text{ g})\left(15.0 \ \frac{\text{cm}}{\text{s}}\right)^2 = 337.5 \ \frac{\text{g cm}^2}{\text{s}^2}$$

Rounding gives 338 g cm²/s²; thus, the more massive particle has more kinetic energy.

Learning Check 6.3	Calculate the kinetic energy of two 3.00-g particles if one has a velocity of 10.0 cm/s and the other has a velocity of 20.0 cm/s.

Potential energy
The energy a particle has as a result of attractive or repulsive forces acting on it.

Potential energy results from attractions or repulsions of particles for one another. A number of these interactions are familiar, such as the gravitational attraction of Earth and the behavior of the poles of two magnets brought near each other. In each of these examples, the size of the force and the potential energy depend on the separation distance. The same behavior is found for the potential energy of atomic-sized particles. The potential energy of attraction increases as separation increases, whereas the potential energy of repulsion decreases with increasing separation.

The kinetic molecular theory provides reasonable explanations for many of the observed properties of matter. An important factor in these explanations is the relative influence of cohesive forces and disruptive forces. **Cohesive forces** are the attractive forces associated with potential energy, and **disruptive forces** result from particle motion (kinetic energy). Disruptive forces tend to scatter particles and make them independent of each other; cohesive forces have the opposite effect. Thus, the state of a substance depends on the relative strengths of the cohesive forces that hold the particles together and the disruptive forces tending to separate them. Cohesive forces are essentially temperature-independent because they involve interparticle attractions of the type described in Chapter 4. Disruptive forces increase with temperature because they arise from particle motion, which increases with temperature (Postulate 4). This explains why temperature plays such an important role in determining the state in which matter is found.

Cohesive force
The attractive force between particles; it is associated with potential energy.

Disruptive force
The force resulting from particle motion; it is associated with kinetic energy.

➤ 6.3 The Solid State

In the solid state, the cohesive forces are stronger than the disruptive forces (see ■ Figure 6.3a). Each particle of a crystalline solid occupies a fixed position in the crystal lattice (see Section 4.11). Disruptive kinetic energy causes the particles to vibrate about their fixed positions, but the strong cohesive forces prevent the lattice from breaking down. The properties of solids in Table 6.1 are explained by the kinetic theory as follows:

High density. The particles of solids are located as closely together as possible. Therefore, large numbers of particles are contained in a small volume, resulting in a high density.

Definite shape. The strong cohesive forces hold the particles of solids in essentially fixed positions, resulting in a definite shape.

Small compressibility. Because there is very little space between particles of solids, increased pressure cannot push them closer together, and it will have little effect on the volume.

Very small thermal expansion. Increased temperature increases the vibrational motion of the particles and the disruptive forces acting on them. Each particle vibrates with an increased amplitude and "occupies" a

(a) Solid state: The particles are close together and held in fixed positions; they do not need a container.

(b) Liquid state: The particles are close together but not held in fixed positions; they take the shape of the container.

(c) Gaseous state: The particles are far apart and completely fill the container.

slightly larger volume. However, there is only a slight expansion of the solid because the strong cohesive forces prevent this effect from becoming very large.

➤ 6.4 The Liquid State

Particles in the liquid state are randomly packed and relatively close to each other (see Figure 6.3b). They are in constant, random motion, sliding freely over one another, but without sufficient kinetic energy to separate completely from each other. The liquid state is a situation in which cohesive forces dominate slightly. The characteristic properties of liquids are also explained by the kinetic theory:

High density. The particles of liquids are not widely separated; they essentially touch each other. There will, therefore, be a large number of particles per unit volume and a high density.

Indefinite shape. Although not completely independent of each other, the particles in a liquid are free to move over and around each other in a random manner, limited only by the container walls and the extent to which the container is filled.

Small compressibility. Because the particles in a liquid essentially touch each other, there is very little space between them. Therefore, increased pressure cannot squeeze the particles much more closely together.

Small thermal expansion. Most of the particle movement in a liquid is vibrational because the particles can move only a short distance before colliding with a neighbor. Therefore, the increased particle velocity that accompanies a temperature increase results only in increased vibration. The net effect is that the particles push away from each other a little more, thereby causing a slight volume increase in the liquid.

➤ 6.5 The Gaseous State

Disruptive forces completely overcome cohesive forces between particles in the gaseous or vapor state. As a result, the particles of a gas move essentially independently of one another in a totally random way (see Figure 6.3c). Under ordinary pressure, the particles are relatively far apart except when they collide with each other. Between collisions with each other or with the container walls, gas particles travel in straight lines. The particle velocities and resultant collision frequencies are quite high for gases, as shown in ■ Table 6.2.

TABLE 6.2 Some numerical data related to the gaseous state

Gas	Average speed at 0°C	Average distance traveled between collisions at 1 atm of pressure	Number of collisions of 1 molecule in 1 s at 1 atm of pressure
Hydrogen (H_2)	169,000 cm/s (3700 mi/h)	1.12×10^{-5} cm	1.6×10^{10}
Nitrogen (N_2)	45,400 cm/s (1015 mi/h)	0.60×10^{-5} cm	0.80×10^{10}
Carbon dioxide (CO_2)	36,300 cm/s (811 mi/h)	0.40×10^{-5} cm	0.95×10^{10}

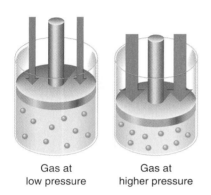

Gas at low pressure   Gas at higher pressure

➤ **FIGURE 6.4** The compression of a gas.

The kinetic theory explanation of gaseous-state properties follows the same pattern seen earlier for solids and liquids:

Low density. The particles of a gas are widely separated. There are relatively few of them in a given volume, which means there is little mass per unit volume.

Indefinite shape. The forces of attraction between particles have been overcome by kinetic energy, and the particles are free to travel in all directions. The particles, therefore, completely fill the container and assume its inner shape.

Large compressibility. The gas particles are widely separated, so that a gas sample is mostly empty space. When pressure is applied, the particles are easily pushed closer together, decreasing the amount of empty space and the gas volume, as shown by ■ Figure 6.4.

Moderate thermal expansion. Gas particles move in straight lines except when they collide with each other or with container walls. An increase in temperature causes the particles to collide with more energy. Thus, they push each other away more strongly, and at constant pressure this causes the gas itself to occupy a significantly larger volume.

It must be understood that the size of the particles is not changed during expansion or compression of gases, liquids, or solids. The particles are merely moving farther apart or closer together, and the space between them is changed.

➤ 6.6 The Gas Laws

The states of matter have not yet been discussed in quantitative terms. We have pointed out that solids, liquids, and gases expand when heated, but the amounts by which they expand have not been calculated. Such calculations for liquids and solids are beyond the intended scope of this text, but gases obey relatively simple quantitative relationships that were discovered during the 17th, 18th, and 19th centuries. These relationships, called **gas laws**, describe in mathematical terms the behavior of gases as they are mixed, subjected to pressure or temperature changes, or allowed to diffuse.

To use the gas laws, you must clearly understand the units used to express pressure, and you must utilize the Kelvin temperature scale introduced in Section 1.6. **Pressure** is defined as force per unit area. However, most of the units commonly used to express pressure reflect a relationship to barometric measurements of atmospheric pressure. The mercury barometer was invented by

Gas law
A mathematical relationship that describes the behavior of gases as they are mixed, subjected to pressure or temperature changes, or allowed to diffuse.

Pressure
A force per unit area of surface on which the force acts. In measurements and calculations involving gases, it is often expressed in units related to measurements of atmospheric pressure.

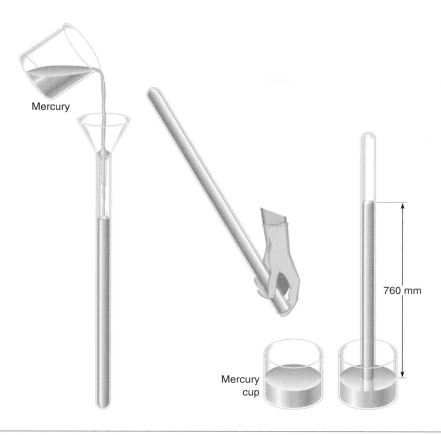

Mercury

760 mm

Mercury
cup

> **FIGURE 6.5** Setting up a simple mercury barometer. Standard atmospheric pressure is 760 mm of mercury.

Italian physicist Evangelista Torricelli (1608–1647). Its essential components are shown in ■ Figure 6.5. A glass tube, sealed at one end, is filled with mercury, stoppered, and inverted so the stoppered end is under the surface of a pool of mercury. When the stopper is removed, the mercury in the tube falls until its weight is just balanced by the weight of air pressing on the mercury pool. The pressure of the atmosphere is then expressed in terms of the height of the supported mercury column.

Despite attempts to standardize measurement units, you will probably encounter a number of different units of pressure in your future studies and employment. Some of the more common are standard atmosphere (atm), torr, millimeters of mercury (mmHg), inches of mercury (in. Hg), pounds per square inch (psi), bar, and kilopascals (kPa). One **standard atmosphere** is the pressure needed to support a 760-mm column of mercury in a barometer tube, and 1 **torr** (named in honor of Torricelli) is the pressure needed to support a 1-mm column of mercury in a barometer tube. The relationships of the various units to the standard atmosphere are given in ■ Table 6.3. Note that the values of 1 atm and 760 torr (or 760 mmHg) are exact numbers based on definitions and do not limit the number of significant figures in calculated numbers.

Standard atmosphere
The pressure needed to support a 760-mm column of mercury in a barometer tube.

Torr
The pressure needed to support a 1-mm column of mercury in a barometer tube.

Example 6.4

The gauge on a cylinder of compressed oxygen gas reads 1500 psi. Express this pressure in terms of (a) atm, (b) torr, and (c) mmHg.

TABLE 6.3	Units of pressure	
Unit	Relationship to standard atmosphere	Typical application
Atmosphere	—	Gas laws
Torr	760 torr = 1 atm	Gas laws
Millimeters of mercury	760 mmHg = 1 atm	Gas laws
Pounds per square inch	14.7 psi = 1 atm	Compressed gases
Bar	1.01 bar = 29.9 in. Hg = 1 atm	Meteorology
Kilopascal	101 kPa = 1 atm	Gas laws

Solution

We will use the factor-unit method of calculation from Section 1.9, and the necessary factors from Table 6.3.

a. It is seen from Table 6.3 that 14.7 psi = 1 atm. Therefore, the known quantity, 1500 psi, is multiplied by the factor atm/psi in order to generate units of atm. The result is

$$1500 \ \text{psi} \left(\frac{1 \ \text{atm}}{14.7 \ \text{psi}} \right) = 102 \ \text{atm}$$

b. Table 6.3 does not give a direct relationship between psi and torr. However, both psi and torr are related to atm. Therefore, the atm is used somewhat like a "bridge" between psi and torr:

$$(1500 \ \text{psi}) \left(\frac{1 \ \text{atm}}{14.7 \ \text{psi}} \right)\left(\frac{760 \ \text{torr}}{1 \ \text{atm}} \right) = 77{,}551 \ \text{torr} = 7.76 \times 10^4 \ \text{torr}$$

Note how the "bridge" term canceled out.

c. Because the torr and mmHg are identical, the problem is worked as it was in part b:

$$(1500 \ \text{psi}) \left(\frac{1 \ \text{atm}}{14.7 \ \text{psi}} \right)\left(\frac{760 \ \text{mmHg}}{1 \ \text{atm}} \right) = 7.76 \times 10^4 \ \text{mmHg}$$

Learning Check 6.4

A barometer has a pressure reading of 670 torr. Convert this reading into (a) atm and (b) psi.

According to the kinetic molecular theory, a gas expands when it is heated at constant pressure because the gaseous particles move faster at the higher temperature. It makes no difference to the particles what temperature scale is used to describe the heating. However, gas law calculations are based on the Kelvin temperature scale (Section 1.6) rather than the Celsius or Fahrenheit scales. The only apparent difference between the Kelvin and Celsius scales is the location of the zero reading (see Figure 1.14). However, a 0 reading on the Kelvin scale has a great deal of significance. It is the lowest possible temperature and is called **absolute zero**. It represents the temperature at which particles have no kinetic energy because all motion stops. The Kelvin and Celsius temperature scales have the same size degree, but the 0 of the Kelvin scale is 273 degrees below the freezing point of water, which is the 0 of the Celsius

Absolute zero
The temperature at which all motion stops; a value of 0 on the Kelvin scale.

CHEMISTRY AND YOUR HEALTH • 6.1

HOW TO LOWER YOUR BLOOD PRESSURE

As you read the title of this feature, you might think it has little to do with you. However, unless your blood pressure has been measured recently by a health professional and you know it is normal, you should continue reading. It is estimated that nearly one-third of the U.S. population has high blood pressure (hypertension), and many individuals with the problem are not aware of it. Blood pressure is reported as the ratio of two numbers, such as 120/80. The numbers represent pressures in terms of the height of a column of mercury in mm that the pressure can support. The larger number (systolic pressure) is the pressure exerted by the blood when the heart contracts, and the smaller number (diastolic pressure) is the pressure of the blood when the heart relaxes between contractions. In young adults the normal systolic range is 100–120, and the normal diastolic range is 60–80. In older people the corresponding normal ranges are 115–135 and 75–85 mm Hg.

From a health perspective, the focus is on the diastolic pressure, and studies have shown that even mild elevations of this indicator to values between 90 and 104 bring increased risk of cardiovascular disease that can lead to heart attack, stroke, and kidney failure. High blood pressure can be reduced and controlled by making changes in lifestyle factors such as diet, smoking, and exercise and by the administration of drugs. The following lifestyle recommendations are generally recognized as ways to reduce blood pressure and maintain it at a healthful level:

• Lose weight if you are overweight.
• Limit daily alcohol intake to no more than two drinks daily.
• Exercise regularly.
• Don't smoke.
• Keep your sodium intake under 2400 mg per day (the amount in about a teaspoon of table salt).
• Maintain adequate dietary intake of potassium, calcium, magnesium, and dietary fiber.
• Reduce your intake of saturated fat and cholesterol.

The drugs used to treat high blood pressure work in many different ways. The drugs most commonly used are diuretics that increase urine excretion and in the process decrease the blood volume as well as the concentration of Na^+. A number of non-diuretic drugs work by affecting the contraction and relaxation of smooth muscles that surround blood vessels, while others influence the rate at which the heart beats. The variety of ways that drugs can influence and control high blood pressure coupled with the large number of individuals requiring treatment has led to a great deal of research activity in the pharmaceutical industry to develop drugs that are more effective and have fewer side effects than those currently on the market.

Individuals suffering from high blood pressure do not generally feel ill. The only way to properly diagnose the problem is by regular blood pressure measurements taken as a part of a physical examination. It is important for everyone to have such measurements taken, but it is especially important for those who suspect they might be at risk because of serious deviations from the seven lifestyle recommendations given here.

scale. Thus, a Celsius reading can be converted to a Kelvin reading simply by adding 273.

Example 6.5

Refer to Figure 1.14 for the necessary conversion factors, and make the following temperature conversions (remember, Kelvin temperatures are expressed in kelvins, K):

a. 37°C (body temperature) to K   **b.** −50°C to K   **c.** 400 K to °C

Solution

a. Celsius is converted to Kelvin by adding 273, K = °C + 273. Therefore,

$$K = 37°C + 273 = 310 \text{ K (body temperature)}$$

b. Again, 273 is added to the Celsius reading. However, in this case the Celsius reading is negative, so the addition must be done with the signs in mind. Therefore,

$$K = −50°C + 273 = 223 \text{ K}$$

c. Because K = °C + 273, an algebraic rearrangement shows that °C = K − 273. Therefore,

$$°C = 400 \text{ K} − 273 = 127°C$$

Convert the following Celsius temperatures into kelvins, and the Kelvin (K) temperatures into degrees Celsius:

a. 27°C   **b.** 0°C   **c.** 0 K   **d.** 100 K

➤ 6.7 Pressure, Temperature, and Volume Relationships

Experimental investigations into the behavior of gases as they were subjected to changes in temperature and pressure led to several gas laws that could be expressed by simple mathematical equations. In 1662 Robert Boyle, an Irish chemist, reported his discovery of a relationship between the pressure and volume of a gas sample kept at constant temperature. This relationship, known as **Boyle's law,** is

Boyle's law
A gas law that describes the pressure and volume behavior of a gas sample kept at constant temperature. Mathematically, it is $PV = k$.

$$P = \frac{k}{V} \tag{6.3}$$

or

$$PV = k \tag{6.4}$$

where k is an experimentally determined constant, and (remember) the measurements are made without changing the temperature of the gas. Mathematically, the pressure and volume are said to be related by an inverse relationship because the change in one is in a direction opposite (inverse) to the change in the other. That is, as the pressure on a gas sample is increased, the volume of the gas sample decreases.

Another gas law was discovered in 1787 by Jacques Charles, a French scientist. He studied the volume behavior of gas samples kept at constant pressure as they were heated. He found that at constant pressure, the volume of a gas sample was directly proportional to its temperature expressed in kelvins. In other words, if the temperature was doubled, the sample volume doubled as long as the pressure was kept constant (see ■ Figure 6.6). This behavior, known as **Charles's law,** is represented mathematically as

Charles's law
A gas law that describes the temperature and volume behavior of a gas sample kept at constant pressure. Mathematically, it is $V/T = k'$.

$$V = k'T \tag{6.5}$$

or

$$\frac{V}{T} = k' \tag{6.6}$$

where k' is another experimentally determined constant.

Boyle's law and Charles's law can be combined to give a single law called the combined gas law, which provides a relationship between the pressure, volume, and temperature of gases. The **combined gas law** is written as

Combined gas law
A gas law that describes the pressure, volume, and temperature behavior of a gas sample. Mathematically, it is $PV/T = k''$.

$$\frac{PV}{T} = k'' \tag{6.7}$$

where k'' is another experimentally determined constant.

Equation 6.7 can be put into a useful form by the following line of reasoning. Suppose a gas sample is initially at a pressure and temperature of P_i and T_i and has a volume of V_i. Now suppose that the pressure and temperature are changed to some new (final) values represented by P_f and T_f and that the volume changes to a new value of V_f. According to Equation 6.7,

$$\frac{P_iV_i}{T_i} = k''$$

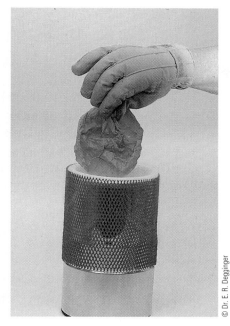

© Dr. E. R. Degginger

© Dr. E. R. Degginger

➤ **FIGURE 6.6** A balloon collapses when the gas it contains is cooled in liquid nitrogen (temperature = −196°C, or 77 K). Assume the inflated balloon had a volume of 4.0 L at room temperature (25°C) and the prevailing pressure. Assume the pressure remained constant and calculate the volume of the gas in the balloon at the temperature of liquid nitrogen.

and

$$\frac{P_f V_f}{T_f} = k''$$

Since both quotients on the left side are equal to the same constant, they can be set equal to each other, and we get

$$\frac{P_i V_i}{T_i} = \frac{P_f V_f}{T_f} \tag{6.8}$$

Example 6.6

a. A sample of helium gas has a volume of 5.00 L at 25°C and a pressure of 0.951 atm. What volume will the sample have if the temperature and pressure are changed to 50°C and 1.41 atm?

b. A sample of gas has a volume of 3.75 L at a temperature of 25°C and a pressure of 1.15 atm. What will the volume be at a temperature of 35°C and a pressure of 620 torr?

Solution

a. Equation 6.8 will be used to solve this problem. An important step is to identify the quantities that are related and will thus have the same subscripts in Equation 6.8. In this problem, the 5.00-L volume, 25°C temperature, and 0.951 atm pressure are all related and will be used as the initial conditions. The final conditions are the 50°C temperature, the 1.41 atm pressure, and the final volume V_f, which is to be calculated. Substitution of these quantities into Equation 6.8 gives

$$\frac{(0.951 \text{ atm})(5.00 \text{ L})}{(298 \text{ K})} = \frac{(1.41 \text{ atm})V_f}{(323 \text{ K})}$$

Note that the Celsius temperatures have been changed to kelvins. The desired quantity, V_f, can be isolated by multiplying both sides of the equation by 323 K and dividing both sides by 1.41 atm:

$$\frac{(323 \; \cancel{K})(0.951 \; \cancel{atm})(5.00 \; L)}{(298 \; K)(1.41 \; \cancel{atm})} = \frac{(\cancel{323 \; K})(\cancel{1.41 \; atm})V_f}{(\cancel{323 \; K})(\cancel{1.41 \; atm})}$$

or

$$V_f = \frac{(323 \; \cancel{K})(0.951 \; \cancel{atm})(5.00 \; L)}{(298 \; \cancel{K})(1.41 \; \cancel{atm})} = 3.66 \; L$$

b. This is the same as the problem in part a, except that the initial and final pressures are given in different units. As we saw in part a, the pressure units must cancel in the final calculation, so the initial and final units must be the same. The units of pressure used make no difference as long as they are the same, so we will convert the initial pressure into torr. The necessary conversion factor was obtained from Table 6.3:

$$P_i = (1.15 \; \cancel{atm})\left(\frac{760 \; torr}{1 \; \cancel{atm}}\right) = 874 \; torr$$

The quantities are substituted into Equation 6.8 to give

$$\frac{(874 \; torr)(3.75 \; L)}{(298 \; K)} = \frac{(620 \; torr)(V_f)}{(308 \; K)}$$

or

$$V_f = \frac{(308 \; K)(874 \; \cancel{torr})(3.75 \; L)}{(298 \; K)(620 \; \cancel{torr})} = 5.46 \; L$$

Example 6.7

a. A steel cylinder has a bursting point of 10,000 psi. It is filled with gas at a pressure of 2500 psi when the temperature is 20°C. During transport, the cylinder is allowed to sit in the hot sun, and its temperature reaches 100°C. Will the cylinder burst?
b. A sample of gas has a volume of 3.00 ft³ at a temperature of 30°C. If the pressure on the sample remains constant, at what Celsius temperature will the volume be half the volume at 30°C?

Solution

a. The volume of the gas is constant (unless the cylinder bursts), so $V_i = V_f$. With this in mind, Equation 6.8 is used with the following values: $P_i = 2500$ psi, $V_i = V_f$, $T_i = 20°C + 273$ K, $P_f =$ unknown pressure, and $T_f = 100°C + 273$ K. Substitution into Equation 6.8 gives

$$\frac{(2500 \; psi)(V_i)}{293 \; K} = \frac{(P_f)(V_f)}{373 \; K}$$

or

$$P_f = \frac{(2500 \; psi)(\cancel{V_f})(373 \; \cancel{K})}{(\cancel{V_f})(293 \; \cancel{K})} = 3.18 \times 10^3 \; psi$$

The cylinder will not burst.
b. In this case, $P_i = P_f$ because the pressure remains constant. Also, $V_i = 3.00$ ft³; $V_f = 1/2V_i$, or 1.50 ft³; $T_i = 30°C + 273 = 303$ K; and T_f is to be determined. Substitution into Equation 6.8 gives

$$\frac{(P_i)(3.00 \; ft^3)}{(303 \; K)} = \frac{(P_f)(1.50 \; ft^3)}{T_f}$$

OVER THE COUNTER • 6.1

BRONCHODILATORS

The body's lungs are located in the thoracic cavity along with the heart. Ribs form the outer wall of the cavity and provide protection for the lungs and heart. The floor of the thoracic cavity consists of the diaphragm, a large dome-shaped sheet of muscle that completely separates the thoracic cavity from the abdominal cavity. When a person inhales, the diaphragm moves downward, increasing the volume of the thoracic cavity and the contained lungs. As a result, the air pressure inside the lungs is reduced to a value less than atmospheric pressure in accordance with Boyle's law. Air outside the lungs at atmospheric pressure then flows into the lungs. When a person exhales, the diaphragm moves upward, decreasing the volume of the lungs and increasing the pressure of the contained air to a value greater than atmospheric pressure. As a result, air flows out of the lungs. For most people, the act of breathing just described is done automatically, with little thought or effort. However, during an asthma attack, breathing can require a great deal of effort.

The respiratory airway through which air travels during breathing is a tube that begins at the back of the throat and leads into the thoracic cavity where a separate branch enters each lung. Once inside the lungs, the tubes continue to branch, eventually forming a network of many extremely small tubes called bronchioles. Each bronchiole terminates in a tiny air sac. An asthma attack occurs when the bronchioles and larger tubes leading to them (bronchi) are partially obstructed because of a contraction of muscles located in their walls. Most asthma attacks are triggered by allergic reactions, but infections, vigorous exercise, and psychological stress have also been identified as causes.

Most products used to treat asthma are available only by prescription, but a limited number are available over the counter.

The best selling OTC product is available in two forms: a spray device that produces an aerosol that is inhaled, and tablets that are swallowed. The active ingredient in the spray is epinephrine, a hormone normally produced by the body. The active ingredient in the tablets is ephedrine, a drug obtained from the twigs of a Chinese plant, mahuang. Both of these active ingredients provide relief from asthma by relaxing the muscles in the walls of the bronchi and bronchioles, allowing them to open and permitting the unobstructed flow of air into and out of the lungs.

Over-the-counter relief for asthma sufferers.

The quantity to be calculated, T_f, is in the denominator of the right side. We put it into the numerator by inverting both sides of the equation:

$$\frac{(303 \text{ K})}{(P_i)(3.00 \text{ ft}^3)} = \frac{T_f}{(P_f)(1.50 \text{ ft}^3)}$$

The desired quantity, T_f, is now isolated by multiplying both sides of the equation by $(P_f)(1.50 \text{ ft}^3)$:

$$\frac{(P_f)(1.50 \text{ ft}^3)(303 \text{ K})}{(P_i)(3.00 \text{ ft}^3)} = \frac{(P_f)(1.50 \text{ ft}^3)(T_f)}{(P_f)(1.50 \text{ ft}^3)}$$

or

$$T_f = \frac{(P_f)(1.50 \text{ ft}^3)(303 \text{ K})}{(P_i)(3.00 \text{ ft}^3)} = 152 \text{ K}$$

Remember, P_f and P_i canceled because they were equal (the pressure was kept constant). The answer is given in kelvins, but we want it in degrees Celsius:

$$K = °C + 273$$

or

$$°C = K - 273 = 152 - 273 = -121°C$$

Learning Check 6.6

a. A sample of argon gas is confined in a 10.0-L container at a pressure of 1.90 atm and a temperature of 30°C. What volume would the sample have at 1.00 atm and −10.2°C?

b. A sample of gas has a volume of 500 mL at a temperature and pressure of 300 K and 800 torr. It is desired to compress the sample to a volume of 250 mL at a pressure of 900 torr. What temperature in both kelvins and Celsius degrees will be required?

GOB
Chemistry∙⚛∙Now™

Go to GOB Now and click to learn about the relationship between different gas properties for a gas in a syringe.

Avogadro's law
Equal volumes of gases measured at the same temperature and pressure contain equal numbers of molecules.

Standard conditions (STP)
A set of specific temperature and pressure values used for gas measurements.

Ideal gas law
A gas law that relates the pressure, volume, temperature, and number of moles in a gas sample. Mathematically, it is $PV = nRT$.

Universal gas constant
The constant that relates pressure, volume, temperature, and number of moles of gas in the ideal gas law.

➤ **FIGURE 6.7** The box has a 22.4-L volume, the volume of 1 mol of any gas at STP. The basketball has a volume of 7.4 L. How many moles of gas would it contain at STP?

➤ 6.8 The Ideal Gas Law

The combined gas law (Equation 6.8) works for samples only in which the mass of gas remains constant during changes in temperature, pressure, and volume. However, it is often useful to work with situations in which the amount of gas varies. The foundation for this kind of work was proposed in 1811 by Amadeo Avogadro, an Italian scientist. According to his proposal, which is now known as **Avogadro's law,** equal volumes of different gases measured at the same temperature and pressure contain equal numbers of gas molecules. According to this idea, two identical compressed gas cylinders of helium and oxygen at the same pressure and temperature would contain identical numbers of molecules of the respective gases. However, the mass of gas in the cylinders would *not* be the same because the molecules of the two gases have different molecular weights.

The actual temperature and pressure used do not influence the validity of Avogadro's law, but it is convenient to specify a standard set of values. Chemists have chosen 0°C (273 K) and 1.00 atm to represent what are called **standard conditions** for gas measurements. These conditions are often abbreviated STP (standard temperature and pressure).

As we have seen, the mole (defined in Section 2.6) is a convenient quantity of matter to work with. What volume does 1 mol of a gas occupy according to Avogadro's law? Experiments show that 1 mol of any gas molecules has a volume of 22.4 L at STP (see ■ Figure 6.7).

A combination of Boyle's, Charles's, and Avogadro's laws leads to another gas law that includes the quantity of gas in a sample as well as the temperature, pressure, and volume of the sample. This law, known as the **ideal gas law,** is written as

$$PV = nRT \qquad (6.9)$$

In this equation, *P*, *V*, and *T* are defined as they were in the gas laws given earlier. The symbol *n* stands for the number of moles of gas in the sample being used, and *R* is a constant known as the **universal gas constant.** The measured value for the volume of 1 mol of gas at STP allows *R* to be evaluated by substituting the values into Equation 6.9 after rearrangement to isolate *R*:

$$R = \frac{PV}{nT} = \frac{(1.00 \text{ atm})(22.4 \text{ L})}{(1 \text{ mol})(273 \text{ K})} = 0.0821 \frac{\text{L atm}}{\text{mol K}}$$

This value of *R* is the same for all gases under any conditions of temperature, pressure, and volume.

Example 6.8

Use the ideal gas law to calculate the volume of 0.413 mol hydrogen gas at a temperature of 20°C and a pressure of 1200 torr.

Solution

To use the ideal gas law (Equation 6.9), it is necessary that all units match those of R. Thus, the temperature will have to be expressed in kelvins and the pressure in atmospheres:

$$K = °C + 273 = 20°C + 273 = 293 \text{ K}$$

$$P = (1200 \text{ torr})\left(\frac{1 \text{ atm}}{760 \text{ torr}}\right) = 1.58 \text{ atm}$$

Because volume is the desired quantity, we isolate it by dividing both sides of the ideal gas law by P:

$$\frac{PV}{P} = \frac{nRT}{P} \qquad \text{or} \qquad V = \frac{nRT}{P}$$

Substitution of quantities gives

$$V = \frac{(0.413 \text{ mol})\left(0.0821 \dfrac{\text{L atm}}{\text{mol K}}\right)(293 \text{ K})}{(1.59 \text{ atm})} = 6.29 \text{ L}$$

Learning Check 6.7

A 2.15-mol sample of sulfur dioxide gas (SO_2) occupies a volume of 12.6 L at 30°C. What is the pressure of the gas?

Because R is a constant for all gases, it follows that if any three of the quantities P, V, T, or n are known for a gas sample, the fourth quantity can be calculated by using Equation 6.9. An interesting application of this concept makes it possible to determine the molecular weights of gaseous substances. If the mass of a sample is known, the number of moles in the sample is the mass in grams divided by the molecular weight. This fact is represented by Equation 6.10, where n is the number of moles in a sample that has a mass in grams of m and a molecular weight of MW:

$$n = \frac{m}{\text{MW}} \tag{6.10}$$

When $\dfrac{m}{\text{MW}}$ is substituted for n in Equation 6.9, the result is

$$PV = \frac{mRT}{\text{MW}} \tag{6.11}$$

GOB
Chemistry⋅⋅Now™
Go to GOB Now and click to study the control of gas composition, temperature, and pressure on gas density.

GOB
Chemistry⋅⋅Now™
Go to GOB Now and click to learn how to calculate gas pressure from mass, temperature, and volume measurements.

GOB
Chemistry⋅⋅Now™
Go to GOB Now and click to discover how to use gas properties to determine the molecular weight of a gas.

Example 6.9

A gas sample has a volume of 2.74 L and a mass of 16.12 g and is stored at a pressure of 4.80 atm and a temperature of 25°C. The gas might be CH_4, C_2H_6, or C_3H_{10}. Which gas is it?

Solution

The molecular weight of the gas is determined using Equation 6.11 and compared with the molecular weights of the three possibilities calculated from their formulas. Rearrangement of Equation 6.11 to isolate molecular weight gives

$$\text{MW} = \frac{mRT}{PV}$$

Substitution of the known quantities after conversion to proper units gives

$$MW = \frac{mRT}{PV} = \frac{(16.12 \text{ g})\left(0.0821 \frac{L \text{ atm}}{\text{mol } K}\right)(298 \text{ K})}{(4.80 \text{ atm})(2.74 \text{ L})} = 30.0 \text{ g/mol}$$

Thus, the molecular weight is 30.0 u. This matches the molecular weight of C_2H_6 calculated from atomic weights and the molecular formula.

Learning Check 6.8

A sample of unknown gas has a mass of 3.35 g and occupies 2.00 L at 1.21 atm and 27°C. What is the molecular weight of the gas? Is the gas H_2S, HBr, or NH_3?

The gas laws discussed in this chapter apply only to gases that are ideal, but interestingly, no ideal gases actually exist. If they did exist, ideal gases would behave exactly as predicted by the gas laws at all temperatures and pressures. Real gases deviate from the behavior predicted by the gas laws, but under normally encountered temperatures and pressures, the deviations are small for many real gases. This fact allows the gas laws to be used for real gases. Interparticle attractions tend to make gases behave less ideally. Thus, the gas laws work best for gases in which such forces are weak, that is, those made up of single atoms (the noble gases) or nonpolar molecules (O_2, N_2, etc.). Highly polar molecules such as water vapor, hydrogen chloride, and ammonia deviate significantly from ideal behavior.

➤ 6.9 Dalton's Law

Dalton's law of partial pressures
The total pressure exerted by a mixture of gases is equal to the sum of the partial pressures of the gases in the mixture.

Partial pressure
The pressure an individual gas of a mixture would exert if it were in the container alone at the same temperature as the mixture.

John Dalton (1766–1844), an English schoolteacher, made a number of important contributions to chemistry. Some of his experiments led to the law of partial pressures, also called **Dalton's law.** According to this law, the total pressure exerted by a mixture of different gases kept at constant volume and temperature is equal to the sum of the partial pressures of the gases in the mixture. The **partial pressure** of each gas in such mixtures is the pressure each gas would exert if it were confined alone under the same temperature and volume conditions as the mixture.

Imagine you have four identical gas containers, as shown in ■ Figure 6.8. Place samples of three different gases (represented by △, ○, and □) into three of the containers, one to a container, and measure the pressure exerted by each sample. Then place all three samples into the fourth container and measure the total pressure (P_t) exerted. The result is a statement of Dalton's law:

$$P_t = P_\triangle + P_\bigcirc + P_\square \qquad (6.12)$$

where $P_\triangle$, $P_\bigcirc$, and $P_\square$ are the partial pressures of gases △, ○, and □, respectively.

➤ **FIGURE 6.8** Dalton's law of partial pressures.

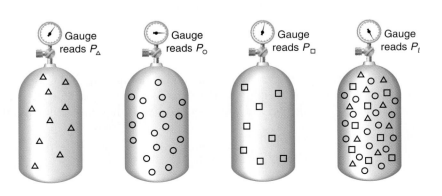

Example 6.10

A sample of air is collected when the atmospheric pressure is 742 torr. The partial pressures of nitrogen and oxygen in the sample are found to be 581 torr and 141 torr, respectively. Assume water vapor to be the only other gas present in the air sample, and calculate its partial pressure.

Solution

Dalton's law says $P_t = P_{O_2} + P_{N_2} + P_{H_2O}$. The total pressure of the sample is the atmospheric pressure of 742 torr. Therefore,

$$742 \text{ torr} = 141 \text{ torr} + 581 \text{ torr} + P_{H_2O}$$

and

$$P_{H_2O} = 742 - 141 - 581 = 20 \text{ torr}$$

GOB
Chemistry·Now™

Go to GOB Now and click to see how to calculate partial pressures and the total pressure for a mixture of three gases.

A mixture is made of helium, nitrogen, and oxygen. Their partial pressures are, respectively, 310 torr, 0.200 atm, and 7.35 psi. What is the total pressure of the mixture in torr?

Learning Check 6.9

➤ 6.10 Graham's Law

Effusion and diffusion are processes by which gases move and mix. **Effusion** is the escape of a gas through a small hole in its container. **Diffusion** is the process in which two or more gases spontaneously intermingle when brought together. Even though the processes appear to be different, they are related, and both follow a law proposed in 1828 by Thomas Graham to describe effusion (see ■ Figure 6.9). For two gases, represented by A and B, **Graham's law** is

$$\frac{\text{effusion rate } A}{\text{effusion rate } B} = \sqrt{\frac{\text{molecular mass of } B}{\text{molecular mass of } A}} \qquad (6.13)$$

Effusion
A process in which a gas escapes from a container through a small hole.

Diffusion
A process that causes gases to spontaneously intermingle when they are brought together.

Graham's law
A mathematical expression that relates rates of effusion or diffusion of two gases to the masses of the molecules of the two gases.

Example 6.11

Oxygen molecules weigh 16 times as much as hydrogen molecules. Which molecule will diffuse faster and how much faster?

A balloon freshly filled with helium.

The same balloon the next morning.

➤ **FIGURE 6.9** Graham's law in action. Would a balloon filled with hydrogen gas deflate more slowly or faster than the helium-filled balloon?

GOB
Chemistry·Now™

Go to GOB Now and click to learn how to use Graham's law to determine the molecular weight of an unknown gas.

Solution

Applying Equation 6.13 but recognizing that diffusion follows the equation as well as effusion, we get

$$\frac{\text{rate } H_2}{\text{rate } O_2} = \sqrt{\frac{\text{mass } O_2}{\text{mass } H_2}} = \sqrt{\frac{32}{2}} = \sqrt{16} = 4$$

Therefore, we conclude that hydrogen will diffuse four times faster than oxygen.

Learning Check 6.10 | Which will diffuse faster, He molecules or Ne molecules? How much faster?

➤ 6.11 Changes in State

Matter can be changed from one state into another by processes such as heating, cooling, or changing the pressure. Heating and cooling are the processes most often used, and a change in state that requires an input of heat is called endothermic, whereas one in which heat is given up (or removed) is called exothermic (see Section 5.8). (These terms come from the Greek *endo* = in and *exo* = out). Endothermic changes are those in which particles are moved farther apart as cohesive forces are being overcome, such as the change of a solid to a liquid or gas. Exothermic changes are those in the opposite direction. See ■ Figure 6.10.

Learning Check 6.11 | Classify the following processes as endothermic or exothermic:

a. Alcohol evaporates.
b. Water freezes.
c. Iron melts in a furnace.

➤ 6.12 Evaporation and Vapor Pressure

Evaporation or vaporization
An endothermic process in which a liquid is changed to a gas.

The evaporation of liquids is a familiar process. Water in a container will soon disappear (evaporate) if the container is left uncovered. **Evaporation**, or **vaporization**, is an endothermic process that takes place as a result of molecules leaving the surface of a liquid. The rate of evaporation depends on the temperature of the liquid and the surface area from which the molecules can escape. Temperature is an important factor because it is related directly to the speed and kinetic energy of the molecules and hence to their ability to break away from the attractive forces present at the liquid surface.

Evaporating molecules carry significant amounts of kinetic energy away from the liquid, and as a result, the temperature of the remaining liquid will drop unless heat flows in from the surroundings. This principle is involved in all evaporative cooling processes, including evaporative coolers for homes,

➤ **FIGURE 6.10** Endothermic and exothermic changes of state.

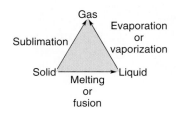

Endothermic

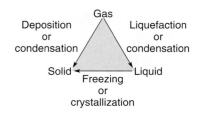

Exothermic

CHEMISTRY AROUND US • 6.1

SWEATING IT OUT

Even though the cosmetics industry sells millions of dollars worth of products each year to prevent it, sweating is as natural a process for the human body as is breathing. It is also an essential process that helps control body temperature. Because of hydrogen bonding between its molecules, liquid water has a surprisingly large heat of vaporization of 540 cal/g. This means that 540 calories of heat must be absorbed by each gram of water that changes from the liquid to the vapor state. When excess body heat is generated by metabolic activity in the body or by exposure to surroundings that are warmer than the body, sweat, a dilute solution that contains about 99% water, is actively released onto the surface of the skin by millions of sweat glands. At an air temperature of 20°C (68°F), an adult loses an average of about 100 mL of water daily through sweating. However, as the air temperature and degree of activity increase, the rate of sweating increases dramatically, and in some cases gets as high as 2 liters per hour.

If 1 liter (or 1000 g) of sweat evaporated per hour from an individual, 540 thousand cal of heat energy would be removed from that person's body during that hour. Usually, however, when sweat is produced in such abundance it does not all evaporate; a significant amount drips off or runs off the skin and does not contribute to the cooling process. The amount that evaporates also depends on the relative humidity of the surrounding air. Relative humidity is the percentage of water vapor present in the air compared to the maximum amount the air can hold at

that temperature. For example, if the relative humidity is 70%, the air already has in it 70% of the maximum amount of water vapor it can possibly hold. Such high-humidity air does not accept as much evaporated water as less humid air and therefore slows the evaporation process. Thus, a profusely sweating person on a hot, humid day is likely to be getting less-efficient body cooling than a person on a hot, dry day who shows only a slight amount of moisture on the skin.

Sweat is a part of the body's cooling system.

© Rori Adamski Peek/Tony Stone Worldwide

---

the cooling of the human body by perspiration, and the cooling of a panting dog by the evaporation of saliva from the mucous membranes of its mouth.

Compare evaporation in a closed container (■ Figure 6.11) with that in an open container. Evaporation occurs in both containers, as indicated by a drop in liquid level. But, unlike that in an open container, the liquid level in the closed container eventually stops dropping and becomes constant.

What would explain this behavior? In a closed container the molecules of liquid that go into the vapor (gaseous) state are unable to move completely away from the liquid surface as they do in an open container. Instead, the vapor molecules are confined to a space immediately above the liquid, where they have many random collisions with the container walls, other vapor molecules, and the liquid surface. Occasionally, their collisions with the liquid result in condensation, and they are recaptured by the liquid. **Condensation** is an exothermic process in which a gas or vapor is converted to either a liquid or a solid. Thus, two processes—evaporation (escape) and condensation (recapture)—actually take place in the closed container. Initially, the rate of evaporation exceeds that of condensation, and the liquid level drops. However, the rates of the two processes eventually become equal, and the liquid level stops dropping because the number of molecules that escape in a given time is the same as the number recaptured.

A system in which two opposite processes take place at equal rates is said to be in equilibrium (see Chapter 8). Under the equilibrium conditions just

GOB
Chemistry Now™

Go to GOB Now and click to examine the effect of temperature and molecular composition on vapor pressure and boiling point.

Condensation

An exothermic process in which a gas or vapor is changed to a liquid or solid.

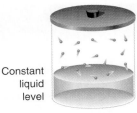

➤ **FIGURE 6.11** Liquid evaporation in a closed container. The drop in liquid level is greatly exaggerated for emphasis.

Initially   After some time   Constant liquid level

At equilibrium

Vapor pressure
The pressure exerted by vapor that is in equilibrium with its liquid.

GOB
Chemistry⚛Now™
Go to GOB Now and click to view a simulation demonstrating the effect of molecular size and ability to hydrogen bond on boiling points of liquids.

described, the number of molecules in the vapor state remains constant. This constant number of molecules will exert a constant pressure on the liquid surface and the container walls. This pressure exerted by a vapor in equilibrium with a liquid is called the **vapor pressure** of the liquid. The magnitude of a vapor pressure depends on the nature of the liquid (molecular polarity, mass, etc.) and the temperature of the liquid. These dependencies are illustrated in
■ Tables 6.4 and 6.5.

The effect of molecular mass on vapor pressure is seen in both the pentane–heptane and ethanol–1-butanol series of compounds in Table 6.4. Within each series, the molecular polarities are the same for each member, but vapor pressure decreases as molecular weight increases. This shows the effect of stronger dispersion forces between heavier molecules. A comparison of pentane and 1-butanol shows the effects of polar attractions between molecules. Molecules of these liquids have similar molecular weights, but the hydrogen-bonded 1-butanol has a much lower vapor pressure.

The effect of increasing the kinetic energy of molecules by heating is clearly evident by the vapor pressure behavior of water shown in Table 6.5.

| Learning Check 6.12 | Decide which member of each of the following pairs of compounds would have the higher vapor pressure. Explain your choice in each case. |

a. Methyl alcohol (CH_3OH) and propyl alcohol (C_3H_7OH)
b. Liquid helium (He) and liquid nitrogen (N_2)
c. Liquid HF and liquid neon (Ne)

➤ 6.13 Boiling and the Boiling Point

Boiling point
The temperature at which the vapor pressure of a liquid is equal to the prevailing atmospheric pressure.

Normal or standard boiling point
The temperature at which the vapor pressure of a liquid is equal to 1 standard atmosphere (760 torr).

As a liquid is heated, its vapor pressure increases, as shown for water in Table 6.5. If the liquid's temperature is increased enough, its vapor pressure will reach a value equal to that of the prevailing atmospheric pressure. Up to that temperature, all vaporization appears to take place at the liquid surface. However, when the vapor pressure becomes equal to atmospheric pressure, vaporization begins to occur beneath the surface as well. When bubbles of vapor form and rise rapidly to the surface, where the vapor escapes, the liquid is boiling. The **boiling point** of a liquid is the temperature at which the vapor pressure of the liquid is equal to the atmospheric pressure above the liquid. The **normal** or **standard boiling point** of a liquid is the temperature at which the vapor pressure is equal to 1 standard atmosphere (760 torr). Normal values were used in all examples of boiling points given earlier in this book.

Liquids boil at temperatures higher than their normal boiling points when external pressures are greater than 760 torr. When external pressures

TABLE 6.4 The vapor pressure of various liquids at 20°C

Liquid	Molecular weight (u)	Polarity	Vapor pressure (torr)
pentane (C_5H_{12})	72	Nonpolar	414.5
hexane (C_6H_{14})	86	Nonpolar	113.9
heptane (C_7H_{16})	100	Nonpolar	37.2
ethanol (C_2H_5—OH)	46	Polar (hydrogen bonds)	43.9
1-propanol (C_3H_7—OH)	60	Polar (hydrogen bonds)	17.3
1-butanol (C_4H_9—OH)	74	Polar (hydrogen bonds)	7.1

TABLE 6.5 Vapor pressure of water at various temperatures

Temperature (°C)	Vapor pressure (torr)
0	4.6
20	17.5
40	55.3
60	149.2
80	355.5
100	760.0

are less than 760 torr, liquids boil at temperatures below their normal boiling points. Thus, the boiling point of water fluctuates with changes in atmospheric pressure. Such fluctuations seldom exceed 2°C at a specific location, but there can be striking variations at different elevations, as shown by ■ Table 6.6. (This is why cooking directions are sometimes different for different altitudes.)

The increase in boiling point caused by an increase in pressure is the principle used in the ordinary pressure cooker. Increasing the pressure inside the cooker causes the boiling point of the water in it to rise, as shown in ■ Table 6.7. It then becomes possible to increase the temperature of the food-plus-water in the cooker above 100°C. Such an increase of just 10°C will make the food cook approximately twice as fast (see ■ Figure 6.12).

➤ 6.14 Sublimation and Melting

Solids, like liquids, have vapor pressures. Although the motion of particles is much more restricted in solids, particles at the surface can escape into the vapor state if they acquire sufficient energy. However, the strong cohesive forces characteristic of the solid state usually cause the vapor pressures of solids to be quite low.

As expected, vapor pressures of solids increase with temperature. When the vapor pressure of a solid is high enough to allow escaping molecules to go directly into the vapor state without passing through the liquid state, the

TABLE 6.6 Variations in the boiling point of water with elevation

Location	Elevation (feet above sea level)	Boiling point of water (°C)
San Francisco, CA	Sea level	100.0
Salt Lake City, UT	4,390	95.6
Denver, CO	5,280	95.0
La Paz, Bolivia	12,795	91.4
Mount Everest	20,028	76.5

© Phil Degginger

➤ **FIGURE 6.12** A pressure cooker shortens the time required to cook food. A carrot cooks completely in boiling water (100°C) in 8 minutes. About how long would it take to cook in a pressure cooker that raised the boiling point of water to 110°C?

TABLE 6.7 The boiling point of water in a pressure cooker		
Pressure above atmospheric		**Boiling point of water (°C)**
psi	torr	
5	259	108
10	517	116
15	776	121

Sublimation
The endothermic process in which a solid is changed directly to a gas without first becoming a liquid.

Melting point
The temperature at which a solid changes to a liquid; the solid and liquid have the same vapor pressure.

Decomposition
A change in chemical composition that can result from heating.

process is called **sublimation** (see ■ Figure 6.13). Sublimation is characteristic of materials such as solid carbon dioxide (dry ice) and naphthalene (moth crystals). Frozen water also sublimes under appropriate conditions. Wet laundry hung out in freezing weather eventually dries as the frozen water sublimes. Freeze drying, a technique based on this process, is used to remove water from materials that would be damaged by heating (e.g., freeze-dried foods).

Even though solids have vapor pressures, most pure substances in the solid state melt before appreciable sublimation takes place. Melting involves the breakdown of a rigid, orderly, solid structure into a mobile, disorderly liquid state. This collapse of the solid structure occurs at a characteristic temperature called the **melting point.** At the melting point, the kinetic energy of solid particles is large enough to partially overcome the strong cohesive forces holding the particles together, and the solid and liquid states have the same vapor pressure.

In some instances, solids cannot be changed into liquids, or liquids into gases, by heating. The atoms making up the molecules of some solids acquire enough kinetic energy on heating to cause bonds within the molecules to break before the solid (or liquid) can change into another state. This breaking of bonds within molecules changes the composition of the original substance. When this **decomposition** occurs, the original substance is said to have decomposed. This is why cotton and paper, when heated, char rather than melt.

➤ 6.15 Energy and the States of Matter

A pure substance in the gaseous state contains more energy than in the liquid state, which in turn contains more energy than in the solid state. Before we look at this, note the following relationships. Kinetic energy, the energy of particle motion, is related to heat. In fact, temperature is a measurement of the average kinetic energy of the particles in a system. Potential energy, in contrast, is related to particle separation distances rather than motion. Thus, we conclude that an increase in temperature on adding heat corresponds to an increase in kinetic energy of the particles, whereas no increase in temperature on adding heat corresponds to an increase in the potential energy of the particles.

Now let's look at a system composed of 1 g of ice at an initial temperature of −20°C. Heat is added at a constant rate until the ice is converted into 1 g of steam at 120°C. The atmospheric pressure is assumed to be 760 torr throughout the experiment. The changes in the system take place in several steps, as shown in ■ Figure 6.14.

The solid is first heated from −20°C to the melting point of 0°C (line AB). The temperature increase indicates that most of the added heat causes an increase in the kinetic energy of the molecules. Along line BC, the temperature remains constant at 0°C while the solid melts. The constant temperature during melting reveals that the added heat has increased the poten-

➤ **FIGURE 6.13** Solid CO_2 (left) and solid H_2O (right) both sublime. Contrast the endothermic processes involved when each is used in the most common way to keep things cold. Are the same processes involved?

CHEMISTRY AROUND US • 6.2

THERAPEUTIC USES OF OXYGEN GAS

A steady supply of oxygen is essential for the human body to function properly. The most common source of this gas is inhaled air, in which the partial pressure of oxygen is about 160 torr. In a healthy individual, this partial pressure is high enough to allow sufficient oxygen needed for body processes to be transported into the blood and distributed throughout the body. Patients suffering from a lung disease, such as pneumonia or emphysema, often cannot transport sufficient oxygen to the blood from the air they breathe unless the partial pressure of oxygen in the air is increased. This is done by mixing an appropriate amount of oxygen with the air by using an oxygen mask or nasal cannula.

Oxygen at very high partial pressures is used in other clinical applications based on a technique called *hyperbaric oxygena-*

tion. In one application, patients infected by anaerobic bacteria, such as those that cause tetanus and gangrene, are placed in a hyperbaric chamber in which the partial pressure of oxygen is 3 to 4 standard atmospheres (2.2×10^3 to 3.0×10^3 torr). Because of the high partial pressure, body tissues pick up large amounts of oxygen, and the bacteria are killed.

Hyperbaric oxygenation may also be used to treat other abnormal conditions or injuries, such as certain heart disorders, carbon monoxide poisoning, crush injuries, certain hard-to-treat bone infections, smoke inhalation, near-drowning, asphyxia, and burns.

---

tial energy of the molecules without increasing their kinetic energy; the molecules are moved farther apart, but their motion is not increased. The addition of more heat warms the liquid water from 0°C to the normal boiling point of 100°C (line *CD*). At 100°C, another change of state occurs as heat is added, and the liquid is converted into vapor (steam) at 100°C (line *DE*). The constant temperature of this process again indicates an increase in potential energy. Line *EF* represents heating the steam from 100°C to 120°C by adding more heat.

The addition of heat to this system resulted in two obviously different results: The temperature of water in a specific state was increased, or the water was changed from one state to another at constant temperature. The amount of heat required to change the temperature of a specified amount of a substance by 1°C is called the **specific heat** of the substance. In scientific work, this is often given in units of calories or joules (J) per gram degree. Thus, the specific heat of a substance is the number of calories or joules required to raise the temperature of 1 g of the substance by 1°C. (NOTE: 1 cal = 4.184 J.) The specific heat is related to the amount of heat required to increase the temperature of a sample of substance by Equation 6.14.

$$\text{Heat} = (\text{sample mass})(\text{specific heat})(\text{temp. change}) \qquad (6.14)$$

Specific heats for a number of substances in various states are listed in ■ Table 6.8. A substance with a high specific heat is capable of absorbing more

Specific heat
The amount of heat energy required to raise the temperature of exactly 1 g of a substance by exactly 1°C.

GOB
Chemistry ⚛ Now™
Go to GOB Now and click to learn how energy addition to a substance leads to a specific temperature change.

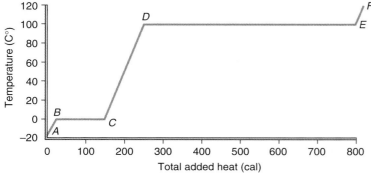

> **FIGURE 6.14** The temperature behavior of a system during changes in state.

GOB
Chemistry ⚛ Now™
Go to GOB Now and click to see how to calculate the mass of ice that the energy released by a match can melt.

TABLE 6.8	Specific heats for selected substances	
	Specific heat	
Substance and state	**cal/g degree**	**J/g degree**
Aluminum (solid)	0.24	1.0
Copper (solid)	0.093	0.39
Ethylene glycol (liquid)	0.57	2.4
Helium (gas)	1.25	5.23
Hydrogen (gas)	3.39	14.2
Lead (solid)	0.031	0.13
Mercury (liquid)	0.033	0.14
Nitrogen (gas)	0.25	1.1
Oxygen (gas)	0.22	0.92
Sodium (solid)	0.29	1.2
Sodium (liquid)	0.32	1.3
Water (solid)	0.51	2.1
Water (liquid)	1.00	4.18
Water (gas)	0.48	2.0

heat with a small temperature change than substances with lower specific heats. Thus, substances to be used as heat transporters (e.g., in cooling and heating systems) should have high specific heats—note the value for liquid water.

Example 6.12

You notice that your car is running hot on a summer day. Someone tells you to drain out the ethylene glycol antifreeze and replace it with water. Will this allow the engine to run cooler?

Solution

The engine will run cooler because a given mass of water has a greater ability to carry heat from the engine to the radiator than does an equal mass of ethylene glycol. The coolant in your car is a mixture of ethylene glycol and water, but to simplify the following comparison, let us imagine it is pure ethylene glycol. We calculate the amount of heat absorbed by 1000 g each of pure ethylene glycol and pure water as the temperature changes from 20°C to 80°C.

$$\text{Heat absorbed} = (\text{mass})(\text{specific heat})(\text{temp. change})$$

ethylene glycol: $\quad \text{Heat absorbed} = (1000 \text{ g})\left(\dfrac{0.57 \text{ cal}}{\text{g} \,°C}\right)(60°\,C)$

$$= 34{,}200 \text{ cal} = 34.2 \text{ kcal}$$

water: $\quad \text{Heat absorbed} = (1000 \text{ g})\left(\dfrac{1.00 \text{ cal}}{\text{g} \,°C}\right)(60°\,C)$

$$= 60{,}000 \text{ cal} = 60.0 \text{ kcal}$$

Thus, the same amount of water will absorb (and transport) nearly twice as much heat for the same temperature increase.

Learning Check 6.13

Some nuclear reactors are cooled by gases. Calculate the number of calories that 1.00 kg of helium gas will absorb when it is heated from 25°C to 700°C. See Table 6.8 for the specific heat of He.

The amount of heat required to change the state of 1 g of a substance at constant temperature is called the **heat of fusion** (for melting) and the **heat of vaporization** (for boiling). The units used for these quantities are just calories or joules per gram because no temperature changes are involved. These heats represent the amount of energy required to change 1 g of a substance to the liquid or vapor state at the characteristic melting or boiling point. The heats of fusion and vaporization for water are 80 and 540 cal/g, respectively. This explains why a burn caused by steam at 100°C is more severe than one caused by water at 100°C. Vaporization has added 540 cal to each gram of steam, and each gram will release these 540 cal when it condenses on the skin. Liquid water at 100°C would not have this extra heat with which to burn the skin (see ■ Figure 6.15).

Heat of fusion
The amount of heat energy required to melt exactly 1 g of a solid substance at constant temperature.

Heat of vaporization
The amount of heat energy required to vaporize exactly 1 g of a liquid substance at constant temperature.

Example 6.13

Calculate the heat released when 5.0 kg of steam at 120°C condenses to water at 100°C in a radiator of a steam heating system.

Solution

Consider the process as taking place in two steps. The steam must cool from 120°C to 100°C and then condense to liquid water at 100°C. When the steam cools to 100°C,

$$\text{Heat released} = (\text{mass})(\text{specific heat})(\text{temp. change})$$

$$= (5.0 \times 10^3 \text{ g})\left(\frac{0.48 \text{ cal}}{\text{g }°\text{C}}\right)(20° \text{C})$$

$$= 4.8 \times 10^4 \text{ cal} = 48 \text{ kcal}$$

A paper cup filled with water does not reach the ignition temperature when heated by a burner, but the water boils.

An empty paper cup burns when heated by a burner because the heat from the burner increases the temperature of the cup to the ignition point.

➤ **FIGURE 6.15** Water can be boiled in a paper cup. Explain why heat from the burner does not increase the temperature of the water-containing cup to the ignition temperature.

WHICH GAS LAW TO USE

For many students, the biggest challenge of this chapter is solving problems using the gas laws. Once you recognize that you are faced with a gas law problem, you must decide which of the six gas laws will work to solve the problem. One aid to selecting the appropriate law is to look for key words, phrases, or ideas that are often associated with specific laws. Some of these are given here:

Gas law	Equation	Key
Boyle's law	$PV = k$	T is constant
Charles's law	$\dfrac{V}{T} = k'$	P is constant
Combined gas law	$\dfrac{P_iV_i}{T_i} = \dfrac{P_fV_f}{T_f}$	P, V, and T all change
Dalton's law	$P_t = P_\triangle + P_\bigcirc + P_\square$	Two or more different gases
Graham's law	$\dfrac{\text{effusion rate } A}{\text{effusion rate } B} = \sqrt{\dfrac{\text{molecular mass of } B}{\text{molecular mass of } A}}$	Effusion and diffusion
Ideal gas law	$PV = nRT$	Moles of gas

For example, if you see the key word *diffusion,* it is likely that you are dealing with a Graham's law problem.

The first three gas laws (Boyle's, Charles's, and the combined) are very similar in that they all use the symbols P, V, and T. If you have narrowed down a gas law problem to one of these three, a simple approach is to use the combined gas law. It works in all three cases. If T is constant,

$$\frac{P_iV_i}{T_i} = \frac{P_fV_f}{T_f}$$

simplifies to $P_iV_i = P_fV_f$, a form of Boyle's law. If P is constant, it simplifies to

$$\frac{V_i}{T_i} = \frac{V_f}{T_f}$$

a form of Charles's law. If V is constant, the combined law becomes

$$\frac{P_i}{T_i} = \frac{P_f}{T_f}$$

an equation that is not named but is useful for certain problems.

A final point to remember is that temperatures must be expressed in kelvins to get correct answers using any of the gas laws that involve temperature.

When the steam condenses to liquid water at 100°C,

$$\text{Heat released} = (\text{mass})(\text{heat of vaporization})$$

$$= (5.0 \times 10^3 \text{ g})\left(\frac{540 \text{ cal}}{\text{g}}\right)$$

$$= 2.7 \times 10^6 \text{ cal} = 2.7 \times 10^3 \text{ kcal}$$

Note that the heat of vaporization was used even though the water was changing from the vapor to the liquid state. The only difference between vaporization and condensation is the direction of heat flow; the amount of heat involved remains the same for a specific quantity of material. (Accordingly, 1 g of liquid water will release 80 cal when it freezes.)

These results make it clear that most of the transported heat (98%) was carried in the form of potential energy, which was released when the steam condensed.

Suppose the radiator of your car overheats and begins to boil. Calculate the number of calories absorbed by 5.00 kg of water (about 1 gallon) as it boils and changes to steam.

Learning Check 6.14

> **FOR FUTURE REFERENCE** DRUGS THAT MAY TRIGGER AN ASTHMA ATTACK

An asthma attack occurs when the bronchioles and larger tubes leading to them become partially obstructed by the contraction of muscles in their walls (see Over the Counter 6.1). Such attacks are most often triggered by allergic reactions, and sometimes drugs taken to treat other ailments can be the cause. The following list contains the names of drugs that may cause asthma attacks:

acetaminophen
aspirin
beta-adrenergic-
   blocking drugs
carbachol
cephalosporins
chloramphenicol
deanol
demeclocycline

erythromycin
griseofulvin
maprotiline
methacholine
methoxypsoralen
metoclopramide
morphine
neomycin
neostigmine

nitrofurantoin
nonsteroidal anti-inflammatory
   drugs (NSAIDs)
penicillins
pilocarpine
propafenone
pyridostigmine
streptomycin
tartrazine (a coloring agent)

SOURCE: Adapted from Rybacki, J. J.; Long, J. W. *The Essential Guide to Prescription Drugs.* New York: Harper Perennial, 1999.

CONCEPT SUMMARY

Observed Properties of Matter. Matter in the solid, liquid, or gaseous state shows differences in physical properties such as density, shape, compressibility, and thermal expansion.

The Kinetic Molecular Theory of Matter. Much of the behavior of matter in different states can be explained by the kinetic molecular theory, according to which all matter is composed of tiny molecules that are in constant motion and are attracted or repelled by each other.

The Solid State. In the solid state, cohesive forces between particles of matter are stronger than disruptive forces. As a result, the particles of solids are held in rigid three-dimensional lattices in which the particle's kinetic energy takes the form of vibrations about each lattice site.

The Liquid State. In the liquid state, cohesive forces between particles slightly dominate disruptive forces. As a result, particles of liquids are randomly arranged but relatively close to each other and are in constant random motion, sliding freely over each other but without enough kinetic energy to become separated.

The Gaseous State. In the gaseous state, disruptive forces dominate and particles move randomly, essentially independent of each other. Under ordinary pressure, the particles are separated from each other by relatively large distances except when they collide.

The Gas Laws. Mathematical relationships, called gas laws, describe the observed behavior of gases when they are mixed, sub-

jected to pressure or temperature changes, or allowed to diffuse. When these relationships are used, it is necessary to express the volume and pressure in consistent units and the temperature in kelvins.

Pressure, Temperature, and Volume Relationships. Gas laws discovered by Robert Boyle and Jacques Charles led to the development of the combined gas law. This law allows calculations to be made that relate temperature, pressure, and volume changes for gases.

The Ideal Gas Law. The application of a gas law discovered by Amadeo Avogadro to the combined gas law led to the ideal gas law, which allows calculations to be made that account for the number of moles of gas in a sample.

Dalton's Law. John Dalton discovered experimentally that the total pressure exerted by a mixture of gases is equal to the sum of the partial pressures of the individual gases in the mixture.

Graham's Law. Thomas Graham proposed a relationship that mathematically relates the rate of effusion or diffusion of a gas to the mass of the gas molecules:

$$\frac{\text{effusion rate } A}{\text{effusion rate } B} = \sqrt{\frac{\text{molecular mass of } B}{\text{molecular mass of } A}}$$

Changes in State. Most matter can be changed from one state to another by heating, cooling, or changing pressure. State changes that give up heat are called exothermic, and those that absorb heat are called endothermic.

Evaporation and Vapor Pressure. The evaporation of a liquid is an endothermic process and as a result is involved in many cooling processes. In a closed container, evaporation takes place only until the rate of escape of molecules from the liquid is equal to the rate at which they return to the liquid. The pressure exerted by the vapor, which is in equilibrium with the liquid, is called the vapor pressure of the liquid. Liquid vapor pressures increase as the liquid temperature increases.

Boiling and the Boiling Point. At the boiling point of a liquid, its vapor pressure equals the prevailing atmospheric pressure, and bubbles of vapor form within the liquid and rise to the surface as the liquid boils. The boiling point of a liquid changes as the prevailing atmospheric pressure changes.

Sublimatiom and Melting. Solids, like liquids, have vapor pressures that increase with temperature. Some solids have high enough vapor pressures to allow them to change to vapor without first becoming a liquid, a process called sublimation. Most solids change to the liquid state before they change to the vapor state. The temperature at which solids change to liquids is called the melting point.

Energy and the States of Matter. Energy is absorbed or released when matter is changed in temperature or changed from one state to another. The amount of heat energy required to produce temperature changes is called the specific heat of the matter involved. For phase changes, the amount of heat required is called the heat of fusion, or vaporization.

LEARNING OBJECTIVES ASSESSMENT

You can get an approximate but quick idea of how well you have met the learning objectives given at the beginning of this chapter by working the selected end-of-chapter exercises given below. The answer to each exercise is given in Appendix B of the book.

Objective 1 (Section 6.1): Exercise 6.2

Objective 2 (Sections 6.3, 6.4, and 6.5): Exercises 6.12 and 6.16

Objective 3 (Sections 6.6, 6.7, 6.9, and 6.10): Exercises 6.20, 6.28, 6.58, and 6.60

Objective 4 (Section 6.8): Exercise 6.48

Objective 5 (Sections 6.11, 6.12, 6.13, and 6.15): Exercises 6.64, 6.68, 6.70, and 6.74

KEY TERMS AND CONCEPTS

Absolute zero (6.6)
Avogadro's law (6.8)
Boiling point (6.13)
Boyle's law (6.7)
Charles's law (6.7)
Cohesive force (6.2)
Combined gas law (6.7)
Compressibility (6.1)
Condensation (6.12)
Dalton's law of partial pressures (6.9)
Decomposition (6.14)
Diffusion (6.10)

Disruptive force (6.2)
Effusion (6.10)
Evaporation or vaporization (6.12)
Gas law (6.6)
Graham's law (6.10)
Heat of fusion (6.15)
Heat of vaporization (6.15)
Ideal gas law (6.8)
Kinetic energy (6.2)
Melting point (6.14)
Normal or standard boiling point (6.13)
Partial pressure (6.9)

Potential energy (6.2)
Pressure (6.6)
Shape (6.1)
Specific heat (6.15)
Standard atmosphere (6.6)
Standard conditions (STP) (6.8)
Sublimation (6.14)
Thermal expansion (6.1)
Torr (6.6)
Universal gas constant (6.8)
Vapor pressure (6.12)

KEY EQUATIONS

1. Calculation of volume from mass and density (Section 6.1):

$$V = \frac{m}{d}$$

Equation 6.1

2. Calculation of kinetic energy of particles in motion (Section 6.2):

$$KE = \frac{1}{2}mv^2$$

Equation 6.2

3. Boyle's law (Section 6.7):

$$P = \frac{k}{V}$$

Equation 6.3

$$PV = k$$

Equation 6.4

4. Charles's law (Section 6.7):

$$V = k'T$$

Equation 6.5

$$\frac{V}{T} = k'$$

Equation 6.6

5. Combined gas law (Section 6.7):

$$\frac{PV}{T} = k''$$

Equation 6.7

$$\frac{P_i V_i}{T_i} = \frac{P_f V_f}{T_f}$$

Equation 6.8

6. Ideal gas law (Section 6.8):

$$PV = nRT$$

Equation 6.9

7. Molecular weight determination (Section 6.8):

$$PV = \frac{mRT}{\text{MW}}$$

Equation 6.11

8. Dalton's law (Section 6.9):

$$P_t = P_\triangle + P_\bigcirc + P_\square$$

Equation 6.12

9. Graham's law of effusion (Section 6.10):

$$\frac{\text{effusion rate } A}{\text{effusion rate } B} = \sqrt{\frac{\text{molecular mass of } B}{\text{molecular mass of } A}}$$

Equation 6.13

10. Heat calculation (Section 6.15):

$$\text{Heat} = (\text{sample mass})(\text{specific heat})(\text{temp. change})$$

Equation 6.14

······································ **EXERCISES** ······································

LEGEND: 1 = straightforward, 2 = intermediate, 3 = challenging. All even-numbered exercises are answered in Appendix B.

OBSERVED PROPERTIES OF MATTER (SECTION 6.1)

6.1 Calculate the volume of 25.0 g of the following liquids:
   a. Acetone (d = 0.792 g/mL)
   b. Olive oil (d = 0.918 g/mL)
   c. Chloroform (d = 1.49 g/mL)

6.2 Calculate the volume of 85.0 mL of the following liquids:
   a. Sea water (d = 1.03 g/mL)
   b. Methyl alcohol (d = 0.792 g/mL)
   c. Concentrated sulfuric acid (d = 1.84 g/mL)

6.3 Copper metal has a density of 8.92 g/cm^3 at 20.0°C and 8.83 g/cm^3 at 100.0°C. Calculate the change in volume that occurs when a 10.0-cm^3 piece of copper is heated from 20.0°C to 100.0°C.

6.4 Liquid water has a density of 1.00 g/mL at 10.0°C and 0.996 g/mL at 30.0°C. Calculate the change in volume that occurs when 500 mL of water is heated from 10.0°C to 30.0°C.

6.5 Gallium metal melts at 29.8°C. At the melting point, the density of the solid is 5.90 g/mL, and that of the liquid is 6.10 g/mL.
   a. Does solid gallium expand or contract when it is melted? Explain.

   b. What is the change in volume when 5.00 mL (cm^3) of solid gallium is melted?

6.6 A 1.50-L rubber balloon is filled with carbon dioxide gas at a temperature of 0.00°C and a pressure of 1.00 atm. The density of the carbon dioxide gas under these conditions is 1.98 g/L.
   a. Will the density of the carbon dioxide gas increase or decrease when the balloon is heated?
   b. At 50.0°C, the balloon has a volume of 1.78 L. Calculate the carbon dioxide density at this temperature.

THE KINETIC MOLECULAR THEORY OF MATTER (SECTION 6.2)

6.7 Describe the changes in form of energy (kinetic changes to potential, etc.) that occur for the energy of a rock dropped to the ground from a cliff. What form or forms do you suppose the energy takes when the rock hits the ground?

6.8 Suppose a toy ball is thrown into the air such that it goes straight up, then falls and is caught by the person who threw it. Describe the changes in form of energy that occur for the ball from the time it is thrown until it is caught.

6.9 Suppose a 180-lb (81.8-kg) halfback running at a speed of 8.0 m/s collides head-on with a 260-lb (118.2-kg) tackle running at 3.0 m/s. Which one will be pushed back? That is, which one has more kinetic energy? If you're not familiar with football, check with someone who is for definition of terms.

6.10 At 25.0°C, He molecules (He) have an average velocity of 1.26×10^5 cm/s, and methane molecules (CH_4) have an average velocity of 6.30×10^4 cm/s. Calculate the kinetic energy of each type of molecule at 25.0°C and determine which is greater. Express molecular masses in u for this calculation.

6.11 Which have the greater kinetic energy, hydrogen molecules traveling with a velocity of $2v$, or helium molecules traveling with a velocity of v? Express molecular masses in u.

THE SOLID, LIQUID, AND GASEOUS STATES (SECTIONS 6.3–6.5)

6.12 Explain each of the following observations using the kinetic molecular theory of matter:
a. A liquid takes the shape, but not necessarily the volume, of its container.
b. Solids and liquids are practically incompressible.
c. A gas always exerts uniform pressure on all walls of its container.

6.13 Explain each of the following observations using the kinetic molecular theory of matter:
a. Gases have low densities.
b. The densities of a substance in the solid and liquid states are nearly identical.
c. Solids, liquids, and gases all expand when heated.

6.14 Discuss differences in kinetic and potential energy of the constituent particles for a substance in the solid, liquid, and gaseous states.

6.15 The following statements are best associated with the solid, liquid, or gaseous states of matter. Match the statements to the appropriate state of matter.
a. This state is characterized by the lowest density of the three.
b. This state is characterized by an indefinite shape and a high density.
c. In this state, disruptive forces prevail over cohesive forces.
d. In this state, cohesive forces are most dominant.

6.16 The following statements are best associated with the solid, liquid, or gaseous states of matter. Match the statements to the appropriate state of matter.
a. Temperature changes influence the volume of this state substantially.
b. In this state, constituent particles are less free to move about than in other states.
c. Pressure changes influence the volume of this state more than that of the other two states.
d. This state is characterized by an indefinite shape and a low density.

THE GAS LAWS (SECTION 6.6)

6.17 What is a gas law?

6.18 A weather reporter on TV reports the barometer pressure as 27.7 inches of mercury. Calculate this pressure in the following units:
a. atm
b. torr
c. psi
d. bars

6.19 The pressure of a gas sample is recorded as 615 torr. Calculate this pressure in the following units:
a. atm
b. in. Hg
c. psi
d. bars

6.20 An engineer reads the pressure gauge of a boiler as 190 psi. Calculate this pressure in the following units:
a. atm
b. bars
c. mmHg
d. in. Hg

6.21 A chemist reads a pressure from a manometer attached to an experiment as 17.6 cmHg. Calculate this pressure in the following units:
a. atm
b. mmHg
c. torr
d. psi

6.22 Convert each of the following temperatures from the unit given to the unit indicated:
a. The melting point of lithium metal, 180.5°C, to kelvins.
b. The freezing point of liquid helium, 0.9 K, to degrees Celsius.
c. The boiling point of liquid hydrogen, −252.8°C, to kelvins.

6.23 Convert each of the following temperatures from the unit given to the unit indicated:
a. The melting point of gold, 1337.4 K, to degrees Celsius.
b. The melting point of tungsten, 3410°C, to kelvins.
c. The melting point of tin, 505 K, to degrees Celsius.

PRESSURE, TEMPERATURE, AND VOLUME RELATIONSHIPS (SECTION 6.7)

6.24 Use the combined gas law (Equation 6.8) to calculate the unknown quantity for each gas sample described in the following table.

	Sample		
	A	**B**	**C**
P_i	1.50 atm	2.35 atm	9.86 atm
V_i	2.00 L	1.97 L	11.7 L
T_i	300 K	293 K	500 K
P_f	?	1.09 atm	5.14 atm
V_f	3.00 L	?	9.90 L
T_f	450 K	310 K	?

6.25 A 200-mL sample of oxygen gas is collected at 26.0°C and a pressure of 690 torr. What volume will the gas occupy at STP (0°C and 760 torr)?

6.26 A 250-mL sample of nitrogen gas is collected at 35.0°C and a pressure of 650 torr. What volume in mL will the gas sample occupy at STP (0.0°C and 760 torr)?

6.27 A 3.00-L sample of helium at 0.00°C and 1.00 atm is compressed into a 0.50-L cylinder. What pressure will the gas exert in the cylinder at 50°C?

6.28 A 3.00-L sample of neon gas at 0.00°C and 1.25 atm is compressed into a 1.00-L cylinder. What pressure will the gas exert in the cylinder at 25.0°C?

6.29 What volume (in liters) of air measured at 1.00 atm would have to be put into a bicycle tire with a 1.00-L volume if the pressure in the bike tire is to be 65.0 psi? Assume the temperature of the gas remains constant.

6.30 What volume (in liters) of air measured at 1.00 atm would have to be put into a car tire with a volume of 12.9 L if the pressure in the car tire is to be 30.0 psi? Assume the temperature of the gas remains constant.

6.31 A sample of gas has a volume of 500 mL at a pressure of 640 torr. What volume will the gas occupy at the same temperature but at the standard atmospheric pressure, 760 torr?

6.32 A sample of gas has a volume of 800 mL at a pressure of 650 torr. What volume in mL will the gas occupy at the same temperature, but at standard atmospheric pressure, 760 torr?

6.33 A 3.0-L sample of gas at 1.0 atm and 0.0°C is heated to 85°C. Calculate the gas volume at the higher temperature if the pressure remains at 1.0 atm.

6.34 A 4.2-L sample of gas at 1.0 atm and 25°C is heated to 85°C. Calculate the gas volume in liters at the higher temperature if the pressure remains constant at 1.0 atm.

6.35 A sample of gas has a volume of 350 mL at 27°C. The gas is heated at a constant pressure until the volume is 500 mL. What is the new temperature of the gas in degrees Celsius?

6.36 What volume of gas in liters at 150°C must be cooled to 40°C if the gas volume at constant pressure and 40°C is to be 2.0 L?

6.37 A 5.00-L gas sample is collected at a temperature and pressure of 27.0°C and 1.20 atm. The gas is to be transferred to a 3.00-L container at a pressure of 1.00 atm. What must the Celsius temperature of the gas in the 3.00-L container be?

6.38 A 2000-L sample of oxygen gas is produced at 1.00 atm pressure. It is to be compressed and stored in a 25.0-L steel cylinder. Assume it is produced and stored at the same temperature and calculate the pressure in atm of the stored oxygen in the cylinder.

6.39 A steel tank with a volume of 8.00 L is full of gas at a pressure of 3.25 atm. What volume in liters would the gas occupy at a pressure of 0.250 atm if its temperature did not change?

6.40 A helium balloon was partially filled with 8000 ft³ of helium when the atmospheric pressure was 0.98 atm and the temperature was 23°C. The balloon rose to an altitude where the atmospheric pressure was 400 torr and the temperature was 5.3°C. What volume did the helium occupy at this altitude?

6.41 You have a 1.50-L balloon full of air at 30°C. To what Celsius temperature would you have to heat the balloon to double its volume if the pressure remained unchanged?

6.42 A gas has a volume of 200 mL at a pressure of 2.50 atm. What volume in mL would it occupy at the same temperature and a pressure of 60.0 kPa?

6.43 What minimum pressure in psi would a 250-mL aerosol can have to withstand if it were to contain 2.50 L of gas measured at 700 torr? Assume the temperature remains constant.

6.44 A 2.00-L sample of nitrogen gas at 760 torr and 0.0°C weighs 2.50 g. The pressure on the gas is increased to 4.00 atm at 0.0°C. Calculate the gas density at the new pressure in g/L.

THE IDEAL GAS LAW (SECTION 6.8)

6.45 Use the ideal gas law and calculate the following:
   a. The pressure exerted by 2.00 mol of oxygen confined to a volume of 500 mL at 20.0°C.
   b. The volume of hydrogen gas in a steel cylinder if 0.525 mol of the gas exerts a pressure of 3.00 atm at a temperature of 10.0°C.
   c. The temperature (in degrees Celsius) of a nitrogen gas sample that has a volume of 2.50 L and a pressure of 300 torr and contains 0.100 mol.

6.46 Use the ideal gas law and calculate the following:
   a. The number of moles of argon in a gas sample that occupies a volume of 500 mL at a temperature of 85.5°C and a pressure of 800 torr.
   b. The pressure in atm exerted by 0.855 mol of helium gas confined to a volume of 2.00 L at 50.0°C.
   c. The volume in liters of a tank of nitrogen that contains 2.15 mol of the gas at a pressure of 3.85 atm and a temperature of 30.0°C.

6.47 Suppose 0.246 mol of SO_2 gas was compressed into a 1.25-L steel cylinder at a temperature of 38.0°C. What pressure in atm would the gas exert on the walls of the cylinder?

6.48 Suppose 12.5 g of SO_2 gas was compressed into a 1.25-L steel cylinder at a temperature of 38.0°C. What pressure in atm would the gas exert on the walls of the cylinder?

6.49 Calculate the volume occupied by 8.75 g of oxygen gas (O_2) at a pressure of 0.890 atm and a temperature of 35.0°C.

6.50 The pressure gauge of a steel cylinder of methane gas (CH_4) reads 400 psi. The cylinder has a volume of 1.50 L and is at a temperature of 30.0°C. How many grams of methane does the cylinder contain?

6.51 Suppose 10.0 g of dry ice (solid CO_2) was placed in an empty 400-mL steel cylinder. What pressure would develop if all the solid sublimed at a temperature of 35.0°C?

6.52 An experimental chamber has a volume of 85 L. How many moles of oxygen gas will be required to fill the chamber at STP?

6.53 How many molecules of nitrogen (N_2) are present in a sample that fills a 10.0-L tank at STP?

6.54 A sample of methyl ether has a mass of 8.12 g and occupies a volume of 3.96 L at STP. What is the molecular weight of methyl ether?

6.55 A sample of a gaseous nitrogen oxide is found to weigh 0.525 g. The sample has a volume of 300 mL at a pressure of 708 torr and a temperature of 25.7°C. Is the gas NO or NO_2?

6.56 A sample of gas weighs 0.176 g and has a volume of 114.0 mL at a pressure and temperature of 640 torr and 20°C. Determine the molecular weight of the gas, and identify it as CO, CO_2, or O_2.

6.57 A 2.00-g sample of gas has a volume of 1.12 L at STP. Calculate its molecular weight and identify it as He, Ne, or Ar.

DALTON'S LAW (SECTION 6.9)

6.58 A steel cylinder contains a mixture of nitrogen, oxygen, and carbon dioxide gases. The total pressure in the tank is 2100 torr. The pressure exerted by the nitrogen and oxygen is 810 and 920 torr, respectively. What is the partial pressure in torr of the carbon dioxide in the mixture?

6.59 A 250-mL sample of oxygen gas is collected by water displacement. As a result, the oxygen is saturated with water vapor. The partial pressure of water vapor at the prevailing temperature is 22 torr. Calculate the partial pressure of the oxygen if the total pressure of the sample is 720 torr.

GRAHAM'S LAW (SECTION 6.10)

6.60 Hydrogen gas (H_2) is found to diffuse approximately four times as fast as oxygen gas (O_2). Using this information, determine how the masses of hydrogen molecules and oxygen molecules compare. How do they compare based on information in the periodic table?

6.61 The mass of a bromine molecule is 160 u, and the mass of an argon molecule is 40 u. Compare the rates at which these gases will diffuse.

6.62 Two identical rubber balloons were filled with gas, one with helium and the other with nitrogen. After a time, it was noted that one of the balloons appeared to be going "flat." Which one do you think it was? Explain.

6.63 Assume the balloon in Exercise 6.62 that went flat first showed signs of "flatness" 12 hours after it was filled. How long would it take for the other balloon to begin to show signs of going flat?

CHANGES IN STATE (SECTION 6.11)

6.64 Classify each of the following processes as endothermic or exothermic:
  a. Condensation
  b. Liquefaction
  c. Boiling

6.65 Classify each of the following processes as endothermic or exothermic:
  a. Freezing
  b. Sublimation
  c. Vaporization

6.66 Discuss what is meant by a change in state.

EVAPORATION AND VAPOR PRESSURE (SECTION 6.12)

6.67 The following are all nonpolar liquid hydrocarbon compounds derived from petroleum: butane (C_4H_{10}), pentane (C_5H_{12}), hexane (C_6H_{14}), and heptane (C_7H_{16}). Arrange these compounds in order of increasing vapor pressure (lowest first, highest last) and explain how you arrived at your answer.

6.68 Methylene chloride (CH_2Cl_2) was used at one time as a local anesthetic by dentists. It was sprayed onto the area to be anesthetized. Propose an explanation for how it worked.

6.69 Suppose a drop of methyl ether (C_2H_6O) was put on the back of one of your hands and a drop of ethyl ether ($C_4H_{10}O$) was put on your other hand. Propose a way you could tell which compound was which without smelling them.

BOILING AND THE BOILING POINT (SECTION 6.13)

6.70 Each of two glass containers contains a clear, colorless, odorless liquid that has been heated until it is boiling. One liquid is water (H_2O) and the other is ethylene glycol ($C_2H_6O_2$). Explain how you could make one measurement of each boiling liquid, using the same device, and tell which liquid was which.

6.71 Suppose a liquid in an open container was heated to a temperature just 1 or 2 degrees below its boiling point, then insulated so it stayed at that temperature. Describe how the liquid would behave (what you would see happen) if the hot sample was suspended beneath a helium balloon and taken rapidly to higher altitudes.

6.72 Suppose you were on top of Mount Everest and wanted to cook a potato as quickly as possible. You left your microwave oven home, so you could either boil the potato in water or throw it into a campfire. Explain which method you would use and why.

ENERGY AND THE STATES OF MATTER (SECTION 6.15)

6.73 Using the specific heat data of Table 6.8, calculate the amount of heat (in calories) needed to increase the temperature of the following:
  a. 50 g of aluminum from 25°C to 55°C
  b. 2500 g of ethylene glycol from 80°C to 85°C
  c. 500 g of steam from 110°C to 120°C

6.74 Using the specific heat data of Table 6.8, calculate the amount of heat (in calories) needed to increase the temperature of the following:
  a. 210 g of copper from 40°C to 95°C
  b. 150 g of mercury from 120°C to 300°C
  c. 2500 g of helium gas from 250°C to 900°C

6.75 For solar energy to be effective, collected heat must be stored for use during periods of decreased sunshine. One proposal suggests that heat can be stored by melting solids that, upon solidification, would release the heat. Calculate the heat that could be stored by melting 1000 kg of each of the following solids. (NOTE: The water in each formula is included in the molecular weight.)
  a. Calcium chloride ($CaCl_2 \cdot 6H_2O$): melting point = 30.2°C, heat of fusion = 40.7 cal/g
  b. Lithium nitrate ($LiNO_3 \cdot 3H_2O$): melting point = 29.9°C, heat of fusion = 70.7 cal/g
  c. Sodium sulfate ($Na_2SO_4 \cdot 10H_2O$): melting point = 32.4°C, heat of fusion = 57.1 cal/g

6.76 Why wouldn't a solid such as K_2SO_4 (melting point = 1069°C, heat of fusion = 50.3 cal/g) be suitable for use in a solar heat storage system? (See Exercise 6.75.)

6.77 Liquid Freon (CCl_2F_2) is used as a refrigerant. It is circulated inside the cooling coils of older refrigerators or freezers. As it vaporizes, it absorbs heat. How much heat can be removed by 2.00 kg of Freon as it vaporizes inside the coils of a refrigerator? The heat of vaporization of Freon is 38.6 cal/g.

6.78 Calculate the total amount of heat needed to change 500 g of ice at −10°C into 500 g of steam at 120°C. Do this by cal-culating the heat required for each of the following steps and adding to get the total:

Step 1. Ice (−10°C) → ice (0°C)
Step 2. Ice (0°C) → water (0°C)
Step 3. Water (0°C) → water (100°C)
Step 4. Water (100°C) → steam (100°C)
Step 5. Steam (100°C) → steam (120°C)

CHEMISTRY FOR THOUGHT

6.1 As solids are heated, they melt. Explain this in terms of the effect of temperature on cohesive and disruptive forces.

6.2 Explain how a hot-air balloon works in terms of gas densities.

6.3 Which of the following gases would you expect to behave most ideally: He, Ar, or HCl? Explain.

6.4 Refer to Figure 6.6 and do the calculation. When a gas is heated, it expands. Explain how the gas in a hot-air balloon remains at constant pressure as the gas is heated. Hot-air balloons do *not* stretch.

6.5 Refer to Figure 6.7 and answer the question. Would a basketball that was inflated for use actually contain the number of moles you calculated? Explain.

6.6 Refer to Figure 6.9 and answer the question. Calculate the actual time factor for the rate of deflation for the two gases, that is, twice as fast and so on.

6.7 Refer to Figure 6.12 and answer the question. Water is sometimes made safe to drink by boiling. Explain why this might not work if you attempted to do it in an open pan on the summit of Mount Everest.

6.8 Refer to Figure 6.15 and answer the question. Suppose a sample of water was heated to the boiling point in a glass beaker, using a single burner. What would happen to the temperature of the boiling water if a second burner was added to help with the heating? Explain.

6.9 Suppose you put four 250-mL bottles of water into an ice chest filled with crushed ice. If the bottles were initially at a temperature of 23°C, calculate the number of grams of ice that would have to melt in order to cool the water in the bottles to 5°C. Assume the density of water is 1.00g/mL.

InfoTrac College Edition Readings

"Carbon copy," *Scientific American*, Feb 2003, 288(2):82(2). Record number A97920051.

"Caffeine does not increase blood pressure," *Cardiovascular Device Update*, March 2003, 9(3):10(1). Record number A98922695.

"LCDs or plasmas?" *EMedia, The Digital Studio Magazine*, Feb 2003, 16(2):49(1). Record number A98171816.

Chemistry·Now™

Assess your understanding of this chapter's topics with additional quizzing and conceptual-based problems at http://chemistry.brookscole.com/seager5e

CHAPTER 7

Solutions and Colloids

Urine, a solution formed in the kidneys, is an excellent indicator of the body's state of health. Urinalysis is an essential part of a physical examination or a diagnosis of disease. Here, a clinical laboratory technician examines a urine sample before preparing it for testing. In this chapter you will learn many of the important characteristics of solutions.

© Spencer Rowell

LEARNING OBJECTIVES

When you have completed your study of this chapter, you should be able to:

1. Classify mixtures as solutions or non-solutions based on their appearance. (Section 7.1)

2. Predict in a general way the solubility of solutes in solvents on the basis of molecular polarity. (Section 7.3)

3. Calculate solution concentrations in units of molarity, weight/weight percent, weight/volume percent, and volume/volume percent. (Section 7.4)

4. Describe how to prepare solutions of specific concentration using pure solutes and solvent, or solutions of greater concentration than the one desired. (Section 7.5)

5. Do stoichiometric calculations based on solution concentrations. (Section 7.6)

6. Do calculations based on the colligative solution properties of boiling point, freezing point, and osmotic pressure. (Section 7.7)

7. Describe the characteristics of colloids. (Sections 7.8 and 7.9)

8. Describe the process of dialysis and compare it to the process of osmosis. (Section 7.10)

arlier, homogeneous matter was classified into two categories—pure substances and mixtures (see Figure 1.5). Since then, the discussion has been limited to pure substances. We now look at homogeneous mixtures called solutions and their distant relatives, colloidal suspensions.

Solutions and colloidal suspensions are very important in our world. They bring nutrients to the cells of our bodies and carry away waste products. The ocean is a solution of water, sodium chloride, and many other substances (even gold). Many chemical reactions take place in solution—including most of those discussed in this book.

➤ 7.1 Physical States of Solutions

Solutions are homogeneous mixtures of two or more substances in which the components are present as atoms, molecules, or ions. These uniformly distributed particles are too small to reflect light, and as a result solutions are transparent (clear); light passes through them. In addition, some solutions are colored. The component particles are in constant motion (remember the kinetic theory, Section 6.2) and do not settle under the influence of gravity.

In most solutions, a larger amount of one substance is present compared to the other components. This most abundant substance in a solution is called the **solvent,** and any other components are called **solutes.** Most people normally think of solutions as liquids, but solutions in solid and gaseous forms are known as well. The state of a solution is often the same as the state of the solvent. This is illustrated by ■ Table 7.1, which lists examples of solutions in various states. The original states of the solvents and solutes are given in parentheses. Solution formation takes place when one or more solutes **dissolve** in a solvent.

Solution
A homogeneous mixture of two or more substances in which the components are present as atoms, molecules, or ions.

Solvent
The substance present in a solution in the largest amount.

Solute
One or more substances present in a solution in amounts less than that of the solvent.

Dissolving
A term used to describe the process of solution formation when one or more solutes are dispersed in a solvent to form a homogeneous mixture.

Example 7.1

Identify the solvent and solute(s) in each of the following solutions:

a. A sample of natural gas contains 97% methane (CH_4), 1.5% ethane (C_2H_6), 1% carbon dioxide (CO_2), and 0.5% nitrogen (N_2).
b. The label on a bottle of Scotch whiskey says, among other things, 86 proof. This means the contents contain 43% ethyl alcohol (C_2H_5OH). Assume water to be the only other component.
c. The physiological saline solution used in hospitals contains 0.9 g NaCl for each 100 g water.

Solution

a. Methane is present in the largest amount; it is the solvent. All other components are solutes.
b. The whiskey is 43% alcohol and 57% water. Thus, water is the solvent and alcohol the solute. The whiskey actually contains small amounts of many other components, which impart flavor and color. These components are also solutes.
c. Water is the component present in the larger amount and is therefore the solvent.

TABLE 7.1 Solutions in various states

Solution	Solution state	Solvent	Solute
Salt water	Liquid	Water (liquid)	Sodium chloride (solid)
Alcoholic beverage	Liquid	Water (liquid)	Alcohol (liquid)
Carbonated water	Liquid	Water (liquid)	Carbon dioxide (gas)
Gold alloy (jewelry)	Solid	Gold (solid)	Copper (solid)
Gold amalgam	Solid	Gold (solid)	Mercury (liquid)
Hydrogen in palladium	Solid	Palladium (solid)	Hydrogen (gas)
Air	Gaseous	Nitrogen (gas)	Oxygen (gas)
Humid oxygen	Gaseous	Oxygen (gas)	Water (liquid)
Camphor in nitrogen	Gaseous	Nitrogen (gas)	Camphor (solid)

Identify the solvent and solute(s) in the following solutions:

Learning Check 7.1

a. White gold is a solid solution containing 58.3% gold, 17.0% copper, 7.7% zinc, and 17.0% nickel.
b. Clean, dry air contains about 78.1% nitrogen, 21.0% oxygen, 0.9% argon, and very small amounts of other gases.

➤ 7.2 Solubility

A few experiments with water, sugar, cooking oil, and rubbing alcohol (isopropyl alcohol) illustrate some important concepts associated with solution formation. Imagine you have three drinking glasses, each containing 100 mL (100 g) of pure water. You begin by adding small amounts of the other three substances to the glasses. The mixtures are stirred well, with results shown in ■ Figure 7.1a. The sugar and alcohol form homogeneous mixtures (solutions) with the water, while the oil forms a two-layer (heterogeneous) mixture. **Soluble substances** such as the sugar and alcohol dissolve completely in the solvent and form solutions. **Insoluble substances** do not dissolve in the solvent. the term **immiscible** is used to describe a liquid solute that does not dissolve in a liquid solvent.

The experiments are continued by adding more sugar, alcohol, and oil to the water samples. This ultimately leads to the situation shown in Figure 7.1b. Regardless of the amount of alcohol added, a solution forms. In fact, other experiments could be done that show that water and isopropyl alcohol are completely soluble in each other and will mix in any proportion. Sugar behaves differently. About 204 g can be dissolved in the 100 mL of water (at 20°C), but any additional sugar simply sinks to the bottom of the glass and remains undissolved. Thus, the sugar has a solubility in water of 204 g/100 g H_2O. The term **solubility** refers to the maximum amount of solute that can dissolve in a specific amount of solvent at a specific temperature. With the oil–water mixture, any additional oil simply floats on the surface of the water along with the oil added initially.

Soluble substance
A substance that dissolves to a significant extent in a solvent.

Insoluble substance
A substance that does not dissolve to a significant extent in a solvent.

Immiscible
A term used to describe liquids that are insoluble in each other.

Solubility
The maximum amount of solute that can be dissolved in a specific amount of solvent under specific conditions of temperature and pressure.

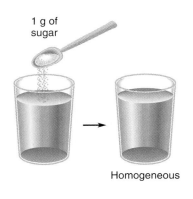

1 g of
sugar

Homogeneous

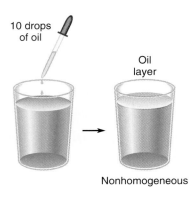

10 drops
of oil

Oil
layer

Nonhomogeneous

10 drops
of alcohol

Homogeneous

(a) Small amounts of substance added

300 g of
sugar

Nonhomogeneous

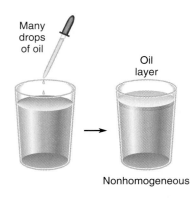

Many
drops
of oil

Oil
layer

Nonhomogeneous

Many drops
of alcohol

Homogeneous

(b) Larger amounts of substance added

➤ **FIGURE 7.1** The homogeneity of solutions and mixtures of sugar (limited solubility), cooking oil (immiscible), and alcohol (completely soluble).

Saturated solution
A solution that contains the maximum amount possible of dissolved solute in a stable situation under the prevailing conditions of temperature and pressure.

Supersaturated solution
An unstable solution that contains an amount of solute greater than the solute solubility under the prevailing conditions of temperature and pressure.

The solubilities of a number of solutes in water are given in ■ Table 7.2. The use of specific units, such as grams of solute per 100 g of water, makes it possible to compare solubilities precisely. However, such precision is often unnecessary, and when this is the case, the approximate terms defined in ■ Table 7.3 will be used.

A solution in which the maximum amount of solute has been dissolved in a quantity of solvent is called a **saturated solution.** The final sugar solution described in the earlier experiments (204 g in 100 g H_2O) was a saturated solution. Solutions in which the amount of solute dissolved is greater than the solute solubility are called **supersaturated solutions.** Supersaturated solutions are usually prepared by forming a nearly saturated solution at a high temperature and then cooling the solution to a lower temperature at which the solubility is lower. Such solutions are not stable. The addition of a small amount of solid solute (or even a dust particle) will usually cause the excess solute to crystallize out of solution until the solution becomes saturated (see ■ Figure 7.2). The temperature dependence of solute solubility is illustrated in ■ Figure 7.3.

Whereas the solubility of most liquids and solids in water increases with temperature, the solubility of most gases in water decreases as the temperature increases (see the SO_2 curve in Figure 7.3). This is easily demonstrated for gaseous CO_2 by opening both a cold and a warm carbonated beverage. The solubility of gaseous solutes is also influenced significantly by pressure; the effect on liquid or solid solutes is minimal. It has been found that the solubility of many gases is directly proportional to the pressure of the gas above the solution at constant temperature. Thus, if the gas pressure is doubled, the solubility doubles.

TABLE 7.2	Examples of solute solubilities in water (0°C)	

Name	Formula	Solubility (g solute/100 g H_2O)
Ammonium chloride	NH_4Cl	29.7
Ammonium nitrate	NH_4NO_3	118.3
Ammonium orthophosphate	$NH_4H_2PO_4$	22.7
Ammonium sulfate	$(NH_4)_2SO_4$	70.6
Calcium carbonate	$CaCO_3$	0.0012
Calcium chloride	$CaCl_2$	53.3
Calcium sulfate	$CaSO_4$	0.23
Potassium carbonate	K_2CO_3	101
Potassium chloride	KCl	29.2
Sodium bicarbonate	$NaHCO_3$	6.9
Sodium bromide	$NaBr$	111
Sodium carbonate	Na_2CO_3	7.1
Sodium chloride	$NaCl$	35.7
Sodium iodide	NaI	144.6
Ascorbic acid (vitamin C)	$C_6H_8O_6$	33
Ethyl alcohol	C_2H_5OH	∞[a]
Ethylene glycol (antifreeze)	$C_2H_4(OH)_2$	∞
Glycerin	$C_3H_5(OH)_3$	∞
Sucrose (table sugar)	$C_{12}H_{22}O_{11}$	179.2

[a]Soluble in all proportions.

TABLE 7.3	Approximate solubility terms

Solute solubility (g solute/100 g H_2O)	Solubility term
Less than 0.1	Insoluble
0.1–1	Slightly soluble
1–10	Soluble
Greater than 10	Very soluble

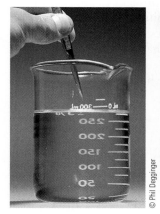

A supersaturated solution.

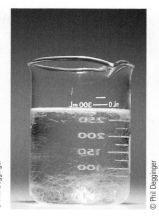

Seed crystal is added and induces rapid crystallization.

After excess solute is crystallized, the remaining solution is saturated.

➤ **FIGURE 7.2** Crystallization converts a supersaturated solution to a saturated solution.

➤ **FIGURE 7.3** The effect of temperature on solute solubility.

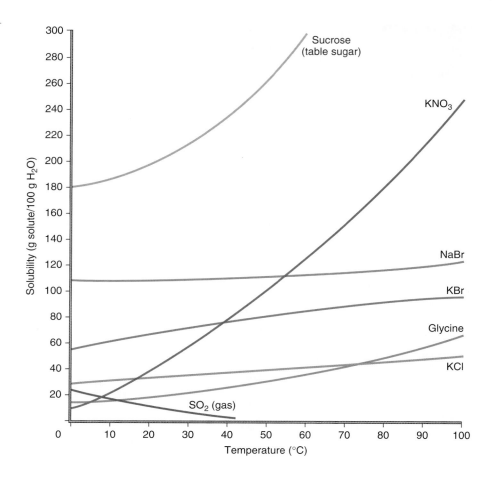

The pressure dependence of gas solubility provides the "sparkle" for carbonated beverages. The cold beverage is saturated with CO_2 and capped under pressure. When the bottle is opened, the pressure is relieved, and the gas, now less soluble, comes out of solution as fine bubbles. A similar effect sometimes takes place in the bloodstream of deep-sea divers. While submerged, they inhale air under pressure that causes nitrogen to be more soluble in the blood than it is under normal atmospheric pressure. If the diver is brought to the lower pressure on the surface too quickly, the excess dissolved nitrogen comes out of solution and forms bubbles in the blood and joints. The result, called decompression sickness or the bends, is painful and dangerous. The chances of getting the bends are decreased by breathing a mixture of oxygen and helium rather than air (oxygen and nitrogen) because helium is less soluble in the body fluids than nitrogen.

Example 7.2

A 260-g sample of sugar is added to 100 g of water at 70°C. The sugar dissolves completely. The resulting solution is allowed to cool slowly. At 30°C, crystals of sugar form rapidly and increase in size as the solution cools to 20°C. Refer to Figure 7.3 and describe the nature of the solution at 70°, 60°, 50°, 40°, 30°, and 20°C.

Solution

The solubility of sugar is greater than 260 g/100 g H_2O at all temperatures above 50°C. Therefore, the solution is unsaturated at 70°C and 60°C. At 50°C, the solubility is equal to the amount dissolved, so the solution is saturated. At 40°C, the solution contains more dissolved sugar than it should on the basis of solubility, and the solution is supersaturated. At 30°C, the excess sugar crystalizes from solution, and the resulting solution becomes saturated. From that point, excess sugar continues to crystallize from the solution, and the solution remains saturated to 20°C.

Learning Check 7.2

Refer to Figure 7.3. Saturated solutions of potassium nitrate (KNO₃) and sodium bromide (NaBr) are made at 80°C. Which solution contains more solute per 100 g of H₂O? The solutions are cooled to 50°C, and excess solute crystalizes out of each solution. Which solution at 50°C contains more solute per 100 g H₂O? Are the solutions saturated at 50°C?

➤ 7.3 The Solution Process

The how and why of solution formation are the topics of this section. How are solute particles removed from the bulk solute and uniformly distributed throughout the solvent? Why are some solutes very soluble whereas others are not?

Consider the formation of a saltwater solution. Earlier we pointed out (Section 4.2) that solid ionic compounds are collections of ions held together by attractions between the opposite charges. When an ionic compound dissolves, the orderly ionic arrangement is destroyed as the interionic attractions are overcome. Thus, the attractive forces between water molecules and ions must be stronger than the interionic attractions within the crystal.

The solution-forming process for an ionic solute is represented in ■ Figure 7.4. When the solid ionic crystal is placed in water, the polar water molecules become oriented so that the negative oxygen portion points toward positive sodium ions, and the positive hydrogen portion points toward negative chloride ions. As the polar water molecules begin to surround ions on the crystal surface, they tend to create a shielding effect that reduces the attraction

➤ **FIGURE 7.4** The dissolving of an ionic substance in water.

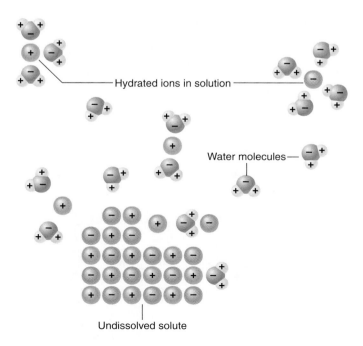

Hydrated ions in solution

Water molecules

Undissolved solute

Hydrated ion
An ion in solution that is surrounded by water molecules.

between the ion and the remainder of the crystal. As a result, the ion breaks away from the crystal surface and is surrounded by water molecules. Ions surrounded by water molecules in solution are called **hydrated ions.** As each ion leaves the surface, others are exposed to the water, and the crystal is picked apart ion by ion. Once in solution, the hydrated ions are uniformly distributed by stirring or by random collisions with other molecules or ions.

The random motion of solute ions in solution causes them to collide with one another, with solvent molecules, and occasionally with the surface of any undissolved solute. Ions undergoing such collisions occasionally stick to the solid surface and thus leave the solution. When the number of ions in solution is low, the chances for collision with the undissolved solute are low. However, as the number of ions in solution increases, so do the chances for such collisions, and more ions leave the solution and become attached once again to the solid. Eventually, the number of ions in solution reaches a level at which ions return to the undissolved solute at the same rate as other ions leave. At this point the solution is saturated, an equilibrium condition. Even though the processes of leaving and returning continue, no net changes in the number of ions in solution or the amount of undissolved solute can be detected as time passes, and an experimenter would observe that no more solid solute dissolves.

Supersaturated solutions form when there are no sites with which the excess solute particles can collide. The addition of such sites in the form of a seed crystal of solute causes the excess solute to crystallize from solution very quickly.

Polar but nonionic solutes such as sugar dissolve in water in much the same way as ionic solutes. The only difference is the attraction of polar water molecules to both poles of the solute molecules. The process is represented in ■ Figure 7.5.

This solution-forming process also explains the low solubility of some solutes. A solute will not dissolve in a solvent if (1) the forces between solute particles are too strong to be overcome by interactions with solvent particles or (2) the solvent particles are more strongly attracted to each other than to solute particles.

The cooking oil of the earlier experiment did not dissolve in water because the polar water molecules were attracted to each other more strongly than

➤ **FIGURE 7.5** The dissolving of a polar solute in water.

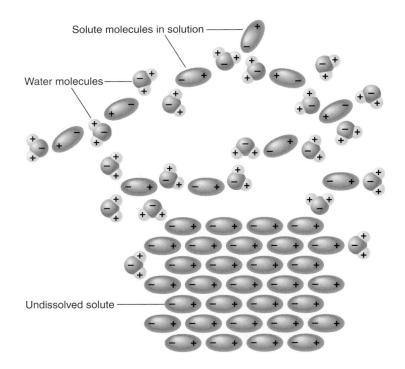

Solute molecules in solution

Water molecules

Undissolved solute

they were to the nonpolar oil molecules. The cooking oil would dissolve in a nonpolar solvent such as gasoline or carbon tetrachloride (CCl_4). In these solvents, the weak forces between oil molecules and solvent molecules are no stronger than the weak forces between nonpolar solvent molecules. A good rule of thumb is "like dissolves like." Thus, polar solvents will dissolve polar or ionic solutes, and nonpolar solvents will dissolve nonpolar or nonionic solutes.

These generalizations apply best to nonionic compounds. Some ionic compounds (such as $CaCO_3$ and $CaSO_4$) have very low solubilities in water (see Table 7.2). Both the attractive forces between ions and the attraction of polar solvent molecules for ions are electrical and depend on such characteristics as ionic charge and size. Changes in ionic compounds that increase the attractive forces between ions also increase the forces between ions and polar solvent molecules, but not always by the same amount. Thus, simple rules for quantitatively predicting the water solubility of ionic compounds are not available. (We can't easily predict how many grams will dissolve in a specific quantity of water.) However, ionic compounds are generally insoluble in nonpolar solvents; they usually follow the solubility guidelines given in ■ Table 7.4 when water is the solvent.

GOB
Chemistry·᠅·Now™
Go to GOB Now and click to view a simulation exploring the relationship between the composition of ionic compounds and their solubility.

GOB
Chemistry·᠅·Now™
Go to GOB Now and click to see how to use solubility guidelines to predict ionic compound solubilities.

OVER THE COUNTER • 7.1

ORAL REHYDRATION THERAPY

Dehydration, the excessive loss of water from the body, can result from a number of causes, including severe episodes of diarrhea or vomiting, and excessive sweating without proper fluid intake. The condition can be life threatening when it is severe enough and goes untreated. Dehydration is serious because the body loses electrolytes such as sodium, potassium, and chloride ions along with water. When the electrolyte balance in the body is upset, many organs, including the heart, cannot function properly. Small children are especially susceptible to dehydration caused by diarrhea because their small bodies do not have much of a fluid reserve, and it doesn't take much fluid loss to get their electrolytes out of balance.

Oral rehydration therapy (ORT) is a simple and effective way to treat or prevent dehydration and the accompanying electrolyte loss, especially if the dehydration is caused by diarrhea. Oral rehydration therapy was developed in the 1950s for use in developing countries where diarrhea-producing diseases like cholera, combined with unsanitary water and food, cause the death of an estimated 4 million children annually. The threat to children in developed countries is not nearly as great; an estimated 500 children die annually from diarrhea in the United States.

The materials used for ORT are simple mixtures of water, salts, and carbohydrates. These materials are regulated by the FDA as a medical food, and are available in most grocery and drug stores. A list of ingredients from the label of the liquid form of a popular ORT product includes water; the two carbohydrate sugars dextrose and fructose; citric acid; and the salts potassium citrate, sodium citrate, and sodium chloride. This material is available in an unflavored form, and in several flavors, such as berry and bubblegum, to appeal to children. The product also comes in the form of a powder that has to be dissolved in water before use. A third form consists of liquid sealed in plastic sleeves that can be frozen and then eaten like popular frozen treats.

The appropriate dosage for ORT products depends on the weight of the child; the directions included on the product labels should be read and followed. Many pediatricians suggest that parents of young children should include at least one bottle of oral rehydration fluid in the family medicine chest. The low cost of the commercially available materials makes it possible for most families to follow this prudent advice.

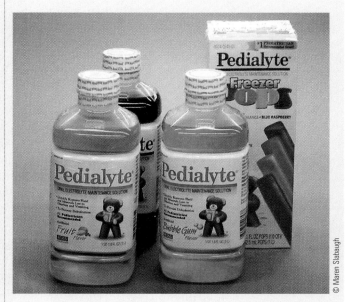

Oral rehydration therapy products are available in numerous forms and flavors.

TABLE 7.4 General solubilities of ionic compounds in water

Compounds	Solubility	Exceptions
Group IA (Na^+, K^+, etc.) and NH_4^+	Soluble	
Nitrates (NO_3^-)	Soluble	
Acetates ($C_2H_3O_2^-$)	Soluble	
Chlorides (Cl^-)	Soluble	Chlorides of Ag^+, Pb^{2+}, Hg^+ (Hg_2^{2+})
Sulfates (SO_4^{2-})	Soluble	Sulfates of Ba^{2+}, Sr^{2+}, Pb^{2+}, Hg^+ (Hg_2^{2+})
Carbonates (CO_3^{2-})	Insoluble[a]	Carbonates of group IA and NH_4^+
Phosphates (PO_4^{3-})	Insoluble[a]	Phosphates of group IA and NH_4^+

[a]Many hydrogen carbonates (HCO_3^-) and phosphates (HPO_4^{2-}, $H_2PO_4^-$) are soluble.

Example 7.3

Predict the solubility of the following solutes in the solvent indicated:

a. ammonia gas (NH_3) in water
b. oxygen gas (O_2) in water
c. $Ca(NO_3)_2$ in water
d. $Mg_3(PO_4)_2$ in water
e. paraffin wax (nonpolar) in CCl_4

Solution

a. Soluble: NH_3 is polar—like dissolves like. The actual solubility at 20°C is 51.8 g/100 g H_2O.
b. Insoluble: O_2 is nonpolar. The actual solubility at 20°C is 4.3×10^{-3} g/100 g H_2O.
c. Soluble: nitrates are soluble in water (Table 7.4).
d. Insoluble: phosphates are insoluble except those of group IA(1) and NH_4^+ (Table 7.4).
e. Soluble: CCl_4 is nonpolar—like dissolves like.

Learning Check 7.3

a. Refer to Table 7.4 and decide which of the following salts you would use as a solute if you wanted to prepare a solution that contained as many barium ions (Ba^{2+}) as possible: $BaSO_4$, $Ba(NO_3)_2$, and $BaCO_3$. Explain your answer.
b. A common sight on the evening TV news since the late 1960s is sea birds soaked with crude oil being cleaned by concerned people. Which of the following do you think is used to clean the oil from the birds? Explain your answer. Light mineral oil, purified water, seawater, or gasoline.

Be careful not to confuse solute solubility with the rate at which a solute dissolves. Under some conditions, even a very soluble solute will dissolve slowly—for example, a lump of rock candy (sugar) dissolves much more slowly than an equal weight of granulated sugar. The dissolving rate can be increased in a number of ways (see ■ Figure 7.6):

1. Crushing or grinding the solute—small particles provide more surface area for solvent attack and dissolve more rapidly than larger particles.

2. Heating the solvent—solvent molecules move faster and have more frequent collisions with solute at higher temperatures.
3. Stirring or agitating the solution—stirring removes locally saturated solution from the vicinity of the solute and allows unsaturated solvent to take its place.

Heat is usually absorbed or released when a solute dissolves in a solvent. When heat is absorbed, the process is endothermic, and the solution becomes cooler. The fact that heat absorption leads to cooling might sound strange, and it needs to be explained. The heat is absorbed by the interacting solvent and solute molecules, and it is removed from the solvent molecules that are not involved in the actual attack on the solute. Since most of the solvent is in the latter category, the entire solution becomes cooled.

When heat is released, the process is exothermic, and the solution temperature increases, this time because the heat released by the interacting molecules is absorbed by the uninvolved solvent. This behavior is the more common one. These processes can be represented by equations:

$$\text{Endothermic:} \quad \text{Solute} + \text{solvent} + \text{heat} \rightarrow \text{solution} \quad (7.1)$$

$$\text{Exothermic:} \quad \text{Solute} + \text{solvent} \rightarrow \text{solution} + \text{heat} \quad (7.2)$$

An endothermic solution process is the basis for commercially available instant cold packs. Water is sealed in a thin plastic bag and placed inside a larger, stronger bag together with a quantity of solid solute (NH_4Cl or NH_4NO_3). When the inner bag is broken by squeezing, the solid dissolves in the water, heat is absorbed, and the mixture becomes quite cold (see ■ Figure 7.7).

➤ 7.4 Solution Concentrations

In Chapter 5, quantitative calculations were done using equations for reactions such as the following:

$$CH_4(g) + 2O_2(g) \rightarrow CO_2(g) + 2H_2O(g) \quad (7.3)$$

➤ **FIGURE 7.6** Heat and agitation increase the rate at which solutes dissolve. Will sugar dissolve faster in hot tea or iced tea?

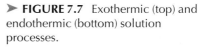

➤ **FIGURE 7.7** Exothermic (top) and endothermic (bottom) solution processes.

Solid NaOH not yet added to water.

Solution becomes hot when NaOH dissolves in water.

Solid NH_4NO_3 not yet added to water.

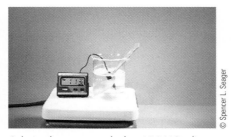

Solution becomes cool when NH_4NO_3 dissolves in water.

The reaction represented by this equation, and most of the reactions we have discussed to this point in the book, involve pure substances as reactants and products. However, many of the reactions done in laboratories, and most of those that go on in our bodies, take place between substances dissolved in a solvent to form solutions. In our bodies, the solvent is almost always water. A double-replacement reaction of this type done in laboratories is represented by the following equation:

$$2AgNO_3(aq) + Na_2CO_3(aq) \rightarrow Ag_2CO_3(s) + 2NaNO_3(aq) \qquad (7.4)$$

We remember from Chapter 5 that the coefficients in equations such as Equation 7.3 allow the relative number of moles of pure reactants and products involved in the reaction to be determined. These relationships coupled with the mole definition in terms of masses then yield factors that can be used to solve stoichiometric problems involving the reactants and products. Similar calculations can be done for reactions that take place between the solutes of solutions if the amount of solute contained in a specific quantity of the reacting solutions is known. Such relationships are known as solution **concentrations**. Solution concentrations may be expressed in a variety of units, but only two, molarity and percentage, will be discussed at this time.

The **molarity (M)** of a solution expresses the number of moles of solute contained in exactly 1 L of the solution:

$$M = \frac{\text{moles of solute}}{\text{liters of solution}} \qquad (7.5)$$

It is important to note that even though a concentration in molarity expresses the number of moles contained in 1 L of solution, the molarity of solutions that have total volumes different from 1 L can be calculated using Equation 7.5. We simply determine the number of moles of solute contained in a specified volume of solution, then express that volume in liters before substituting the values into Equation 7.5.

Concentration
The relationship between the amount of solute and the specific amount of solution in which it is contained.

Molarity (M)
A solution concentration expressed in terms of the number of moles of solute contained in a liter of solution.

Example 7.4

Express the concentration of each of the following solutions in terms of molarity:

a. 2.00 L of solution contains 1.50 mol of solute.
b. 150 mL of solution contains 0.210 mol of solute.
c. 315 mL of solution contains 10.3 g of isopropyl alcohol, C_3H_7OH.

Solution

a. Because the data are given in terms of moles of solute and liters of solution, the data may be substituted directly into Equation 7.5:

$$M = \frac{1.50 \text{ mol solute}}{2.00 \text{ L solution}} = 0.750 \frac{\text{mol solute}}{\text{L solution}}$$

The solution is 0.750 molar, or 0.750 M.

b. In this problem, the number of moles of solute is given, but the volume of solution is given in milliliters rather than liters. The volume must first be converted into liters, then the data may be substituted into Equation 7.5:

$$(150 \text{ mL solution})\left(\frac{1 \text{ L}}{1000 \text{ mL}}\right) = 0.150 \text{ L solution}$$

$$M = \frac{0.210 \text{ mol solute}}{0.150 \text{ L solution}} = 1.40 \frac{\text{mol solute}}{\text{L solution}}$$

The solution is 1.40 molar, or 1.40 M.

c. Before Equation 7.5 can be used, the number of moles of solute (isopropyl alcohol) must be determined. This is done as follows, where the factor

$$\frac{1 \text{ mol alcohol}}{60.1 \text{ g alcohol}}$$

comes from the calculated molecular weight of 60.1 u for isopropyl alcohol:

$$(10.3 \text{ g alcohol})\left(\frac{1 \text{ mol alcohol}}{60.1 \text{ g alcohol}}\right) = 0.171 \text{ mol alcohol}$$

Next, this number of moles of solute and the solution volume expressed in liters are substituted into Equation 7.5:

$$M = \frac{0.171 \text{ mol alcohol}}{0.315 \text{ L solution}} = 0.543 \frac{\text{mol alcohol}}{\text{L solution}}$$

The solution is 0.543 molar, or 0.543 M.

Express the concentrations of each of the following solutions in terms of molarity:

a. 2.50 L of solution contains 1.25 mol of solute.
b. 225 mL of solution contains 0.486 mol of solute.
c. 100 mL of solution contains 2.60 g of NaCl solute.

Learning Check 7.4

Sometimes a detailed knowledge of the actual stoichiometry of a process involving solutions is not needed, but some information about the solution concentrations would be useful. When this is true, solution concentrations are often expressed as percentages. In general, a concentration in **percent** gives the number of parts of solute contained in 100 parts of solution. You saw in Section 1.10 that it is convenient to use a formula for percentage calculations because we seldom work with exactly 100 units of anything. The general formula used is:

Percent
A solution concentration that expresses the amount of solute in 100 parts of solution.

$$\% = \frac{\text{part}}{\text{total}} \times 100$$

Three different percent concentrations for solutions are used. A **weight/weight percent** (abbreviated w/w) is the mass of solute contained in 100 mass units of solution. Thus, a 12.0% (w/w) sugar solution contains 12.0 grams of sugar in each 100 g of solution. In terms of this concentration, the general formula for percent calculations becomes

Weight/weight percent
A concentration that expresses the mass of solute contained in 100 mass units of solution.

$$\%(\text{w/w}) = \frac{\text{solute mass}}{\text{solution mass}} \times 100 \qquad (7.6)$$

Any mass units may be used, but the mass of solute and solution must be expressed in the same units.

A more commonly used percent concentration is **weight/volume percent** (abbreviated w/v), which is the grams of solute contained in 100 mL of solution. In these units, a 12.0% (w/v) sugar solution would contain 12.0 g of sugar in each 100 mL of solution. This percent concentration is normally used when the solute is a solid and the solvent and resulting solutions are liquids. The general formula for the calculation of percent concentrations in these units is

Weight/volume percent
A concentration that expresses the grams of solute contained in 100 mL of solution.

$$\%(\text{w/v}) = \frac{\text{grams of solute}}{\text{milliliters of solution}} \times 100 \qquad (7.7)$$

In weight/volume percent calculations, the solute amount is always given in grams, and the solution volume is always given in milliliters.

Volume/volume percent
A concentration that expresses the volume of solute contained in 100 volumes of solution.

A percent concentration that is useful when the solute and solvent are either both liquids or both gases is **volume/volume percent** (abbreviated v/v). Concentrations given in these units express the number of volumes of solute found in 100 volumes of solution. For these units, the general percentage equation becomes:

$$\%(v/v) = \frac{\text{solute volume}}{\text{solution volume}} \times 100 \qquad (7.8)$$

Any volume units may be used, but they must be the same for both the solute and the solution.

Example 7.5

a. A solution contains 100 g of water and 1.20 g of solute. What is the %(w/w) concentration?
b. A solution is made by mixing 90.0 mL of alcohol with enough water to give 250 mL of solution. What is the %(v/v) concentration of alcohol in the solution?
c. A 150-mL sample of saltwater is evaporated to dryness. A residue of salt weighing 27.9 g is left behind. Calculate the %(w/v) of the original saltwater.

Solution

a. Equation 7.6 is used:

$$\%(w/w) = \frac{\text{solute mass}}{\text{solution mass}} \times 100 = \frac{1.20 \text{ g}}{101.2 \text{ g}} \times 100 = 1.19\% \ (w/w)$$

Note that the solution mass is the sum of the solvent (water) and solute masses.

b. Equation 7.8 is used:

$$\%(v/v) = \frac{\text{solute volume}}{\text{solution volume}} \times 100 = \frac{90 \text{ mL}}{250 \text{ mL}} \times 100 = 36.0\% \ (v/v)$$

c. Equation 7.7 is used after checking to make certain the amount of solute is given in grams and the solution volume is in mL:

$$\%(w/v) = \frac{\text{grams of solute}}{\text{milliliters of solution}} \times 100$$

$$= \frac{27.9 \text{ g}}{150 \text{ mL}} \times 100$$

$$= 18.6\% \ (w/v)$$

Learning Check 7.5

a. A solution is made by dissolving 0.900 g of salt in 100.0 mL of water. Assume that each milliliter of water weighs 1.00 g and that the final solution volume is 100.0 mL. Calculate the %(w/w) and %(w/v) for the solution using the assumptions as necessary.
b. An alcoholic beverage is labeled 90 proof, which means the alcohol concentration is 45% (v/v). How many milliliters of pure alcohol would be present in 1 oz (30 mL) of the beverage?

CHEMISTRY AND YOUR HEALTH • 7.1

WATER, WATER EVERYWHERE, BUT HOW SAFE IS IT TO DRINK?

Because of its polarity and ability to hydrogen bond, water is a good solvent for a wide variety of substances, including ionic compounds and covalent compounds, especially those containing highly electronegative elements such as nitrogen and oxygen. It is also a medium in which many microorganisms can live and multiply. These characteristics lead to a great deal of concern and a great number of opinions about the safety of drinking water. Some information provided by those marketing bottled or purified water or water purification systems creates confusion about the safety of public water supplies. On the other hand, the information provided by those administering public water supplies sometimes contains technical words and abbreviations that also add to confusion about water safety.

Scientific evidence clearly indicates that one of the great achievements of the 20th century was the development of safe drinking water supplies in the United States, Canada, and most other industrialized nations. The safety of the water contributes to the increased longevity enjoyed by residents of these countries. In spite of these accomplishments, individuals still need to be vigilant concerning their water supplies. Local water suppliers are required by law to notify water users of any unsafe conditions, and to provide an annual report of the sources and quality of the water they supply. This report must include a list of contaminant levels found during required testing. Municipal water is tested annually for microorganisms, various organic and inorganic chemicals, and radioactive substances.

A number of potential drinking water contaminants get a lot of attention in advertisements for products such as purification systems. These include lead, Pb, arsenic, As, and byproducts of chlorination. Lead is especially dangerous for children and pregnant women. Lead can get into water from lead-containing materials such as solder, pipes, and faucets. Lead tests are relatively inexpensive and sometimes provided free by local water suppliers. Arsenic can leach into water from the ground and from some industrial wastes. Constant exposure to low arsenic levels has been related to increased risks of bladder, lung, and skin cancer. Most water supplies in the United States are chlorinated to kill harmful microorganisms. During the process of killing the microorganisms, chlorine is converted into chlorine-containing organic byproducts. An increased risk of cancer has been associated with exposure to high levels of some of these byproducts. The effects of exposure to the very low levels found in chlorinated drinking water has been studied for many years with only inconclusive results.

A number of options exist to deal with drinking water thought to contain harmful amounts of these or other contaminants. Water treatment systems based on ion exchange, reverse osmosis, or distillation are available. The purchase price is not necessarily the only cost for most of these systems. Some require filters or other parts to be replaced on a regular schedule. The true annual cost to operate such systems varies over a wide range and should be carefully investigated before a system is purchased. Another option is to purchase bottled water for drinking. However, the sources of bottled water should be carefully checked, or the water itself should be tested. It has been found that some bottled water is just tap water in a fancy container.

➤ 7.5 Solution Preparation

Solutions are usually prepared by mixing together proper amounts of solute and solvent or by diluting a concentrated solution with solvent to produce a solution of lower concentration. In the first method, the solute is measured out and placed in a container, and the correct amount of solvent is added. When the concentration is based on solution volume [%(v/v), %(w/v), and M], a volumetric flask or other container is used that, when filled to a specific mark, holds an accurately known volume (see ■ Figure 7.8). When the concentration is based on solvent mass [%(w/w)], the correct mass of solvent is added. This mass is usually converted to a volume by using the density, so that a volume of solvent can be measured rather than an amount weighed on a balance.

Example 7.6

Describe how you would prepare the following solutions from pure solute and water:

a. 1.00 L of 1.50 M $CoCl_2$ solution
b. 250 mL of 0.900% (w/v) NaCl solution
c. 500 mL of 8.00% (v/v) methyl alcohol solution

GOB
Chemistry⚛Now™

Go to GOB Now and click to examine what mass of solute to use to create a solution of a specific molarity.

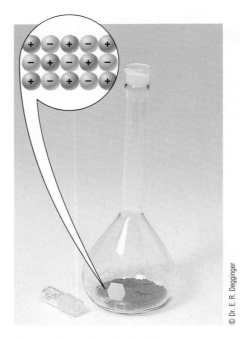 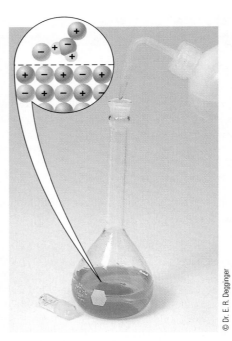 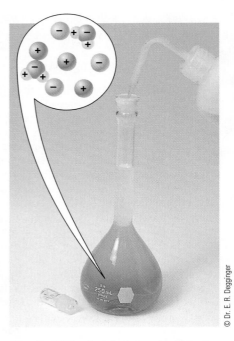

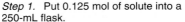

© Dr. E. R. Degginger

Step 1. Put 0.125 mol of solute into a 250-mL flask.

Step 2. Add some water and dissolve the solute.

Step 3. Fill the flask to the mark with water. Mix thoroughly.

➤ **FIGURE 7.8** Preparation of a 0.500 M solution. Use the data given above and show by a calculation that the resulting solution is 0.500 M.

Solution

a. According to Equation 7.5, molarity is the moles of solute per (divided by) the liters of solution containing the solute:

$$M = \frac{\text{moles of solute}}{\text{liters of solution}}$$

This equation can be rearranged to calculate the moles of solute required:

$$(M)(\text{liters of solution}) = \text{moles of solute}$$

By substituting the molarity and volume of the desired solution, the required number of moles of solute is determined:

$$\left(1.50 \frac{\text{mol}}{\text{L}}\right)(1.00 \text{ L}) = 1.50 \text{ mol}$$

The formula weight of $CoCl_2$ is 129.9 u. The mass of solute is calculated:

$$(1.50 \text{ mol } CoCl_2)\left(\frac{129.9 \text{ g } CoCl_2}{1 \text{ mol } CoCl_2}\right) = 195 \text{ g } CoCl_2$$

The solution is prepared by weighing out 195 g $CoCl_2$, putting it into a 1-L volumetric flask, and adding enough water to fill the flask to the mark (see Figure 7.8). It is a good practice to let the solute dissolve completely in part of the solvent before filling the flask to the mark.

b. Similarly, rearrange Equation 7.7, and use the result to calculate the amount of solute needed:

$$\%(\text{w/v}) = \frac{\text{grams of solute}}{\text{milliliters of solution}} \times 100$$

Therefore,

$$\frac{(\%)(\text{milliliters of solution})}{100} = \text{grams of solute}$$

Substitute percentage and solution volume:

$$\frac{(0.900\%)(250\ \text{mL})}{100} = 2.25\ \text{g NaCl}$$

Thus, the solution is prepared by putting 2.25 g NaCl into a 250-mL volumetric flask and adding water to the mark.

c. Rearrange Equation 7.7, and use the result to calculate the volume of solute needed:

$$\%(\text{v/v}) = \frac{\text{solute volume}}{\text{solution volume}} \times 100$$

Therefore,

$$\frac{(\%)(\text{solution volume})}{100} = \text{solute volume}$$

Substitute percentage and solution volume:

$$\frac{(8.00\%)(500\ \text{mL})}{100} = 40.0\ \text{mL}$$

Thus, 40.0 mL methyl alcohol is put into a 500-mL volumetric flask, and water is added up to the mark.

Describe how you would prepare the following, using pure solute and water:

Learning Check 7.6

a. 500 mL of 1.00 M $MgCl_2$ solution
b. 100 mL of 12.0% (w/v) $MgCl_2$ solution
c. 1.00 L of 20.0% (v/v) ethylene glycol solution

Solutions are often prepared by diluting a more concentrated solution with solvent (usually water) to produce a solution of lower concentration. Suppose, for example, that you want to prepare 250 mL of 0.100 M NaCl solution using a 2.00 M NaCl solution as the source of NaCl. First, calculate the number of moles of NaCl that would be contained in 250 mL of 0.100 M solution. Remember Equation 7.5,

$$M = \frac{\text{moles of solute}}{\text{liters of solution}}$$

which can be rearranged to

$$(M)(\text{liters of solution}) = \text{moles of solute}$$

So, the number of moles of NaCl contained in the desired solution is

$$\left(0.100\ \frac{\text{mol}}{\text{L}}\right)(0.250\ \text{L}) = 0.0250\ \text{mol}$$

These moles of NaCl must be obtained from the 2.00 M NaCl solution. The volume of this solution that contains the desired number of moles can be obtained by rearranging Equation 7.5 a different way:

$$\text{liters of solution} = \frac{\text{moles of solute}}{M}$$

GOB Chemistry Now™

Go to GOB Now and click to engage in a tutorial examining what volume of a stock solution is needed to prepare a solution of specific molarity by dilution.

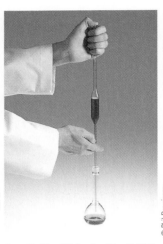

Step 1. 20.0 mL of 0.200 M $K_2Cr_2O_7$ solution is withdrawn from a beaker using a pipette.

Step 2. The 20.0 mL of the 0.200 M $K_2Cr_2O_7$ solution is put into a 100-mL flask.

Step 3. Water is added (while the solution is swirled) to fill the flask to the mark.

Step 4. The new solution is transferred to a bottle and labeled.

➤ **FIGURE 7.9** Preparation of a 0.0400 M $K_2Cr_2O_7$ solution by dilution of a 0.200 M $K_2Cr_2O_7$ solution. Use the data and show by a calculation that the new solution is 0.0400 M.

or

$$\text{liters of solution} = \frac{0.0250 \text{ mol}}{2.00 \text{ mol/L}} = 0.0125 \text{ L} = 12.5 \text{ mL}$$

Thus, the solution is prepared by putting 12.5 mL of 2.00 M NaCl solution into a 250-mL volumetric flask and adding water up to the mark. ■ Figure 7.9 will help you understand this process.

The same result can be obtained by using a simplified calculation. Note that the product of concentration in molarity (M) and solution volume in liters will give the number of moles of solute in a sample of solution. In a dilution such as the one done above, the number of moles of solute taken from the concentrated solution and diluted with water is the same as the number of moles of solute in the resulting more dilute solution (see Figure 7.9). Thus, the following can be written:

$$M_c V_c = \begin{array}{c} \text{solute moles in} \\ \text{concentrated solution} \end{array} = \begin{array}{c} \text{solute moles in} \\ \text{dilute solution} \end{array} = M_d V_d$$

or

$$(C_c)(V_c) = (C_d)(V_d) \tag{7.9}$$

where the subscripts c and d refer to the more concentrated and dilute solutions, respectively, and C is used to represent any appropriate concentration. An advantage of using this equation is that any volume units may be used, as long as the same one is used for both V_c and V_d. Also, Equation 7.9 is true for any solution concentration based on volume. This means that M, %(v/v), or %(w/v) concentrations can be used just as long as C_c and C_d are expressed in the same units.

Example 7.7

Use Equation 7.9 and describe how to prepare 250 mL of 0.100 M NaCl solution using a 2.00 M NaCl solution as the source of NaCl.

Solution

Equation 7.9 can be used to determine the volume of 2.00 M NaCl needed. The relationships between the terms in Equation 7.9 and the quantities of the problem are C_c = 2.00 M, V_c = volume of 2.00 M NaCl needed, C_d = 0.100 M, and V_d = 250 mL. Substitution gives

$$(2.00 \text{ M})(V_c) = (0.100 \text{ M})(250 \text{ mL})$$

or

$$V_c = \frac{(0.100 \text{ M})(250 \text{ mL})}{(2.00 \text{ M})} = 12.5 \text{ mL}$$

Thus, we quickly find the volume of 2.00 M NaCl needs to be 12.5 mL, the same result found earlier. This volume of solution is put into a 250-mL flask, and water is added to the mark as before.

Use Equation 7.9 and describe how to prepare 500 mL of 0.250 M NaOH solution from 6.00 M NaOH solution.

Learning Check 7.7

➤ 7.6 Solution Stoichiometry

The stoichiometry calculations of solution reactions can be done by extending Figure 5.12 to include solutions and using the molarity of solutions as a source of factors. This useful extension of Figure 5.12 is shown in ■ Figure 7.10. In the examples that follow, note that the primary difference between the calculations of Chapter 5 and this chapter is that solutions are the source of the reactants, and the molarity of the solutions is used to calculate the number of moles of reactants and products involved in the reactions.

GOB
Chemistry.⚛.Now™

Go to GOB Now and click to learn how to calculate the theoretical yield of a reaction performed by mixing two solutions of known concentration.

GOB
Chemistry.⚛.Now™

Go to GOB Now and click to engage in a tutorial examining the volume of a solution needed to produce a desired mass of product.

Example 7.8

NaOH and HCl solutions react as follows:

$$\text{NaOH(aq)} + \text{HCl(aq)} \rightarrow \text{NaCl(aq)} + \text{H}_2\text{O}(\ell)$$

a. What volume of 0.250 M NaOH solution contains exactly 0.110 mol NaOH?

b. What volume of 0.200 M NaOH solution is required to exactly react with 0.150 mol HCl?

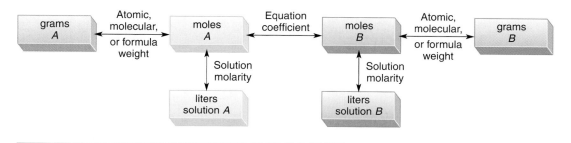

➤ **FIGURE 7.10** Relationships for problem solving based on balanced equations and solution molarities.

c. What volume of 0.185 M NaOH solution is required to exactly react with 25.0 mL of 0.225 M HCl solution?

Solution

a. According to Figure 7.10, this problem has the pattern mol A → liters solution A, and the pathway is mol NaOH → liters NaOH solution. The necessary factor, according to Figure 7.10, comes from the molarity of the NaOH solution and will be either

$$\frac{0.250 \text{ mol NaOH}}{1 \text{ L NaOH solution}} \quad \text{or} \quad \frac{1 \text{ L NaOH solution}}{0.250 \text{ mol NaOH}}$$

The known quantity is 0.110 mol NaOH, and the unit of the unknown quantity is L NaOH solution. The steps in the factor-unit method are as follows:

Step 1. 0.110 mol NaOH

Step 2. 0.110 mol NaOH = L NaOH solution

Step 3. 0.110 mol NaOH × $\dfrac{1 \text{ L NaOH solution}}{0.250 \text{ mol NaOH}}$ = L NaOH solution

Step 4. $(0.110)\left(\dfrac{1 \text{ L NaOH solution}}{0.250}\right)$ = 0.440 L NaOH solution

The volume can also be expressed as 440 mL NaOH solution.

b. According to Figure 7.10, the pattern for this problem is mol A → liters solution B, and the required pathway is mol HCl → mol NaOH → liters NaOH solution. The factor for the first step comes from the coefficients of the balanced equation and will be either

$$\frac{1 \text{ mol HCl}}{1 \text{ mol NaOH}} \quad \text{or} \quad \frac{1 \text{ mol NaOH}}{1 \text{ mol HCl}}$$

The factor for the second step comes from the molarity of the NaOH solution and will be either

$$\frac{0.200 \text{ mol NaOH}}{1 \text{ L NaOH solution}} \quad \text{or} \quad \frac{1 \text{ L NaOH solution}}{0.200 \text{ mol NaOH}}$$

The problem will be solved in combined form. The steps in the factor-unit method are as follows:

Step 1. 0.150 mol HCl

Step 2. 0.150 mol HCl = L NaOH solution

Step 3. 0.150 mol HCl × $\dfrac{1 \text{ mol NaOH}}{1 \text{ mol HCl}}$ × $\dfrac{1 \text{ L NaOH solution}}{0.200 \text{ mol NaOH}}$ = L NaOH solution

Step 4. $(0.150)\left(\dfrac{1}{1}\right)\left(\dfrac{1 \text{ L NaOH solution}}{0.200}\right)$ = 0.750 L NaOH solution

The answer can also be expressed as 750 mL.

c. The pattern is liters solution A → liters solution B, and the required pathway is liters HCl solution → mol HCl → mol NaOH → liters NaOH solution. The factors come from the molarities of the two solutions and the coefficients of the balanced equation. The combined steps in the factor-unit method are given on the next page. Note that the known quantity, 25.0 mL HCl solution, is changed to liters so that units will cancel properly.

Step 1. 0.0250 L HCl solution

Step 2. 0.0250 L HCl solution   = L NaOH solution

Step 3. $0.0250 \;\cancel{L}\; \cancel{HCl}\; \cancel{solution} \times \dfrac{0.225 \;\cancel{mol}\; \cancel{HCl}}{1 \;\cancel{L}\; \cancel{HCl}\; \cancel{solution}}$

$\times \dfrac{1 \;\cancel{mol}\; \cancel{NaOH}}{1 \;\cancel{mol}\; \cancel{HCl}} \times \dfrac{1 \text{ L NaOH solution}}{0.185 \;\cancel{mol}\; \cancel{NaOH}} = \text{L NaOH solution}$

Step 4. $(0.0250)\left(\dfrac{0.225}{1}\right)\left(\dfrac{1}{1}\right)\left(\dfrac{1 \text{ L NaOH solution}}{0.185}\right) = 0.0304 \text{ L NaOH solution}$

The answer can also be expressed as 30.4 mL NaOH solution.

NaOH and H_2SO_4 solutions react as follows:

$$2NaOH(aq) + H_2SO_4(aq) \rightarrow Na_2SO_4(aq) + 2H_2O(\ell)$$

a. What volume of 0.200 M H_2SO_4 solution contains exactly 0.150 mol H_2SO_4?

b. What volume of 0.255 M NaOH solution is required to exactly react with 0.125 mol H_2SO_4?

c. What volume of 0.250 M NaOH solution is required to exactly react with 25.0 mL of 0.120 M H_2SO_4 solution?

➤ 7.7 Solution Properties

The experiment represented in ■ Figure 7.11 reveals an interesting property of some solutions. As shown by the figure, the circuit is completed and the

A solution of a strong electrolyte (salt) conducts electricity well.

© Dr. E. R. Degginger

A solution of a weak electrolyte (acetic acid) conducts electricity poorly.

© Dr. E. R. Degginger

A solution of a nonelectrolyte (sugar) does not conduct electricity.

© Dr. E. R. Degginger

➤ **FIGURE 7.11** Electrical conductivity of solutions.

Electrolyte
A solute that when dissolved in water forms a solution that conducts electricity.

Nonelectrolyte
A solute that when dissolved in water forms a solution that does not conduct electricity.

Colligative property
A solution property that depends only on the concentration of solute particles in solution.

light goes on only when the solution conducts electricity. Experiments of this type demonstrate that conductive solutions are formed when soluble ionic materials (Section 4.3), such as sodium chloride, or highly polar covalent materials (Section 4.9), such as hydrogen chloride, dissolve in water. Solutes that form conductive water solutions are called **electrolytes,** whereas solutes that form nonconductive solutions are called **nonelectrolytes.** Conductive solutions result when the dissolved solute dissociates, or breaks apart, to form ions, as shown in the following equations representing the dissociation into ions of polar covalent hydrogen chloride and ionic calcium nitrate:

$$HCl(aq) \rightarrow H^+(aq) + Cl^-(aq) \tag{7.10}$$

$$Ca(NO_3)_2(aq) \rightarrow Ca^{2+}(aq) + 2NO_3^-(aq) \tag{7.11}$$

Some electrolytes, such as HCl and $Ca(NO_3)_2$, dissociate essentially completely in solution and are called *strong electrolytes*. They form strongly conducting solutions. Other electrolytes, such as acetic acid (the acid found in vinegar), dissociate only slightly and form weakly conductive solutions. These solutes are classified as *weak electrolytes*. The solutes in nonconductive solutions dissolve in the solvent but remain in the form of uncharged molecules.

Besides electrical conductivity (or the lack of it), all solutions have properties that depend only on the concentration of solute particles present and not on the actual identity of the solute. Thus, these properties, called **colligative properties,** would be identical for water solutions containing 1 mol of sugar or 1 mol of alcohol per liter.

Three closely related colligative properties are vapor pressure, boiling point, and freezing point. Experiments demonstrate that the vapor pressure of water (solvent) above a solution is lower than the vapor pressure of pure water (see ■ Figure 7.12). This behavior causes the boiling point of solutions to be higher than the boiling point of the pure solvent used in the solutions and the freezing point to be lower (see ■ Table 7.5).

➤ **FIGURE 7.12** The vapor pressure of water above a solution is lower than the vapor pressure of pure water. Remember that pressure is a force per unit area. Think of this force as "pushing" water molecules in the vapor state, and explain the process shown in the photos.

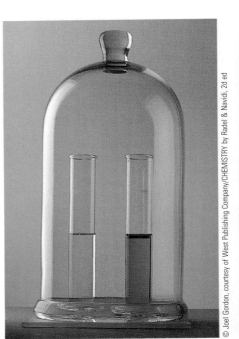

Initially, equal volumes of pure water and copper sulfate solution (blue) are placed under a glass dome that prevents water vapor from escaping to the outside.

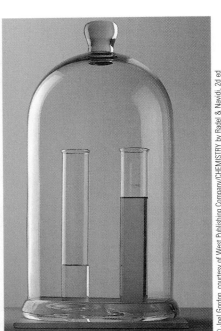

After some time has passed, the volume of pure water has decreased, while that of copper sulfate solution has increased.

TABLE 7.5	A comparison of colligative properties of pure solvent and solutions	
Property	Pure solvent	Solution
Vapor pressure	Normal	Lower than pure solvent
Boiling point	Normal	Higher than pure solvent
Freezing point	Normal	Lower than pure solvent

The boiling and freezing point differences between pure solvent and solutions can be calculated by using an equation of the general form

$$\Delta t = nkm \qquad (7.12)$$

In this equation, m is the solution concentration expressed as a molality, a unit we have not yet discussed. For very dilute solutions, the molality m and molarity M are essentially equal, and M can be used in Equation 7.12 instead of m. This approximation will be used for calculating colligative properties of solutions in this book. The symbol Δt is the boiling point or freezing point difference between pure solvent and solution. The specific equations used to calculate Δt for boiling and freezing points are

$$\Delta t_b = nK_b M \qquad (7.13)$$

$$\Delta t_f = nK_f M \qquad (7.14)$$

The subscripts b and f refer to boiling or freezing, and K_b and K_f are constants characteristic of the solvent used in the solution (remember, to this point we have focused on water as the solvent). Values for K_b and K_f are given in ■ Table 7.6 for a number of solvents. In Equations 7.13 and 7.14, M is the molarity of solute in solution, and n is the number of moles of solute particles put into solution when 1 mol of solute dissolves.

GOB Chemistry Now™
Go to GOB Now and click to learn how to calculate the boiling point of a solution.

GOB Chemistry Now™
Go to GOB Now and click to learn how to calculate the freezing point of a solution.

Example 7.9

Calculate the boiling and freezing points of the following solutions:

a. 171 g of sugar ($C_{12}H_{22}O_{11}$) is dissolved in enough water to give 1.00 L of solution.
b. 13.4 g of NH_4Cl is dissolved in water to form 500 mL of solution.

TABLE 7.6	Boiling and freezing point constants for various solvents			
Solvent	Normal boiling point (°C)	K_b (°C/M)	Normal freezing point (°C)	K_f (°C/M)
Benzene	80.1	2.53	5.5	4.90
Camphor			174.0	40.0
Carbon tetrachloride	76.8	5.03		
Chloroform	61.2	3.63		
Cyclohexane	81.0	2.79	6.5	20.0
Ethyl alcohol	78.5	1.22		
Water	100.0	0.52	0.0	1.86

GETTING STARTED WITH MOLARITY CALCULATIONS

Knowing where to start or what to do first is a critical part of working any math-type problem. This might be a factor if molarity calculations are difficult for you. First, of course, you must identify a problem as a molarity problem. This can be done by looking for a key word or phrase (molar, molarity, or moles/liter) or the abbreviation M. Second, remember that you have a formula for molarity

$$M = \frac{\text{moles of solute}}{\text{liters of solution}}$$

and the problem might be treated as a formula-type problem. Look for the given numbers and their units, and see if they match the units of the formula. If they do, put them into the formula and do the calculations. For example, if you are asked to calculate the molarity of a solution that contains 0.0856 mol NaCl dissolved in enough water to give 0.100 L of solution, you could put the numbers and their units directly into the formula:

$$M = \frac{0.0856 \text{ mol NaCl}}{0.100 \text{ L solution}} = 0.856 \text{ mol NaCl/L solution}$$

However, if you must calculate the molarity of a solution that contains 5.00 g of NaCl in enough water to give 100 mL of solution, the units of the numbers do not match those of the formula. The factor-unit method can be used to convert each quantity into the units needed by the formula

$$500 \text{ g NaCl} \times \frac{1 \text{ mol NaCl}}{58.4 \text{ g NaCl}} = 0.0856 \text{ mol NaCl}$$

$$100 \text{ mL solution} \times \frac{1 \text{ L solution}}{1000 \text{ mL solution}} = 0.100 \text{ L solution}$$

These numbers and units can then be put into the formula

$$M = \frac{0.0856 \text{ mol NaCl}}{0.100 \text{ L solution}} = 0.856 \text{ mol NaCl/L solution}$$

Alternatively, the problem can be solved by remembering the units of the answer (molarity would have the units mol NaCl/L solution), noting the numbers and units of the given quantities (5.00 g NaCl/100 mL solution), and using the factor-unit method to convert the units of the given quantity to those of the answer:

Step 1. $\dfrac{5.00 \text{ g NaCl}}{100 \text{ mL solution}}$

Step 2. $\dfrac{5.00 \text{ g NaCl}}{100 \text{ mL solution}} \qquad = \dfrac{\text{mol NaCl}}{\text{L solution}}$

Step 3. $\dfrac{5.00 \text{ g NaCl}}{100 \text{ mL solution}} \times \dfrac{1 \text{ mol NaCl}}{58.4 \text{ g NaCl}} \times \dfrac{1000 \text{ mL solution}}{1 \text{ L solution}} = \dfrac{\text{mol NaCl}}{\text{L solution}}$

Step 4. $\dfrac{(5.00)(1 \text{ mol NaCl})(1000)}{(100)(58.4)(1 \text{ L solution})} = 0.856 \dfrac{\text{mol NaCl}}{\text{L solution}}$

The final answer, which has the correct units for a molarity, has been rounded to three significant figures.

Solution

In each case, Equations 7.13 and 7.14 are used to calculate the difference between the normal boiling and freezing point of water and the solution. To use these equations, K_b and K_f are obtained from Table 7.6; the solution molarity, M, is calculated; and n is determined.

a. To find the boiling point, calculate solution molarity.

$$(171 \text{ g } C_{12}H_{22}O_{11})\left(\frac{1 \text{ mol } C_{12}H_{22}O_{11}}{342.0 \text{ g } C_{12}H_{22}O_{11}}\right) = 0.500 \text{ mol } C_{12}H_{22}O_{11}$$

$$M = \frac{\text{moles of solute}}{\text{liters of solution}} = \frac{0.500 \text{ mol}}{1.00 \text{ L}} = 0.500 \frac{\text{mol}}{\text{L}}$$

Determine n: Because sugar does not dissociate on dissolving, $n = 1$. Therefore,

$$\Delta t_b = nK_bM = (1)(0.52°C/M)(0.500 \text{ M}) = 0.26°C$$

Because boiling points are higher in solutions, we add Δt_b to the normal boiling point of water:

$$\text{solution boiling point} = 100.00°C + 0.26°C = 100.26°C$$

To find the freezing point, calculate Δt_f:

$$\Delta t_f = nK_fM = (1)(1.86°C/M)(0.500 \text{ M}) = 0.93°C$$

Because freezing points are lower in solutions, we subtract Δt_f from the normal freezing point of water:

$$\text{solution freezing point} = 0.00°C - 0.93°C = -0.93°C$$

b. Similarly,

$$(13.4 \text{ g } NH_4Cl)\left(\frac{1 \text{ mol } NH_4Cl}{53.5 \text{ g } NH_4Cl}\right) = 0.250 \text{ mol } NH_4Cl$$

$$M = \frac{\text{moles of solute}}{\text{liters of solution}} = \frac{0.250 \text{ mol}}{0.500 \text{ L}} = 0.500 \frac{\text{mol}}{\text{L}}$$

Because NH_4Cl dissociates in water, 1 mol of solute gives 2 mol of particles (ions):

$$NH_4Cl(aq) \rightarrow NH_4^+(aq) + Cl^-(aq)$$

Thus, we conclude that $n = 2$. Therefore,

$$\Delta t_b = nK_bM = (2)(0.52°C/M)(0.500 \text{ M}) = 0.52°C$$

$$\text{solution boiling point} = 100.00°C + 0.52°C = 100.52°C$$

$$\Delta t_f = nK_fM = (2)(1.86°C/M)(0.500 \text{ M}) = 1.86°C$$

$$\text{solution freezing point} = 0.00°C - 1.86°C = -1.86°C$$

Example 7.9 demonstrates the influence of solute dissociation on colligative properties. Even though the solutions in parts a and b have the same molarity, the NH_4Cl dissociates and produces twice as many solute particles in solution. Hence, it causes twice as much change in the colligative properties.

Osmotic pressure, another important colligative property, can be illustrated by some hypothetical experiments. Consider ■ Figure 7.13, where a sugar solution is separated from pure water by a barrier (a). The barrier is removed, but the mixture is not stirred (b). After a day or so, the barrier is replaced, with the results shown in (c). These results are not surprising; the sugar has diffused throughout the mixture uniformly. It would have been very surprising if the process had not taken place.

Now consider the experiment shown in ■ Figure 7.14, in which two solutions similar to those used before are separated by a semipermeable membrane, which has pores large enough to allow small molecules such as water to pass through but small enough to prevent passage by larger molecules or hydrated

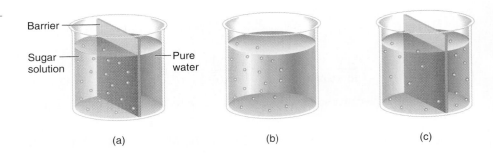

➤ **FIGURE 7.13** Diffusion eliminates concentration gradients.

ions. In Figure 7.14, the membrane allows water molecules to pass but not sugar molecules.

In this experiment, a concentration difference between the liquids has been created, but diffusion is prevented from taking place as it did before. Water molecules move through the membrane in both directions, but sugar molecules cannot move into the pure water to equalize the concentration. As a result, the net flow of water through the membrane is into the sugar solution. This has the effect of increasing the volume of the sugar solution and decreasing its concentration to a value closer to that of the pure water. The movement of water creates a difference in liquid levels *(h)*, which causes hydrostatic pressure against the membrane. This pressure increases until it becomes high enough to balance the tendency for net water to flow through the membrane into the sugar solution. From that time on, the flow of water in both directions through the membrane is equal, and the volume of liquid on each side of the membrane no longer changes. The hydrostatic pressure required to prevent the net flow of water through a semipermeable membrane into a solution is called the **osmotic pressure** of the solution. The process in which solvent molecules move through semipermeable membranes is called **osmosis**. Although our experiment doesn't illustrate it, solvent will also flow osmotically through semipermeable membranes separating solutions of different concentrations. The net flow of solvent is always from the more dilute solution into the more concentrated (see ■ Figure 7.15).

The osmotic pressure that will develop across a semipermeable membrane separating pure solvent from a solution of molarity M is given by Equation 7.15, which is similar to the ideal gas law given earlier (Section 6.8):

$$\pi = nMRT \qquad (7.15)$$

In this equation, π is the osmotic pressure in units that are the same as the pressure units used in the ideal gas constant R. M is the solution molarity, T is the temperature in kelvins, and n, as before, is the number of moles of solute

Osmotic pressure
The hydrostatic pressure required to prevent the net flow of solvent through a semipermeable membrane into a solution.

Osmosis
The process in which solvent flows through a semipermeable membrane into a solution.

➤ **FIGURE 7.14** Osmosis.

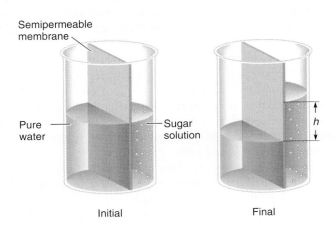

Initial   Final

➤ **FIGURE 7.15** Osmosis through carrot membranes. Look at the photos carefully. What evidence is there that some molasses moved out of the carrot into the water?

A hollowed-out carrot is filled with molasses (a concentrated sugar solution) and immersed in pure water.

The carrot acts as a semipermeable osmotic membrane through which water flows and increases the volume of the molasses. The volume increase forces the molasses up the tube.

particles obtained when 1 mol of solute dissolves. Scientists in biological and medical fields often call the product of *n* and M the **osmolarity** of the solution.

Osmolarity
The product of *n* and M in the equation $\pi = nMRT$.

Example 7.10

Calculate the osmolarity and the osmotic pressure that would develop across a semipermeable membrane if the solutions of Example 7.9 were separated from pure water by the membrane. Assume a solution temperature of 27°C and use an *R* value of 62.4 L torr/K mol.

Solution

a. The molarity of the sugar solution was found to be 0.500 mol/L. Because sugar does not dissociate, *n* is equal to 1, and the osmolarity *n*M is also equal to 0.500 mol/L.

CHEMISTRY AROUND US • 7.1

CARBON DIOXIDE AND DRY CLEANING

The dry cleaning process is not really dry since liquids are used, but the liquids do not contain water. It is based on the solubility rule "like dissolves like," and uses nonpolar liquid solvents to dissolve nonpolar soil and stains from fabrics. At one time, gasoline was a popular dry cleaning solvent for both commercial and home use. However, the flammability of gasoline led to the development of less hazardous nonpolar cleaning solvents by chemically substituting chlorine in place of some hydrogen atoms in petroleum-based solvent molecules. Today, the most widely used product to come out of this development is perchloroethylene (called "perc" in the dry cleaning industry) with the formula $CCl_2{=}CCl_2$.

Perchloroethylene is not without its problems, such as being a suspected human carcinogen and an EPA-classified hazardous waste. A new cleaning technology that uses carbon dioxide, CO_2, has none of the problems associated with perc. The technology is based on the fact that CO_2, which is a gas at room temperature under normal atmospheric pressure (14.7 psi) can be liquified by putting it under a pressure of about 1100 psi and keeping it at a temperature of less than 31°C. A mixture of liquid CO_2 and appropriate detergents is able to dissolve and wash nonpolar oily, greasy soil out of fabrics. Nontoxic, environmentally friendly detergents and washing equipment capable of withstanding the necessary high pressure have been developed and are being marketed in the form of dry cleaning franchises. The future of the dry cleaning industry looks bright (and clean).

$$\pi = nMRT = (1)\left(0.500\ \frac{\text{mol}}{\text{L}}\right)\left(\frac{62.4\ \text{L torr}}{\text{K mol}}\right)(300\ \text{K})$$

$$= 9.36 \times 10^3\ \text{torr}$$

Thus, this solution would develop a pressure sufficient to support a column of mercury 9.36×10^3 mm high. This is equal to about 12.3 standard atmospheres of pressure (see Table 6.3 for pressure units).

b. The molarity of this solution was also found to be 0.500. But, because this solute dissociates into two ions, $n = 2$. This makes the osmolarity equal to $(2)(0.500\ \text{mol/L}) = 1.00\ \text{mol/L}$:

$$\pi = nMRT = (2)\left(\frac{0.500\ \text{mol}}{\text{L}}\right)\left(\frac{62.4\ \text{L torr}}{\text{K mol}}\right)(300\ \text{K})$$

$$= 1.87 \times 10^4\ \text{torr, or about 24.6 standard atmospheres}$$

Learning Check 7.9

Calculate the boiling point, freezing point, and osmotic pressure of the following solutions. Assume that the osmotic pressure is measured at 27°C.

a. A 0.100 M solution of $CaCl_2$ in water. The $CaCl_2$ is a strong electrolyte.
b. A 0.100 M solution of ethylene glycol in water. Ethylene glycol does not dissociate in solution.

➤ 7.8 Colloids

Colloid
A homogeneous mixture of two or more substances in which the dispersed substances are present as larger particles than are found in solutions.

Dispersing medium
The substance present in a colloidal dispersion in the largest amount.

Dispersed phase
The substance present in a colloidal dispersion in amounts less than the amount of dispersing medium.

Tyndall effect
A property of colloids in which the path of a beam of light through the colloid is visible because the light is scattered.

Like solutions, **colloids** (or colloidal dispersions) are homogeneous mixtures of two or more components in which there is more of one component than of the others. In solutions, the terms *solvent* and *solute* are used for the components, but in colloids, the terms **dispersing medium** (for solvent) and **dispersed phase** (for solute) are used.

Solute particles (in solutions) and dispersed phase particles (in colloids) cannot be seen and do not settle under the influence of gravity. However, solute particles do not scatter or reflect light, whereas dispersed phase particles do. This and other variations in properties result from the principal difference between solutions and colloids, the size of the particles making up the solute or dispersed phase. The dissolved solute in a solution is present in the form of tiny particles (small molecules or ions) that are less than about 10^{-7} cm (0.1 μm) in diameter. The dispersed phase of colloids is made up of much larger particles (very large molecules or small pieces of matter) with diameters of 10^{-7} to 10^{-5} cm (0.1–10 μm). As a result of light scattering, colloids often appear to be cloudy. When a beam of light passes through them, they demonstrate the **Tyndall effect** in which the path of the light becomes visible (■ Figure 7.16).

The word *colloidal* means "gluelike," and some colloids, including some glues, fit this description quite well. However, many, including smoke, shaving cream, and cheese, do not. Colloids are usually differentiated according to the states of the dispersing medium and dispersed phase. Some colloid types are listed in ■ Table 7.7, together with examples and specific names.

Sols that become viscous and semisolid are called *gels*. In these colloids, the solid dispersed phase has a very high affinity for the dispersing medium. The gel "sets" by forming a three-dimensional network of solid and dispersing medium. Other examples of gels are fruit jellies and "canned heat" (jellied alcohol).

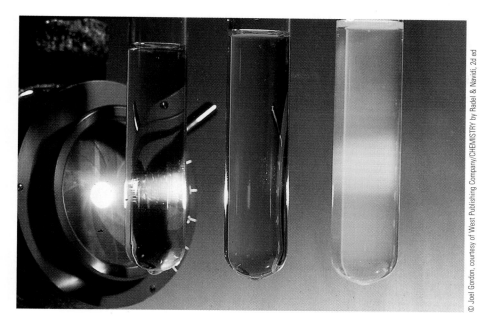

➤ **FIGURE 7.16** The Tyndall effect. The light beam passes from left to right through a purple gold sol (a colloid), a blue copper sulfate solution, and a colloidal iron (III) hydroxide. The light path can be seen in both colloids but not in the copper sulfate solution.

➤ 7.9 Colloid Formation and Destruction

Much of the interest in colloids is related to their formation or destruction. Ions that are present in the dispersing medium are attracted to colloid particles and stick on their surfaces. The charge (+ or −) of the ions depends on the nature of the colloid, but all colloid particles within a particular system will attract ions of only one charge or the other. In this way, the colloid particles all acquire the same charge and repel each other. This repulsion helps prevent the particles from coalescing into aggregates large enough to settle out.

TABLE 7.7 Types of colloids

Colloid type			
Dispersing medium	Dispersed phase	Name	Examples
Gas	Liquid	Aerosol	Fog, aerosol sprays, some air pollutants
Gas	Solid		Smoke, some air pollutants
Liquid	Gas	Foam	Whipped cream, shaving cream
Liquid	Liquid	Emulsion	Milk, mayonnaise
Liquid	Solid	Sol	Paint, ink, gelatin dessert
Solid	Gas	Solid foam	Marshmallow, pumice stone, foam rubber
Solid	Liquid		Butter, cheese
Solid	Solid		Pearls, opals, colored glass, some metal alloys

In the Cottrel precipitator, colloidal solids are removed from gaseous smokestack wastes before they are released into the atmosphere. The precipitator contains a number of highly charged plates or electrodes. As smoke passes over the charged surfaces, the colloid particles lose their charges. The particles then coalesce into larger particles that settle out and are collected for disposal.

Emulsifying agent (stabilizing agent)
A substance that when added to colloids prevents them from coalescing and setting.

Some colloids are stabilized (prevented from coalescing) by substances known as **emulsifying agents** or **stabilizing agents**. Mayonnaise-like salad dressing is a colloid of oil in water, with compounds from egg yolk acting as the emulsifying agent. These compounds form a coating around the oil droplets, which keeps them separated and suspended in the water.

The cleaning action of soaps and detergents comes from their activity as emulsifying agents. Both soaps and detergents contain long molecules with structures like the following:

$$H-\overset{\overset{\displaystyle H}{|}}{\underset{\underset{\displaystyle H}{|}}{C}}-\overset{\overset{\displaystyle H}{|}}{\underset{\underset{\displaystyle H}{|}}{C}}-\overset{\overset{\displaystyle H}{|}}{\underset{\underset{\displaystyle H}{|}}{C}}-\overset{\overset{\displaystyle H}{|}}{\underset{\underset{\displaystyle H}{|}}{C}}-\overset{\overset{\displaystyle H}{|}}{\underset{\underset{\displaystyle H}{|}}{C}}-\overset{\overset{\displaystyle H}{|}}{\underset{\underset{\displaystyle H}{|}}{C}}-\overset{\overset{\displaystyle H}{|}}{\underset{\underset{\displaystyle H}{|}}{C}}-\overset{\overset{\displaystyle H}{|}}{\underset{\underset{\displaystyle H}{|}}{C}}-\overset{\overset{\displaystyle H}{|}}{\underset{\underset{\displaystyle H}{|}}{C}}-\overset{\overset{\displaystyle H}{|}}{\underset{\underset{\displaystyle H}{|}}{C}}-\overset{\overset{\displaystyle H}{|}}{\underset{\underset{\displaystyle H}{|}}{C}}-\overset{\overset{\displaystyle O}{||}}{C}-O^-Na^+$$

Soap

$$H-\overset{\overset{\displaystyle H}{|}}{\underset{\underset{\displaystyle H}{|}}{C}}-\overset{\overset{\displaystyle H}{|}}{\underset{\underset{\displaystyle H}{|}}{C}}-\overset{\overset{\displaystyle H}{|}}{\underset{\underset{\displaystyle H}{|}}{C}}-\overset{\overset{\displaystyle H}{|}}{\underset{\underset{\displaystyle H}{|}}{C}}-\overset{\overset{\displaystyle H}{|}}{\underset{\underset{\displaystyle H}{|}}{C}}-\overset{\overset{\displaystyle H}{|}}{\underset{\underset{\displaystyle H}{|}}{C}}-\overset{\overset{\displaystyle H}{|}}{\underset{\underset{\displaystyle H}{|}}{C}}-\overset{\overset{\displaystyle H}{|}}{\underset{\underset{\displaystyle H}{|}}{C}}-\overset{\overset{\displaystyle H}{|}}{\underset{\underset{\displaystyle H}{|}}{C}}-\overset{\overset{\displaystyle H}{|}}{\underset{\underset{\displaystyle H}{|}}{C}}-\overset{\overset{\displaystyle H}{|}}{\underset{\underset{\displaystyle H}{|}}{C}}-O-\overset{\overset{\displaystyle O}{||}}{\underset{\underset{\displaystyle O}{||}}{S}}-O^-Na^+$$

Detergent

When placed in water, soaps and detergents dissociate to form ions, as shown in Equations 7.16 and 7.17, where the long carbon chains are represented by a wavy line.

$$\sim\sim\sim\sim\overset{\overset{\displaystyle O}{||}}{C}-O^-Na^+ \longrightarrow \sim\sim\sim\sim\overset{\overset{\displaystyle O}{||}}{C}-O^- + Na^+ \tag{7.16}$$

$$\sim\sim\sim\sim O-\overset{\overset{\displaystyle O}{||}}{\underset{\underset{\displaystyle O}{||}}{S}}-O^-Na^+ \longrightarrow \sim\sim\sim\sim O-\overset{\overset{\displaystyle O}{||}}{\underset{\underset{\displaystyle O}{||}}{S}}-O^- + Na^+ \tag{7.17}$$

Nonpolar oils and greases are not soluble in water, but they are attracted to the uncharged ends of the soap or detergent ions. As a result of this attraction, the soap or detergent forms a charged layer around the oil droplets, which keeps them separated and suspended (see ■ Figure 7.17). Certain compounds (lecithins) of egg yolk act in much the same way to stabilize the mayonnaise-like dressing discussed earlier.

➤ 7.10 Dialysis

Dialyzing membrane
A semipermeable membrane with pores large enough to allow solvent molecules, other small molecules, and hydrated ions to pass through.

Earlier we discussed semipermeable membranes that selectively allow solvent to pass but retain dissolved solutes during osmosis. **Dialyzing membranes** are semipermeable membranes with larger pores than osmotic membranes. They hold back colloid particles and large molecules but allow solvent, hydrated ions, and small molecules to pass through. The passage of these ions and small

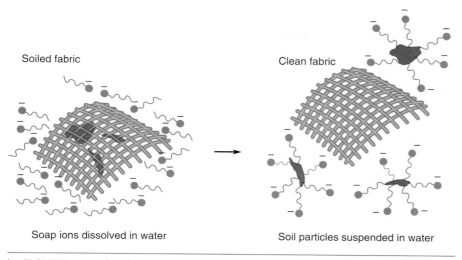

Soiled fabric

Clean fabric

Soap ions dissolved in water

Soil particles suspended in water

➤ **FIGURE 7.17** The cleaning action of soaps and detergents.

molecules through such membranes is called **dialysis**. Dialysis can be used to separate small particles from colloids, as shown in ■ Figure 7.18.

A mixture containing water, ions, small molecules, and colloid particles is placed inside a bag made from a dialyzing membrane. Water flowing around the bag carries away ions and small molecules that pass through the membrane. Water molecules move through the membrane in both directions, but the colloid particles remain inside the bag.

A similar technique is used to clean the blood of people suffering from kidney malfunction. The blood is pumped through tubing made of a dialyzing membrane. The tubing passes through a bath in which impurities collect after passing out of the blood. Blood proteins and other important large molecules remain in the blood.

Dialysis
A process in which solvent molecules, other small molecules, and hydrated ions pass from a solution through a membrane.

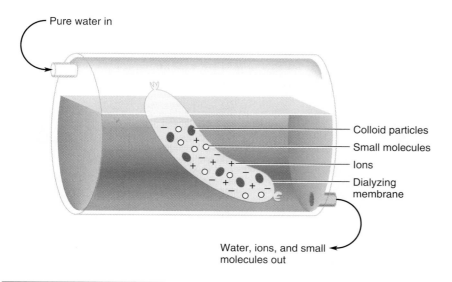

Pure water in

Colloid particles

Small molecules

Ions

Dialyzing membrane

Water, ions, and small molecules out

➤ **FIGURE 7.18** Dialysis. This is one method of dialysis used to purify proteins.

CHEMISTRY AROUND US • 7.2

DIALYSIS AND RED BLOOD CELLS

Each of the millions of red blood cells (RBCs) circulating in the bloodstream is enclosed by a semipermeable membrane through which *dialysis* can take place. The direction of liquid flow through the membranes is determined by the difference in osmolarity of the fluid inside and outside the cells. For example, when RBCs are placed in pure water, the osmolarity inside is higher than the pure water outside. Water flows into the cells to such an extent that they burst. The rupture of RBCs this way is called *hemolysis*. In contrast, fluid flows out of RBCs placed in a solution with an osmolarity higher than the fluid inside the cells. Because they lose fluid, such cells shrivel and shrink and are said to have undergone *crenation*.

It is sometimes necessary to administer body fluids or nutrients to patients by an intravenous drip technique. Care must be taken to ensure that the fluids added to the bloodstream do not change the osmolarity of the liquid in which the RBCs are suspended to an extent that would result in hemolysis or crenation.

Solutions that have the same osmolarity are called *isotonic* solutions. When two solutions have different osmolarities, the one of higher osmolarity is said to be *hypertonic* compared with the other. The solution of lower osmolarity is said to be *hypotonic* to the other. Thus, RBCs undergo hemolysis when surrounded by a hypotonic solution (compared with the interior of the RBC). They undergo crenation in hypertonic environments. A solution that is 0.89% (w/w) in NaCl is isotonic with the interior of RBC. This solution is called physiological, or normal, saline solution.

➤ **FOR FUTURE REFERENCE**   DRUG CONCENTRATIONS IN BLOOD

In Section 7.6, we demonstrated that the amount of chemical reaction that can take place between two solutions depends on the concentration of reactants in the solutions. The therapeutic effect of most drugs in the body results from chemical reactions of the drugs, and thus depends on the amount of the drugs present. A convenient way for a physician to monitor the amount of a drug present in the body is to measure the concentration of the drug in the blood. In order to be effective, most drugs must be maintained within a specified concentration range or level in the blood. Some drugs work best when their maximum level (peak level) falls within a specified range of values, and their minimum level (trough level) falls within a lower specified range of values.

The table below lists drugs that are most suitable for blood-level monitoring, and their therapeutic concentration ranges. Individuals using any of these regularly should ask their doctor about periodically checking the blood for drug levels. The concentrations in the table are given using several units: micrograms/mL (μg/mL), nanograms/mL (ng/mL), milliequivalents/L (meq/L), and micromoles/L (μmol/L).

Generic name (Trade name example)	Therapeutic blood level range
acetaminophen (Tylenol)	10–20 μg/mL
amikacin (Amikin)	12–25 μg/mL (peak)
	5–10 μg/mL (trough)
amitriptyline (Elavil) combined with nortriptyline	120–250 ng/mL
amoxapine (Asendin)	200–500 ng/mL
aspirin and other salicylates	100–250 μg/mL
carbamazepine (Tegretol)	5–10 μg/mL
chloramphenicol (Chloromycetin)	10–25 μg/mL
chlorpromazine (Thorazine)	50–300 ng/mL
ciprofloxacin (Cipro)	0.94–3.4 μg/mL
clonazepam (Klonopin)	10–50 ng/mL
cyclosporine (Sandimmune)	100–150 ng/mL
desipramine (Norpramin, Pertofrane)	150–300 ng/mL

(continued)

digitoxin (Crystodigin)	15–30 ng/mL
digoxin (Lanoxin)	0.5–2.0 ng/mL
diltiazem (Cardizem)	100–200 ng/mL
disopyramide (Norpace)	2.0–4.5 µg/mL
doxepin (Adapin, Sinequan)	100–275 ng/mL
ethosuximide (Zarontin)	40–100 µg/mL
flecainide (Tambocor)	0.2–1.0 µg/mL
flucytosine (Ancobon)	50–100 µg/mL
gentamicin (Garamycin)	4–10 µg/mL (peak)
	less than 2 µg/mL (trough)
gold salts (Auranofin)	1–2.0 µg/mL
imipramine (Janimine, Tofranil)	150–300 ng/mL
kanamycin (Kantres)	25–35 µg/mL
lidocaine (Xylocaine)	2–5 µg/mL
lithium (Lithobid, Lithotabs)	0.3–1.3 meq/L
mephobarbital (Mebaral)	1–7 µg/mL
methotrexate (Mexate)	up to 0.1 µmol/L
methsuximide (Celontin)	up to 1 µg/mL
metoprolol (Lopressor)	20–200 ng/mL
mexiletine (Mexitil)	0.75–2.0 µg/mL
nifedipine (Procardia)	25–100 ng/mL
nortriptyline (Aventyl, Pamelor)	50–150 ng/mL
combined with amitriptyline	120–250 ng/mL
phenobarbital (Luminal)	10–25 µg/mL
phenytoin (Dilantin)	10–20 µg/mL
primidone (Mysoline)	6–12 µg/mL
procainamide (Pronestyl)	4–10 µg/mL
propranolol (Inderal)	50–100 ng/mL
protriptyline (Vivactil)	70–250 ng/mL
quinidine (Quinaglute)	1–4 µg/mL
specific to quinidine test method	
sulfadiazine (Microsulfon)	100–120 µg/mL
sulfamethoxazole (Gantanol)	90–100 µg/mL
theophylline (Aminophylline)	10–20 µg/mL
thioridazine (Mellaril)	50–300 ng/mL
tobramycin (Nebcin)	4–10 µg/mL (peak)
	less than 2 µg/mL (trough)
tocainide (Tonocard)	5–12 µg/mL
trimethadione (Tridione)	10–30 µg/mL
trimethoprim (Proloprim)	1–3 µg/mL
valproic acid (Depakene)	50–100 µg/mL
vancomycin (Vancocin)	30–40 µg/mL (peak)
	5–10 µg/mL (trough)
verapamil (Calan)	50–200 ng/mL

SOURCE: Adapted from Rybacki, J.J.; Long, J. W. *The Essential Guide to Prescription Drugs*. New York: Harper Perennial, 1999.

CONCEPT SUMMARY

Physical States of Solutions. Solutions, homogeneous mixtures of a solvent and one or more solutes, can be found in any of the three states of matter: solid, liquid, or gas. The physical state of the solution is often the same as the physical state of the solvent.

Solubility. The amount of solute that will dissolve in a quantity of solvent to form a saturated solution is the solubility of the solute. Solubility depends on similarities in polarities of the solvent and solute; it generally increases with temperature for solid and liquid solutes, but decreases with temperature for gaseous solutes.

The Solution Process. The solution process of solid solute in a liquid solvent can be thought of in terms of the solvent molecules attracting the solute particles away from the solute crystal lattice. The solute is picked apart and solute particles hydrated as a result of attractions between water molecules and solute particles. Heat is generally absorbed or released when a solute dissolves in a solvent. When heat is released, the solution process is called exothermic, and the solution temperature increases. When heat is absorbed, the solution process is endothermic, and the solution cools as solute dissolves.

Solution Concentrations. Relationships between the amount of solute and the amount of solution containing the solute are called concentrations. They may be expressed as molarity, weight/weight percent, weight/volume percent, or volume/volume percent.

Solution Preparation. Solutions of specific concentration can be prepared by mixing appropriate amounts of solute and solvent or by diluting a more concentrated solution with solvent.

Solution Stoichiometry. When solution concentrations are expressed as molarities, the mole concept can be applied to reactions taking place between substances that are solutes in the solutions.

Solution Properties. A number of solution properties differ from the properties of pure solvent. Ionic or highly polar solutes that dissociate in solution result in solutions that conduct electricity. Colligative solution properties depend on the concentration of solute particles in the solution and include vapor pressure, boiling point, freezing point, and osmotic pressure.

Colloids. Homogeneous mixtures called colloids, or colloidal dispersions, differ from solutions in terms of the size of the dispersed phase particles. In colloids, the particles are large enough to scatter light and thus show the Tyndall effect. Colloids, like solutions, occur in all three physical states, depending primarily on the physical state of the dispersing medium.

Colloid Formation and Destruction. In true colloids, the suspension is permanent, and dispersed materials do not settle out. This happens because dispersed particles acquire similar charges by adsorbing ions from the dispersing medium and repel each other or because emulsifying agents function to keep the dispersed particles from coalescing.

Dialysis. Dialyzing membranes are semipermeable but with pores large enough to allow solvent molecules, hydrated ions, and small molecules to pass through in a process called dialysis. The process is important in the removal of impurities from the blood and is applied artificially for people suffering from kidney malfunction.

LEARNING OBJECTIVES ASSESSMENT

You can get an approximate but quick idea of how well you have met the learning objectives given at the beginning of this chapter by working the selected end-of-chapter exercises given below. The answer to each exercise is given in Appendix B of the book.

Objective 1 (Section 7.1): Exercise 7.4

Objective 2 (Section 7.3): Exercise 7.16

Objective 3 (Section 7.4): Exercises 7.22 b, 7.30 c, 7.34 a, and 7.38 c

Objective 4 (Section 7.5): Exercises 7.46 and 7.48 b

Objective 5 (Section 7.6): Exercise 7.56

Objective 6 (Section 7.7): Exercises 7.64 a & c and 7.74

Objective 7 (Sections 7.8 and 7.9): Exercise 7.80

Objective 8 (Section 7.10): Exercise 7.82

KEY TERMS AND CONCEPTS

Colligative property (7.7)
Colloid (7.8)
Concentration (7.4)
Dialysis (7.10)
Dialyzing membrane (7.10)
Dispersed phase (7.8)
Dispersing medium (7.8)
Dissolving (7.1)
Electrolyte (7.7)
Emulsifying agent (stabilizing agent) (7.9)

Hydrated ion (7.3)
Immiscible (7.2)
Insoluble substance (7.2)
Molarity (M) (7.4)
Nonelectrolyte (7.7)
Osmolarity (7.7)
Osmosis (7.7)
Osmotic pressue (7.7)
Percent (7.4)
Saturated solution (7.2)

Solubility (7.2)
Soluble substance (7.2)
Solute (7.1)
Solution (7.1)
Solvent (7.1)
Supersaturated solution (7.2)
Tyndall effect (7.8)
Volume/volume percent (7.4)
Weight/volume percent (7.4)
Weight/weight percent (7.4)

.... **KEY EQUATIONS** ....

1. Solution concentration in terms of molarity (Section 7.4):

$$M = \frac{\text{moles of solute}}{\text{liters of solution}}$$

Equation 7.5

2. Solution concentration in terms of weight/weight % (Section 7.4):

$$\%(w/w) = \frac{\text{solute mass}}{\text{solution mass}} \times 100$$

Equation 7.6

3. Solution concentration in terms of weight/volume % (Section 7.4):

$$\%(w/v) = \frac{\text{grams of solute}}{\text{milliliters of solution}} \times 100$$

Equation 7.7

4. Solution concentration in terms of volume/volume % (Section 7.4):

$$\%(v/v) = \frac{\text{solute volume}}{\text{solution volume}} \times 100$$

Equation 7.8

5. Dilution of concentrated solution to make less-concentrated solution (Section 7.5):

$$(C_c)(V_c) = (C_d)(V_d)$$

Equation 7.9

6. Boiling point elevation of a solution (Section 7.7):

$$\Delta t_b = nK_b M$$

Equation 7.13

7. Freezing point depression of a solution (Section 7.7):

$$\Delta t_f = nK_f M$$

Equation 7.14

8. Osmotic pressure of a solution (Section 7.7):

$$\pi = nMRT$$

Equation 7.15

.... **EXERCISES** ....

LEGEND: 1 = straightforward, 2 = intermediate, 3 = challenging. All even-numbered exercises are answered in Appendix B.

PHYSICAL STATES OF SOLUTIONS (SECTION 7.1)

7.1 Many solutions are found in the home. Some are listed below, with the composition as printed on the label. When no percentage is indicated, components are usually given in order of decreasing amount. When water is present, it is often not mentioned on the label or it is included in the inert ingredients. Identify the solvent and solutes of the following solutions:
 a. Antiseptic mouthwash: alcohol 25%, thymol, eucalyptol, methyl salicylate, menthol, benzoic acid, boric acid
 b. Paregoric: alcohol 45%, opium 0.4%
 c. Baby oil: mineral oil, lanolin (there happens to be no water in this solution—why?)
 d. Distilled vinegar: acetic acid 5%

7.2 Many solutions are found in the home. Some are listed below, with the composition as printed on the label. When no percentage is indicated, components are usually given in order of decreasing amount. When water is present, it is often not mentioned on the label or it is included in the inert ingredients. Identify the solvent and solutes of the following solutions:
 a. Liquid laundry bleach: sodium hypochlorite 5.25%, inert ingredients 94.75%

 b. Rubbing alcohol: isopropyl alcohol 70%
 c. Hydrogen peroxide: 3% hydrogen peroxide
 d. Aftershave: SD alcohol, water, glycerin, fragrance, menthol, benzophenone-1, coloring

7.3 Classify the following as being a solution or not a solution. Explain your reasons when you classify one as *not* a solution. For the ones classified as solutions, identify the solvent and solute(s).
 a. Maple syrup
 b. Milk
 c. Eyedrops
 d. Tomato juice
 e. Tap water

7.4 Classify the following as being a solution or not a solution. Explain your reasons when you classify one as *not* a solution. For the ones classified as solutions, identify the solvent and solute(s).
 a. Foggy air
 b. Tears
 c. Freshly squeezed orange juice
 d. Strained tea
 e. Creamy hand lotion

SOLUBILITY (SECTION 7.2)

7.5 Use the term *soluble, insoluble,* or *immiscible* to describe the behavior of the following pairs of substances when they are shaken together:
   a. 25 mL of water and 1 g of salt—the resulting mixture is clear and colorless.
   b. 25 mL of water and 1 g of solid silver chloride—the resulting mixture is cloudy and solid settles out.
   c. 25 mL of water and 5 mL of mineral oil—the resulting mixture is cloudy and gradually separates into two layers.

7.6 Use the term *soluble, insoluble,* or *immiscible* to describe the behavior of the following pairs of substances when they are shaken together:
   a. 25 mL of cooking oil and 25 mL of vinegar—the resulting mixture is cloudy and gradually separates into two layers.
   b. 25 mL of water and 10 mL of rubbing alcohol—the resulting mixture is clear and colorless.
   c. 25 mL of chloroform and 1 g of roofing tar—the resulting mixture is clear but dark brown in color.

7.7 Define the term *miscible.* It is *not* defined in the text.

7.8 Classify the following solutions as unsaturated, saturated, or supersaturated:
   a. A solution to which a small piece of solute is added, and it dissolves.   ~~solvent~~ unsaturated
   b. A solution to which a small piece of solute is added, and much more solute comes out of solution.   super
   c. The final solution resulting from the process in part (b).   satur

7.9 Suppose you put 35.8 g of ammonium sulfate into a flask and add 100 g of water at 0°C. After stirring to dissolve as much solute as possible, will you have a saturated or unsaturated solution? Explain your answer. See Table 7.2.

7.10 Suppose you have a saturated solution that is at room temperature. Discuss how it could be changed into a supersaturated solution without using any additional solute.

7.11 Classify each of the following solutes into the approximate solubility categories of Table 7.3. The numbers in parentheses are the grams of solute that will dissolve in 100 g of water at the temperature indicated.
   a. boric acid, H_3BO_3 (6.35 g at 30°C)
   b. calcium hydroxide, $Ca(OH)_2$ (5.35 g at 30°C)
   c. antimony(III) sulfide, Sb_2S_3 (1.75×10^{-4} g at 18°C)
   d. copper(II) chloride, $CuCl_2$ (70.6 g at 0°C)
   e. iron(II) bromide, $FeBr_2$ (109 g at 10°C)

7.12 Classify each of the following solutes into the approximate solubility categories of Table 7.3. The numbers in parentheses are the grams of solute that will dissolve in 100 g of water at the temperature indicated.
   a. barium nitrate, $Ba(NO_3)_2$ (8.7 g at 20°C)
   b. aluminum oxide, Al_2O_3 (9.8×10^{-5} g at 29°C)
   c. calcium sulfate, $CaSO_4$ (0.21 g at 30°C)
   d. manganese chloride, $MnCl_2$ (72.3 g at 25°C)
   e. lead bromide, $PbBr_2$ (0.46 g at 0°C)

THE SOLUTION PROCESS (SECTION 7.3)

7.13 What is the difference between a nonhydrated ion and a hydrated ion? Draw a sketch using the Cl^- ion to help illustrate your answer.

7.14 Suppose you had a sample of white crystalline solid that was a mixture of calcium carbonate ($CaCO_3$) and calcium chloride ($CaCl_2$). Describe how you could treat the sample to isolate one of the solids in the pure state. Which solid would it be?

7.15 Ground-up limestone ($CaCO_3$) is used as a gentle abrasive in some powdered cleansers. Why is this a better choice than ground-up soda ash (Na_2CO_3)?

7.16 Indicate which of the following substances (with geometries as given) would be soluble in water (a polar solvent) and in benzene (a nonpolar solvent):
   a.

   $\begin{array}{c} H \\ | \\ H - C - H \\ | \\ H \end{array}$ (tetrahedral)

   b. Ne

   c.

   $\begin{array}{c} N \\ H - | - H \\ H \end{array}$ (triangular-based pyramid)

   d.

   $\begin{array}{c} F \\ | \\ B \\ F \nearrow \searrow F \end{array}$ (flat triangle)

7.17 Indicate which of the following substances (with geometries as given) would be soluble in water (a polar solvent) and in benzene (a nonpolar solvent):
   a. $H - S \diagdown_H$

   b. $H - Cl$

   c. $\begin{array}{c} O - O \\ H \diagup \quad \diagdown H \end{array}$

   d. $N \equiv N$

7.18 Freons are compounds formerly used in a variety of ways. Explain why freon-114 was useful as a degreasing agent. The molecular structure is

$$\begin{array}{ccc} F & F \\ | & | \\ Cl - C - C - Cl \\ | & | \\ F & F \end{array}$$

7.19 Suppose you put a piece of a solid into a beaker that contains water and stir the mixture briefly. You find that the solid does not immediately dissolve completely. Describe three things you might do to try to get the solid to dissolve.

SOLUTION CONCENTRATIONS (SECTION 7.4)

7.20 Calculate the molarity of the following solutions:
   a. 1.25 L of solution contains 0.455 mol of solute.
   b. 250 mL of solution contains 0.215 mol of solute.
   c. 0.175 mol of solute is put into a container and enough distilled water is added to give 100 mL of solution.

7.21 Calculate the molarity of the following solutions:
   a. 2.00 L of solution contains 0.860 mol of solute.
   b. 500 mL of solution contains 0.304 mol of solute.
   c. 0.115 mol of solute is put into a container and enough distilled water is added to give 250 mL of solution.

7.22 Calculate the molarity of the following solutions:
   a. A sample of solid KBr weighing 11.9 g is put in enough distilled water to give 200 mL of solution.
   b. A 14.2-g sample of solid Na_2SO_4 is dissolved in enough water to give 500 mL of solution.
   c. A 10.0-mL sample of solution is evaporated to dryness and leaves 0.29 g of solid residue that is identified as Li_2SO_4.

7.23 Calculate the molarity of the following solutions:
   a. A sample of solid Na_2SO_4 weighing 0.140 g is dissolved in enough water to make 10.0 mL of solution.
   b. A 4.50-g sample of glucose ($C_6H_{12}O_6$) is dissolved in enough water to give 150 mL of solution.
   c. A 43.5-g sample of K_2SO_4 is dissolved in a quantity of water, and the solution is stirred well. A 25.0-mL sample of the resulting solution is evaporated to dryness and leaves behind 2.18 g of solid K_2SO_4.

7.24 Calculate:
   a. How many moles of solute is contained in 1.75 L of 0.215 M solution?
   b. How many moles of solute is contained in 250 mL of 0.300 M solution?
   c. How many mL of 0.350 M solution contains 0.200 mol of solute?

7.25 Calculate:
   a. How many moles of solute is contained in 1.25 L of 0.350 M solution?
   b. How many moles of solute is contained in 200 mL of 0.750 M solution?
   c. How many mL of 0.415 M solution contains 0.500 mol of solute?

7.26 Calculate:
   a. How many grams of solid would be left behind if 25.0 mL of 0.400 M NaCl solution was evaporated to dryness?
   b. How many liters of 0.255 M HCl solution is needed to provide 0.0400 mol of HCl?
   c. How many mL of 0.500 M $NaNO_3$ solution contains 50.0 g of solute?

7.27 Calculate:
   a. How many grams of solid $AgNO_3$ will be needed to prepare 200 mL of 0.200 M solution?
   b. How many grams of vitamin C ($C_6H_8O_6$) would be contained in 25.0 mL of 1.00 M solution?
   c. How many moles of HCl is contained in 250 mL of 6.0 M solution?

7.28 Calculate the concentration in %(w/w) of the following solutions. Assume water has a density of 1.00 g/mL.
   a. 5.3 g of sugar and 100 mL of water
   b. 5.3 g of any solute and 100 mL of water
   c. 5.3 g of any solute and 100 g of any solvent

7.29 Calculate the concentration in %(w/w) of the following solutions. Assume water has a density of 1.00 g/mL.
   a. 7.5 g of table salt and 100 mL of water
   b. 7.5 g of any solute and 100 mL of water
   c. 7.5 g of any solute and 100 g of any solvent

7.30 Calculate the concentration in %(w/w) of the following solutions. Assume water has a density of 1.00 g/mL.
   a. 15.0 g of salt is dissolved in 200 mL of water.
   b. 0.150 mol of solid NH_4Cl is dissolved in 150 mL of water.
   c. 75.0 g of solid is dissolved in 500 mL of water.
   d. 20.0 mL of ethyl alcohol (density = 0.789 g/mL) is mixed with 15.0 mL of water.

7.31 Calculate the concentration in %(w/w) of the following solutions. Assume water has a density of 1.00 g/mL.
   a. 5.20 g of $CaCl_2$ is dissolved in 125 mL of water.
   b. 0.200 mol of solid KBr is dissolved in 200 mL of water.
   c. 50.0 g of solid is dissolved in 250 mL of water.
   d. 10.0 mL of ethyl alcohol (density = 0.789 g/mL) is mixed with 10.0 mL of ethylene glycol (density = 1.11 g/mL).

7.32 Calculate the concentration in %(w/w) of the following solutions:
   a. 15.0 g of solute is dissolved in enough water to give 100 mL of solution. The density of the resulting solution is 1.20 g/mL.
   b. A 10.0-mL solution sample with a density of 1.15 g/mL leaves 1.54 g of solid residue when evaporated.
   c. A 50.0-g sample of solution on evaporation leaves a 3.12-g residue of $MgCl_2$.

7.33 Calculate the concentration in %(w/w) of the following solutions:
   a. 424 g of solute is dissolved in enough water to give 1.00 L of solution. The density of the resulting solution is 1.18 g/mL.
   b. A 50.0-mL solution sample with a density of 0.898 g/mL leaves 12.6 g of solid residue when evaporated.
   c. A 25.0-g sample of solution on evaporation leaves a 2.32-g residue of NH_4Cl.

7.34 Calculate the concentration in %(v/v) of the following solutions:
   a. 150 mL of solution contains 12.0 mL of alcohol.
   b. 150 mL of solution contains 12.0 mL of any soluble liquid solute.
   c. 6.0 fluid ounces of oil is added to 2.0 gallons (256 fluid ounces) of gasoline.
   d. A solution of alcohol and water is separated by distillation. A 150-mL solution sample gives 51.3 mL of alcohol.

7.35 Calculate the concentration in %(v/v) of the following solutions:
   a. 250 mL of solution contains 20.0 mL of acetone.
   b. 250 mL of solution contains 20.0 mL of any soluble liquid solute.

c. 1.0 quart of acetic acid is put into a 5-gallon container, and enough water is added to fill the container.

d. A solution of acetone and water is separated by distillation. A 300-mL sample gives 109 mL of acetone.

7.36 Consider the blood volume of an adult to be 5.0 L. A blood alcohol level of 0.50% (v/v) can cause a coma. What volume of pure ethyl alcohol, if consumed in one long drink and assumed to be absorbed completely into the blood, would result in this critical blood alcohol level?

7.37 The blood serum acetone level for a person is determined to be 1.8 mg of acetone per 100 mL of serum. Express this concentration as %(v/v) if liquid acetone has a density of 0.79 g/mL.

7.38 Calculate the concentration in %(w/v) of the following solutions:

a. 150 mL of solution contains 7.50 g of dissolved solid Na_2SO_4.

b. 150 mL of solution contains 7.50 g of any dissolved solid solute.

c. 350 mL of solution contains 30.7 g of dissolved solid solute.

7.39 Calculate the concentration in %(w/v) of the following solutions:

a. 28.0 g of solute is dissolved in 200 mL of water to give a solution with a density of 1.10 g/mL.

b. A 25.0-mL solution sample on evaporation leaves a solid residue of 0.38 g.

c. On analysis for total protein, a blood serum sample of 15.0 mL is found to contain 1.02 g of total protein.

7.40 A saturated solution of KBr in water is formed at 20.0°C. Consult Figure 7.3 and calculate the concentration of the solution in %(w/w).

7.41 Assume the density of the solution prepared in Exercise 7.40 is 1.18 g/mL and express the concentraton in %(w/v).

SOLUTION PREPARATION (SECTION 7.5)

7.42 Explain how you would prepare the following solutions using pure solute and water. Assume water has a density of 1.00 g/mL.

a. 200 mL of 0.150 M Na_2SO_4 solution

b. 250 mL of 0.250 M $Zn(NO_3)_2$ solution

c. 150 g of 2.25% (w/w) NaCl solution

d. 125 mL of 0.75% (w/v) KCl solution

7.43 Explain how you would prepare the following solutions using pure solute and water. Assume water has a density of 1.00 g/mL.

a. 500 mL of 2.00 M NaOH solution

b. 250 mL of 40.0% (v/v) alcohol solution (C_2H_5OH)

c. 100 mL of 10.0% (w/v) glycerol solution. Glycerol is a liquid with a density of 1.26 g/mL. Describe two ways to measure out the amount of glycerol needed.

d. Approximately 100 mL of normal saline solution, 0.89% (w/w) NaCl

7.44 A solution is prepared by mixing 45.0 g of water and 15.0 g of ethyl alcohol. The resulting solution has a density of 0.952 g/mL. Express the solution concentration in %(w/w) ethyl alcohol and %(w/v) ethyl alcohol.

7.45 Calculate the following:

a. The number of moles of NaI in 50.0 mL of 0.400 M solution

b. The number of grams of KBr in 120 mL of 0.720 M solution

c. The number of grams of NaCl in 20.0 mL of 1.20% (w/v) NaCl solution

d. The number of milliliters of alcohol in 250 mL of 20.0% (v/v) solution

7.46 Calculate the following:

a. The number of grams of Li_2CO_3 in 250 mL of 1.75 M Li_2CO_3 solution

b. The number of moles of NH_3 in 200 mL of 3.50 M NH_3 solution

c. The number of mL of alcohol in 250 mL of 12.5% (v/v) solution

d. The number of grams of $CaCl_2$ in 50.0 mL of 4.20% (w/v) $CaCl_2$ solution

7.47 Explain how you would prepare the following dilute solutions from the more concentrated ones:

a. 200 mL of 0.500 M HCl from 6.00 M HCl solution

b. 50 mL of 2.00 M H_2SO_4 from 6.00 M H_2SO_4 solution

c. 100 mL of normal saline solution, 0.89% (w/v) NaCl, from 5.0% (w/v) NaCl solution

d. 250 mL of 5.00% (v/v) acetone from 20.5% (v/v) acetone.

7.48 Explain how you would prepare the following dilute solutions from the more concentrated ones:

a. 5.00 L of 6.00 M H_2SO_4 from 18.0 M H_2SO_4 solution

b. 250 mL of 0.500 M $CaCl_2$ from 3.00 M $CaCl_2$ solution

c. 200 mL of 1.50% (w/v) KBr from 10.0% (w/v) KBr solution

d. 500 mL of 10.0% (v/v) alcohol from 50.0% (v/v) alcohol

7.49 What is the molarity of the solution prepared by diluting 25.0 mL of 0.412 M $Mg(NO_3)_2$ solution to each of the following final volumes?

a. 50.0 mL

b. 120 mL

c. 1.50 L

d. 475 mL

7.50 What is the molarity of the solution prepared by diluting 100 mL of 0.225 M KBr solution to each of the following volumes?

a. 2.00 L

b. 225 mL

c. 600 mL

d. 850 mL

SOLUTION STOICHIOMETRY (SECTION 7.6)

7.51 How many milliliters of 6.00 M HCl solution would be needed to react exactly with 25.0 g of pure solid NaOH?

$$HCl(aq) + NaOH(s) \rightarrow NaCl(aq) + H_2O(\ell)$$

7.52 How many grams of solid Na_2CO_3 will exactly react with 250 mL of 1.25 M HCl solution?

$$Na_2CO_3(s) + 2HCl(aq) \rightarrow 2NaCl(aq) + CO_2(g) + H_2O(\ell)$$

7.53 How many milliliters of 0.250 M HCl would be needed to react exactly with 10.5 g of solid $NaHCO_3$?

$$NaHCO_3(s) + HCl(aq) \rightarrow NaCl(aq) + CO_2(g) + H_2O(\ell)$$

7.54 How many milliliters of 0.250 M $AgNO_3$ solution will exactly react with 25.0 mL of 0.200 M NaCl solution?

$$NaCl(aq) + AgNO_3(aq) \rightarrow NaNO_3(aq) + AgCl(s)$$

7.55 How many milliliters of 0.115 M Na_2S solution will exactly react with 35.0 mL of 0.150 M $AgNO_3$ solution?

$$2AgNO_3(aq) + Na_2S(aq) \rightarrow Ag_2S(s) + 2NaNO_3(aq)$$

7.56 How many milliliters of 0.225 M NH_3 solution will exactly react with 30.0 mL of 0.190 M H_2SO_4 solution?

$$2NH_3(aq) + H_2SO_4(aq) \rightarrow (NH_4)_2SO_4(aq)$$

7.57 How many milliliters of 0.124 M NaOH solution will exactly react with 25.0 mL of 0.210 M H_3PO_4 solution?

$$3NaOH(aq) + H_3PO_4(aq) \rightarrow Na_3PO_4(aq) + 3H_2O(\ell)$$

7.58 How many milliliters of 0.135 M NaOH solution will exactly react with 30.0 mL of 0.125 M HCl solution?

$$NaOH(aq) + HCl(aq) \rightarrow NaCl(aq) + H_2O(\ell)$$

7.59 How many milliliters of 0.124 M NaOH solution will exactly react with 25.0 mL of 0.210 M H_2SO_4 solution?

$$2NaOH(aq) + H_2SO_4(aq) \rightarrow Na_2SO_4(aq) + 2 H_2O(\ell)$$

7.60 Stomach acid is essentially 0.10 M HCl. An active ingredient found in a number of popular antacids is calcium carbonate, $CaCO_3$. Calculate the number of grams of $CaCO_3$ needed to exactly react with 250 mL of stomach acid.

$$CaCO_3(s) + 2HCl(aq) \rightarrow CO_2(g) + CaCl_2(aq) + H_2O(\ell)$$

7.61 An ingredient found in some antacids is magnesium hydroxide, $Mg(OH)_2$. Calculate the number of grams of $Mg(OH)_2$ needed to exactly react with 250 mL of stomach acid (see Exercise 7.60).

$$Mg(OH)_2(s) + 2HCl(aq) \rightarrow MgCl_2(aq) + 2H_2O(\ell)$$

SOLUTION PROPERTIES (SECTION 7.7)

7.62 Before it is frozen, ice cream is essentially a solution of sugar, flavorings, etc., dissolved in water. Use the idea of colligative solution properties and explain why a mixture of ice (and water) and salt is used to freeze homemade ice cream. Why won't just ice work?

7.63 If you look at the labels of automotive products used to prevent radiator freezing (antifreeze) and radiator boiling, you will find the same ingredient listed, ethylene glycol. Use the idea of colligative properties to explain how the same material can prevent an automobile cooling system from freezing and boiling.

7.64 Calculate the boiling and freezing points of water solutions that are 1.50 M in the following solutes:
a. KBr, a strong electrolyte

b. glycerol, a nonelectrolyte
c. $(NH_4)_2SO_4$, a strong electrolyte
d. $Al(NO_3)_3$, a strong electrolyte

7.65 Calculate the boiling and freezing points of water solutions that are 1.00 M in the following solutes:
a. KBr, a strong electrolyte
b. ethylene glycol, a nonelectrolyte
c. $(NH_4)_2CO_3$, a strong electrolyte
d. $Al_2(SO_4)_3$, a strong electrolyte

7.66 Calculate the boiling and freezing points of the following solutions. Water is the solvent unless otherwise indicated.
a. A 0.750 M solution of urea, a nonelectrolyte
b. A 0.150 M solution of $CaCl_2$, a strong electrolyte
c. A solution containing 120 g of ethylene glycol ($C_2H_6O_2$) per 400 mL of solution

7.67 Calculate the boiling and freezing points of the following solutions. Water is the solvent, unless otherwise indicated.
a. A solution containing 50.0 g of H_2SO_4, a strong electrolyte (both Hs dissociate), per 250 mL
b. A solution containing 200 g of table sugar ($C_{12}H_{22}O_{11}$), a nonelectrolyte, per 250 mL
c. A solution containing 75.0 g of octanoic acid ($C_8H_{16}O_2$), a nonelectrolyte, in enough benzene to give 250 mL of solution

7.68 Calculate the osmolarity for the following solutions:
a. A 0.20 M solution of glycerol, a nonelectrolyte
b. A 0.20 M solution of $(NH_4)_2SO_4$, a strong electrolyte
c. A solution containing 33.7 g of LiCl (a strong electrolyte) per 500 mL

7.69 Calculate the osmolarity for the following solutions:
a. A 0.25 M solution of KCl, a strong electrolyte
b. A solution containing 15.0 g of urea (CH_4N_2O), a nonelectrolyte, per 500 mL
c. A solution containing 50.0 mL of ethylene glycol ($C_2H_6O_2$), a nonelectrolyte with a density of 1.11 g/mL, per 250 mL

NOTE: In Exercises 7.70–7.79, assume the temperature is 25.0°C, and express your answer in torr, mmHg, and atm.

7.70 Calculate the osmotic pressure of any solution with an osmolarity of 0.300.

7.71 Calculate the osmotic pressure of any solution with an osmolarity of 0.250.

7.72 Calculate the osmotic pressure of a 0.200 M solution of Na_2SO_4, a strong electrolyte.

7.73 Calculate the osmotic pressure of a 0.300 M solution of methanol, a nonelectrolyte.

7.74 Calculate the osmotic pressure of a solution that contains 95.0 g of the nonelectrolyte urea, CH_4N_2O, per 500 mL of solution.

7.75 Calculate the osmotic pressure of a solution that contains 1.20 mol of $CaCl_2$ in 1500 mL.

7.76 Calculate the osmotic pressure of a solution that has a freezing point of −0.35°C.

7.77 Calculate the osmotic pressure of a solution that is 0.122 M in solute and has a boiling point 0.19°C above that of pure water.

7.78 Calculate the osmotic pressure of a solution that contains 5.30 g of NaCl and 8.20 g of KCl per 750 mL.

7.79 Calculate the osmotic pressure of a solution that contains 245.0 g of ethylene glycol ($C_2H_6O_2$), a nonelectrolyte, per liter.

COLLOIDS AND DIALYSIS (SECTIONS 7.8–7.10)

7.80 Suppose you have a bag made of a membrane like that in Figure 7.18. Inside the bag is a solution containing water and dissolved small molecules. Describe the behavior of the

system when the bag functions as an osmotic membrane and when it functions as a dialysis membrane.

7.81 Explain why detergents or soaps are needed if water is to be used as a solvent for cleaning clothes and dishes.

7.82 Explain how the following behave in a colloidal suspension: dispersing medium, dispersed phase, and colloid emulsifying agent.

7.83 Suppose an osmotic membrane separates a 5.00% sugar solution from a 10.0% sugar solution. In which direction will water flow? Which solution will become diluted as osmosis takes place?

CHEMISTRY FOR THOUGHT

7.1 When a patient has blood cleansed by hemodialysis, the blood is circulated through dialysis tubing submerged in a bath that contains the following solutes in water: 0.6% NaCl, 0.04% KCl, 0.2% NaHCO₃, and 0.72% glucose (all percentages are w/v). Suggest one or more reasons why the dialysis tubing is not submerged in pure water.

7.2 Can the terms *saturated* and *supersaturated* be used to describe solutions made of liquids that are soluble in all proportions? Explain.

7.3 Refer to Figure 7.3 and propose a reason why fish sometimes die when the temperature of the water in which they live increases.

7.4 Small souvenir salt-covered objects are made by forming the object out of wire mesh and suspending the mesh object in a container of water from a salt lake such as the Dead Sea or Great Salt Lake. As the water evaporates, the wire mesh becomes coated with salt crystals. Describe this process using the key terms introduced in Section 7.2.

7.5 Refer to Figure 7.6 and answer the question. How would the solubility of sugar compare in equal amounts of hot and iced tea?

7.6 Refer to Figure 7.12 and explain the process as requested. Draw simple diagrams showing the initial appearance and appearance after some time for a similar experiment in which the two liquids are 0.20 M copper sulfate solution and 2.0 M copper sulfate solution. Explain your reasoning.

7.7 Refer to Figure 7.15 and answer the question. Propose at least two explanations for the movement of molasses out of the carrot into the water.

7.8 Strips of fresh meat can be preserved by drying. In one process, the strips are coated with table salt and exposed to the air. Use a process discussed in this chapter and one discussed in Chapter 6 to explain how the drying takes place.

InfoTrac College Edition Readings

"Biomedical & biomembrane: delivery method for low solubility drugs," *Membrane & Separation Technology News*, July 2002, 20(10):0. Record number A89841213.

"Drug therapy attempts to avoid dialysis," *Biotech Week*, Oct 2, 2002:46. Record number A92071493.

"Beyond Jell-O: new ideas gel in the lab," *Science News*, May 25, 2002, 161(21):323(2). Record number A87353950.

GOB Chemistry⋅Now™

Assess your understanding of this chapter's topics with additional quizzing and conceptual-based problems at http://chemistry.brookscole.com/seager5e

CHAPTER 8

Reaction Rates and Equilibrium

Medical laboratory technicians perform diagnostic tests on all types of samples from the body. Here, a technician is collecting blood samples to be tested for several substances, including glucose. The rate at which blood glucose is used by the body is a valuable indicator in the diagnosis and treatment of diabetes. This chapter will introduce you to the important areas of reaction rates.

© Roger Ressmeyeer/CORBIS

LEARNING OBJECTIVES

When you have completed your study of this chapter, you should be able to:

1. Use the concepts of energy and entropy to predict the spontaneity of processes and reactions. (Section 8.1)

2. Calculate reaction rates from experimental data. (Section 8.2)

3. Use the concept of molecular collisions to explain reaction characteristics. (Section 8.3)

4. Represent and interpret the energy relationships for reactions by using energy diagrams. (Section 8.4)

5. Explain how factors such as reactant concentrations, temperature, and catalysts influence reaction rates. (Section 8.5)

6. Write equilibrium expressions based on reaction equations. (Section 8.7)

7. Do calculations based on equilibrium expressions. (Section 8.7)

8. Use Le Châtelier's principle to predict the influence of changes in concentration and reaction temperature on the position of equilibrium for a reaction. (Section 8.8)

According to calculations, carbon in the form of a diamond will spontaneously change into graphite. Owners of diamonds seem unconcerned by this fact; they know from experience that their investments in gems are secure. In reality, the change does take place, but so slowly that it is not detectable over many human lifetimes. The low rate of the reaction makes the difference.

The element barium is very poisonous when it is taken into the body as barium ions (Ba^{2+}). When barium sulfate ($BaSO_4$) dissolves, Ba^{2+} and SO_4^{2-} ions are formed. However, suspensions of solid $BaSO_4$ are routinely swallowed by patients undergoing diagnostic X-ray photography of the stomach and intestinal tract. The patients are not affected because very little $BaSO_4$ dissolves in the body. This lack of solubility is an equilibrium property of $BaSO_4$.

➤ 8.1 Spontaneous and Nonspontaneous Processes

Years ago, tourists in Salt Lake City visited "gravity hill," where an optical illusion made a stream of water appear to flow uphill. The fascination with the scene came from the apparent violation of natural laws—everyone knew that water does not flow uphill. Processes that take place naturally with no apparent cause or stimulus are called **spontaneous processes**. Nonspontaneous processes take place only as the result of some cause or stimulus. For example, imagine you are in the positions depicted in ■ Figure 8.1 and want to move a boulder.

In (a), the boulder will roll down the hill as soon as you release it; the process begins and takes place spontaneously. In (b), you must push the boulder a little to get it over the hump, but once started, the boulder spontaneously rolls downhill. Situation (c) is very different. You must continually push on the boulder, or it will not move up the hill; at no time can you stop pushing and expect the boulder to continue moving up. The process is nonspontaneous; it takes place only because of the continuous application of an external stimulus.

As the boulder rolls downhill, it gives up energy; it moves from a state of high potential energy to a lower one. As you push the boulder uphill, it gains energy (from you). Processes that give up energy are called **exergonic** (energy out), whereas those that gain energy are called **endergonic** (energy in). Very often, energy changes in chemical processes involve heat. Those changes that do are referred to as either exothermic (heat out) or endothermic (heat in), two terms you have encountered before (Sections 5.8 and 6.11).

Spontaneous process
A process that takes place naturally with no apparent cause or stimulus.

Exergonic process
A process that gives up energy as it takes place.

Endergonic process
A process that gains or accepts energy as it takes place.

➤ **FIGURE 8.1** The problem of moving a boulder.

(a) Spontaneous process

(b) Spontaneous process once started

(c) Nonspontaneous process

Many spontaneous processes give up energy. A piece of wood, once ignited, spontaneously burns and liberates energy as heat and light. At normal room temperature, steam spontaneously condenses to water and releases heat. However, some spontaneous processes take place and either give up no energy or actually gain energy. These spontaneous processes are always accompanied by another change called an entropy increase. **Entropy** describes the disorder or mixed-up character (randomness) of a system (see ■ Figure 8.2). Thus, an entropy increase accompanies all spontaneous processes in which energy remains constant or is gained. A drop of ink placed in a glass of water will eventually become uniformly distributed throughout the water, even though the water is not stirred. No energy change accompanies the diffusion of the ink, but the distribution of ink throughout the water is a more disorderly (higher-entropy) state than when the ink is all together in a single drop. Ice melts spontaneously at 20°C and in the process absorbs heat. The process takes place because the random distribution of water molecules in the liquid is a higher-entropy state than the orderly molecular arrangement found in crystalline ice.

Energy and entropy changes influence the spontaneity of processes in several ways:

1. A process will always be spontaneous if the energy decreases and the entropy increases. An example of such a process is the burning of a piece of wood in which heat is given up and the gaseous products of combustion are distributed throughout the surroundings, providing the entropy increase.
2. When a spontaneous process is accompanied by an energy increase, a large entropy increase must also occur. Thus, when ice spontaneously melts at 20°C, the increase in entropy is large enough to compensate for the increase in energy that also takes place.
3. A spontaneous process accompanied by an entropy decrease must also be accompanied by a compensating energy decrease. Thus, when water spontaneously freezes at 0°C, the energy loss compensates for the entropy decrease that occurs as the water molecules assume the well-ordered arrangement in ice.

It is useful to think of energy and entropy changes in terms of the directions favoring spontaneity and the directions not favoring spontaneity. It is apparent from the preceding discussion that spontaneity is favored by an energy decrease and/or an entropy increase. Processes in which one of these factors changes in a nonspontaneous direction will be spontaneous only if the other factor changes in a spontaneous direction by a large enough amount to compensate for the nonspontaneous change. The influences listed under 2 and 3 above are examples of this.

Entropy
A measurement or indication of the disorder or randomness of a system. The more disorderly a system is, the higher its entropy.

➤ **FIGURE 8.2** Entropy is an indicator of randomness. The mixture on the right has higher entropy (more randomness, or mixed-up character) than the mixture on the left.

Stable substance
A substance that does not undergo spontaneous changes under the surrounding conditions.

Substances that do not undergo spontaneous changes are said to be **stable**. However, stability depends on the surrounding conditions, and a change in those conditions might cause a nonspontaneous process to become spontaneous. Ice is a stable solid at $-10°C$ and 760 torr pressure but spontaneously melts to a liquid at $5°C$ and 760 torr. Wood is stable at room temperature but spontaneously burns when its temperature equals or exceeds the ignition temperature. In the following discussions, the surrounding conditions are assumed to be those normally encountered. Otherwise, the differences are specified.

➤ 8.2 Reaction Rates

Reaction rate
The speed of a reaction.

The speed of a reaction is called the **reaction rate**. It is determined experimentally as the change in concentration of a reactant or product divided by the time required for the change to occur. This average rate is represented for changes in product concentration by Equation 8.1:

$$\text{Rate} = \frac{\Delta C}{\Delta t} = \frac{C_t - C_0}{\Delta t} \tag{8.1}$$

where the delta symbol (Δ) stands for change. C_t and C_0 are the concentrations of a product at the end and beginning, respectively, of the measured time change, Δt. The time can be measured in any convenient unit.

OVER THE COUNTER • 8.1

TIMED-RELEASE MEDICATIONS

Most of us have experienced the inconvenience of having to take repeated doses of a medication at specific intervals; this dosing maintains an effective level of the medication in the body throughout the day. In 1961, the first commercial attempt was made to overcome this inconvenience when the decongestant Contac® became available in the form of timed-release capsules. The approach used to time the release was simple. Each capsule contained many tiny beads of medication. Each bead was coated with a slow-dissolving polymer, but the thickness of the polymer coating was not the same on all beads. Beads with a thin coating dissolved rapidly, while those with a thicker coating dissolved more slowly. By including an appropriate mix of beads with coatings of different thickness in each capsule, a gradual, steady release of medication over a specific time period was achieved.

This same method for timed-release is still used in some products, but other methods have also been developed. The modern technology is used not only to provide a steady release of medication, but also to control the point of release in the digestive tract. Some medications are known to irritate and damage the stomach lining, or to be destroyed by the acidic environment of the stomach. An *enteric coating* that is stable in acid but that dissolves in the more basic environment of the small intestines is used to allow tablets or caplets of such materials to pass through the stomach and release their medication in the intestines.

In a pharmacy, OTC products in the following categories can be found with enteric coatings or in timed-release forms that claim to provide effective levels of medication for periods ranging from 6 to 24 hours: analgesics (pain relievers), allergy treatments, cold/flu remedies, laxatives, sleeping aids, gas-relief medications, appetite-control materials, diuretics (water pills), and tablets to prevent sleeping.

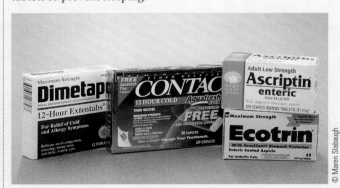

A wide variety of products is available in timed-release form.

Example 8.1

The gases NO_2 and CO react as follows:

$$NO_2(g) + CO(g) \rightarrow NO(g) + CO_2(g)$$

Calculate the average rate of the reaction if pure NO_2 and CO are mixed and after 50.0 seconds the concentration of CO_2 is found to be 0.0160 mol/L.

Solution

Because the reaction was started by mixing pure NO_2 and CO, the initial concentration C_0 of CO_2 was 0.000 mol/L. After 50.0 seconds, the concentration C_t of CO_2 was 0.0160 mol/L.

$$\text{Rate} = \frac{C_t - C_0}{\Delta t} = \frac{0.0160 \text{ mol/L} - 0.000 \text{ mol/L}}{50.0 \text{ s}} = 3.2 \times 10^{-4} \frac{\text{mol/L}}{\text{s}}$$

The answer is read 3.2×10^{-4} mole per liter per second. It means that during the 50.0-second time interval an average of 3.2×10^{-4} mol CO_2 was formed in each liter of reacting mixture each second.

Learning Check 8.1

The Ce^{4+} and Fe^{2+} ions react in solutions as follows:

$$Ce^{4+}(aq) + Fe^{2+}(aq) \rightarrow Ce^{3+}(aq) + Fe^{3+}(aq)$$

Solutions of Ce^{4+} and Fe^{2+} are mixed and allowed to react. After 75.0 seconds, the concentration of Ce^{3+} is found to be 1.50×10^{-5} mol/L. Calculate the average rate of the reaction.

➤ 8.3 Molecular Collisions

Chemical equations such as those in Example 8.1 and Learning Check 8.1 are useful in identifying reactants and products and in representing the stoichiometry of a reaction. However, they indicate nothing about how a reaction takes place—how reactants get together (or break apart) to form products. The explanation of how a reaction occurs is called a **reaction mechanism.** Reaction mechanisms are not discussed much in this book, but most are based on the following assumptions:

Reaction mechanism
A detailed explanation of how a reaction actually takes place.

1. Reactant particles must collide with one another in order for a reaction to occur.
2. Particles must collide with at least a certain minimum total amount of energy if the collision is to result in a reaction.
3. In some cases, colliding reactants must be oriented in a specific way if a reaction is to occur.

The validity of the first assumption is fairly obvious. Two molecules cannot react with each other if they are far apart. In order to break bonds, exchange atoms, and form new bonds, they must come in contact. There are, however, some exceptions, such as decompositions in which molecules break into fragments and processes in which molecules react by an internal rearrangement of their atoms.

In general, reactions take place more rapidly in the gaseous or liquid state than in the solid state. This observation verifies the first assumption because molecules of gases and liquids move about freely and can undergo many more collisions than can the rigidly held molecules of solids. Reactions involving solids usually take place only on the solid surface and therefore involve only a small fraction of the total molecules present in the solid. As the reaction proceeds and the products dissolve, diffuse, or fall from the surface, fresh solid is exposed. In this way, the reaction proceeds into the solid. The rusting of iron is an example of such a process.

If collisions were the only factor, however, most gaseous and liquid state reactions would take place almost instantaneously if every collision resulted in a reaction. Such high reaction rates are not observed, a fact that brings us to assumptions 2 and 3.

One of the ways to speed up a chemical reaction is to add energy in the form of heat. The added heat increases both the average speed (kinetic energy) and the internal energy of the molecules. **Internal energy** is the energy associated with molecular vibrations. An increase in internal energy increases the amplitude of molecular vibrations and, if large enough, breaks bonds (see ■ Figure 8.3). Internal energy is also increased by the conversion of some kinetic energy into internal energy during collisions. When heat is added, both the kinetic and the internal energy of molecules is increased. This increases the frequency and speed of collisions and the chances that a collision will cause a sufficient increase in internal energy to break bonds and bring about a reaction (see ■ Figure 8.4).

In some reaction mixtures, the average total energy of the molecules is too low at the prevailing temperature for a reaction to proceed at a detectable rate; the reaction mixture is stable. However, some reactions can be started by providing **activation energy.** Once the reaction is started, enough energy is released to activate other molecules and keep the reaction going at a good rate. The energy needed to get the boulder over the small hump in Figure 8.1b is a kind of activation energy. In many chemical reactions, activation energy causes bonds in reactant molecules to break. When the broken bonds react to form the new bonds of the products, energy is released that can cause bonds in more reactant molecules to break and the reaction to continue. The striking of a kitchen match is a good example. Activation energy is provided by rubbing the match head against a rough surface. Once started, however, the match continues to burn spontaneously.

A number of gases are routinely used in hospitals. Some of these form very flammable or even potentially explosive mixtures. Cyclopropane, a formerly used anesthetic, will burn vigorously in the presence of the oxygen in air. However, a mixture of the two will not react unless activation energy is provided in the form of an open flame or a spark. This is the reason that

Internal energy
The energy associated with vibrations within molecules.

Activation energy
Energy needed to start some spontaneous processes. Once started, the processes continue without further stimulus or energy from an outside source.

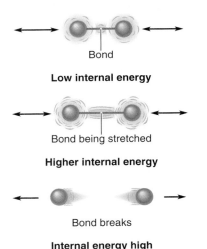

Bond

Low internal energy

Bond being stretched

Higher internal energy

Bond breaks

Internal energy high enough to break bonds

➤ **FIGURE 8.3** The internal energy of molecules.

© Richard Megna/Fundamental Photographs

➤ **FIGURE 8.4** The reaction that makes a Cyalume™ light stick glow takes place more rapidly in hot water (left) than in ice water (right). Which stick would glow for a longer time? Explain your reasoning.

extreme precautions are taken to avoid open flames and sparks in a hospital operating room. Even sparks from static electricity can set off such gaseous mixtures, so special materials are used in clothing, floor coverings, and so on to prevent static electricity from building up.

Oxygen gas does not burn, but it reacts with anything that is combustible. Oxygen is often used in hospitals in concentrations much higher than the 20% found in air. Thus, precautions must be taken to prevent fire or sparks in any areas where oxygen gas is used in high concentrations.

Orientation effects are related to which side or end of a particle actually hits another particle in collision. Orientation effects are unimportant in many reactions. For example, the orientation of silver ions (Ag^+) and chloride ions (Cl^-) toward each other during collision has no effect on the rate of forming AgCl:

$$Ag^+(aq) + Cl^-(aq) \rightarrow AgCl(s) \qquad (8.2)$$

The reason is that Ag^+ and Cl^- are both essentially spherical charged particles. However, collision orientation may be important in reactions that involve non-spherical molecules (assumption 3). Consider the following hypothetical reaction:

$$A—B + C—D \rightarrow A—C + B—D \qquad (8.3)$$

It is clear that the collision orientation of $A—B$ and $C—D$ shown in ■ Figure 8.5a is more favorable to reaction than the orientations shown in (b) and (c).

(a) Favorable orientation: A is near C, and B is near D during collision.

(b) Unfavorable orientation: A is not near C, and B is not near D during collision.

(c) Unfavorable orientation: B is near D, but A and C are far removed during collision.

➤ **FIGURE 8.5** Molecular orientations during collisions.

➤ 8.4 Energy Diagrams

Energy relationships for reactions can be represented by energy diagrams like the one in ■ Figure 8.6. Notice the similarity to the earlier example of rolling boulders.

The energy diagrams for most reactions look generally alike, but there are some differences. Typical diagrams for exothermic (exergonic) and endothermic (endergonic) reactions are given in ■ Figure 8.7. It is clear that the products of exothermic reactions have lower energy than the reactants and that the products of endothermic reactions have higher energy than the reactants. When exothermic reactions occur, the energy difference between reactants and products is released. This energy often appears as heat, so heat is given up to the surroundings. When endothermic reactions occur, the energy (heat) added to the products is absorbed from the surroundings. Also, different reactions generally have different activation energies, a fact easily represented by energy diagrams.

White phosphorus, a nonmetallic element, spontaneously bursts into flame if left in a rather warm room (34°C):

$$4P(s) + 5O_2(g) \rightarrow P_4O_{10}(s) \qquad (8.4)$$

➤ **FIGURE 8.6** A typical energy diagram for chemical reactions.

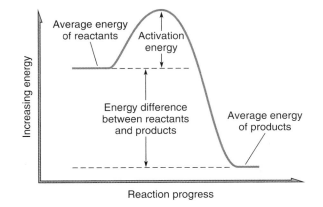

Reaction progress

➤ **FIGURE 8.7** Energy diagrams for exothermic and endothermic reactions.

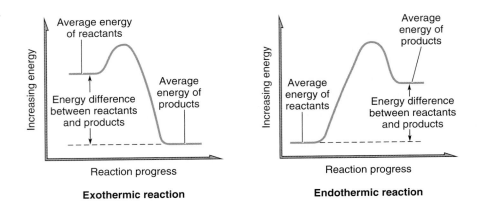

Exothermic reaction

Endothermic reaction

➤ **FIGURE 8.7** Energy diagrams for exothermic and endothermic reactions.

Sulfur, another nonmetallic element, will also burn, but it does not ignite until heated to about 232°C:

$$S(s) + O_2(g) \rightarrow SO_2(g) \qquad (8.5)$$

The different activation energies for these two exothermic reactions are represented in ■ Figure 8.8.

It should now be clear why we stressed the notion that some substances that are "stable" at normal living conditions undergo spontaneous changes at other conditions. It is simply that a higher temperature would provide the necessary activation energy. In a room hotter than 232°C, for example, both white phosphorus and sulfur would be "unstable" in the presence of oxygen.

➤ 8.5 Factors That Influence Reaction Rates

Reaction rates are influenced by a number of different factors. Four factors that affect the rates of all reactions are:

1. The nature of the reactants.
2. The concentration of the reactants.
3. The temperature of the reactants.
4. The presence of catalysts.

The formation of insoluble AgCl by mixing together solutions containing Ag^+ and Cl^- ions is represented by Equation 8.2, given earlier. The white solid AgCl forms the instant the two solutions are mixed. This behavior is typical

➤ **FIGURE 8.8** Differences in activation energies.

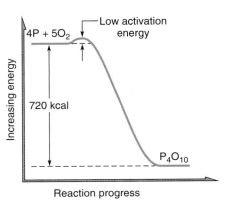

Low activation energy, high heat of reaction

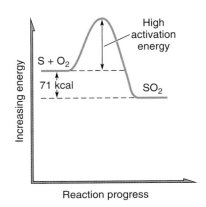

High activation energy, low heat of reaction

of reactions involving ionic reactants. The high reaction rate results from the attraction of the charged reactants to each other. In contrast, reactions that require covalent bonds to be broken or formed often proceed slowly. The production of water gas, a mixture of hydrogen (H_2) and carbon monoxide (CO), is a reaction involving covalent bonds:

$$H_2O(g) + C(s) \xrightarrow{\text{Heat}} CO(g) + H_2(g) \tag{8.6}$$

In addition, other structural characteristics of reactants, such as bond polarity or molecular size, may also be important factors in reaction rates.

The influence of reactant concentration on reaction rates can be illustrated by using the concept of molecular collisions. Suppose a reaction takes place between the hypothetical molecules A and B, which are mixed in a 1:1 ratio. The proposed reaction is

$$A + B \rightarrow \text{products} \tag{8.7}$$

Collisions with the capability to cause a reaction to occur are called **effective collisions**. Only collisions between A and B molecules can be classified as effective, since collisions between two A molecules or between two B molecules cannot possibly yield products. Imagine the reaction is begun with two A molecules and two B molecules as shown in ■ Figure 8.9a. The example is simplified by looking only at the collisions of a single A molecule. Initially, two of every three collisions of the A molecule will be effective (a). On doubling the number (concentration) of B molecules, the number of effective collisions also doubles, and four of every five collisions is effective (b).

When larger numbers of molecules are used, the results approach more closely those actually observed experimentally for simple chemical reactions. Imagine a mixture containing 1000 A molecules and 1000 B molecules. On the average, 1000 out of every 1999 collisions of a single A molecule will involve a B molecule. (Each A molecule has almost a 50–50 chance of bumping into a B molecule.) This gives essentially a 1:1 ratio of effective to noneffective collisions. When the number of B molecules is doubled to 2000, the ratio becomes 2:1, because 2000 of every 2999 collisions will be effective. Therefore, the reaction rate should double when the concentration of one reactant is doubled. This result has been verified in numerous chemical reactions. Thus, higher concentrations produce a larger number of effective collisions in a given period of time, and this increases the reaction rate.

Gas-phase reactions are easily visualized this way, but reacting liquids and solids must be looked at differently. A large piece of solid contains a large number of molecules, but, as described before, only those on the surface can react with the molecules of another substance. For this reason, the total amount of solid in a sample is not as important as the surface area of solid in contact with other reactants. The effective concentration of a solid therefore depends on its surface area and state of division. A 100-pound sack of flour is difficult to burn when it is in a single pile. The same 100 pounds, dispersed in the air as a fine dust, burns very rapidly, and a dust explosion results (see ■ Figure 8.10). The effective concentrations of reacting liquids must be thought of in a similar way unless the reactants are completely miscible.

The effect of reactant temperature on reaction rates can also be explained by using the molecular collision concept. As was noted earlier, an increase in the temperature of a system increases the average kinetic energy and internal energy of the reacting molecules. The increased molecular speed (kinetic energy) causes more collisions to take place in a given time. Also, because the kinetic and internal energies of the colliding molecules are greater, a larger fraction of the collisions will be effective because they will provide the necessary activation energy.

Effective collision
A collision that causes a reaction to occur between the colliding molecules.

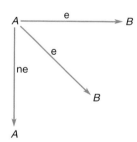

(a) Low concentration of B

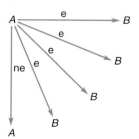

(b) Higher concentration of B

e = possibly effective collisions
ne = noneffective collisions

➤ **FIGURE 8.9** The effect of concentration on reaction rates.

A compact pile of lycopodium powder burns sluggishly.

A similar sample of lycopodium powder in a blowpipe.

The blown powder burns instantly with a bright flash.

➤ **FIGURE 8.10** Surface area influences reaction rates of solids. What is the other reactant in this reaction? How finely divided is it?

As a rough rule of thumb, it has been found that the rate of a chemical reaction doubles for every 10°C increase in temperature. The chemical reactions of cooking take place faster in a pressure cooker because of a higher cooking temperature (Section 6.13). Also, cooling or freezing is used to slow the chemical reactions involved in the spoiling of food, souring of milk, and ripening of fruit.

Catalysts are substances that change reaction rates without being used up in the reactions. Usually the term *catalyst* is used to describe substances that speed up reactions. Substances that slow reactions are known as **inhibitors**. Catalysts are used in a number of different forms. Those dispersed uniformly throughout a reaction mixture in the form of individual ions or molecules are called **homogeneous catalysts. Heterogeneous** or **surface catalysts** are used in the form of solids; usually they have large surface areas on which reactions take place readily.

Catalysts enhance a reaction rate by providing an alternate reaction pathway that requires less activation energy than the normal pathway. This effect is represented by ■ Figure 8.11. According to some theories, a catalyst provides the lower-energy pathway by entering into a reaction and forming an intermediate structure, which then breaks up to produce the final products and regenerate the catalyst:

Uncatalyzed: $\qquad A + B \rightarrow \text{products}$ (8.8)

Catalyzed: $\quad A + B + \text{catalyst} \rightarrow A\!-\!B \rightarrow \text{products} + \text{catalyst}$ (8.9)
$$\underset{\text{catalyst}}{|\quad|}$$

Intermediate structure

According to another proposed mechanism, solid catalysts provide a surface to which reactant molecules become attached with a particular orientation. Reactants attached to these surfaces are sufficiently close to one another and oriented favorably enough to allow the reaction to take place. The products of the reaction then leave the surface and make the attachment sites available for catalyzing other reactant molecules.

Catalyst
A substance that changes (usually increases) reaction rates without being used up in the reaction.

Inhibitor
A substance that decreases reaction rates.

Homogeneous catalyst
A catalytic substance that is distributed uniformly throughout the reaction mixture.

Heterogeneous or surface catalyst
A catalytic substance normally used in the form of a solid with a large surface area on which reactions take place.

➤ **FIGURE 8.11** The effect of catalysts on activation energy.

➤ 8.6 Chemical Equilibrium

So far, the focus has been only on the reactants of a reaction. However, all reactions can (in principle) go in both directions, and the products can be looked on as reactants. Suppose equal amounts of gaseous H_2 and I_2 are placed

CHEMISTRY AND YOUR HEALTH • 8.1

PROTECTING THE ELDERLY FROM WINTER'S COLD

An individual is said to be in a state of hypothermia when his or her body temperature is lower than 35°C (95°F). Chemical reactions essential to life occur in the organs of the body. Because the speed of chemical reactions decreases as temperature decreases, the life-supporting functions of the body's organs also slow as the temperature drops, creating a very serious health threat. Hypothermia occurs more often in men than in women, and the risk is greater for those who do not eat a balanced diet, suffer from diabetes or heart disease, and have liver or hypothyroid conditions. Certain drugs also increase an older person's risk, including drugs used to treat anxiety, depression, and nausea, and even some over-the-counter cold remedies.

Elderly people are quite susceptible to hypothermia because they often fail to realize how cold they are, and their bodies do not generate and distribute as much heat as they did at a younger age. Some symptoms of hypothermia include sluggish movement, confusion, dizziness, slurred speech, slow breathing, drowsiness, irregular or slowed heartbeat, blue fingers and toes, rigid muscles, very cold skin, and coma. The ways to avoid hypothermia seem obvious: Dress warmly, use appropriate clothing, keep living quarters warm, and sleep warmly. While such measures seem obvious, some elderly people, especially those living alone, need to be checked on a regular basis during cold weather and, where necessary, helped to take adequate precautions.

---

in a closed container and allowed to react and form HI. Initially, no HI is present, so the only possible reaction is

$$H_2(g) + I_2(g) \rightarrow 2HI(g) \tag{8.10}$$

However, after a short time, some HI molecules are produced. They can collide with one another in a way that causes the reverse reaction to occur:

$$2HI(g) \rightarrow H_2(g) + I_2(g) \tag{8.11}$$

The low concentration of HI makes this reaction slow at first, but as the concentration increases, so does the reaction rate. The rate of Reaction 8.10 decreases as the concentrations of H_2 and I_2 decrease. Eventually, the concentrations of H_2, I_2, and HI in the reaction mixture reach levels at which the rates of the forward (Equation 8.10) and reverse (Equation 8.11) reactions are equal. From that time on, the concentrations of H_2, I_2, and HI in the mixture remain constant, since both reactions take place at the same rate and each substance is being produced as fast as it is used up. When the forward and reverse reaction rates are equal, the reaction is in a **state of equilibrium**, and the concentrations are called **equilibrium concentrations.**

The behavior of reaction rates and reactant concentrations for both the forward and reverse reactions is shown graphically in ■ Figure 8.12.

Chemistry·⚛·Now™

Go to GOB Now and click to see how systems starting with differing amounts of reactants and products can each attain a state of equilibrium.

State of equilibrium
A condition in a reaction system when the rates of the forward and reverse reactions are equal.

Equilibrium concentrations
The unchanging concentrations of reactants and products in a reaction system that is in a state of equilibrium.

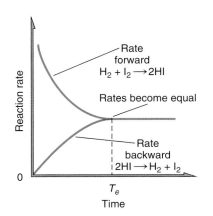 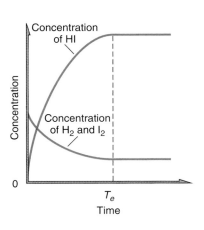

> ➤ **FIGURE 8.12** Variation of reaction rates and reactant concentrations as equilibrium is established (T_e is the time needed to reach equilibrium).

Instead of writing separate equations for both the forward and reverse reactions, the usual practice is to represent reversibility by double arrows. Thus, the equation for the reaction between H_2 and I_2 is written

$$H_2(g) + I_2(g) \rightleftarrows 2HI(g) \qquad (8.12)$$

▶ 8.7 The Position of Equilibrium

Position of equilibrium

An indication of the relative amounts of reactants and products present at equilibrium.

The **position of equilibrium** for a reaction indicates the relative amounts of reactants and products present at equilibrium. When the position is described as being far to the right, it means that at equilibrium the concentration of products is much higher than the concentration of reactants. A position far to the left means the concentration of reactants is much higher than that of products.

The position of equilibrium can be represented numerically by using the concept of an equilibrium constant. Any reaction that establishes an equilibrium can be represented by a general equation:

$$aA + bB + \cdots \rightleftarrows wW + xX + \cdots \qquad (8.13)$$

GOB
Chemistry·⚛·Now™

Go to GOB Now and click to engage in a simulation exploring the nature of the equilibrium constant.

In this equation, the capital letters stand for substances such as the H_2, I_2, and HI of Equation 8.12, and the lowercase letters are the coefficients in the balanced equation such as the 1, 1, and 2 of Equation 8.12. The dots in Equation 8.13 indicate that any number of reactants and products can be involved in the reaction, but we will limit ourselves to just the four shown.

As we implied earlier, a reaction at equilibrium can be recognized because the concentrations of reactants and products remain constant. That is, they do not change as time passes. For a reaction at equilibrium, the following equation is valid:

$$K = \frac{[W]^w[X]^x}{[A]^a[B]^b} \qquad (8.14)$$

Equilibrium expression

An equation relating the equilibrium constant and reactant and product equilibrium concentrations.

Equilibrium constant

A numerical relationship between reactant and product concentrations in a reaction at equilibrium.

In this **equilibrium expression**, the brackets [ ] represent molar concentrations of the reactants (A and B) and the products (W and X). The K is a constant called the **equilibrium constant,** and the powers on each bracket are the coefficients from the balanced equation for the reaction. According to this equation, the product of equilibrium concentrations of products (raised to appropriate powers) divided by the product of the equilibrium concentration of reactants (also raised to appropriate powers) gives a number that does not change with time (the equilibrium constant). The reason K does not change with time is that at equilibrium the concentrations used to calculate K do not change with time. We could get a different constant by dividing the reactant concentrations by the product concentrations, but it is the accepted practice to calculate K values as we have shown in Equation 8.14.

Example 8.2

GOB
Chemistry·⚛·Now™

Go to GOB Now and click to determine how to construct equilibrium constant expressions.

Write equilibrium expressions for the reactions represented by the following equations:

a. $H_2(g) + I_2(g) \rightleftarrows 2HI(g)$
b. $2SO_2(g) + O_2(g) \rightleftarrows 2SO_3(g)$

Solution

a. The concentration of the product HI goes on top and is raised to the power 2. The concentrations of the reactants H_2 and I_2 go on the bottom, and each is raised to the power 1, which is not written but understood:

$$K = \frac{[HI]^2}{[H_2][I_2]}$$

b. The product is SO_3, and the power is 2. The reactants are SO_2 (power = 2) and O_2 (power = 1).

$$K = \frac{[SO_3]^2}{[SO_2]^2[O_2]}$$

Write equilibrium expressions for the reactions represented by the following equations:

a. $N_2O_4(g) \rightleftarrows 2NO_2(g)$
b. $3NO(g) \rightleftarrows N_2O(g) + NO_2(g)$

Learning Check 8.2

For a reaction with an equilibrium position far to the right, the product concentrations [W] and [X] will be much higher than the reactant concentrations [A] and [B]. According to Equation 8.14, such a situation should result in a large K value. For reactions with equilibrium positions far to the left, a similar line of reasoning leads to the conclusion that K values should be small. Thus, the value of the equilibrium constant gives an indication of the position of equilibrium for a reaction.

Example 8.3

For each of the following equations, the equilibrium molarity concentration is given below each reactant and product. Calculate K for each reaction and comment on the position of equilibrium.

a. $2NO_2(g) \rightleftarrows N_2O_4(g)$
     8.23×10^{-3} M 1.46×10^{-2} M at 25°C

b. $CH_4(g) + H_2O(g) \rightleftarrows CO(g) + 3H_2(g)$
     0.200 M   0.150 M 1.37×10^{-2} M 4.11×10^{-2} M at 900 K

Solution

a. $K = \dfrac{[N_2O_4]}{[NO_2]^2} = \dfrac{(1.46 \times 10^{-2} \text{ mol/L})}{(8.23 \times 10^{-3} \text{ mol/L})^2} = \dfrac{2.16 \times 10^2}{\text{mol/L}}$

We see that K is fairly large, which tells us the equilibrium position is toward the right, or product, side of the reaction. The peculiar sounding "per mole per liter" unit is the result of the way concentration terms are arranged in the equilibrium expression. The unit will not be the same for the K of all reactions. These units are not ordinarily shown, a practice we will follow for the remainder of this chapter.

b. $K = \dfrac{[CO][H_2]^3}{[CH_4][H_2O]}$

$= \dfrac{(1.37 \times 10^{-2} \text{ mol/L})(4.11 \times 10^{-2} \text{ mol/L})^3}{(0.200 \text{ mol/L})(0.150 \text{ mol/L})}$

$= 3.17 \times 10^{-5}$

The small value of K indicates the equilibrium position is toward the left.

The molar concentrations are given below the reactants and products in the equations for the following reactions. Calculate K and comment on the position of each equilibrium.

a.
$$Br_2(g) \quad + \quad I_2(g) \quad \rightleftarrows \quad 2IBr(g)$$
1.50×10^{-1} M $\quad 5.00 \times 10^{-2}$ M $\quad 1.96 \times 10^{-2}$ M at 25°C

b.
$$N_2(g) \quad + \quad 3H_2(g) \quad \rightleftarrows \quad 2NH_3(g)$$
9.23×10^{-3} M $\quad 2.77 \times 10^{-2}$ M $\quad 9.00$ M at 25°C

Example 8.3 illustrates that equilibrium constants for different reactions may be small or large. In fact, the two reactions used are by no means even close to the extremes encountered for K values. Some are so small (such as 1.1×10^{-36}) that for all practical purposes no products are present at equilibrium. Others are so large (such as 1.2×10^{40}) that the reaction can be considered to go completely to products. For reactions with K values between 10^{-3} and 10^3, the position of equilibrium is not extremely favorable to either side, and significant concentrations of both reactants and products can be detected in the equilibrium mixtures.

You might also have noticed in Example 8.3 that the equilibrium concentrations were given for specific temperatures. The reason for this is that K values are constant for a reaction as long as the temperature remains constant. However, K values will change as the temperature is changed. This is a demonstration of the effect of temperature on the position of equilibrium described in the next section.

➤ 8.8 Factors That Influence Equilibrium Position

Le Châtelier's principle
The position of an equilibrium shifts in response to changes made in factors of the equilibrium.

A number of factors can change the position of an established equilibrium. The influence of such factors can be predicted by using a concept known as Le Châtelier's principle, in honor of its originator. According to **Le Châtelier's principle**, when a change is made in any factor of an established equilibrium, the position of equilibrium will shift in a direction that will minimize or oppose the change. The factors we will be most concerned with are concentrations of reactants and products, and reaction temperature.

The effect of concentration changes can be illustrated by the reaction of H_2 with I_2 (Equation 8.12):

$$H_2(g) + I_2(g) \rightleftarrows 2HI(g)$$

Suppose an equilibrium mixture of H_2, I_2, and HI is formed. According to the molecular collision concept, favorable collisions are occurring between H_2 and I_2 molecules to form HI molecules, and favorable collisions are also taking place between HI molecules that cause them to form H_2 and I_2 molecules. Now, suppose some additional I_2 is added to the equilibrium mixture. The chances for favorable collisions between H_2 and I_2 molecules are increased; the rate of formation of HI is increased, and more HI is formed than disappears. The rate of reaction of HI to give H_2 and I_2 increases as the concentration of HI increases, and eventually a new equilibrium position and a new set of equilibrium concentrations will be established. The new equilibrium mixture will contain more HI than the original mixture, but the amounts of the reactants will also be different such that the equilibrium constant remains unchanged. The original equilibrium of the reaction has been shifted toward the right.

CHEMISTRY AROUND US • 8.1

HYPERTHERMIA: A HOT-WEATHER HAZARD

In Chemistry and Your Health 8.1, some of the dangers associated with low body temperature or hypothermia were described. On the other end of the temperature spectrum is a range of conditions collectively called hyperthermia, in which the body suffers ill effects from exposure to high temperatures related to the weather or living conditions. A person with symptoms such as headache, nausea, and fatigue after exposure to heat is probably suffering from some degree of heat-related illness. The two most common forms of hyperthermia are heat exhaustion and life-threatening heat stroke. It is important to recognize the difference between heat stroke and other heat-related illnesses. The severity of such illnesses is indicated by the following terminology and associated symptoms, which are given in order of increasing seriousness.

Heat Stress: A feeling of slight discomfort brought on by exposure to hot weather or surroundings.

Heat Fatigue: A feeling of weakness with symptoms of cool, moist skin and a weakened pulse resulting from exposure to high temperatures.

Heat Syncope: A sudden dizziness occurring after exercising in a hot environment. The skin appears pale and sweaty but is generally moist and cool. Body temperature is normal, but the pulse rate is usually rapid.

Heat Cramps: Painful muscle spasms in the abdomen, arms, or legs that occur after strenuous physical activity in a hot environment. The skin is usually moist and cool, the pulse rate is normal or slightly elevated, and body temperature is near normal.

Heat Exhaustion: This condition is an indication that the body is getting too hot even though a measured body temperature is near normal. The person may be thirsty, weak, uncoordinated, and nauseous, while sweating profusely. The pulse rate may be normal or elevated, and the skin is cold and clammy.

Heat Stroke: This condition is life threatening, so immediate medical attention is required. A person with heat stroke will have a measured body temperature greater than 40°C (104°F). Other symptoms include confusion; bizarre behavior; combativeness; fainting; staggering; a strong, rapid pulse; dry, flushed skin; a lack of sweating; and delirium or coma.

The longer a person is exposed to high temperatures, the greater the severity of the resulting hyperthermia. For this reason, it is important that symptoms of even less-severe hyperthermia conditions be taken seriously and appropriate steps taken to treat the condition, always remembering that if heat stroke is indicated immediate medical attention is needed. Heat exhaustion and other less-severe conditions may be treated in a number of ways, including the following:

1. Get the victim out of the sun and into a cool place, preferably one that is air conditioned.
2. Give the victim fluids to drink, but avoid alcohol or caffeine-containing beverages. Water, fruit juices, and vegetable juices are best.
3. If possible, get the victim to shower or bathe. If that is not possible, sponge the victim with cool water.
4. Encourage the victim to lie down and rest, preferably in a cool place.

Le Châtelier's principle can also be used to predict the influence of temperature on an equilibrium by treating heat as a product or reactant. Consider the following equation for a hypothetical exothermic reaction:

$$A + B \rightleftarrows \text{products} + \text{heat} \qquad (8.15)$$

If the temperature is increased (by adding more heat, which appears on the right side of the equation), the equilibrium shifts to the left in an attempt to use up the added heat. In the new equilibrium position, the concentrations of A and B will be higher, whereas the chemical product concentrations will be lower than those in the original equilibrium. In this case the value of the equilibrium constant is changed by the change in temperature (see ■ Figure 8.13).

Example 8.4

Using Le Châtelier's principle, answer the following questions:

a. Ammonia is made from hydrogen and atmospheric nitrogen:

$$N_2(g) + 3H_2(g) \rightleftarrows 2NH_3(g) + \text{heat}$$

What effect will cooling have on an equilibrium mixture?

GOB
Chemistry ⚛ Now™
Go to GOB Now and click to view a tutorial showing how to determine the effect a change in temperature has on the position of an equilibrium.

➤ **FIGURE 8.13** The effect of temperature on the position of equilibrium for the reaction $N_2O_4 \rightleftarrows 2NO_2$. On which side of the equation will heat appear? Is the reaction exothermic or endothermic?

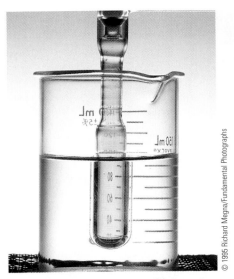

A sealed tube containing an equilibrium mixture of red-brown NO_2 and colorless N_2O_4 is cooled in an ice bath.

The same sealed tube is heated in a hot water bath.

b. What effect will removing H_2 have on the equilibrium mixture described in part (a)?

c. Consider the following equation for a reaction that takes place in solution:

$$Fe^{3+}(aq) + 6SCN^-(aq) \rightleftarrows Fe(SCN)_6{}^{3-}(aq)$$

light brown   colorless   dark red

What effect will the addition of colorless NaSCN solid have on the color of the solution?

Solution

a. Heat is removed by cooling. Heat is replenished when the equilibrium shifts to the right. Thus, more NH_3 will be present at equilibrium in the cooler mixture.

b. The equilibrium will shift to the left in an attempt to replenish the H_2. At the new equilibrium position, less NH_3 will be present.

c. When solid NaSCN is added, it will dissolve and form ions:

$$NaSCN(s) \rightarrow Na^+(aq) + SCN^-(aq)$$

The SCN^- concentration in the solution will therefore be increased. In an attempt to use up the added SCN^-, the equilibrium will shift to the right. This shift generates more of the dark red $Fe(SCN)_6{}^{3-}$. Therefore, the new position of equilibrium will be characterized by a darker red color.

| Learning Check 8.4 | Use Le Châtelier's principle and answer the following: |

a. A saturated solution of ammonium nitrate is formed as follows. The solution process is endothermic:

$$heat + NH_4NO_3(s) \rightleftarrows NH_4{}^+(aq) + NO_3{}^-(aq)$$

Which way will the equilibrium shift when heat is added? What does this shift mean in terms of the solubility of NH_4NO_3 at the higher temperature?

b. Sulfur trioxide gas is formed when oxygen and sulfur dioxide gas react:

$$2SO_2(g) + O_2(g) \rightleftarrows 2SO_3(g)$$

Which way would the equilibrium shift if O_2 were removed from the system? How would the new equilibrium concentration of SO_3 compare with the earlier equilibrium concentration?

Catalysts cannot change the position of an equilibrium. This fact becomes clear when you remember that a catalyst functions by lowering the activation energy for a reaction. A lowering of the energy barrier for the forward reaction also lowers the barrier for the reverse reaction (see ■ Figure 8.14). Hence, a catalyst speeds up both the forward and reverse reactions and cannot change the position of equilibrium. However, the lowered activation energy allows equilibrium to be established more quickly than if the catalyst were absent.

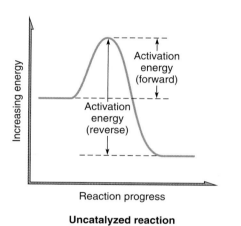

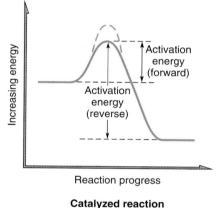

➤ **FIGURE 8.14** The influence of catalysts on forward and reverse activation energies.

Uncatalyzed reaction

Catalyzed reaction

LE CHÂTELIER'S PRINCIPLE IN EVERYDAY LIFE

Le Châtelier's principle is extremely important in laboratory work and in the chemical industry when the goal is to obtain the maximum amount of product from a reaction. In such cases, increasing reactant concentrations, or adjusting the reaction temperature or pressure to shift the equilibrium position to the product side of the reaction, is a common practice.

To help you understand Le Châtelier's principle, it is often useful to observe familiar situations and events and interpret them in terms of shifting equilibria. For example, when you stand in an upright position, you are in equilibrium with the force of gravity. If a friend leans against you, your equilibrium is upset, and you will lose your balance unless you respond by leaning toward your friend such that the forces on you are once again balanced.

Children playing on a seesaw provide a similar example of forces in equilibrium. If one child slides toward the center of the seesaw, the force acting on that side is reduced, the equilibrium is upset, and the other child will be permanently on the ground. However, if the second child also shifts toward the center, the forces on each side can be made equal again, and play can resume.

The preceding examples involved equilibrium between forces, but we are not limited to only those types of examples. Imagine you are enjoying a warm shower when someone turns on a nearby hot water faucet. Suddenly, your shower turns cold. You respond to this stress by adjusting the shower controls (after shrieking loudly) to let in more hot water in an attempt to restore the pleasant temperature you were enjoying. (However, be aware of what will happen when the other faucet is turned off.)

Be mindful of your surroundings and activities, and you will see many more examples that will remind you of responses to stress that are attempts to restore a previous situation.

> **FOR FUTURE REFERENCE** DRUG INTERACTIONS WITH ALCOHOL

The effectiveness of many drugs depends on the concentration of the drug in the blood or other body fluids. The concentration of a drug is one factor that can influence the rate of chemical reactions involving the drug (see Section 8.5), and thus may affect the rate at which the drug acts therapeutically. The concentration of a drug and hence the rate of the reaction of the drug may also be changed by equilibrium shifts that involve the drug (see Section 8.7). These are just two examples of numerous mechanisms by which the effect of drugs on the body could be influenced by interactions with other materials.

Alcohol is a substance that is known to interact with a wide variety of drugs and influence their behavior in the body. The table below does not describe the mechanism for the interactions, but it includes drugs for which the interactions are serious enough to suggest that alcohol should be completely avoided. Individuals using any of the drugs should consult a physician for advice about using alcohol.

DRUG NAME OR CLASS	POSSIBLE INTERACTION WITH ALCOHOL
amphetamines	Excessive increase in blood pressure with alcoholic beverages containing tyramine (unpasteurized beer, Chianti wine, sherry wine, and vermouth)
antidepressants	Excessive sedation, increased intoxication
barbiturates	Excessive sedation
bromides	Confusion, delirium, increased intoxication
calcium carbimide	Disulfiram-like reaction*
carbamazepine	Excessive sedation
chlorprothixene	Excessive sedation
chlorzoxazone	Excessive sedation
disulfiram	Disulfiram-like reaction*
ergotamine	Reduced effectiveness of ergotamine
furazolidone	Disulfiram-like reaction*
haloperidol	Excessive sedation
monoamine oxidase (MAO) inhibitor drugs	Excessive increase in blood pressure with alcoholic beverages containing tyramine (unpasteurized beer, Chianti wine, sherry, and vermouth)
meperidine	Excessive sedation
meprobamate	Excessive sedation
methotrexate	Increased liver toxicity and excessive sedation
metronidazole	Disulfiram-like reaction*
narcotic drugs	Excessive sedation
oxyphenbutazone	Increased stomach irritation and/or bleeding
pentazocine	Excessive sedation
pethidine	Excessive sedation
phenothiazines	Excessive sedation
phenylbutazone	Increased stomach irritation and/or bleeding
procarbazine	Disulfiram-like reaction*
propoxyphene	Excessive sedation
reserpine	Excessive sedation, orthostatic hypotension**
sleep-inducing drugs (hypnotics)	Excessive sedation
—carbromal	
—chloral hydrate	
—ethchlorvynol	
—ethinamate	
—glutethimide	
—flurazepam	
—methaqualone	

(continued)

—methyprylon
—temazepam
—triazolam
thiothixene   Excessive sedation
tricyclic antidepressants   Excessive sedation, increased intoxication
trimethobenzamide   Excessive sedation

*Symptoms include intense facial flushing, severe headache, shortness of breath, chest pains, nausea, repeated vomiting, sweating, and weakness. With large amounts of alcohol, symptoms include blurred vision, vertigo, decrease in blood pressure, and loss of consciousness. Severe reactions may lead to convulsions and death.
**Low blood pressure related to body position or posture. People with symptoms may have normal blood pressure lying down, but on sitting upright or standing will feel light-headed, dizzy, and like they are going to faint. The symptoms result from inadequate blood flow to the brain.

SOURCE: Adapted from Rybacki, J.J.; Long, J.W. *The Essential Guide to Prescription Drugs.* New York: Harper Perennial, 1999.

CONCEPT SUMMARY

Spontaneous and Nonspontaneous Processes. Spontaneous processes take place naturally with no apparent cause or stimulus. Process spontaneity depends on the energy and entropy changes that accompany the process. Energy decreases and entropy increases favor spontaneity. However, a nonspontaneous change in one of these factors can be compensated for by a large spontaneous change in the other to cause processes to be spontaneous.

Reaction Rates. The speed of a reaction is called a reaction rate, which can be determined by measuring how fast reactants are used up or products are formed.

Molecular Collisions. Explanations of how reactions take place are called reaction mechanisms. Most mechanisms are based on three assumptions: (1) Molecules must collide with one another, (2) the collision must involve a certain minimum of energy, and (3) some colliding molecules must be oriented in a specific way during collision in order to react.

Energy Diagrams. Energy relationships for reactions can be represented by energy diagrams, in which energy is plotted versus the reaction progress. The concepts of exothermic and endothermic reactions and activation energy are clearly represented by such diagrams.

Factors That Influence Reaction Rates. Four factors affect the rates of all reactions: (1) the nature of the reactants, (2) reactant concentrations, (3) reactant temperature, and (4) the presence of catalysts.

Chemical Equilibrium. Reactions are in equilibrium when the rate of the forward reaction is equal to the rate of the reverse reaction. Equilibrium is emphasized in equations for reactions by writing double arrows pointing in both directions between reactants and products.

The Position of Equilibrium. The relative amounts of reactants and products present in a system at equilibrium define the position of equilibrium. The equilibrium position is toward the right when a large amount of product is present and toward the left when a large amount of reactant is present. The position is indicated by the value of the equilibrium constant.

Factors That Influence Equilibrium Position. Factors known to influence the position of equilibrium include changes in amount of reactants and/or products and changes in temperature. The influence of such factors can be predicted by using Le Châtelier's principle. Catalysts cannot change the position of equilibrium.

LEARNING OBJECTIVES ASSESSMENT

You can get an approximate but quick idea of how well you have met the learning objectives given at the beginning of this chapter by working the selected end-of-chapter exercises given below. The answer to each exercise is given in Appendix B of the book.

Objective 1 (Section 8.1): Exercise 8.6

Objective 2 (Section 8.2): Exercise 8.14

Objective 3 (Section 8.3): Exercise 8.20

Objective 4 (Section 8.4): Exercise 8.26

Objective 5 (Section 8.5): Exercise 8.30

Objective 6 (Section 8.7): Exercise 8.38

Objective 7 (Section 8.7): Exercise 8.44

Objective 8 (Section 8.8): Exercise 8.50

Activation energy (8.3)
Catalyst (8.5)
Effective collision (8.5)
Endergonic process (8.1)
Entropy (8.1)
Equilibrium concentrations (8.6)
Equilibrium constant (8.7)

Equilibrium expression (8.7)
Exergonic process (8.1)
Heterogeneous (surface) catalyst (8.5)
Homogeneous catalyst (8.5)
Inhibitor (8.5)
Internal energy (8.3)
Le Châtelier's principle (8.8)

Position of equilibrium (8.7)
Reaction mechanism (8.3)
Reaction rate (8.2)
Spontaneous process (8.1)
Stable substance (8.1)
State of equilibrium (8.6)

KEY EQUATIONS

1. Calculation of reaction rate (Section 8.2):

$$\text{Rate} = \frac{\Delta C}{\Delta t} = \frac{C_t - C_0}{\Delta t}$$

Equation 8.1

2. Equilibrium expression for general reaction (Section 8.7):

$$aA + bB + \cdots \rightleftharpoons wW + xX + \cdots$$

Equation 8.13

$$K = \frac{[W]^w[X]^x}{[A]^a[B]^b}$$

Equation 8.14

EXERCISES

LEGEND: 1 = straightforward, 2 = intermediate, 3 = challenging. All even-numbered exercises are answered in Appendix B.

SPONTANEOUS AND NONSPONTANEOUS PROCESSES (SECTION 8.1)

8.1 Classify the following processes as spontaneous or non-spontaneous. Explain your answers in terms of whether energy must be continually supplied to keep the process going.
   a. Water is decomposed into hydrogen and oxygen gas by passing electricity through the liquid.
   b. An explosive detonates after being struck by a falling rock.
   c. A coating of magnesium oxide forms on a clean piece of magnesium exposed to air.
   d. A light bulb emits light when an electric current is passed through it.
   e. A cube of sugar dissolves in a cup of hot coffee.

8.2 Classify the following processes as spontaneous or non-spontaneous. Explain your answers in terms of whether energy must be continually supplied to keep the process going.
   a. The space shuttle leaves its pad and goes into orbit.
   b. The fuel in a booster rocket of the space shuttle burns.
   c. Water boils at 100°C and 1 atm pressure.
   d. Water temperature increases to 100°C at 1 atm pressure.
   e. Your bedroom becomes orderly.

8.3 Classify the following processes as exergonic or endergonic. Explain your answers.
   a. Any combustion process
   b. Perspiration evaporating from the skin
   c. Melted lead solidifying
   d. An explosive detonating
   e. An automobile being pushed up a slight hill (from point of view of the automobile)

8.4 Classify the following processes as exergonic or endergonic. Explain your answers.
   a. An automobile being pushed up a slight hill (from point of view of the one pushing)
   b. Ice melting (from point of view of the ice)
   c. Ice melting (from point of view of surroundings of the ice)
   d. Steam condensing to liquid water (from point of view of the steam)
   e. Steam condensing to liquid water (from point of view of surroundings of the steam)

8.5 Describe the energy and entropy changes that occur in the following processes, and indicate whether the processes are spontaneous under the conditions stated:
   a. Lumber becomes a house.
   b. A seed grows into a tree.
   c. On a hot day, water evaporates from a lake.

8.6 Describe the energy and entropy changes that occur in the following processes, and indicate whether the processes are spontaneous under the conditions stated:
   a. On a cold day, water freezes.
   b. A container of water at 40°C cools to room temperature.
   c. The odor from an open bottle of perfume spreads throughout a room.

8.7 Pick the example with the highest entropy from each of the following sets. Explain your answers.
   a. Solid ice, liquid water, or steam
   b. Leaves on a tree, fallen leaves blown about on the ground, fallen leaves raked and placed in a basket
   c. A stack of sheets of paper, a wastebasket containing

sheets of paper, a wastebasket containing torn and crumpled sheets of paper

d. A 0.10 M sugar solution, a 1.0 M sugar solution, a 10.0 M sugar solution

e. A banquet table set for dinner, a banquet table during dinner, a banquet table immediately after dinner

8.8 Pick the example with the highest entropy from each of the following sets. Explain your answers.

a. Two opposing football teams just before the ball is snapped, two opposing football teams 1 second after the ball is snapped, two opposing football teams when the whistle is blown, ending the play

b. A 10% copper/gold alloy, a 2% copper/gold alloy, pure gold

c. A purse on which the strap just broke, a purse just hitting the ground, a purse on the ground with contents scattered

d. Coins in a piggy bank, coins in piles containing same type of coins, coins in stacks of same type of coins

e. A dozen loose pearls in a box, a dozen pearls randomly strung on a string, a dozen pearls strung on a string in order of decreasing size

8.9 You probably know that, on exposure to air, silver tarnishes and iron rusts; but gold, stainless steel, and chromium do not change. Explain these facts, using the concept of stability.

REACTION RATES (SECTION 8.2)

8.10 Classify the following processes according to their rates as very slow, slow, or fast:

a. The cooking of a pizza in a hot oven

b. The change in leaf color from green to red or orange in the autumn

c. The drying of a wet swimsuit hung out in the sun

d. The burning of a match after it has been started

e. The growing of grass during a warm summer

8.11 Classify the following processes according to their rates as very slow, slow, or fast:

a. The melting of butter put into a hot pan

b. The ripening of a piece of fruit stored at room temperture

c. The cooking of a raw potato in a hot oven

d. The melting of an ice cube in a glass of cool water

e. The combustion of gasoline in the engine of a car

8.12 Describe the observations or measurements that could be made to allow you to follow the rate of the following processes:

a. The melting of a block of ice

b. The setting (hardening) of concrete

c. The burning of a candle

8.13 Describe the observations or measurements that could be made to allow you to follow the rate of the following processes:

a. The diffusion of ink from a drop placed in a pan of quiet, undisturbed water

b. The loss of water from a pan of boiling water

c. The growth of a corn plant

8.14 Consider the following hypothetical reaction:

$$A + B \rightarrow C$$

Calculate the average rate of the reaction on the basis of the following information:

a. Pure A and B are mixed, and after 12.0 minutes the measured concentration of C is 0.396 mol/L.

b. Pure A, B, and C are mixed together at equal concentrations of 0.300 M. After 8.00 minutes, the concentration of C is found to be 0.455 M.

8.15 Consider the following reaction:

$$A + B \rightarrow C$$

Calculate the average rate of the reaction on the basis of the following information:

a. Pure A, B, and C are mixed together at concentrations of $A = B = 0.400$ M, $C = 0.150$ M. After 6.00 minutes, the concentration of C is 0.418 M.

b. Pure A and B are mixed together at the same concentration of 0.361 M. After 7.00 minutes, the concentration of A is found to be 0.048 M.

8.16 A reaction generates chlorine gas (Cl_2) as a product. The reactants are mixed and sealed in a 250-mL container. After 30.0 minutes, 2.97×10^{-2} mol of Cl_2 has been generated. Calculate the average rate of the reaction.

8.17 A reaction generates hydrogen gas (H_2) as a product. The reactants are mixed in a sealed 250-mL vessel. After 20.0 minutes, 3.91×10^{-2} mol H_2 has been generated. Calculate the average rate of the reaction.

8.18 Ammonium and nitrite ions react in solution to form nitrogen gas:

$$NH_4^+(aq) + NO_2^-(aq) \rightarrow N_2(g) + 2H_2O(\ell)$$

A reaction is run, and the liberated N_2 gas is collected in a previously evacuated 500-mL container. After the reaction has gone on for 750 seconds, the pressure of N_2 in the 500-mL container is 2.77×10^{-2} atm, and the temperature of the N_2 is 25.0°C. Use the ideal gas law (Equation 6.9) to calculate the number of moles of N_2 liberated. Then calculate the average rate of the reaction.

8.19 Suppose a small lake is contaminated with an insecticide that decomposes with time. An analysis done in June shows the decomposition product concentration to be 7.8×10^{-4} mol/L. An analysis done 35 days later shows the concentration of decomposition product to be 9.9×10^{-4} mol/L. Assume the lake volume remains constant and calculate the average rate of decomposition of the insecticide.

MOLECULAR COLLISIONS (SECTION 8.3)

8.20 In each of the following, which reaction mechanism assumption is apparently being violated? Explain your answers.

a. A reaction takes place more rapidly when the concentration of reactants is decreased.

b. A reaction takes place more rapidly when the reaction mixture is cooled.

c. The reaction rate of $A + B \rightarrow A—B$ increases as the concentration of A is increased but does not change as the concentration of B is increased.

8.21 Which reaction mechanism assumptions are unimportant in describing simple ionic reactions between cations and anions? Why?

8.22 Describe two ways by which an increase in temperature increases a reaction rate.

ENERGY DIAGRAMS (SECTION 8.4)

8.23 Sketch energy diagrams to represent each of the following. Label the diagrams completely and tell how they are similar to each other and how they are different.
   a. Exothermic (exergonic) reaction with activation energy
   b. Exothermic (exergonic) reaction without activation energy

8.24 Sketch energy diagrams to represent each of the following. Label the diagrams completely and tell how they are similar to each other and how they are different.
   a. Endothermic (endergonic) reaction with activation energy
   b. Endothermic (endergonic) reaction without activation energy

8.25 Use energy diagrams to compare catalyzed and uncatalyzed reactions.

8.26 One reaction occurs at room temperature and liberates 500 kJ/mol of reactant. Another reaction does not take place until the reaction mixture is heated to 150°C. However, it also liberates 500 kJ/mol of reactant. Draw an energy diagram for each reaction and indicate the similarities and differences between the two.

FACTORS THAT INFLUENCE REACTION RATES (SECTION 8.5)

8.27 The following reactions are proposed. Make a rough estimate of the rate of each one—rapid, slow, won't react. Explain each answer.
   a. $H_2O(\ell) + H^+(aq) \rightarrow H_3O^+(aq)$
   b. $H_3O^+(aq) + H^+(aq) \rightarrow H_4O^{2+}(aq)$
   c. $3H_2(g) + N_2(g) \rightarrow 2NH_3(g)$
   d. $Ba^{2+}(aq) + SO_4^{2-}(aq) \rightarrow BaSO_4(s)$

8.28 The following reactions are proposed. Make a rough estimate of the rate of each one—rapid, slow, won't react. Explain each answer.
   a. $2I^-(aq) + Pb^{2+}(aq) \rightarrow PbI_2(s)$
   b. $Br^-(aq) + I^-(aq) \rightarrow BrI^{2-}(aq)$
   c. $NH_3(g) + HCl(g) \rightarrow NH_4Cl(s)$
   d. $CaO(s) + CO_2(g) \rightarrow CaCO_3(s)$

8.29 Which reaction mechanism assumption is most important in explaining the following? Why?
   a. The influence of concentration on reaction rates
   b. The influence of catalysts on reaction rates

8.30 Suppose you are running a reaction and you want to speed it up. Describe three things you might try to do this.

8.31 A reaction is started by mixing reactants. As time passes, the rate decreases. Explain this behavior that is characteristic of most reactions.

8.32 A reaction is run at 20°C and takes 30 minutes to go to completion. How long would it take to complete the same reaction at a temperature of 40°C?

8.33 What factor is more important than simply the amount of solid reactant present in determining the rate of a reaction? Explain.

8.34 Describe two ways catalysts might speed up a reaction.

CHEMICAL EQUILIBRIUM (SECTION 8.6)

8.35 Describe the establishment of equilibrium in a system represented by a shopper walking up the "down" escalator.

8.36 Describe the observation or measurement result that would indicate when each of the following had reached equilibrium:
   a. $\quad H_2 \quad + \quad I_2 \quad \rightleftarrows \quad 2HI$
      colorless gas   violet gas   colorless gas
   b. solid sugar + water $\rightleftarrows$ sugar solution
   c. $\quad N_2 \quad + \quad 2O_2 \quad \rightleftarrows \quad 2NO_2$
      colorless gas   colorless gas   red-brown gas

8.37 Describe the observation or measurement result that would indicate when each of the following had reached equilibrium:
   a. $\quad 2CO \quad + \quad O_2 \quad \rightleftarrows \quad 2CO_2$
      colorless gas   colorless gas   colorless gas
      (apply Dalton's law)
   b. $\quad LiOH \quad + \quad CO_2 \quad \rightleftarrows \quad LiHCO_3$
      colorless solid   colorless gas   colorless solid
   c. paycheck $\rightarrow$ checking account $\rightarrow$ checks to pay bills

THE POSITION OF EQUILIBRIUM (SECTION 8.7)

8.38 Write an equilibrium expression for each of the following gaseous reactions:
   a. $2CO + O_2 \leftrightarrows 2CO_2$
   b. $N_2O_4 \leftrightarrows 2NO_2$
   c. $2C_2H_6 + 7O_2 \leftrightarrows 4CO_2 + 6H_2O$
   d. $2NOCl \leftrightarrows 2NO + Cl_2$
   e. $2Cl_2O_5 \leftrightarrows O_2 + 4ClO_2$

8.39 Write an equilibrium expression for each of the following gaseous reactions:
   a. $H_2 + Br_2 \rightleftarrows 2HBr$
   b. $2H_2S + 3O_2 \rightleftarrows 2H_2O + 2SO_2$
   c. $3NO_2 \rightleftarrows N_2O_5 + NO$
   d. $4NH_3 + 3O_2 \rightleftarrows 2N_2 + 6H_2O$
   e. $2NO + 2H_2 \rightleftarrows N_2 + 2H_2O$

8.40 The following equilibria are established in water solutions. Write an equilibrium expression for each reaction.
   a. $Fe^{3+} + 6CN^- \leftrightarrows Fe(CN)_6^{3-}$
   b. $Ag^+ + 2NH_3 \leftrightarrows Ag(NH_3)_2^+$
   c. $Au^{3+} + 4Cl^- \leftrightarrows AuCl_4^-$

8.41 The following equilibria are established in water solutions. Write an equilibrium expression for each reaction.
   a. $Ni^{2+} + 6NH_3 \rightleftarrows Ni(NH_3)_6^{2+}$
   b. $Sn^{2+} + 2Fe^{3+} \rightleftarrows Sn^{4+} + 2Fe^{2+}$
   c. $F_2 + 2Cl^- \rightleftarrows 2F^- + Cl_2$

8.42 Write a reaction equation that corresponds to each of the following equilibrium expressions:
   a. $K = \dfrac{[CO_2][H_2O]^2}{[CH_4][O_2]^2}$
   b. $K = \dfrac{[CH_4][H_2O]}{[H_2]^3[CO]}$

c. $K = \dfrac{[O_2]^3}{[O_3]^2}$

d. $K = \dfrac{[NH_3]^4[O_2]^7}{[NO_2]^4[H_2O]^6}$

8.43 Write an equation that corresponds to each of the following equilibrium expressions:

a. $K = \dfrac{[PH_3][F_2]^3}{[HF]^3[PF_3]}$

b. $K = \dfrac{[O_2]^7[NH_3]^4}{[NO_2]^4[H_2O]^6}$

c. $K = \dfrac{[O_2][ClO_2]^4}{[Cl_2O_5]^2}$

d. $K = \dfrac{[N_2][H_2O]^2}{[NO]^2[H_2]^2}$

8.44 A sample of gaseous BrCl is allowed to decompose in a closed container at 25°C:

$$2BrCl(g) \rightleftarrows Br_2(g) + Cl_2(g)$$

When the reaction reaches equilibrium, the following concentrations are measured: [BrCl] = 0.38 M, [Cl$_2$] = 0.26 M, [Br$_2$] = 0.26 M. Evaluate the equilibrium constant for the reaction at 25°C.

8.45 At 600°C, gaseous CO and Cl$_2$ are mixed together in a closed container. At the instant they are mixed, their concentrations are CO = 0.79 mol/L and Cl$_2$ = 0.69 mol/L. After equilibrium is established, their concentrations are CO = 0.25 mol/L and Cl$_2$ = 0.15 mol/L. Evaluate the equilibrium constant for the reaction

$$CO(g) + Cl_2(g) \rightleftarrows COCl_2(g)$$

8.46 A mixture of the gases NOCl, Cl$_2$, and NO is allowed to reach equilibrium at 25°C. The measured equilibrium concentrations are [NO] = 0.92 mol/L, [NOCl] = 1.31 mol/L, and [Cl$_2$] = 0.20 mol/L. What is the value of the equilibrium constant at 25°C for the reaction

$$2NOCl(g) \rightleftarrows 2NO(g) + Cl_2(g)$$

8.47 Consider the following equilibrium constants. Describe how you would expect the equilibrium concentrations of reactants and products to compare with each other (larger than, smaller than, etc.) for each case.

a. $K = 2.1 \times 10^{-6}$ c. $K = 1.2 \times 10^8$
b. $K = 0.15$ d. $K = 0.00036$

8.48 Consider the following equilibrium constants. Describe how you would expect the equilibrium concentrations of reactants and products to compare with each other (larger than, smaller than, etc.) for each case.

a. $K = 5.9$ c. $K = 2.7 \times 10^{-4}$
b. $K = 3.3 \times 10^6$ d. $K = 0.0000558$

FACTORS THAT INFLUENCE EQUILIBRIUM POSITION (SECTION 8.8)

8.49 Use Le Châtelier's principle to predict the direction of equilibrium shift in the following equilibria when the indicated stress is applied:

a. $2A + B + heat \rightleftarrows D$; some B is removed.

b. $2A + B + heat \rightleftarrows C + D$; the system is heated.
c. $N_2O_4 \rightleftarrows 2NO_2$; some NO_2 is added.

8.50 Use Le Châtelier's principle to predict the direction of equilibrium shift in the following equilibria when the indicated stress is applied:

a. $Ag^+(aq) + Cl^-(aq) \rightleftarrows AgCl(s)$; some Ag^+ is removed.
b. $2HI(g) + heat \rightleftarrows H_2(g) + I_2(g)$; the system is heated.
c. $6Cu(s) + N_2(g) + heat \rightleftarrows 2Cu_3N(s)$; the system is cooled and some N_2 is removed.

8.51 Using Le Châtelier's principle, predict the direction of equilibrium shift and the changes that will be observed (color, amount of precipitate, etc.) in the following equilibria when the indicated stress is applied:

a. heat + $Co^{2+}(aq)$ + $4Cl^-(aq)$ $\rightleftarrows$ $CoCl_4^{2-}(aq)$;
    pink   colorless   blue
  the equilibrium mixture is cooled.

b. heat + $Co^{2+}(aq)$ + $4Cl^-(aq)$ $\rightleftarrows$ $CoCl_4^{2-}(aq)$;
    pink   colorless   blue
  Cl^- is added to the equilibrium mixture.

c. $Fe^{3+}(aq)$ + $6SCN^-(aq)$ $\rightleftarrows$ $Fe(SCN)_6^{3-}(aq)$;
  brown   colorless   red
  Fe^{3+} is added to the equilibrium mixture.

d. $Pb^{2+}(aq)$ + $2Cl^-(aq)$ $\rightleftarrows$ $PbCl_2(s)$ + heat;
  colorless   colorless white solid
  Cl^- is added to the equilibrium mixture.

e. C_2H_4 + I_2 $\rightleftarrows$ $C_2H_4I_2$ + heat;
  colorless gas violet gas colorless gas
  a catalyst is added to the equilibrium mixture.

8.52 Using Le Châtelier's principle, predict the direction of equilibrium shift and the changes that will be observed (color, amount of precipitate, etc.) in the following equilibria when the indicated stress is applied:

a. $Cu^{2+}(aq)$ + $4NH_3(aq)$ $\rightleftarrows$ $Cu(NH_3)_4^{2+}(aq)$;
  blue   colorless   dark purple
  some NH_3 is added to the equilibrium mixture.

b. $Pb^{2+}(aq)$ + $2Cl^-(aq)$ $\rightleftarrows$ $PbCl_2(s)$ + heat;
  colorless colorless white solid
  the equilibrium mixture is cooled.

c. C_2H_4 + I_2 $\rightleftarrows$ $C_2H_4I_2$ + heat;
  colorless gas violet gas colorless gas
  some $C_2H_4I_2$ is removed from the equilibrium mixture.

d. C_2H_4 + I_2 $\rightleftarrows$ $C_2H_4I_2$ + heat;
  colorless gas violet gas colorless gas
  the equilibrium mixture is cooled.

e. heat + $4NO_2$ + $6H_2O$ $\rightleftarrows$ $7O_2$ + $4NH_3$;
    brown colorless colorless colorless
     gas   gas   gas   gas
  a catalyst is added, and NH_3 is added to the equilibrium mixture.

8.53 Tell what will happen to each equilibrium concentration in the following when the indicated stress is applied and a new equilibrium position is established:

a. $H^+(aq) + HCO_3^-(aq) \rightleftarrows H_2O(\ell) + CO_2(g)$;
  HCO_3^- is added.

b. $CO_2(g) + H_2O(\ell) \rightleftarrows H_2CO_3(aq)$ + heat;
  CO_2 is removed.

c. $CO_2(g) + H_2O(\ell) \rightleftharpoons H_2CO_3(aq) + heat$; the system is cooled.

8.54 Tell what will happen to each equilibrium concentration in the following when the indicated stress is applied and a new equilibrium position is established:
a. $2N_2O(g) + 3O_2(g) \leftrightharpoons 4NO_2(g) + heat$; O_2 is added.
b. $2SO_3(g) + heat \leftrightharpoons 2SO_2(g) + O_2(g)$; the system is heated.
c. $2CO(g) + O_2(g) \leftrightharpoons 2CO_2(g) + heat$; the system is cooled.

8.55 The gaseous reaction $2HBr(g) \rightleftharpoons H_2(g) + Br_2(g)$ is endothermic. Tell which direction the equilibrium will shift for each of the following:
a. Some H_2 is removed.
b. The temperature is decreased.

c. Some Br_2 is added.
d. A catalyst is added.
e. Some HBr is added.
f. The temperature is decreased, and some HBr is removed.

8.56 The gaseous reaction $N_2(g) + 3H_2(g) \rightleftharpoons 2NH_3(g)$ is exothermic. Tell which direction the equilibrium will shift for each of the following:
a. Some N_2 is added.
b. The temperature is increased.
c. Some NH_3 is removed.
d. Some H_2 is removed.
e. A catalyst is added.
f. The temperature is increased, and some H_2 is removed.

CHEMISTRY FOR THOUGHT

8.1 Refer to Figure 8.4 and answer the question. How would the total energy released as light for each light stick compare when they have both stopped glowing?

8.2 A mixture of the gases NOCl, NO, and Cl_2 is allowed to come to equilibrium at 400°C in a 1.50-L reaction container. Analysis shows the following number of moles of each substance to be present at equilibrium: NOCl = 1.80 mol, NO = 0.70 mol, and Cl_2 = 0.35 mol. Calculate the value of the equilibrium constant for the reaction at 400°C. The equation for the reaction is

$$2NOCl(g) \rightleftharpoons 2NO(g) + Cl_2(g)$$

8.3 The equilibrium constant for the reaction $PCl_5 \rightleftharpoons PCl_3 + Cl_2$ is 0.0245 at 250°C. What molar concentration of PCl_5 would be present at equilibrium if the concentrations of PCl_3 and Cl_2 were both 0.250 M?

8.4 At 448°C, the equilibrium constant for the reaction $H_2 + I_2 \rightleftharpoons 2HI$ is 50.5. What concentration of I_2 would be found in an equilibrium mixture in which the concentrations of HI and H_2 were 0.500 M and 0.050 M, respectively?

8.5 Refer to Figure 8.10 and answer the questions. Would you expect a crushed antacid tablet (like Alka-Seltzer) to dissolve faster or slower than a whole tablet? Explain.

8.6 Refer to Figure 8.13 and answer the questions. Would the presence of a catalyst in the tube influence the equilibrium concentrations of the two gases? Explain.

8.7 In the blood, both oxygen (O_2) and poisonous carbon monoxide (CO) can react with hemoglobin (Hb). The respective compounds are HbO_2 and HbCO. When both gases are present in the blood, the following equilibrium is established.

$$HbCO + O_2 \rightleftharpoons HbO_2(g) + CO$$

Use Le Châtelier's principle to explain why pure oxygen is often administered to victims of CO poisoning.

8.8 Use the concept of reaction rates to explain why no smoking is allowed in hospital areas where patients are being administered oxygen gas.

8.9 Suppose you have two identical unopened bottles of carbonated beverage. The contents of both bottles appear to be perfectly clear. You loosen the cap of one of the bottles and hear a hiss as gas escapes, and at the same time gas bubbles appear in the liquid. The liquid in the unopened bottle still appears to be perfectly clear. Explain these observations using the concept of equilibrium and LeChâtelier's principle. Remember, a carbonated beverage contains carbon dioxide gas dissolved in a liquid under pressure.

8.10 Someone once suggested that it is impossible to unscramble a scrambled egg. Describe an unscrambled and a scrambled egg in terms of the concept of entropy.

InfoTrac College Edition Readings

"At physiological temperature, peptide binding is endothermic and entropy-driven," *Immunotherapy Weekly*, Feb 26, 2003:92. Record number A98044525.

"Material could halt catalyst waste," *Science News*, July 27, 2002, 162(4):61(1). Record number A90870743.

"EPA OKs three halon alternatives for fires and explosions," *Pesticide & Toxic Chemical News*, Feb 3, 2003, 31(15):18. Record number A97592470.

CHAPTER 9

Acids, Bases, and Salts

This respiratory therapist is administering oxygen-enriched air to a patient in order to increase the concentration of oxygen in his blood. Oxygen and carbon dioxide concentrations in the blood exert a significant influence on the acid–base balance of the body. This chapter introduces you to the concepts of acids and bases and their characteristic reactions.

LEARNING OBJECTIVES

When you have completed your study of this chapter, you should be able to:

1. Write reaction equations that illustrate the Arrhenius and Brønsted acid–base behavior. (Sections 9.1 and 9.2)

2. Identify Brønsted acids and bases from written reaction equations. (Section 9.2)

3. Name common acids. (Section 9.3)

4. Do calculations using the concept of the self-ionizaton of water. (Section 9.4)

5. Do calculations using the pH concept. (Section 9.5)

6. Write reaction equations that illustrate the characteristic reactions of acids. (Section 9.6)

7. Write reaction equations that illustrate various ways to prepare salts. (Sections 9.6, 9.7, and 9.8)

8. Write reaction equations that represent neutralization reactions between acids and bases. (Section 9.7)

9. Do calculations using the concept of an equivalent of a salt. (Section 9.8)

10. Explain the difference between the words *strong* and *weak* as applied to acids and bases. (Section 9.9)

11. Do calculations related to the analysis of acids and bases by titration. (Sections 9.10 and 9.11)

12. Explain the concept of salt hydrolysis, and write equations to illustrate the concept. (Section 9.12)

13. Explain how buffers work, and write equations to illustrate their action. (Section 9.13)

cids, bases, and salts are among the most common and important solutes found in solutions. Until late in the 19th century, these substances were characterized by such properties as taste and color changes induced in certain dyes. Acids taste sour; bases, bitter; and salts, salty. Litmus, a dye, is red in the presence of acids and blue in the presence of bases. These and other observations led to the correct conclusions that acids and bases are chemical opposites, and that salts are produced when acids and bases react with each other. Today, acids and bases are defined in more precise ways that are useful in studying their characteristics.

➤ 9.1 The Arrhenius Theory

Arrhenius acid
Any substance that provides H^+ ions when dissolved in water.

Arrhenius base
Any substance that provides OH^- ions when dissolved in water.

In 1887, Swedish chemist Svante Arrhenius proposed a theory dealing with electrolytic dissociation. He defined **acids** as substances that dissociate when dissolved in water and produce hydrogen ions (H^+). Similarly, **bases** are substances that dissociate and release hydroxide ions (OH^-) into the solution. Hydrogen chloride (HCl) and sodium hydroxide (NaOH) are examples of an Arrhenius acid and base, respectively. They dissociate in water as follows:

$$HCl(aq) \rightarrow H^+(aq) + Cl^-(aq) \tag{9.1}$$

$$NaOH(aq) \rightarrow Na^+(aq) + OH^-(aq) \tag{9.2}$$

Note that the hydrogen ion is a bare proton, the nucleus of a hydrogen atom.

➤ 9.2 The Brønsted Theory

Arrhenius did not know that free hydrogen ions cannot exist in water. They covalently bond with water molecules to form hydronium ions (H_3O^+). The water molecules provide both electrons used to form the covalent bond:

$$H^+ + :\ddot{O}\!-\!H \longrightarrow \left[H\!-\!\ddot{O}\!-\!H \right]^+ \tag{9.3}$$

new bond

Or, more simply,

$$H^+(aq) + H_2O(\ell) \rightarrow H_3O^+(aq) \tag{9.4}$$

Brønsted acid
Any hydrogen-containing substance that is capable of donating a proton (H^+) to another substance.

Brønsted base
Any substance capable of accepting a proton from another substance.

In 1923, Johannes Brønsted in Denmark and Thomas Lowry in England proposed an acid–base theory that took into account this behavior of hydrogen ions. They defined an **acid** as any hydrogen-containing substance that donates a proton (hydrogen ion) to another substance and a **base** as any substance that accepts a proton.

In conformity with this theory, the acidic behavior of covalently bonded HCl molecules in water is written

$$HCl(aq) + H_2O(\ell) \rightleftarrows H_3O^+(aq) + Cl^-(aq) \tag{9.5}$$

The HCl behaves as a Brønsted acid by donating a proton to a water molecule. The water molecule, by accepting the proton, behaves as a base.

The double arrows of unequal length in Equation 9.5 indicate that the reaction is reversible with the equilibrium lying far to the right. In actual water solutions, essentially 100% of the dissolved HCl is in the ionic form at equilibrium. Remember, both the forward and the reverse reactions are taking place at equilibrium (Section 8.6). When the reverse reaction occurs, hydronium ions donate protons to chloride ions to form HCl and H_2O molecules. Thus, H_3O^+ behaves as a Brønsted acid, and Cl^- behaves as a Brønsted base. We see from this discussion that when a substance like HCl behaves as a Brønsted acid by donating a proton, the species that remains (Cl^-) is a Brønsted base. The Cl^- is called the **conjugate base** of HCl. Every Brønsted acid and the base formed when it donates a proton is called a **conjugate acid–base pair.** Thus, in Equation 9.5, we see that HCl and Cl^- form a conjugate acid–base pair, as do H_3O^+ and H_2O for the reverse reaction. Notice that the acid and base in a conjugate acid–base pair differ only by a proton, H^+.

Conjugate base
The species remaining when a Brønsted acid donates a proton.

Conjugate acid–base pair
A Brønsted acid and its conjugate base.

Example 9.1

Identify all Brønsted acids, bases, and acid–base conjugate pairs in the reactions represented by the following equations:

a. $HNO_3(aq) + H_2O(\ell) \rightleftharpoons H_3O^+(aq) + NO_3^-(aq)$
b. $HClO_4(aq) + H_2O(\ell) \rightleftharpoons H_3O^+(aq) + ClO_4^-(aq)$
c. $NH_3(aq) + H_2O(\ell) \rightleftharpoons OH^-(aq) + NH_4^+(aq)$

Solution

a. Nitric acid, HNO_3, behaves as a Brønsted acid by donating a proton to H_2O, a base, in the forward reaction. In the reverse reaction, H_3O^+, the conjugate acid of H_2O, donates a proton to NO_3^-, the conjugate base of HNO_3. In summary, the Brønsted acids are HNO_3 and H_3O^+, the Brønsted bases are H_2O and NO_3^-, and the conjugate acid–base pairs are HNO_3/NO_3^- and H_3O^+/H_2O:

$$HNO_3(aq) + H_2O(\ell) \rightleftharpoons H_3O^+(aq) + NO_3^-(aq)$$

acid   base   acid   base
conjugate pair
conjugate pair

b. Similarly, perchloric acid ($HClO_4$) is a Brønsted acid, and H_2O is a Brønsted base (forward reaction). Also, H_3O^+ is a Brønsted acid, and the perchlorate ion (ClO_4^-) is a Brønsted base (reverse reaction). The conjugate acid–base pairs are $HClO_4/ClO_4^-$ and H_3O^+/H_2O:

$$HClO_4(aq) + H_2O(\ell) \rightleftharpoons H_3O^+(aq) + ClO_4^-(aq)$$

acid   base   acid   base
conjugate pair
conjugate pair

c. In this reaction, water donates a proton instead of accepting one. Therefore, H_2O is a Brønsted acid, and ammonia (NH_3) is a Brønsted base (forward reaction). The ammonium ion (NH_4^+) is an acid, and the hydroxide ion (OH^-) is a base (reverse reaction). Note that an Arrhenius base was a substance that released the OH^-, whereas according to Brønsted, the OH^- is a base. The conjugate acid–base pairs are H_2O/OH^- and NH_4^+/NH_3:

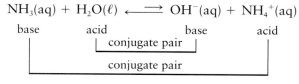

$$NH_3(aq) + H_2O(\ell) \rightleftharpoons OH^-(aq) + NH_4^+(aq)$$

base   acid   base   acid
conjugate pair
conjugate pair

| Learning Check 9.1 | Identify all Brønsted acids, bases, and acid–base conjugate pairs in the reactions represented by the following equations: |

a. $HC_2H_3O_2(aq) + H_2O(\ell) \rightleftarrows H_3O^+(aq) + C_2H_3O_2^-(aq)$
b. $NO_2^-(aq) + H_2O(\ell) \rightleftarrows HNO_2(aq) + OH^-(aq)$
c. $HS^-(aq) + H_2O(\ell) \rightleftarrows H_3O^+(aq) + S^{2-}(aq)$

➤ 9.3 Naming Acids

In Section 4.4, the rules for naming binary ionic compounds were given. The rules for naming binary covalent compounds and ionic compounds that contain polyatomic ions were discussed in Section 4.10. We now conclude our discussion of inorganic nomenclature by giving the rules used to name hydrogen-containing compounds that behave as acids. Examples of the two types of compounds that behave as acids were given in Section 9.2. Acids of the first type, represented by HCl, are compounds in which hydrogen is covalently bonded to a nonmetal. In acids of the second type, hydrogen is covalently bonded to a polyatomic ion. An example of the second type is HNO_3.

You probably recognized that HCl is a binary covalent compound that can be named by the rules given earlier in Section 4.10. According to those rules, HCl should be named hydrogen chloride. In fact, that is the correct name for the compound HCl that has not been dissolved in water and is represented in reaction equations by the notation HCl(g). Such compounds that have not been dissolved in water are said to be anhydrous (without water). However, when the gas is dissolved in water and represented in equations by the notation HCl(aq), it behaves as an acid and is given another name. The following rules are used to name acidic water solutions of such compounds:

1. The word *hydrogen* in the anhydrous compound name is dropped.
2. The prefix *hydro-* is attached to the stem of the name of the nonmetal that is combined with hydrogen.
3. The suffix *-ide* on the stem of the name of the nonmetal that is combined with hydrogen is replaced with the suffix *-ic*.
4. The word *acid* is added to the end of the name as a separate word.

| Example 9.2 | |

Determine the name that would be given to each of the following binary covalent compounds in the anhydrous form and in the form of water solutions:

a. HCl (stomach acid)
b. H_2S (a gas produced when some sulfur-containing foods such as eggs decay)

Solution

a. The name of the anhydrous compound was given above as hydrogen chloride. The name of the water solution is obtained by dropping *hydrogen* from the anhydrous compound name and adding the prefix *hydro-* to the stem *chlor*. The *-ide* suffix on the *chlor* stem is replaced by the suffix *-ic* to give the name *hydrochloric*. The word *acid* is added, giving the final name *hydrochloric acid* for the water solution.
b. According to the rules of Section 4.10, the anhydrous compound name is hydrogen sulfide. The first two steps in obtaining the name of the water solution are to drop *hydrogen* and to add the prefix *hydro-* to the stem *sulf*.

However, in acids involving sulfur as the nonmetal combined with hydrogen, the stem *sulf* is replaced by the entire name *sulfur* for pronunciation reasons. The next steps involve dropping the suffix *-ide*, adding the suffix *-ic*, and adding the word *acid*. The resulting name of the water solution is *hydrosulfuric acid*.

Learning Check 9.2

Determine the name that would be given to each of the following binary covalent compounds in the anhydrous form and in the form of water solutions:

a. HI
b. HBr

Acids of the second type, in which hydrogen is covalently bonded to a polyatomic ion, have the same name in the anhydrous form and in the form of water solutions. The names for these acids are based on the name of the polyatomic ion to which the hydrogen is bonded. The rules are as follows:

1. All hydrogens that are written as the first part of the formula of the acid are removed. The hydrogens are removed in the form of H^+ ions.
2. The polyatomic ion that remains after the H^+ ions are removed is named by referring to sources such as Table 4.7.
3. When the remaining polyatomic ion has a name ending in the suffix *-ate*, the suffix is replaced by the suffix *-ic,* and the word *acid* is added.
4. When the remaining polyatomic ion has a name ending in the suffix *-ite*, the suffix is replaced by the suffix *-ous,* and the word *acid* is added.
5. If the polyatomic ion contains sulfur or phosphorus, the stems *-sulf* or *-phosph* that remain in Steps 3 or 4, when the suffixes *-ate* or *-ite* are replaced, are expanded for pronunciation reasons to *-sulfur* and *-phosphor* before the *-ic* or *-ous* suffixes are added.

Example 9.3

Compounds derived from the acid H_3PO_4 serve numerous important functions in the body, including the control of acidity in urine and body cells and the storage of energy in the form of ATP. Name this important acid.

Solution

The removal of the three H^+ ions leaves behind the PO_4^{3-} polyatomic ion. This ion is named the phosphate ion in Table 4.7. According to Rules 1–4, the *-ate* suffix is replaced by the *-ic* suffix to give the name *phosphic acid.* However, Rule 5 must be used for this phosphorus-containing acid. The stem is expanded to *phosphor* to give the final name, *phosphoric acid.*

Learning Check 9.3

The acid H_2CO_3 is involved in many processes in the body, including the removal of CO_2 gas produced by cellular metabolism and the control of the acidity of various body fluids. Name this important acid.

➤ 9.4 The Self-Ionization of Water

In Examples 9.1a and b, water behaved as a Brønsted base. In Example 9.1c, it was a Brønsted acid. But what happens when only pure water is present?

The answer is that water behaves as both an acid and a base and undergoes a self- or auto-ionization. The equation representing this self-ionization is

$$H_2O(\ell) + H_2O(\ell) \rightleftarrows H_3O^+(aq) + OH^-(aq) \qquad (9.6)$$

or

$$\begin{bmatrix} :\!\ddot{O}\!-\!H \\ | \\ H \end{bmatrix} + \begin{bmatrix} :\!\ddot{O}\!-\!H \\ | \\ H \end{bmatrix} \longrightarrow \begin{bmatrix} H\!-\!\ddot{O}\!-\!H \\ | \\ H \end{bmatrix}^+ + :\!\ddot{O}\!-\!H^-$$

The transfer of a proton from one water molecule (the acid) to another (the base) causes one H_3O^+ and one OH^- to form. Therefore, in pure water the concentrations of H_3O^+ and OH^- must be equal. At 25°C these concentrations are 10^{-7} mol/L (M). Thus, the equilibrium position is far to the left, as indicated by the arrows in Equation 9.6. Unless noted otherwise, all concentrations and related terms in this chapter are given at 25°C.

The term **neutral** is used to describe any water solution in which the concentrations of H_3O^+ and OH^- are equal. Thus, pure water is neutral because each liter of pure water contains 10^{-7} mol H_3O^+ and 10^{-7} mol OH^-, at equilibrium. Although all water solutions are not necessarily neutral, it is true that in any solution that contains water, the product of the molar concentrations of H_3O^+ and OH^- is a constant. This becomes apparent by writing the equilibrium expression for Reaction 9.6:

$$K = \frac{[H_3O^+][OH^-]}{[H_2O][H_2O]} = \frac{[H_3O^+][OH^-]}{[H_2O]^2} \qquad (9.7)$$

This expression contains the square of the concentration of water in the denominator. However, only a tiny amount of water actually reacts to form H_3O^+ and OH^- (10^{-7} mol/L), so the concentration of water remains essentially constant. Rearrangement of Equation 9.7 gives

$$K[H_2O]^2 = [H_3O^+][OH^-] \qquad (9.8)$$

This equation may be written as

$$K[H_2O]^2 = K_w = [H_3O^+][OH^-] \qquad (9.9)$$

where K_w is a new constant called the **ion product of water.** It is a constant because it is equal to the product of two constants, K and $[H_2O]^2$. At 25°C, K_w can be evaluated from the measured values of $[H_3O^+]$ and $[OH^-]$ in pure water.

$$K_w = [H_3O^+][OH^-] = (1.0 \times 10^{-7} \text{ mol/L})(1.0 \times 10^{-7} \text{ mol/L})$$

$$= 1.0 \times 10^{-14} \text{ (mol/L)}^2 \qquad (9.10)$$

Equation 9.10 is valid not only for pure water but for any solution in which water is the solvent. Notice that we are including units for equilibrium constants in this discussion. This makes the calculation of concentrations easier to follow and understand.

A solution is classified as **acidic** when the concentration of H_3O^+ is greater than the concentration of OH^-. In a **basic** or **alkaline solution**, the concentration of OH^- is greater than that of H_3O^+. However, the product of the molar concentrations of H_3O^+ and OH^- will be 1.0×10^{-14} (mol/L)^2 in either case. Many acidic and basic materials are found in the home (see ■ Figure 9.1).

Neutral
A term used to describe any water solution in which the concentrations of H_3O^+ and OH^- are equal. Also, a water solution with pH = 7.

Ion product of water
The equilibrium constant for the dissociation of pure water into H_3O^+ and OH^-.

Acidic solution
A solution in which the concentration of H_3O^+ is greater than the concentration of OH^-. Also, a solution in which pH is less than 7.

Basic or alkaline solution
A solution in which the concentration of OH^- is greater than the concentration of H_3O^+. Also, a solution in which pH is greater than 7.

Acidic materials found in the home.   Basic or alkaline materials found in the home.

➤ **FIGURE 9.1** Acidic and basic materials are common in the home. Look at the photos, and note a single category to which most of the acidic materials belong, and one to which most of the basic materials belong.

Example 9.4

Classify each of the following solutions as acidic, basic, or neutral. Calculate the molar concentration of the ion whose concentration is not given.

a. $[H_3O^+] = 1.0 \times 10^{-4}$ mol/L
b. $[OH^-] = 1.0 \times 10^{-9}$ mol/L
c. $[OH^-] = 1.0 \times 10^{-6}$ mol/L

Solution

a. Because $[H_3O^+] = 1.0 \times 10^{-4}$ mol/L, a rearrangement of Equation 9.10 gives

$$[OH^-] = \frac{1.0 \times 10^{-14} \, (mol/L)^2}{[H_3O^+]} = \frac{1.0 \times 10^{-14} \, (mol/L)^2}{1.0 \times 10^{-4} \, mol/L}$$

$$= 1.0 \times 1.0^{-10} \, mol/L$$

Thus,

$$[H_3O^+] = 1.0 \times 10^{-4} \, mol/L \quad \text{and}$$

$$[OH^-] = 1.0 \times 10^{-10} \, mol/L$$

The solution is acidic because the $[H_3O^+]$ is greater than the $[OH^-]$; 1.0×10^{-4} is greater than 1.0×10^{-10}.

b. Similarly, $[OH^-] = 1.0 \times 10^{-9}$ mol/L; therefore,

$$[H_3O^+] = \frac{1.0 \times 10^{-14} \, (mol/L)^2}{[OH^-]} = \frac{1.0 \times 10^{-14} \, (mol/L)^2}{1.0 \times 10^{-9} \, mol/L}$$

$$= 1.0 \times 1.0^{-5} \, mol/L$$

This solution is also acidic because 1.0×10^{-5} is greater than 1.0×10^{-9}.

c. $[OH^-] = 1.0 \times 10^{-6}$ mol/L; therefore,

$$[H_3O^+] = \frac{1.0 \times 10^{-14} \, (mol/L)^2}{[OH^-]} = \frac{1.0 \times 10^{-14} \, (mol/L)^2}{1.0 \times 10^{-6} \, mol/L}$$

$$= 1.0 \times 1.0^{-8} \, mol/L$$

This solution is basic because the OH^- concentration (1.0×10^{-6} mol/L) is greater than the H_3O^+ concentration (1.0×10^{-8} mol/L).

Learning Check 9.4	Classify each of the following solutions as acidic, basic, or neutral. Calculate the molarity of the ion whose concentration is not given.

a. $[OH^-] = 1.0 \times 10^{-5}$ mol/L
b. $[H_3O^+] = 1.0 \times 10^{-9}$ mol/L
c. $[H_3O^+] = 1.0 \times 10^{-2}$ mol/L

➤ 9.5 The pH Concept

In Section 9.4, you learned that the concentration of H_3O^+ in pure water is 1.0×10^{-7} M. Chemists, technologists, and other laboratory personnel routinely work with solutions in which the H_3O^+ concentration may be anywhere from 10 to 10^{-14} M. Because of the inconvenience of working with numbers that extend over such a wide range, chemists long ago adopted a shortcut notation known as the pH. Mathematically, the **pH** is defined in terms of the negative logarithm (log) of $[H_3O^+]$ by the following two equations, where we now introduce the common practice of substituting H^+ for H_3O^+ to simplify equations:

pH
The negative logarithm of the molar concentration of H^+ (H_3O^+) in a solution.

$$pH = -\log[H^+] \tag{9.11}$$

$$[H^+] = 1 \times 10^{-pH} \tag{9.12}$$

Thus, pH is simply the negative of the exponent used to express the hydrogen-ion concentration in moles per liter.

GOB
Chemistry⚛Now™
Go to GOB Now and click to engage in a simulation showing the relationship between $[H_3O^+]$, $[OH^-]$, and pH.

Example 9.5	Express the following concentrations in terms of pH:

a. $[H^+] = 1 \times 10^{-5}$ mol/L
b. $[OH^-] = 1 \times 10^{-9}$ mol/L
c. $[H^+] = 1 \times 10^{-7}$ mol/L
d. $[H^+] = 1 \times 10^{-11}$ mol/L

Solution

a. pH is the negative of the exponent used to express $[H^+]$. Therefore, pH = $-(-5) = 5.0$

b. Here $[OH^-]$ is given, and $[H^+]$ must be calculated. We remember Equation 9.10 and get

$$[H^+] = \frac{1.0 \times 10^{-14} \, (\text{mol}/\text{L})^2}{[OH^-]} = \frac{1.0 \times 10^{-14} \, (\text{mol}/\text{L})^2}{1 \times 10^{-9} \, \text{mol}/\text{L}} = 1 \times 10^{-5} \, \text{mol}/\text{L}$$

Therefore, pH = $-(-5) = 5.0$

GOB
Chemistry⚛Now™
Go to GOB Now and click to learn how to convert between $[H_3O^+]$, $[OH^-]$, and pH.

c. This pH = $-(-7) = 7.0$. $[H^+] = 1 \times 10^{-7}$ mol/L corresponds to pure water and neutral solutions. Thus, a pH of 7 represents neutrality.

d. pH = $-(-11) = 11.0$

Learning Check 9.5	Express the following concentrations in terms of pH.

a. $[H^+] = 1 \times 10^{-14}$ mol/L
b. $[OH^-] = 1.0$ mol/L
c. $[OH^-] = 1 \times 10^{-8}$ mol/L

TABLE 9.1 Relationships between [H$^+$], [OH$^-$], and pH

	[H$^+$]	[OH$^-$]	pH	Examples (solids are dissolved in water)
	10^0	10^{-14}	0	HCl (1 mol/L)
	10^{-1}	10^{-13}	1	
				Gastric juice
	10^{-2}	10^{-12}	2	
				Lemon juice
	10^{-3}	10^{-11}	3	
				Vinegar, carbonated drink
				Aspirin
	10^{-4}	10^{-10}	4	Orange juice
				Apple juice
	10^{-5}	10^{-9}	5	Black coffee
	10^{-6}	10^{-8}	6	Normal urine (average value)
				Milk, liquid dishwashing detergent
Neutral	10^{-7}	10^{-7}	7	Saliva, pure water
				Blood
	10^{-8}	10^{-6}	8	Soap (not synthetic detergent)
				Baking soda
				Phosphate-containing detergent
	10^{-9}	10^{-5}	9	
				Milk of magnesia
				Powdered household cleanser
	10^{-10}	10^{-4}	10	
				Phosphate-free detergent
	10^{-11}	10^{-3}	11	Household ammonia
				Liquid household cleaner
	10^{-12}	10^{-2}	12	
	10^{-13}	10^{-1}	13	NaOH (0.1 mol/L)
	10^{-14}	10^0	14	NaOH (1 mol/L)

The pH value of 5.0 obtained in Example 9.5a corresponds to a solution in which the H$^+$ concentration is greater than the OH$^-$ concentration. Thus, the solution is acidic. Any solution with a pH less than 7 is classified as acidic. Any solution with a pH greater than 7 is classified as basic or alkaline. The pH values of some familiar solutions are given in ■ Table 9.1.

Example 9.6

Determine the H$^+$ and OH$^-$ molar concentrations that correspond to the following pH values:

a. pH = 9.0
b. pH = 3.0
c. pH = 11.0

Solution

In each case, the relationships [H$^+$] = $1 \times 10^{-\text{pH}}$ and K_w = [H$^+$][OH$^-$] = 1.0×10^{-14} (mol/L)^2 can be used. Note that we have substituted [H$^+$] for [H$_3$O$^+$] in Equation 9.10.

a. Because pH = 9.0, $[H^+] = 1 \times 10^{-pH} = 1 \times 10^{-9}$ mol/L

$$[OH^-] = \frac{K_w}{[H^+]} = \frac{1.0 \times 10^{-14} \, (mol/L)^2}{1 \times 10^{-9} \, mol/L} = 1 \times 10^{-5} \, mol/L$$

b. Because pH = 3.0, $[H^+] = 1 \times 10^{-pH} = 1 \times 10^{-3}$ mol/L

$$[OH^-] = \frac{K_w}{[H^+]} = \frac{1.0 \times 10^{-14} \, (mol/L)^2}{1 \times 10^{-3} \, mol/L} = 1 \times 10^{-11} \, mol/L$$

c. Because pH = 11.0, $[H^+] = 1 \times 10^{-pH} = 1 \times 10^{-11}$ mol/L

$$[OH^-] = \frac{K_w}{[H^+]} = \frac{1.0 \times 10^{-14} \, (mol/L)^2}{1 \times 10^{-11} \, mol/L} = 1 \times 10^{-3} \, mol/L$$

Learning Check 9.6	Determine the $[H^+]$ and $[OH^-]$ values that correspond to the following pH values:

a. pH = 10.0
b. pH = 4.0
c. pH = 5.0

It is apparent from Table 9.1 that not all solutions have pH values that are neat whole numbers. For example, the pH of vinegar is about 3.3. How do we deal with such numbers? The H^+ concentration of vinegar could be written $1 \times 10^{-3.3}$, but it is not convenient to work with exponents that are not whole numbers. When exact values are not needed, the pH, $[H^+]$, and so on can be expressed as a range. Thus, the H^+ concentration of vinegar is between 1×10^{-3} and 1×10^{-4} mol/L. Similarly, a solution with $[H^+] = 2 \times 10^{-5}$ mol/L has a pH between 4 and 5.

When more exact values are needed, we must work with logarithms. This is most easily done by using a hand calculator. A hydrogen-ion concentration is converted to pH by taking the logarithm and changing its sign. ■ Table 9.2 gives the steps of a typical calculator procedure (what button is pushed, etc.) and a typical calculator readout or display for the conversion of $[H^+] = 3.6 \times 10^{-4}$ mol/L into pH. The pH from Table 9.2 would be recorded as 3.44.

Learning Check 9.7	Convert the following $[H^+]$ values into pH:

a. $[H^+] = 4.2 \times 10^{-5}$ mol/L
b. $[H^+] = 8.1 \times 10^{-9}$ mol/L

TABLE 9.2 Calculating pH from molarity with a calculator

Step	Calculator procedure	Calculator display
1. Enter 3.6	Press buttons 3, ., 6	3.6
2. Enter 10^{-4}	Press button that activates exponential mode (EE, Exp, etc.)	3.6 00
	Press 4	3.6 04
	Press change-sign button (±, etc.)	3.6 −04
3. Take logarithm	Press log button (log, etc.)	−3.4437
4. Change sign	Press change-sign button (±, etc.)	3.4437

TABLE 9.3 Calculating molarity from pH with a calculator

Step	Calculator procedure	Calculator display
1. Enter 5.92	Press 5, ., 9, 2	5.92
2. Change sign	Press change-sign button ($\pm$, etc.)	-5.92
3. Take antilog	Press antilog or 10^x button, or (more commonly) press inv. or 2nd function button and then log button	.0000012 or 1.2 -06

Calculators can also be used to convert pH values into corresponding molar concentrations. The steps are given in ■ Table 9.3 for the conversion of a pH value of 5.92 into $[H^+]$. As shown in Table 9.3, a pH of 5.92 corresponds to a $[H^+]$ of 1.2×10^{-6} mol/L. Note that the number of figures to the right of the decimal in a pH value should be the same as the number of significant figures in the $[H^+]$ value. Thus, the two figures to the right of the decimal in pH = 5.92 is reflected in the two significant figures in $[H^+]$ = 1.2×10^{-6} mol/L.

Learning Check 9.8

Convert the following pH values into molar concentrations of H^+:

a. pH = 2.75
b. pH = 8.33

➤ 9.6 Properties of Acids

Acids and bases are used so often in most laboratories that stock solutions are kept readily available at each work space. The common solutions, their concentrations, and label designations are given in ■ Table 9.4.

Example 9.7

Describe how you would make 250 mL of 1 M HNO_3 by using dilute HNO_3 stock solution.

Solution

According to Table 9.4, dilute HNO_3 stock solution is 6 M. The volume of 6 M HNO_3 needed to make 250 mL of 1 M HNO_3 is obtained by using Equation 7.9. It must be remembered that this equation is useful only for dilution problems such as this one. The equation is not useful for calculations involving the volume of one solution that will react with a specific volume of a second solution (Sections 9.10 and 9.11).

$$C_c V_c = C_d V_d$$
$$(6\ M)V_c = (1\ M)(250\ mL)$$
$$V_c = \frac{(1\ M)(250\ mL)}{6\ M} = 41.7\ mL$$

Marble, a naturally occurring form of $CaCO_3$, reacts with hydrochloric acid, HCl. What gas is produced?

Eggshells are also made of $CaCO_3$.

➤ **FIGURE 9.2** The reaction of hydrochloric acid with two natural forms of calcium carbonate ($CaCO_3$).

TABLE 9.4		Common laboratory acids and bases	
Name	Formula	Label concentration	Molarity
Acids			
acetic acid	$HC_2H_3O_2$	Glacial	18
acetic acid	$HC_2H_3O_2$	Dilute	6
hydrochloric acid	HCl	Concentrated	12
hydrochloric acid	HCl	Dilute	6
nitric acid	HNO_3	Concentrated	16
nitric acid	HNO_3	Dilute	6
sulfuric acid	H_2SO_4	Concentrated	18
sulfuric acid	H_2SO_4	Dilute	3
Bases			
aqueous ammonia[a]	NH_3	Concentrated	15
aqueous ammonia	NH_3	None usually given	6
sodium hydroxide	NaOH	None usually given	6

[a]Often erroneously called ammonium hydroxide and given the formula NH_4OH.

Because the molarity is given using only one significant figure, the solution can be made without too much attention to volumetric flasks and the like. Thus, 42 mL of 6 M HNO_3 is measured with a graduated cylinder, and this is added to 208 mL of distilled water that has also been measured with a graduated cylinder.

Learning Check 9.9	Describe how you would prepare 500 mL of 3.0 M aqueous ammonia using concentrated NH_3 stock solution (Table 9.4).

Different acids have different properties that make some more practical than others for specific uses. However, all acids have certain properties in common. Two of these were mentioned earlier—all acids taste sour and produce H_3O^+ ions when dissolved in water. In addition, all acids undergo characteristic double-replacement reactions with solid oxides, hydroxides, carbonates, and bicarbonates (see ■ Figure 9.2).

$$2HCl(aq) + CuO(s) \rightarrow CuCl_2(aq) + H_2O(\ell) \qquad (9.13)$$
copper
oxide

$$2HCl(aq) + Ca(OH)_2(s) \rightarrow CaCl_2(aq) + 2H_2O(\ell) \qquad (9.14)$$
calcium
hydroxide

$$2HCl(aq) + CaCO_3(s) \rightarrow CaCl_2(aq) + CO_2(g) + H_2O(\ell) \qquad (9.15)$$
calcium
carbonate

$$2HCl(aq) + Sr(HCO_3)_2(s) \rightarrow SrCl_2(aq) + 2CO_2(g) + 2H_2O(\ell) \quad (9.16)$$
strontium
bicarbonate

GOB
Chemistry··Now™

Go to GOB Now and click to learn how to write net ionic equations.

Notice that the preceding reactions are written using molecular equations (Section 5.7).

Reactions involving ionic substances can also be written as total ionic equations or net ionic equations. Equation 9.13 is written in total ionic form in Equation 9.17.

$$\underbrace{2H^+(aq) + 2Cl^-(aq)}_{2HCl} + CuO(s) \longrightarrow \underbrace{Cu^{2+}(aq) + 2Cl^-(aq)}_{CuCl_2} + H_2O(\ell) \quad (9.17)$$

We see that the chloride ions are spectator ions, so a net ionic equation can be written as shown in Equation 9.18.

$$2H^+(aq) + CuO(s) \rightarrow Cu^{2+}(aq) + H_2O(\ell) \quad (9.18)$$

The general nature of the reaction is emphasized in the net ionic form because the H^+ could come from any acid.

Remember that correctly written molecular equations must have their atoms balanced (Section 5.1). Total ionic and net ionic equations must also have their atoms balanced, but in addition the total charges on each side of the equation must balance. In Equation 9.18, for example, the two H^+ ions provide two positive charges on the left, which are balanced by the two positive charges of Cu^{2+} on the right.

Example 9.8

Write Equations 9.14 and 9.15 in total ionic and net ionic forms.

Solution

In Equation 9.14, HCl and $CaCl_2$ are soluble and ionizable.

Total ionic: $2H^+(aq) + 2Cl^-(aq) + Ca(OH)_2(s) \rightarrow Ca^{2+}(aq) + 2Cl^-(aq) + 2H_2O(\ell)$

The Cl^- is a spectator ion.

Net ionic: $2H^+(aq) + Ca(OH)_2(s) \rightarrow Ca^{2+}(aq) + 2H_2O(\ell)$

In Equation 9.15, HCl and $CaCl_2$ are soluble and ionizable.

Total ionic: $2H^+(aq) + 2Cl^-(aq) + CaCO_3(s) \rightarrow Ca^{2+}(aq) + 2Cl^-(aq) + CO_2(g) + H_2O(\ell)$

Again, Cl^- is a spectator ion.

Net ionic: $2H^+(aq) + CaCO_3(s) \rightarrow Ca^{2+}(aq) + CO_2(g) + H_2O(\ell)$

Write Equation 9.16 in total and net ionic forms. Consider HCl and $SrCl_2$ as soluble and ionizable.

Learning Check 9.10

Another property of acids is their ability to react with (and dissolve) certain metals to yield hydrogen gas. This is a redox reaction (Section 5.3), as evidenced by the change in oxidation number of hydrogen as the reaction takes place. In compounds such as acids, hydrogen has an oxidation number of $+1$. In hydrogen gas, the oxidation number is 0 (Section 5.3). Thus, hydrogen is

TABLE 9.5 The activity series of the metals

Metal	Symbol	Comments
potassium	K	React violently with cold water
sodium	Na	
calcium	Ca	Reacts slowly with cold water
magnesium	Mg	React very slowly with steam, but quite rapidly in higher H_3O^+ concentrations
aluminum	Al	
zinc	Zn	
chromium	Cr	
iron	Fe	React in moderately high H_3O^+ concentrations
nickel	Ni	
tin	Sn	
lead	Pb	
copper	Cu	Do not react with H_3O^+
mercury	Hg	
silver	Ag	
platinum	Pt	
gold	Au	

reduced during the reaction, and, as shown by Equations 9.19 and 9.20, the metal is oxidized. The ability to reduce hydrogen ions to hydrogen gas is not the same for all metals. Some are such strong reducing agents that they can react with hydrogen ions of very low concentrations such as that found in water. Others are so weak as reducing agents that they cannot react with H^+ at the high concentration found in concentrated acids. These tendencies are represented by the **activity series** of metals shown in ■ Table 9.5. The higher a metal is in the series, the more active it is as a reducing agent. Some typical reactions are given in Equations 9.19 and 9.20 (see ■ Figure 9.3):

Molecular equation: $Zn(s) + 2HCl(aq) \rightarrow ZnCl_2(aq) + H_2(g)$ (9.19)

Net ionic equation: $Zn(s) + 2H^+(aq) \rightarrow Zn^{2+}(aq) + H_2(g)$

Molecular equation: $2K(s) + 2H_2O(\ell) \rightarrow 2KOH(aq) + H_2(g)$ (9.20)

Net ionic equation: $2K(s) + 2H_2O(\ell) \rightarrow 2K^+(aq) + 2OH^-(aq) + H_2(g)$

Activity series
A tabular representation of the tendencies of metals to react with H^+.

➤ **FIGURE 9.3** Metals vary in their ability to reduce hydrogen ions to hydrogen gas. All the metals are in hydrochloric acid (HCl) of the same molarity. Are these results consistent with Table 9.5? Explain.

Iron and HCl

Zinc and HCl

Magnesium and HCl

© Spencer L. Seager

Write molecular, total ionic, and net ionic equations to represent the following reactions:

a. calcium (Ca) with cold water. (NOTE: $Ca(OH)_2$ is not soluble in water.)
b. Mg with H_2SO_4. (NOTE: $MgSO_4$ is soluble and ionizable in water.)

➤ 9.7 Properties of Bases

Solutions containing bases feel soapy or slippery and change the color of litmus from red to blue. Equation 9.14 illustrates their most characteristic chemical property—they react readily with acids. In most of the earliest acid–base reactions studied, the complete reaction of an acid with a base produced a neutral solution. For this reason, such reactions were often called **neutralization reactions** (see below). It is now known that many "neutralization" reactions do not produce neutral solutions (Section 9.12). However, the name for the reactions is still used. More than 20 billion pounds of sodium hydroxide (NaOH) is produced and used in the United States each year. This useful crystalline solid is quite soluble in water and dissociates to form basic solutions.

Neutralization reaction
A reaction in which an acid and base react completely, leaving a solution that contains only a salt and water.

$$NaOH(s) + H_2O(\ell) \rightarrow Na^+(aq) + OH^-(aq) \qquad (9.21)$$

The neutralization reaction between NaOH and HCl is

Molecular equation: $HCl(aq) + NaOH(aq) \rightarrow NaCl(aq) + H_2O(\ell)$ $\qquad$ (9.22)

Total ionic equation: $H^+(aq) + Cl^-(aq) + Na^+(aq) + OH^-(aq) \rightarrow$ $\qquad$ (9.23)
$$Na^+(aq) + Cl^-(aq) + H_2O(\ell)$$

Net ionic equation: $H^+(aq) + OH^-(aq) \rightarrow H_2O(\ell)$ $\qquad$ (9.24)

The molecular equation (9.22) illustrates the following statement, which is a common definition of neutralization: During a neutralization reaction, an acid and a base combine to form a salt and water. (More is said about salts in the next section.) The net ionic form of the equation (9.24) emphasizes the general nature of neutralization reactions: H^+ ions (from any source) react with OH^- (from any source) to form water.

Bases also react with fats and oils and convert them into smaller, soluble molecules. For this reason, most household cleaning products contain basic substances. For example, lye (NaOH) is the active ingredient in numerous drain cleaners, and many liquid household cleaners contain ammonia.

Write molecular, total ionic, and net ionic equations to represent neutralization reactions between the following acids and bases:

a. HNO_3 and NaOH
b. H_2SO_4 and KOH

➤ 9.8 Salts

At room temperature, **salts** are solid ionic compounds that contain the **cation** (positive ion) of a base and the **anion** (negative ion) of an acid. Thus, ordinary table salt (NaCl) contains Na^+, the cation of NaOH, and Cl^-, the anion of HCl (look again at Equation 9.22). Similarly, $CuSO_4$ is a salt containing the cation of $Cu(OH)_2$ and the anion of H_2SO_4. You must be careful to think of the term *salt* in a general way, and not as representing only table salt (NaCl).

Salt
A solid crystalline ionic compound at room temperature that contains the cation of a base and the anion of an acid.

Cation
A positively charged ion.

Anion
A negatively charged ion.

Some acids and bases are not stable enough to be isolated even though their salts are. For example, carbonic acid (H_2CO_3) cannot be isolated in the pure state. When it forms in water, it promptly decomposes:

$$H_2CO_3(aq) \rightleftarrows H_2O(\ell) + CO_2(g) \qquad (9.25)$$

Despite this characteristic, salts of carbonic acid, such as Na_2CO_3 and $NaHCO_3$, are quite stable.

It is not necessary to identify the parent acid and base in order to write correct salt formulas or names. Just remember that the cation of a salt can be any positive ion except H^+, and it will usually be a simple metal ion or NH_4^+. The salt anion can be any negative ion except OH^-. Most of the polyatomic anions you will use were given earlier in Table 4.7. The rules for naming salts were given in Sections 4.4 and 4.10.

As previously observed, salts dissolved in solution are dissociated into ions. The salt can be recovered by evaporating away the water solvent. When this is done carefully, some salts retain specific numbers of water molecules as part of the solid crystalline structure. Such salts are called **hydrates,** and the retained water is called the **water of hydration.** Most hydrates lose all or part of the water of hydration when they are heated to moderate or high temperatures. A number of useful hydrates are given in ■ Table 9.6.

Many salts occur in nature, and some are used as industrial raw materials. Examples are sodium chloride, $NaCl$ (a source of Cl_2 and $NaOH$); calcium carbonate or limestone, $CaCO_3$ (a source of cement and building stone); and calcium phosphate or rock phosphate, $Ca_3(PO_4)_2$ (a source of fertilizer). In the laboratory, salts can be prepared by reacting a solution of an appropriate acid with a metal, a metal oxide, a metal hydroxide, a metal carbonate, or a metal bicarbonate. These reactions, given earlier as examples of acid properties (Equations 9.19, 9.13, 9.14, 9.15, 9.16), are given below in a general form:

$$\text{acid} + \text{metal} \rightarrow \text{salt} + H_2 \qquad (9.26)$$

$$\text{acid} + \text{metal oxide} \rightarrow \text{salt} + H_2O \qquad (9.27)$$

$$\text{acid} + \text{metal hydroxide} \rightarrow \text{salt} + H_2O \qquad (9.28)$$

$$\text{acid} + \text{metal carbonate} \rightarrow \text{salt} + H_2O + CO_2 \qquad (9.29)$$

$$\text{acid} + \text{metal bicarbonate} \rightarrow \text{salt} + H_2O + CO_2 \qquad (9.30)$$

Hydrate

A salt that contains specific numbers of water molecules as part of the solid crystalline structure.

Water of hydration

Water retained as part of the solid crystalline structure of some salts.

TABLE 9.6 Some useful and common hydrates

Formula	Chemical name	Common name	Uses
$CaSO_4 \cdot H_2O$	Calcium sulfate monohydrate	Plaster of Paris	Plaster, casts, molds
$CaSO_4 \cdot 2H_2O$	Calcium sulfate dihydrate	Gypsum	Casts, molds, wallboard
$MgSO_4 \cdot 7H_2O$	Magnesium sulfate heptahydrate	Epsom salts	Cathartic
$Na_2B_4O_7 \cdot 10H_2O$	Sodium tetraborate decahydrate	Borax	Laundry
$Na_2CO_3 \cdot 10H_2O$	Sodium carbonate decahydrate	Washing soda	Water softener
$Na_3PO_4 \cdot 12H_2O$	Sodium phosphate dodecahydrate	Trisodium phosphate (TSP)	Water softener
$Na_2SO_4 \cdot 10H_2O$	Sodium sulfate decahydrate	Glauber's salt	Cathartic
$Na_2S_2O_3 \cdot 5H_2O$	Sodium thiosulfate pentahydrate	Hypo	Photography

Example 9.9

Write equations to represent the preparation of $Mg(NO_3)_2$, using Reactions 9.26 through 9.30. Use molecular equations to emphasize the salt formation.

Solution

In each case, a dilute solution of nitric acid (HNO_3) is reacted with magnesium metal or the appropriate magnesium compound:

$$2HNO_3(aq) + Mg(s) \rightarrow Mg(NO_3)_2(aq) + H_2(g)$$

$$2HNO_3(aq) + MgO(s) \rightarrow Mg(NO_3)_2(aq) + H_2O(\ell)$$

$$2HNO_3(aq) + Mg(OH)_2(s) \rightarrow Mg(NO_3)_2(aq) + 2H_2O(\ell)$$

$$2HNO_3(aq) + MgCO_3(s) \rightarrow Mg(NO_3)_2(aq) + H_2O(\ell) + CO_2(g)$$

$$2HNO_3(aq) + Mg(HCO_3)_2(s) \rightarrow Mg(NO_3)_2(aq) + 2H_2O(\ell) + 2CO_2(g)$$

Learning Check 9.13

Write balanced equations to represent the preparation of $AlCl_3$, using Reactions 9.26 through 9.30. Use molecular equations to emphasize the salt formation.

We have expressed the amounts of materials involved in chemical reactions primarily in terms of mass (grams) or number of moles. However, in certain applications, it is important to express the amount of salt in a solution in terms of the amount of electrical charge represented by the ions of the salt. This is especially true in medical applications, where the levels and balance of electrolytes in various body fluids are extremely important.

A unit that expresses the amount of ionic electrical charge for salts is the equivalent. One **equivalent of salt** is the amount that will produce 1 mol of positive (or negative) charges when dissolved and dissociated. To determine the amount of salt that represents 1 equivalent (eq), you must know what ions are produced when the salt dissociates. For example, potassium chloride (KCl), an electrolyte often administered to patients following surgery, dissociates as follows: $KCl(aq) \rightarrow K^+(aq) + Cl^-(aq)$. Thus, 1 mol, or 74.6 g, of solid KCl provides 1 mol of positive charges (1 mol of K^+ ions) when it dissolves and dissociates. Thus, 1 eq of KCl is equal to 1 mol, or 74.6 g, of KCl.

Equivalent of salt
The amount that will produce 1 mol of positive electrical charge when dissolved and dissociated.

Example 9.10

Determine the number of equivalents and milliequivalents (meq) of salt contained in the following:

a. 0.050 mol KCl
b. 0.050 mol $CaCl_2$

Solution

a. As shown above, 1 mol KCl = 1 eq KCl. Therefore, 0.050 mol KCl = 0.050 eq KCl. Also, because 1 eq = 1000 meq,

$$0.050 \; \cancel{eq} \times \frac{1000 \text{ meq}}{1 \; \cancel{eq}} = 50 \text{ meq}$$

b. The dissociation reaction is

$$CaCl_2(aq) \rightarrow Ca^{2+}(aq) + 2Cl^-(aq)$$

Thus, we see that 1 mol $CaCl_2$ produces 1 mol Ca^{2+} or 2 mol of positive charges. Thus, 1 mol $CaCl_2$ = 2 eq $CaCl_2$. Therefore,

$$0.050 \text{ mol } CaCl_2 \times \frac{2 \text{ eq } CaCl_2}{1 \text{ mol } CaCl_2} = 0.10 \text{ eq } CaCl_2$$

Also, because 1 eq = 1000 meq,

$$0.10 \text{ eq} \times \frac{1000 \text{ meq}}{1 \text{ eq}} = 1.0 \times 10^2 \text{ meq}$$

Notice that we could have focused on the negative charges in either part above and still arrived at the same answers. In part (a), 0.050 mol KCl produces 0.050 mol Cl^-, or 0.050 mol of negative charge. Similarly, in part (b), 0.050 mol $CaCl_2$ produces 2×0.050, or 0.10 mol, Cl^- ions, or 0.10 mol of negative charge. We have arbitrarily chosen to use the positive charges, but either will work. Just remember that you do not count both the negative and positive charges for a salt.

| Learning Check 9.14 | Determine the number of equivalents and milliequivalents in each of the following: |

a. 0.10 mol NaCl
b. 0.10 mol $Mg(NO_3)_2$

CHEMISTRY AROUND US • 9.1

BAKERY CHEMISTRY

We use a characteristic reaction of acids to make light, fluffy pancakes, waffles, cakes, biscuits, and other baked goodies. This useful reaction is the one that takes place between acids and bicarbonates (Equation 9.30).

A leavening agent, used to make batter and dough rise, works by generating carbon dioxide gas, which becomes distributed throughout the batter or dough in little bubbles. When these bubbles are heated during baking, they expand, and the batter or dough rises to give a light product. The baking powder called for in recipes contains the gas-producing materials. In two popular brands of baking powder, the active ingredients are sodium bicarbonate, $NaHCO_3$; calcium hydrogen phosphate, $CaHPO_4$; and sodium aluminum sulfate, $NaAl(SO_4)_2$. Both the $CaHPO_4$ and the $NaAl(SO_4)_2$ behave as acids when they come in contact with water. The liberated H^+ ions then react characteristically with the sodium bicarbonate. The net ionic reaction is

$$H^+(aq) + HCO_3^-(aq) \rightarrow H_2O(\ell) + CO_2(g)$$

The advantage of using baking powder is that it acts rapidly. A slower leavening process is used with most breads. Yeast, a living microorganism, is added to the dough along with sugar. The yeast slowly metabolizes the sugar and produces ethyl alcohol and carbon dioxide as products. The ethyl alcohol evaporates as the bread bakes, and the CO_2 gas, which has been distributed throughout the dough by kneading, expands and causes the dough to rise.

Some recipes don't call for baking powder but include ingredients like baking soda and sour milk or buttermilk. These ingredients do the same thing as baking powder. Baking soda is just good old sodium bicarbonate again, $NaHCO_3$, and sour milk or buttermilk contains lactic acid. Once again, the H^+ ions produced by the acid react with the bicarbonate to give CO_2. Why do you suppose sodium bicarbonate is called baking soda?

Living yeast is the leavening for most breads.

© Mark Slabaugh

Example 9.11

A sample of blood serum contains 0.139 eq/L of Na^+ ion. Assume the Na^+ comes from dissolved NaCl, and calculate the number of equivalents, number of moles, and number of grams of NaCl in 250 mL of the serum.

Solution

The Na^+ ion has a single charge, so we may write

$$1.00 \text{ mol NaCl} = 1.00 \text{ mol } Na^+ = 1.00 \text{ eq } Na^+ = 1.00 \text{ eq NaCl}$$

A slight modification to the solution stoichiometry approach given in Section 7.6 leads to the following pattern and pathway for this problem. The pattern is liters solution $A \rightarrow$ eq A, and the pathway is liters Na^+ solution $\rightarrow$ eq Na^+. The factor for the conversion comes from the concentration given above.

$$(0.250 \text{ L } Na^+ \text{ solution}) \times \frac{0.139 \text{ eq } Na^+}{1.00 \text{ L } Na^+ \text{ solution}} = 0.0347 \text{ eq } Na^+$$

The equivalents of Na^+ are converted to the quantities asked for by using factors from the relationships given above and the formula weight for NaCl of 58.44 u.

$$(0.0347 \text{ eq } Na^+) \times \frac{1.00 \text{ eq NaCl}}{1.00 \text{ eq } Na^+} = 0.0347 \text{ eq NaCl}$$

$$(0.0347 \text{ eq NaCl}) \times \frac{1.00 \text{ mol NaCl}}{1.00 \text{ eq NaCl}} = 0.0347 \text{ mol NaCl}$$

$$(0.0347 \text{ mol NaCl}) \times \frac{58.44 \text{ g NaCl}}{1.00 \text{ mol NaCl}} = 2.03 \text{ g NaCl}$$

A sample of blood serum contains 0.103 eq/L of Cl^- ion. Assume the Cl^- comes from dissolved NaCl, and calculate the number of equivalents, number of moles, and number of grams of NaCl in 250 mL of serum.

Learning Check 9.15

➤ 9.9 The Strengths of Acids and Bases

When salts dissolve in water, they generally dissociate completely, but this is not true for all acids and bases. The acids and bases that do dissociate almost completely are classified as **strong acids** and **strong bases** (they are also strong electrolytes). Those that dissociate to a much smaller extent are called **weak** or **moderately weak,** depending on the degree of dissociation (they are also weak or moderately weak electrolytes). Examples of strong and weak acids are given in ■ Table 9.7.

A 0.10 M solution of hydrochloric acid could be prepared by dissolving 0.10 mol (3.7 g) of HCl gas in enough water to give 1.0 L of solution. According to Table 9.7, 100% of the dissolved gas would dissociate into H^+ and Cl^-. Thus, the concentration of H^+ in a 0.10 M HCl solution is 0.10 mol/L, and the pH is 1.00.

The strength of acids and bases is shown quantitatively by the value of the equilibrium constant for the dissociation reaction in water solutions.

Strong acids and strong bases
Acids and bases that dissociate (ionize) essentially completely when dissolved to form a solution.

Weak (or moderately weak) acids and bases
Acids and bases that dissociate (ionize) less than completely when dissolved to form a solution.

| TABLE 9.7 | Some common strong and weak acids |

Name	Formula	% Dissociation[a]	K_a	Classification
Hydrochloric acid	HCl	100	Very large	Strong
Hydrobromic acid	HBr	100	Very large	Strong
Nitric acid	HNO_3	100	Very large	Strong
Sulfuric acid	H_2SO_4	100	Very large	Strong
Phosphoric acid	H_3PO_4	28	7.5×10^{-3}	Moderately weak
Sulfurous acid[b]	H_2SO_3	34	1.5×10^{-2}	Moderately weak
Acetic acid	$HC_2H_3O_2$	1.3	1.8×10^{-5}	Weak
Boric acid	H_3BO_3	0.01	7.3×10^{-10}	Weak
Carbonic acid[b]	H_2CO_3	0.2	4.3×10^{-7}	Weak
Nitrous acid[b]	HNO_2	6.7	4.6×10^{-4}	Weak

[a]Based on dissociation of one proton in 0.1 M solutions at 25°C.
[b]Unstable acid.

Equation 9.31 represents the dissociation of a general acid, HB, in water, where B represents the conjugate base of the acid:

$$HB(aq) + H_2O(\ell) \leftrightharpoons H_3O^+(aq) + B^-(aq) \tag{9.31}$$

The equilibrium expression for this reaction is

$$K = \frac{[H_3O^+][B^-]}{[HB][H_2O]} \tag{9.32}$$

In Equation 9.32, the brackets, again, represent molar concentrations of the materials in the solution. Only a tiny amount of the water in the solution actually enters into the reaction, so the concentration of water is considered to be constant (see Section 9.4). We can then write Equation 9.33, where K_a is a new constant called the **acid dissociation constant:**

$$K[H_2O] = K_a = \frac{[H_3O^+][B^-]}{[HB]} \tag{9.33}$$

Acid dissociation constant
The equilibrium constant for the dissociation of an acid.

In Equation 9.33, $[H_3O^+]$ and $[B^-]$ are, respectively, the equilibrium concentrations of the hydronium ion and the anion conjugate base that is characteristic of the acid. The [HB] represents the concentration of that part of the dissolved acid that remains undissociated in the equilibrium mixture. In solutions of strong acids, $[H_3O^+]$ and $[B^-]$ values are quite large, while [HB] has a value near 0, so K_a is quite large. In weak acids, [HB] has larger values, while $[H_3O^+]$ and $[B^-]$ are smaller, so K_a is smaller. Thus, the larger a K_a value, the stronger the acid it represents. This is illustrated by the K_a values given in Table 9.7. If we simplify by substituting H^+ for H_3O^+, introduced in Section 9.5, we obtain

$$K_a = \frac{[H^+][B^-]}{[HB]} \tag{9.34}$$

It is important to remember that the terms *weak* and *strong* apply to the extent of dissociation and not to the concentration of an acid or base. For example, gastric juice (0.05% HCl) is a dilute (not weak) solution of a strong acid.

WRITING REACTIONS OF ACIDS

As you study this chapter, you will acquire a knowledge of the characteristic reactions of acids and the ability to write balanced equations for the reactions. In the Key Equations section at the end of the chapter, five characteristic reactions are summarized in item 3. One useful way to remember the reactions is to learn them as general word equations, such as

"An acid plus a metal gives a salt plus hydrogen gas"

rather than as specific equations such as $H_2SO_4(aq) + Zn(s) \rightarrow ZnSO_4(aq) + H_2(g)$. If you learn the general word equations and you recognize the starting materials for a reaction (such as an acid and a metal), you will know what the products will be (a salt and H_2 gas).

A second approach is to remember that all five of the general reactions of acids are either single-replacement

$$\text{acid} + \text{metal} \rightarrow \text{salt} + H_2$$

or double-replacement (the remaining four acid reactions in the Key Equations section). The formulas of the products of double-replacement reactions can be predicted by simply breaking each reactant into its positive and negative parts and recombining the parts in the other possible way (the positive part of one reactant with the negative part of the other reactant). For example, let's determine the products and the balanced equation for a reaction between H_2SO_4 and KOH. If you remember that H_2SO_4 is an acid and KOH is a base, the general word equation says the products will be a salt and water:

$$H_2SO_4(aq) + KOH(aq) \rightarrow \text{salt}(aq) + H_2O(\ell)$$

H_2SO_4 breaks apart to give H^+ and SO_4^{2-}; KOH breaks apart to give K^+ and OH^-. If we combine the positive part of the acid (H^+) with the negative part of the base (OH^-), we get the water (H_2O). A similar combination of the positive part of the base (K^+) with the negative part of the acid (SO_4^{2-}) gives the salt. We must remember that the total charges of the combined parts must add up to 0. Thus, two K^+ will combine with one SO_4^{2-} to give the salt K_2SO_4. The balanced equation is

$$H_2SO_4(aq) + 2KOH(aq) \rightarrow K_2SO_4(aq) + 2H_2O(\ell)$$

As another example, consider a reaction between HBr and KOH. KBr breaks into H^+ and Br^-. KOH breaks into K^+ and OH^-. Now, switch the parts and recombine:

The products are KBr and H_2O, and the balanced equation is

$$HBr(aq) + KOH(aq) \rightarrow KBr(aq) + H_2O(\ell)$$

For a final example, let's try a reaction between HBr and $NaHCO_3$. The word equation predicts that the products should be a salt, water, and carbon dioxide gas. HBr breaks into H^+ and Br^-; $NaHCO_3$ breaks into Na^+ and HCO_3^-. Now, switch the parts and recombine:

The products are NaBr and H_2CO_3. However, H_2CO_3 is not stable; it decomposes to give H_2O and CO_2, the products predicted earlier. The balanced equation is

$$HBr(aq) + NaHCO_3(aq) \rightarrow NaBr(aq) + H_2O(\ell) + CO_2(g)$$

Example 9.12

Write dissociation reactions and expressions for K_a for each of the following weak acids:

a. Hydrocyanic acid (HCN)
b. Phosphoric acid (H_3PO_4) (1st H only)
c. Dihydrogen phosphate ion ($H_2PO_4^-$) (1st H only)

Solution

In each case, H^+ has been substituted for H_3O^+.

a. $HCN \leftrightharpoons H^+ + CN^-$; $\quad K_a = \dfrac{[H^+][CN^-]}{[HCN]}$

b. $H_3PO_4 \leftrightharpoons H^+ + H_2PO_4^-$; $\quad K_a = \dfrac{[H^+][H_2PO_4^-]}{[H_3PO_4]}$

c. $H_2PO_4^- \leftrightharpoons H^+ + HPO_4^{2-}$; $\quad K_a = \dfrac{[H^+][HPO_4^{2-}]}{[H_2PO_4^-]}$

Learning Check 9.16

Write dissociation reactions and K_a expressions for the following weak acids:

a. Hydrogen phosphate ion (HPO_4^{2-})
b. Nitrous acid (HNO_2)
c. Hydrofluoric acid (HF)

Monoprotic acid
An acid that gives up only one proton (H^+) per molecule when dissolved.

Diprotic acid
An acid that gives up two protons (H^+) per molecule when dissolved.

Triprotic acid
An acid that gives up three protons (H^+) per molecule when dissolved.

Acid behavior is linked to the loss of protons. Thus, acids must contain hydrogen atoms that can be removed to form H^+. **Monoprotic acids** can lose only one proton per molecule, whereas **diprotic** and **triprotic** acids can lose two and three, respectively. For example, HCl is monoprotic, H_2SO_4 is diprotic, and H_3PO_4 is triprotic. Di- and triprotic acids dissociate in steps, as shown for H_2SO_4 in Equations 9.35 and 9.36:

$$H_2SO_4(aq) \rightleftharpoons H^+(aq) + HSO_4^-(aq) \tag{9.35}$$

$$HSO_4^-(aq) \leftrightharpoons H^+(aq) + SO_4^{2-}(aq) \tag{9.36}$$

The second proton is not as easily removed as the first because it must be pulled away from a negatively charged particle, HSO_4^-. Accordingly, HSO_4^- is a weaker acid than H_2SO_4.

The number of ionizable hydrogens cannot always be determined from the molecular formula for an acid. For example, acetic acid ($HC_2H_3O_2$) is monoprotic even though the molecule contains four hydrogen atoms. The dissociation of acetic acid is represented by Equation 9.37, where structural formulas are used to emphasize the different H atoms in the molecule:

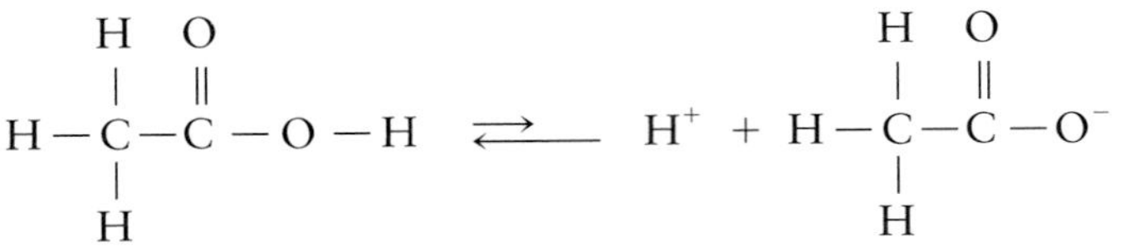

$$\text{(9.37)}$$

Only the hydrogen bound to the oxygen is ionizable. Those hydrogens bound to C are too tightly held to be removed. ■ Table 9.8 contains other examples, with the ionizable hydrogens shown in color.

We have focused our attention on the strength of acids, using the extent of dissociation as a basis. However, all acid dissociations are reversible to some

TABLE 9.8 Examples of monoprotic, diprotic, and triprotic acids

Name	Formula	Structural formula	Classification
Butyric acid	$HC_4H_7O_2$		Monoprotic
Carbonic acid	H_2CO_3		Diprotic
Formic acid	$HCHO_2$		Monoprotic
Nitric acid	HNO_3		Monoprotic
Phosphoric acid	H_3PO_4		Triprotic
Phosphorous acid	H_3PO_3		Diprotic

degree, and in the reverse reactions, anions produced by the forward reaction behave as Brønsted bases. What can be said about the strength of these bases? Because Brønsted acid–base behavior is really just competition for protons, we can answer this question by looking again at the simplified form of the equation for dissociation of a general acid:

$$HB(aq) \leftrightarrows H^+(aq) + B^-(aq) \tag{9.38}$$

We see from this equation that the strength of HB as an acid depends on how tightly the conjugate base B^- holds onto the proton. If HB is a strong acid, the conjugate base holds onto the proton only weakly. If HB is a weak acid, the conjugate base holds on more strongly, depending on the strength of HB. Thus, we have answered our earlier question. If HB is a strong acid, the H^+ is held only weakly by the B^-. We can conclude then that B^- is not strongly attracted to protons—it is a weak base. Conversely, if HB is a weak acid, the H^+ is held tightly by the B^-, and we conclude that B^- is more strongly attracted to protons—it is a stronger base than the B^- from a strong acid.

In general, the conjugate base anions produced by the dissociation of strong Brønsted acids are weak Brønsted bases. The conjugate base anions of weak acids are stronger bases, with their strengths dependent on the strength of the parent acid.

DO YOU HAVE ACID REFLUX DISEASE?

The medical name for acid reflux disease is *gastroe-sophageal reflux disease,* which is often abbreviated and referred to as GERD. The disease is often mistaken for oc-casional heartburn and treated with over-the-counter remedies (See Over the Counter 9.1). GERD is the result of a malfunc-tioning muscle (the LES muscle) located at the bottom of the esophagus, just above the stomach. When operating normally, this muscle relaxes and opens to allow food to pass from the esophagus down into the stomach, then contracts to close the opening and prevent the acidic contents of the stomach from backing up into the esophagus.

When the muscle relaxes at inappropriate times, the acidic stomach contents get into the esophagus and cause the burning chest pain called heartburn. However, when this occurs repeat-edly and frequently, the acidic stomach contents can also erode the lining of the esophagus. GERD is a complex condition with many degrees of severity, ranging from only frequent heartburn symp-toms to erosive esophagitis, in which the esophagus can suffer dif-ferent degrees of damage. In extreme cases of erosive esophagitis, ulcers develop in the esophagus and lead to esophageal bleeding that, if persistent and undetected, can lead to iron deficiency and anemia as well as extreme pain and weight loss. In some cases, se-vere GERD can lead to other serious medical conditions that re-quire hospitalization and even surgery to correct.

Doctors often recommend lifestyle and dietary changes for most GERD patients, including the avoidance of foods and bev-erages that weaken the LES muscle. These foods include choco-late, peppermint, fatty foods, coffee, and alcoholic beverages. The use of foods and beverages that can irritate a damaged esophageal lining, such as acidic fruits and juices, pepper, and tomato products is also discouraged. GERD symptoms in over-weight individuals often diminish when some weight is lost. Smokers who quit also generally gain some relief. Prescription medications are also available that reduce the amount of acid in the stomach. Two types of medication are available. Both types, called H2 blockers and proton (acid) pump inhibitors, decrease the amount of acid secreted into the stomach, but by different mechanisms.

The main symptom that may indicate the presence of GERD is frequent, persistent heartburn that occurs two or more times a week. Other symptoms include difficulty in swallowing and fre-quent belching and regurgitation. Less common symptoms that occur in some people resemble respiratory conditions and in-clude a persistent sore throat, wheezing, chronic coughing, and hoarseness. Individuals who suspect they might be suffering from some degree of GERD should consult a physician to deter-mine the extent of the disease and proper treatment.

Ammonia (NH_3) is the weak base most often encountered in addition to the anions of strong acids. The dissociation reaction of gaseous NH_3 in water, given earlier in Example 9.1, is

$$NH_3(aq) + H_2O(\ell) \rightleftarrows NH_4^+(aq) + OH^-(aq) \qquad (9.39)$$

The most common strong bases are the hydroxides of group IA(1) metals (NaOH, KOH, etc.) and the hydroxides of group IIA(2) metals ($Mg(OH)_2$, $Ca(OH)_2$, etc.).

Example 9.13

Classify each of the following pairs by identifying the stronger of the pair according to the indicated behavior. Information from Tables 9.7 and 9.8 may be used.

a. H_3PO_4 and $H_2PO_4^-$ (acid)
b. $H_2PO_4^-$ and HPO_4^{2-} (base)
c. HNO_3 and HNO_2 (acid)

Solution

a. H_3PO_4 is stronger as an acid. In di- and triprotic acids, each proton in the removal sequence is harder to remove, so the corresponding acid is weaker.
b. The stronger acid produces the weaker anion base. Thus, $H_2PO_4^-$, the anion of the stronger acid H_3PO_4, would be a weaker base than HPO_4^{2-}, the anion of the weaker acid $H_2PO_4^-$. So HPO_4^{2-} is a stronger base than $H_2PO_4^-$.

c. HNO_3 is the stronger acid according to Table 9.7. Generally, when related acids (same atoms, etc.) are compared for strength, the one containing more oxygen atoms will be the stronger.

Classify each of the following according to strength for the indicated behavior. If more than two are compared, list them with the strongest at the top and the weakest at the bottom. Use Tables 9.7 and 9.8 as needed.

a. $HClO$, $HClO_3$, $HClO_2$ (acid)
b. NO_2^- and NO_3^- (base)
c. $HC_2H_3O_2$ and $C_2H_3O_2^-$ (acid)

➤ 9.10 Analyzing Acids and Bases

The analysis of solutions for the total amount of acid or base they contain is a regular activity in many laboratories. The total amount of acid in a solution is indicated by its capacity to neutralize a base. The pH is related to the acidity or concentration of H^+ in solution, while the capacity to neutralize a base depends on the total amount of H^+ available. For example, a 0.10 M acetic acid solution has an H^+ concentration of about 1.3×10^{-3} M (pH = 2.89). However, 1 L of the solution can neutralize 0.10 mol of OH^-, not just 1.3×10^{-3} mol. The reason is that the dissociation equilibrium of acetic acid is

$$HC_2H_3O_2(aq) \rightleftarrows H^+(aq) + C_2H_3O_2^-(aq) \qquad (9.40)$$

As OH^- is added, H^+ reacts to form water (see Equation 9.24). The removal of H^+ causes the equilibrium of Reaction 9.40 to shift right in accordance with Le Châtelier's principle. The continued addition of OH^- and removal of H^+ will eventually cause all of the acetic acid molecules to dissociate and react.

A common procedure often used to analyze acids and bases is called **titration** (■ Figure 9.4). Suppose the total acidity of an unknown acid solution needs to be determined. A known volume of the acidic solution is first measured out by drawing it up to the calibration mark of a pipet. This solution is placed in a container (Step 1). A basic solution of known concentration (a standard solution) is added to the acid solution in the container until the **equivalence point** is reached. This is the point where the unknown acid is completely reacted with base. The volume of base needed to reach the equivalence point is obtained from the buret readings.

To successfully complete a titration, the point at which the reaction is completed must somehow be detected. One way to do this is to add an indicator to the solution being titrated. An indicator is an organic compound that changes to different colors depending on the pH of its surroundings. An indicator is selected that will change color at a pH as close as possible to the pH the solution will have at the equivalence point. If the acid and base are both strong, the pH at that point will be 7. However, for reasons discussed later, the pH is not always 7 at the equivalence point. The point at which the indicator changes color and the titration is stopped is called the titration **endpoint**. ■ Figure 9.5 gives the colors shown by a number of indicators at various pH values.

Indicator papers (litmus paper, pH paper, etc.) are often used to make routine pH measurements. These papers are impregnated with one or more indicators that change to a variety of colors depending on the pH. However, indicators may not be practical under certain conditions. For example, there might not be any indicator that changes color close enough to the equivalence point,

Titration
An analytical procedure in which one solution (often a base) of known concentration is slowly added to a measured volume of an unknown solution (often an acid). The volume of the added solution is measured with a buret.

Equivalence point of a titration
The point at which the unknown solution has exactly reacted with the known solution. Neither is in excess.

Endpoint of a titration
The point at which the titration is stopped on the basis of an indicator color change or pH meter reading.

➤ **FIGURE 9.4** Titration.

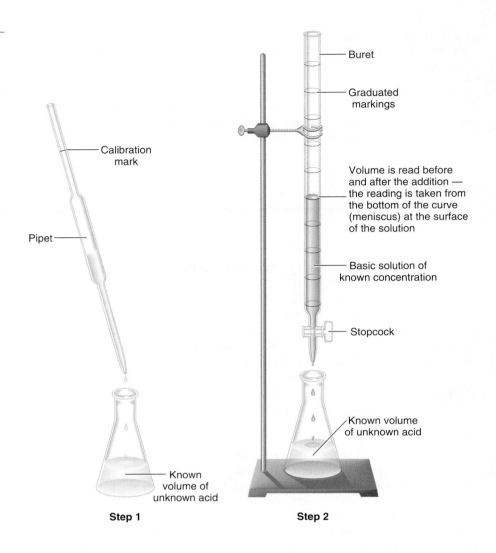

or the solution being titrated might be so highly colored that an indicator color change cannot be detected. Under such circumstances, a pH meter can be used. The electrodes of the meter are placed in the solution being titrated, and the pH is read directly (see ■ Figure 9.6). The titration is continued until the meter reading matches the pH of the equivalence point.

➤ 9.11 Titration Calculations

Equations 9.41 and 9.42 represent the reactions that occur when solutions of nitric and sulfuric acid are titrated with sodium hydroxide:

$$HNO_3(aq) + NaOH(aq) \rightarrow H_2O(\ell) + NaNO_3(aq) \qquad (9.41)$$

$$H_2SO_4(aq) + 2NaOH(aq) \rightarrow 2H_2O(\ell) + Na_2SO_4(aq) \qquad (9.42)$$

One mole of HNO_3 requires 1 mol NaOH for complete reaction, but 1 mol H_2SO_4 requires 2 mol NaOH. Because these are solution reactions, stoichiometric calculations can be done using the methods described in Section 7.6.

Example 9.14

Calculate the molarity of the HNO_3 and H_2SO_4 solutions involved in the following titrations:

phenolphthalein

bromthymol blue

methyl red

a. A 25.0-mL sample of HNO_3 solution requires the addition of 16.3 mL of 0.200 M NaOH solution to reach the equivalence point.

b. A 25.0-mL sample of H_2SO_4 solution requires the addition of 32.6 mL of 0.200 M NaOH to reach the equivalence point.

> **FIGURE 9.5** Indicators change color with changes in pH (the numbers on the tubes). Would phenolphthalein be a useful indicator to differentiate between two solutions with pH values of 5 and 7? Explain.

Solution

In each case we know the volume of acid solution reacted. If we also knew the number of moles of acid in the volume of reacted solution, we could calculate the solution molarity by using Equation 7.5:

$$M = \frac{\text{moles of solute}}{\text{liters of solution}}$$

Therefore, our task will be to calculate the number of moles of acid reacted in each case. We will use the methods described in Section 7.6.

a. According to Figure 7.10, this problem has the pattern liters solution $A \rightarrow$ mol B, and the pathway is liters NaOH solution $\rightarrow$ mol NaOH $\rightarrow$ mol HNO_3. The factors come from the molarity of the NaOH solution and the coefficients of the balanced equation. In combined form, the steps in the factor-unit method are as follows:

Step 1. 0.0163 L NaOH solution

Step 2. 0.0163 L NaOH solution $= \text{mol } HNO_3$

Step 3. $0.0163 \; \cancel{\text{L NaOH solution}} \times \dfrac{0.200 \; \cancel{\text{mol NaOH}}}{1 \; \cancel{\text{L NaOH solution}}} \times \dfrac{1 \text{ mol } HNO_3}{1 \; \cancel{\text{mol NaOH}}} = \text{mol } HNO_3$

Step 4. $(0.0163)\left(\dfrac{0.200}{1}\right)\left(\dfrac{1 \text{ mol } HNO_3}{1}\right) = 0.00326 \text{ mol } HNO_3$

GOB
Chemistry·⚛·Now™
Go to GOB Now and click to view three tutorials showing how to use titration data to determine the concentration of an unknown acid.

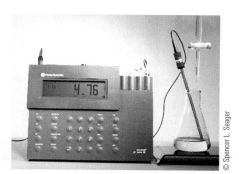
At the beginning, the pH meter gives the pH of the acid solution being titrated.

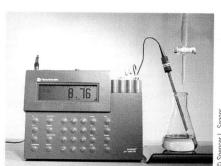
Partway through the titration, the pH meter reading is of a solution of unreacted acid and the salt produced by the reaction.

At the end of the titration, the pH meter gives the pH of the salt solution formed by the complete reaction of acid with base.

> **FIGURE 9.6** An acid–base titration using a pH meter to detect the equivalence point.

Notice that the volume of NaOH solution was changed to liters so that units would cancel properly. The result of this calculation tells us that the 25.0-mL sample of HNO_3 solution contains 0.00326 mol HNO_3. After converting the volume to liters, the molarity can be calculated:

$$M = \frac{\text{moles HNO}_3}{\text{liters HNO}_3 \text{ solution}} = \frac{0.00326 \text{ mol HNO}_3}{0.0250 \text{ L HNO}_3 \text{ solution}} = \frac{0.130 \text{ mol HNO}_3}{\text{L HNO}_3 \text{ solution}}$$

b. The problem of calculating the number of moles of H_2SO_4 in the titrated sample also has the pattern liters solution $A \rightarrow$ mol B; the pathway is liters NaOH solution $\rightarrow$ mol NaOH $\rightarrow$ mol H_2SO_4. Once again, the factors come from the molarity of NaOH solution and the coefficients of the balanced equation. In combined form, the steps in the factor-unit method are as follows:

Step 1.   0.0326 L NaOH solution

Step 2.   0.0326 L NaOH solution $= $ mol H_2SO_4

Step 3.   0.0326 $\cancel{L}$ $\cancel{\text{NaOH}}$ $\cancel{\text{solution}}$ $\times \dfrac{0.200 \cancel{\text{ mol NaOH}}}{1 \cancel{L} \cancel{\text{NaOH}} \cancel{\text{solution}}} \times \dfrac{1 \text{ mol H}_2\text{SO}_4}{2 \cancel{\text{ mol NaOH}}} = $ mol H_2SO_4

Step 4. $(0.0326)\left(\dfrac{0.200}{1}\right)\left(\dfrac{1 \text{ mol H}_2\text{SO}_4}{2}\right)$ $= 0.00326$ mol H_2SO_4

The molarity of the solution is calculated as before:

$$M = \frac{\text{moles H}_2\text{SO}_4}{\text{liters H}_2\text{SO}_4 \text{ solution}} = \frac{0.00326 \text{ mol H}_2\text{SO}_4}{0.0250 \text{ L H}_2\text{SO}_4 \text{ solution}} = \frac{0.130 \text{ mol H}_2\text{SO}_4}{\text{L H}_2\text{SO}_4 \text{ solution}}$$

Thus, we see that the two acid solutions have the same molar concentrations. Why did the second one require twice the volume of NaOH solution for the titration?

| Learning Check 9.18 | Calculate the molarity of a solution of phosphoric acid (H_3PO_4) if a 25.0-mL sample of acid solution requires 14.1 mL of 0.250 M NaOH solution to titrate to the equivalence point. The equation for the reaction is |

$$H_3PO_4(aq) + 3NaOH(aq) \rightarrow 3H_2O(aq) + Na_3PO_4(aq)$$

➤ 9.12 Hydrolysis Reactions of Salts

Hydrolysis reaction
Any reaction with water. For salts it is a reaction of the acidic cation and/or basic anion of the salt with water.

In general, a **hydrolysis reaction** is a reaction with water. Many types of hydrolysis reactions are known, but at this point we will discuss only one, the hydrolysis of salts. In Section 9.10, we said that the pH at the equivalence point is not 7 for all acid–base titrations. However, we pointed out earlier that the only products of acid–base reactions are water and a salt. Therefore, it seems reasonable to conclude that some salts and water must interact (a hydrolysis reaction) and cause the solution pH to differ from that of pure water.

Suppose three solutions are prepared by dissolving equal molar amounts of sodium acetate, ammonium chloride, and sodium chloride in identical volumes of pure water. Measurement of pH shows that the sodium acetate solution is alkaline (pH higher than 7), the ammonium chloride solution is acidic (pH lower than 7), and the sodium chloride solution is neutral (pH = 7). The alkaline sodium acetate solution must contain more OH^- than H^+ ions. However, the only source of both ions in a solution of salt and water is the dissociation of water, so both should be present in equal amounts. The equation for the water dissociation is

$$H_2O(\ell) \rightleftharpoons H^+(aq) + OH^-(aq) \qquad (9.43)$$

The OH^- excess would result if something removed H^+ from this equilibrium, causing it to shift right and produce more OH^-. This is exactly what happens. The acetate anion is the conjugate base of a weak acid (acetic acid), and it is a strong enough Brønsted base to react with the H^+ ions of the above equilibrium:

$$C_2H_3O_2^-(aq) + H^+(aq) \rightleftharpoons HC_2H_3O_2(aq) \qquad (9.44)$$

The overall result of this reaction can be seen by adding Equations 9.43 and 9.44:

$$H_2O(\ell) + C_2H_3O_2^-(aq) + H^+(aq) \rightleftharpoons$$
$$H^+(aq) + OH^-(aq) + HC_2H_3O_2(aq) \qquad (9.45)$$

Net reaction: $H_2O(\ell) + C_2H_3O_2^-(aq) \rightleftharpoons$
$$OH^-(aq) + HC_2H_3O_2^-(aq) \qquad (9.46)$$

Now it can be seen that the excess OH^- comes from a reaction between the acetate ion and water. But what about the Na^+ ion that is also in the solution? Does it react with water? The Na^+ cation is the conjugate acid of the strong base NaOH. As a result, Na^+ is an extremely weak Brønsted acid and makes no contribution to the pH of the solution.

In general, salts influence the pH of water solutions as follows: (1) Salts containing an anion of a weak acid and a cation of a strong base will form alkaline water solutions. (2) Salts containing an anion of a strong acid and a cation of a weak base will form acidic solutions. (3) Salts containing an anion and a cation from equal-strength acids and bases (both weak or both strong) will form neutral solutions (see ■ Figure 9.7).

The acidic ammonium chloride solution is an example of the second category of salts, where the NH_4^+ cation is the conjugate acid of the weak base NH_3, and the Cl^- anion is the conjugate base of the strong acid HCl. The hydrolysis reaction is given in Equation 9.47, where H_3O^+ is used to emphasize the reaction with water:

$$NH_4^+(aq) + H_2O(\ell) \rightleftharpoons H_3O^+(aq) + NH_3(aq) \qquad (9.47)$$

The neutral sodium chloride solution is an example of the third category. Both NaOH and HCl are strong, and an NaCl solution is neutral.

A sample of pure water with a little phenolphthalein indicator added.

A solution of sodium acetate in water with a little phenolphthalein indicator added.

➤ **FIGURE 9.7** The acetate ion hydrolyzes in water to form a basic solution. Why didn't the Na^+ ion from the sodium acetate hydrolyze?

Predict the relative pH value (higher than 7, lower than 7, etc.) for water solutions of the following salts. Table 9.7 may be useful.

a. sodium nitrate ($NaNO_3$)
b. sodium nitrite ($NaNO_2$)
c. potassium borate (K_3BO_3)

Learning Check 9.19

➤ **9.13 Buffers**

Buffers are solutions that have the ability to resist changes in pH when acids or bases are added (see ■ Figure 9.8). Most buffers consist of a pair of compounds, one with the ability to react with H^+ and the other with the ability

Buffer
A solution with the ability to resist changing pH when acids (H^+) or bases (OH^-) are added.

The solution on the left is not buffered; the one on the right is; bromcresol green indicator has been added to each solution.

Sodium hydroxide has been added to each solution.

Hydrochloric acid has been added to two fresh samples that originally looked like the first pair of samples.

➤ **FIGURE 9.8** Buffered solutions resist changing pH when bases and acids are added. What observation of this experiment indicates that the pH of the buffered solutions did not change significantly when base or acid was added?

Buffer capacity
The amount of acid (H^+) or base (OH^-) that can be absorbed by a buffer without causing a significant change in pH.

Go to GOB Now and click to explore how a buffer works.

to react with OH^-. An example is a mixture of acetic acid and its salt, sodium acetate. The acetate ion from sodium acetate is the conjugate base of acetic acid and reacts with any added acid (H^+ ions):

$$C_2H_3O_2^-(aq) + H^+(aq) \rightleftarrows HC_2H_3O_2(aq) \qquad (9.48)$$

Any added base (OH^- ions) reacts with nonionized acetic acid molecules:

$$HC_2H_3O_2(aq) + OH^-(aq) \rightleftarrows C_2H_3O_2^-(aq) + H_2O(\ell) \qquad (9.49)$$

In this way, the buffer solution resists changes in pH. The amount of H^+ or OH^- that a buffer system can absorb without allowing significant pH changes to occur is called its **buffer capacity.**

Buffered solutions are found extensively in the body, where they serve to protect us from the disastrous effects of large deviations in pH. An important buffer system in blood is composed of carbonic acid (H_2CO_3) and bicarbonate salts such as $NaHCO_3$. The unstable carbonic acid results when dissolved CO_2 reacts with water in the blood:

$$H_2O(\ell) + CO_2(aq) \rightleftarrows H_2CO_3(aq) \qquad (9.50)$$

Added H^+ is neutralized by reacting with HCO_3^- from the dissolved bicarbonate salts:

$$HCO_3^-(aq) + H^+(aq) \rightleftarrows H_2CO_3(aq) \qquad (9.51)$$

The carbonic acid protects against added OH^-:

$$H_2CO_3(aq) + OH^-(aq) \rightleftarrows HCO_3^-(aq) + H_2O(\ell) \qquad (9.52)$$

If large amounts of H^+ or OH^- are added to a buffer, the buffer capacity can be exceeded, the buffer system is overwhelmed, and the pH changes. For example, if large amounts of H^+ were added to the bicarbonate ion–carbonic acid buffer, Reaction 9.51 would take place until the HCO_3^- was depleted. The pH would then drop as additional H^+ ions were added. In blood, the concentration of HCO_3^- is ten times the concentration of H_2CO_3. Thus, this buffer has a greater capacity against added acid than against bases. This is consistent with the normal functions of the body that cause larger amounts of acidic than basic substances to enter the blood.

The pH of buffers can be calculated by using a form of Equation 9.34 that was given earlier and is repeated here:

$$K_a = \frac{[H^+][B^-]}{[HB]}$$

This equation can be rearranged to give

$$[H^+] = K_a \frac{[HB]}{[B^-]}$$

(9.53)

Application of the logarithm concept to Equation 9.53 gives

$$pH = pK_a + \log\frac{[B^-]}{[HB]}$$

(9.54)

In this equation, $pH = -\log[H+]$ (Equation 9.11), and $\mathbf{pK_a} = -\log K_a$. Equation 9.54 is known as the **Henderson–Hasselbalch equation**; it is often used by biologists, biochemists, and others who frequently work with buffers. We see from Equation 9.54 that when the concentrations of a weak acid and its conjugate base (anion) are equal in a solution, the pH of the solution is equal to pK_a. When it is desired to produce a buffer with a pH different from the

pK_a
The negative logarithm of K_a.

Henderson–Hasselbalch equation
A relationship between the pH of a buffer, pK_a, and the concentrations of acid and salt in the buffer.

OVER THE COUNTER • 9.1

HEARTBURN REMEDIES: SOMETHING OLD, SOMETHING NEW

Every day, the average person's stomach produces about a quart of hydrochloric acid as a part of the digestive gastric juices. The resulting 0.1 M solution of HCl does not cause any significant problems for someone with a healthy, normally functioning stomach. However, it causes extreme discomfort for anyone suffering from a peptic ulcer or a gastroesophageal reflux condition. This latter condition is characterized by a tendency for the acidic contents of the stomach to be forced back up into the esophagus, causing what is commonly called *heartburn* or, in extreme cases, *gastroesophageal reflux disease (GERD)*.

For many years the traditional nonprescription treatment for heartburn has been antacids, which are stocked by most drugstores in a variety of forms. Some of them fizz when placed in water, others relieve headache as well as heartburn, several will relieve gas pains, and still others provide a laxative action to clean out the intestinal tract. They come in various forms and flavors. Some are soft and chewy, others are crunchy, and at least one is a thick liquid that "coats" a distressed stomach.

Regardless of their form, taste, or added side effects, most of these conventional antacids act on stomach acid by either neutralizing it with a hydroxide-containing base or reacting it with a carbonate or bicarbonate. A quick survey of 12 antacid products resulted in the list of active ingredients below. The number in parentheses is the number of different antacids that list it as an ingredient. Some products contained more than one active ingredient, so the numbers in parentheses will add up to more than 12.

- Sodium bicarbonate (5), $NaHCO_3$
- Potassium bicarbonate (1), $KHCO_3$
- Calcium carbonate (4), $CaCO_3$
- Magnesium carbonate (2), $MgCO_3$
- Magnesium hydroxide (4), $Mg(OH)_2$
- Aluminum hydroxide (2), $Al(OH)_3$

For years, compounds known collectively as H2 blockers had been available by prescription for use as maintenance therapy for peptic ulcers. H2 blockers are named for the fact that they block histamine signals in the body that cause the stomach to produce acid. However, the discovery that most ulcers are caused by a bacterium and can be completely cured with antibiotics essentially eliminated this use for H2 blockers. The manufacturers quickly responded by making their products available in OTC strengths to compete with the traditional antacids. These new products have been heavily advertised in the media under such names as Pepcid AC®, Zantac 75®, Tagamet HB®, and Axid AR®. A disadvantage of H2 blockers is that they must be taken up to an hour before a meal in order to be effective, while traditional antacids may be taken when symptoms first appear. However, H2 blockers are effective for several hours, whereas the effects of regular antacids may wear off in as little as a few minutes. Another advantage of the traditional antacids that use $CaCO_3$ as an ingredient is that they provide a source of calcium for the body.

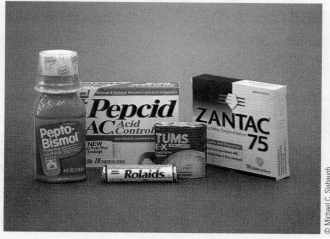

Antacids neutralize stomach acid; H2 blockers limit its production.

TABLE 9.9 K_a and pK_a values for selected weak acids

Name	Formula	K_a	pK_a
Acetic acid[a]	CH_3COOH	1.8×10^{-5}	4.74
Ammonium ion	NH_4^+	5.6×10^{-10}	9.25
Boric acid	H_3BO_3	7.3×10^{-10}	9.14
Dihydrogen borate ion	$H_2BO_3^-$	1.8×10^{-13}	12.75
Hydrogen borate ion	HBO_3^{2-}	1.6×10^{-14}	13.80
Carbonic acid	H_2CO_3	4.3×10^{-7}	6.37
Bicarbonate ion	HCO_3^-	5.6×10^{-11}	10.25
Citric acid[a]	$C_3H_4(OH)(COOH)_3$	8.4×10^{-4}	3.08
Dihydrogen citrate ion[a]	$C_3H_4(OH)(COOH)_2COO^-$	1.8×10^{-5}	4.74
Hydrogen citrate ion[a]	$C_3H_4(OH)(COOH)(COO)_2^{2-}$	4.0×10^{-6}	5.40
Formic acid[a]	$HCOOH$	1.8×10^{-4}	3.74
Lactic acid[a]	$C_2H_4(OH)COOH$	1.4×10^{-4}	3.85
Nitrous acid	HNO_2	4.6×10^{-4}	3.33
Phosphoric acid	H_3PO_4	7.5×10^{-3}	2.12
Dihydrogen phosphate ion	$H_2PO_4^-$	6.2×10^{-8}	7.21
Hydrogen phosphate ion	HPO_4^{2-}	2.2×10^{-13}	12.66
Sulfurous acid	H_2SO_3	1.5×10^{-2}	1.82
Bisulfite ion	HSO_3^-	1.0×10^{-7}	7.00

[a]The hydrogen that ionizes in organic acids and ions is a part of a carboxylic acid group, represented by COOH.

exact pK_a, Equation 9.54 indicates it can be done. An acid with a pK_a near the desired pH is selected, and the ratio of the concentrations of conjugate base (anion of the acid) and acid is adjusted to give the desired pH. When a weak acid and its salt (the source of the acid's conjugate base anion) are mixed in solution, you can generally assume that the amount of acid that dissociates is small and may be neglected. This means that the buffer concentrations of the acid and anion (the salt of acid) are "equal" to the made-up concentrations. ■ Table 9.9 lists some weak acids, together with values of K_a and pK_a.

Example 9.15

a. Calculate the pH of a buffer solution that contains 0.10 mol acetic acid (CH_3COOH) and 0.10 mol sodium acetate (CH_3COONa) per liter.
b. What is the pH of a buffer in which the concentration of NaH_2PO_4 is 0.10 M and that of Na_2HPO_4 is 0.50 M?
c. A buffer system consisting of the ions $H_2PO_4^-$ and HPO_4^{2-} helps control the pH of urine. What value of the ratio $[HPO_4^{2-}]/[H_2PO_4^-]$ would be required to maintain a pH of 6.00, the average value for normal urine?

Solution

a. The acetic acid concentration is 0.10 M and that of its conjugate base, the acetate ion from the sodium acetate salt, is also 0.10 M. The pK_a for acetic acid is 4.74:

$$pH = pK_a + \log\frac{[CH_3COO^-]}{[CH_3COOH]} = 4.74 + \log\left(\frac{0.10 \text{ mol/L}}{0.10 \text{ mol/L}}\right)$$

$$= 4.74 + \log(1) = 4.74 + 0 = 4.74$$

We see that, as mentioned earlier, when the acid and conjugate base concentrations are equal, $pH = pK_a$.

b. In this case, both compounds are salts that produce the $H_2PO_4^-$ and HPO_4^{2-} ions. In such situations, the ion with more hydrogens will be the acid, and the one with fewer will be the anion or conjugate base. Therefore, the acid is $H_2PO_4^-$ at a concentration of 0.10 M, and the conjugate base or anion is HPO_4^{2-} at a concentration of 0.50 M. pK_a for $H_2PO_4^-$ is 7.21:

$$pH = pK_a + \log\frac{[HPO_4^{2-}]}{[H_2PO_4^-]} = 7.21 + \log\left(\frac{0.50 \text{ mol/L}}{0.10 \text{ mol/L}}\right)$$

$$= 7.21 + \log(5.00) = 7.21 + 0.70 = 7.91$$

c. As in part (b), the acid is $H_2PO_4^-$ and the conjugate base is HPO_4^{2-}. The desired pH is 6.00. Substitution into Equation 9.54 gives

$$pH = pK_a + \log\frac{[HPO_4^{2-}]}{[H_2PO_4^-]}$$

$$6.00 = 7.21 + \log\frac{[HPO_4^{2-}]}{[H_2PO_4^-]}$$

$$6.00 - 7.21 = \log\frac{[HPO_4^{2-}]}{[H_2PO_4^-]}$$

$$-1.21 = \log\frac{[HPO_4^{2-}]}{[H_2PO_4^-]}$$

At this point, we know the value of the log of the desired ratio. To get the ratio, we must evaluate the antilog of -1.21. Use Step 3 given in Table 9.3 to get 0.062 as the value of the ratio. This result says the ratio $[HPO_4^{2-}]/[H_2PO_4^-]$ must be 0.062. In other words, the HPO_4^{2-} concentration must be only about 0.06 times the value of the $H_2PO_4^-$ concentration. Any concentrations giving this ratio would be satisfactory. For example, if $[H_2PO_4^-] = 0.50$ M, then $[HPO_4^{2-}]$ would have to equal 0.062 × 0.5 M, or 0.031 M.

Learning Check 9.20

a. What is the pH of a buffer solution in which formic acid (HCOOH) and sodium formate (HCOONa) are both at a concentration of 0.22 M?

b. What is the pH of a buffer solution that is 0.25 M in sulfurous acid (H_2SO_3) and 0.10 M in sodium bisulfite ($NaHSO_3$)?

c. A buffer system in the blood consists of the bicarbonate ion, HCO_3^-, and carbonic acid, H_2CO_3. What value of the ratio $[HCO_3^-]/[H_2CO_3]$ would be required to maintain blood at the average normal pH of 7.40?

FOR FUTURE REFERENCE PHOSPHORUS-RICH FOODS

Phosphorus is the second most abundant mineral in the body (calcium is the most abundant). About 85% of it is found in teeth and bones in the form of calcium salts of the moderately weak phosphoric acid, H_3PO_4 (See Tables 9.7 and 9.9). In addition to their role as components of bones and teeth, phosphoric acid and its salts (phosphates) form an important buffer system inside all body cells. Phosphorus compounds are also important components of genetic material, energy-transfer compounds, cell membranes, some proteins, and lipid transport systems. Dietary deficiencies of phosphorus are unknown for individuals with diets that provide adequate energy and protein. Most foods, including processed foods, contain significant amounts of phosphorus. In processed foods, phosphorus from additives represents a significant source. The chart below contains a selection of foods that are rich in phosphorus. Notice that most of these are foods that are also high in proteins.

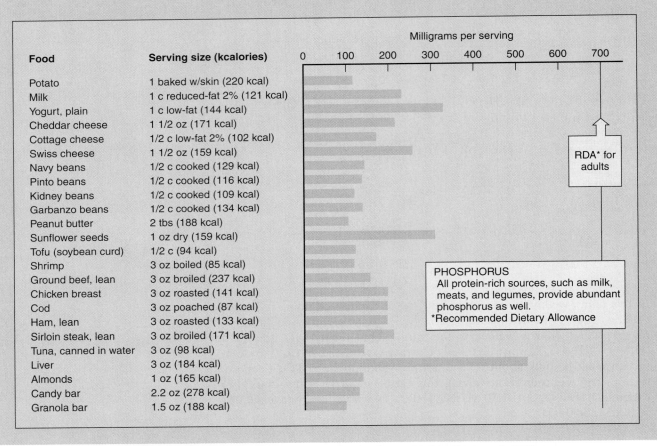

Food	Serving size (kcalories)
Potato	1 baked w/skin (220 kcal)
Milk	1 c reduced-fat 2% (121 kcal)
Yogurt, plain	1 c low-fat (144 kcal)
Cheddar cheese	1 1/2 oz (171 kcal)
Cottage cheese	1/2 c low-fat 2% (102 kcal)
Swiss cheese	1 1/2 oz (159 kcal)
Navy beans	1/2 c cooked (129 kcal)
Pinto beans	1/2 c cooked (116 kcal)
Kidney beans	1/2 c cooked (109 kcal)
Garbanzo beans	1/2 c cooked (134 kcal)
Peanut butter	2 tbs (188 kcal)
Sunflower seeds	1 oz dry (159 kcal)
Tofu (soybean curd)	1/2 c (94 kcal)
Shrimp	3 oz boiled (85 kcal)
Ground beef, lean	3 oz broiled (237 kcal)
Chicken breast	3 oz roasted (141 kcal)
Cod	3 oz poached (87 kcal)
Ham, lean	3 oz roasted (133 kcal)
Sirloin steak, lean	3 oz broiled (171 kcal)
Tuna, canned in water	3 oz (98 kcal)
Liver	3 oz (184 kcal)
Almonds	1 oz (165 kcal)
Candy bar	2.2 oz (278 kcal)
Granola bar	1.5 oz (188 kcal)

Milligrams per serving

RDA* for adults

PHOSPHORUS
All protein-rich sources, such as milk, meats, and legumes, provide abundant phosphorus as well.
*Recommended Dietary Allowance

SOURCE: Adapted from Whitney, E.N.; Rolfes, S.R. *Understanding Nutrition,* 8th ed. Belmont, CA: West/Wadsworth, 1999.

CONCEPT SUMMARY

The Arrhenius Theory. Svante Arrhenius defined acids as substances that dissociate in water to provide hydrogen ions (H^+), and bases as substances that dissociate in water to provide hydroxide ions, (OH^-).

The Brønsted Theory. Johannes Brønsted and Thomas Lowry proposed a theory in which acids are defined as any hydrogen-containing substances capable of donating protons to other substances. Bases are substances that accept and form covalent bonds with protons. When a substance behaves as a Brønsted acid by donating a proton, the substance becomes a conjugate base.

Naming Acids. Two types of acids are named differently. Water solutions of binary covalent compounds containing hydrogen and a nonmetal are named following the pattern hydro(stem)ic acid, where (stem) is the stem of the name of the nonmetal bonded to hydrogen. Acids in which hydrogen is bonded to polyatomic ions have names based on the name of the polyatomic ion to which hydrogen is bonded.

The Self-Ionization of Water. Water, a hydrogen-containing material, is able to behave as both a Brønsted acid and a Brønsted base. In pure water, a small number of water molecules (10^{-7} mol/L) donate protons to other water molecules.

The pH Concept. The pH is the negative logarithm of the molar H^+ concentration of a solution. Solutions with pH values lower than 7 are acidic, those with pH values higher than 7 are basic or alkaline, and those with a pH value of 7 are neutral.

Properties of Acids. All acids have certain characteristic properties that include (1) a sour taste; (2) a reaction with water to produce H_3O^+; (3) reactions with solid metallic oxides, hydroxides, carbonates, and bicarbonates; and (4) a reaction with certain metals to give hydrogen gas.

Properties of Bases. Basic solutions feel soapy or slippery and change the color of litmus from red to blue. Their most characteristic chemical property is a neutralization reaction with acids to produce water and a salt.

Salts. At room temperature, salts are solid crystalline substances

that contain the cation of a base and the anion of an acid. Hydrated salts contain specific numbers of water molecules as a part of their crystalline structures. Salts can be prepared by reacting an appropriate acid with one of a number of other materials.

The Strengths of Acids and Bases. Acids and bases that dissociate completely when dissolved in solution are called strong, and those that do not are called weak or moderately weak depending on the degree of dissociation they undergo. Acid strength is indicated by the value of K_a, the acid dissociation constant. In general, polyprotic acids become weaker as an acid during each successive dissociation reaction. The Brønsted base produced by the dissociation of an acid has a strength opposite that of the acid. The acid strengths of cations produced by the dissociation of bases follow a similar pattern.

Analyzing Acids and Bases. The neutralization reaction of acids and bases is used in a process called titration to analyze acids and bases. During a typical titration, a base solution of known concentration is added slowly to an acid solution of unknown concentration. The titration is stopped at the endpoint when a color change occurs in an indicator. The volumes of acid and base required are used to calculate the acid concentration.

Titration Calculations. Titrations are used to determine the total amount of acid or base in solutions. Data collected from titrations are treated like other stoichiometric data in calculations.

Hydrolysis Reactions of Salts. The cation and anion of a salt may have the same or different Brønsted acid and base strengths. When they are the same, solutions of the salt have a pH of 7. When they are different, salt solutions have H^+ and OH^- concentrations that are unequal and pH values different than 7. The reactions of salt ions with water that cause these results are called hydrolysis reactions.

Buffers. Solutions with the ability to maintain essentially constant pH values when acid (H^+) or base (OH^-) are added are called buffers. All buffers have a limit to the amount of acid or base they can absorb without changing pH. This limit is called the buffer capacity.

LEARNING OBJECTIVES ASSESSMENT

You can get an approximate but quick idea of how well you have met the learning objectives given at the beginning of this chapter by working the selected end-of-chapter exercises given below. The answer to each exercise is given in Appendix B of the book.

Objective 1 (Sections 9.1 and 9.2): Exercises 9.2 and 9.6

Objective 2 (Section 9.2): Exercise 9.8

Objective 3 (Section 9.3): Exercise 9.22

Objective 4 (Section 9.4): Exercises 9.28 a & b and 9.30 a & b

Objective 5 (Section 9.5): Exercises 9.36 and 9.40

Objective 6 (Section 9.6): Exercise 9.50

Objective 7 (Sections 9.6, 9.7, and 9.8): Exercises 9.54, 9.60, and 9.68

Objective 8 (Section 9.7): Exercise 9.58

Objective 9 (Section 9.8): Exercise 9.74

Objective 10 (Section 9.9): Exercise 9.86

Objective 11 (Sections 9.10 and 9.11): Exercises 9.92 and 9.96

Objective 12 (Section 9.12): Exercise 9.108

Objective 13 (Section 9.13): Exercise 9.116

KEY TERMS AND CONCEPTS

Acid dissociation constant (9.9)
Acidic solution (9.4)
Activity series (9.6)
Anion (9.8)
Arrhenius acid (9.1)
Arrhenius base (9.1)
Basic or alkaline solution (9.4)
Brønsted acid (9.2)
Brønsted base (9.2)
Buffer capacity (9.13)
Buffer (9.13)
Cation (9.8)

Conjugate acid–base pair (9.2)
Conjugate base (9.2)
Diprotic acid (9.9)
Endpoint of a titration (9.10)
Equivalence point of a titration (9.10)
Equivalent of salt (9.8)
Henderson–Hasselbalch equation (9.13)
Hydrate (9.8)
Hydrolysis reaction (9.12)
Ion product of water (9.4)
Monoprotic acid (9.9)
Neutral (9.4)

Neutralization reaction (9.7)
pH (9.5)
pK_a (9.13)
Salt (9.8)
Strong acids and strong bases (9.9)
Titration (9.10)
Triprotic acid (9.9)
Water of hydration (9.8)
Weak (or moderately weak) acids and bases (9.9)

KEY EQUATIONS

1. Relationship between $[H^+]$ and $[OH^-]$ in water solutions (Section 9.4):

$$K_w = [H_3O^+][OH^-] = 1.0 \times 10^{-14} \ (mol/L)^2$$

Equation 9.10

2. Relationships between pH and $[H^+]$ (Section 9.5):

$$pH = -\log[H^+]$$

Equation 9.11

$$[H^+] = 1 \times 10^{-pH}$$

Equation 9.12

3. General reactions of acids (Section 9.8):

$$acid + metal \rightarrow salt + H_2$$

Equation 9.26

$$acid + metal \ oxide \rightarrow salt + H_2O$$

Equation 9.27

$$acid + metal \ hydroxide \rightarrow salt + H_2O$$

Equation 9.28

$$acid + metal \ carbonate \rightarrow salt + H_2O + CO_2$$

Equation 9.29

$$acid + metal \ bicarbonate \rightarrow salt + H_2O + CO_2$$

Equation 9.30

4. Dissociation constant for acids (Section 9.9):

$$K_a = \frac{[H^+][B^-]}{[HB]}$$

Equation 9.34

5. Calculation of buffer pH, Henderson–Hasselbalch equation (Section 9.13):

$$pH = pK_a + \log\frac{[B^-]}{[HB]}$$

Equation 9.54

EXERCISES

LEGEND: 1 = straightforward, 2 = intermediate, 3 = challenging. All even-numbered exercises are answered in Appendix B.

THE ARRHENIUS THEORY (SECTION 9.1)

9.1 Write the dissociation equations for the following that emphasize their behavior as Arrhenius acids:
  a. HI
  b. HBrO
  c. HCN
  d. $HClO_2$

9.2 Write the dissociation equations for the following that emphasize their behavior as Arrhenius acids:
  a. $HBrO_2$
  b. HS^-
  c. HBr
  d. $HC_2H_3O_2$ (only the 1st listed H dissociates)

9.3 Each of the following produces a basic solution when dissolved in water. Identify those that behave as Arrhenius bases and write dissociation equations to illustrate that behavior.
a. CsOH
b. CH_3NH_2
c. NH_3
d. $Ca(OH)_2$

9.4 Each of the following produces a basic solution when dissolved in water. Identify those that behave as Arrhenius bases and write dissociation equations to illustrate that behavior.
a. LiOH
b. $C_2H_5NH_2$
c. $Sr(OH)_2$
d. $NaNH_2$

THE BRØNSTED THEORY (SECTION 9.2)

9.5 Identify each Brønsted acid and base in the following equations. Note that the reactions are assumed to be reversible.
a. $HBr(aq) + H_2O(\ell) \rightleftarrows H_3O^+(aq) + Br^-(aq)$
b. $H_2O(\ell) + N_3^-(aq) \rightleftarrows HN_3(aq) + OH^-(aq)$
c. $H_2S(aq) + H_2O(\ell) \rightleftarrows H_3O^+(aq) + HS^-(aq)$
d. $SO_3^{2-}(aq) + H_2O(\ell) \rightleftarrows HSO_3^-(aq) + OH^-(aq)$
e. $HCN(aq) + H_2O(\ell) \rightleftarrows H_3O^+(aq) + CN^-(aq)$

9.6 Identify each Brønsted acid and base in the following equations. Note that the reactions are assumed to be reversible.
a. $HC_2O_4^-(aq) + H_2O(\ell) \rightleftarrows H_3O^+(aq) + C_2O_4^{2-}(aq)$
b. $HNO_2(aq) + H_2O(\ell) \rightleftarrows H_3O^+(aq) + NO_2^-(aq)$
c. $PO_4^{3-}(aq) + H_2O(\ell) \rightleftarrows HPO_4^{2-}(aq) + OH^-(aq)$
d. $H_2SO_3(aq) + H_2O(\ell) \rightleftarrows HSO_3^-(aq) + H_3O^+(\ell)$
e. $F^-(aq) + H_2O(\ell) \rightleftarrows HF(aq) + OH^-(aq)$

9.7 Identify each conjugate acid–base pair in the equations of Exercise 9.5.

9.8 Identify each conjugate acid–base pair in the equations of Exercise 9.6.

9.9 Write equations to represent the Brønsted acid behavior for each of the following acids in water solution. Remember to represent the reactions as being reversible.
a. HI
b. HBrO
c. HCN
d. HSe^-

9.10 Write equations to represent the Brønsted acid behavior for each of the following acids in water solution. Remember to represent the reactions as being reversible.
a. HF
b. $HClO_3$
c. HClO
d. HS^-

9.11 Write a formula for the conjugate base formed when each of the following behaves as a Brønsted acid:
a. HSO_3^-
b. HPO_4^{2-}
c. $HClO_3$
d. $CH_3NH_3^+$
e. $H_2C_2O_4$

9.12 Write a formula for the conjugate base formed when each of the following behaves as a Brønsted acid:
a. HSO_4^-
b. $CH_3NH_3^+$
c. $HClO_4$
d. NH_4^+
e. HCl

9.13 Write a formula for the conjugate acid formed when each of the following behaves as a Brønsted base:
a. NH_2^-
b. CO_3^{2-}
c. OH^-
d. $(CH_3)_2NH$
e. NO_2^-

9.14 Write a formula for the conjugate acid formed when each of the following behaves as a Brønsted base:
a. HCO_3^-
b. S^{2-}
c. HS^-
d. $HC_2O_4^-$
e. $HN_2O_2^-$

9.15 The following reactions illustrate Brønsted acid–base behavior. Complete each equation.
a. $HI(aq) + ? \rightarrow H_3O^+(aq) + I^-(aq)$
b. $NH_3(\ell) + ? \rightarrow NH_4^+ + NH_2^-$
c. $H_2C_2O_4(aq) + H_2O(\ell) \rightarrow ? + HC_2O_4^-(aq)$
d. $H_2N_2O_2(aq) + H_2O(\ell) \rightarrow H_3O^+(aq) + ?$
e. $? + H_2O(\ell) \rightarrow H_3O^+(aq) + CO_3^{2-}(aq)$

9.16 The following reactions illustrate Brønsted acid–base behavior. Complete each equation.
a. $H_2AsO_4^-(aq) + ? \rightarrow NH_4^+(aq) + HAsO_4^{2-}(aq)$
b. $C_6H_5NH_2(aq) + ? \rightarrow C_6H_5NH_3^+(aq) + OH^-(aq)$
c. $S^{2-}(aq) + H_2O(\ell) \rightarrow ? + OH^-(aq)$
d. $(CH_3)_2NH(aq) + ? \rightarrow (CH_3)_2NH_2^+(aq) + Br^-(aq)$
e. $H_2PO_4^-(aq) + CH_3NH_2(aq) \rightarrow HPO_4^{2-}(aq) + ?$

9.17 Write equations to illustrate the acid–base reaction of each of the following pairs of Brønsted acids and bases:

	Acid	Base
a.	HOCl	H_2O
b.	$HClO_4$	NH_3
c.	H_2O	NH_2^-
d.	H_2O	OCl^-
e.	$HC_2O_4^-$	H_2O

9.18 Write equations to illustrate the acid–base reaction of each of the following pairs of Brønsted acids and bases:

	Acid	Base
a.	HS^-	NH_3
b.	H_2O	ClO_3^-
c.	H_2O	NH_2^-
d.	HBO_3^{2-}	H_2O
e.	HNO_2	NH_3

NAMING ACIDS (SECTION 9.3)

9.19 A water solution of HF gas is used to etch glass. Name the water solution as an acid.

9.20 Hydrogen cyanide, HCN, behaves in water solution very much like the binary covalent compounds of hydrogen, but it liberates the cyanide ion, CN^-. Name the acidic water solution by following the rules for binary covalent compounds of hydrogen.

9.21 Name the following acids. Refer to Table 4.7 as needed.
  a. $H_2Se(aq)$
  b. $HClO_3$
  c. H_2SO_4
  d. HNO_3

9.22 Name the following acids. Refer to Table 4.7 as needed.
  a. $H_2Te(aq)$
  b. $HClO$
  c. H_2SO_3
  d. HNO_2

9.23 The acid $H_3C_6H_5O_7$ forms the citrate ion, $C_6H_5O_7^{3-}$, when all three hydrogens are removed. This acid is involved in an important energy-storing process in the body. Name the acid.

9.24 The acid $H_2C_4H_4O_4$ forms the succinate ion, $C_4H_4O_4^{2-}$, when both hydrogens are removed. This acid is involved in the same energy-storing process as the acid of Exercise 9.23. Name $H_2C_4H_4O_4$ as an acid.

9.25 Refer to Table 4.7, and write the formula for chromic acid.

9.26 Refer to Table 4.7, and write the formula for carbonic acid.

THE SELF-IONIZATION OF WATER (SECTION 9.4)

9.27 Calculate the molar concentration of OH^- in water solutions with the following H_3O^+ molar concentrations:
  a. 1.0×10^{-7}
  b. 3.2×10^{-3}
  c. 4.7×10^{-11}
  d. 1.2
  e. 0.043

9.28 Calculate the molar concentration of OH^- in water solutions with the following H_3O^+ molar concentrations:
  a. 0.044
  b. 1.3×10^{-4}
  c. 0.0087
  d. 7.9×10^{-10}
  e. 3.3×10^{-2}

9.29 Calculate the molar concentration of H_3O^+ in water solutions with the following OH^- molar concentrations:
  a. 1.0×10^{-7}
  b. 5.2×10^{-4}
  c. 9.9×10^{-10}
  d. 0.092
  e. 3.7

9.30 Calculate the molar concentration of H_3O^+ in water solutions with the following OH^- molar concentrations:
  a. 6.9×10^{-5}
  b. 0.074
  c. 4.9
  d. 1.7×10^{-3}
  e. 9.2×10^{-9}

9.31 Classify the solutions represented in Exercises 9.27 and 9.29 as acidic, basic, or neutral.

9.32 Classify the solutions represented in Exercises 9.28 and 9.30 as acidic, basic, or neutral.

THE pH CONCEPT (SECTION 9.5)

9.33 Classify solutions with the following characteristics as acidic, basic, or neutral:
  a. pH = 10
  b. pH = 4
  c. pH = 7.3
  d. pH = 6

9.34 Classify solutions with the following characteristics as acidic, basic, or neutral.
  a. pH = 4.7
  b. pH = 8.8
  c. pH = 1.5
  d. pH = 11
  e. pH = 5.9

9.35 Determine the pH of water solutions with the following characteristics. Classify each solution as acidic, basic, or neutral.
  a. $[H^+] = 1.0 \times 10^{-5}$ M
  b. $[OH^-] = 6.0 \times 10^{-3}$ M
  c. $[H^+] = [OH^-]$
  d. $[H^+] = 9.0 \times 10^{-4}$ M
  e. $[OH^-] = 3.0 \times 10^{-9}$ M

9.36 Determine the pH of water solutions with the following characteristics. Classify each solution as acidic, basic, or neutral.
  a. $[H^+] = 4.1 \times 10^{-9}$
  b. $[OH^-] = 9.4 \times 10^{-4}$
  c. $[OH^-] = 10[H^+]$
  d. $[H^+] = 2.3 \times 10^{-2}$
  e. $[OH^-] = 5.1 \times 10^{-10}$

9.37 Determine the pH of water solutions with the following characteristics. Classify each solution as acidic, basic, or neutral.
  a. $[H^+] = 3.7 \times 10^{-4}$ M
  b. $[H^+] = 7.4 \times 10^{-8}$ M
  c. $[H^+] = 1.9 \times 10^{-10}$ M
  d. $[OH^-] = 1.3 \times 10^{-1}$ M
  e. $[OH^-] = 6.8 \times 10^{-7}$ M

9.38 Determine the pH of water solutions with the following characteristics. Classify each solution as acidic, basic, or neutral.
  a. $[H^+] = 2.2 \times 10^{-3}$
  b. $[H^+] = 3.9 \times 10^{-12}$
  c. $[H^+] = 7.5 \times 10^{-6}$
  d. $[OH^-] = 2.5 \times 10^{-4}$
  e. $[OH^-] = 8.6 \times 10^{-10}$

9.39 Determine the $[H^+]$ value for solutions with the following characteristics:
  a. pH = 4.45
  b. pH = 13.12
  c. pH = 7.73

9.40 Determine the $[H^+]$ value for solutions with the following characteristics:
  a. pH = 9.27

b. pH = 2.55

c. pH = 5.42

9.41 Convert the following pH values into both $[H^+]$ and $[OH^-]$ values:

a. pH = 8.00

b. pH = 6.15

c. pH = 1.30

9.42 Convert the following pH values into both $[H^+]$ and $[OH^-]$ values:

a. pH = 3.95

b. pH = 4.00

c. pH = 11.86

9.43 The pH values listed in Table 9.1 are generally the average values for the listed materials. Most natural materials, such as body fluids and fruit juices, have pH values that cover a range for different samples. Some measured pH values for specific body fluid samples are given below. Convert each one to $[H^+]$, and classify the fluid as acidic, basic, or neutral.

a. Blood, pH = 7.41

b. Gastric juice, pH = 1.60

c. Urine, pH = 5.93

d. Saliva, pH = 6.85

e. Pancreatic juice, pH = 7.85

9.44 The pH values listed in Table 9.1 are generally the average values for the listed materials. Most natural materials, such as body fluids and fruit juices, have pH values that cover a range for different samples. Some measured pH values for specific body fluid samples are given below. Convert each one to $[H^+]$, and classify the fluid as acidic, basic, or neutral.

a. Bile, pH = 8.05

b. Vaginal fluid, pH = 3.93

c. Semen, pH = 7.38

d. Cerebrospinal fluid, pH = 7.40

e. Perspiration, pH = 6.23

9.45 The pH values of specific samples of food items are listed below. Convert each value to $[H^+]$, and classify the sample as acidic, basic, or neutral.

a. Milk, pH = 6.39

b. Coffee, pH = 5.10

c. Orange juice, pH = 4.07

d. Vinegar, pH = 2.65

9.46 The pH values of specific samples of food items are listed below. Convert each value to $[H^+]$, and classify the sample as acidic, basic, or neutral.

a. Soft drink, pH = 2.91

b. Tomato juice, pH = 4.11

c. Lemon juice, pH = 2.32

d. Grapefruit juice, pH = 3.07

PROPERTIES OF ACIDS (SECTION 9.6)

9.47 Using the information in Table 9.4, describe how you would prepare each of the following solutions.

a. About 750 mL of 0.5 M H_2SO_4 from dilute sulfuric acid

b. About 200 mL of 0.1 M NaOH from stock sodium hydroxide solution

c. About 1.0 L of 1 M acetic acid from glacial acetic acid

9.48 Use the information in Table 9.4 and describe how you would prepare each of the following solutions.

a. About 2 L of 3.0 M HNO_3 from dilute nitric acid solution

b. About 500 mL of 1.5 M aqueous ammonia from concentrated aqueous ammonia solution

c. About 5 L of 0.2 M HCl from concentrated hydrochloric acid solution

9.49 Write balanced molecular equations to illustrate the following characteristic reactions of acids, using nitric acid (HNO_3).

a. Reaction with water to form hydronium ions

b. Reaction with the solid oxide, CaO

c. Reaction with the solid hydroxide, $Mg(OH)_2$

d. Reaction with the solid carbonate, $CuCO_3$

e. Reaction with the solid bicarbonate, $KHCO_3$

f. Reaction with Mg metal

9.50 Write balanced molecular equations to illustrate the following characteristic reactions of acids, using sulfuric acid (H_2SO_4).

a. Reaction with water to form hydronium ions

b. Reaction with the solid oxide, CaO

c. Reaction with the solid hydroxide, $Mg(OH)_2$

d. Reaction with the solid carbonate, $CuCO_3$

e. Reaction with the solid bicarbonate, $KHCO_3$

f. Reaction with Mg metal

9.51 Write each molecular equation of Exercise 9.49 in total ionic and net ionic form. Use Table 7.4 to decide which products will be soluble.

9.52 Write each molecular equation of Exercise 9.50 in total ionic and net ionic form. Use Table 7.4 to decide which products will be soluble.

9.53 Write balanced molecular equations to illustrate five different reactions that could be used to prepare $BaCl_2$ from hydrochloric acid (HCl) and other appropriate substances.

9.54 Write balanced molecular equations to illustrate five different reactions that could be used to prepare $MgCl_2$ from hydrochloric acid (HCl) and other appropriate substances.

9.55 Write balanced molecular, total ionic, and net ionic equations to illustrate each of the following reactions. All the metals form 2+ ions.

a. zinc with H_2SO_4

b. magnesium with HCl

c. calcium with $HC_2H_3O_2$

9.56 Write balanced molecular, total ionic, and net ionic equations to illustrate each of the following reactions. All the metals form 2+ ions.

a. tin with H_2SO_3

b. magnesium with H_3PO_4

c. calcium with HBr

PROPERTIES OF BASES (SECTION 9.7)

9.57 Write balanced molecular, total ionic, and net ionic equations to represent neutralization reactions between RbOH and the following acids. Use all H's possible for each acid.
   a. HCl
   b. HNO_3
   c. H_2SO_4

9.58 Write balanced molecular, total ionic, and net ionic equations to represent the neutralization reactions between RbOH and the following acids. Use all H's possible for each acid.
   a. HBr
   b. H_2SO_3
   c. H_3PO_3 (only two H's react)

9.59 Some polyprotic acids can form more than one salt depending on the number of H's that react with base. Write balanced molecular, total ionic, and net ionic equations to represent the following neutralization reactions between KOH and
   a. H_2SO_4 (only react one H)
   b. H_2SO_4 (react both H's)
   c. H_3PO_4 (only react one H)

9.60 Some polyprotic acids can form more than one salt depending on the number of H's that react with base. Write balanced molecular, total ionic, and net ionic equations to represent the following neutralization reactions between KOH and
   a. H_3PO_4 (react two H's)
   b. H_3PO_4 (react three H's)
   c. $H_2C_2O_4$ (react one H)

SALTS (SECTION 9.8)

9.61 Identify with ionic formulas the cations and anions of the following salts:
   a. LiCl
   b. $Cu(NO_3)_2$
   c. $SrSO_4$
   d. K_3PO_4
   e. K_2HPO_4
   f. $CaCO_3$

9.62 Identify with ionic formulas the cations and anions of the following salts:
   a. $CuCl_2$
   b. $(NH_4)_2SO_4$
   c. Li_3PO_4
   d. $MgCO_3$
   e. $Ca(C_2H_3O_2)_2$
   f. KNO_3

9.63 Identify with formulas the acid and base from which the anion and cation of each salt in Exercise 9.61 was derived. Pay special attention to salts derived from polyprotic acids and be sure to list the acid formula with all H's.

9.64 Identify with formulas the acid and base from which the anion and cation of each salt in Exercise 9.62 was derived. Pay special attention to salts derived from polyprotic acids and be sure to list the acid formula with all H's.

9.65 Calculate the mass of water that would be released if the water of hydration were completely driven off 1.0 mol of (a) plaster of Paris and (b) gypsum (see Table 9.6). How would the products of these reactions compare?

9.66 Calculate the mass of water that would be released if the water of hydration were completely driven off 1.0 mol of (a) Epsom salts and (b) borax (see Table 9.6). How would the products of these reactions compare?

9.67 Write formulas for the acid and indicated solid that could be used to prepare each of the following salts:
   a. $CuCl_2$ (solid is an oxide)
   b. $MgSO_4$ (solid is a carbonate)
   c. LiBr (solid is a hydroxide)

9.68 Write formulas for the acid and indicated solid that could be used to prepare each of the following salts:
   a. $Mg(NO_3)_2$ (solid is a carbonate)
   b. $CaCl_2$ (solid is an oxide)
   c. Rb_2SO_4 (solid is a bicarbonate)

9.69 Write balanced molecular equations to illustrate each salt preparation described in Exercise 9.67.

9.70 Write balanced molecular equations to illustrate each salt preparation described in Exercise 9.68.

9.71 Determine the number of moles of each of the following salts that would equal 1 eq of salt:
   a. KNO_3
   b. Li_2CO_3
   c. $SrCl_2$

9.72 Determine the number of moles of each of the following salts that would equal 1 eq of salt:
   a. $CaCl_2$
   b. $Cu(NO_3)_2$
   c. $CrCl_3$

9.73 Determine the number of equivalents and milliequivalents in each of the following:
   a. 0.10 mol KI
   b. 0.25 mol $MgCl_2$
   c. 4.73×10^{-2} mol $AgNO_3$

9.74 Determine the number of equivalents and milliequivalents in each of the following:
   a. 0.22 mol $ZnCl_2$
   b. 0.45 mol CsCl
   c. 3.12×10^{-2} mol $Fe(NO_3)_2$

9.75 Determine the number of equivalents and milliequivalents in 5.00 g of each of the following salts. Include any waters of hydration given in the salt formula when you calculate salt formula weights.
   a. NaCl
   b. $NaNO_3$
   c. Na_3PO_4
   d. $MgSO_4 \cdot 7H_2O$

9.76 Determine the number of equivalents and milliequivalents in 5.00 g of each of the following salts. Include any waters of hydration given in the salt formula when you calculate salt formula weights.
   a. $Na_2CO_3 \cdot 10H_2O$
   b. $CuSO_4 \cdot 5H_2O$
   c. Li_2CO_3
   d. NaH_2PO_4

9.77 A sample of intracellular fluid contains 45.1 meq/L of Mg^{2+} ion. Assume the Mg^{2+} comes from dissolved $MgCl_2$, and calculate the number of moles and number of grams of $MgCl_2$ that would be found in 250 mL of the intracellular fluid.

9.78 A sample of intracellular fluid contains 133 meq/L of K^+ ion. Assume the K^+ comes from dissolved K_2SO_4, and calculate the number of moles and number of grams of K_2SO_4 that would be found in 150 mL of the intracellular fluid.

THE STRENGTHS OF ACIDS AND BASES (SECTION 9.9)

9.79 Illustrate the difference between weak, moderately strong, and strong acids by writing dissociation reactions for the hypothetical acid HB, using arrows of various lengths.

9.80 The K_a values have been determined for four acids and are listed below. Arrange the acids in order of increasing acid strength (weakest first, strongest last).

$$\text{acid } A \ (K_a = 5.6 \times 10^{-5})$$
$$\text{acid } B \ (K_a = 1.8 \times 10^{-5})$$
$$\text{acid } C \ (K_a = 1.3 \times 10^{-4})$$
$$\text{acid } D \ (K_a = 1.1 \times 10^{-3})$$

9.81 Arrange the four acids classified as weak in Table 9.7 in order of increasing strength (weakest first, strongest last).

9.82 K_a values for four weak acids are given below:

$$\text{acid } A \ (K_a = 2.6 \times 10^{-4})$$
$$\text{acid } B \ (K_a = 3.7 \times 10^{-5})$$
$$\text{acid } C \ (K_a = 5.8 \times 10^{-4})$$
$$\text{acid } D \ (K_a = 1.5 \times 10^{-3})$$

a. Arrange the four acids in order of increasing acid strength (weakest first, strongest last).
b. Arrange the conjugate bases of the acids (identify as base A, etc.) in order of increasing base strength (weakest base first, strongest last).

9.83 Write dissociation reactions and K_a expressions for the following weak acids:
a. hypobromous acid, HBrO
b. sulfurous acid, H_2SO_3 (1st H only)
c. hydrogen sulfite ion, HSO_3^-
d. hydroselenic acid, H_2Se (1st H only)
e. arsenic acid, H_3AsO_4 (1st H only)

9.84 Write dissociation reactions and K_a expressions for the following weak acids:
a. nitrous acid, HNO_2
b. hydrogen carbonate ion, HCO_3^-
c. dihydrogen phosphate ion, $H_2PO_4^-$ (1st H only)
d. hydrogen sulfide ion, HS^-
e. hypobromous acid, HBrO

9.85 Equal molar solutions are made of three monoprotic acids HA, HB, and HC. The pHs of the solutions are, respectively, 4.82, 3.16, and 5.47. Rank the acids in order of increasing acid strength and explain your reasoning.

9.86 If someone asked you for a weak acid solution, which of the following would you provide according to definitions in this chapter?

a. 0.05 M HCl
b. 20% acetic acid

If the individual really wanted the other solution, what term should have been used instead of *weak*?

9.87 Arsenic acid (H_3AsO_4) is a moderately weak triprotic acid. Write equations showing its stepwise dissociation. Which of the three anions formed in these reactions will be the strongest Brønsted base? Which will be the weakest Brønsted base? Explain your answers.

ANALYZING ACIDS AND BASES (SECTION 9.10)

9.88 Explain the purpose of doing a titration.

9.89 Describe the difference between the information obtained by measuring the pH of an acid solution and by titrating the solution with base.

9.90 Suppose a student is going to titrate an acidic solution with a base and just picks an indicator at random. Under what circumstances will (a) the endpoint and equivalence point be the same? (b) The endpoint and equivalence point be different?

9.91 Determine the number of moles of NaOH that could be neutralized by each of the following:
a. 1.00 L of 0.200 M HCl
b. 500 mL of 0.150 M HNO_3

9.92 Determine the number of moles of NaOH that could be neutralized by each of the following:
a. 250 mL of 0.400 M HBr
b. 750 mL of 0.300 M $HClO_4$

TITRATION CALCULATIONS (SECTION 9.11)

9.93 Write a balanced molecular equation to represent the neutralization reaction between NaOH and each of the following acids. React all of the acid H's.
a. molybdic acid, H_2MoO_4
b. permanganic acid, $HMnO_4$
c. phosphoric acid, H_3PO_4

9.94 Write a balanced molecular equation to represent the neutralization reaction between NaOH and each of the following acids. React all of the acid H's.
a. chromic acid, H_2CrO_4
b. chloric acid, $HClO_3$
c. arsenous acid, H_3AsO_3

9.95 Write a balanced molecular equation to represent the neutralization reaction between HCl and each of the following bases:
a. $Cd(OH)_2$
b. $Cr(OH)_3$
c. $Fe(OH)_2$

9.96 Write a balanced molecular equation to represent the neutralization reaction between HCl and the following bases:
a. $Sr(OH)_2$
b. $Ni(OH)_3$
c. $Zn(OH)_2$

9.97 A 25.00-mL sample of gastric juice is titrated with 0.0210 M NaOH solution. The titration to the equivalence point requires 29.8 mL of NaOH solution. If the equation for the reaction is

$$HCl(aq) + NaOH(aq) \rightarrow NaCl(aq) + H_2O(\ell)$$

what is the molarity of HCl in the gastric juice?

9.98 A 25.00-mL sample of $H_2C_2O_4$ solution required 43.88 mL of 0.1891 M NaOH solution to titrate it to the equivalence point. Calculate the molarity of the $H_2C_2O_4$ solution.

9.99 A 20.00-mL sample of each of the following acid solutions is to be titrated to the equivalence point using 0.120 M NaOH solution. Determine the number of milliliters of NaOH solution that will be needed for each acid sample.
   a. 0.180 M HCl
   b. 0.180 M H_2SO_4
   c. 0.100 M HCl
   d. 10.00 g of H_3PO_4 in 250 mL of solution
   e. 0.150 mol H_2MoO_4 in 500 mL of solution
   f. 0.215 mol H_2MoO_4 in 700 mL of solution

9.100 A 20.00-mL sample of each of the following acid solutions is to be titrated to the equivalence point using 0.120 M NaOH solution. Determine the number of milliliters of NaOH solution that will be needed for each acid sample.
   a. 0.200 M $HClO_4$
   b. 0.125 M H_2SO_4
   c. 0.150 M $H_4P_2O_6$
   d. 0.120 mol H_3PO_4 in 500 mL of solution
   e. 6.25 g of H_2SO_4 in 250 mL of solution
   f. 0.500 mol $HClO_3$ in 1.00 L of solution

9.101 The following acid solutions were titrated to the equivalence point with the base listed. Use the titration data to calculate the molarity of each acid solution.
   a. 25.00 mL of HI solution required 27.15 mL of 0.250 M NaOH solution.
   b. 20.00 mL of H_2SO_4 solution required 11.12 mL of 0.109 M KOH solution.
   c. 25.00 mL of gastric juice (HCl) required 18.40 mL of 0.0250 M NaOH solution.

9.102 The following acid solutions were titrated to the equivalence point with the base listed. Use the titration data to calculate the molarity of each acid solution.
   a. 5.00 mL of dilute H_2SO_4 required 29.88 mL of 1.17 M NaOH solution.
   b. 10.00 mL of vinegar (acetic acid) required 35.62 mL of 0.250 M KOH solution.
   c. 10.00 mL of muriatic acid (HCl) used to clean brick and cement required 20.63 mL of 6.00 M NaOH solution.

9.103 A 20.00-mL sample of diprotic oxalic acid ($H_2C_2O_4$) solution is titrated with 0.250 M NaOH solution. A total of 27.86 mL of NaOH is required. Calculate:
   a. The number of moles of oxalic acid in the 20.00-mL sample.
   b. The molarity of the oxalic acid solution.
   c. The number of grams of oxalic acid in the 20.00-mL sample.

9.104 A sample of monoprotic benzoic acid weighing 0.5823 g is dissolved in about 25 mL of water. The solution is titrated to the equivalence point using 0.1021 M NaOH. The volume of base required is 46.75 mL. Calculate the molecular weight of the solid acid.

HYDROLYSIS REACTIONS OF SALTS (SECTION 9.12)

9.105 A solution of solid NH_4Cl dissolved in pure water is acidic (the pH is less than 7). Explain.

9.106 A solution of solid Na_2CO_3 dissolved in pure water has a pH significantly higher than the pH of pure water. Explain why.

9.107 Predict the relative pH (greater than 7, less than 7, etc.) for water solutions of the following salts. Table 9.9 may be useful. For each solution in which the pH is greater or less than 7, explain why and write a net ionic equation to justify your answer.
   a. potassium sulfite, K_2SO_3
   b. lithium nitrite, $LiNO_2$
   c. sodium carbonate, Na_2CO_3
   d. methylammonium chloride, CH_3NH_3Cl (CH_3NH_2 is a weak base)

9.108 Predict the relative pH (greater than 7, less than 7, etc.) for water solutions of the following salts. Table 9.9 may be useful. For each solution in which the pH is greater or less than 7, explain why and write a net ionic equation to justify your answer.
   a. sodium hypochlorite, NaOCl (HOCl is a weak acid)
   b. sodium formate, $NaCHO_2$
   c. potassium nitrate, KNO_3
   d. sodium phosphate, Na_3PO_4

9.109 A chemist has 20.00-mL samples of 0.100 M acid A and 0.100 M acid B in separate flasks. Both acids are monoprotic. Unfortunately, the flasks were not labeled, so the chemist doesn't know which sample is in which flask. But fortunately, it is known that acid A is strong and acid B is weak. Before thinking about the problem, the chemist adds 20.00 mL of 0.100 M NaOH solution to each flask. Explain how the chemist could use a pH meter (or pH paper) to determine which flask originally contained which acid.

9.110 Explain why the hydrolysis of salts makes it necessary to have available in a laboratory more than one acid–base indicator for use in titrations.

9.111 How would the pH values of equal molar solutions of the following salts compare (highest, lowest, etc.)? NaH_2PO_4, Na_2HPO_4, and Na_3PO_4.

BUFFERS (SECTION 9.13)

9.112 Write equations similar to Equations 9.48 and 9.49 of the text to illustrate how a mixture of sodium hydrogen phosphate (Na_2HPO_4) and sodium dihydrogen phosphate (NaH_2PO_4) could function as a buffer when dissolved in water. Remember that phosphoric acid (H_3PO_4) ionizes in three steps.

9.113 Could a mixture of ammonia (NH_3), a weak base, and ammonium chloride (NH_4Cl) behave as a buffer when dissolved in water? Use reaction equations to justify your answer.

9.114 Some illnesses lead to a condition of excess acid (acidosis) in the body fluids. An accepted treatment is to inject solutions containing bicarbonate ions (HCO_3^-) directly into the bloodstream. Write an equation to show how this treatment would help combat the acidosis.

9.115 Calculate the pH of a buffer made by dissolving 1 mol formic acid (HCOOH) and 1 mol sodium formate (HCOONa) in 1 L of solution (see Table 9.9).

9.116 a. Calculate the pH of a buffer that is 0.1 M in lactic acid ($C_2H_4(OH)COOH$) and 0.1 M in sodium lactate, $C_2H_4(OH)COONa$.
   b. What is the pH of a buffer that is 1 M in lactic acid and 1 M in sodium lactate?
   c. What is the difference between the buffers described in parts a and b?

9.117 Which of the following weak acids and its conjugate base would you use to make a buffer with a pH of 5.00? Explain your reasons: acetic acid, nitrous acid, or formic acid

9.118 Calculate the pH of buffers that contain the acid and conjugate base concentrations listed below.
   a. $[CH_3COOH] = 0.40$ M, $[CH_3COO^-] = 0.25$ M
   b. $[H_2PO_4^-] = 0.10$ M, $[HPO_4^{2-}] = 0.40$ M
   c. $[HSO_3^-] = 1.50$ M, $[SO_3^{2-}] = 0.20$ M

9.119 Calculate the pH of buffers that contain the acid and conjugate base concentrations listed below.
   a. $[HPO_4^{2-}] = 0.33$ M, $[PO_4^{3-}] = 0.52$ M
   b. $[HNO_2] = 0.029$ M, $[NO_2^-] = 0.065$ M
   c. $[HCO_3^-] = 0.50$ M, $[CO_3^{2-}] = 0.15$ M

9.120 What ratio of concentrations of NaH_2PO_4 and Na_2HPO_4 in solution would give a buffer with pH = 7.65?

9.121 A citric acid–citrate buffer has a pH of 3.20. You want to increase the pH to a value of 3.35. Would you add citric acid or sodium citrate to the solution? Explain.

CHEMISTRY FOR THOUGHT

9.1 In an early industrial method, H_2SO_4 was manufactured in lead-lined chambers. Propose an explanation for this.

9.2 A saturated solution of solid $Ca(OH)_2$ in water has a $[OH^-]$ of only 2.50×10^{-2} M, and yet $Ca(OH)_2$ is a strong base. Explain this apparent contradiction.

9.3 Imagine a solution of weak acid is being titrated with a strong base. Describe the substances present in the solution being titrated when it has been titrated halfway to the equivalence point. How is the pH of this half-titrated solution related to pK_a for the weak acid?

9.4 Calculate K_a for the following weak acids based on the equilibrium concentrations and pH values given:

   a. Benzoic acid, represented by HBz:

   pH = 2.61, $[Bz^-] = 2.48 \times 10^{-3}$, $[HBz] = 0.0975$

   b. Abietic acid, represented by HAb:

   pH = 4.16, $[Ab^-] = 6.93 \times 10^{-5}$, $[HAb] = 0.200$

   c. Cacodylic acid, represented by HCc:

   pH = 3.55, $[Cc^-] = 0.000284$, $[HCc] = 0.150$

9.5 Bottles of ketchup are routinely left on the counters of cafés, yet the ketchup does not spoil. Why not?

9.6 Refer to Figure 9.2 and answer the question. What implications does this reaction have for the long-term durability of marble structures when exposed to acid rain?

9.7 Refer to Figure 9.3 and answer the question. Do you think the results would be the same if sulfuric acid (H_2SO_4) were substituted for the HCl? Explain.

9.8 Refer to Figure 9.5 and answer the question. Answer the same question for the other two indicators.

9.9 Refer to Figure 9.6 and explain why the pH reading of the meter might not be 7.00 at the equivalence point in a titration. Would the pH at the equivalence point be greater or less than 7 if hydrochloric acid were titrated with aqueous ammonia?

9.10 Refer to Figure 9.7 and answer the question. List at least two other ions that would behave like the Na^+ ion.

9.11 Refer to Figure 9.8 and answer the question. What color is bromcresol green in acidic solutions? In basic solutions? Explain how you arrived at your conclusions.

InfoTrac College Edition Readings

"Unless you balance acidity, your muscles may become tense," *Better Nutrition,* March 1996, 58(3):26(1). Record number A18169206.

"The risk related to alcohol is unaltered by folate, methionine, or vitamin use," *Cancer Weekly,* April 15, 2003:29. Record number A99874585.

"Study: Antacids may be more crucial than calcium in building healthy bones," *Obesity, Fitness & Wellness Week,* June 29, 2002:16. Record number A87506174.

CHAPTER 10 Radioactivity and Nuclear Processes

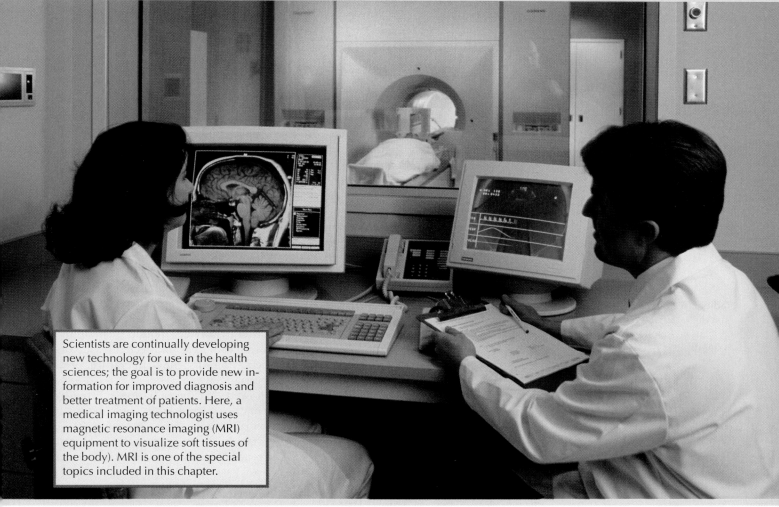

Scientists are continually developing new technology for use in the health sciences; the goal is to provide new information for improved diagnosis and better treatment of patients. Here, a medical imaging technologist uses magnetic resonance imaging (MRI) equipment to visualize soft tissues of the body). MRI is one of the special topics included in this chapter.

LEARNING OBJECTIVES

When you have completed your study of this chapter, you should be able to:

1. Describe and characterize the common forms of radiation emitted during radioactive decay and other nuclear processes. (Section 10.1)

2. Write balanced equations for nuclear reactions. (Section 10.2)

3. Solve problems using the half-life concept. (Section 10.3)

4. Describe the effects of radiation on health. (Section 10.4)

5. Describe and compare the units used to measure quantities of radiation. (Section 10.5)

6. Describe, with examples, medical and nonmedical uses of radioisotopes. (Sections 10.6 and 10.7)

7. Show that you understand the concept of induced nuclear reactions. (Section 10.8)

8. Describe the differences between nuclear fission and nuclear fusion reactions. (Section 10.9)

In Chapters 3 and 4, we learned that the arrangements of electrons around the nuclei of atoms determine the chemical behavior of elements and compounds. We saw that chemical reactions take place between atoms and molecules as electrons rearrange to achieve noble gas configurations for the atoms involved. These noble gas configurations represent a stable arrangement for the electrons. In this chapter, we will see that the achievement of stability is also a driving force that causes certain atomic nuclei to undergo changes that release small subnuclear particles and energy.

➤ 10.1 Radioactive Nuclei

In 1896 Henri Becquerel, a French physicist, discovered that uranium compounds emitted rays that could expose and fog photographic plates wrapped in lightproof paper. Research conducted since that time has shown that the penetrating rays originate from changes that occur in the nuclei of some atoms. We saw earlier (Section 2.3) that isotopes of the same element differ only in the number of neutrons present in their nuclei. Thus, each of the three isotopes of hydrogen, represented by the symbols ^1_1H, ^2_1H, and ^3_1H, contains one proton in the nucleus and one electron outside the nucleus. However, besides the one nuclear proton, ^2_1H and ^3_1H contain one and two nuclear neutrons, respectively. It has been found that some combinations of neutrons and protons in the nucleus are stable and do not change spontaneously, but some combinations are not stable. It is these unstable nuclei that emit radiation as they become more stable. For example, both ^1_1H and ^2_1H are stable, but ^3_1H emits radiation. Nuclei that emit radiation are said to be **radioactive nuclei.**

Henri Becquerel discovered natural radioactivity by chance. Through research he found that radiation was emitted by any compound of uranium. He further found that the intensity of the radiation was unaffected by factors that normally influence the rates of chemical reactions: the temperature, pressure, and type of uranium compound used.

Later studies by other investigators showed that the radiation emitted by uranium, and by other radioactive elements discovered later, could be separated into three types by an electrical or magnetic field (see ■ Figure 10.1). The three types had different electrical charges: One was positive, one was negative, and one carried no charge. The types of radiation were given names that are still used today: alpha rays (positive), beta rays (negative), and gamma rays (uncharged).

Today it is known that other types of radiation, such as neutrons and positrons, are also emitted by radioactive nuclei. However, alpha, beta, and gamma are the most common. ■ Table 10.1 summarizes the characteristics of these forms. In this book, we usually use the symbol given first for each form; the symbols in parentheses are alternatives you might find elsewhere. Except for gamma radiation, which is very high-energy radiation somewhat like X rays, the radiation emitted by radioactive nuclei consists of streams of par-

---

Radioactive nuclei
Nuclei that undergo spontaneous changes and emit energy in the form of radiation.

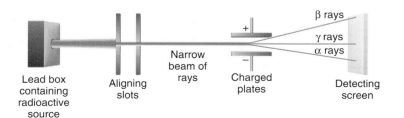

➤ **FIGURE 10.1** The three types of emitted radiation.

TABLE 10.1 Characteristics of nuclear radiation

Type of radiation	Symbols	Mass number	Charge	Composition
Alpha	$_2^4\alpha$ (α, $_2^4$He, He^{2+})	4	+2	Helium nuclei, 2 protons + 2 neutrons
Beta	$_{-1}^{0}\beta$ (β, β^-, $_{-1}^{0}e$, e^-)	0	−1	Electrons produced in nucleus and ejected
Gamma	γ ($_0^0\gamma$)	0	0	Electromagnetic radiation
Neutron	$_0^1 n$ (n)	1	0	Neutrons
Positron	$_1^0\beta$ (β^+, $_1^0 e$, e^+)	0	+1	Positive electrons

ticles. The emission of radiation by unstable nuclei is often called **radioactive decay.**

We note from Table 10.1 that the particles that make up alpha rays are identical to the nuclei of helium atoms; that is, they are clusters containing two protons and two neutrons. Because they are positively charged, alpha particles are attracted toward the negatively charged plate in Figure 10.1. **Alpha particles** move at a speed of nearly one-tenth the speed of light. They are the most massive particles emitted by radioactive materials, and they have the highest charge. Because of their mass and charge, they collide often with molecules of any matter through which they travel. As a result of these collisions, their energy is quickly dissipated, and they do not travel far. They cannot even penetrate through a few sheets of writing paper or the outer cells of the skin. However, exposure to intense alpha radiation can result in severe burns of the skin.

Beta particles are actually electrons, but they are produced within the nucleus and then emitted. They are not part of the group of electrons moving around the nucleus that are responsible for chemical characteristics of the atoms. Beta particles have a smaller charge than alpha particles and are also 7000 times less massive than alpha particles. They undergo far fewer collisions with molecules of matter through which they pass and therefore are much more penetrating than alpha particles. Because of their negative charge and tiny mass, beta particles are attracted toward the positive plate of Figure 10.1, and are deflected more than alpha particles.

Gamma rays are not streams of particles but rays of electromagnetic radiation similar to X rays, but with higher energies. Like X rays, gamma radiation is very penetrating and dangerous to living organisms. Adequate shielding from gamma rays is provided only by thick layers of heavy materials like metallic lead or concrete. Because they are not charged, gamma rays are not attracted to either charged plate in Figure 10.1.

➤ 10.2 Equations for Nuclear Reactions

Isotopes of elements that emit nuclear radiation are called **radioisotopes.** In nuclear reactions, a specific isotope of an element may behave differently from another isotope of the same element. Thus, all particles involved in nuclear reactions are designated by a symbol, a mass number (the sum of protons and neutrons in the particle), and an atomic number (or charge for electrons or positrons). The symbolism used was introduced in Section 2.3. It is $_Z^A X$, where X is the symbol for the particle, A is the mass number, and Z is the atomic

Radioactive decay
A process in which an unstable nucleus changes energy states and in the process emits radiation.

Alpha particle
The particle that makes up alpha rays. It is identical to the helium nucleus and is composed of two protons and two neutrons.

Beta particle
The particle that makes up beta rays. It is identical to an electron but is produced in the nucleus when a neutron is changed into a proton and an electron.

Gamma ray
A high-energy ray that is like an X ray, but with a higher energy.

Radioisotope
An isotope of an element that emits nuclear radiation.

number or charge. When X represents the nucleus of an isotope of an element, the chemical symbol for the element is generally used.

Example 10.1

Write appropriate symbols for the following particles using the $^A_Z X$ symbolism:

a. A lead-214 nucleus
b. An alpha particle
c. A deuteron, a particle containing one proton and one neutron

Solution

a. Lead-214 is an isotope of lead, so the symbol X is replaced by Pb. The mass number is 214 according to rules learned earlier for designating isotopes, so $A = 214$. The atomic number of lead obtained from the periodic table is 82, so $Z = 82$, and the symbol is $^{214}_{82}\text{Pb}$.
b. An alpha particle is represented by the symbol α. Since it contains 2 protons and 2 neutrons, $A = 4$ and $Z = 2$. The symbol is $^4_2\alpha$, as shown in Table 10.1.
c. A deuteron is not familiar to you, so we will replace its symbol X with D. It contains one proton and one neutron, so $A = 2$ and $Z = 1$. The symbol is $^2_1 D$. This particle is the nucleus of a deuterium or heavy hydrogen atom and so could also be represented as ^2_1H.

Learning Check 10.1

Write appropriate symbols for the following particles using the $^A_Z X$ symbolism:

a. An iodine-131 nucleus
b. A beta particle
c. A hypothetical particle composed of four protons
d. A hypothetical particle composed of four electrons

A nuclear equation is balanced when the sum of the atomic numbers on the left side equals the sum of the atomic numbers on the right side, and the sum of the mass numbers on the left equals the sum of the mass numbers on the right. Thus, we focus on mass numbers and atomic numbers in nuclear reactions. As shown in the following example, a nucleus of an element that undergoes a nuclear reaction by emitting a particle is changed to a nucleus of a different element.

Example 10.2

Uranium-238 nuclei undergo radioactive decay by emitting an alpha particle and changing to the nucleus of another element. Write a balanced nuclear equation for the process.

Solution

Because the uranium-238 nucleus gives up an alpha particle, $^4_2\alpha$, the mass number of the uranium-238 must decrease by 4 and the atomic number must decrease by 2. As a result, the new nucleus, called a **daughter nucleus,** has a mass number of 234 and an atomic number of 90 (remember, the atomic number of uranium can be obtained from a periodic table). The balanced nuclear equation is

$$^{238}_{92}\text{U} \rightarrow {}^4_2\alpha + {}^{234}_{90}\text{Th}$$

Daughter nuclei
The new nuclei produced when unstable nuclei undergo radioactive decay.

Note that this equation is balanced because

$$238 = 4 + 234 \quad \text{and} \quad 92 = 2 + 90$$

The symbol for the daughter (thorium) was obtained by locating element number 90 in the periodic table.

Processes such as this where one alpha particle is ejected can be represented by the general equation

$$^{A}_{Z}X \rightarrow \, ^{4}_{2}\alpha + \, ^{A-4}_{Z-2}Y$$

where Y is the symbol of the daughter nucleus.

Thorium, element number 90, exists in a number of isotopic forms. One form, thorium-234, decays by emitting a beta particle. Write a balanced nuclear equation for the process and write a general equation for decays in which one beta particle is ejected.

Learning Check 10.2

Because gamma rays have no mass or atomic numbers, they do not enter into the balancing process of nuclear reactions. However, they do represent energy and should be included in balanced equations when they are known to be emitted.

As described earlier, beta particles are electrons ejected from the nucleus of a radioactive atom, and yet, we know from our discussions of atomic theory in Chapter 2 that atomic nuclei contain only protons and neutrons. Where then do the beta particles originate? According to a simplified theory of nuclear behavior, a beta particle (or electron) is produced in the nucleus when a neutron changes into a proton. The balanced equation for the nuclear reaction is

$$^{1}_{0}n \rightarrow \, ^{1}_{1}p + \, ^{0}_{-1}\beta \qquad (10.1)$$

Note that the criteria for balance in this equation are satisfied; that is, $1 = 1 + 0$ and $0 = 1 + (-1)$. Thus, a nuclear neutron is converted to a proton during beta emission, and the daughter has an atomic number higher by 1 than the decaying nucleus. You found this result if you did Learning Check 10.2 correctly. Similar changes in nuclear particles occur during two other types of nuclear reactions. Some nuclei decay by positron emission. A **positron** is a positively charged electron with the symbol $^{0}_{1}\beta$ (see Table 10.1). When a nucleus emits a positron, a nuclear proton is changed to a neutron. Positron-emitting isotopes are used in diagnostic PET scans of the brain (see ■ Figure 10.2).

Positron
A positively charged electron.

Write a general balanced equation for the nuclear process that occurs during positron emission. How do the atomic numbers of daughter nuclei compare with those of the decaying nucleus?

Learning Check 10.3

Electron capture
A mode of decay for some unstable nuclei in which an electron from outside the nucleus is drawn into the nucleus, where it combines with a proton to form a neutron.

Certain nuclear changes occur when an electron from outside the nucleus is drawn into the nucleus where it reacts with a proton and converts it to a neutron. This process, called **electron capture,** is not as common as other decay processes.

Write a general balanced equation for the nuclear process that occurs during electron capture. How do the atomic numbers of daughter nuclei compare with those of the decaying nucleus?

Learning Check 10.4

➤ **FIGURE 10.2** Positron emission tomography (PET) scan shows normal brain activity during sleep. A radioactive isotope that emits positrons is made into a chemical compound that is absorbed by active areas of the brain. The emitted positrons collide with nearby electrons and produce gamma rays that pass through the skull to detectors surrounding the patient's head. A computer uses the detector data to construct the image.

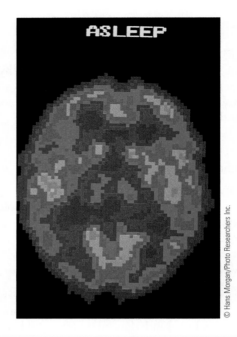

Example 10.3

Write a balanced nuclear equation for the decay of each of the following. The mode of decay is indicated in parentheses.

a. $^{14}_{6}C$ (beta emission)
b. $^{122}_{53}I$ (positron emission)
c. $^{55}_{26}Fe$ (electron capture)
d. $^{212}_{84}Po$ (alpha emission)

Solution

In each case, the atomic number of the daughter is obtained by balancing atomic numbers in the equation. The symbol for the daughter is obtained by matching atomic numbers to symbols in the periodic table.

a. $^{14}_{6}C \rightarrow {}^{0}_{-1}\beta + {}^{14}_{7}N$
b. $^{122}_{53}I \rightarrow {}^{0}_{1}\beta + {}^{122}_{52}Te$
c. $^{55}_{26}Fe + {}^{0}_{-1}e \rightarrow {}^{55}_{25}Mn$
d. $^{212}_{84}Po \rightarrow {}^{4}_{2}\alpha + {}^{208}_{82}Pb$

Learning Check 10.5

Write a balanced nuclear equation for the decay of each of the following. The mode of decay is indicated in parentheses.

a. $^{50}_{25}Mn$ (positron emission)
b. $^{54}_{25}Mn$ (electron capture)
c. $^{56}_{25}Mn$ (beta emission)
d. $^{224}_{88}Ra$ (alpha emission)

➤ 10.3 Isotope Half-Life

Some radioactive isotopes are more stable than others. The more stable isotopes undergo radioactive decay more slowly than the less stable isotopes. The

TABLE 10.2 Examples of half-lives

Isotope	Half-life	Source
$^{238}_{92}U$	4.5×10^9 years	Naturally occurring
$^{40}_{19}K$	1.3×10^9 years	Naturally occurring
$^{226}_{88}Ra$	1600 years	Naturally occurring
$^{14}_{6}C$	5600 years	Naturally occurring
$^{239}_{94}Pu$	24,000 years	Synthetically produced
$^{90}_{38}Sr$	28 years	Synthetically produced
$^{131}_{53}I$	8 days	Synthetically produced
$^{24}_{11}Na$	15 hours	Synthetically produced
$^{15}_{8}O$	2 minutes	Synthetically produced
$^{5}_{3}Li$	10^{-21} seconds	Synthetically produced

half-life of an isotope is used to indicate stability, and it is equal to the time required for one-half (50%) of the atoms of a sample of the isotope to decay. ■ Table 10.2 contains examples showing the wide range of half-lives that have been determined.

Half-life
The time required for one-half the unstable nuclei in a sample to undergo radioactive decay.

Example 10.4

Rubidium-79 decays by positron emission and forms krypton-79, which is a gas. A weighed 100.00-mg sample of solid rubidium-79 was allowed to decay for 42 minutes, then weighed again. Its mass was 25.00 mg. What is the half-life of rubidium-79?

Solution

We assume all the gaseous krypton-79 that was formed escaped into the surrounding air and so was not weighed. In the first half-life, one-half the original 100.00-mg sample of rubidium would have been lost, so the sample would have a mass of 50.00 mg. During the second half-life, this 50.00-mg sample would again lose half its mass and would be reduced to 25.00 mg. Thus, we can conclude that two half-lives passed during the 42 minutes that elapsed between the weighings. Because two half-lives equal 42 minutes, one half-life equals 21 minutes.

Learning Check 10.6

Potassium-38 decays by positron emission to argon-38, a gas. Potassium-38 has a half-life of 7.7 minutes. How long will it take a 200.00-mg sample of potassium-38 to be reduced to 25.00 mg?

Because radioisotopes are continuously decaying, any found in nature must belong to one of three categories. They may have very long half-lives and decay very slowly ($^{238}_{92}U$), they may be daughters produced by the decay of long-lived isotopes ($^{226}_{88}Ra$), or they may result from natural processes such as cosmic-ray bombardment of stable nuclei ($^{14}_{6}C$).

The fraction of original radioactive atoms remaining in a sample after a

➤ **FIGURE 10.3** A radioactive decay curve (bars give values at 1, 2, and 3 half-lives).

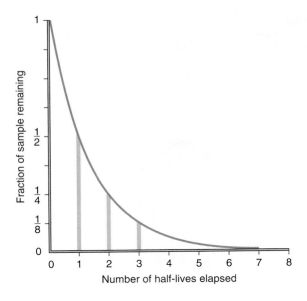

specific time has passed can be determined from the half-life. After one half-life has passed, $\frac{1}{2}$ the original number have decayed, so $\frac{1}{2}$ remain. During the next half-life, $\frac{1}{2}$ the remaining $\frac{1}{2}$ decay, so $\frac{1}{4}$ of the original atoms remain. After three half-lives, $\frac{1}{2} \times \frac{1}{2} \times \frac{1}{2} = \frac{1}{8}$ of the original atoms remain undecayed (see ■ Figure 10.3).

The amount of radiation given off by a sample of radioactive material is proportional to the number of radioactive atoms present in the sample. The amount of radiation given off is called the *activity* of the sample and is measured by devices such as the Geiger–Müller counter described in Section 10.5. Data such as the sample fractions shown in Figure 10.3 are obtained quite easily. For example, suppose a sample gave an initial reading of 120 units of activity, and it took 35 minutes for the reading to drop to 60 units. The reading of 60 units means that the sample contained just half as many radioactive atoms as it contained when the reading was 120 units. Thus, the 35 minutes represents one half-life, and the fraction remaining from the time of the initial measurement is $\frac{1}{2}$.

Example 10.5

The activity (amount of radiation emitted) of a radioactive substance is measured at 9 A.M. At 5 P.M. the activity is found to be only $\frac{1}{16}$ the original (9 A.M.) value. What is the half-life of the radioactive material?

Solution

Because only $\frac{1}{16}$ the original radioactive material is present, four half-lives have passed: $\frac{1}{2} \times \frac{1}{2} \times \frac{1}{2} \times \frac{1}{2} = \frac{1}{16}$. The time between measurements is 8 hours (9 A.M. to 5 P.M.). Therefore, 8 hours is four half-lives, and the half-life is 2 hours.

Learning Check 10.7

Iodine-123 is a radioisotope used to diagnose the function of the thyroid gland. It has a half-life of 13.3 hours. What fraction of a diagnostic dose of iodine-123 would be present in a patient 79.8 hours (a little over 3 days) after it was administered?

➤ 10.4 The Health Effects of Radiation

You are probably aware of the fact that radiation is hazardous to living organisms. But why? What does radiation do that is so dangerous? The greatest danger to living organisms results from the ability of high-energy or ionizing radiation to knock electrons out of compounds and generate highly reactive particles called **radicals** or **free radicals** in tissue through which the radiation passes. These electron-deficient radicals are very reactive. They cause reactions to occur among more stable materials in the cells of living organisms. When these reactions involve genetic materials such as genes and chromosomes, the changes might lead to genetic mutations, cancer, or other serious consequences.

Long-term exposure to low-level radiation is more likely to cause the problems just described than are short exposures to intense radiation. Short-term intense radiation tends to destroy tissue rapidly in the area exposed and can cause symptoms of so-called **acute radiation syndrome.** Some of these symptoms are given in ■ Table 10.3. It is this rapid destruction of tissue that makes relatively intense radiation a useful tool in the treatment of some cancers.

The health hazards presented by radiation make it imperative that exposures be minimized. This is especially true for individuals with occupations that continually present opportunities for exposure. One approach is to absorb radiation by shielding. As shown in ■ Table 10.4, alpha and beta rays are the easiest to stop. Gamma radiation and X rays require the use of very dense materials such as lead to provide protection. If the type of radiation is known, a careful choice of shielding materials can provide effective protection.

Another protection against radiation is distance. Because radiation spreads in all directions from a source, the amount falling on a given area decreases the farther that area is from the source. Imagine a light bulb at the center of a balloon that is being filled slowly with air. The intensity of light on each

Radical or free radical
An electron-deficient particle that is very reactive.

Acute radiation syndrome
The condition associated with and following short-term exposure to intense radiation.

TABLE 10.3	The effects on humans of short-term, whole-body radiation exposure

Dose (rems)[a]	Effects
0–25	No detectable clinical effects.
25–100	Slight short-term reduction in number of some blood cells, disabling sickness not common.
100–200	Nausea and fatigue, vomiting if dose is greater than 125 rems, longer-term reduction in number of some blood cells.
200–300	Nausea and vomiting first day of exposure, up to a 2-week latent period followed by appetite loss, general malaise, sore throat, pallor, diarrhea, and moderate emaciation. Recovery in about 3 months unless complicated by infection or injury.
300–600	Nausea, vomiting, and diarrhea in first few hours. Up to a 1-week latent period followed by loss of appetite, fever, and general malaise in the second week, followed by hemorrhage, inflammation of mouth and throat, diarrhea, and emaciation. Some deaths in 2–6 weeks. Eventual death for 50% if exposure is above 450 rems; others recover in about 6 months.
600 or more	Nausea, vomiting, and diarrhea in first few hours. Rapid emaciation and death as early as second week. Eventual death of nearly 100%.

[a]The rem, a biological unit of radiation, is defined in Section 10.5.
SOURCE: U.S. Atomic Energy Commission.

TABLE 10.4	Penetrating abilities of alpha, beta, and gamma rays		
Type of radiation	Depth of penetration into		
	Dry air	Tissue	Lead
alpha	4 cm	0.05 mm	0
beta[a]	6–300 cm	0.06–4 mm	0.005–0.3 mm
gamma[b]	400 m	50 cm	300 mm

[a]Depth depends on energy of rays.
[b]Depth listed is that necessary to reduce initial intensity by 10%.

Inverse square law of radiation
A mathematical way of saying that the intensity of radiation is inversely proportional to the square of the distance from the source of the radiation.

square centimeter of the balloon will decrease as the surface of the balloon moves away from the light. It has been shown that the intensity is inversely proportional to the square of the distance of the surface from the radiation source. This **inverse square law of radiation** is true only for radiation traveling in a vacuum, but it gives pretty good results when the radiation travels through air. The law is represented by Equation 10.2, where I_x is the intensity at distance d_x and I_y is the intensity at distance d_y:

$$\frac{I_x}{I_y} = \frac{d_y^2}{d_x^2}$$

(10.2)

Example 10.6

The intensity of radiation is 18.5 units at a distance of 50 cm from a source. What is the intensity at a distance of 100 cm?

Solution

Let $I_x = 18.5$ units and $d_x = 50$ cm. Then I_y is wanted at $d_y = 100$ cm. Substitution into Equation 10.2 gives

$$\frac{18.5}{I_y} = \frac{(100 \text{ cm})^2}{(50 \text{ cm})^2}$$

or

$$I_y = \frac{(50 \text{ cm})^2(18.5)}{(100 \text{ cm})^2} = 4.63 \text{ units}$$

Thus, doubling the distance cut the intensity to one-fourth the initial intensity.

Learning Check 10.8

The intensity of a radioactive source is measured at a distance of 25 feet and equals 10.0 units. What is the intensity 5 feet from the source?

Physical unit of radiation
A radiation measurement unit indicating the activity of the source of the radiation; for example, the number of nuclear decays per minute.

Biological unit of radiation
A radiation measurement unit indicating the damage caused by radiation in living tissue.

► 10.5 Measurement Units for Radiation

Two methods, physical and biological, are used to describe quantities of radiation. **Physical units** indicate the activity of a source of radiation, typically in terms of the number of nuclei that decay per unit of time. **Biological units** are related to the damage caused by radiation and account for the fact that a given

CHEMISTRY AROUND US • 10.1

MEDICAL IMAGING

Through the first half of the 20th century, physicians often used exploratory surgery to diagnose abnormalities of internal organs. The technique allowed them to locate and diagnose tumors or malfunctioning organs. The development of various imaging techniques has greatly reduced the number of such surgeries done each year. Medical imaging has its roots in a discovery made in 1895 when Wilhelm Roentgen devised a method to generate invisible rays that could pass through solid matter. Because their nature was unknown at the time, these rays were called *X rays*. Today, X-ray-generated images are indispensable aids in the practice of medicine and dentistry. However, the penetrating ability of X rays makes them most effective at imaging dense materials such as bones or teeth, and their usefulness in making soft tissues visible is limited.

The usefulness of X-ray imaging was greatly enhanced in the early 1970s by integrating X rays with computers. In the resulting technique, called *computed tomography* or CT scanning, the intensity of X rays that pass through the body is analyzed by a powerful computer. The X-ray intensity depends on the density and opacity of the material through which the X rays pass. In soft tissues, these differences are very small and cannot be detected by ordinary X-ray techniques. However, the computer is able to distinguish between these small differences and construct an image of the structures through which the X rays pass. A cross-sectional image is created by moving the X-ray source around the body in a circle. The computer interprets the intensities of the X rays that pass through and creates a cross-sectional image of organs, bones, and tissues that were in the path of the X rays. The image is projected onto a TV-like screen where it can be photographed or studied. By moving the X-ray source along the length of the body, an image can be obtained at any point from the head to the feet. The resulting images are like a series of orange slices, where each one shows a cross section of the orange at a different point.

A discovery made in 1946 supplied the means for another step forward in medical imaging. E. M. Purcell and Felix Bloch independently developed a method of detecting hydrogen atoms in matter. They discovered that when atoms of specific isotopes of certain elements (including hydrogen) were placed in a strong magnetic field, the atoms became capable of absorbing, then re-emitting, radio waves. Many years passed before this discovery was applied to medicine, but today *magnetic resonance imaging* (MRI) scans are routinely used to observe soft tissues without exposing patients to X rays.

MRI scans are produced using a method similar to CT scans. The patient is placed in a strong magnetic field, which causes the hydrogen atoms in water and other molecules of the body to become radio-wave absorbers. Radio waves are then used like the X rays of a CT scan. The radio waves that are absorbed and re-emitted by the hydrogen atoms are detected, then analyzed by a computer. The intensity of the radio waves allows the computer to construct an image of the tissue from which the radio waves come. All soft tissue and organs contain large amounts of hydrogen, which makes it possible to obtain MRI scans of essentially any organ in the body. This makes the detection of soft-tissue disorders such as brain tumors much safer and much more comfortable for the patient than was true when exploratory surgery was the only way to get a look inside.

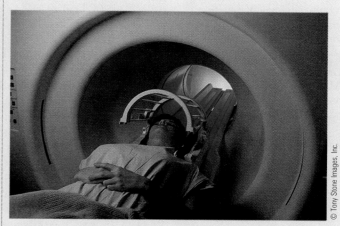

Magnetic resonance imaging (MRI) scans enable medical personnel to view soft tissues in detail.

quantity (or number of particles) of one type of radiation does not have the same damaging effect on tissue as the same quantity of another type of radiation.

The curie and the becquerel are physical units. The **curie** is most widely used at the present time and is equal to 3.7×10^{10} nuclear disintegrations per second. One curie (Ci) represents a large unit, so fractions such as the millicurie (mCi), the microcurie (μCi), and picocurie (pCi) are often used. The **becquerel** (Bq) is a relatively new unit that has achieved little use among U.S. scientists, even though it is smaller and more manageable than the curie. One becquerel is equal to one disintegration per second.

A number of biological units are used, including the roentgen (R), rad (D), gray (Gy), and rem. The **roentgen** is a biological unit used with X rays or gamma rays. One roentgen is the quantity of radiation that will generate

Curie
A physical unit of radiation measurement corresponding to 3.7×10^{10} nuclear disintegrations per second.

Becquerel
A physical unit of radiation measurement corresponding to one nuclear disintegration per second.

Roentgen
A biological unit of radiation measurement used with X rays and gamma rays; the quantity of radiation that generates 2.1×10^9 ion pairs per 1 cm³ of dry air or 1.8×10^{12} ion pairs per 1 g of tissue.

TABLE 10.5 Units for measuring radiation

Unit	Type of unit	Relationships between units
Curie (Ci)	Physical	$1 \text{ Ci} = 3.7 \times 10^{10} \text{ Bq}$
Becquerel (Bq)	Physical	$1 \text{ Bq} = 2.7 \times 10^{-11} \text{ Ci}$
Roentgen (R)	Biological	$1 \text{ R} = 0.96 \text{ D (tissue)}$
Rad (D)	Biological	$1 \text{ D} = 1 \times 10^{-2} \text{ Gy}$
Gray (Gy)	Biological	$1 \text{ Gy} = 1 \times 10^{2} \text{ D}$
Rem	Biological	1 rem = 1 R X ray or gamma ray in health effect

Rad

A biological unit of radiation measurement corresponding to the transfer of 1×10^{-2} J or 2.4×10^{-3} cal of energy to 1 kg of tissue.

Gray

A biological unit of radiation measurement corresponding to the transfer of 1 J of energy to 1 kg of tissue.

Rem

A biological unit of radiation measurement corresponding to the health effect produced by 1 roentgen of gamma or X rays regardless of the type of radiation involved.

Scintillation counter

A radiation-detection device operating on the principle that phosphors give off light when struck by radiation.

Geiger–Müller tube

A radiation-detection device operating on the principle that ions form when radiation passes through a tube filled with low-pressure gas.

2.1×10^9 ion pairs in 1 cm^3 of dry air at normal temperature and pressure. The ionization effect of X rays and gamma rays is greater when the radiation passes through tissue; one roentgen generates about 1.8×10^{12} pairs per gram of tissue.

The rad and the gray are biological units used to describe the effects of radiation in terms of the amount of energy transferred from the radiation to the tissue through which it passes. One **rad** transfers 1×10^{-2} J or 2.4×10^{-3} cal of energy to 1 kg of tissue. One **gray**, a larger unit, transfers 1 J per 1 kg of tissue. From a health point of view, the roentgen and rad are essentially equal. One roentgen of X rays or gamma rays equals 0.96 rad.

The rem is a biological unit devised to account for health differences in various types of radiation. A 1-rad dose of gamma radiation does not produce the same health effects as a 1-rad dose of alpha radiation. An additive unit was needed so that one unit of alpha had the same health effects as one unit of gamma. The devised unit is the rem, which stands for roentgen equivalent in man. One **rem** of any type of radiation has the same health effect as 1 roentgen of gamma rays or X rays. The advantage of such a unit is that individuals exposed to low levels of various types of radiation have to keep track only of the rems absorbed from each type and add them up to get the total health effect. If other units were used, the number of units from each type of radiation would have to be determined and the health effects calculated for each type. The units used for measuring radiation are summarized in ■ Table 10.5, together with relationships among the units.

Radiation can be detected in a number of ways. Film badges are commonly used by people working in environments where they might be exposed to radiation. The badges contain photographic film that becomes exposed when subjected to radiation. The extent of exposure of the film increases with the amount of radiation. The film is developed after a specific amount of time, and the degree of exposure indicates the radiation dose that has been absorbed during that time (see ■ Figure 10.4).

Radiation causes certain substances, called phosphors, to give off visible light. The light appears as brief flashes when radiation strikes a surface coated with a thin layer of phosphors. The number of flashes (called *scintillations*) per unit time is proportional to the amount of radiation striking the surface. Devices employing this approach are called **scintillation counters.**

A third type of detector, called a **Geiger–Müller tube**, is based on the fact that electricity is conducted through a gas that contains charged particles. A Geiger–Müller tube is represented in ■ Figure 10.5. The tube essentially consists of two electrodes charged to high voltages of opposite sign. The tube is filled with a gas at very low pressure. When radiation passes through the tube, it creates a path of charged particles (ions) by knocking electrons off gas molecules. They conduct a brief pulse of current between the electrodes. This brief current is amplified and triggers a counting circuit to record one "count." The number of counts per unit time is proportional to the amount of radiation passing through the Geiger–Müller tube.

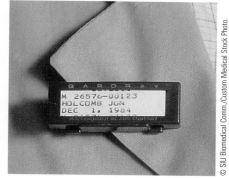

> **FIGURE 10.4** Film badges provide a convenient way to monitor the total amount of radiation received during a specific time period.

► 10.6 Medical Uses of Radioisotopes

Chemically, radioisotopes undergo the same reactions as nonradioactive isotopes of the same element. For example, radioactive iodine-123 is absorbed by the thyroid gland and used to produce the hormone thyroxine just as if it were ordinary nonradioactive iodine. This characteristic, together with the fact that the location of radioisotopes in the body can be readily detected, makes them useful for diagnostic and therapeutic medical applications. In diagnostic applications, radioisotopes are used as **tracers** whose progress through the body or localization in specific organs can be followed.

To minimize the risks to patients from exposure to radiation, radioisotopes used as diagnostic tracers should have as many of the following characteristics as possible:

1. Tracers should have short half-lives so they will decay while the diagnosis is being done but will give off as little radiation as possible after the diagnostic procedure is completed.
2. The daughter produced by the decaying radioisotope should be nontoxic and give off little or no radiation of its own. Ideally, it will be stable.
3. The radioisotope should have a long enough half-life to allow it to be prepared and administered conveniently.
4. The radiation given off by the radioisotope should be penetrating gamma rays, if possible, to ensure that they can be detected readily by detectors located outside the body.
5. The radioisotope should have chemical properties that make it possible for the tissue being studied to either concentrate it in diseased areas and form a **hot spot** or essentially reject it from diseased areas to form a **cold spot**.

Iodine-123 is used diagnostically to determine whether a thyroid gland is properly functioning. As mentioned earlier, the thyroid gland concentrates iodine and uses it in the production of thyroxine, an iodine-containing hormone. Iodine is an excellent tracer to use because the thyroid is the only user of iodine in the body. Iodine-123 is a good tracer because it is a gamma-emitter with a half-life of 13.3 hours. Iodine-131 is also radioactive and has a short half-life (8 days), but it emits gamma and beta radiation. The beta radiation cannot penetrate through the tissue to be detected diagnostically and therefore is simply an added radiation risk to the patient. For this reason, iodine-123 is

Tracer
A radioisotope used medically because its progress through the body or localization in specific organs can be followed.

Hot spot
Tissue in which a radioactive tracer concentrates.

Cold spot
Tissue from which a radioactive tracer is excluded or rejected.

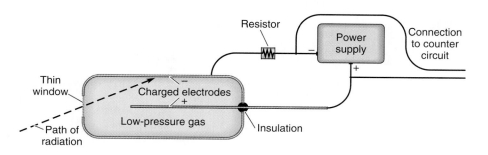

> **FIGURE 10.5** A cross section of a detection tube used in Geiger–Müller counters.

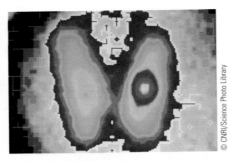

> **FIGURE 10.6** A thyroid scan produced after the administration of a radioactive iodine isotope. Is either a hot spot or a cold spot present?

preferred. An overactive thyroid absorbs more iodine and forms a hot spot, whereas an underactive or nonactive gland absorbs less iodine and becomes a cold spot (see ■ Figure 10.6).

Radioisotopes administered internally for therapeutic use should ideally have the following characteristics:

1. The radioisotope should emit less penetrating alpha or beta radiation to restrict the extent of damage to the desired tissue.
2. The half-life should be long enough to allow sufficient time for the desired therapy to be accomplished.
3. The decay products should be nontoxic and give off little or no radiation.
4. The target tissue should concentrate the radioisotope to restrict the radiation damage to the target tissue.

Iodine-131 is used therapeutically to treat thyroid cancer and hyperthyroidism. The primary advantage of this radioisotope is the ability of thyroid tissue to absorb it and localize its effects.

Until the mid-1950s, when improvements in the generation of penetrating X rays made them the therapy of choice, cobalt-60 was widely used to treat cancer. This radioisotope emits beta and gamma radiation and has a half-life of 5.3 years. When used therapeutically today, a sample of cobalt-60 is placed in a heavy lead container with a window aimed at the cancerous site. The beam of gamma radiation that exits through the window is focused on a small area of the body where the tumor is located. ■ Table 10.6 contains other examples of radioisotopes used diagnostically and therapeutically.

TABLE 10.6   Examples of medically useful radioisotopes

Isotope	Emission	Half-life	Applications
$^{3}_{1}H$	beta	12.3 years	To measure water content of body
$^{32}_{15}P$	beta	14.3 days	Detection of tumors, treatment of a form of leukemia
$^{51}_{24}Cr$	gamma	27.8 days	Diagnosis: size and shape of spleen, gastrointestinal disorders
$^{59}_{26}Fe$	beta	45.1 days	Diagnosis: anemia, bone marrow function
$^{60}_{27}Co$	beta, gamma	5.3 years	Therapy: cancer treatment
$^{67}_{31}Ga$	gamma	78.1 hours	Diagnosis: various tumors
$^{75}_{34}Se$	beta	120.4 days	Diagnosis: pancreatic scan
$^{81}_{36}Kr$	gamma	2.1×10^5 years	Diagnosis: lung ventilation scan
$^{85}_{38}Sr$	gamma	64 days	Diagnosis: bone scan
$^{99}_{43}Tc$	gamma	6 hours	Diagnosis: brain, liver, kidney, bone, and heart muscle scans
$^{123}_{53}I$	gamma	13.3 hours	Diagnosis: thyroid cancer
$^{131}_{53}I$	beta, gamma	8.1 days	Diagnosis and treatment: thyroid cancer
$^{197}_{80}Hg$	gamma	65 hours	Diagnosis: kidney scan

© CNRI/Science Photo Library

➤ 10.7 Nonmedical Uses of Radioisotopes

The study of the photosynthetic process in plants has been made much easier because of the use of radioactive carbon-14 as a tracer. The overall photosynthetic process is represented by Equation 10.3:

$$6CO_2(g) + 6H_2O(\ell) \rightarrow C_6H_{12}O_6(aq) + 6O_2(g) \qquad (10.3)$$

However, the conversion of water (H_2O) and carbon dioxide (CO_2) into carbohydrates ($C_6H_{12}O_6$) involves numerous chemical reactions and carbon-containing intermediate compounds. By using CO_2 that contains radioactive carbon-14, scientists have been able to follow the chemical path of CO_2 through the various intermediate compounds.

Interesting uses for radioactive tracers are found in industry. In the petroleum industry, for example, radioisotopes are used in pipelines to indicate the boundary between different products moving through the lines (see ■ Figure 10.7). In another application, tracer added to fluids moving in pipes makes it easy to detect leaks. Also, the effectiveness of lubricants has been studied by the use of radioactive metal isotopes as components of metal parts; as the metal part wears, the isotope shows up in the lubricant, and the amount of wear can be determined.

Radioisotopes are also useful in forms other than tracers. The thickness of manufactured materials, such as metal foils or sheets, can be determined continuously by the use of a radioisotope. The amount of radiation that penetrates the material to the detector is related to the thickness of the material.

An interesting application of radioisotopes involves the use of naturally occurring radioactive materials to determine the ages of artifacts and rocks. Perhaps the most widely known example of **radioactive dating** is the use of carbon-14, which is produced in the atmosphere when cosmic-ray neutrons strike nitrogen nuclei. The equation for the reaction is

$$^{14}_{7}N + ^{1}_{0}n \rightarrow ^{14}_{6}C + ^{1}_{1}p \qquad (10.4)$$

The resulting $^{14}_{6}C$ is converted to carbon dioxide $^{14}_{6}CO_2$ through a chemical reaction with oxygen. In this form, the radioactive carbon is absorbed by plants and converted into cellulose or other carbohydrates during photosynthesis. As long as the plant lives, it takes in $^{14}_{6}CO_2$. As a result, an equilibrium

Radioactive dating
A process for determining the age of artifacts and rocks, based on the amount and half-life of radioisotopes contained in the object.

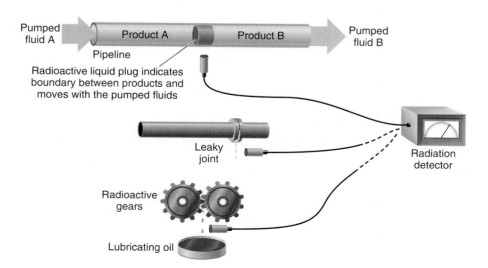

Pumped fluid A

Product A   Product B

Pumped fluid B

Pipeline

Radioactive liquid plug indicates boundary between products and moves with the pumped fluids

Leaky joint

Radiation detector

Radioactive gears

Lubricating oil

➤ **FIGURE 10.7** Radioactive tracers used in industry.

CHEMISTRY AROUND US • 10.2

RADON: A CHEMICALLY INERT HEALTH RISK

Substances that are chemically unreactive are generally not thought to be dangerous to one's health. However, radon, a colorless, odorless, chemically inert noble gas, is considered to be a health risk because it is radioactive. It is believed to be the leading cause of lung cancer among nonsmokers. It has been estimated by U.S. officials that up to 8% of the country's annual lung cancer deaths can be attributed to indoor exposure to radon-222.

Radon-222 is formed during a multistep process in which naturally occurring uranium-238 decays to stable lead-206. During one step of the process, radium-226, with a half-life of 1600 years, emits alpha and gamma radiation to produce radon-222:

$$^{226}_{88}\text{Ra} \rightarrow {}^{4}_{2}\alpha + {}^{222}_{86}\text{Rn} + \gamma$$

Radon-222, with a half-life of 4 days, is produced in rocks and soil that contain even minute amounts of uranium-238. Because it is a gas, radon migrates readily from the soil into the surrounding air. It seeps into houses and other buildings through openings around pipes, and through cracks in basement floors and walls.

The average level of radon in homes is lower than 1 picocurie/L of air. A picocurie (10^{-12} curie) is equivalent to the radioactive decay of 2 radon nuclei per minute. The EPA recommends an upper limit of 4 picocuries/L for indoor air, and it has been estimated that 99% of the homes in the United States have levels below this value. However, much higher levels have been found. The highest recorded level of radioactivity from radon in a home was 2700 picocuries/L of air. This was found in a home in Pennsylvania that was built on soil rich in uranium-238. The level of radon in home air can be minimized by sealing the entry points and providing good ventilation.

The hazards associated with breathing radon-222 come not only from the alpha radiation given off as the radon decays but also from the daughter products of that decay. None of these decay products is a gas, but all are radioactive. Because they are not gases, they are not exhaled and become lodged in the lungs, where their radiation can cause serious damage.

It is difficult to decide just how much concern people should have about radon in their homes. Some experts in the field agree with the EPA recommendations, but others question the data on which those recommendations are based. It is quite easy and relatively inexpensive to test for radon in a home, and homes built in locations where the gas has been identified as a problem should be tested.

Radon detectors are available for home use.

is established in which the plant contains the same fraction of carbon-14 as the surrounding air. Thus, as long as the plant lives, a constant fraction of the total carbon present is carbon-14. When the plant dies, $^{14}_{6}\text{CO}_2$ intake stops, and the fractional amount of carbon-14 begins to decrease. From a measurement of this amount in, say, a wooden object, and a knowledge of the half-life of carbon-14 (5600 years), it is possible to calculate the age of the object—how long since the tree was cut down to make the object. An object containing only about one-eighth as much carbon-14 as a fresh wood sample from a living tree would be about three half-lives, or 16,800 years old.

This method is limited to objects less than about 50,000 years old because of difficulties in measuring the small amount of carbon-14 present. For this reason, other dating methods have been developed. One method involves potassium-40 and argon-40. The half-life of potassium-40, which absorbs one of its own inner electrons to produce argon-40, is 1.3×10^9 years. By determining the amount of argon-40 in a potassium-containing mineral, it is possible to estimate the age of the mineral. The equation for the reaction is

$$^{40}_{19}\text{K} + {}^{0}_{-1}e \rightarrow {}^{40}_{18}\text{Ar} \tag{10.5}$$

➤ 10.8 Induced Nuclear Reactions

Before 1934, the study of radioactivity was limited to reactions of the relatively few radioisotopes found in nature. In that year, Irene and Frederic Joliot-Curie, French physical chemists, found that radioactivity could be induced in nonradioactive nuclei by bombarding them with small, subatomic particles. They produced an artificial radioactive isotope, nitrogen-13, by bombarding boron-10 with alpha particles from a natural radioisotope:

$$^{10}_{5}B + {}^{4}_{2}\alpha \rightarrow {}^{13}_{7}N + {}^{1}_{0}n \qquad (10.6)$$

Notice that a neutron, ${}^{1}_{0}n$, is produced in addition to ${}^{13}_{7}N$. Fifteen years earlier, British physicist Ernest Rutherford had used a similar reaction to produce nonradioactive oxygen-17 from nitrogen-14. A small particle, a proton, was produced in addition to oxygen-17:

$$^{14}_{7}N + {}^{4}_{2}\alpha \rightarrow {}^{17}_{8}O + {}^{1}_{1}p \qquad (10.7)$$

A great deal of research has been done since these first experiments were performed, and today it is known that a variety of nuclear reactions can be induced by bombarding either stable or naturally radioactive nuclei with high-energy particles. These artificially induced reactions may lead to one of four results. The first possibility is that the bombarded nucleus may be changed into a different nucleus that is stable. A second possibility is that a new nucleus may be formed that is unstable and undergoes radioactive decay. The other two possibilities are that nuclear fission or nuclear fusion will take place. These are discussed in more detail in Section 10.9. The result of bombardment depends on the kind of nuclei bombarded and the nature and energy of the bombarding particles.

Charged or uncharged particles may be used to bombard nuclei, and each type presents unique problems to overcome. It is believed that a bombarding particle will be drawn into the nucleus of an atom by strong nuclear forces if it can get to within 10^{-12} cm of the nucleus. Neutrons are captured readily if their speeds are reduced enough to allow the nuclear forces to work. Neutrons are slowed by passing them through materials known as **moderators**. Graphite is an example of a neutron moderator. Charged particles present a different problem. If they are negatively charged (${}^{0}_{-1}\beta$), they are repelled by the electron cloud around the nucleus; for a reaction to take place they must have sufficient energy to overcome this repulsion. Similarly, positive particles must be energetic enough to overcome repulsion by the positively charged nucleus.

Neutrons for bombardment reactions are obtained from other nuclear reactions and, together with proper moderators, are used as they come from the sources. Charged particles are obtained from radioactive materials or from ionization reactions. For example, alpha particles are produced by an ionization reaction in which the two electrons are removed from helium atoms. These charged particles are generally accelerated to high speeds (and energies) before bombarding the intended target nuclei. Two types of particle accelerators, cyclic and linear, are used. The **cyclotron**, one type of cyclic accelerator, is represented in ■ Figure 10.8.

The charged particles to be accelerated enter the evacuated chamber at the center of the cyclotron and, because of the magnetic field, move in a circular path toward the gap between the Dees. Just as the particles reach the gap, the electrical charge on the Dees is adjusted so the particles are repelled by the Dee they are leaving and attracted to the other one. The particles then coast inside the Dee until they again reach a gap, at which point the charges are again adjusted to cause acceleration. This process continues, and the particles are accelerated each time they pass through the gap. As the speed and energy of the particles increase, so does the radius of their circular path, until they finally leave the Dees and strike the target.

Moderators
Materials capable of slowing down neutrons that pass through them.

Cyclotron
A cyclic particle accelerator that works by changing electrical polarities as charged particles cross a gap. The particles are kept moving in a spiral path by a strong magnetic field.

➤ **FIGURE 10.8** The cyclotron. Early versions of the cyclotron were about the size of a dinner plate.

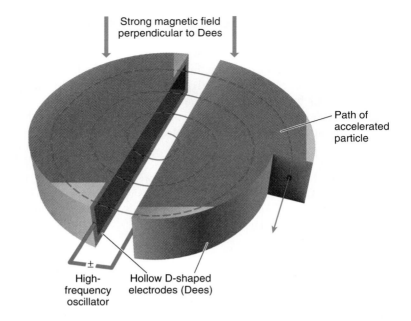

Strong magnetic field perpendicular to Dees

Path of accelerated particle

±

High-frequency oscillator

Hollow D-shaped electrodes (Dees)

Linear accelerator

A particle accelerator that works by changing electrical polarities as charged particles cross gaps between segments of a long tube.

In a **linear accelerator,** the particles are accelerated through a series of axially aligned, charged tubes located in an evacuated chamber (see ■ Figure 10.9). Each time a particle passes through a gap and goes from one tube to another, it is accelerated in the same way as particles passing between the Dees of a cyclotron. To make the particle accelerate, the charges on the tubes are reversed at the proper times. The tubes get successively longer to allow the particles the same residence or coasting time as they move toward the end at ever-increasing speeds. After acceleration, the particles exit and strike the target.

One of the most interesting results of bombardment is the creation of completely new elements. Four of these elements, produced between 1937 and 1941, filled gaps in the periodic table for which no naturally occurring element had been found. These four are technetium (Tc, number 43), promethium (Pm, number 61), astatine (At, number 85), and francium (Fr, number 87). The equations for the reactions for their production are:

$$^{96}_{42}\text{Mo} + {}^{2}_{1}\text{H} \rightarrow {}^{1}_{0}n + {}^{97}_{43}\text{Tc} \text{ (half-life: } 2.6 \times 10^{6} \text{ years)} \tag{10.8}$$

$$^{142}_{60}\text{Nd} + {}^{1}_{0}n \rightarrow {}^{0}_{-1}\beta + {}^{143}_{61}\text{Pm} \text{ (half-life: 265 days)} \tag{10.9}$$

$$^{209}_{83}\text{Bi} + {}^{4}_{2}\alpha \rightarrow 3{}^{1}_{0}n + {}^{210}_{85}\text{At} \text{ (half-life: 8.3 hours)} \tag{10.10}$$

$$^{230}_{90}\text{Th} + {}^{1}_{1}p \rightarrow 2{}^{4}_{2}\alpha + {}^{223}_{87}\text{Fr} \text{ (half-life: 22 minutes)} \tag{10.11}$$

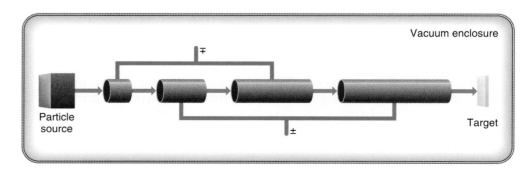

Vacuum enclosure

∓

Particle source

±

Target

➤ **FIGURE 10.9** A linear particle accelerator.

CHEMISTRY AND YOUR HEALTH • 10.1

IS IRRADIATED FOOD SAFE?

In 1993, four deaths and hundreds of illnesses were linked to the eating of undercooked ground beef that was contaminated with *E. coli* bacteria. In 1997 the largest recall of meat products in U.S. history up to that time took place when a producer voluntarily recalled 25 million pounds of ground beef suspected of being contaminated with the same type of bacteria. In 2002, a new meat product recall record was established when 27.4 million pounds of poultry products was recalled because of possible contamination by *Listeria,* a potentially deadly type of bacteria.

It is a well-established fact that treating food by exposing it to radiation destroys disease-causing organisms, kills insects, delays maturation and sprouting of seeds, and slows spoiling. But is irradiated food safe? After years of scientific study, the U.S. government began answering this question with a somewhat-qualified *yes* in 1963, when the Food and Drug Administration (FDA) first allowed radiation-treated wheat and wheat flour to be marketed. Ongoing studies have focused on such areas as the chemical effects of radiation on food, the impact of radiation on nutrient content, and toxicity concerns. Food irradiation for the control of disease-causing organisms has received endorsements and praise from numerous food industry and health organizations.

Since 1963, the list of foods approved for radiation treatment has grown steadily. The effect of the radiation on specific foods is determined by controlling the dose of radiation used. The following table shows the types of foods approved for radiation at this time along with the approved dose and the desired effect the dose has on the food. The doses are measured in units of kilograys (kGys), which were defined in Table 10.5.

Food	Effect	Dose (kGys)
Spices and dry vegetable seasonings	Decontaminates, controls insects and harmful bacteria	30
Dry or dehydrated enzymes	Controls insects and harmful bacteria	10
All foods	Controls insects	1
Fresh foods	Delays maturation	1
Poultry	Controls harmful bacteria	3
Red meat	Controls spoilage and harmful bacteria	4.5 (fresh) 7 (frozen)

In spite of scientific studies, endorsements from health organizations, and repeated assurances about the safety of irradiated foods, most consumers do not intentionally purchase and use them. Some irradiated foods are used unintentionally as a result of the labeling rules. For example, irradiated spices or fresh strawberries sold in a grocery store must have labels that indicate they have been irradiated. However, when such irradiated foods are used as ingredients in other foods, the other foods are not required to have irradiation labels. Also, irradiation labeling does not apply to foods served in restaurants.

Two factors that currently contribute to low consumer acceptance of irradiated food are continuing misconceptions about the effects of radiation on food and the low availability of irradiated foods for sale. Some misconceptions that are difficult to eliminate despite reassuring results from many studies are that irradiation causes food to become radioactive, irradiation produces harmful compounds in food, and irradiation reduces the amounts of nutrients in food. Studies indicate that continued education about the safety of irradiated food will eventually overcome this problem. The availability problem stems primarily from the lack of food-irradiation facilities. However, this factor is rapidly becoming less important as more facilities are being built in response to the acceptance of irradiation as a protection against harmful bacteria in food, especially in meat products. Some large meat-packing companies are adding irradiation facilities to existing packing plants, and including irradiation capabilities in plans for new plants, in order to eliminate the costs of shipping their products away from the packing plants to be irradiated.

It appears that food irradiation is becoming more and more acceptable to consumers, and the acceptance will continue to grow. It is likely that in the future food irradiation will be considered by most consumers to be a normal practice in providing a safe food supply to the consuming public.

Synthetic elements heavier than uranium, the heaviest naturally occurring element, have also been produced. All these elements, called **transuranium elements,** are radioactive. The first, neptunium (Np, number 93), was produced in 1940; since then at least 19 more (numbers 94 through 112) have been synthesized.

Significant quantities of only a few of these elements have been produced. Plutonium is synthesized in quantities large enough to make it available for use in atomic weapons and reactors. In 1968 the entire world's supply of californium

Transuranium elements
Synthetic elements with atomic numbers greater than that of uranium.

$(10^{-4}\,g)$ was gathered and made into a target for bombardment by heavy, accelerated particles. These experiments led to the discovery of several elements heavier than californium. Usually, the amounts of the new elements produced are extremely small—sometimes only hundreds of atoms. When the half-life is short, this small amount quickly disappears. Thus, these elements have to be identified not by chemical properties but by the use of instruments that analyze the characteristic radiation emitted by each new element.

Example 10.7

An isotope of gallium (Ga) useful in medical diagnostic work is formed when zinc-66 is bombarded with accelerated protons. Write the equation for this nuclear reaction and determine what isotope of gallium is produced.

Solution

The equation for the reaction is written

$$^{66}_{30}\text{Zn} + ^{1}_{1}p \rightarrow ?$$

where the product is to be determined. By balancing the mass numbers and atomic numbers we see that the product has a mass number of 67 and an atomic number of 31. Reference to the periodic table shows element number 31 to be gallium, Ga. Therefore, the isotope produced is gallium-67, and the equation for the reaction is:

$$^{66}_{30}\text{Zn} + ^{1}_{1}p \rightarrow ^{67}_{31}\text{Ga}$$

Learning Check 10.9

When manganese-55 is bombarded by protons, a neutron is one product. Write the equation for the process and identify the other product.

➤ 10.9 Nuclear Energy

In 1903, Ernest Rutherford and Frederick Soddy made observations that led them to conclude that the nuclei of all atoms, not just radioactive ones, must contain large quantities of energy. However, the form it took and the means for releasing it remained a mystery. In 1905, Albert Einstein, who was then working as a patent clerk, provided a partial answer. According to his theory of relativity, matter and energy were equivalent and one could be converted into the other. This fact was expressed in his now-famous equation,

$$E = mc^2 \qquad (10.12)$$

According to his equation, a small amount of mass, m, would, upon conversion, yield a huge amount of energy, E. The amount of energy liberated is huge because of the size of the multiplier c^2, the velocity of light squared. The velocity of light is a very large number (3.0×10^8 m/s, or 1.9×10^5 mi/s). The verification of Einstein's theory was not made until 1939, when the work of Otto Hahn, Lise Meitner, Fritz Strassman, and Enrico Fermi led to the discovery of **nuclear fission**. Attempts to produce new heavier transuranium elements by bombarding uranium-235 with neutrons led instead to the production of elements much less massive than uranium-235. The uranium-235 nuclei had undergone fission and split into smaller fragments. In the process a small amount of matter was converted into energy, just as Einstein had predicted.

Nuclear fission
A process in which large nuclei split into smaller, approximately equal-sized nuclei when bombarded by neutrons.

Today, we know that uranium-235 is the only naturally occurring isotope that will undergo fission, and a number of different products can result. Equations 10.13–10.16 represent some of the numerous fission reactions that can occur.

$$\begin{aligned}
&\nearrow \; {}^{135}_{53}\text{I} + {}^{97}_{39}\text{Y} + 4{}^{1}_{0}n &\quad(10.13)\\
&\nearrow \; {}^{139}_{56}\text{Ba} + {}^{94}_{36}\text{Kr} + 3{}^{1}_{0}n &\quad(10.14)\\
{}^{235}_{92}\text{U} + {}^{1}_{0}n &\longrightarrow {}^{103}_{42}\text{Mo} + {}^{132}_{50}\text{Sn} + 2{}^{1}_{0}n &\quad(10.15)\\
&\searrow \; {}^{139}_{54}\text{Xe} + {}^{95}_{38}\text{Sr} + 2{}^{1}_{0}n &\quad(10.16)
\end{aligned}$$

For every uranium-235 nucleus undergoing fission, more than one neutron is generated, which opens up the possibility for a chain reaction to take place. If only one of the neutrons produced each time reacts with another uranium-235 nucleus, the process will become a **chain reaction** that continues at a constant rate. If more than one of the neutrons generated per fission reaction produces another reaction, the process becomes an **expanding** or **branching chain reaction** that leads to an explosion. A constant-rate chain reaction is called a **critical reaction.** A branching chain reaction that will lead to an explosion is called **supercritical** (see ■ Figure 10.10).

History's first critical chain reaction was carried out in an atomic pile, which was literally a stack or pile of uranium-containing graphite blocks and blocks of pure graphite that served as a neutron moderator (see ■ Figure 10.11). The pile was constructed in a squash court beneath the stands of an athletic fieldhouse at the University of Chicago, and the experiments were carried out by a team of scientists led by Italian physicist Enrico Fermi. In December 1942, the team observed that nuclear reactions in the pile had become self-sustaining, or critical. The control rods, composed of strong neutron absorbers, were pushed into the pile to stop the reaction.

Once it had been demonstrated that uranium-235 would undergo self-sustaining fission, the construction of an atomic bomb was fairly simple in principle, but it depended on the concept of critical mass. Neutrons released inside a small piece of fissionable material might escape to the outside before they collide with and split another atom. If the majority of neutrons are lost this way, a chain reaction will not take place. However, as the size of the piece of fissionable material is increased, the chance for neutrons to escape without hitting other atoms decreases. The amount of fissionable material

Chain reaction
A nuclear reaction in which the products of one reaction cause a repeat of the reaction to take place. In the case of uranium fission, neutrons from fission reactions cause other fission reactions to occur.

Expanding or branching chain reaction
A reaction in which the products of one reaction cause more than one more reaction to occur.

Critical reaction
A constant-rate chain reaction.

Supercritical reaction
A branching chain reaction.

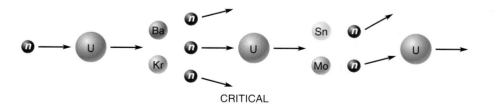

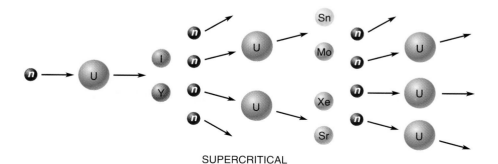

➤ **FIGURE 10.10** Nuclear chain reactions.

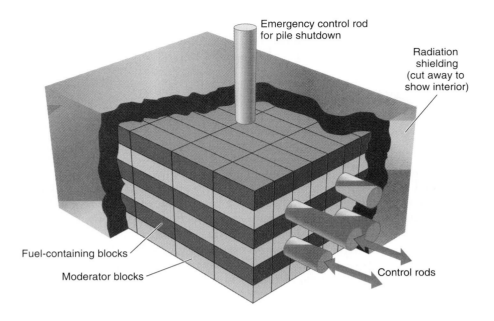

➤ **FIGURE 10.11** An atomic pile, the world's first nuclear reactor.

Emergency control rod for pile shutdown

Radiation shielding (cut away to show interior)

Fuel-containing blocks

Moderator blocks

Control rods

Critical mass
The minimum amount of fissionable material needed to sustain a critical chain reaction at a constant rate.

Supercritical mass
The minimum amount of fissionable material that must be present to cause a branching chain reaction to occur.

needed to cause a critical reaction to occur is the **critical mass.** The amount needed to cause a branching chain reaction and explosion to occur is called a **supercritical mass.**

During World War II, one major difficulty encountered in building a nuclear bomb was the size of device needed to bring together subcritical masses to form a supercritical mass. Any bomb used in combat would have to fit into a B-29, the largest bomber available at the time, and could not be heavier than the load capacity of the airplane. The configurations of the first two devices actually used are represented in ■ Figure 10.12. The spherical configuration resulted in a short, wide bomb, while the tubular configuration gave rise to a long, narrow bomb. In each configuration, explosive charges were used to drive subcritical masses together to form a supercritical mass. These nuclear bombs were the first applications of the concepts of subcritical and supercritical mass.

➤ **FIGURE 10.12** Configurations used in fission-type nuclear bombs.

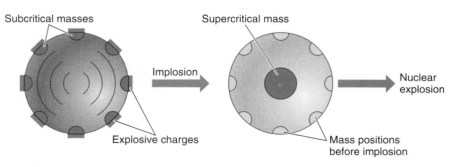

Subcritical masses

Supercritical mass

Implosion

Nuclear explosion

Explosive charges

Mass positions before implosion

SPHERICAL CONFIGURATION

Explosive charges

Subcritical masses

Supercritical mass

Nuclear explosion

TUBULAR CONFIGURATION

OVER THE COUNTER • 10.1

DIETARY SUPPLEMENTS

Every year, Americans purchase billions of dollars' worth of the dietary supplements known as vitamins and minerals. For many years, these were the only dietary supplements commonly available in drugstores. However, today many products with names such as *dong quai root, echinacea,* and *saw palmetto extract* are found on the same shelves as conventional vitamins and minerals. These herbal products represent a growing and somewhat controversial segment of the dietary supplements industry, with annual sales approaching those of conventional products. The sales increase is attributed to a growing fascination with alternative medicine in the general population, while the controversy is primarily related to a lack of FDA authority to regulate the products.

Factors contributing to the growing interest in alternative medicine are the "natural-is-better" perception, an increasing percentage of older people in the population, and some disenchantment with and concern about conventional medical health care. One source of concern for some is related to a knowledge of the health effects of radiation, such as those given in Table 10.3, and the medical use of radioisotopes (see Section 10.6), X rays, and other types of radiation (see Chemistry Around Us 10.1). The well-known drugs quinine, aspirin, and digitalis were all originally discovered in plants that were being used medicinally, a fact not lost on those who look to "natural" sources for their medicines. Older people who have had experience with modern medicine might have grown a little cynical and are willing to try alternatives. Modern worldwide communication has made it possible for these same people to become informed about practices in other parts of the world, where herbal medicines are more widely used and accepted by medical practitioners than they are in the United States.

The controversy surrounding dietary supplements such as herbal medicines took a dramatic jump in 1994. Prior to that time, the U.S. Food and Drug Administration did not quite know how to deal with products like herbal medicines. Neither the FDA rules governing the safety and effectiveness of prescription and OTC drugs, nor the rules that set manufacturing and safety standards for foods, applied to most plant products that claimed medicinal effects. Herbal medicines fell somewhere between foods and drugs, and the FDA permitted their sale as long as they were labeled as nutritional supplements, and no medicinal uses were mentioned.

In 1993, the FDA gave advance notice of proposed guidelines for dietary supplements. As a result of the announcement, a multimillion-dollar campaign was launched by the dietary supplements industry in opposition to the proposed stricter regulation of its products. Deluged with millions of responses from the voting public, Congress responded in 1994 with the Dietary Supplement Health and Education Act. This law created a new category of product that is distinct from food or drugs, and that is nearly exempt from the rules the FDA uses against questionable products. The new category includes vitamins, minerals, herbs, amino acids, and just about everything that had been sold as a supplement before the new law was passed.

Provisions of the new law severely limited the authority of the FDA to regulate dietary supplements. For example, the FDA cannot review evidence used to support claims that a supplement is safe or effective. Thus, dietary supplements can be manufactured and marketed with little regulatory control concerning their safety or the validity of the health benefits claimed for them. The law does require that as of 1999, the label of any dietary supplement must contain specific information, including the name and quantity of each ingredient, appropriate dosage, and the nutrients it contains. The label must also contain a statement indicating that any claims made for the product have not been reviewed or approved by the FDA.

In spite of the lack of FDA regulations, some manufacturers of herbal products are beginning to establish manufacturing standards, and some valid research results support the health claims of a few herbs. Research done in the United States, Germany, and other parts of Europe and Asia support claims of beneficial effects from products that represent about half the total of herbal medicines sold in the United States each year. These include echinacea, garlic, ginseng, gingko biloba, saw palmetto, aloe gel, ephedra, eleuthero, and cranberry. However, in some cases valid research results have disproved the health claims of certain herbs, and some herbs have been shown to be quite dangerous. Those planning to use an herbal product or other dietary supplement would be well advised to check first with their physician, and to obtain as much information about the product as possible from sources other than the product label.

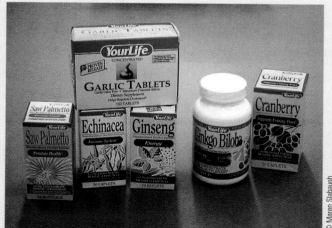

Research supports claims of beneficial effects for some herbal medicine.

On August 6, 1945, the United States became the first nation to use a nuclear weapon in combat. This bomb used a tubular configuration. Approximately 70,000 inhabitants of Hiroshima, Japan, were killed in the explosion; an equal number were seriously injured. A few days later a second bomb, with a spherical configuration, was dropped on the city of Nagasaki. The casualty figures were similar, and World War II came to an end. The Nagasaki bomb was made of plutonium, an artificial fissionable material produced in atomic piles by breeding reactions.

Breeder reactions are induced reactions in which the product nuclei undergo fission reactions. The plutonium used in the Nagasaki bomb was made by bombarding uranium-238 with neutrons in the first atomic pile. Uranium-238 is the most abundant isotope of uranium found in nature, but it will not undergo fission, as will the much less abundant uranium-235. However, uranium-238 reacts as follows when placed in a neutron-rich environment.

$$^{238}_{92}U + {}^1_0n \rightarrow {}^{239}_{92}U \tag{10.17}$$

$$^{239}_{92}U \rightarrow {}^{239}_{93}Np + {}^{\;\;0}_{-1}\beta \tag{10.18}$$

$$^{239}_{93}Np \rightarrow {}^{239}_{94}Pu + {}^{\;\;0}_{-1}\beta \tag{10.19}$$

The plutonium-239 will undergo fission and, when collected into a supercritical mass, will cause an explosion. Reactions such as 10.17–10.19 are called breeding reactions because a fissionable material is "bred" from a nonfissionable material.

Curiosity about the seemingly unlimited energy supply of the stars (especially the sun) eventually led scientists to an understanding of a second nuclear process that releases vast amounts of energy. In 1920, Sir Arthur Eddington suggested that the energy of the stars is a byproduct of a reaction in which hydrogen is changed into helium. This was followed in 1929 by the concept of **thermonuclear reactions**, nuclear reactions that are started by very high temperatures. According to this idea, at high temperatures the nuclei of very light-weight elements will combine to form nuclei of heavier elements, and in the process some mass is converted into energy. This idea was consistent with Eddington's suggestion that hydrogen is converted to helium in the sun and other stars. It was not until nuclear fission reactions became available that high enough temperatures could be achieved to test the theory of **nuclear fusion**. The theory proved to be correct, and today most nuclear weapons in the world are based on fusion reactions.

The fusion reactions responsible for the energy output of the sun are thought to be

$$2{}^1_1H \rightarrow {}^2_1H + {}^0_1\beta \tag{10.20}$$

$$^1_1H + {}^2_1H \rightarrow {}^3_2He + \gamma \tag{10.21}$$

$$^3_2He + {}^3_2He \rightarrow {}^4_2He + 2{}^1_1H \tag{10.22}$$

The net overall reaction is

$$4{}^1_1H \rightarrow {}^4_2He + 2{}^0_1\beta + 2\gamma \tag{10.23}$$

which is what Eddington proposed; hydrogen (${}^1_1H$) is converted to helium (${}^4_2He$).

A goal of many of the early developers of nuclear energy was to harness it and make it a virtually unlimited source of useful energy. Today, that goal has been partially achieved. Fission reactors generate a small percentage of the electricity used in the world, and researchers are making progress in their attempts to harness fusion reactions.

Breeder reaction
A nuclear reaction in which isotopes that will not undergo spontaneous fission are changed into isotopes that will.

Thermonuclear reactions
Nuclear fusion reactions that require a very high temperature to start.

Nuclear fusion
A process in which small nuclei combine or fuse to form larger nuclei.

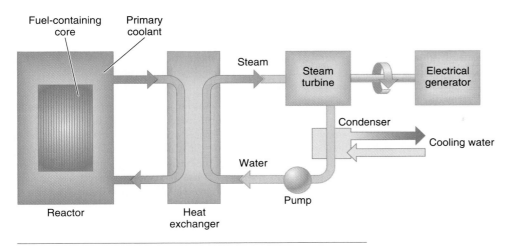

➤ **FIGURE 10.13** Components of a nuclear electricity power plant.

The essential components of a fission-powered generating plant are shown in ■ Figure 10.13. The use of nuclear power plants has created much controversy. The plants release large amounts of waste heat and thus cause thermal pollution of natural waters. In addition, radioactive wastes create disposal problems—where do you dump radioactive materials that will be a health hazard for thousands of years? Electricity consumers and producers have lost some of their confidence in the safety of nuclear power plants as a result of accidents such as the well-known incidents at Three-Mile Island (1979) and Chernobyl (1986). Another problem is the relatively short supply of fissionable uranium-235. Breeder reactors could solve this problem, since they can convert abundant supplies of uranium-238 and thorium-232 into fissionable fuels, but the breeder reactor program in the United States is at a standstill. As a result of these factors, the future of fission power generation is uncertain at the present time.

In principle, fusion reactors can be used as controlled energy sources just as fission reactors are. Fusion reactors would have many advantages over fission reactors. The deuterium fuel (hydrogen-2) is abundant in water, no significant amounts of radioactive wastes would be produced during operation, and the possibility of a serious accident is much smaller. Unfortunately, the feasibility of fusion reactors has not yet been established, although scientists are getting closer through continuing research.

CONCEPT SUMMARY

Radioactive Nuclei. Some nuclei are unstable and undergo radioactive decay. The common types of radiation emitted during decay processes are alpha, beta, and gamma, which can be characterized by mass and charge values.

Equations for Nuclear Reactions. Nuclear reactions can be represented by balanced equations in which the focus is on mass number and atomic number balance on each side. Symbols are used in nuclear equations that make it convenient to balance mass and atomic numbers.

Isotope Half-Life. Different radioisotopes generally decay at different rates, which are indicated by half-lives. One half-life is the time required for one-half of the unstable nuclei in a sample to undergo radioactive decay.

The Health Effects of Radiation. Radiation generates free radicals in tissue as it passes through. Radiation is hazardous even at low intensity if there is long-term exposure. Radiation sickness is caused by short-term intense radiation. Those working around radioactive sources can minimize exposure by using shielding or distance as a protection.

Measurement Units for Radiation. Two systems, physical and biological, are used to describe quantities of radiation. Physical units indicate the number of nuclei of radioactive material that decay per unit of time. Biological units are related to the damage caused by radiation in living tissue. The common physical units are the curie and its fractions and the becquerel. Biological units include the roentgen (for gamma and X rays), the rad, the gray, and the rem.

Medical Uses of Radioisotopes. Radioisotopes behave chemically like nonradioisotopes of the same element and can be used diagnostically and therapeutically. Diagnostically, radioisotopes are used as tracers whose movement or localization in the body can be followed. Therapeutic radioisotopes localize in diseased areas of the body, where their radiation can destroy diseased tissue.

Nonmedical Uses of Radioisotopes. Nonmedical uses of radioisotopes include (1) tracers in chemical reactions, (2) boundary markers between liquids in pipelines, (3) tracers to detect leaks, (4) metal-wear indicators, and (5) thickness indicators for foils and sheet metal. An especially interesting application is the determination of the ages of artifacts and rocks.

Induced Nuclear Reactions. The bombardment of nuclei with small high-energy particles causes the bombarded nuclei to (1) change into stable nuclei, (2) change into radioactive nuclei, (3) break into smaller pieces (fission), or (4) fuse to form larger nuclei (fusion). Most bombarding particles are accelerated by using cyclic or linear accelerators. Bombardment reactions have been used to produce elements that are not found in nature.

Nuclear Energy. During spontaneous nuclear fission reactions, heavy nuclei split when bombarded by neutrons, and release large amounts of energy. This process was used to produce two atomic bombs whose use ended World War II. A second energy-releasing nuclear process, fusion, is the basis for today's hydrogen bombs. Nuclear fission is in limited use as a source of electrical power; however, this use is controversial. Nuclear fusion has not yet proved feasible as a controlled source of power, but research toward this end continues.

LEARNING OBJECTIVES ASSESSMENT

You can get an approximate but quick idea of how well you have met the learning objectives given at the beginning of this chapter by working the selected end-of-chapter exercises given below. The answer to each exercise is given in Appendix B of the book.

Objective 1 (Section 10.1): Exercise 10.2

Objective 2 (Section 10.2): Exercise 10.12

Objective 3 (Section 10.3): Exercise 10.16

Objective 4 (Section 10.4): Exercise 10.22

Objective 5 (Section 10.5): Exercise 10.24

Objective 6 (Sections 10.6 and 10.7): Exercises 10.30 and 10.36

Objective 7 (Section 10.8): Exercise 10.38

Objective 8 (Section 10.9): Exercise 10.46

KEY TERMS AND CONCEPTS

Alpha particle (10.1)
Becquerel (10.5)
Beta particle (10.1)
Biological unit of radiation (10.5)
Breeding reaction (10.9)
Chain reaction (10.9)
Cold spot (10.6)
Critical mass (10.9)
Critical reaction (10.9)
Curie (10.5)
Cyclotron (10.8)
Daughter nuclei (10.2)
Electron capture (10.2)
Expanding or branching chain reaction (10.9)

Gamma ray (10.1)
Geiger–Müller tube (10.5)
Gray (10.5)
Half-life (10.3)
Hot spot (10.6)
Inverse square law of radiation (10.4)
Linear accelerator (10.8)
Moderators (10.8)
Nuclear fission (10.9)
Nuclear fusion (10.9)
Physical unit of radiation (10.5)
Positron (10.2)
Rad (10.5)
Radiation sickness (10.4)
Radical or free radical (10.4)

Radioactive dating (10.7)
Radioactive decay (10.1)
Radioactive nuclei (10.1)
Rem (10.5)
Roentgen (10.5)
Scintillation counter (10.5)
Supercritical mass (10.9)
Supercritical reaction (10.9)
Thermonuclear reaction (10.9)
Tracer (10.6)
Transuranium elements (10.8)

KEY EQUATIONS

1. Symbol for isotopes (Section 10.2): $_Z^A X$
2. Conversion of a neutron to a proton and electron in the nucleus, leading to beta emission (Section 10.2):

$$_0^1 n \rightarrow {}_1^1 p + {}_{-1}^0 \beta$$

Equation 10.1

3. Variation of radiation intensity with distance (Section 10.4):

$$\frac{I_x}{I_y} = \frac{d_y^2}{d_x^2}$$

Equation 10.2

4. Fission of uranium-235 (Section 10.9):

$$^{235}_{92}U + ^{1}_{0}n \rightarrow ^{135}_{53}I + ^{97}_{39}Y + 4^{1}_{0}n$$

Equation 10.13

See also Equations 10.14, 10.15, and 10.16.

5. Hydrogen fusion reaction of the sun (Section 10.9):

$$4^{1}_{1}H \rightarrow ^{4}_{2}He + 2^{0}_{1}\beta + 2\gamma$$

Equation 10.23

EXERCISES

LEGEND: 1 = straightforward, 2 = intermediate, 3 = challenging. All even-numbered exercises are answered in Appendix B.

RADIOACTIVE NUCLEI (SECTION 10.1)

10.1 Define the term *radioactive;* then criticize the following statements:
  a. Beta rays are radioactive.
  b. Radon is a stable radioactive element.

10.2 Group the common nuclear radiations (Table 10.1) into the following categories:
  a. Those with a mass number of 0
  b. Those with a positive charge
  c. Those with a charge of 0

10.3 Group the common nuclear radiations (Table 10.1) into the following categories:
  a. Those with a negative charge
  b. Those with a mass number greater than 0
  c. Those that consist of particles

10.4 Characterize the following nuclear particles in terms of the fundamental particles—protons, neutrons, and electrons:
  a. A beta particle
  b. An alpha particle
  c. A positron

10.5 Discuss how the charge and mass of particles that radiation comprises influence the range or ability of the radiation to penetrate matter.

EQUATIONS FOR NUCLEAR REACTIONS (SECTION 10.2)

10.6 Summarize how the atomic number and mass number of daughter nuclei compare with the original nuclei after
  a. An alpha particle is emitted.
  b. A beta particle is emitted.
  c. An electron is captured.
  d. A gamma ray is emitted.
  e. A positron is emitted.

10.7 Write appropriate symbols for the following particles using the $^{A}_{Z}X$ symbolism:
  a. A tin-117 nucleus
  b. A nucleus of the chromium (Cr) isotope containing 26 neutrons
  c. A nucleus of element number 20 that contains 24 neutrons

10.8 Write appropriate symbols for the following particles using the $^{A}_{Z}X$ symbolism:
  a. An iron-54 nucleus
  b. A nucleus of element number 48 with a mass number of 110
  c. A nucleus of the titanium isotope that contains 26 neutrons

10.9 Complete the following equations, using appropriate notations and formulas:
  a. $^{10}_{4}Be \rightarrow ? + ^{10}_{5}B$
  b. $^{210}_{83}Bi \rightarrow ^{4}_{2}\alpha + ?$
  c. $^{15}_{8}O \rightarrow ? + ^{15}_{7}N$
  d. $^{44}_{22}Ti + ^{0}_{-1}e \rightarrow ?$
  e. $^{8}_{4}Be \rightarrow ? + ^{4}_{2}He$
  f. $^{46}_{23}V \rightarrow ? + ^{46}_{22}Ti$

10.10 Complete the following equations, using appropriate notations and formulas:
  a. $^{234}_{90}Th \rightarrow ? + ^{234}_{91}Pa$
  b. $^{35}_{16}S \rightarrow ? + ^{0}_{-1}\beta$
  c. $? \rightarrow ^{4}_{2}\alpha + ^{244}_{98}Cf$
  d. $^{226}_{88}Ra \rightarrow ? + ^{4}_{2}\alpha$
  e. $^{113}_{47}Ag \rightarrow ^{113}_{48}Cd + ?$
  f. $^{41}_{18}Ar + ^{0}_{-1}e \rightarrow ?$

10.11 Write balanced equations to represent decay reactions of the following isotopes. The decay process or daughter isotope is given in parentheses.
  a. $^{66}_{29}Cu$ (beta emission)
  b. $^{22}_{11}Na$ (positron emission)
  c. $^{19}_{8}O$ (daughter = $^{19}_{9}F$)
  d. $^{192}_{78}Pt$ (alpha emission)
  e. $^{108}_{50}Sn$ (electron capture)
  f. $^{67}_{28}Ni$ (beta emission)

10.12 Write balanced equations to represent decay reactions of the following isotopes. The decay process or daughter isotope is given in parentheses.
  a. $^{121}_{50}Sn$ (beta emission)
  b. $^{190}_{78}Pt$ (daughter = osmium-186)
  c. $^{55}_{26}Fe$ (electron capture)

    d. $^{72}_{31}Ga$ (daughter = Ge-72)

    e. $^{10}_{4}Be$ (beta emission)

    f. $^{238}_{92}U$ (alpha emission)

ISOTOPE HALF-LIFE (SECTION 10.3)

10.13 What is meant by a half-life?

10.14 Describe half-life in terms of something familiar, such as a cake or cookies or your checking account.

10.15 An isotope of lead, $^{194}_{82}Pb$ has a half-life of 11 minutes. What fraction of the lead-194 atoms in a sample would remain after 44 minutes had elapsed?

10.16 Technetium-99 has a half-life of 6 hours. This isotope is used diagnostically to perform brain scans. A patient is given a 9.0-ng dose. How many nanograms will be present in the patient 24 hours later?

10.17 An archaeologist sometime in the future analyzes the iron used in an old building. The iron contains tiny amounts of nickel-63, with a half-life of 92 years. On the basis of the amount of nickel-63 and its decay products found, it is estimated that about 0.78% (1/128) of the original nickel-63 remains. If the building was constructed in 1980, in what year did the archaeologist make the discovery?

10.18 An archaeologist unearths the remains of a wooden box, analyzes them for the carbon-14 content, and finds that about 12.5 % of the carbon-14 initially present remains. Estimate the age of the box. The half-life of carbon-14 is 5600 years.

10.19 Germanium-66 decays by positron emission, with a half-life of 2.5 hours. What mass of germanium-66 remains in a sample after 10.0 hours if the sample originally weighed 50.0 mg?

10.20 A grain sample was found in a cave. The amount of carbon-14 in the sample was 1/16 the amount found in a fresh sample of grain. How old was the grain in the cave? The half-life of carbon-14 is 5600 years.

THE HEALTH EFFECTS OF RADIATION (SECTION 10.4)

10.21 A source of radiation has an intensity of 130 units at a distance of 10 feet. How far away from the source would you have to be to reduce the intensity to 15 units?

10.22 Compare and contrast the general health effects of long-term exposure to low-level radiation and short-term exposure to intense radiation.

10.23 A radiologist found that in a 30-minute period the dose from a radioactive source was 75 units at a distance of 12.0 m from the source. What would the dose be in the same amount of time at a distance of 2.00 m from the source?

MEASUREMENT UNITS FOR RADIATION (SECTION 10.5)

10.24 Explain the difference between physical and biological units of radiation and give an example of each type.

10.25 Explain why the rem is the best unit to use when evaluating the radiation received by a person working in an area where exposure to several types of radiation is possible.

10.26 An individual receives a short-term whole-body dose of 3.1 rads of beta radiation. How many roentgens of X rays would represent the same health hazard?

10.27 Describe how scintillation counters and Geiger–Müller counters detect radiation.

10.28 One Ci corresponds to 3.7×10^{10} nuclear disintegrations per second. How many disintegrations per second would take place in a sample containing 4.6 μCi of radioisotope?

MEDICAL USES OF RADIOISOTOPES (SECTION 10.6)

10.29 Explain what a diagnostic tracer is and list the ideal characteristics one should have.

10.30 Describe the importance of hot and cold spots in diagnostic work using tracers.

10.31 List the ideal characteristics of a radioisotope that is to be administered internally for therapeutic use.

10.32 Chromium-51 is used medically to monitor kidney activity. Chromium-51 decays by electron capture. Write a balanced equation for the decay process and identify the daughter that is produced.

10.33 Gold-198 is a β emitter used to treat leukemia. It has a half-life of 2.7 days. The dosage is 1.0 mCi/kg body weight. How long would it take for a 70-mCi dose to decay to an activity of 2.2 mCi?

NONMEDICAL USES OF RADIOISOTOPES (SECTION 10.7)

10.34 A mixture of water (H_2O) and hydrogen peroxide (H_2O_2) will give off oxygen gas when solid manganese dioxide is added as a catalyst. Describe how you could use a tracer to determine if the oxygen comes from the water or the peroxide.

10.35 Suppose you are planning a TV commercial for motor oil. Describe how you would set it up to show your oil prevents engine wear better than a competing brand.

10.36 Propose a method for measuring the volume of water in an irregular-shaped swimming pool. You have 1 gallon of water that contains a radioisotope with a long half-life, and a Geiger–Müller counter.

10.37 Explain why carbon-14, with a half-life of about 5600 years, is not a good radioisotope to use if you want to determine the age of a coal bed thought to be several million years old.

INDUCED NUCLEAR REACTIONS (SECTION 10.8)

10.38 Write a balanced equation to represent the synthesis of silicon-27 that takes place when magnesium-24 reacts with an accelerated alpha particle. A neutron is also produced.

10.39 Americium-241 (Am) results when plutonium-239 captures two neutrons. A beta particle is also produced. Write a balanced equation for the process.

10.40 Write a balanced equation to represent the net breeding reaction that occurs when uranium-238 reacts with a neutron to form plutonium-239 and two beta particles.

10.41 Explain why neutrons cannot be accelerated in a cyclotron or linear accelerator.

10.42 Describe the role of a moderator in nuclear reactions involving neutrons.

10.43 To make indium-111 for diagnostic work, silver-109 is bombarded with α particles. What isotope forms when a silver-109 nucleus captures an α particle? Write an equation to represent the process. What two particles must be emitted by the isotope to form indium-111?

10.44 To make gallium-67 for diagnostic work, zinc-66 is bombarded with accelerated protons. Write a balanced equation to represent the process when a zinc-66 nucleus captures a single proton.

NUCLEAR ENERGY (SECTION 10.9)

10.45 Write one balanced equation that illustrates nuclear fission.

10.46 Write one balanced equation that illustrates nuclear fusion.

10.47 Describe the difference between nuclear fission and nuclear fusion processes. Use nuclear equations to illustrate the difference.

10.48 Explain why critical mass is an important concept in neutron-induced fission chain reactions.

10.49 Complete the following equations, which represent two additional ways uranium-235 can undergo nuclear fission.

a. $^{235}_{92}U + ^{1}_{0}n \rightarrow ^{160}_{62}Sm + ? + 4^{1}_{0}n$

b. $^{235}_{92}U + ^{1}_{0}n \rightarrow ^{87}_{35}Br + ? + 3^{1}_{0}n$

10.50 Plutonium-238 is used to power batteries for pacemakers. It decays by α emission. Write a balanced equation for the process.

> ## CHEMISTRY FOR THOUGHT

10.1 One (unrealized) goal of ancient alchemists was to change one element into another (such as lead to gold). Do such changes occur naturally? Explain your reasoning.

10.2 Refer to Figure 10.2. Some isotopes used as positron emitters in PET scans are fluorine-18, oxygen-15, and nitrogen-13. What element results in each case when a positron is emitted?

10.3 Consider the concept of half-life and decide if, in principle, a radioactive isotope ever completely disappears by radioactive decay. Explain your reasoning.

10.4 Do you think Earth is more or less radioactive than it was when it first formed? Explain your answer.

10.5 Nuclear wastes typically have to be stored for at least 20 half-lives before they are considered safe. This can be a time of hundreds of years for some isotopes. With that in mind, would you consider sending such wastes into outer space a responsible solution to the nuclear waste disposal problem? Explain your answer.

10.6 Read Chemistry Around Us 10.2 and write an equation to represent the radioactive decay of radon-222. Then, write an equation to represent the decay of the daughter produced by the radon decay. The daughter decays by alpha emission. Then, write an equation for the decay of the daughter produced by this second reaction, which decays by beta emission. In each of the three reactions, assume only one particle in addition to the daughter is produced. What element is radon converted into by this series of three decays?

10.7 Uranium-238 is the most abundant naturally occurring isotope of uranium. It undergoes radioactive decay to form other isotopes. In the first three steps of this decay, uranium-238 is converted to thorium-234, which is converted to protactinium-234, which is converted to uranium-234. Assume only one particle in addition to the daughter is produced in each step, and use equations for the decay processes to determine the type of radiation emitted during each step.

---

nfoTrac College Edition Readings

"Use and abuse of isotopic exchange data in soil chemistry," *Australian Journal of Soil Research,* Dec 15, 2002, 40(8):1371(11). Record number A96237880.

"Aiming for superheavy elements," *Science News,* Sept 24, 1994, 146(13):206(1). Record number A15757837.

"Smashing atoms," *Poptronics,* Sept 2000, 1(9):17. Record number A64693119.

GOB
Chemistry·⚛·Now™

Assess your understanding of this chapter's topics with additional quizzing and conceptual-based problems at http://chemistry.brookscole.com/seager5e

CHAPTER 11

Organic Compounds: Alkanes

Pharmacists are responsible for the appropriate dispensing of drugs and other medications. Pharmacists must have a broad knowledge of the many drugs and medications available, as well as the effects they have on the human body when administered individually and in combinations. The great majority of drugs are organic compounds. This and the next several chapters provide the basis for understanding the structure, properties, and physiological effects of organic compounds.

© Royalty Free/CORBIS

LEARNING OBJECTIVES

When you have completed your study of this chapter, you should be able to:

1. Show that you understand the general importance of organic chemical compounds. (Section 11.1)

2. Be able to recognize the molecular formulas of organic and inorganic compounds. (Section 11.1)

3. Explain some general differences between inorganic and organic compounds. (Section 11.2)

4. Be able to use structural formulas to identify compounds that are isomers of each other. (Section 11.3)

5. Write condensed or expanded structural formulas for compounds. (Section 11.4)

6. Classify alkanes as normal or branched. (Section 11.5)

7. Use structural formulas to determine whether compounds are structural isomers. (Section 11.6)

8. Assign IUPAC names and draw structural formulas for alkanes. (Section 11.7)

9. Assign IUPAC names and draw structural formulas for cycloalkanes. (Section 11.8)

10. Name and draw structural formulas for geometric isomers of cycloalkanes. (Section 11.9)

11. Describe the key physical properties of alkanes. (Section 11.10)

12. Write alkane combustion reactions. (Section 11.11)

The word *organic* is used in several different contexts. Scientists of the 18th and 19th centuries studied compounds extracted from plants and animals and labeled them "organic" because they had been obtained from organized (living) systems. Organic fertilizer is organic in the original sense that it comes from a living organism. There is no universal definition of organic foods, but the term is generally taken to mean foods grown without the application of pesticides or synthetic fertilizers. When referring to organic chemistry, however, we mean the chemistry of carbon-containing compounds.

➤ 11.1 Carbon: The Element of Organic Compounds

Early chemists thought organic compounds could be produced only through the action of a "vital force," a special force active only in living organisms. This idea was central to the study of organic chemistry until 1828, because up to that time, no one had been able to synthesize an organic compound from its elements or from naturally occurring minerals. In that year, Friedrich Wöhler, a German chemist, heated an inorganic salt called ammonium cyanate and produced urea. This compound, normally found in blood and urine, was unquestionably organic, and it had come from an inorganic source. The reaction is

$$NH_4NCO \xrightarrow{\text{Heat}} \underset{\text{urea}}{H-N-C-N-H} \tag{11.1}$$

ammonium
cyanate

Wöhler's urea synthesis discredited the "vital force" theory, and his success prompted other chemists to attempt to synthesize organic compounds. Today, organic compounds are being synthesized in thousands of laboratories, and most of the synthetics have never been isolated from natural sources.

Organic compounds share one unique feature: They all contain carbon. Therefore, **organic chemistry** is defined as the study of carbon-containing compounds. There are a few exceptions to this definition; a small number of carbon compounds—such as CO, CO_2, carbonates, and cyanides—were studied before Wöhler's urea synthesis. These were classified as inorganic because they were obtained from nonliving systems, and even though they contain carbon, we still consider them to be a part of **inorganic chemistry.**

The importance of carbon compounds to life on Earth cannot be overemphasized. If all carbon compounds were removed from Earth, its surface would be somewhat like the barren surface of the moon (see ■ Figure 11.1). There

Organic compound
A compound that contains the element carbon.

Organic chemistry
The study of carbon-containing compounds.

Inorganic chemistry
The study of the elements and all noncarbon compounds.

➤ **FIGURE 11.1** Organic chemistry makes a tremendous difference between Earth and the moon.

© NASA

would be no animals, plants, or any other form of life. If carbon-containing compounds were removed from the human body, all that would remain would be water, a very brittle skeleton, and a small residue of minerals. Many of the essential constituents of living matter—such as carbohydrates, fats, proteins, nucleic acids, enzymes, and hormones—are organic chemicals.

The essential needs of daily human life are food, fuel, shelter, and clothing. The principal components of food (with the exception of water) are organic. The fuels we use (e.g., wood, coal, and natural gas) are mixtures of organic compounds. Our homes typically involve wood construction, and our clothing, whether made of natural or synthetic fibers, is organic.

Besides the major essentials, many of the smaller everyday things often taken for granted are also derived from carbon and its compounds. Consider an ordinary pencil. The "lead" (actually graphite), the wood, the rubber eraser, and the paint on the surface are all either carbon or carbon compounds. The paper in this book, the ink on its pages, and the glue holding it all together are also made of carbon compounds.

➤ 11.2 Organic and Inorganic Compounds Compared

It is interesting that the subdivision of chemistry into its organic and inorganic branches results in one branch that deals with compounds composed mainly of one element and another branch that deals with compounds formed by the more than 100 remaining elements. However, this classification seems more reasonable when we recognize that known organic compounds are much more numerous than inorganic compounds. An estimated 250,000 inorganic compounds have been identified, but more than 6 million organic compounds are known, and thousands of new ones are synthesized or isolated each year.

One of the reasons for the large number of organic compounds is the unique ability of carbon atoms to form stable covalent bonds with other carbon atoms and with atoms of other elements. The resulting covalently bonded molecules may contain as few as one or more than a million carbon atoms.

In contrast, inorganic compounds are often characterized by the presence of ionic bonding. Covalent bonding also may be present, but it is less common. These differences generally cause organic and inorganic compounds to differ physically (see ■ Figure 11.2) and chemically, as shown in ■ Table 11.1.

TABLE 11.1 Properties of typical organic and inorganic compounds

Property	Organic compounds	Inorganic compounds
Bonding within molecules	Usually covalent	Often ionic
Forces between molecules	Generally weak	Quite strong
Normal physical state	Gases, liquids, or low-melting-point solids	Usually high-melting-point solids
Flammability	Often flammable	Usually nonflammable
Solubility in water	Often low	Often high
Conductivity of water solutions	Nonconductor	Conductor
Rate of chemical reactions	Usually slow	Usually fast

> **FIGURE 11.2** Many organic compounds, such as ski wax, have relatively low melting points. What does this fact reveal about the forces between organic molecules?

© Clark Taylor

Learning Check 11.1

Classify each of the following compounds as organic or inorganic:

a. NaCl   **d.** NaOH
b. CH_4 **e.** CH_3OH
c. C_6H_6 **f.** $Mg(NO_3)_2$

Learning Check 11.2

Decide whether each of the following characteristics most likely describes an organic or inorganic compound:

a. Flammable   **b.** Low boiling point   **c.** Soluble in water

---

> **STUDY SKILLS 11.1**

CHANGING GEARS FOR ORGANIC CHEMISTRY

You will find that organic chemistry is very different from general or inorganic chemistry. By quickly picking up on the changes, you will help yourself prepare for quizzes and exams.

There is almost no math in these next six chapters or in the biochemistry section. Very few mathematical formulas need to be memorized. The problems you will encounter fall mainly into four categories: naming compounds and drawing structures, describing physical properties of substances, writing reactions, and identifying typical uses of compounds. This pattern holds true for all six of the organic chemistry chapters.

The naming of compounds is introduced in this chapter, and the rules developed here will serve as a starting point in the next five chapters. Therefore, it is important to master naming in this chapter. A well-developed skill in naming will help you do well on exams covering the coming chapters.

Only a few reactions are introduced in this chapter, but many more will be in future chapters. Writing organic reactions is just as important (and challenging) as naming, and Study Skills 12.1 will help you. Identifying the uses of compounds can best be handled by making a list as you read the chapter or by highlighting compounds and their uses so that they are easy to review. All four categories of problems are covered by numerous end-of-chapter exercises to give you practice.

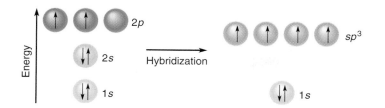

➤ 11.3 Bonding Characteristics and Isomerism

There are two major reasons for the astonishing number of organic compounds: the bonding characteristics of carbon atoms, and the isomerism of carbon-containing molecules. As a group IVA(14) element, a carbon atom has four valence electrons. Two of these outermost-shell electrons are in an s orbital, and two are in p orbitals (see Section 3.4):

With only two unpaired electrons, we might predict that carbon would form just two covalent bonds with other atoms. Yet, we know from the formula of methane (CH_4) that carbon forms four bonds.

Linus Pauling (1901–1994), winner of the Nobel Prize in chemistry (1954) and Nobel Peace Prize (1963), developed a useful model to explain the bonding characteristics of carbon. Pauling found that a mathematical mixing of the $2s$ and three $2p$ orbitals could produce four new, equivalent orbitals (see ■ Figure 11.3). Each of these **hybrid orbitals** has the same energy and is designated sp^3. An sp^3 orbital has a two-lobed shape, similar to the shape of a p orbital but with different-sized lobes (see ■ Figure 11.4). Each of the four sp^3 hybrid orbitals contains a single unpaired electron available for covalent bond formation. Thus, carbon forms four bonds.

Each carbon–hydrogen bond in methane arises from an overlap of a C (sp^3) and an H ($1s$) orbital. The sharing of two electrons in this overlap region creates a sigma (σ) bond. The four equivalent sp^3 orbitals point toward the corners of a regular tetrahedron (see ■ Figure 11.5).

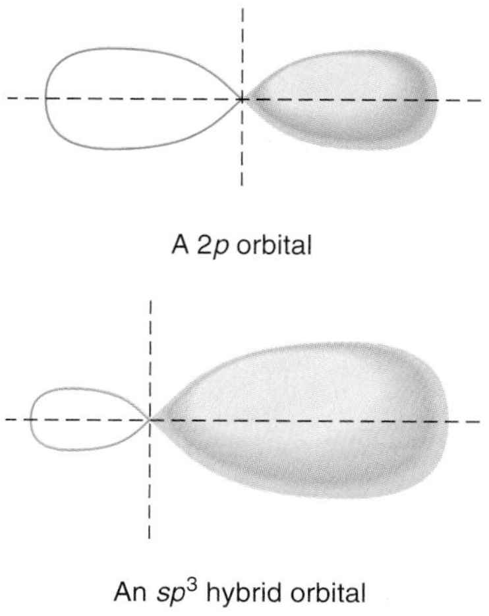

Hybrid orbital
An orbital produced from the combination of two or more nonequivalent orbitals of an atom.

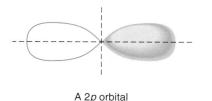

A 2p orbital

➤ **FIGURE 11.4** A comparison of unhybridized p and sp^3 hybridized orbital shapes. The atomic nucleus is at the junction of the lobes in each case.

An sp^3 hybrid orbital

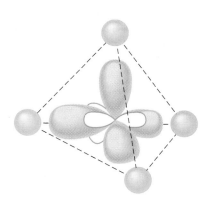

➤ **FIGURE 11.5** Directional characteristics of sp^3 hybrid orbitals of carbon and the formation of C—H bonds in methane (CH_4). The hybrid orbitals point toward the corners of a regular tetrahedron. Hydrogen $1s$ orbitals are illustrated in position to form bonds by overlap with the major lobes of the hybrid orbitals.

Carbon atoms also have the ability to bond covalently to other carbon atoms to form chains and networks. This means that two carbon atoms can join by sharing two electrons to form a single covalent bond:

$$\cdot \overset{}{\underset{}{C}} \cdot \; + \; \cdot \overset{}{\underset{}{C}} \cdot \; \longrightarrow \; \cdot \overset{}{\underset{}{C}} \text{:} \overset{}{\underset{}{C}} \cdot \qquad \text{or} \qquad \cdot \overset{}{\underset{}{C}} - \overset{}{\underset{}{C}} \cdot \qquad (11.2)$$

A third carbon atom can join the end of this chain:

$$\cdot \overset{}{\underset{}{C}} - \overset{}{\underset{}{C}} \cdot \; + \; \cdot \overset{}{\underset{}{C}} \cdot \; \longrightarrow \; \cdot \overset{}{\underset{}{C}} - \overset{}{\underset{}{C}} - \overset{}{\underset{}{C}} \cdot \qquad (11.3)$$

This process can continue and form carbon chains of almost any length, such as

$$\cdot \overset{}{\underset{}{C}} - \overset{}{\underset{}{C}} - \overset{}{\underset{}{C}} - \overset{}{\underset{}{C}} - \overset{}{\underset{}{C}} - \overset{}{\underset{}{C}} - \overset{}{\underset{}{C}} - \overset{}{\underset{}{C}} - \overset{}{\underset{}{C}} \cdot$$

The electrons not involved in forming the chain can be shared with electrons of other carbon atoms (to form chain branches) or with electrons of other elements such as hydrogen, oxygen, or nitrogen. Carbon atoms may also share more than one pair of electrons to form multiple bonds:

$$\cdot \overset{}{\underset{}{C}} - \overset{}{\underset{}{C}} = \overset{}{\underset{}{C}} - \overset{}{\underset{}{C}} \cdot \qquad\qquad \cdot \overset{}{\underset{}{C}} - C \equiv C - \overset{}{\underset{}{C}} \cdot$$

Chain with
double bond

Chain with
triple bond

In principle, there is no limit to the number of carbon atoms that can bond covalently. Thus, organic molecules range from the simple molecules such as methane (CH_4) to very complicated molecules containing over a million carbon atoms.

The variety of possible carbon atom arrangements is even more important than the size range of the resulting molecules. The carbon atoms in all but the very simplest organic molecules can bond in more than one arrangement, giving rise to different compounds with different structures and properties. This property, called **isomerism**, is characterized by compounds that have identical molecular formulas but different arrangements of atoms. One type of isomerism is characterized by compounds called **structural isomers.** Other types of isomerism are covered in Chapters 12 and 17.

Isomerism
A property in which two or more compounds have the same molecular formula but different arrangements of atoms.

Structural isomers
Compounds that have the same molecular formula but in which the atoms bond in different patterns.

Example 11.1

Use the usual rules for covalent bonding to show that a compound with the molecular formula C_2H_6O demonstrates the property of isomerism. Draw formulas for the isomers, showing all covalent bonds.

Solution

Carbon forms four covalent bonds by sharing its four valence-shell electrons. Similarly, oxygen should form two covalent bonds, and hydrogen a single bond. On the basis of these bonding relationships, two structural isomers are possible:

ethyl alcohol   dimethyl ether

Which one of the structures below represents a structural isomer of

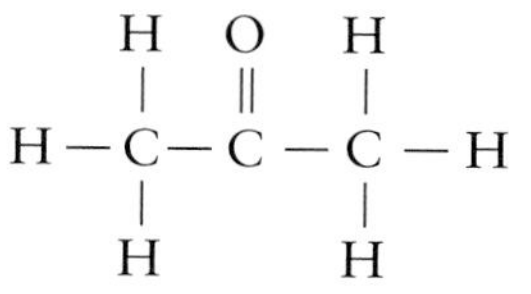

$$
\begin{array}{c}
\quad\; H \quad O \quad H \\
\quad\; | \quad\;\; || \quad\; | \\
H - C - C - C - H \quad ? \\
\quad\; | \qquad\quad | \\
\quad\; H \qquad\quad H
\end{array}
$$

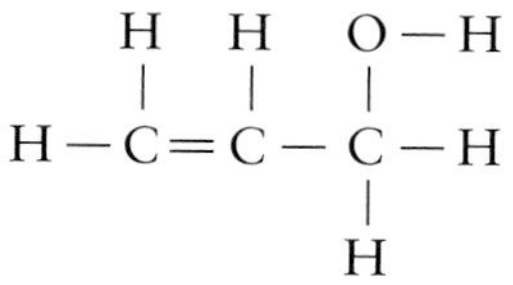

a.
$$
\begin{array}{c}
H \quad H \quad O-H \\
| \quad\; | \quad\; | \\
H - C = C - C - H \\
\qquad\qquad | \\
\qquad\qquad H
\end{array}
$$

c.
$$
\begin{array}{c}
H \quad H \quad O \\
| \quad\; | \quad\; || \\
H - C - C - C - O - H \\
| \quad\; | \\
H \quad H
\end{array}
$$

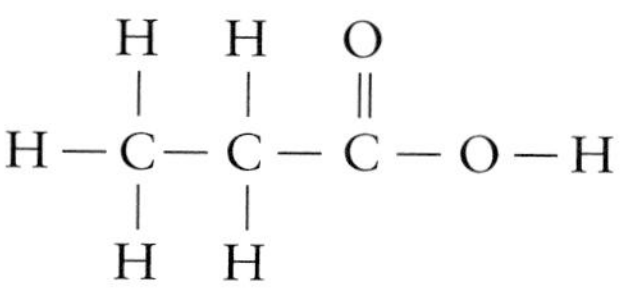

b.
$$
\begin{array}{c}
H \quad H \quad O \\
| \quad\; | \quad\; || \\
H - C = C - C - H
\end{array}
$$

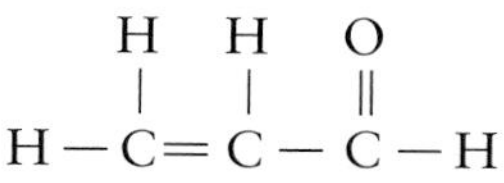

The two isomers of Example 11.1 are quite different. Ethyl alcohol (grain alcohol) is a liquid at room temperature, whereas dimethyl ether is a gas. Thus, the structural differences exert a significant influence on properties. From this example, we can see that molecular formulas such as C_2H_6O provide much less information about a compound than do structural formulas. ■ Figure 11.6 shows ball-and-stick models of these two molecules.

As the number of carbon atoms in the molecular formula increases, the number of possible isomers increases dramatically. For example, 366,319 different isomers are possible for a molecular formula of $C_{20}H_{42}$. No one has prepared all these isomers or even drawn their structural formulas, but the number helps us understand why so many organic compounds have been either isolated from natural sources or synthesized.

➤ 11.4 Functional Groups: The Organization of Organic Chemistry

GOB
Chemistry ⚛ Now™
Go to GOB Now and click to learn how to identify organic functional groups.

Because of the enormous number of possible compounds, the study of organic chemistry might appear to be hopelessly difficult. However, the arrangement of organic compounds into a relatively small number of classes can simplify the study a great deal. This organization is done on the basis of characteristic structural features called **functional groups**. For example, compounds with a carbon–carbon double bond

Functional group
A unique reactive combination of atoms that differentiates molecules of organic compounds of one class from those of another.

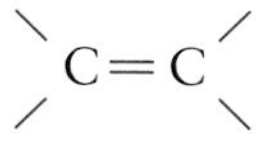

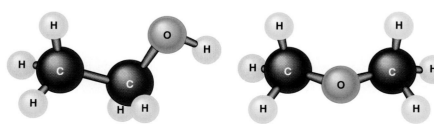

ethyl alcohol   dimethyl ether

➤ **FIGURE 11.6** Ball-and-stick models of the isomers of C_2H_6O. Ethyl alcohol is a liquid at room temperature and completely soluble in water, whereas dimethyl ether is a gas at room temperature and only partially soluble in water.

TABLE 11.2 Classes and functional groups of organic compounds

Class	Functional group	Example of expanded structural formula	Example of condensed structural formula	Common name
Alkane	None	$H-\underset{\underset{H}{\mid}}{\overset{\overset{H}{\mid}}{C}}-\underset{\underset{H}{\mid}}{\overset{\overset{H}{\mid}}{C}}-H$	CH_3CH_3	ethane
Alkene	$\diagup C = C \diagdown$	$\underset{H}{\overset{H}{\diagdown}} C = C \underset{H}{\overset{H}{\diagup}}$	$H_2C = CH_2$	ethylene
Alkyne	$-C \equiv C-$	$H - C \equiv C - H$	$HC \equiv CH$	acetylene
Aromatic				benzene
Alcohol	$-\overset{\mid}{\underset{\mid}{C}}-O-H$	$H-\underset{\underset{H}{\mid}}{\overset{\overset{H}{\mid}}{C}}-\underset{\underset{H}{\mid}}{\overset{\overset{H}{\mid}}{C}}-O-H$	CH_3CH_2-OH	ethyl alcohol
Ether	$-\overset{\mid}{\underset{\mid}{C}}-O-\overset{\mid}{\underset{\mid}{C}}-$	$H-\underset{\underset{H}{\mid}}{\overset{\overset{H}{\mid}}{C}}-O-\underset{\underset{H}{\mid}}{\overset{\overset{H}{\mid}}{C}}-H$	CH_3-O-CH_3	dimethyl ether
Amine	$-\overset{\overset{H}{\mid}}{N}-H$	$H-\underset{\underset{H}{\mid}}{\overset{\overset{H}{\mid}}{C}}-\overset{\overset{H}{\mid}}{N}-H$	CH_3-NH_2	methylamine
Aldehyde	$-\overset{\overset{O}{\parallel}}{C}-H$	$H-\underset{\underset{H}{\mid}}{\overset{\overset{H}{\mid}}{C}}-\overset{\overset{O}{\parallel}}{C}-H$	$CH_3-\overset{\overset{O}{\parallel}}{C}-H$	acetaldehyde
Ketone	$-\overset{\mid}{\underset{\mid}{C}}-\overset{\overset{O}{\parallel}}{C}-\overset{\mid}{\underset{\mid}{C}}-$	$H-\underset{\underset{H}{\mid}}{\overset{\overset{H}{\mid}}{C}}-\overset{\overset{O}{\parallel}}{C}-\underset{\underset{H}{\mid}}{\overset{\overset{H}{\mid}}{C}}-H$	$CH_3-\overset{\overset{O}{\parallel}}{C}-CH_3$	acetone
Carboxylic acid	$-\overset{\overset{O}{\parallel}}{C}-O-H$	$H-\underset{\underset{H}{\mid}}{\overset{\overset{H}{\mid}}{C}}-\overset{\overset{O}{\parallel}}{C}-O-H$	$CH_3-\overset{\overset{O}{\parallel}}{C}-OH$	acetic acid
Ester	$-\overset{\overset{O}{\parallel}}{C}-O-\overset{\mid}{\underset{\mid}{C}}-$	$H-\underset{\underset{H}{\mid}}{\overset{\overset{H}{\mid}}{C}}-\overset{\overset{O}{\parallel}}{C}-O-\underset{\underset{H}{\mid}}{\overset{\overset{H}{\mid}}{C}}-H$	$CH_3-\overset{\overset{O}{\parallel}}{C}-O-CH_3$	methyl acetate
Amide	$-\overset{\overset{O}{\parallel}}{C}-\overset{\overset{H}{\mid}}{N}-H$	$H-\underset{\underset{H}{\mid}}{\overset{\overset{H}{\mid}}{C}}-\overset{\overset{O}{\parallel}}{C}-\overset{\overset{H}{\mid}}{N}-H$	$CH_3-\overset{\overset{O}{\parallel}}{C}-NH_2$	acetamide

are classified as alkenes. The major classes and functional groups are given in ■ Table 11.2. Notice that each functional group in Table 11.2 (except for alkanes) contains a multiple bond or at least one oxygen or nitrogen atom.

In Table 11.2, we have used both expanded and condensed structural formulas for the compounds. **Expanded structural formulas** show all covalent bonds, whereas **condensed structural formulas** show only specific bonds. You should become familiar with both types, but especially with condensed formulas because they will be used often.

Expanded structural formula
A structural molecular formula showing all the covalent bonds.

Condensed structural formula
A structural molecular formula showing the general arrangement of atoms but without showing all the covalent bonds.

Example 11.2

Write a condensed structural formula for each of the following compounds:

a.
$$
\begin{array}{cccc}
\text{H} & \text{H} & \text{H} & \text{H} \\
| & | & | & | \\
\text{H}-\text{C}-\text{C}-\text{C}-\text{C}-\text{H} \\
| & | & | & | \\
\text{H} & \text{H} & \text{H} & \text{H}
\end{array}
$$

b.
$$
\begin{array}{c}
\text{H} \\
| \\
\text{H}-\text{C}-\text{H} \\
\end{array}
$$

$$
\text{H}-\text{C}\text{——}\text{C}\text{——}\text{C}-\text{C}-\text{C}-\text{H}
$$

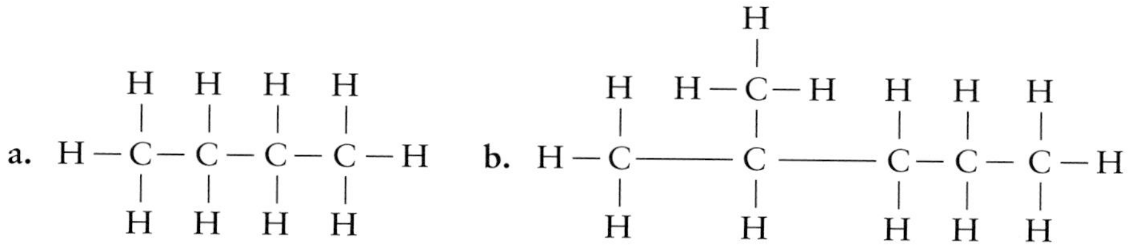

Solution

a. Usually the hydrogens belonging to a carbon are grouped to the right. Thus, the group

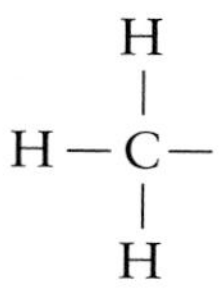

$$
\begin{array}{c}
\text{H} \\
| \\
\text{H}-\text{C}- \\
| \\
\text{H}
\end{array}
$$

condenses to CH_3—, and

$$
\begin{array}{c}
\text{H} \\
| \\
-\text{C}- \\
| \\
\text{H}
\end{array}
$$

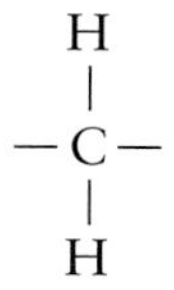

condenses to $-CH_2-$. Thus, the formula condenses to

$$CH_3-CH_2-CH_2-CH_3$$

Other acceptable condensations are

$$CH_3CH_2CH_2CH_3 \quad \text{and} \quad CH_3(CH_2)_2CH_3$$

Parentheses are used here to denote a series of two $-CH_2-$ groups.

b. The group

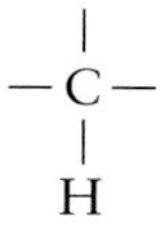

$$
\begin{array}{c}
| \\
-\text{C}- \\
| \\
\text{H}
\end{array}
$$

condenses to $-\overset{|}{\text{C}}\text{H}-$. The condensed formula is therefore

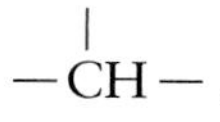

$$
\begin{array}{c}
\text{CH}_3 \\
| \\
\text{CH}_3-\text{CH}-\text{CH}_2-\text{CH}_2-\text{CH}_3
\end{array}
\quad \text{or} \quad
\begin{array}{c}
\text{CH}_3 \\
| \\
\text{CH}_3\text{CHCH}_2\text{CH}_2\text{CH}_3
\end{array}
$$

or

$$(CH_3)_2CH(CH_2)_2CH_3$$

In the last form, the first set of parentheses indicates two identical CH_3 groups attached to the same carbon atom, and the second set of parentheses denotes two CH_2 groups.

Learning Check 11.4   Write a condensed structural formula for each of the following compounds. Retain the bonds to and within the functional groups.

a.
$$\begin{array}{ccccc} & H & OH & H & H & H \\ & | & | & | & | & | \\ H- & C- & C- & C- & C- & C-H \\ & | & | & | & | & | \\ & H & H & H & H & H \end{array}$$

b.
$$\begin{array}{cccc} & H & H & H\ O \\ & | & | & | \ \| \\ H- & C- & C- & C-C-OH \\ & | & | & | \\ & H & H-C-H & H \\ & & | \\ & & H \end{array}$$

CHEMISTRY AND YOUR HEALTH • 11.1

ARE ORGANIC FOODS BETTER FOR YOU?

The answer to the question posed by the title is mixed. The United States Department of Agriculture (USDA) makes no claim that organically produced food is safer or more nutritious than conventionally produced food. Experts in the organic food producing industry seem to agree with the USDA. They point out that the word *organic* refers only to a method of food production. In December 2001, the USDA standardized the way the word *organic* can be used in food labeling. According to the USDA, a food product labeled *100% organic* must contain only ingredients that meet the following requirements. No genetic engineering, ionizing radiation, sewage-sludge fertilizer, or synthesized antibiotics, pesticides, hormones, or fertilizers can be used in their production. In order for a food product to be labeled *95% organic,* at least 95% of the ingredients must meet this definition, and the label *made with organic ingredients* can only be used on food products that contain a minimum of 70% organic ingredients.

Consumers of organic foods have a different answer to the question. Some say they use organic foods because of a combination of environmental and personal health concerns, while a larger number use good flavor as their primary reason. Those who are concerned about their health feel that organic products are better for them because no pesticides, growth hormones, antibiotics, or other synthesized chemicals were used in the food production and so cannot remain in the food as a residue.

While it is true that conventionally produced food may contain residues of such things as pesticides that are known to be toxic in high doses, there is no scientific evidence that they cause health problems when ingested in the quantities found on conventional food products. Some researchers feel that the concern over pesticide residues is misplaced because food-borne bacteria are a much greater health hazard than pesticide residues, and organic farming techniques that use no antibiotics are more likely to produce food carrying disease-causing organisms than are conventional techniques.

There is supporting evidence for those who say organic foods taste better. Organically grown fruits and vegetables are allowed to ripen naturally on the tree or vine, a practice generally recognized to improve flavor over produce that is picked green and ripened artificially. Also, such produce must be transported to market quickly to avoid spoiling, and so tends to be fresher when consumed. Proponents of organic food also point out that the lack of pesticide and antibiotic use in organically grown foods helps slow down the development of resistant strains of bacteria, weeds, and insects. One characteristic of organic foods on which everyone agrees is that they are generally more expensive than conventional foods.

It appears that the answer to the original question about organic foods versus conventional foods is going to continue to be based on who is answering, but it is important to note that all foods have to meet the same USDA standards of safety and quality. As a result, all consumers can be confident that they are benefitting from a safe, high-quality food supply.

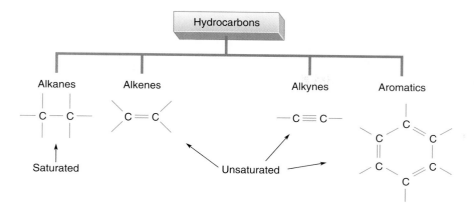

> **FIGURE 11.7** Classification of hydrocarbons.

➤ 11.5 Alkane Structures

Hydrocarbons, the simplest of all organic compounds, contain only two elements, carbon and hydrogen. **Saturated hydrocarbons** or **alkanes** are organic compounds in which carbon is bonded to four other atoms by single bonds; there are no double or triple bonds in the molecule. Unsaturated hydrocarbons, studied later, are called alkenes, alkynes, and aromatics and contain double bonds, triple bonds, or six-carbon rings, as shown in ■ Figure 11.7.

Most life processes are based on the reactions of functional groups. Since alkanes have no functional group, they are not abundant in the human body. However, most compounds in human cells contain parts consisting solely of carbon and hydrogen that behave very much like hydrocarbons. Thus, to understand the chemical properties of the more complex biomolecules, it is useful to have some understanding of the structure, physical properties, and chemical behavior of hydrocarbons.

Another important reason for becoming familiar with the characteristics of hydrocarbons is the crucial role they play in modern industrial society. We use naturally occurring hydrocarbons as primary sources of energy and as important sources of raw materials for the manufacture of plastics, synthetic fibers, drugs, and hundreds of other compounds used daily (see ■ Figure 11.8).

Alkanes can be represented by the general formula C_nH_{2n+2}, where n is the number of carbon atoms in the molecule. The simplest alkane, methane, contains one carbon atom and therefore has the molecular formula CH_4. The carbon atom is at the center, and the four bonds of the carbon atom are directed toward the hydrogen atoms at the corners of a regular tetrahedron; each hydrogen atom is geometrically equivalent to the other three in the molecule (see ■ Figure 11.9). A tetrahedral orientation of bonds with bond angles

Hydrocarbon
An organic compound that contains only carbon and hydrogen.

Saturated hydrocarbon
Another name for an alkane.

Alkane
A hydrocarbon that contains only single bonds.

> **FIGURE 11.8** This racquet has the strength and light weight of aluminum yet is made from graphite (carbon) fibers reinforced with plastic. What other sports equipment is made from graphite fibers?

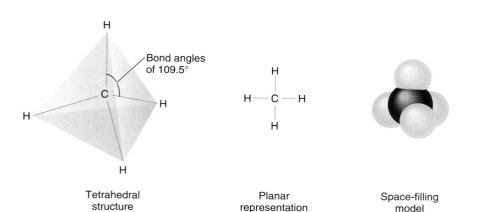

Tetrahedral structure — Planar representation — Space-filling model

> **FIGURE 11.9** Structural representations of methane, CH_4.

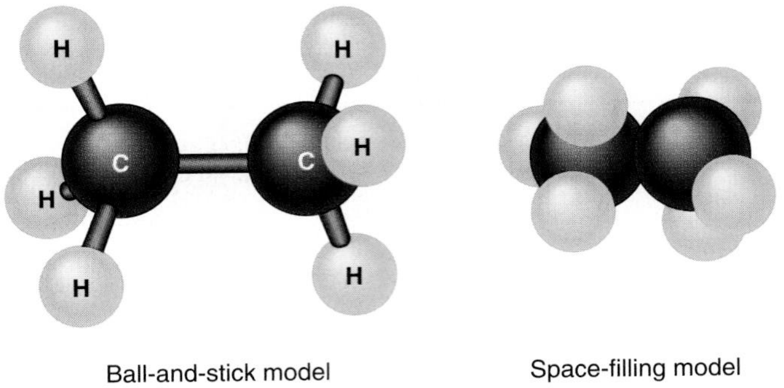

➤ **FIGURE 11.10** Perspective models of the ethane molecule CH_3CH_3.

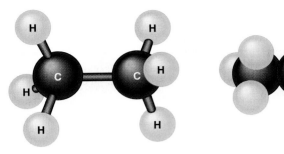

Ball-and-stick model   Space-filling model

of 109.5° is typical for carbon atoms that form four single bonds. Methane is the primary compound in natural gas. Tremendous quantities of natural gas are consumed worldwide because methane is an efficient, clean-burning fuel. It is used to heat homes, cook food, and power factories.

The next alkane is ethane, which has the molecular formula C_2H_6 and the structural formula $CH_3—CH_3$. This molecule may be thought of as a methane molecule with one hydrogen removed and a $—CH_3$ put in its place. Again, the carbon bonds have a tetrahedral geometry as shown in ■ Figure 11.10. Ethane is a minor component of natural gas.

Propane, the third alkane, has the molecular formula of C_3H_8 and the structural formula $CH_3—CH_2—CH_3$. Again, we can produce this molecule by removing a hydrogen atom from the preceding compound (ethane) and substituting a $—CH_3$ in its place (see ■ Figure 11.11). Since all six hydrogen atoms of ethane are equivalent, it makes no difference which one is replaced. Propane is used extensively as an industrial fuel, as well as for home heating and cooking (see ■ Figure 11.12).

The fourth member of the series, butane, with molecular formula C_4H_{10}, can also be produced by removing a hydrogen atom (this time from propane) and adding a $—CH_3$. However, all the hydrogen atoms of propane are not geometrically equivalent, and more than one position is available for substitution. Replacing a hydrogen atom on one of the end carbons of propane with $—CH_3$ produces a butane molecule that has the structural formula $CH_3—CH_2—CH_2—CH_3$. If, however, substitution is made on the central carbon atom of propane, the butane produced is

$$CH_3—\underset{\underset{\displaystyle CH_3}{|}}{CH}—CH_3$$

Notice that both butanes have the same molecular formula, C_4H_{10}. These two possible butanes are structural isomers because they have the same molecular

➤ **FIGURE 11.11** Perspective models of the propane molecule $CH_3CH_2CH_3$.

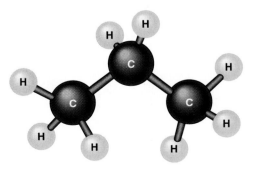

Ball-and-stick model   Space-filling model

➤ FIGURE 11.12 Propane is a common fuel for gas grills.

formulas, but they have different atom-to-atom bonding sequences. The straight-chain isomer is called a **normal alkane** and the other is a **branched alkane** (■ Figure 11.13).

The number of possible structural isomers increases dramatically with the number of carbon atoms in an alkane, as shown in ■ Table 11.3.

Normal alkane
Any alkane in which all the carbon atoms are aligned in a continuous chain.

Branched alkane
An alkane in which at least one carbon atom is not part of a continuous chain.

a. Determine the molecular formula of the alkane containing eight carbon atoms.
b. Draw the condensed structural formula of the normal isomer of the compound in part a.

Learning Check 11.5

➤ 11.6 Conformations of Alkanes

Remember that planar representations such as $CH_3—CH_2—CH_3$ or

$$
\begin{array}{ccc}
H & H & H \\
| & | & | \\
H—C—C—C—H \\
| & | & | \\
H & H & H
\end{array}
$$

are given with no attempt to accurately portray correct bond angles or molecular geometries. Structural formulas are usually written horizontally simply because it is convenient. It is also important to know that actual organic

TABLE 11.3 Molecular formulas and possible structural isomers of alkanes

Molecular formula	Number of possible structural isomers
C_4H_{10}	2
C_5H_{12}	3
C_6H_{14}	5
$C_{10}H_{22}$	75
$C_{20}H_{42}$	366,319
$C_{30}H_{62}$	4,111,846,763

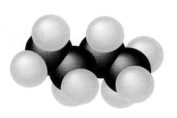

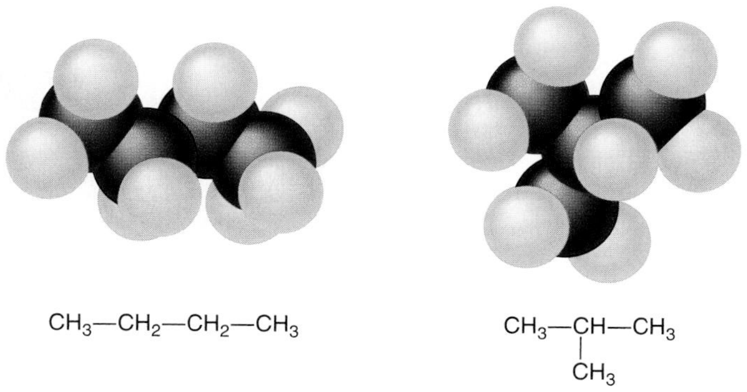

$CH_3—CH_2—CH_2—CH_3$

n-butane

$CH_3—CH—CH_3$
$\quad\quad\quad |$
$\quad\quad CH_3$

isobutane

➤ **FIGURE 11.13** Space-filling models of the isomeric butanes.

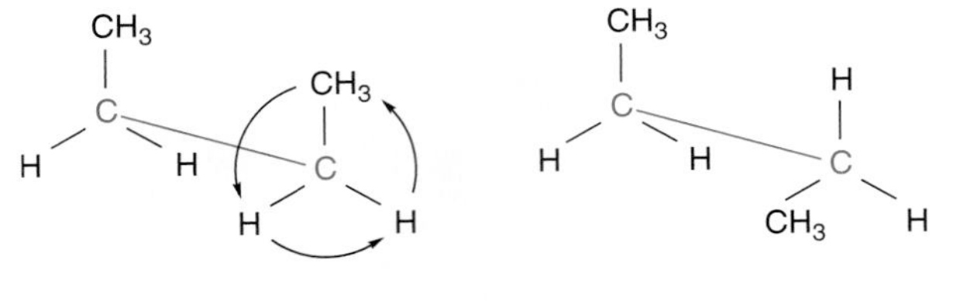

➤ **FIGURE 11.14** Rotation about single bonds (*n*-butane molecule).

molecules are in constant motion—twisting, turning, vibrating, and bending. Groups connected by a single bond are capable of rotating about that bond much like a wheel rotates around an axle (see ■ Figure 11.14). As a result of such rotation about single bonds, a molecule can exist in many different orientations, called **conformations**. In a sample of butane containing billions of identical molecules, there are countless conformations present at any instant, and each conformation is rapidly changing into another. Two of the possible conformations of butane are shown in ■ Figure 11.15. We must be sure to recognize that these different conformations do not represent different structural isomers. In each case, the four carbon atoms are bonded in a continuous (unbranched) chain. Since the order of bonding is not changed, the conformations correspond to the same molecule. Two structures would be structural isomers only if bonds had to be broken and remade to convert one into the other.

Conformations
The different arrangements of atoms in space achieved by rotation about single bonds.

Example 11.3

Which of the following pairs are structural isomers, and which are simply different representations of the same molecule?

a. $CH_3-CH_2-CH_2-CH_3$

$$CH_2-CH_2$$ with CH_3 above on CH_2 and CH_3 below on CH_2

b. $CH_3-\overset{\overset{\displaystyle CH_3}{|}}{CH}-CH_2-CH_3$ $CH_3-CH_2-\overset{\overset{\displaystyle CH_3}{|}}{CH}-CH_3$

c. $CH_3-\overset{\overset{\displaystyle CH_3}{|}}{CH}-CH_2-CH_2-CH_3$ $CH_3-CH_2-\overset{\overset{\displaystyle CH_3}{|}}{CH}-CH_2-CH_3$

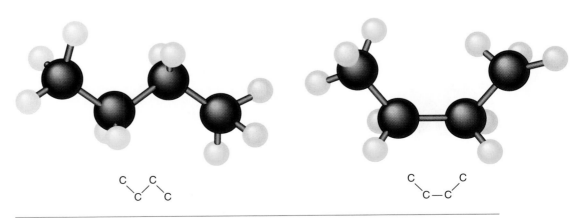

➤ **FIGURE 11.15** Perspective models and carbon skeletons of two conformations of *n*-butane.

Solution

a. Same molecule: In both molecules, the four carbons are bonded in a continuous chain.

$$CH_3 \overset{1}{} - CH_2 \overset{2}{} - CH_2 \overset{3}{} - CH_3 \overset{4}{} \qquad \overset{4}{C}H_3$$

$$\overset{2}{C}H_2 - \overset{}{C}H_2 \overset{3}{}$$

$$\overset{1}{C}H_3$$

b. Same molecule: In both molecules, there is a continuous chain of four carbons with a branch at position 2. The molecule has simply been turned around.

$$\begin{array}{c} CH_3 \\ | \\ CH_3 - CH - CH_2 - CH_3 \\ 1 \quad 2 \quad 3 \quad 4 \end{array} \qquad \begin{array}{c} CH_3 \\ | \\ CH_3 - CH_2 - CH - CH_3 \\ 4 \quad 3 \quad 2 \quad 1 \end{array}$$

c. Structural isomers: Both molecules have a continuous chain of five carbons, but the branch is located at different positions.

$$\begin{array}{c} CH_3 \\ | \\ CH_3 - CH - CH_2 - CH_2 - CH_3 \\ 1 \quad 2 \quad 3 \quad 4 \quad 5 \end{array} \qquad \begin{array}{c} CH_3 \\ | \\ CH_3 - CH_2 - CH - CH_2 - CH_3 \\ 1 \quad 2 \quad 3 \quad 4 \quad 5 \end{array}$$

Which of the following pairs represent structural isomers, and which are simply the same compound?

Learning Check 11.6

a. $CH_3 - CH_2 - CH_2$ $CH_3 - CH_2$
                    | |
              $CH_2 - CH_3$ $CH_2 - CH_2$
                                                 |
                                               CH_3

b. $\begin{array}{c} CH_3 \\ | \\ CH_3 - CH - CH_2 - CH_3 \end{array}$ $\begin{array}{c} CH_2 - CH_3 \\ | \\ CH_3 - CH - CH_3 \end{array}$

c. $\begin{array}{c} CH_3 \\ | \\ CH_3 - CH_2 - CH - CH_2 - CH_3 \end{array}$ $\begin{array}{c} CH_3 \\ | \\ CH_3 - CH_2 - CH_2 - CH - CH_3 \end{array}$

➤ 11.7 Alkane Nomenclature

When only a relatively few organic compounds were known, chemists gave them what are today called trivial or common names, such as methane, ethane, propane, and butane. The names for the larger alkanes were derived from the Greek prefixes that indicate the number of carbon atoms in the molecule. Thus, *pent*ane contains five carbons, *hex*ane has six, *hept*ane has seven, and so forth, as shown in ■ Table 11.4.

As more compounds and isomers were discovered, however, it became increasingly difficult to devise unique names and much more difficult to commit them to memory. Obviously, a systematic method was needed. Such a method is now in use, but so are several common methods.

The names for the two isomeric butanes (*n*-butane and isobutane) illustrate the important features of the common nomenclature system used for alkanes. The stem *but*- indicates that four carbons are present in the molecule.

GOB
Chemistry ⚛ Now ™

Go to GOB Now and click to explore the relationship between the structures of alkanes and their names.

TABLE 11.4	Names of alkanes		
Number of carbon atoms	Name	Molecular formula	Structure of normal isomer
1	methane	CH_4	CH_4
2	ethane	C_2H_6	CH_3CH_3
3	propane	C_3H_8	$CH_3CH_2CH_3$
4	butane	C_4H_{10}	$CH_3CH_2CH_2CH_3$
5	pentane	C_5H_{12}	$CH_3CH_2CH_2CH_2CH_3$
6	hexane	C_6H_{14}	$CH_3CH_2CH_2CH_2CH_2CH_3$
7	heptane	C_7H_{16}	$CH_3CH_2CH_2CH_2CH_2CH_2CH_3$
8	octane	C_8H_{18}	$CH_3CH_2CH_2CH_2CH_2CH_2CH_2CH_3$
9	nonane	C_9H_{20}	$CH_3CH_2CH_2CH_2CH_2CH_2CH_2CH_2CH_3$
10	decane	$C_{10}H_{22}$	$CH_3CH_2CH_2CH_2CH_2CH_2CH_2CH_2CH_2CH_3$

The *-ane* ending signifies the alkane family. The prefix *n-* indicates that all carbons form an unbranched chain. The prefix *iso-* refers to compounds in which all carbons except one are in a continuous chain and in which that one carbon is branched from a next-to-the-end carbon, as shown:

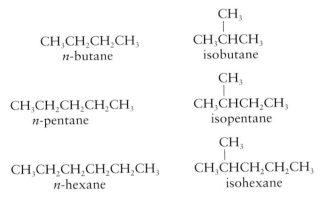

This common naming system has limitations. Pentane has three isomers, and hexane has five. The more complicated the compound, the greater the number of isomers, and the greater the number of special prefixes needed to name all the isomers. It would be extremely difficult and time-consuming to try to identify each of the 75 isomeric alkanes containing 10 carbon atoms by a unique prefix or name.

To devise a system of nomenclature that could be used for even the most complicated compounds, committees of chemists have met periodically since 1892. The system resulting from these meetings is called the IUPAC (International Union of Pure and Applied Chemistry) system. This system is much the same for all classes of organic compounds. The IUPAC name for an organic compound consists of three component parts:

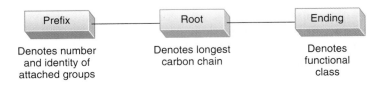

The *root* of the IUPAC name specifies the longest continuous chain of carbon atoms in the compound. The roots for the first 10 normal hydrocarbons are based on the names given in Table 11.4: C_1 *meth-*, C_2 *eth-*, C_3 *prop-*, C_4 *but-*, C_5 *pent-*, C_6 *hex-*, C_7 *hept-*, C_8 *oct-*, C_9 *non-*, C_{10} *dec-*.

The *ending* of an IUPAC name specifies the functional class or the major functional group of the compound. The ending *-ane* specifies an alkane. Each of the other functional classes has a characteristic ending; for example, *-ene* is the ending for alkenes, and the *-ol* ending designates alcohols.

Prefixes are used to specify the identity, number, and location of atoms or groups of atoms that are attached to the longest carbon chain. ■ Table 11.5 lists several common carbon-containing groups referred to as **alkyl groups**. Each alkyl group is a collection of atoms that can be thought of as an alkane minus one hydrogen atom. Alkyl groups are named simply by dropping *-ane* from the name of the corresponding alkane and replacing it with *-yl*. For example, CH_3— is called a methyl group and CH_3—CH_2— an ethyl group:

Alkyl group
A group differing by one hydrogen from an alkane.

CH_4 CH_3—CH_3 CH_3— CH_3—CH_2—
methane   ethane   methyl group   ethyl group

Two different alkyl groups can be derived from propane, depending on which hydrogen is removed. Removal of a hydrogen from an end carbon results in a propyl group:

CH_3—CH_2—CH_3 CH_3—CH_2—CH_2—
propane   propyl group

Removal of a hydrogen from the center carbon results in an isopropyl group:

CH_3—CH_2—CH_3 CH_3—$\overset{|}{CH}$—CH_3
propane   isopropyl group

TABLE 11.5   Common alkyl groups

Parent alkane	Structure of parent alkane	Structure of alkyl group	Name of alkyl group
methane	CH_4	CH_3—	methyl
ethane	CH_3CH_3	CH_3CH_2—	ethyl
propane	$CH_3CH_2CH_3$	$CH_3CH_2CH_2$—	propyl
		$CH_3\overset{|}{CH}CH_3$	isopropyl
n-butane	$CH_3CH_2CH_2CH_3$	$CH_3CH_2CH_2CH_2$—	butyl
		$CH_3CH_2\overset{|}{CH}CH_3$	*sec*-butyl (secondary-butyl)[a]
isobutane	$\overset{\displaystyle CH_3}{\underset{\displaystyle }{CH_3\overset{|}{CH}CH_3}}$	$CH_3\overset{\overset{\textstyle CH_3}{|}}{CH}CH_2$—	isobutyl
		$CH_3\overset{\overset{\textstyle CH_3}{|}}{\underset{|}{C}}CH_3$	*t*-butyl (tertiary-butyl)[a]

[a]For an explanation of *secondary* and *tertiary,* see Section 13.2.

TABLE 11.6	Common nonalkyl groups
Group	**Name**
—F	fluoro
—Cl	chloro
—Br	bromo
—I	iodo
—NO_2	nitro
—NH_2	amino

An isopropyl group also can be represented by $(CH_3)_2CH$—. As shown in Table 11.5, four butyl groups can be derived from butane, two from the straight-chain, or normal, butane, and two from the branched-chain isobutane. A number of nonalkyl groups are also commonly used in naming organic compounds (see ■ Table 11.6).

The following steps are useful when the IUPAC name of an alkane is written on the basis of its structural formula:

Step 1. Name the longest chain. The longest continuous carbon-atom chain is chosen as the basis for the name. The names are those given in Table 11.4.

Example	Longest Chain	Comments
$CH_3 - CH_2 - CH - CH_3$ with CH_3 branch	C — C — C — C with C branch	This compound is a butane.
$CH_3 - CH - CH_2$ with CH_2-CH_3 above and CH_3 below	C — C — C with C — C above and C below	This compound is a pentane.
$CH_3 - CH_2 - C - CH_3$ with CH_2-CH_3 above and CH_2, CH_3-CH_2 below	C — C — C — C with branches	Often there is more than one way to designate the longest chain. Either way, in this case, the compound is a hexane.

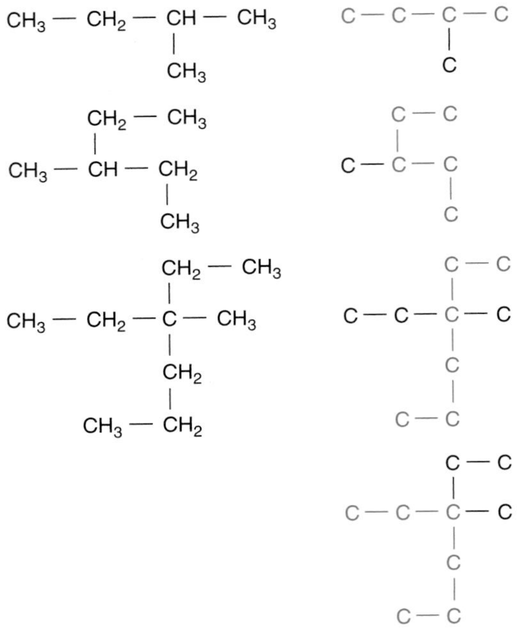

GOB
Chemistry·⚛·Now™
Go to GOB Now and click to learn about the naming of alkane structures.

| Learning Check 11.7 | Identify the longest carbon chain in the following: |

a. 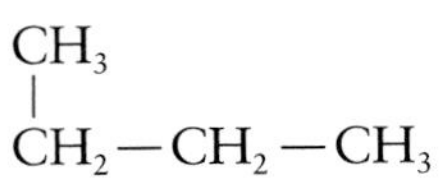 CH_3
    |
    $CH_2 - CH_2 - CH_3$

b. 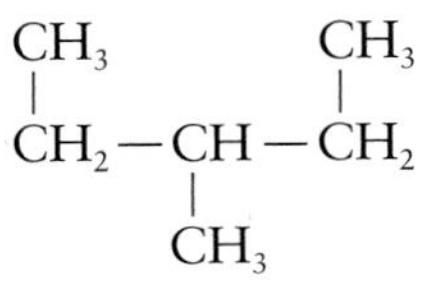 CH_3 CH_3
      | |
      $CH_2 - CH - CH_2$
              |
              CH_3

c. 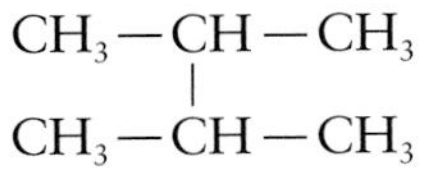 $CH_3 - CH - CH_3$
            |
       $CH_3 - CH - CH_3$

Step 2. Number the longest chain. The carbon atoms in the longest chain are numbered consecutively from the end that will give the lowest possible number to any carbon to which a group is attached.

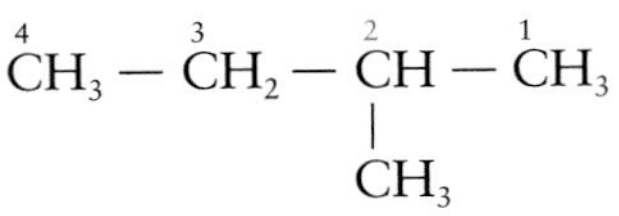

$\overset{4}{C}H_3 - \overset{3}{C}H_2 - \overset{2}{C}H - \overset{1}{C}H_3$ with CH_3 branch   and not $\overset{1}{C}H_3 - \overset{2}{C}H_2 - \overset{3}{C}H - \overset{4}{C}H_3$ with CH_3 branch

$\overset{2}{C}H_2 - \overset{1}{C}H_3$
$\overset{3}{C}H_3 - CH - \overset{4}{C}H_2$ with $\overset{5}{C}H_3$ below   the chain may also be numbered

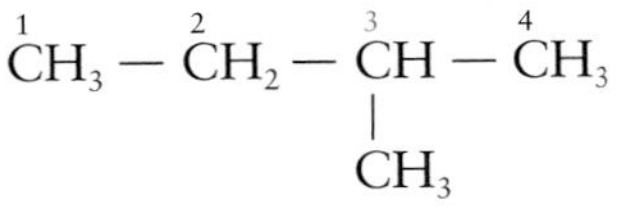

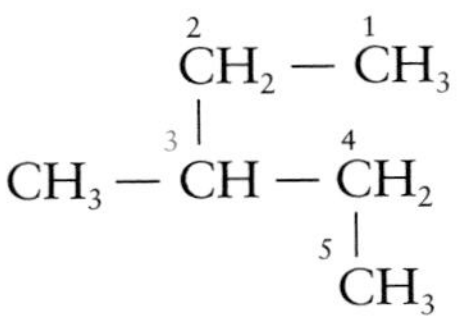

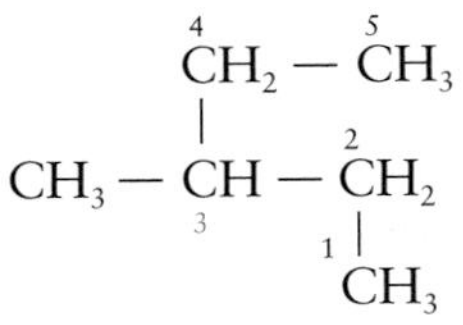

$\overset{4}{C}H_2 - \overset{5}{C}H_3$
$CH_3 - \overset{2}{C}H - \overset{3}{C}H_2$ with $\overset{1}{C}H_3$ below

If two or more alkyl groups are attached to the longest chain and more than one numbering sequence is possible, the chain is numbered to get the lowest series of numbers. An easy way to follow this rule is to number from the end of the chain nearest a branch:

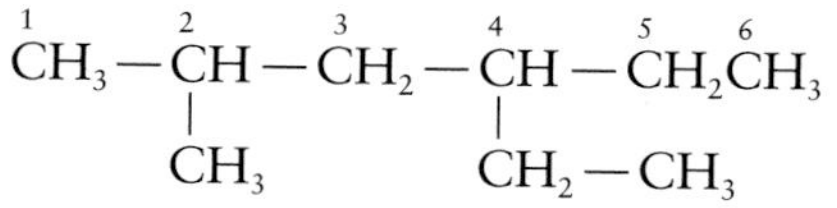

$$\overset{1}{C}H_3-\overset{2}{C}H-\overset{3}{C}H_2-\overset{4}{C}H-\overset{5}{C}H_2\overset{6}{C}H_3$$
$$\qquad\quad |\qquad\qquad\quad |$$
$$\qquad\quad CH_3\qquad\quad CH_2-CH_3$$

and not

$$\overset{6}{C}H_3-\overset{5}{C}H-\overset{4}{C}H_2-\overset{3}{C}H-\overset{2}{C}H_2-\overset{1}{C}H_3$$
$$\qquad\quad |\qquad\qquad\quad |$$
$$\qquad\quad CH_3\qquad\quad CH_2-CH_3$$

Groups are located at positions 2 and 4

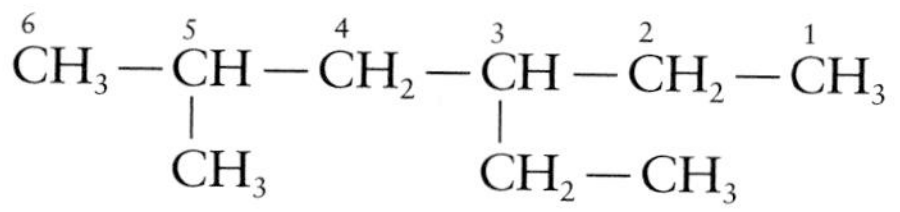

Groups are located at positions 3 and 5

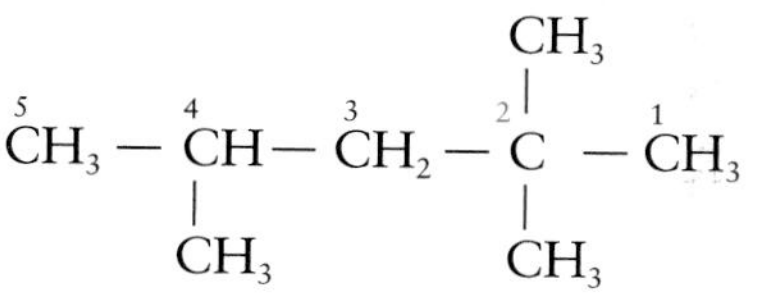

$$\qquad\qquad\qquad CH_3$$
$$\qquad\qquad\qquad |$$
$$\overset{5}{C}H_3-\overset{4}{C}H-\overset{3}{C}H_2-\overset{2}{C}-\overset{1}{C}H_3$$
$$\qquad\quad |\qquad\qquad |$$
$$\qquad\quad CH_3\qquad\quad CH_3$$

and not

$$\qquad\qquad\qquad CH_3$$
$$\qquad\qquad\qquad |$$
$$\overset{1}{C}H_3-\overset{2}{C}H-\overset{3}{C}H_2-\overset{4}{C}-\overset{5}{C}H_3$$
$$\qquad\quad |\qquad\qquad |$$
$$\qquad\quad CH_3\qquad\quad CH_3$$

Groups are located at positions 2,2,4

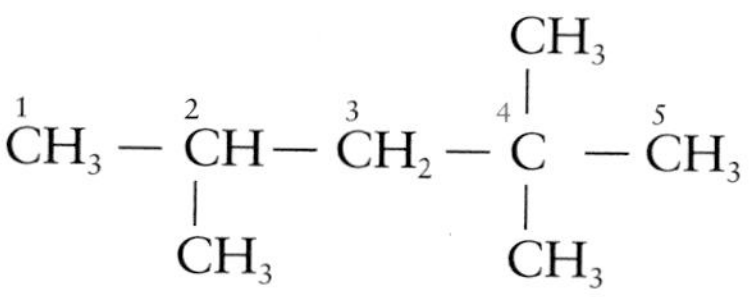

Groups are located at positions 2,4,4

Here, a difference occurs with the second number in the sequence, so positions 2,2,4 is the lowest series and is used rather than positions 2,4,4.

Decide how to correctly number the longest chain in the following according to IUPAC rules:

Learning Check 11.8

a. 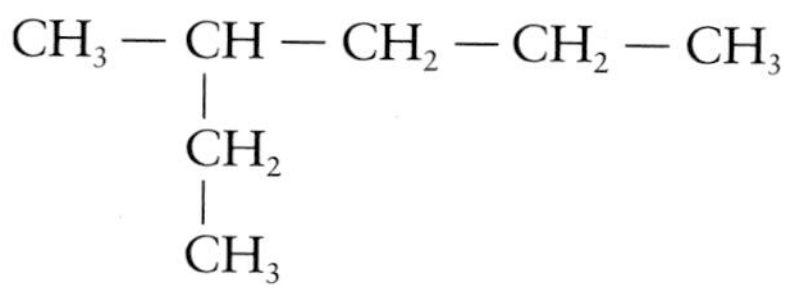 $CH_3-CH-CH_2-CH_2-CH_3$
$\qquad\quad |$
$\qquad\quad CH_2$
$\qquad\quad |$
$\qquad\quad CH_3$

b. 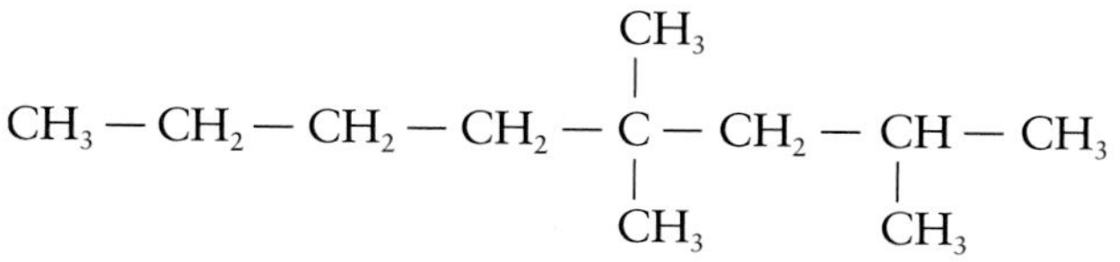 $CH_3-CH_2-CH_2-CH_2-\overset{\overset{\textstyle CH_3}{\textstyle |}}{C}-CH_2-\overset{\overset{\textstyle }{\textstyle |}}{CH}-CH_3$
$\qquad\qquad\qquad\qquad\qquad\quad |\qquad\qquad |$
$\qquad\qquad\qquad\qquad\qquad\quad CH_3\qquad\quad CH_3$

Step 3. Locate and name the attached alkyl groups. Each group is located by the number of the carbon atom to which it is attached on the chain.

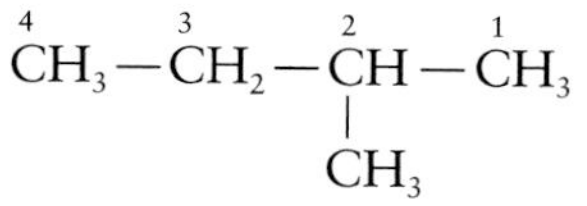

$$\overset{4}{C}H_3-\overset{3}{C}H_2-\overset{2}{C}H-\overset{1}{C}H_3$$
$$\qquad\qquad\quad |$$
$$\qquad\qquad\quad CH_3$$

The attached group is located on carbon 2 of the chain, and it is a methyl group.

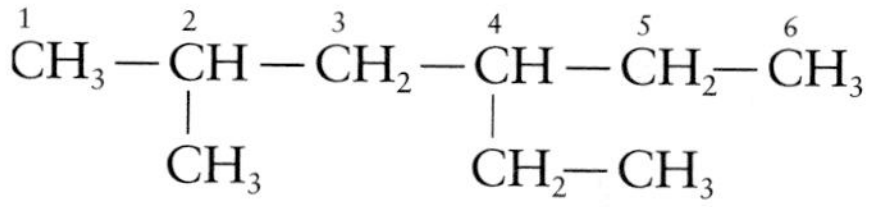

$$\overset{1}{C}H_3-\overset{2}{C}H-\overset{3}{C}H_2-\overset{4}{C}H-\overset{5}{C}H_2\overset{6}{C}H_3$$
$$\qquad\quad |\qquad\qquad\quad |$$
$$\qquad\quad CH_3\qquad\quad CH_2-CH_3$$

The one-carbon group at position 2 is a methyl group. The two-carbon group at position 4 is an ethyl group.

Identify the alkyl groups attached to the dashed line, which symbolizes a long carbon chain. Refer to Table 11.5 if necessary.

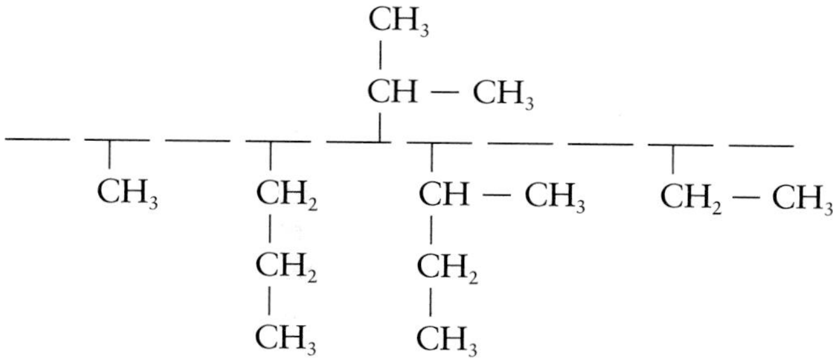

Learning Check 11.9

$$CH_3$$
$$|$$
$$CH-CH_3$$
$$|$$

—T—T——T——T—— —T—

$$CH_3\qquad CH_2\qquad CH-CH_3\qquad CH_2-CH_3$$
$$\qquad\qquad |\qquad\qquad |$$
$$\qquad\qquad CH_2\qquad\quad CH_2$$
$$\qquad\qquad |\qquad\qquad |$$
$$\qquad\qquad CH_3\qquad\quad CH_3$$

Step 4. Combine the longest chain and the branches into the name. The position and the name of the attached alkyl group are added to the name of the longest chain and written as one word:

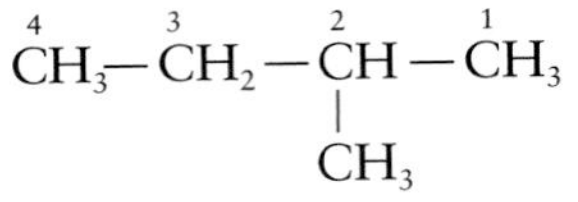

$$\overset{4}{C}H_3-\overset{3}{C}H_2-\overset{2}{C}H-\overset{1}{C}H_3$$
$$|$$
$$CH_3$$

2-methylbutane

Additional steps are needed when more than one alkyl group is attached to the longest chain.

Step 5. Indicate the number and position of attached alkyl groups. If two or more of the same alkyl group occur as branches, the number of them is indicated by the prefixes *di-, tri-, tetra-, penta-*, etc., and the location of each is again indicated by a number. These position numbers, separated by commas, are put just before the name of the group, with hyphens before and after the numbers when necessary:

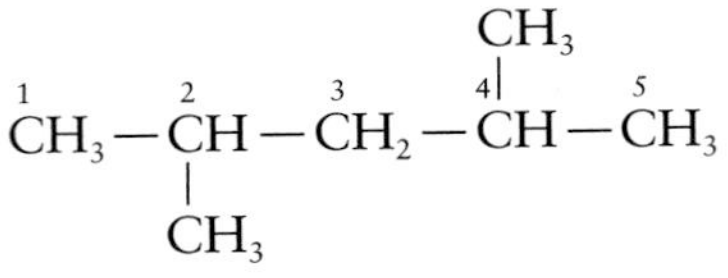

$$\overset{1}{C}H_3-\overset{2}{C}H-\overset{3}{C}H_2-\overset{4}{C}H-\overset{5}{C}H_3$$
$$|\qquad\qquad\quad|$$
$$CH_3\qquad\quad CH_3$$

2,4-dimethylpentane

$$\overset{1}{C}H_3-\overset{2}{C}H_2-\overset{3}{C}-\overset{4}{C}H_2-\overset{5}{C}H_3$$
$$CH_3$$
$$|$$
$$CH_3$$

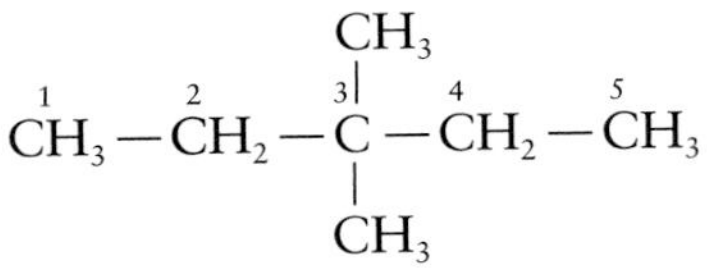

3,3-dimethylpentane

If two or more *different* alkyl groups are present, their names are alphabetized and added to the name of the basic alkane, again as one word. For purposes of alphabetizing, the prefixes *di-, tri-*, and so on are ignored, as are the italicized prefixes secondary *(sec)* and tertiary *(t)*. The prefix *iso-* is an exception and is used for alphabetizing:

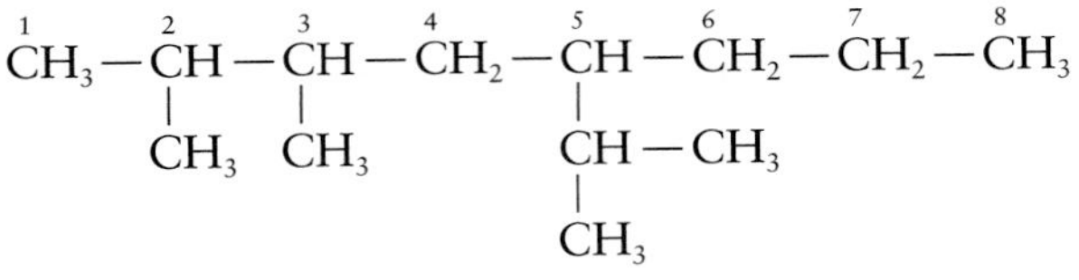

$$\overset{1}{C}H_3-\overset{2}{C}H-\overset{3}{C}H-\overset{4}{C}H_2-\overset{5}{C}H-\overset{6}{C}H_2-\overset{7}{C}H_2-\overset{8}{C}H_3$$
$$|\qquad|\qquad\qquad\qquad|$$
$$CH_3\ \ CH_3\qquad\qquad CH-CH_3$$
$$|$$
$$CH_3$$

5-isopropyl-2,3-dimethyloctane

Learning Check 11.10	Give the correct IUPAC name to each of the following:

a. 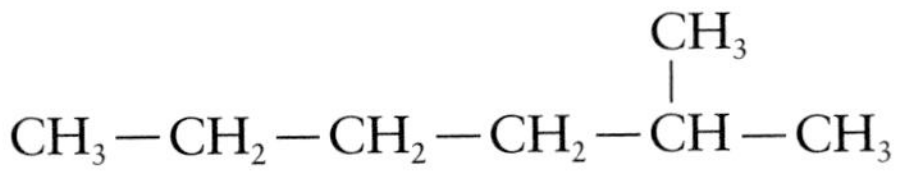
$$CH_3$$
$$|$$
$$CH_3-CH_2-CH_2-CH_2-CH-CH_3$$

b. 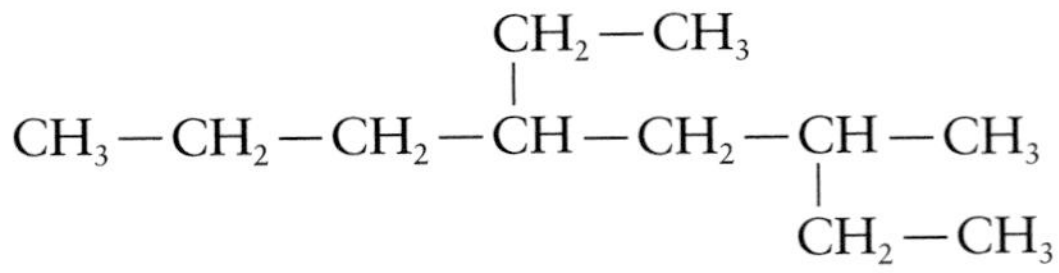
$$CH_2-CH_3$$
$$|$$
$$CH_3-CH_2-CH_2-CH-CH_2-CH-CH_3$$
$$|$$
$$CH_2-CH_3$$

c. 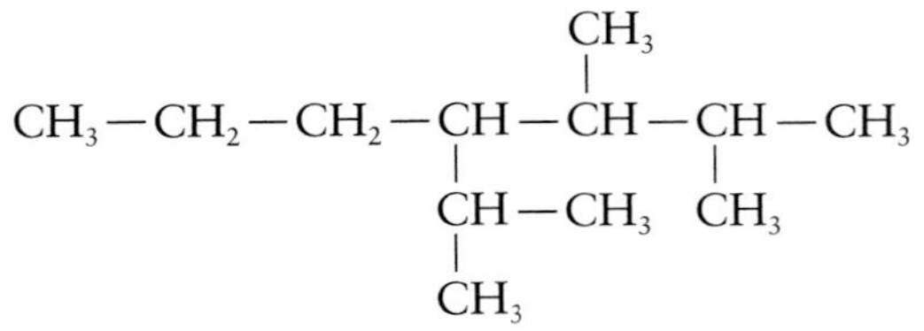
$$CH_3$$
$$|$$
$$CH_3-CH_2-CH_2-CH-CH-CH-CH_3$$
$$|\qquad\quad|$$
$$CH-CH_3\ \ CH_3$$
$$|$$
$$CH_3$$

Naming compounds is a very important skill, as is the reverse process of using IUPAC nomenclature to specify a structural formula. The two processes are very similar. To obtain a formula from a name, determine the longest chain, number the chain, and add any attached groups.

Example 11.4

Draw a condensed structural formula for 3-ethyl-2-methylhexane.

Solution

Use the last part of the name to determine the longest chain. Draw a chain of six carbons. Then, number the carbon atoms.

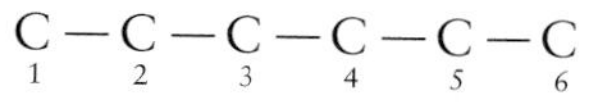

$$\underset{1}{C}-\underset{2}{C}-\underset{3}{C}-\underset{4}{C}-\underset{5}{C}-\underset{6}{C}$$

Attach a methyl group at position 2 and an ethyl group at position 3.

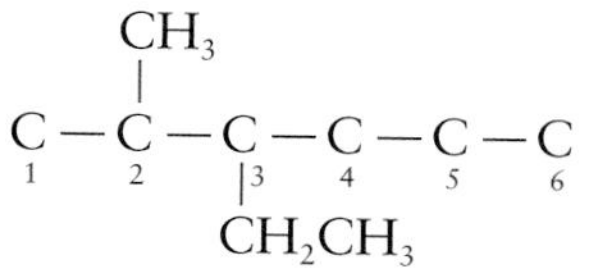

$$\underset{1}{C}-\underset{2}{\overset{\displaystyle CH_3}{\overset{\displaystyle |}{C}}}-\underset{3}{\overset{\displaystyle |}{C}}-\underset{4}{C}-\underset{5}{C}-\underset{6}{C}$$
$$CH_2CH_3$$

Complete the structure by adding enough hydrogen atoms so that each carbon has four bonds.

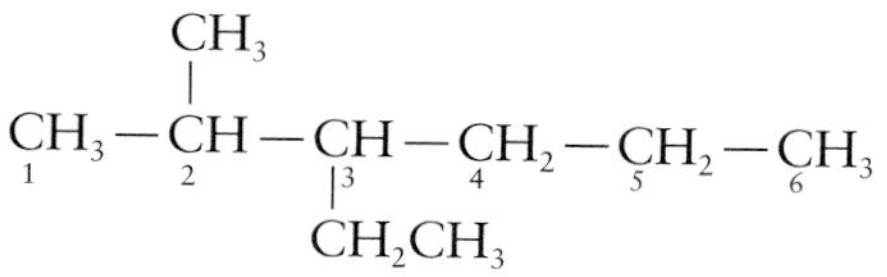

$$\underset{1}{CH_3}-\underset{2}{CH}-\underset{3}{\overset{\displaystyle CH_3}{\overset{\displaystyle |}{CH}}}-\underset{4}{CH_2}-\underset{5}{CH_2}-\underset{6}{CH_3}$$
$$CH_2CH_3$$

Draw a condensed structural formula for each of the following compounds:

Learning Check 11.11

a. 2,2,4-trimethylpentane
b. 3-isopropylhexane
c. 3-ethyl-2,4-dimethylheptane

▶ 11.8 Cycloalkanes

From what we have said so far, the formula C_3H_6 cannot represent an alkane. Not enough hydrogens are present to allow each carbon to form four bonds, unless there are multiple bonds. For example, the structural formula $CH_3-CH=CH_2$ fits the molecular formula but cannot represent an alkane because of the double bond. The C_3H_6 formula does become acceptable for an alkane if the carbon atoms form a ring, or cyclic, structure rather than the open-chain structure shown:

$$C-C-C \qquad \overset{\displaystyle C}{\overset{\displaystyle \diagup\diagdown}{C-C}}$$

Open chain Cyclic

The resulting saturated cyclic compound, called cyclopropane, has the structural formula

$$\overset{\displaystyle CH_2}{\overset{\displaystyle \diagup\diagdown}{CH_2-CH_2}}$$

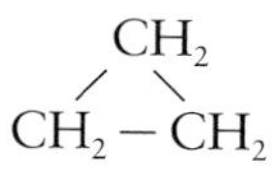

Alkanes containing rings of carbon atoms are called **cycloalkanes.** Like the other alkanes, cycloalkanes are not found in human cells. However, several important molecules in human cells do contain rings of five or six atoms, and the study of cycloalkanes will help you better understand the chemical behavior of these complex molecules.

Cycloalkane
An alkane in which carbon atoms form a ring.

TABLE 11.7   Structural formulas and symbols for common cycloalkanes

Name	Structural formula	Condensed formula
cyclopropane	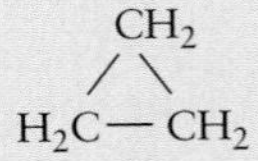 CH_2 / \ H_2C — CH_2	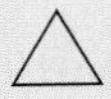 △
cyclobutane	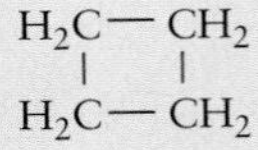 H_2C — CH_2 | | H_2C — CH_2	□
cyclopentane	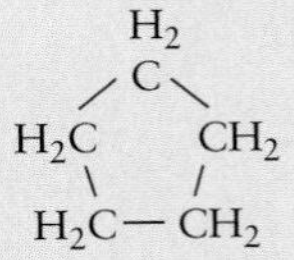 H_2 C H_2C CH_2 \ / H_2C — CH_2	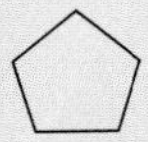 ⬠
cyclohexane	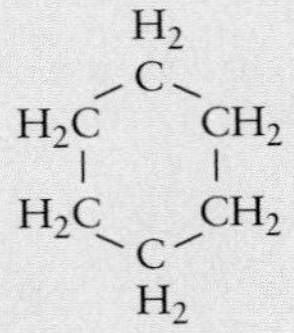 H_2 C H_2C CH_2 | | H_2C CH_2 C H_2	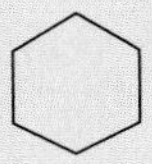 ⬡

According to IUPAC rules, cycloalkanes are named by placing the prefix *cyclo-* before the name of the corresponding alkane with the same number of carbon atoms. Chemists often abbreviate the structural formulas for cycloalkanes and draw them as geometric figures (triangles, squares, etc.) in which each corner represents a carbon atom. The hydrogens are omitted (see ■ Table 11.7). It is important to remember that each carbon atom still possesses four bonds, and that hydrogen is assumed to be bonded to the carbon atoms unless something else is indicated. When substituted cycloalkanes (those with attached groups) are named, the position of a single attached group does not need to be specified in the name because all positions in the ring are equivalent. However, when two or more groups are attached, their positions of attachment are indicated by numbers, just as they were for alkanes. The ring numbering begins with the carbon attached to the first group alphabetically and proceeds around the ring in the direction that will give the lowest numbers for the locations of the other attached groups.

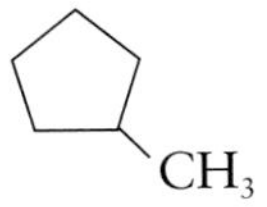

methylcyclopentane

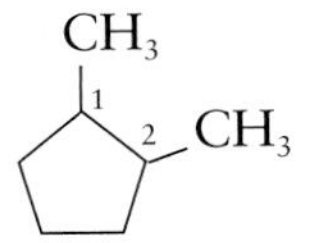

1,2-dimethylcyclopentane
not 1,5-dimethylcyclopentane

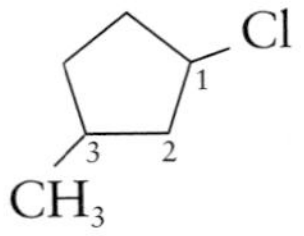

1-chloro-3-methylcyclopentane
not 1-chloro-4-methylcyclopentane
not 3-chloro-1-methylcyclopentane

Example 11.5

Represent each of the following cycloalkanes by a geometric figure, and name each compound:

a.

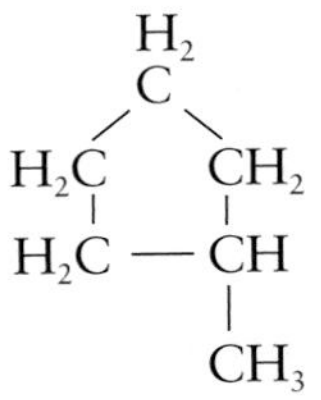

H_2
C
H_2C CH_2
H_2C — CH
CH_3

b.

H_2
C
H_2C CH — CH_3
H_2C CH_2
CH
CH_3

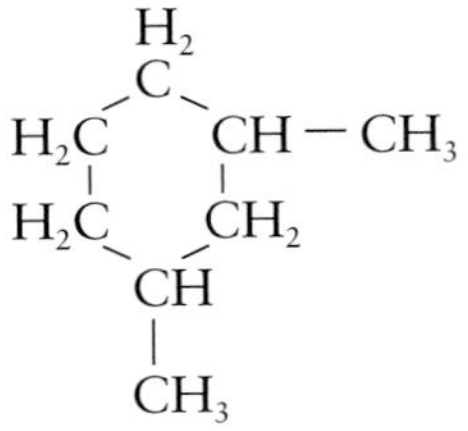

Solution

a. A pentagon represents a five-membered ring, which is called cyclopentane. This compound has a methyl group attached, so the name is methylcyclopentane. The position of a single alkyl group is not indicated by a number because the positions of all carbons in the ring are equivalent.

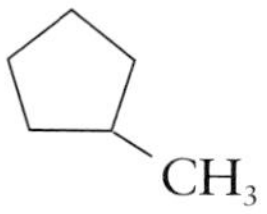

b. A hexagon represents a six-carbon ring, which is called cyclohexane. Two methyl groups are attached; thus we have a dimethylcyclohexane.

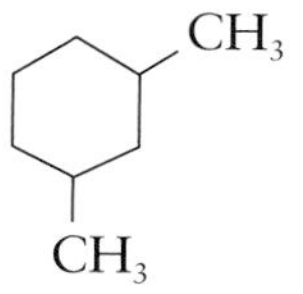

However, the positions of the two alkyl groups must be indicated. The ring is numbered beginning with a carbon to which a methyl group is attached, counting in the direction giving the lowest numbers. The correct name is 1,3-dimethylcyclohexane. Notice that a reverse numbering beginning at the same carbon would have given 1,5-dimethylcyclohexane. The number 3 in the correct name is lower than the 5 in the incorrect name.

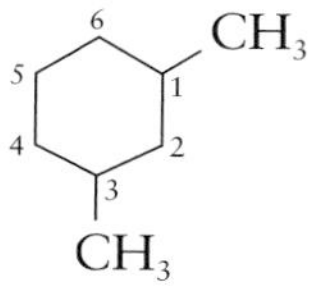

Give each of the following compounds the correct IUPAC name:

a.  CH₃

b. 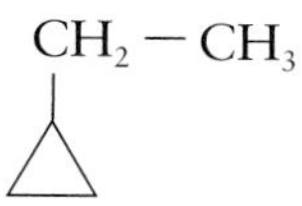 CH₂ — CH₃

c. 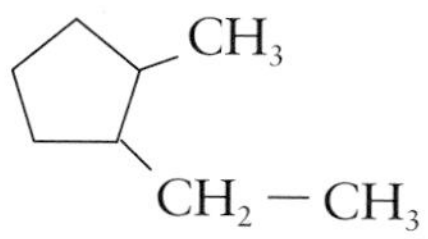 CH₃ / CH₂ — CH₃

➤ 11.9 The Shape of Cycloalkanes

Recall from Section 11.5 that a tetrahedral orientation of bonds with bond angles of 109.5° is characteristic of carbon atoms that form four single bonds. A tetrahedral arrangement is the most stable because it results in the least crowding of the atoms. In certain cycloalkanes, however, a tetrahedral arrangement for all carbon-to-carbon bonds is not possible. For example, the bond angles between adjacent carbon–carbon bonds in planar cyclopropane molecules must be 60° (see ■ Figure 11.16). In cyclobutane, they are close to 90°. As a result, cyclopropane and cyclobutane rings are much less stable than compounds with bond angles of about 109°. Both cyclobutane and cyclopentane bend slightly from a planar structure to reduce the crowding of hydrogen atoms (Figure 11.16). In larger cycloalkanes, the bonds to carbon atoms can be tetrahedrally arranged only when the carbon atoms do not lie in the same plane. For example, cyclohexane can assume several nonplanar shapes. The chair and boat forms, where all the bond angles are 109.5°, are shown in Figure 11.16.

GOB
Chemistry⚛Now™
Go to GOB Now and click to see how to identify axial and equatorial positions in cycloalkanes.

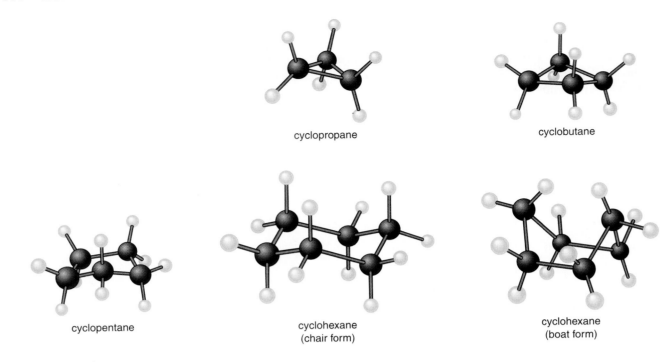

➤ **FIGURE 11.16** Ball-and-stick models for common cycloalkanes.

➤ **FIGURE 11.17** Rotation about C—C single bonds occurs in open-chain compounds but not within rings.

Stereoisomers
Compounds with the same structural formula but different spatial arrangements of atoms.

Geometric isomers
Molecules with restricted rotation around C—C bonds that differ in the three-dimensional arrangements of their atoms in space and not in the order of linkage of atoms.

The free rotation that can take place around C—C single bonds in alkanes (Section 11.6) is not possible for the C—C bonds of cycloalkanes. The ring structure allows bending or puckering but prevents free rotation (see ■ Figure 11.17). Any rotation of one carbon atom 180° relative to an adjacent carbon atom in a cycloalkane would require a single carbon–carbon bond to be broken somewhere in the ring. The breaking of such bonds would require a large amount of energy.

The lack of free rotation around C—C bonds in disubstituted cycloalkanes leads to an extremely important kind of isomerism called *stereoisomerism*. Two different compounds that have the same molecular formula and the same structural formula but different spatial arrangements of atoms are called **stereoisomers**. For example, consider a molecule of 1,2-dimethylcyclopentane. The cyclopentane ring is drawn in ■ Figure 11.18 as a planar pentagon with the heavy lines indicating that two of the carbons are in front as one views the structure. The groups attached to the ring project above or below the plane of the ring. Two stereoisomers are possible: Either both groups may project in the same direction from the plane, or they may project in opposite directions from the plane of the ring. Since the methyl groups cannot rotate from one side of the ring to the other, molecules of the two compounds represented in Figure 11.18 are distinct.

These two compounds have physical and chemical properties that are quite different and therefore can be separated from each other. Stereoisomers of this type, in which the spatial arrangement or geometry of their groups is maintained by rings, are called **geometric isomers** or cis-trans isomers. The

➤ **FIGURE 11.18** Two geometric isomers of 1,2-dimethylcyclopentane.

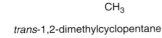

cis-1,2-dimethylcyclopentane   *trans*-1,2-dimethylcyclopentane prefix *cis-* denotes the isomer in which both groups are on the same side of the ring, and *trans-* denotes the isomer in which they are on opposite sides. To exist as geometric isomers, a disubstituted cycloalkane must be bound to groups at two different carbons of the ring. For example, there are no geometric isomers of 1,1-dimethylcyclohexane:

cis-
On the same side (as applied to geometric isomers).

trans-
On opposite sides (as applied to geometric isomers).

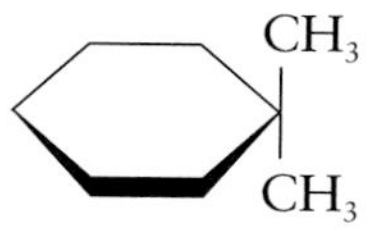

Example 11.6

Name and draw structural formulas for all the isomers of dimethylcyclobutane. Indicate which ones are geometric isomers.

Solution

There are three possible locations for the two methyl groups: positions 1,1, positions 1,2, and positions 1,3.

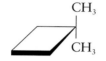

Geometric isomerism is not possible in this case with the two groups bound to the same carbon of the ring

1,1-dimethylcyclobutane

GOB
Chemistry ⚛ Now™
Go to GOB Now and click to discover how to identify *cis-* and *trans-* isomers in cycloalkanes.

Two groups on opposite sides of the planar ring

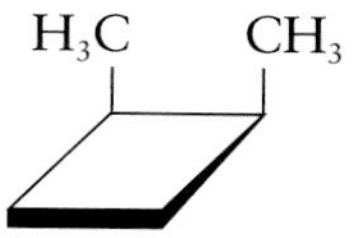

trans-1,2-dimethylcyclobutane

Two groups on the same side of the ring

cis-1,2-dimethylcyclobutane

trans-1,3-dimethylcyclobutane

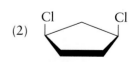

cis-1,3-dimethylcyclobutane

a. Identify each of the following cycloalkanes as a *cis-* or *trans-* compound:

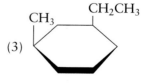

(1)

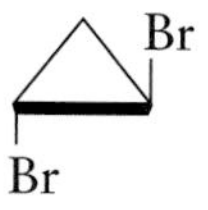

(2)

(3)

b. Draw the structural formula for *cis*-1,2-dichlorocyclobutane.

➤ 11.10 Physical Properties of Alkanes

Since alkanes are composed of nonpolar carbon–carbon and carbon–hydrogen bonds, alkanes are nonpolar molecules. Alkanes have lower melting and boiling points than other organic compounds of comparable molecular weight (see

TABLE 11.8		Physical properties of some normal alkanes				
Number of carbon atoms	IUPAC name	Condensed structural formula	Melting point (°C)	Boiling point (°C)	Density (g/mL)
1	methane	CH_4	−182.5	−164.0	0.55
2	ethane	CH_3CH_3	−183.2	−88.6	0.57
3	propane	$CH_3CH_2CH_3$	−189.7	−42.1	0.58
4	butane	$CH_3CH_2CH_2CH_3$	−133.4	−0.5	0.60
5	pentane	$CH_3CH_2CH_2CH_2CH_3$	−129.7	36.1	0.63
6	hexane	$CH_3CH_2CH_2CH_2CH_2CH_3$	−95.3	68.9	0.66
7	heptane	$CH_3CH_2CH_2CH_2CH_2CH_2CH_3$	−90.6	98.4	0.68
8	octane	$CH_3CH_2CH_2CH_2CH_2CH_2CH_2CH_3$	−56.8	125.7	0.70
9	nonane	$CH_3CH_2CH_2CH_2CH_2CH_2CH_2CH_2CH_3$	−53.5	150.8	0.72
10	decane	$CH_3CH_2CH_2CH_2CH_2CH_2CH_2CH_2CH_2CH_3$	−29.7	174.1	0.73

Homologous series
Compounds of the same functional class that differ by a —CH₂— group.

GOB
Chemistry Now™

Go to GOB Now and click to explore the relationship between alkane structure and boiling point.

■ Table 11.8). This is because their nonpolar molecules exert very weak attractions for each other. Alkanes are odorless compounds.

The normal, or straight-chain, alkanes make up what is called a **homologous series.** This term describes any series of compounds in which each member differs from a previous member only by having an additional —CH₂— unit. The physical and chemical properties of compounds making up a homologous series are usually closely related and vary in a systematic and predictable way. For example, the boiling points of normal alkanes increase smoothly as the length of the carbon chain increases (see ■ Figure 11.19). This pattern results from increasing dispersion forces as molecular weight increases. At ordinary temperatures and pressures, normal alkanes with 1 to 4 carbon atoms are gases, those with 5 to 20 carbon atoms are liquids, and those with more than 20 carbon atoms are waxy solids.

Because they are nonpolar, alkanes and other hydrocarbons are insoluble in water, which is a highly polar solvent. They are also less dense than water and thus float on it. These two properties of hydrocarbons are partly responsible for the well-known serious effects of oil spills from ships (see ■ Figure 11.20).

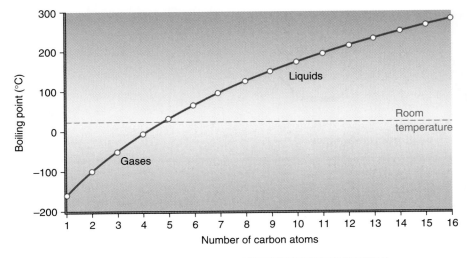

➤ **FIGURE 11.19** Normal alkane boiling points depend on chain length.

© Simon Fraser/Science Photo Library/Photo Researchers Inc.

➤ **FIGURE 11.20** Oil spills can have serious and long-lasting effects on the environment because of the insolubility of hydrocarbons in water.

CHEMISTRY AROUND US • 11.1

PETROLEUM

Petroleum, the most important of the fossil fuels used today, is sometimes called "black gold" in recognition of its importance in the 20th century. At times, the need for petroleum to keep society fueled has seemed second only to our need for food, shelter, and clothing.

It is generally believed that this complex mixture of hydrocarbons was formed over eons through the gradual decay of ocean-dwelling microscopic animals. The resulting crude oil, a viscous black liquid, collects in vast underground pockets in sedimentary rock. It is brought to the surface via drilling and pumping.

Useful products are obtained from crude oil by heating it to high temperatures to produce various fractions according to boiling point (see table). Most petroleum products are eventually burned as a fuel, but about 2% are used to synthesize organic compounds. This seemingly small amount is quite large in actual tonnage because of the huge volume of petroleum that is refined annually. In fact, more than half of all industrial synthetic organic compounds are made from this source. These industrial chemicals are eventually converted into dyes, drugs, plastics, artificial fibers, detergents, insecticides, and other materials deemed indispensable by many in industrialized nations.

Fraction	Boiling point range (°C)	Molecular size range	Typical uses
Gas	−164–30	C_1–C_4	Heating, cooking
Gasoline	30–200	C_5–C_{12}	Motor fuel
Kerosene	175–275	C_{12}–C_{16}	Fuel for stoves and diesel and jet engines
Heating oil	Up to 375	C_{15}–C_{18}	Furnace oil
Lubricating oils	350 and up	C_{16}–C_{20}	Lubrication, mineral oil
Greases	Semisolid	C_{18}–up	Lubrication, petroleum jelly
Paraffin (wax)	Melts at 52–57	C_{20}–up	Candles, toiletries
Pitch and tar	Residue in boiler	High	Roofing, asphalt paving

Asphalt for paving roads is a petroleum product.

© Michael C. Slabaugh

Liquid alkanes of higher molecular weight behave as emollients (skin softeners) when applied to the skin. An alkane mixture known as mineral oil is sometimes used to replace natural skin oils washed away by frequent bathing or swimming. Petroleum jelly (Vaseline is a well-known brand name) is a semisolid mixture of alkanes that is used as both an emollient and a protective film. Water and water solutions such as urine don't dissolve or penetrate the film, and the underlying skin is protected. Many cases of diaper rash have been prevented or treated this way.

The word **hydrophobic** (literally "water fearing") is often used to refer to molecules or parts of molecules that are insoluble in water. Many biomolecules, the large organic molecules associated with living organisms, contain nonpolar (hydrophobic) parts. Thus, such molecules are not water-soluble. Palmitic acid, for example, contains a large nonpolar hydrophobic portion and is insoluble in water.

Hydrophobic
Molecules or parts of molecules that repel (are insoluble in) water.

Nonpolar portion

$$CH_3 - CH_2 - CH_2 - CH_2 - CH_2 - CH_2 - CH_2 - CH_2 - CH_2 - CH_2 - CH_2 - CH_2 - CH_2 - CH_2 - CH_2 - \overset{\displaystyle O}{\overset{\|}{C}} - OH$$

palmitic acid

HYDRATING THE SKIN

Healthy, normal human skin is kept moist by natural body oils contained in sebum, a secretion of the skin's sebaceous glands. The oily sebum helps the epidermis, or outer layer of the skin, retain the 10–30% of water it normally contains. However, some people suffer from skin that is naturally dry, or dry because of aging or contact with materials like paint thinner that dissolve and remove the sebum. Such individuals may find some relief by using OTC products called *moisturizers*.

Two types of skin moisturizers are commonly available. One type, which behaves like natural sebum, contains nonpolar oily substances that form a barrier and prevent water from passing through and evaporating from the skin. These barrier-forming products often contain combinations of materials from various sources, including mineral oil and petroleum jelly from petroleum, vegetable oils such as apricot oil, sesame seed oil, palm kernel oil, olive oil, and safflower oil, and lanolin, an animal fat from sheep oil glands and wool. While they effectively keep water from leaving the skin, these products are somewhat messy to use and leave the skin feeling greasy.

A second, more popular type of moisturizer works by attracting water from the air and skin. These products form a water-rich layer that adheres to the skin without giving it a greasy feel. The substances that attract water are called humectants, and are compounds capable of forming hydrogen bonds with water. Some examples of humectants used in products of this type are glycerol, urea, lactic acid, and propylene glycol.

If you are shopping for a moisturizer, remember that the main characteristic you should look for in a product is the ability to form a barrier to prevent water evaporation, or the ability to act as a humectant. Some expensive products advertise that in addition to moisturizing, they also beautify the skin and even reverse aging because they contain proteins such as collagen and elastin, vitamins, hormones, or even DNA. It is unlikely that such substances can pass through the epidermis of the skin in sufficient amounts to provide the advertised benefits.

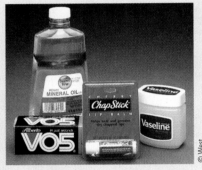

Alkane products marketed to soften and protect the skin.

➤ 11.11 Alkane Reactions

The alkanes are the least reactive of all organic compounds. In general, they do not react with strong acids (such as sulfuric acid), strong bases (such as sodium hydroxide), most oxidizing agents (such as potassium dichromate), and most reducing agents (such as sodium metal). This unreactive nature is reflected in the name *paraffins*, sometimes used to identify alkanes (from Latin words that mean "little affinity"). Paraffin wax, sometimes used to seal jars of homemade preserves, is a mixture of solid alkanes. Paraffin wax is also used in the preparation of milk cartons and wax paper. The inertness of the compounds in the wax makes it ideal for these uses. Alkanes do undergo reactions with halogens such as chlorine and bromine, but these are not important for our purposes.

Perhaps the most significant reaction of alkanes is the rapid oxidation called combustion. In the presence of ample oxygen, alkanes burn to form carbon dioxide and water, liberating large quantities of heat:

$$CH_4 + 2O_2 \rightarrow CO_2 + 2H_2O + 212.8 \text{ kcal/mol} \qquad (11.4)$$

It is this type of reaction that accounts for the wide use of hydrocarbons as fuels.

Natural gas contains methane (80–95%), some ethane, and small amounts of other hydrocarbons. Propane and butane are extracted from natural gas and sold in pressurized metal containers (bottled gas). In this form, they are used for heating and cooking in campers, trailers, boats, and rural homes.

Gasoline is a mixture of hydrocarbons (primarily alkanes) that contain 5 to 12 carbon atoms per molecule. Diesel fuel is a similar mixture, except the molecules contain 12 to 16 carbon atoms. The hot CO_2 and water vapor gen-

CHEMISTRY AROUND US • 11.2

CARBON MONOXIDE: SILENT AND DEADLY

The incomplete combustion of alkanes and other carbon-containing fuels may produce carbon monoxide, CO. It is a common component of furnace, stove, and automobile exhaust. Normally, these fumes are emitted into the air and dispersed. However, whenever CO gas is released into closed spaces, lethal concentrations may develop. All of the following are known to have caused CO poisoning deaths: poor ventilation of a furnace or stove exhaust, running the engine of an automobile for a period of time in a closed garage, idling the engine of a snowbound car to keep the heater working, and using a barbecue grill as a source of heat in a house or camper (burning charcoal generates a large amount of carbon monoxide). Recently, the exhaust systems of snow-blocked vehicles have been identified as another cause of carbon monoxide poisoning. The problem is created when the end of a vehicle's exhaust pipe is obstructed or plugged by snow. CO-containing exhaust can leak through cracks in the exhaust system, penetrate the floorboard, and enter the passenger compartment of the vehicle. An estimated 1000 Americans die each year from unintentional CO poisoning, and as many as 10,000 require medical treatment.

Carbon monoxide is dangerous because of its ability to bind strongly with hemoglobin in red blood cells. When this occurs, the ability of the blood to transport oxygen (O_2) is reduced because the CO molecules occupy sites on hemoglobin molecules that are normally occupied by O_2 molecules.

Because it is colorless and odorless, carbon monoxide gas is a silent and stealthy potential killer. The following symptoms of CO poisoning are given in order of increasing severity and seriousness:

1. Slight headache, dizziness, drowsiness
2. Headache, throbbing temples
3. Weakness, mental confusion, nausea
4. Rapid pulse and respiration, fainting
5. Possibly fatal coma

All except the most severe cases of CO poisoning are reversible. The most important first aid treatment is to get the victim fresh air. Any person who feels ill and suspects CO poisoning might be the cause should immediately evacuate the building, get fresh air, and summon medical assistance. Victims are often treated with 100% oxygen delivered from a mask.

Consumer safety experts stress that combustion appliances should be inspected regularly for leaks or other malfunctions. Also, it is recommended that drivers should inspect exhaust pipes and clear any obstructions before starting vehicles that have been parked in snow. The Consumer Product Safety Commission recommends that CO detectors be installed near bedrooms in residences. These detectors sound an alarm before the gas reaches potentially lethal concentrations in the air.

Fireplaces must be properly constructed to carry away carbon monoxide fumes.

---

erated during combustion in an internal combustion engine have a much greater volume than the air and fuel mixture. It is this sudden increase in gaseous volume and pressure that pushes the pistons and delivers power to the crankshaft.

If there is not enough oxygen available, incomplete combustion of hydrocarbons occurs, and some carbon monoxide (CO) or even carbon (see ■ Figure 11.21) may be produced (Reactions 11.5 and 11.6):

$$2CH_4 + 3O_2 \rightarrow 2CO + 4H_2O \qquad (11.5)$$

$$CH_4 + O_2 \rightarrow C + 2H_2O \qquad (11.6)$$

These reactions are usually undesirable because CO is toxic, and carbon deposits hinder engine performance. Occasionally, however, incomplete combustion is deliberately caused; specific carbon blacks (particularly lampblack, a pigment for ink) are produced by the incomplete combustion of natural gas.

➤ **FIGURE 11.21** A luminous yellow flame from a laboratory burner produces a deposit of carbon when insufficient air (O_2) is mixed with the gaseous fuel.

> **FOR FUTURE REFERENCE**   MEDICINES REMOVED FROM THE MARKET

People generally recognize the Food and Drug Administration (FDA) as the federal agency with the authority to require that drugs be proved safe and effective before they are approved for sale and marketed. However, it is not always understood that the FDA also has the authority to withdraw approved drugs from the marketplace if the use of the drugs shows them to be ineffective or hazardous. This authority, while not used very often, provides an added protection for consumers.

The table below lists drugs that were removed from the market during 1998. In each case the general category to which the drug belongs is given along with its generic name, at least one example of a trade name, and the reason for its removal.

Category	Generic name	Trade name	Reason for removal
Analgesic	bromfenac	Duract	Caused serious liver damage if taken for more than 10 days.
Psoriasis treatment	etretinate	Tegison	Newer drugs became available.
Cholesterol-reduction	probucol	Lorelco	More effective drugs became available.
High blood pressure treatment	mibefradil	Posicor	Serious drug interactions with multiple drugs.
Minimally sedating antihistamine	terfenadine	Seldane	Voluntarily withdrawn because of serious drug interactions with multiple drugs.
Weight-loss agents	dexfenfluramine fenfluramine	Redux Pondimin	Caused heart valve damage. Caused heart valve damage.

SOURCE: Adapted from Rybacki, J.J.; Long, J.W. *The Essential Guide to Prescription Drugs.* New York: Harper Perennial, 1999.

········   **CONCEPT SUMMARY**   ········

Carbon: The Element of Organic Compounds. Organic compounds contain carbon, and organic chemistry is the study of those compounds. Inorganic chemistry is the study of the elements and all noncarbon compounds. Carbon compounds are of tremendous everyday importance to life on Earth and are the basis of all life processes.

Organic and Inorganic Compounds Compared. The properties of organic and inorganic compounds often differ, largely as a result of bonding differences. Organic compounds contain primarily covalent bonds, whereas ionic bonding is more prevalent in inorganic compounds.

Bonding Characteristics and Isomerism. Large numbers of organic compounds are possible because carbon atoms link to form chains and networks. An additional reason for the existence of so many organic compounds is the phenomenon of isomerism. Isomers are compounds that have the same molecular formula but different arrangements of atoms.

Functional Groups: The Organization of Organic Chemistry. All organic compounds are grouped into classes based on characteristic features called functional groups. Compounds with their functional groups are represented by two types of structural formulas. Expanded structural formulas show all covalent bonds, whereas condensed structural formulas show no covalent bonds or only selected bonds.

Alkane Structures. Alkanes are hydrocarbons that contain only single covalent bonds and can be represented by the formula C_nH_{2n+2}. Alkanes possess a three-dimensional geometry in which each carbon is surrounded by four bonds directed to the corners of a tetrahedron. Methane, the simplest alkane, is an important fuel (natural gas) and a chemical feedstock for the preparation of other organic compounds. The number of structural isomers possible for an alkane increases dramatically with the number of carbon atoms present in the molecule.

Conformations of Alkanes. Rotation about the single bonds between carbon atoms allows alkanes to exist in many different conformations. When an alkane is drawn using only two dimensions, the structure can be represented in a variety of ways as long as the order of bonding is not changed.

Alkane Nomenclature. Some simple alkanes are known by common names. More complex compounds are usually named using the IUPAC system. The characteristic IUPAC ending for alkanes is *-ane*.

Cycloalkanes. These are alkanes in which the carbon atoms form a ring. The prefix *cyclo-* is used in the names of these compounds to indicate their cyclic nature.

The Shape of Cycloalkanes. The carbon atom rings of cycloalkanes are usually shown as planar, although only cyclopropane is planar. Because rotation about the single bonds in the ring is restricted, certain disubstituted cycloalkanes can exist as geometric (cis-trans) isomers.

Physical Properties of Alkanes. The physical properties of alkanes are typical of all hydrocarbons: nonpolar, insoluble in water, less dense than water, and increasing melting and boiling points with increasing molecular weight.

Alkane Reactions. Alkanes are relatively unreactive and remain unchanged by most reagents. The reaction that is most significant is combustion.

········· **LEARNING OBJECTIVES ASSESSMENT** ·········

You can get an approximate but quick idea of how well you have met the learning objectives given at the beginning of this chapter by working the selected end-of-chapter exercises given below. The answer to each exercise is given in Appendix B of the book.

Objective 1 (Section 11.1): Exercise 11.2

Objective 2 (Section 11.1): Exercise 11.4

Objective 3 (Section 11.2): Exercise 11.8

Objective 4 (Section 11.3): Exercise 11.20

Objective 5 (Section 11.4): Exercise 11.24

Objective 6 (Section 11.5): Exercise 11.28

Objective 7 (Section 11.6): Exercise 11.30

Objective 8 (Section 11.7): Exercise 11.34

Objective 9 (Section 11.8): Exercise 11.44

Objective 10 (Section 11.9): Exercise 11.54

Objective 11 (Section 11.10): Exercise 11.56

Objective 12 (Section 11.11): Exercise 11.60

········· **KEY TERMS AND CONCEPTS** ·········

Alkane (11.5)
Alkyl group (11.7)
Branched alkane (11.5)
cis- (11.9)
Condensed structural formula (11.4)
Conformations (11.6)
Cycloalkane (11.8)
Expanded structural formula (11.4)

Functional group (11.4)
Geometric isomers (11.9)
Homologous series (11.10)
Hybrid orbital (11.3)
Hydrocarbon (11.5)
Hydrophobic (11.10)
Inorganic chemistry (11.1)
Isomerism (11.3)

Normal alkane (11.5)
Organic chemistry (11.1)
Organic compound (11.1)
Saturated hydrocarbon (11.5)
Stereoisomers (11.9)
Structural isomers (11.3)
trans- (11.9)

········· **KEY REACTIONS** ·········

1. Complete combustion of alkanes (Section 11.11):

$$\text{Alkane} + O_2 \rightarrow CO_2 + H_2O$$

2. Incomplete combustion of alkanes (Section 11.11):

$$\text{Alkane} + O_2 \rightarrow CO \text{ (or C)} + H_2O$$

LEGEND: 1 = straightforward, 2 = intermediate, 3 = challenging. All even-numbered exercises are answered in Appendix B.

CARBON: THE ELEMENT OF ORGANIC COMPOUNDS (SECTION 11.1)

11.1 Why were the compounds of carbon originally called organic compounds?

11.2 Name at least six items you recognize to be composed of organic compounds.

11.3 Describe what Wöhler did that made the vital force theory highly questionable.

11.4 What is the unique structural feature shared by all organic compounds?

11.5 Classify each of the following compounds as organic or inorganic:
a. KBr
b. H_2O
c. $H-C\equiv C-H$
d. LiOH
e. CH_3-NH_2

ORGANIC AND INORGANIC COMPOUNDS COMPARED (SECTION 11.2)

11.6 What kind of bond between atoms is most prevalent among organic compounds?

11.7 Are the majority of all compounds that are insoluble in water organic or inorganic? Why?

11.8 Indicate for each of the following characteristics whether it more likely describes an inorganic or an organic compound. Give one reason for your answer.
a. This compound is a liquid that readily burns.
b. A white solid upon heating is found to melt at 735°C.
c. A liquid added to water floats on the surface and does not dissolve.
d. This compound exists as a gas at room temperature and ignites easily.
e. A solid substance melts at 65°C.

11.9 Devise a test, based on the general properties in Table 11.1, that you could use to quickly distinguish between the substances in each of the following pairs:
a. Gasoline (liquid, organic) and water (liquid, inorganic)
b. Naphthalene (solid, organic) and sodium chloride (solid, inorganic)
c. Methane (gaseous, organic) and hydrogen chloride (gaseous, inorganic)

11.10 Explain why organic compounds are nonconductors of electricity.

11.11 Explain why the rate of chemical reactions is generally slow for organic compounds and usually fast for inorganic compounds.

BONDING CHARACTERISTICS AND ISOMERISM (SECTION 11.3)

11.12 Give two reasons for the existence of the tremendous number of organic compounds.

11.13 How many of carbon's electrons are unpaired and available for bonding according to an sp^3 hybridization model?

11.14 Describe what atomic orbitals overlap to produce a carbon–hydrogen bond in CH_4.

11.15 What molecular geometry exists when a central carbon atom bonds to four other atoms?

11.16 Compare the shapes of unhybridized p and hybridized sp^3 orbitals.

11.17 Use Example 11.1 and Tables 11.2 and 11.6 to determine the number of covalent bonds formed by atoms of the following elements: carbon, hydrogen, oxygen, nitrogen, and bromine.

11.18 Complete the following structures by adding hydrogen atoms where needed:

a. $C-C-C$

b. $C-C=C$

c. $C-\overset{\overset{\displaystyle O}{\|}}{C}$

d. $C-C-N$

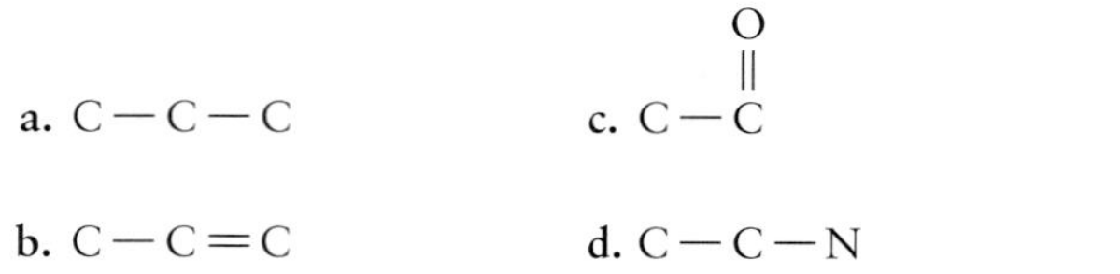

11.19 Complete the following structures by adding hydrogen atoms where needed.

a. $C-\underset{\underset{\displaystyle C}{|}}{C}-C$

b. $C-\underset{\underset{\displaystyle C=O}{|}}{C}-C$

c. $N-C-C-N$

d. $C=C=C$

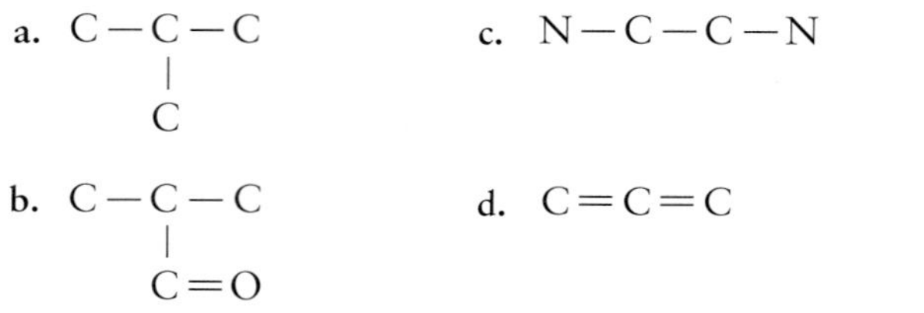

11.20 Which of the following pairs of compounds are structural isomers?

a. $CH_3-CH=CH-CH_3$ and $CH_3-CH_2-CH_2-CH_3$

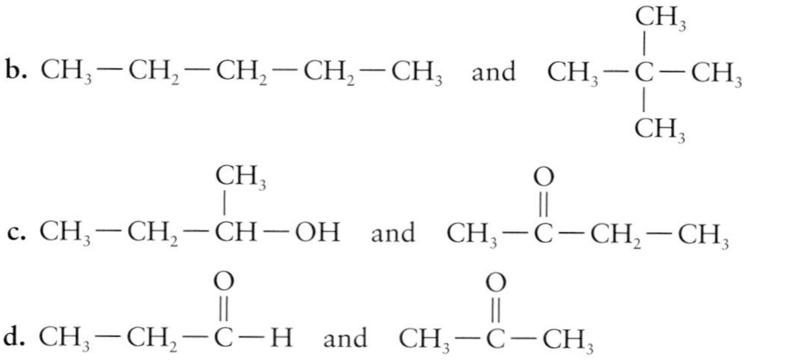

b. $CH_3-CH_2-CH_2-CH_2-CH_3$ and $CH_3-\underset{\underset{\displaystyle CH_3}{|}}{\overset{\overset{\displaystyle CH_3}{|}}{C}}-CH_3$

c. $CH_3-CH_2-\underset{\underset{\displaystyle OH}{}}{\overset{\overset{\displaystyle CH_3}{|}}{CH}}$ and $CH_3-\overset{\overset{\displaystyle O}{\|}}{C}-CH_2-CH_3$

d. $CH_3-CH_2-\overset{\overset{\displaystyle O}{\|}}{C}-H$ and $CH_3-\overset{\overset{\displaystyle O}{\|}}{C}-CH_3$

e. $CH_3-CH_2-CH_2-NH_2$ and $CH_3-CH_2-NH-CH_3$

11.21 Group all the following compounds together that represent structural isomers of each other:

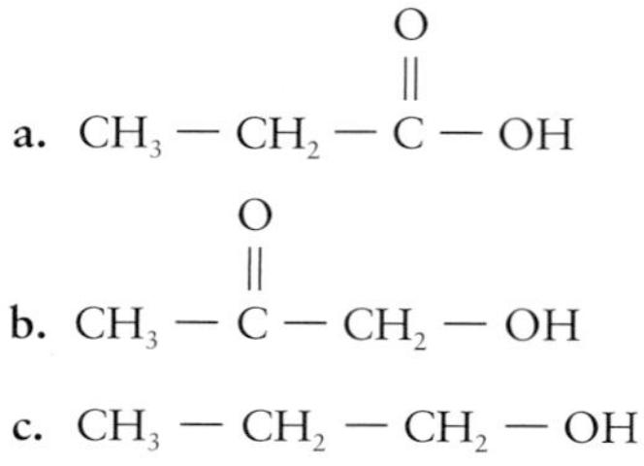

a. $CH_3-CH_2-\overset{\overset{\displaystyle O}{\|}}{C}-OH$

b. $CH_3-\overset{\overset{\displaystyle O}{\|}}{C}-CH_2-OH$

c. $CH_3-CH_2-CH_2-OH$

d.
$$
\begin{array}{c}
\quad\quad\; O \\
\quad\quad\; \| \\
CH_3 - C - CH_2 - CH_3
\end{array}
$$

e.
$$
\begin{array}{c}
\;\; OH\;\;\; O \\
\;\; |\quad\quad \| \\
CH_3 - CH - C - H
\end{array}
$$

f.
$$
\begin{array}{c}
\quad\quad\; O \\
\quad\quad\; \| \\
CH_3 - C - O - CH_3
\end{array}
$$

g. $HO - CH_2 - CH_2 - CH_2 - OH$

11.22 On the basis of the number of covalent bonds possible for each atom, determine which of the following structural formulas are correct. Explain what is wrong with the incorrect structures.

a.
$$
\begin{array}{c}
H\quad\quad H \\
|\quad\quad | \\
H - C - H - C - H \\
|\quad\quad | \\
H\quad\quad H
\end{array}
$$

b.
$$
\begin{array}{c}
\quad\quad H\;\; H \\
\quad\quad |\;\; | \\
H - N - C - C - H \\
\quad\quad |\;\; |\;\; | \\
\quad\quad H\;\; H\;\; H
\end{array}
$$

c.
$$
\begin{array}{c}
H\quad\quad O\quad\quad H \\
|\quad\quad \|\quad\quad | \\
H - C \text{——} C \text{——} C - H \\
|\quad\quad |\quad\quad | \\
H\quad\; H - C - H\;\; H \\
\quad\quad\quad | \\
\quad\quad\quad H
\end{array}
$$

d. $CH_3 - C = C - CH_3$

e.
$$
\begin{array}{c}
\;\;\; OH\;\;\; O \\
\;\;\; |\quad\quad \| \\
CH_3 - CH - CH - C - H \\
\;\;\; | \\
\;\;\; CH_3
\end{array}
$$

FUNCTIONAL GROUPS: THE ORGANIZATION OF ORGANIC CHEMISTRY (SECTION 11.4)

11.23 Identify each of the following as a condensed structural formula, expanded structural formula, or molecular formula:

a.
$$
\begin{array}{c}
H\;\; H\;\; H \\
|\;\; |\;\; | \\
H - C - C = C \\
|\quad\quad | \\
H\quad\quad H
\end{array}
$$

c. $C_6H_{12}O_6$

b. $CH_3CH_2CH_2 - OH$

d.
$$
\begin{array}{c}
\quad\quad\quad O \\
\quad\quad\quad \| \\
H_2N - C - CH_2CH_3
\end{array}
$$

11.24 Write a condensed structural formula for the following compounds:

a.
$$
\begin{array}{c}
H\;\; H\;\; H\quad\quad H\;\; H\;\; H \\
|\;\; |\;\; |\quad\quad |\;\; |\;\; | \\
H - C - C - C - O - C - C - C - H \\
|\;\; |\;\; |\quad\quad |\;\; |\;\; | \\
H\;\; H\;\; H\quad\quad H\;\; H\;\; H
\end{array}
$$

b.
$$
\begin{array}{c}
H\quad O\;\; H\quad\quad\quad H \\
|\quad \|\;\; |\quad\quad\quad | \\
H - C - C - C \text{——} C - H \\
|\quad\quad |\quad\quad\quad | \\
H\quad\; H - C - H\;\; H \\
\quad\quad\quad | \\
\quad\quad\quad H
\end{array}
$$

11.25 Write a condensed structural formula for the following compounds:

a.
$$
\begin{array}{c}
H\;\; H\;\; H\;\; O\;\; H \\
|\;\; |\;\; |\;\; \|\;\; | \\
H - C - C - C - C - N - H \\
|\;\; |\;\; | \\
H\;\; H\;\; H
\end{array}
$$

b.
$$
\begin{array}{c}
\quad\quad\quad H \\
\quad\quad\quad | \\
H\quad\; H - C - H\quad H\;\; O \\
|\quad\quad\quad\quad\quad |\;\; \| \\
H - C \text{——} C \text{——} C - C - OH \\
|\quad\quad |\quad\quad | \\
H\quad\; H - C - H\;\; H \\
\quad\quad\quad | \\
\quad\quad\quad H
\end{array}
$$

11.26 Write an expanded structural formula for the following:

a. $CH_3 - CH_2 - CH_2 - NH_2$

b.
$$
\begin{array}{c}
\quad\quad\quad\quad O \\
\quad\quad\quad\quad \| \\
CH_3 - CH_2 - C - CH_3
\end{array}
$$

ALKANE STRUCTURES (SECTION 11.5)

11.27 The name of the normal alkane containing 9 carbon atoms is nonane. What are the molecular and condensed structural formulas for nonane?

11.28 Classify each of the following compounds as a normal alkane or a branched alkane:

a.
$$
\begin{array}{c}
CH_3 - CH - CH_3 \\
|\\
CH_3
\end{array}
$$

b.
$$
\begin{array}{c}
CH_3 - CH_2 - CH_2 \\
\quad\quad\quad\quad | \\
\quad\quad\quad CH_2 - CH_3
\end{array}
$$

c.
$$
\begin{array}{c}
\quad\quad\quad CH_3 \\
\quad\quad\quad | \\
CH_2 - CH_2 \\
| \\
CH_3
\end{array}
$$

d.
$$
\begin{array}{c}
CH_3 - CH_2 - CH_2 \\
\quad\quad\quad\quad | \\
\quad\quad CH_3 - CH_2
\end{array}
$$

e.
$$
\begin{array}{c}
\quad\quad CH_3\quad\quad CH_3 \\
\quad\quad |\quad\quad\quad | \\
CH_2 - CH_2 - CH \\
\quad\quad\quad\quad\quad | \\
\quad\quad\quad\quad\; CH_3
\end{array}
$$

f.
$$
\begin{array}{c}
\quad\quad\quad\quad CH_3 \\
\quad\quad\quad\quad | \\
CH_3 - CH_2 - C - CH_3 \\
\quad\quad\quad\quad | \\
\quad\quad\quad\quad CH_3
\end{array}
$$

CONFORMATIONS OF ALKANES (SECTION 11.6)

11.29 Why are different conformations of an alkane not considered structural isomers?

11.30 Which of the following pairs represent structural isomers, and which are simply the same compound?

a. 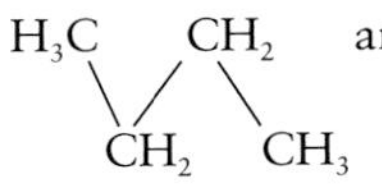 H_3C CH_2 and $CH_3CH_2CH_2CH_3$
    CH_2 CH_3

b. 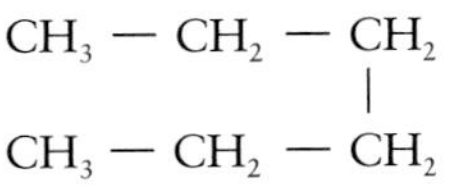 $CH_3 — CH_2 — CH_2$ and $CH_3CH_2CH_2CH_2CH_2CH_3$
            |
  $CH_3 — CH_2 — CH_2$

c. 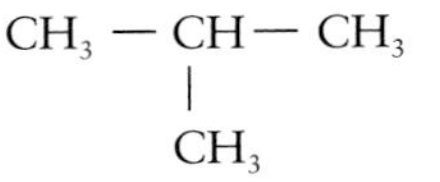 $CH_3 — CH — CH_3$ and $CH_3CH_2CH_2CH_3$
        |
      CH_3

   CH_3
    |
d. 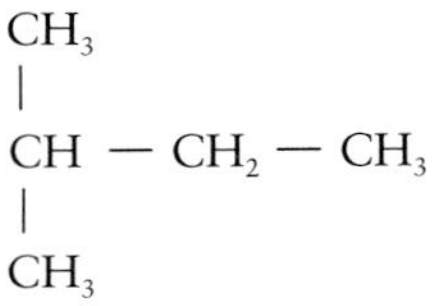 $CH — CH_2 — CH_3$ and $CH_3 — CH_2 — CH — CH_3$
    | |
   CH_3 CH_3

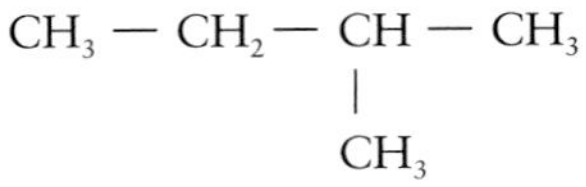

ALKANE NOMENCLATURE (SECTION 11.7)

11.31 For each of the following carbon skeletons, give the number of carbon atoms in the longest continuous chain:

    C   C
    | |
    C   C   C—C   C—C
    | | | |
a. 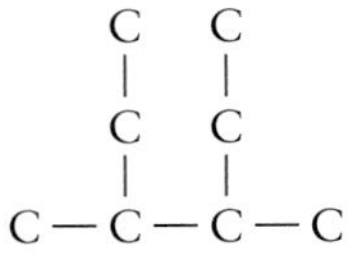 C—C—C—C   **c.** C—C—C—C—C

    C—C
      |
b. 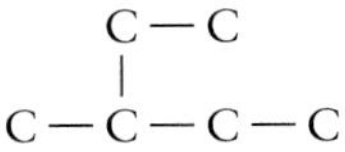 C—C—C—C

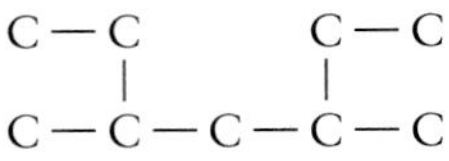

11.32 For each of the following carbon skeletons, give the number of carbon atoms in the longest continuous chain:

     C—C   C—C
       | | |
a. 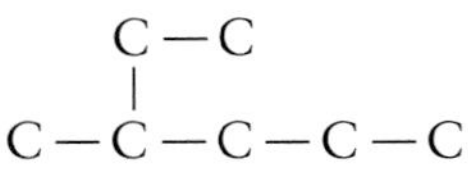   C—C—C—C—C   C   C—C
                            | |
       C   **c.**   C—C   C—C
       |
b. 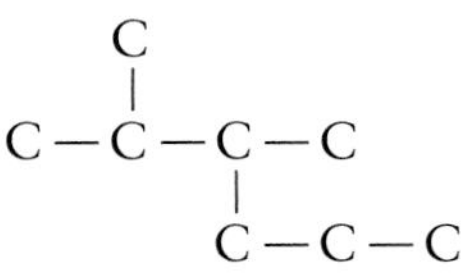   C—C—C—C
         |
        C—C—C

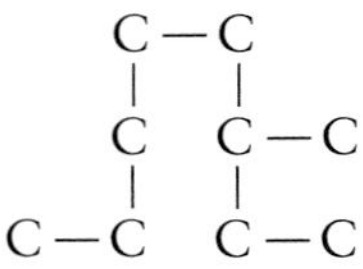

11.33 Identify the following alkyl groups:

a. 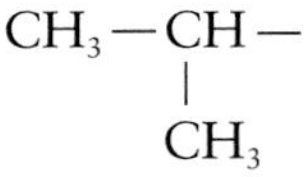 $CH_3 — CH —$
          |
        CH_3

b. $CH_3 — CH_2 — CH_2 — CH_2 —$

c. 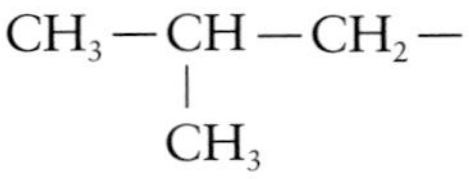 $CH_3 — CH — CH_2 —$
             |
          CH_3

d. $CH_3 —$

11.34 Give the correct IUPAC name for each of the following alkanes:

    $CH_3 — CH_2$
             |
a. 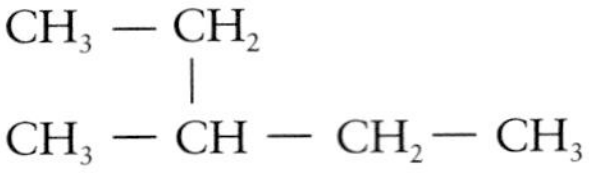 $CH_3 — CH — CH_2 — CH_3$

         CH_3
          |
b. 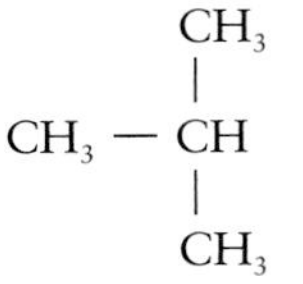 $CH_3 — CH$
          |
         CH_3

               $CH_2 — CH_3$
               |
c. 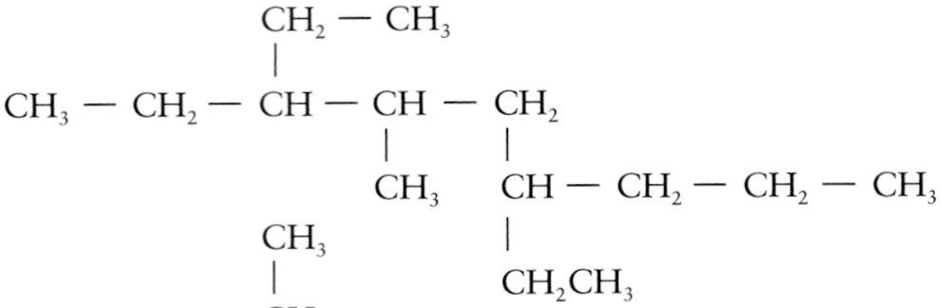 $CH_3 — CH_2 — CH — CH — CH_2$
                   | |
              CH_3 $CH — CH_2 — CH_2 — CH_3$
        CH_3 |
        | CH_2CH_3
        CH_2
        |
        CH_2
        |
d. 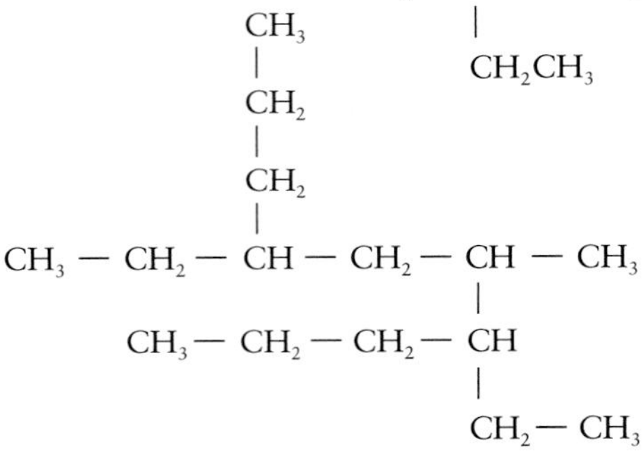 $CH_3 — CH_2 — CH — CH_2 — CH — CH_3$
                              |
      $CH_3 — CH_2 — CH_2 — CH$
                          |
                         $CH_2 — CH_3$

        $CH_2 — CH_2 — CH_2 — CH_3$
        |
e. 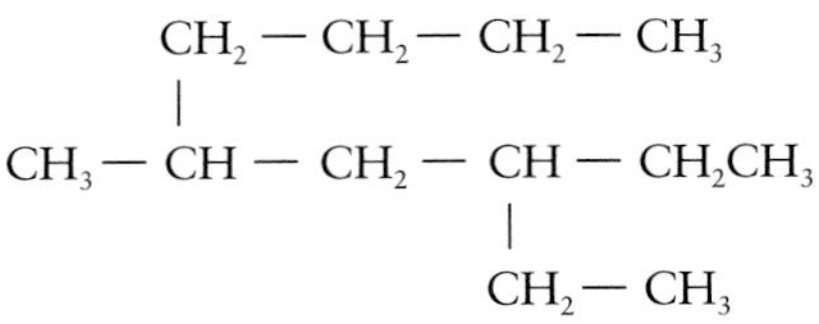 $CH_3 — CH — CH_2 — CH — CH_2CH_3$
                       |
                   $CH_2 — CH_3$

11.35 Give the correct IUPAC name for each of the following alkanes:

a. 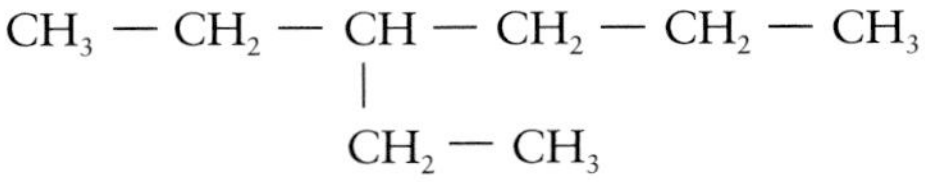 $CH_3 — CH_2 — CH — CH_2 — CH_2 — CH_3$
                    |
            $CH_2 — CH_3$

              CH_3
              |
b. 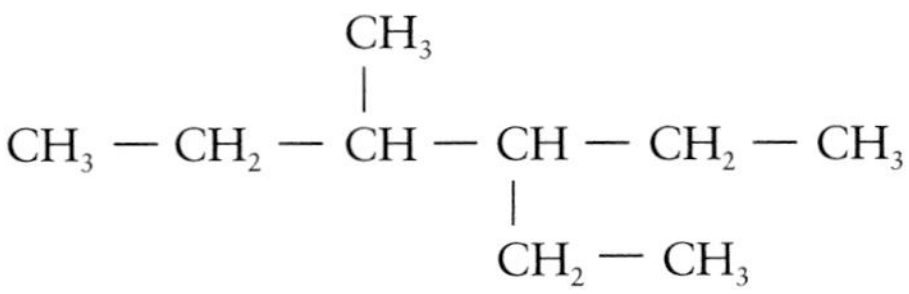 $CH_3 — CH_2 — CH — CH — CH_2 — CH_3$
                    |
            $CH_2 — CH_3$

c. 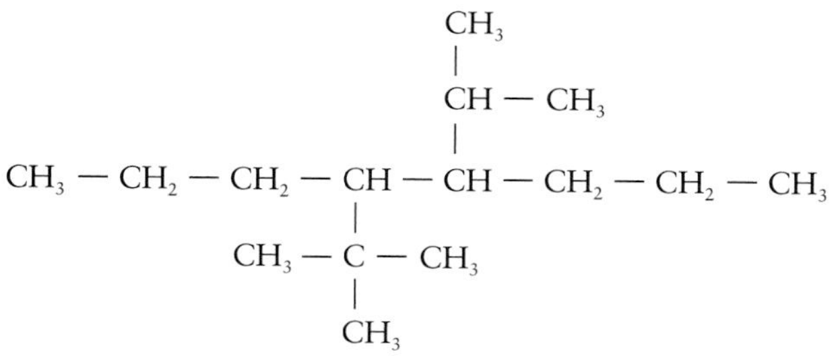
$$CH_3 - CH_2 - CH_2 - CH - CH - CH_2 - CH_2 - CH_3$$

with the substituents:
$$\overset{CH_3}{\underset{\displaystyle CH-CH_3}{|}}$$ on one carbon and $$CH_3 - \overset{CH_3}{\underset{\displaystyle CH_3}{\overset{|}{C}}} - CH_3$$ below

d. 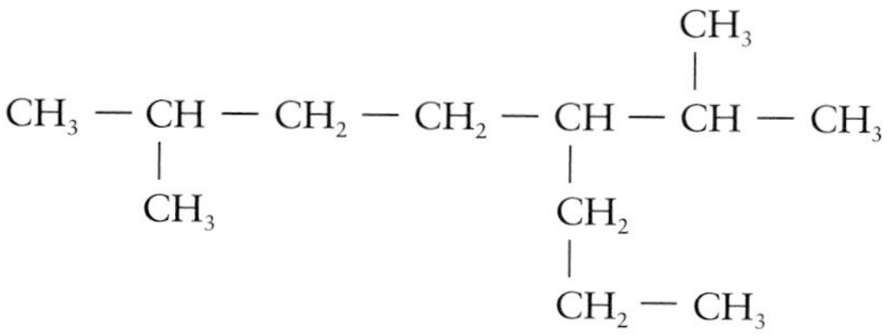
$$CH_3 - CH - CH_2 - CH_2 - CH - CH - CH_3$$
with CH_3 on the second carbon, and $\overset{\displaystyle CH_3}{|}$ above the CH, with CH_2 and $CH_2 - CH_3$ below.

e. 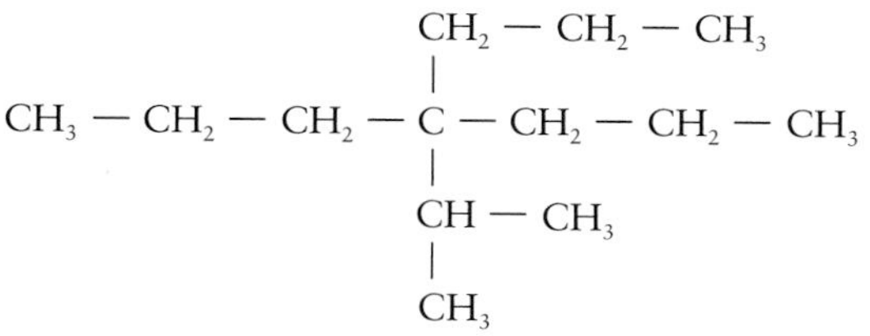
$$CH_3 - CH_2 - CH_2 - C - CH_2 - CH_2 - CH_3$$
with $CH_2 - CH_2 - CH_3$ above, and $CH - CH_3$ / CH_3 below.

11.36 Draw a condensed structural formula for each of the following compounds:
   a. 3-ethylpentane
   b. 2,2-dimethylbutane
   c. 4-ethyl-3,3-dimethyl-5-propyldecane
   d. 5-*sec*-butyldecane

11.37 Draw a condensed structural formula for each of the following compounds:
   a. 2,2,4-trimethylpentane
   b. 4-isopropyloctane
   c. 3,3-diethylhexane
   d. 5-*t*-butyl-2-methylnonane

11.38 Draw the condensed structural formula for each of the three structural isomers of C_5H_{12}, and give the correct IUPAC names.

11.39 Isooctane is 2,2,4-trimethylpentane. Draw structural formulas for and name a branched heptane, hexane, pentane, and butane that are structural isomers of isooctane.

11.40 Draw structural formulas for the compounds and give correct IUPAC names for the five structural isomers of C_6H_{14}.

11.41 The following names are incorrect, according to IUPAC rules. Draw the structural formulas and tell why each name is incorrect. Write the correct name for each compound.
   a. 2,2-methylbutane
   b. 4-ethyl-5-methylheptane
   c. 2-ethyl-1,5-dimethylhexane

11.42 The following names are incorrect, according to IUPAC rules. Draw the structural formulas and tell why each name is incorrect. Write the correct name for each compound.
   a. 1,2-dimethylpropane
   b. 3,4-dimethylpentane
   c. 2-ethyl-4-methylpentane
   d. 2-bromo-3-ethylbutane

CYCLOALKANES (SECTION 11.8)

11.43 The general formula for alkanes is C_nH_{2n+2}. Write a general formula for cycloalkanes.

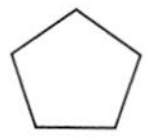

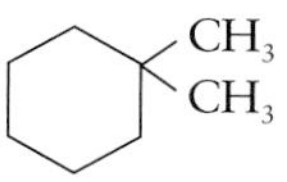

11.44 Write the correct IUPAC name for each of the following:

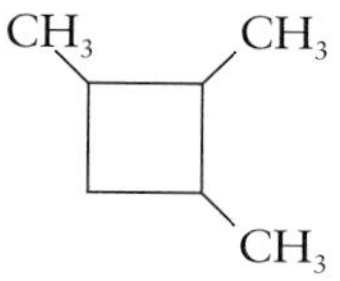

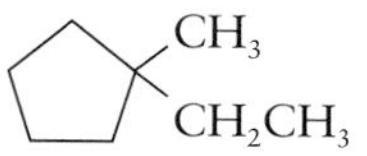

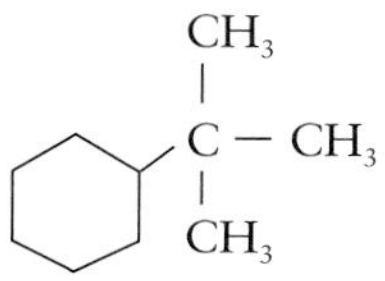

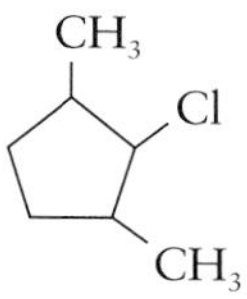

11.45 Write the correct IUPAC name for each of the following:

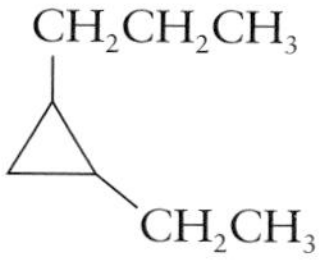

11.46 Draw the structural formulas corresponding to each of the following IUPAC names:
   a. isopropylcyclopentane
   b. 1,1-dimethylcyclobutane
   c. 1-isobutyl-3-isopropylcyclohexane

11.47 Draw the structural formulas corresponding to each of the following IUPAC names:
   a. 1,2-diethylcyclopentane
   b. 1,2,4-trimethylcyclohexane
   c. propylcyclobutane

11.48 Which of the following pairs of cycloalkanes represent structural isomers?

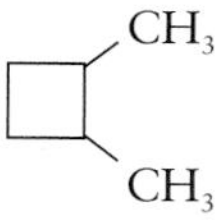

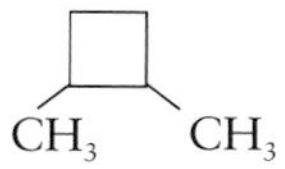

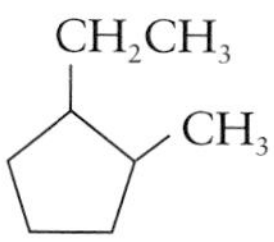

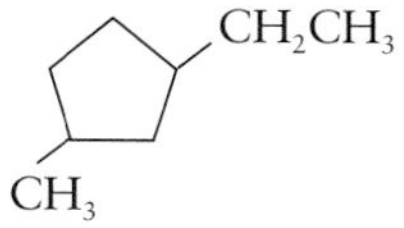

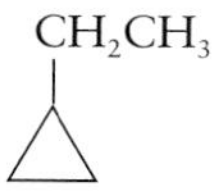

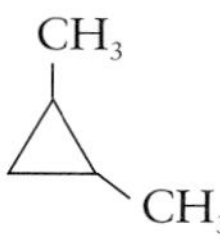

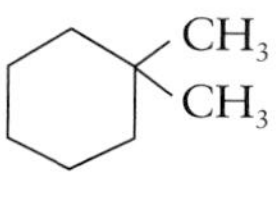

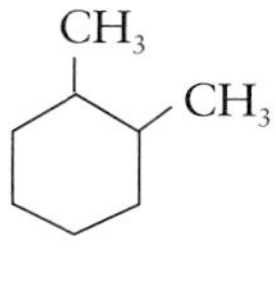

11.49 Draw structural formulas for the five structural isomers of C_5H_{10} that are cycloalkanes.

THE SHAPE OF CYCLOALKANES (SECTION 11.9)

11.50 Why does cyclohexane assume a chair form rather than a planar hexagon?

11.51 Explain the difference between geometric and structural isomers.

11.52 Which of the following cycloalkanes could show geometric isomerism? For each that could, draw structural formulas, and name both the cis and the trans isomers.

a.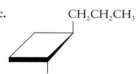

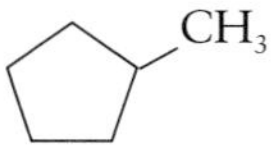

b.

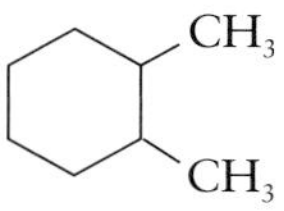

c.

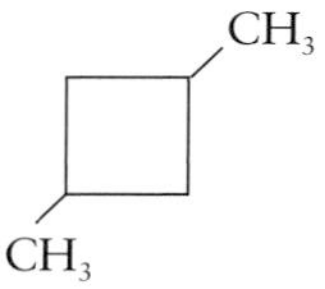

d.

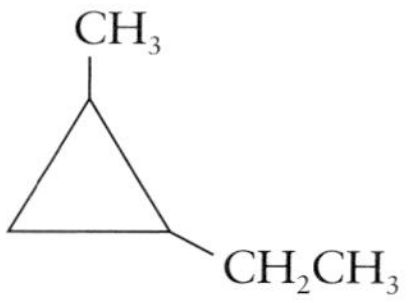

11.53 Draw structural formulas for *cis-* and *trans-* 1,3-dibromocyclobutane.

11.54 Using the prefix *cis-* or *trans-*, name each of the following:

a. 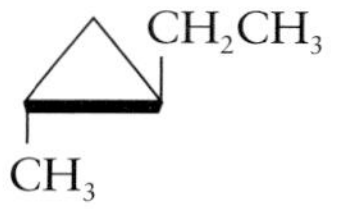   c.

b. 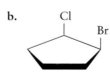   d.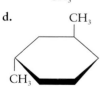

11.55 For each of the following molecular formulas, give the structural formulas requested. In most cases, there are several possible structures.
a. C_6H_{14}, a normal alkane
b. C_6H_{14}, a branched alkane
c. C_5H_{12}, a pair of conformations
d. C_5H_{12}, a pair of structural isomers
e. C_5H_{10}, a cyclic hydrocarbon
f. C_6H_{12}, two cycloalkane geometric isomers
g. C_6H_{12}, a cycloalkane that has no geometric isomers

PHYSICAL PROPERTIES OF ALKANES (SECTION 11.10)

11.56 The compound decane is a straight-chain alkane. Predict the following:
a. Is decane a solid, liquid, or gas at room temperature?
b. Is it soluble in water?
c. Is it soluble in hexane?
d. Is it more or less dense than water?

11.57 Explain why alkanes of low molecular weight have lower melting and boiling points than water.

11.58 Suppose you have a sample of 2-methylhexane and a sample of 2-methylheptane. Which sample would you expect to have the higher melting point? Boiling point?

11.59 Identify (circle) the alkanelike portions of the following molecules:

a. 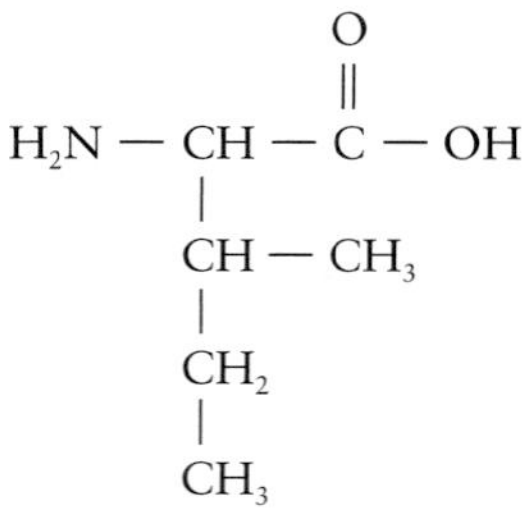
$$H_2N - \underset{\underset{\underset{CH_3}{|}}{\underset{|}{CH_2}}}{\underset{|}{\underset{|}{CH - CH_3}}}{CH} - \overset{\overset{O}{\|}}{C} - OH$$

isoleucine (an amino acid)

b. 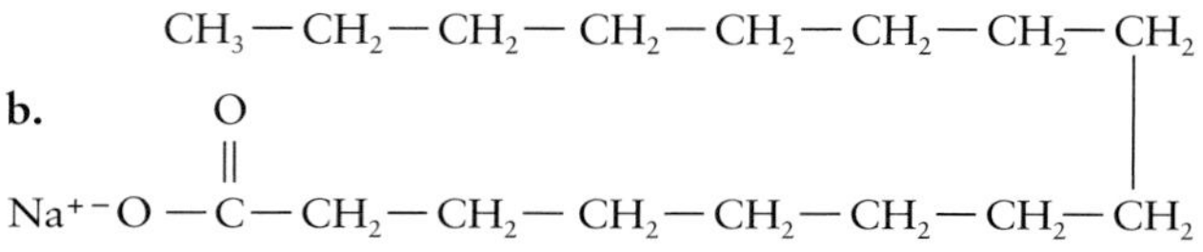
$$CH_3-CH_2-CH_2-CH_2-CH_2-CH_2-CH_2-CH_2$$
$$Na^{+-}O - \overset{\overset{O}{\|}}{C}-CH_2-CH_2-CH_2-CH_2-CH_2-CH_2-CH_2$$
sodium palmitate (a soap)

c. 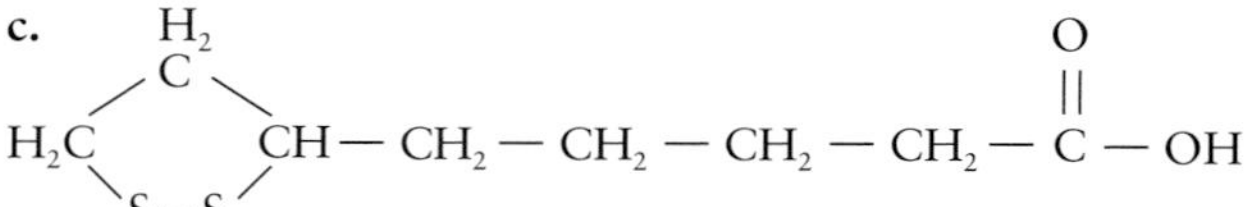 lipoic acid (a component of a coenzyme)

d. 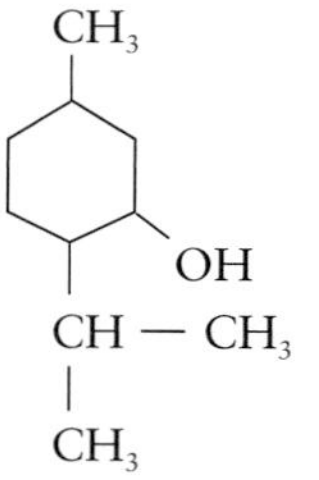
menthol (a flavoring)

ALKANE REACTIONS (SECTION 11.11)

11.60 Write a balanced equation to represent the complete combustion of each of the following:

a. propane   c.

b. CH_3CHCH_3
       $|$
       CH_3

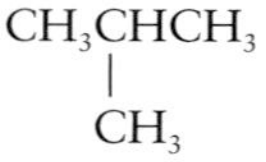

11.61 Write a balanced equation to represent the complete combustion of each of the following:

    a. ethane

    b. $CH_3 — CH — CH_2 — CH_3$
                |
                CH_3

    c.

11.62 Write a balanced equation for the *incomplete* combustion of hexane, assuming the formation of carbon monoxide and water as the products.

11.63 Why is it dangerous to relight a furnace when a foul odor is present?

CHEMISTRY AND YOUR HEALTH, CHEMISTRY AROUND US, AND OVER THE COUNTER

11.64 Why might some organically grown foods be more flavorful than conventionally grown produce?

11.65 What general name is given to substances within skin moisturizers that attract water from the air?

11.66 What petroleum products are sometimes found in skin moisturizers?

11.67 Give a major use for each of the following petroleum fractions:

    a. $C_1–C_4$
    b. $C_5–C_{12}$
    c. $C_{12}–C_{16}$
    d. $C_{16}–C_{20}$
    e. C_{20}–up

11.68 Give another reason why petroleum is so valuable to our society besides serving as a source of fuel.

11.69 In what way does carbon monoxide interact with the body that makes the gas dangerous?

11.70 What is the antidote for CO poisoning?

▷ CHEMISTRY FOR THOUGHT

11.1 Would you expect a molecule of urea produced in the body to have any different physical or chemical properties from a molecule of urea prepared in a laboratory?

11.2 Why might the study of organic compounds be important to someone interested in the health or life sciences?

11.3 Why do very few aqueous solutions of organic compounds conduct electricity?

11.4 The ski wax being examined in Figure 11.2 has a relatively low melting point. What does that fact reveal about the forces between molecules?

11.5 Charcoal briquettes sometimes burn with incomplete combustion when the air supply is limited. Why would it be hazardous to place a charcoal grill inside a home or a camper in an attempt to keep warm?

11.6 If carbon did not form hybridized orbitals, what would you expect to be the formula of the simplest compound of carbon and hydrogen?

11.7 What types of sports equipment are made from graphite fibers besides that shown in Figure 11.8?

11.8 A semi truck loaded with cyclohexane overturns during a rainstorm, spilling its contents over the road embankment. If the rain continues, what will be the fate of the cyclohexane?

11.9 On the way home from school, you drove through a construction zone, resulting in several tar deposits on the car's fender. What substances commonly found in the kitchen might help in removing the tar deposits?

11.10 Oil spills along coastal shores can be disastrous to the environment. What physical and chemical properties of alkanes contribute to the consequences of an oil spill?

11.11 Why might some farmers hesitate to grow and sell organic produce?

InfoTrac College Edition Readings

"See not, hear not, smell not (methane emissions from abandoned mines)," *The Ecologist*, March 2001, 31(2):12. Record number A18169206.

"Oil without end? Revisionists say oil isn't a fossil fuel. That could mean there's lots more of it," *Fortune*, Feb 17, 2003, 147(3):46. Record number A97222343.

"Oil to burn?" *Natural History*, April 2003, 112(3):73(1). Record number A99818081.

CHAPTER 12 Unsaturated Hydrocarbons

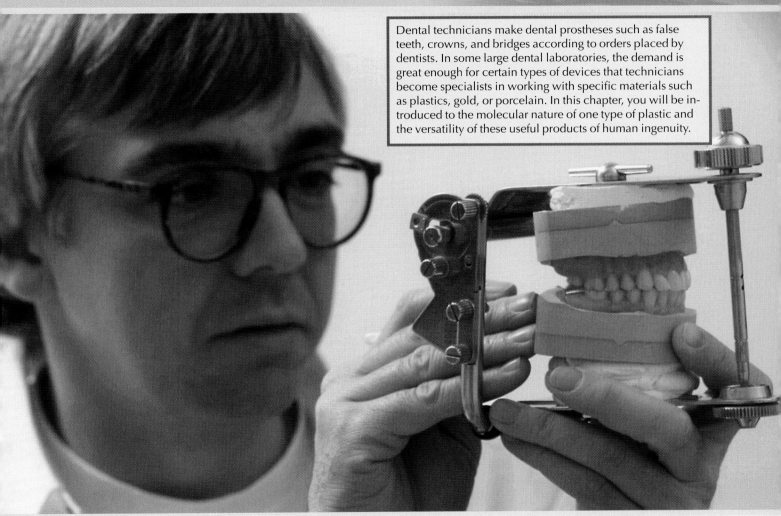

Dental technicians make dental prostheses such as false teeth, crowns, and bridges according to orders placed by dentists. In some large dental laboratories, the demand is great enough for certain types of devices that technicians become specialists in working with specific materials such as plastics, gold, or porcelain. In this chapter, you will be introduced to the molecular nature of one type of plastic and the versatility of these useful products of human ingenuity.

LEARNING OBJECTIVES

When you have completed your study of this chapter, you should be able to:

1. Classify unsaturated hydrocarbons as alkenes, alkynes, or aromatics. (Section 12.1)

2. Write the IUPAC names of alkenes from their molecular structures. (Section 12.1)

3. Predict the existence of geometric (cis-trans) isomers from formulas of compounds. (Section 12.2)

4. Write the names and structural formulas for geometric isomers. (Section 12.2)

5. Write equations for addition reactions of alkenes, and use Markovnikov's rule to predict the major products of certain reactions. (Section 12.3)

6. Write equations for addition polymerization, and list uses for addition polymers. (Section 12.4)

7. Write the IUPAC names of alkynes from their molecular structures. (Section 12.5)

8. Classify organic compounds as aliphatic or aromatic. (Section 12.6)

9. Name and draw structural formulas for aromatic compounds. (Section 12.7)

10. Recognize uses for specific aromatic compounds. (Section 12.8)

Unsaturated hydrocarbon
A hydrocarbon containing one or more multiple bonds.

Alkene
A hydrocarbon containing one or more double bonds.

Alkyne
A hydrocarbon containing one or more triple bonds.

Aromatic hydrocarbons
Any organic compound that contains the characteristic benzene ring or similar feature.

Unsaturated hydrocarbons contain one or more double or triple bonds between carbon atoms and belong to one of three classes: alkenes, alkynes, or aromatic hydrocarbons. **Alkenes** contain one or more double bonds, **alkynes** contain one or more triple bonds, and **aromatic hydrocarbons** contain three double bonds alternating with three single bonds in a six-carbon ring. Ethylene (the simplest alkene), acetylene (the simplest alkyne), and benzene (the simplest aromatic) are represented by the following structural formulas:

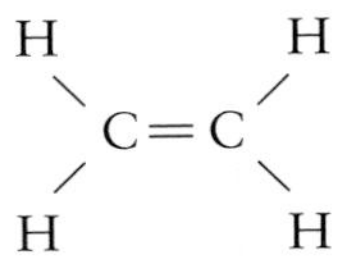

ethylene
(an alkene)

acetylene
(an alkyne)

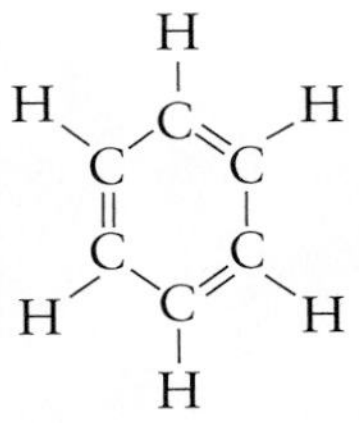

benzene
(an aromatic)

Alkenes and alkynes are called *unsaturated* because more hydrogen atoms can be added in somewhat the same sense that more solute can be added to an unsaturated solution. Benzene and other aromatic hydrocarbons also react to add hydrogen atoms; in general, however, they have chemical properties very different from those of alkenes and alkynes.

➤ 12.1 The Nomenclature of Alkenes

The general formula for alkenes is C_nH_{2n} (the same as that for cycloalkanes). The simplest members are well known by their common names, ethylene and propylene:

$$CH_2=CH_2 \qquad CH_3-CH=CH_2$$
ethylene, C_2H_4 \qquad propylene, C_3H_6

Three structural isomers have the formula C_4H_8:

$$CH_3-CH_2-CH=CH_2 \qquad CH_3-CH=CH-CH_3 \qquad CH_3-C=CH_2$$
$$|$$
$$CH_3$$

The number of structural isomers increases rapidly as the number of carbons increases because, besides variations in chain length and branching, variations occur in the position of the double bond. IUPAC nomenclature is extremely useful in differentiating among these many alkene compounds.

The IUPAC rules for naming alkenes are similar to those used for the alkanes, with a few additions to indicate the presence and location of double bonds.

Step 1. Name the longest chain that contains the double bond. The characteristic name ending is *-ene*.

Step 2. Number the longest chain of carbon atoms so that the carbon atoms joined by the double bond have numbers as low as possible.

Step 3. Locate the double bond by the lower-numbered carbon atom bound by the double bond.

Step 4. Locate and name attached groups.

Step 5. Combine the names for the attached groups and the longest chain into the name.

GOB
Chemistry•⚛•Now™

Go to GOB Now and click to learn the naming system used for alkenes.

Example 12.1

Name the following alkenes:

a. $CH_3-CH=CH-CH_3$

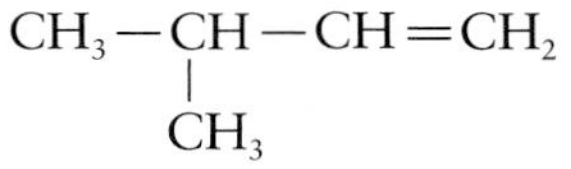

c. 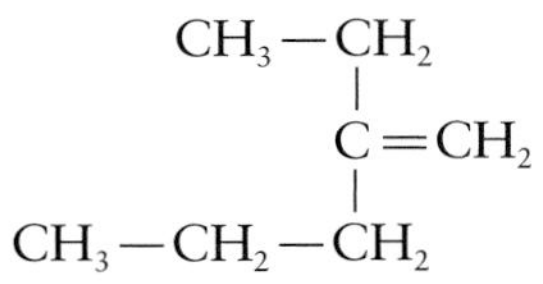
$$CH_3-CH_2$$
$$|$$
$$C=CH_2$$
$$|$$
$$CH_3-CH_2-CH_2$$

b.
$$CH_3-CH-CH=CH_2$$
$$|$$
$$CH_3$$

d.

$$CH_2$$
$$|$$
$$CH_3$$

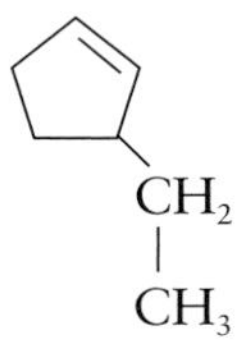

Solution

a. The longest chain containing a double bond has four carbon atoms. The four-carbon alkane is butane. Thus, the compound is a butene:

$$\overset{1}{C}H_3-\overset{2}{C}H=\overset{3}{C}H-\overset{4}{C}H_3$$

The chain can be numbered from either end because the double bond will be between carbons 2 and 3 either way. The position of the double bond is indicated by the lower-numbered carbon atom that is double bonded, carbon 2 in this case. The name is 2-butene.

b. To give lower numbers to the carbons bound by the double bond, the chain is numbered from the right:

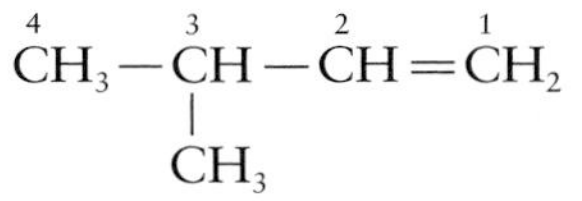

$$\overset{4}{C}H_3-\overset{3}{C}H-\overset{2}{C}H=\overset{1}{C}H_2 \qquad \text{not} \qquad \overset{1}{C}H_3-\overset{2}{C}H-\overset{3}{C}H=\overset{4}{C}H_2$$
$$\qquad\quad |\qquad\qquad\qquad\qquad\qquad\qquad\quad |$$
$$\qquad\quad CH_3\qquad\qquad\qquad\qquad\qquad\qquad CH_3$$

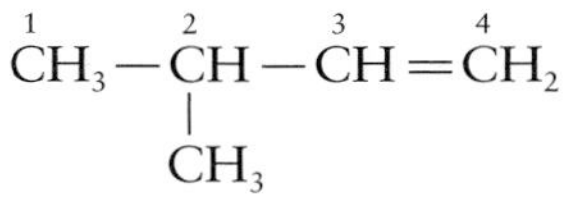

Thus, the compound is a 1-butene with an attached methyl group on carbon 3. Therefore, the name is 3-methyl-1-butene.

c. Care must be taken to select the longest chain containing the double bond. This compound is named as a pentene and not as a hexene because the double bond is not contained in the six-carbon chain:

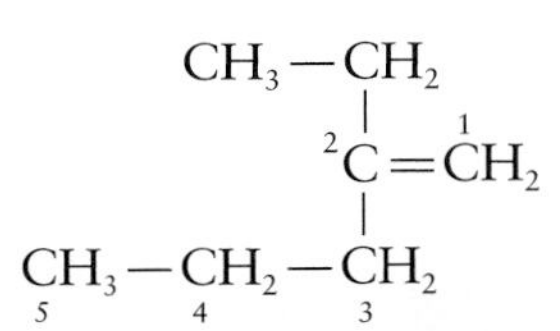

$$CH_3-CH_2 \qquad\qquad\qquad CH_3-\overset{2}{C}H_2$$
$$|\qquad\qquad\qquad\qquad\qquad\qquad\qquad\quad |$$
$$\overset{2}{C}=\overset{1}{C}H_2 \qquad \text{not} \qquad \overset{3}{C}=CH_2$$
$$|\qquad\qquad\qquad\qquad\qquad\qquad\qquad\quad |$$
$$CH_3-CH_2-CH_2 \qquad\qquad CH_3-CH_2-CH_2$$
$$\,\,\,5\qquad\,\,4\qquad\,\,3\qquad\qquad\qquad\quad 6\qquad\,\,5\qquad\,\,4$$

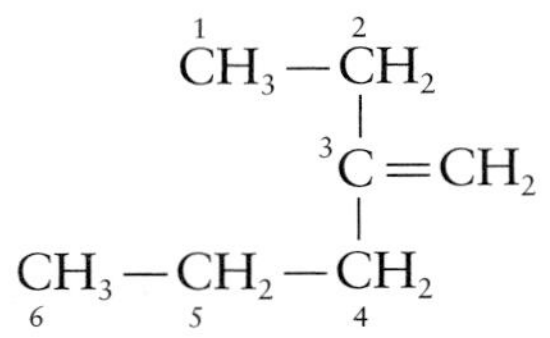

The compound is a 1-pentene with an ethyl group at position 2. Therefore, the name is 2-ethyl-1-pentene.

d. In cyclic alkenes, the ring is numbered so as to give the lowest possible numbers to the double-bonded carbons (they become carbons 1 and 2). The numbering direction around the ring is chosen so that attached groups are located on the lowest-numbered carbon atoms possible. Thus, the name is 3-ethylcyclopentene:

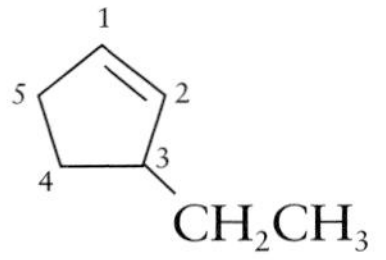

$$CH_2CH_3$$

Notice that it is not called 3-ethyl-1-cyclopentene because the double bond is always between carbons 1 and 2, and therefore its position need not be indicated.

Learning Check 12.1

Give the IUPAC name for each of the following:

a.
$$\underset{\text{Br}}{\overset{|}{\text{CH}_2}}-\text{CH}=\text{CH}_2$$

b.
$$\text{CH}_2=\text{C}-\text{CH}_2-\text{CH}_2-\text{CH}_3$$
$$\overset{|}{\underset{|}{\text{CH}_2}}$$
$$\text{CH}_3$$

c.

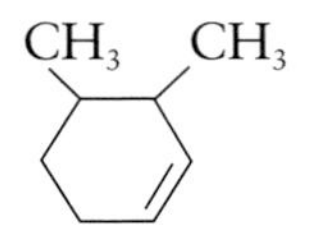

Some compounds contain more than one double bond per molecule. Molecules of this type are important components of natural and synthetic rubber and other useful materials. The nomenclature of these compounds is the same as for the alkenes with one double bond, except that the endings *-diene*, *-triene*, and the like are used to denote the number of double bonds. Also, the locations of all the multiple bonds must be indicated in all molecules, including those with rings:

$$\overset{1}{\text{CH}_2}=\overset{2}{\text{CH}}-\overset{3}{\text{CH}}=\overset{4}{\text{CH}_2}$$

1,3-butadiene

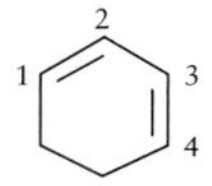

1,3-cyclohexadiene

CHEMISTRY AROUND US • 12.1

SEEING THE LIGHT

Cis-trans isomerism is important in several biological processes, one of which is vision. When light strikes the retina, a cis double bond in the compound *retinal* (structurally related to vitamin A) is converted to a trans double bond. The conversion triggers a chain of events that finally results in our being able to see.

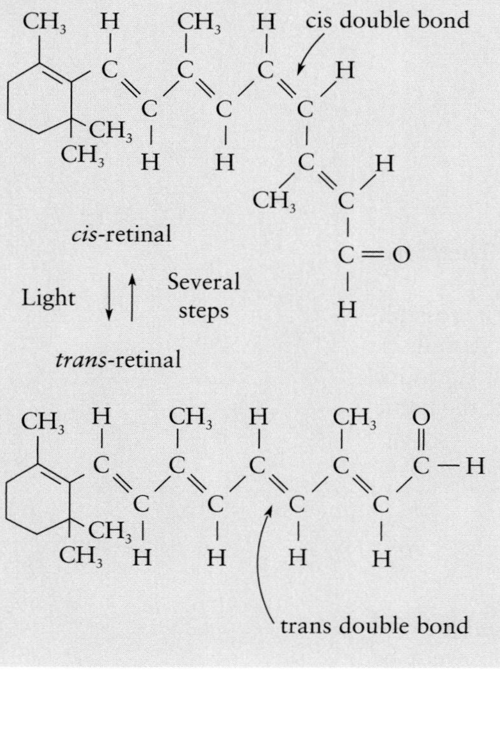

In a series of steps, *trans*-retinal is enzymatically converted back to *cis*-retinal so that the cycle can be repeated. Bright light temporarily destroys our ability to see in dim light because large quantities of *cis*-retinal are rapidly converted to the trans isomer by the bright light. It takes time for conversion of the *trans*-retinal back to *cis*-retinal.

cis-retinal

Light ↓↑ Several steps

trans-retinal

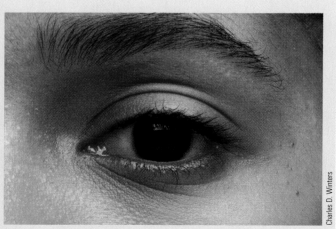

The vision process depends on a cis-trans reaction.

Charles D. Winters

Example 12.2

Name the following compounds:

a. 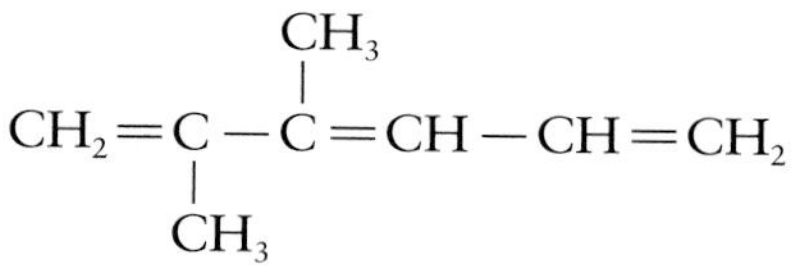 CH$_2$=C—C=CH—CH=CH$_2$ (with CH$_3$ above and CH$_3$ below the second carbon)

b. 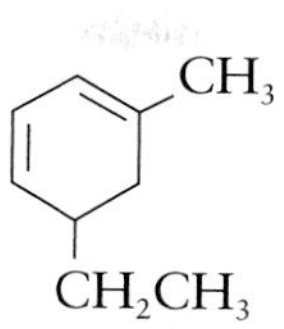 (cyclohexadiene ring with CH$_3$ and CH$_2$CH$_3$ substituents)

Solution

a. This compound is a methyl-substituted hexatriene. The chain is numbered from the end nearest the branch because the direction of numbering, again, makes no difference in locating the double bonds correctly. The name is 2,3-dimethyl-1,3,5-hexatriene:

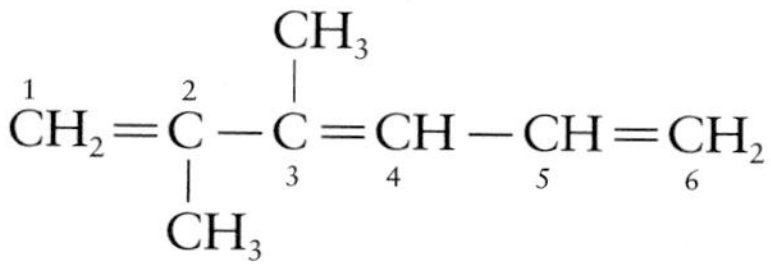

$$\overset{1}{C}H_2=\overset{2}{C}-\overset{3}{C}=\overset{4}{C}H-\overset{5}{C}H=\overset{6}{C}H_2 \quad \text{and not} \quad \overset{6}{C}H_2=\overset{5}{C}-\overset{4}{C}=\overset{3}{C}H-\overset{2}{C}H=\overset{1}{C}H_2$$

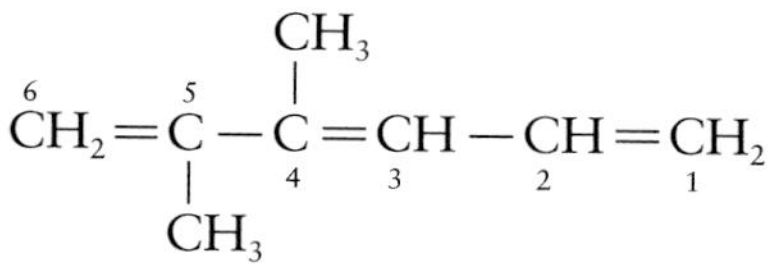

2,3-dimethyl-1,3,5-hexatriene   4,5-dimethyl-1,3,5-hexatriene

b. This compound is a substituted cyclohexadiene. The ring is numbered as shown. The name is 5-ethyl-1-methyl-1,3-cyclohexadiene:

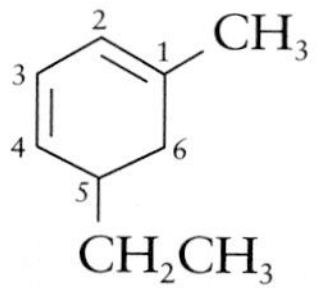

(ring structure with CH$_3$ and CH$_2$CH$_3$)   and not   (ring structure with CH$_3$ and CH$_2$CH$_3$)

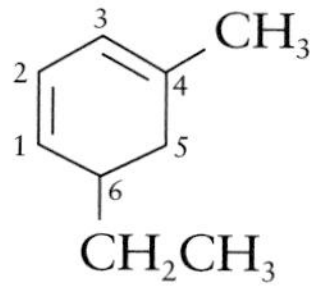

5-ethyl-1-methyl-1,3-cyclohexadiene   6-ethyl-4-methyl-1,3-cyclohexadiene

---

Give the IUPAC name for each of the following:

a. 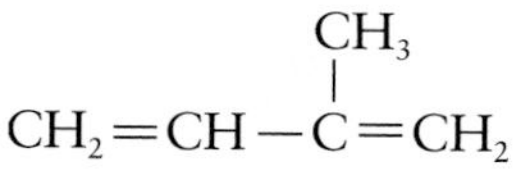 CH$_2$=CH—C=CH$_2$ (with CH$_3$ above the third carbon)

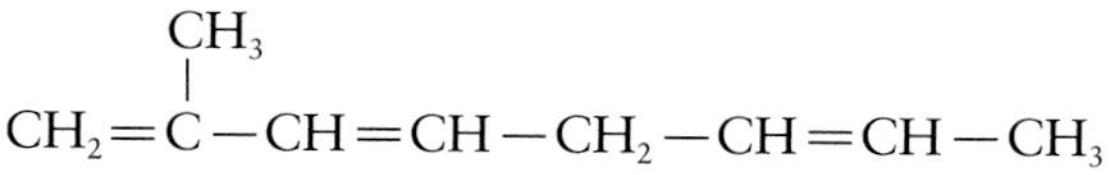

b. CH$_2$=C—CH=CH—CH$_2$—CH=CH—CH$_3$ (with CH$_3$ above the second carbon)

c. Br

(seven-membered ring with double bonds and Br substituent)

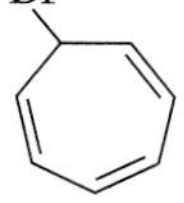

► 12.2 The Geometry of Alkenes

The hybridization of atomic orbitals discussed in Section 11.3 to explain the bonding characteristics of carbon atoms bonded to four other atoms can also be used to describe alkenes, compounds in which some carbon atoms are bonded to only three atoms. This hybridization involves mixing a 2s orbital and two 2p orbitals of a carbon atom to form three hybrid sp^2 orbitals (see ■ Figure 12.1 on page 385).

CHEMISTRY AROUND US • 12.2

WATERMELON: A SOURCE OF LYCOPENE

Summer conjures up images of picnic tables and cool, fresh slices of watermelon. What most watermelon eaters don't realize is that in addition to enjoying a cool summer snack, they are also benefitting from a great source of lycopene, a red substance with the following highly unsaturated molecular structure:

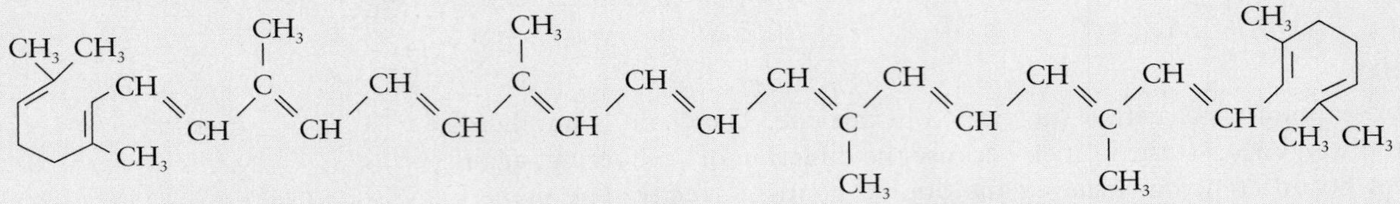

Lycopene is known to help prevent certain types of cancer as well as heart disease, and watermelon is one of a small number of foods that contain this useful compound in large quantities. Other good food sources of lycopene are tomatoes, guava, and pink grapefruit.

The anti-cancer characteristics of lycopene are attributed to its antioxidant properties. It reacts with highly reactive oxygen-containing molecules that can oxidize cell components and cause the cells to malfunction. Research results indicate that tomatoes in the diet—especially cooked tomatoes, which contain concentrated amounts of lycopene—reduce the incidence of prostate cancer. In a study conducted at Harvard University, the incidence of prostate cancer was one-third lower in men who ate a lycopene-rich diet compared to a group who ate a low-lycopene diet.

It was long thought that heat-processed tomatoes represented the best source of lycopene in the diet. This was based on the large amounts of lycopene found in small servings of tomato juice. However, recent studies have found that red seedless watermelons contain as much lycopene as cooked tomatoes and in some cases more, depending on the variety of melon and the growing conditions.

Another factor to consider when looking for a good lycopene source is the ability of the body to digest and use the compound (bioavailability) when a lycopene-containing food is eaten. For example, the lycopene in cooked tomatoes is absorbed more readily during digestion than the lycopene in raw tomatoes. Also, it has been found that the absorption rate of lycopene goes up if the food containing it is eaten with food that contains fat. This characteristic is related to the nonpolar nature of lycopene molecules and their resulting significant solubility in fats. The level of lycopene in body fat is used as an indication of how much lycopene has been consumed in the diet. Increased lycopene levels in fat tissue have been linked to reduced risk of heart attack. The bioavailability of lycopene from raw watermelon has been found to be equal to that of lycopene obtained from cooked tomatoes. Studies are now being conducted to determine if the bioavailability of watermelon lycopene increases if the watermelon is eaten in a diet that also includes fat-containing foods.

In addition to its ability to satisfy a sweet tooth and serve as a good source of lycopene, watermelon is also an excellent source of the vitamins A, B_6, C, and thiamin. It is also fat free and low in calories, so why not include a big juicy slice of it as one of the five servings of fruits and vegetables recommended by the American Institute of Cancer Research as a way to reduce your risk of cancer? Enjoy yourself and try not to drip juice on your clothes.

Watermelon is an excellent source of lycopene.

© Rick Engenbright/CORBIS

The three sp^2 hybrid orbitals lie in the same plane and are separated by angles of 120°. The unhybridized $2p$ orbital of the carbon atom is located perpendicular to the plane of the sp^2 hybrid orbitals (see ■ Figure 12.2).

The bonding between the carbon atoms of ethylene results partially from the overlap of one sp^2 hybrid orbital of each carbon to form a sigma (σ) bond. The second carbon–carbon bond is formed when the unhybridized $2p$ orbitals of the carbons overlap sideways to form what is called a pi (π) bond. The

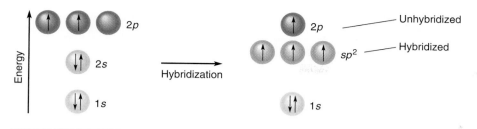

➤ **FIGURE 12.1** A representation of sp^2 hybridization of carbon. During hybridization, two of the $2p$ orbitals mix with the single $2s$ orbital to produce three sp^2 hybrid orbitals. One $2p$ orbital is not hybridized and remains unchanged.

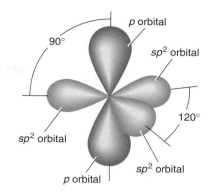

➤ **FIGURE 12.2** The unhybridized p orbital is perpendicular to the plane of the three sp^2 hybridized orbitals.

^{GOB}
Chemistry◦◦Now™

Go to GOB Now and click to see how to determine which type of hybrid orbitals an atom uses.

remaining two sp^2 hybrid orbitals of each carbon overlap with the $1s$ orbitals of the hydrogen atoms to form sigma bonds. Thus, each ethylene molecule contains five sigma bonds (one carbon–carbon and four carbon–hydrogen) and one pi bond (carbon–carbon), as shown in ■ Figure 12.3.

Experimental data support this hybridization model. Ethylene has been found to be a planar molecule with bond angles close to 120° between the atoms (see ■ Figure 12.4).

In addition to geometry, alkenes also differ from open-chain alkanes in that the double bonds prevent the relatively free rotation that is characteristic of carbon atoms bonded by single bonds. As a result, alkenes can exhibit geometric isomerism, the same type of stereoisomerism seen earlier for the cycloalkanes (Section 11.9). There are two geometric isomers of 2-butene:

$$\underset{\text{cis-2-butene}}{\overset{\displaystyle \begin{array}{ccc} H & & H \\ \diagdown & & \diagup \\ & C=C & \\ \diagup & & \diagdown \\ CH_3 & & CH_3 \end{array}}{}}
\qquad
\underset{\text{trans-2-butene}}{\overset{\displaystyle \begin{array}{ccc} CH_3 & & H \\ \diagdown & & \diagup \\ & C=C & \\ \diagup & & \diagdown \\ H & & CH_3 \end{array}}{}}$$

Once again, the prefix *cis-* is used for the isomer in which the two similar or identical groups are on the same side of the double bond and *trans-* for the one in which they are on opposite sides. The two isomers *cis-* and *trans-*2-butene represent distinctly different compounds with different physical properties (see ■ Table 12.1).

Not all double-bonded compounds show cis-trans stereoisomerism. Cis-trans stereoisomerism is found only in alkenes that have two different groups

^{GOB}
Chemistry◦◦Now™

Go to GOB Now and click to examine how pi bonds form.

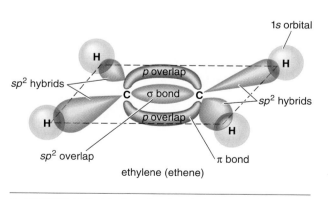

➤ **FIGURE 12.3** The bonding in ethylene is explained by combining two sp^2 hybridized carbon atoms. The C—H sigma bonds all lie in the same plane. The unhybridized p orbitals of the carbon atoms overlap to form the pi bond.

➤ **FIGURE 12.4** All six atoms of ethylene lie in the same plane.

TABLE 12.1 Physical properties of a pair of geometric isomers

Isomer	Melting point (°C)	Boiling point (°C)	Density (g/mL)
cis-2-butene	−139.9	3.7	0.62
trans-2-butene	−105.6	0.9	0.60

attached to each double-bonded carbon atom. In 2-butene, the two different groups are a methyl and a hydrogen for each double-bonded carbon:

two groups are different

cis-2-butene

two groups are different

two groups are different

trans-2-butene

If either double-bonded carbon is attached to identical groups, no cis-trans isomers are possible. Thus, there are no geometric isomers of ethene or propene:

two groups are the same

ethene

two groups are the same

propene

Chemistry⠂Now™

Go to GOB Now and click to learn how to name alkenes.

To see why this is so, let's try to draw geometric isomers of propene:

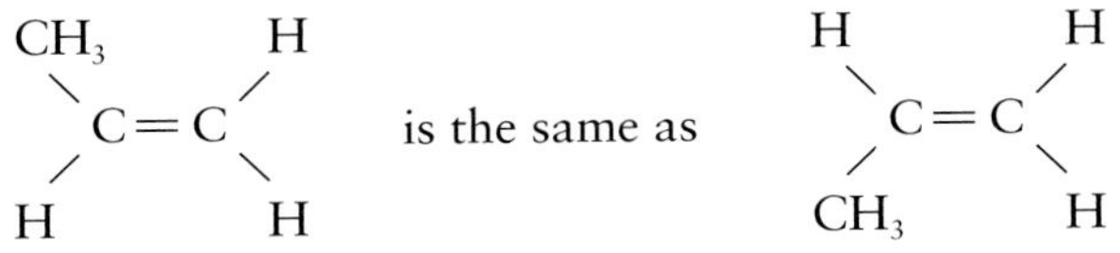

is the same as

(a)   (b)

Notice that structure (b) can be converted into (a) by just flipping it over. Thus, they are identical and not isomers.

Example 12.3

Determine which of the following molecules can exhibit geometric isomerism, and draw structural formulas to illustrate your conclusions:

a. $Cl—CH{=}CH—Cl$
b. $CH_2{=}CH—Cl$
c. $Cl—CH{=}CH—CH_3$

Solution

a. Begin by drawing the carbon–carbon double bond with bond angles of 120° about each carbon atom:

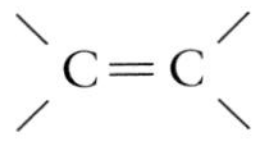

Next, complete the structure and analyze each carbon of the double bond to see if it is attached to two different groups:

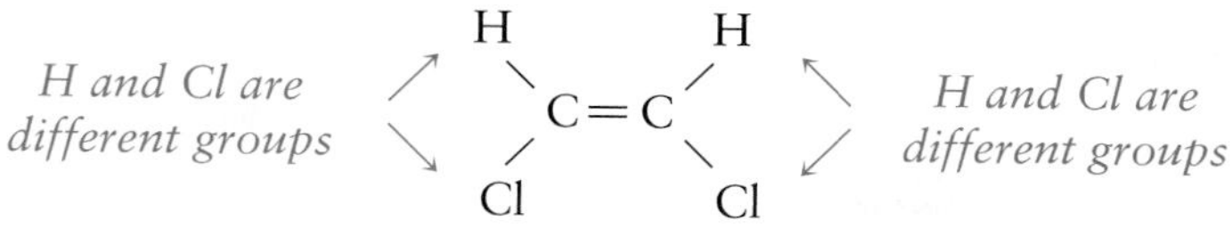

$$
\begin{array}{c}
\text{H}\qquad\text{H}\\
\diagup\quad\diagdown\;\diagup\quad\\
\text{C}=\text{C}\\
\diagup\quad\diagdown\\
\text{Cl}\qquad\text{Cl}
\end{array}
$$

H and Cl are different groups → ← *H and Cl are different groups*

In this case, each carbon is attached to two different groups and geometric isomers are possible:

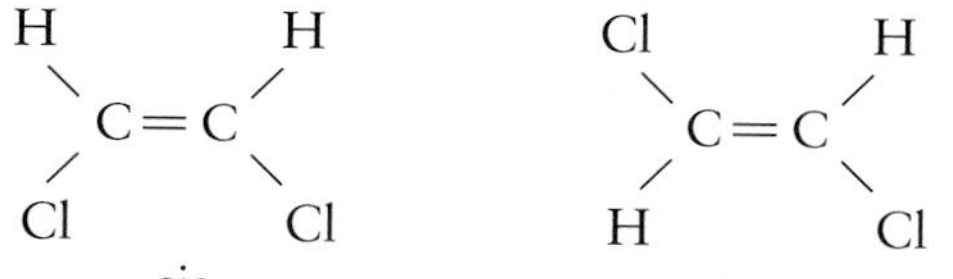

$$
\begin{array}{cc}
\text{H}\qquad\text{H} & \text{Cl}\qquad\text{H}\\
\diagdown\quad\diagup & \diagdown\quad\diagup\\
\text{C}=\text{C} & \text{C}=\text{C}\\
\diagup\quad\diagdown & \diagup\quad\diagdown\\
\text{Cl}\qquad\text{Cl} & \text{H}\qquad\text{Cl}\\
\text{cis} & \text{trans}
\end{array}
$$

b. No geometric isomers are possible because one carbon contains two identical groups:

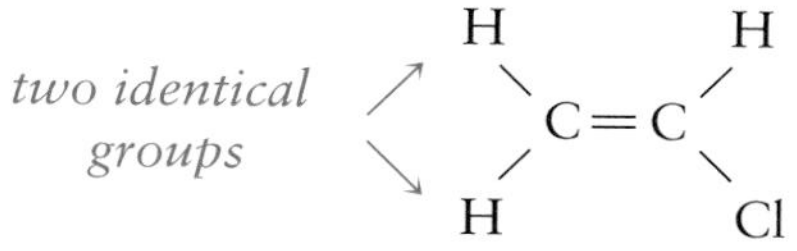

two identical groups →

$$
\begin{array}{c}
\text{H}\qquad\text{H}\\
\diagdown\quad\diagup\\
\text{C}=\text{C}\\
\diagup\quad\diagdown\\
\text{H}\qquad\text{Cl}
\end{array}
$$

c. Geometric isomers are possible because there are two different groups on each carbon:

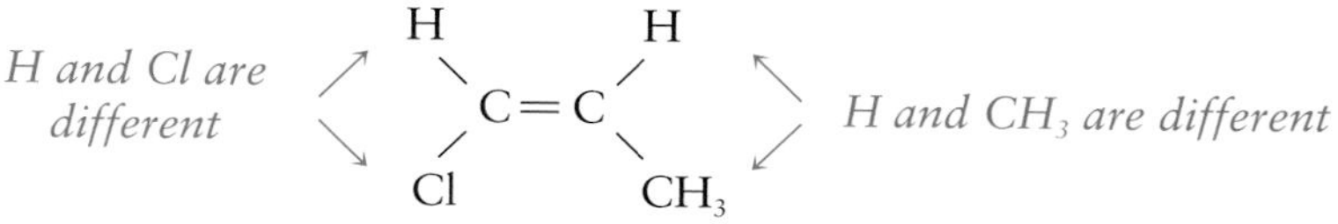

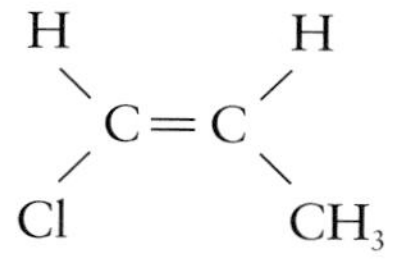

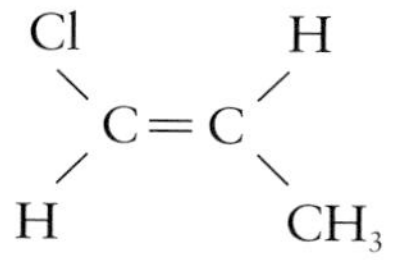

H and Cl are different →

$$
\begin{array}{c}
\text{H}\qquad\text{H}\\
\diagdown\quad\diagup\\
\text{C}=\text{C}\\
\diagup\quad\diagdown\\
\text{Cl}\qquad\text{CH}_3
\end{array}
$$

← *H and CH_3 are different*

$$
\begin{array}{cc}
\text{H}\qquad\text{H} & \text{Cl}\qquad\text{H}\\
\diagdown\quad\diagup & \diagdown\quad\diagup\\
\text{C}=\text{C} & \text{C}=\text{C}\\
\diagup\quad\diagdown & \diagup\quad\diagdown\\
\text{Cl}\qquad\text{CH}_3 & \text{H}\qquad\text{CH}_3\\
\textit{cis}\text{-1-chloropropene} & \textit{trans}\text{-1-chloropropene}
\end{array}
$$

Learning Check 12.3

Determine which of the following can exhibit geometric isomerism, and draw structural formulas for the cis and trans isomers of those that can:

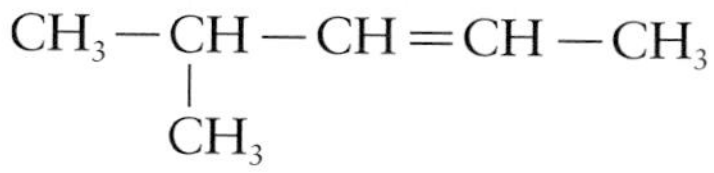

a. 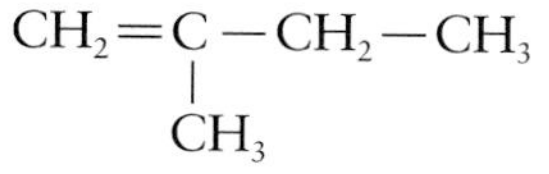 $CH_2{=}C{-}CH_2{-}CH_3$
         |
         CH_3

b. $CH_3{-}CH{-}CH{=}CH{-}CH_3$
             |
             CH_3

c. $CH_3{-}C{=}CH{-}CH_3$
       |
       CH_3

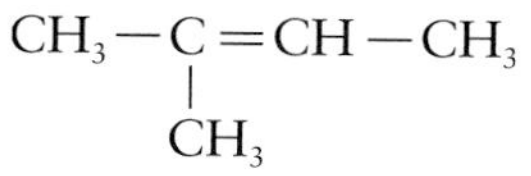

➤ 12.3 Properties of Alkenes

Physical Properties

The alkenes are similar to the alkanes in physical properties—they are non-polar compounds that are insoluble in water, soluble in nonpolar solvents, and less dense than water (see ■ Table 12.2). Alkenes containing 4 carbon atoms or fewer are gases under ordinary conditions. Those containing 5 to 17 carbon atoms are liquids, and those with 18 or more carbon atoms are solids. Low molecular weight alkenes have somewhat unpleasant, gasoline-like odors.

TABLE 12.2 Physical properties of some alkenes

IUPAC name	Structural formula	Boiling point (°C)	Melting point (°C)	Density (g/mL)
ethene	$CH_2{=}CH_2$	−104	−169	0.38
propene	$CH_2{=}CHCH_3$	−47	−185	0.52
1-butene	$CH_2{=}CHCH_2CH_3$	−6	−185	0.60
1-pentene	$CH_2{=}CHCH_2CH_2CH_3$	30	−138	0.64
1-hexene	$CH_2{=}CHCH_2CH_2CH_2CH_3$	63	−140	0.67
cyclohexene	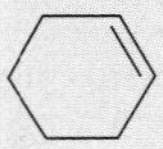	83	−104	0.81

GOB
Chemistry·⚛·Now™

Go to GOB Now and click to explore the reactivity of alkenes.

Chemical Properties

In contrast to alkanes, which are inert to almost all chemical reagents (Section 11.11), alkenes are quite reactive chemically. Since the only difference between an alkane and an alkene is the double bond, it is not surprising to learn that most of the reactions of alkenes take place at the double bond. These reactions follow the pattern

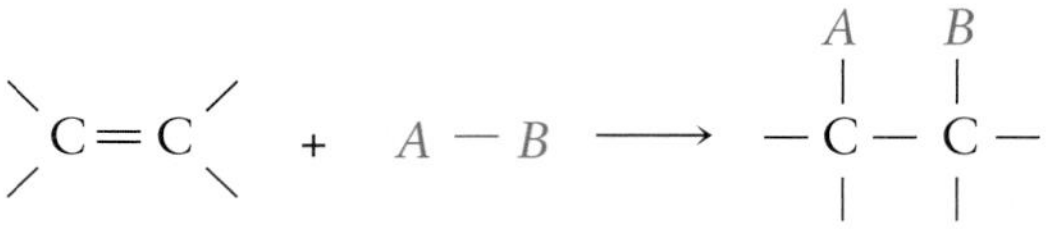

$$\overset{\diagdown}{\underset{\diagup}{}}C=C\overset{\diagup}{\underset{\diagdown}{}} \quad + \quad A-B \quad \longrightarrow \quad -\overset{\overset{\displaystyle A}{\mid}}{C}-\overset{\overset{\displaystyle B}{\mid}}{C}- \tag{12.1}$$

Addition reaction
A reaction in which a compound adds to a multiple bond.

Haloalkane or alkyl halide
A derivative of an alkane in which one or more hydrogens are replaced by halogens.

They are called **addition reactions** because a substance is added to the double bond. Addition reactions are characterized by two reactants that combine to form one product.

Halogenation is one of the most common addition reactions. For example, when bromine, a halogen, is added to an alkene, the double bond reacts, and only a carbon–carbon single bond remains in the product. Without a double bond present, the product is referred to as a **haloalkane** or **alkyl halide.** Reaction 12.2 gives the general reaction, and Reaction 12.3 is a specific example, the bromination of 1-butene.

$$\text{General reaction:} \quad \overset{\diagdown}{\underset{\diagup}{}}C=C\overset{\diagup}{\underset{\diagdown}{}} \quad + \quad Br-Br \quad \longrightarrow \quad -\overset{\mid}{\underset{\underset{\displaystyle Br}{\mid}}{C}}-\overset{\mid}{\underset{\underset{\displaystyle Br}{\mid}}{C}}- \tag{12.2}$$

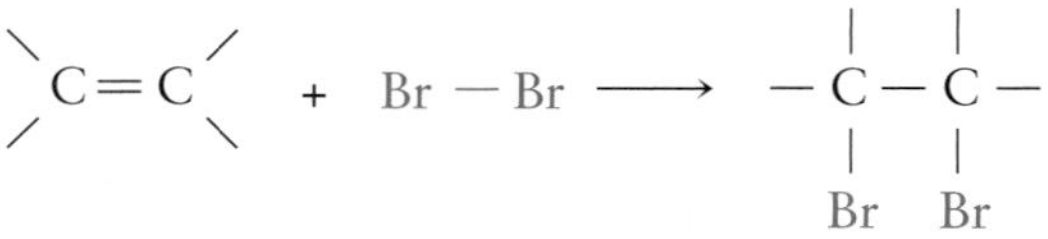

$$\text{Specific example:} \quad CH_3-CH_2-CH=CH_2 \; + \; Br-Br \; \longrightarrow \; CH_3-CH_2-\underset{\underset{\displaystyle Br}{\mid}}{CH}-\underset{\underset{\displaystyle Br}{\mid}}{CH_2} \tag{12.3}$$

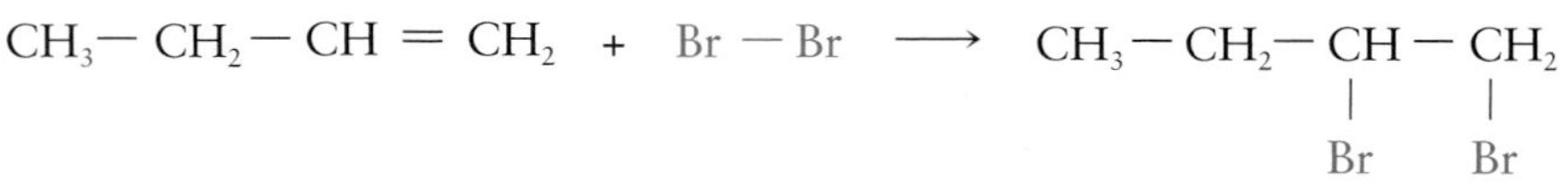

1-butene   1,2-dibromobutane

The addition of Br_2 to double bonds provides a simple laboratory test for unsaturation (see ■ Figure 12.5). As the addition takes place, the characteristic red-brown color of the added bromine fades as it is used up, and the colorless dibromoalkane product forms.

The addition of halogens is also used to quantitatively determine the degree of unsaturation in vegetable oils, margarines, and shortenings (Section 18.4). Chlorine reacts with alkenes to give dichloro products in an addition reaction

Dilute bromine solution added to 1-hexene loses its red-brown color immediately.

The remainder of the bromine solution is added. The last drops react as quickly as the first.

➤ **FIGURE 12.5** The reaction of bromine with an unsaturated hydrocarbon.

similar to that of bromine. However, it is not used as a test for unsaturation because it is difficult to see the pale green color of chlorine in solution.

Learning Check 12.4

Write the structural formula for the product of each of the following reactions:

a. $CH_3 - C = CH_2 + Br_2 \longrightarrow$
         $\quad\quad\;\; | $
         $\quad\quad CH_3$

b. 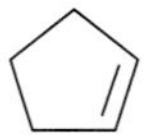 ⬠ + $Cl_2 \longrightarrow$

In the presence of an appropriate catalyst (such as platinum, palladium, or nickel), hydrogen adds to alkenes and converts them into the corresponding alkanes. This reaction, which is called **hydrogenation,** is illustrated in Reactions 12.4 and 12.5.

Hydrogenation
A reaction in which the addition of hydrogen takes place.

General reaction:
$$\overset{\diagdown}{\diagup}C = C\overset{\diagup}{\diagdown} \;+\; H-H \;\overset{Pt}{\longrightarrow}\; \underset{\underset{H}{|}}{-}\overset{|}{C}-\overset{|}{\underset{\underset{H}{|}}{C}}-$$
$$\text{an alkane}$$
(12.4)

Specific example:
$$CH_3CH = CHCH_3 + H-H \overset{Pt}{\longrightarrow} CH_3\underset{\underset{H}{|}}{C}H-\underset{\underset{H}{|}}{C}HCH_3$$
(12.5)

$\quad\quad\quad$ 2-butene $\quad\quad\quad\quad\quad\quad\quad\quad\quad$ butane

The hydrogenation of vegetable oils is a very important commercial process. Vegetable oils, such as soybean and cottonseed oil, are composed of long-chain organic molecules that contain many alkene bonds. The high degree of unsaturation characteristic of these oils gave rise to the term **polyunsaturated.** Upon hydrogenation, the melting point of the oils is raised, and the oils become low-melting-point solids. These products are used in the form of margarine and shortening.

Polyunsaturated
A term usually applied to molecules with several double bonds.

Learning Check 12.5

Write the structural formula for the product of each of the following reactions:

a.
$$CH_3 - \underset{\underset{CH_3}{|}}{C} = CH - CH_3 + H_2 \overset{Pt}{\longrightarrow}$$

b. 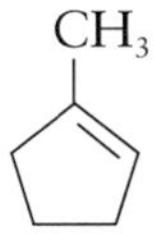
$$\overset{CH_3}{\underset{}{⬠}} + H_2 \overset{Pt}{\longrightarrow}$$

STUDY SKILLS 12.1

KEEPING A REACTION CARD FILE

Remembering organic reactions for exams is challenging for most students. Because the number of reactions being studied increases rapidly, it is a good idea to develop a systematic way to organize them for easy and effective review.

One way to do this is to focus on the functional group concept. When an exam question asks you to complete a reaction by identifying the product, your first step should be to identify the functional group of the reactant. Usually, only the functional group portion of a molecule undergoes reaction; in addition, a particular functional group usually undergoes the same characteristic reactions regardless of the other features of the organic molecule to which it is bound. Thus, by remembering the behavior of a functional group under specific conditions, you can predict the reactions of many compounds, no matter how complex the structures look, as long as they contain the same functional group. For example, any structure containing a C=C will undergo reactions typical of alkenes. Other functional groups will be introduced in later chapters.

Keeping a reaction card file based on the functional group concept is a good way to organize reactions for review. Write the structures and names of the reactants on one side of an index card with an arrow showing any catalyst or special conditions. Write the product structure and name on the back of the card. We recommend that you do this for the general reaction (like those in the Key Reactions section at the end of most chapters) and for a specific example. Review your cards every day (this can even be done while waiting for a bus, etc.), and add to them as new reactions are studied. As an exam approaches, put aside the reactions you know well, and concentrate on the others in what should be a dwindling deck. This is an effective way to focus on learning what you don't know.

A number of acidic compounds, such as the hydrogen halides—HF, HCl, HBr, and HI—also add to alkenes to give the corresponding alkyl halide. The reaction with HCl is illustrated as follows:

General reaction:

$$\begin{array}{c}\diagdown\\ /\end{array}C=C\begin{array}{c}\diagup\\ \diagdown\end{array} + \quad H-Cl \longrightarrow \quad \underset{\underset{H}{|}}{-C}-\underset{\underset{Cl}{|}}{C}- \tag{12.6}$$

Specific example:

$$CH_3CH=CHCH_3 \quad + \quad H-Cl \longrightarrow \quad CH_3\underset{\underset{H}{|}}{C}H\underset{\underset{Cl}{|}}{C}HCH_3 \tag{12.7}$$

The addition reactions involving H_2, Cl_2, and Br_2 yield only one product because the same group (H and H or Br and Br) adds to each double-bonded carbon. However, with H—X, a different group adds to each carbon, and for certain alkenes, there are two possible products. For example, in the reaction of HBr with propene, two products might be expected: 1-bromopropane or 2-bromopropane,

$$CH_2=CH-CH_3 + H-Br \longrightarrow \underset{\underset{\text{1-bromopropane}}{}}{\overset{Br \quad H}{\underset{|\quad\;|}{CH_2-CH-CH_3}}} \text{ or } \underset{\underset{\text{2-bromopropane}}{}}{\overset{H \quad Br}{\underset{|\quad\;|}{CH_2-CH-CH_3}}} \tag{12.8}$$

Markovnikov's rule
In the addition of H—X to an alkene, the hydrogen becomes attached to the carbon atom that is already bonded to more hydrogens.

It turns out that only one product, 2-bromopropane, is formed in significant amounts. This fact, first reported in 1869 by Russian chemist Vladimir Markovnikov, gave rise to a rule for predicting which product will be exclusively or predominantly formed. According to **Markovnikov's rule,** when a molecule of the form H—X adds to an alkene, the hydrogen becomes attached to the carbon atom that is already bonded to more hydrogens. A phrase to help you remember this rule is "the rich get richer." Applying this rule to propene, we find

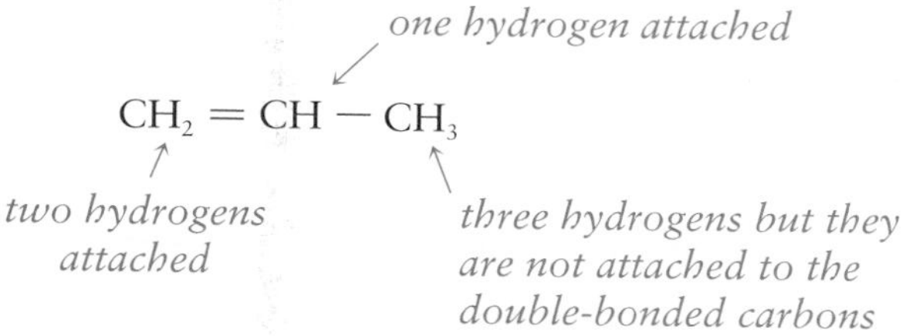

Chemistry·Now™

Go to GOB Now and click to view a simulation exploring Markovnikov's rule.

$$CH_2 = CH - CH_3$$

one hydrogen attached

two hydrogens attached

three hydrogens but they are not attached to the double-bonded carbons

Therefore, H attaches to the end carbon of the double bond and Br attaches to the second carbon, and 2-bromopropane is the major product.

Example 12.4

Use Markovnikov's rule to predict the major product in the following reactions:

a. $CH_3—C=CH_2 + H—Cl \longrightarrow$ (with CH_3 on the central C)

b. (cyclopentene with CH_3) + $H—Cl \longrightarrow$

Solution

a. Analyze the C=C to see which carbon atom has more hydrogens attached:

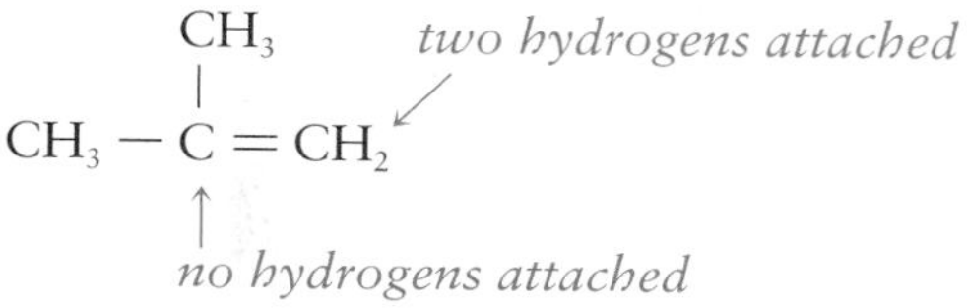

$$CH_3 - C = CH_2$$

two hydrogens attached

no hydrogens attached

The H of H—Cl will attach to the position that has more hydrogens. Thus, 2-chloro-2-methylpropane is the major product:

$$CH_3 - C = CH_2$$

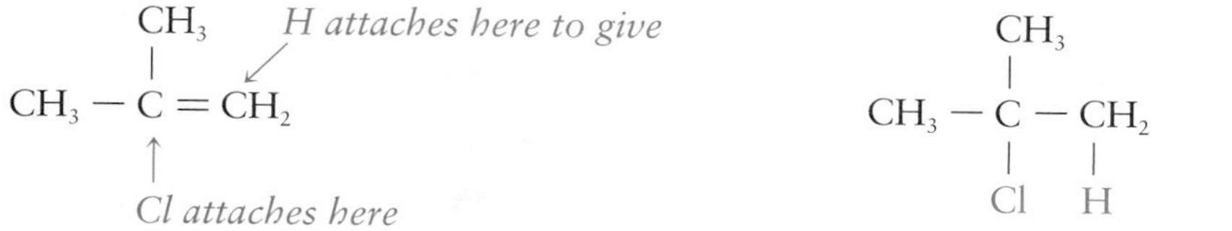

H attaches here to give

Cl attaches here

$$CH_3 - C - CH_2$$
with Cl and H below

2-chloro-2-methylpropane

b. The challenge with a cyclic alkene is to remember that the hydrogens are not shown. Thus,

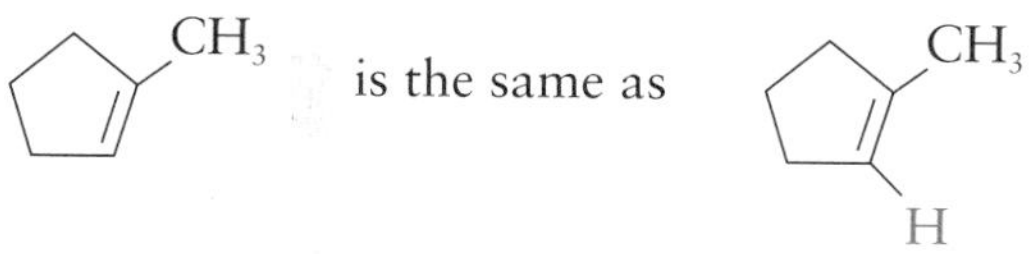

(cyclopentene with CH_3) is the same as (cyclopentene with CH_3 and H shown)

As before, the H of H—Cl will attach to the double-bonded carbon that has more hydrogens:

(cyclopentene, Cl attaches here, H attaches here, H) $\longrightarrow$ (product with CH_3, Cl, H, H) or (product with CH_3, Cl)

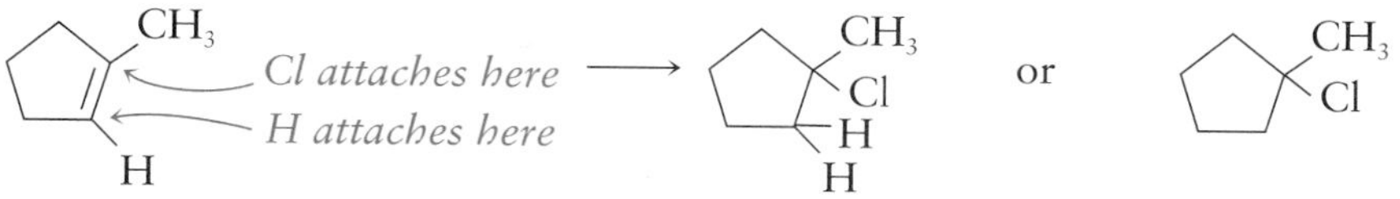

Learning Check 12.6   Use Markovnikov's rule to predict the major product in the following reactions:

$$
\text{a. } \underset{\quad}{CH_3-CH}=\underset{\underset{CH_3}{|}}{C}-CH_2-CH_3 + HBr
$$

b.

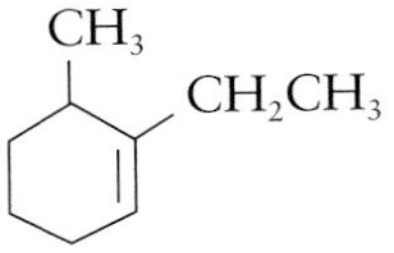

+ HBr $\longrightarrow$

Hydration
The addition of water to a multiple bond.

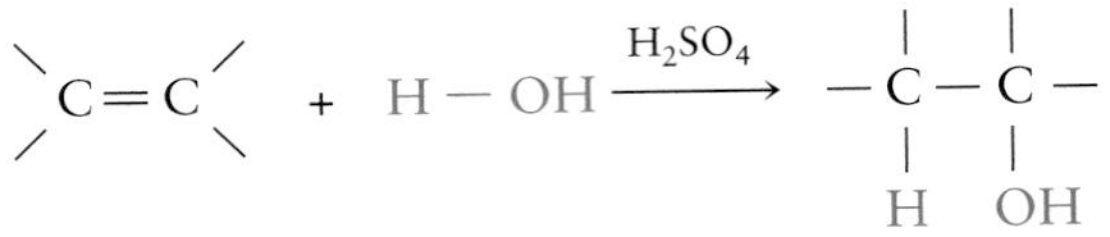

In the absence of a catalyst, water does not react with alkenes. But, if an acid catalyst such as sulfuric acid is added, water adds to carbon–carbon double bonds to give alcohols. In this reaction, which is called **hydration**, a water molecule is split in such a way that —H attaches to one carbon of the double bond, and —OH attaches to the other carbon. In Reactions 12.9–12.11, H_2O is written H—OH to emphasize the portions that add to the double bond. Notice that the addition follows Markovnikov's rule:

$$
\overset{\diagdown}{\underset{\diagup}{C}}=\overset{\diagup}{\underset{\diagdown}{C}} + H-OH \xrightarrow{H_2SO_4} \underset{\underset{H}{|}}{-\overset{|}{C}}-\underset{\underset{OH}{|}}{\overset{|}{C}}- \qquad (12.9)
$$

an alcohol

$$
CH_3CH=CHCH_3 + H-OH \xrightarrow{H_2SO_4} \underset{\underset{H}{|}\quad\underset{OH}{|}}{CH_3CH-CHCH_3} \qquad (12.10)
$$

2-butene   2-butanol

$$
CH_3CH=CH_2 + H-OH \xrightarrow{H_2SO_4} \underset{\underset{OH}{|}\quad\underset{H}{|}}{CH_3CH-CH_2} \qquad (12.11)
$$

propene   2-propanol

The hydration of alkenes provides a convenient method for preparing alcohols on a large scale. The reaction is also important in living organisms, but the catalyst is an enzyme rather than sulfuric acid. For example, one of the steps in the body's utilization of carbohydrates for energy involves the hydration of fumaric acid, which is catalyzed by the enzyme fumarase:

$$
\underset{\text{fumaric acid}}{\overset{\overset{O}{\|}}{HO-C}-CH=CH-\overset{\overset{O}{\|}}{C}-OH} + H_2O \xrightarrow{\text{fumarase}} \underset{\text{malic acid}}{\overset{\overset{O}{\|}\;\;\overset{H}{|}\;\;\;\overset{OH}{|}\;\;\overset{O}{\|}}{HO-C-CH-CH-C-OH}} \qquad (12.12)
$$

Learning Check 12.7   Draw structural formulas for the major organic product of each of the following reactions:

$$
\text{a. } CH_3CH_2CH_2CH=CH_2 + H_2O \xrightarrow{H_2SO_4}
$$

b.

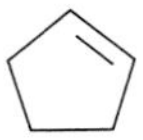

+ H_2O $\xrightarrow{H_2SO_4}$

► 12.4 Addition Polymers

Certain alkenes undergo a very important reaction in the presence of specific catalysts. In this reaction, alkene molecules undergo an addition reaction with one another. The double bonds of the reacting alkenes are converted to single bonds as hundreds or thousands of molecules bond and form long chains. For example, several ethylene molecules react as follows:

$$CH_2 = CH_2 + CH_2 = CH_2 + CH_2 = CH_2 + CH_2 = CH_2 \xrightarrow[\text{catalysts}]{\text{Heat, pressure,}}$$

ethylene molecules

(12.13)

$$-CH_2 - CH_2 - CH_2 - CH_2 - CH_2 - CH_2 - CH_2 - CH_2 -$$

polyethylene

The product is commonly called polyethylene even though there are no longer any double bonds present. The newly formed bonds in this long chain are shown in color. This type of reaction is called a **polymerization**, and the long-chain product made up of repeating units is a **polymer** (*poly* = many, *mer* = parts). The trade names of many polymers such as Orlon®, Plexiglas®, Lucite®, and Teflon® are familiar (see ■ Figure 12.6). These products are referred to as **addition polymers** because of the addition reaction between double-bonded compounds that is used to produce them. The starting materials that make up the repeating units of polymers are called **monomers** (*mono* = one, *mer* = part). Quite often, common names are used for both the polymer and the monomer.

It is not possible to give an exact formula for a polymer produced by a polymerization reaction because the individual polymer molecules vary in size. We could represent polymerization reactions as in Reaction 12.13. However, since this type of reaction is inconvenient, we adopt a commonly used approach: The polymer is represented by a simple repeating unit based on the monomer. For polyethylene, the unit is $+CH_2-CH_2+$. The large number of units making up the polymer is denoted by n, a whole number that varies from several hundred to several thousand. The polymerization reaction of ethylene is then written as

$$n CH_2 = CH_2 \xrightarrow[\text{catalysts}]{\text{Heat, pressure,}} +CH_2 - CH_2 +_n$$

ethylene polyethylene

(12.14)

► **FIGURE 12.6** Gore-Tex® is a thin, membranous material made by stretching Teflon fibers. Fabrics layered with Gore-Tex repel wind and rain but allow body perspiration to escape, making it an excellent fabric for sportswear.

Polymerization
A reaction that produces a polymer.

Polymer
A very large molecule made up of repeating units.

Addition polymer
A polymer formed by the linking together of many alkene molecules through addition reactions.

Monomer
The starting material that becomes the repeating units of polymers.

► STUDY SKILLS 12.2

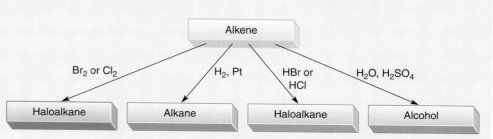

A REACTION MAP FOR ALKENES

A diagram may help you visualize and remember the four common addition reactions in this section. In each case, the alkene double bond reacts, and an alkane or alkane derivative is produced. The specific reagent determines the outcome of the reaction.

Alkene

Br₂ or Cl₂ → Haloalkane

H₂, Pt → Alkane

HBr or HCl → Haloalkane

H₂O, H₂SO₄ → Alcohol

➤ HOW REACTIONS OCCUR

THE HYDRATION OF ALKENES: AN ADDITION REACTION

The mechanism for the hydration of alkenes is believed to begin when H^+ from the acid catalyst is attracted to the electrons of the carbon–carbon double bond. The H^+ becomes bonded to one of the carbons by a sharing of electrons:

Step 1.

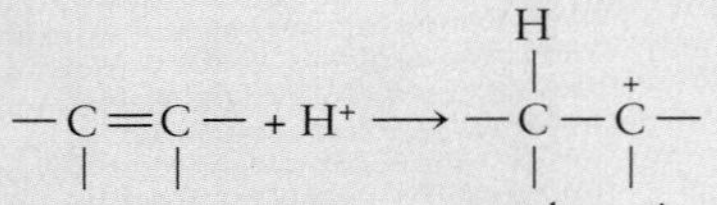

$$-C=C- \ + \ H^+ \longrightarrow -\overset{\displaystyle H}{\underset{\displaystyle |}{C}}-\overset{+}{C}-$$

a carbocation

This process leaves the second carbon with only three bonds about it and thus a positive charge. Such ions, referred to as **carbocations,** are extremely reactive.

As soon as it forms, the positive carbocation attracts any species that has readily available nonbonding electrons, whether it is an anion or a neutral molecule. In the case of water, the oxygen atom has two unshared pairs of electrons:

$$\overset{\displaystyle \cdot\cdot}{\underset{\displaystyle H \qquad H}{O}}$$

One pair of oxygen electrons forms a covalent bond with the carbocation:

Step 2.

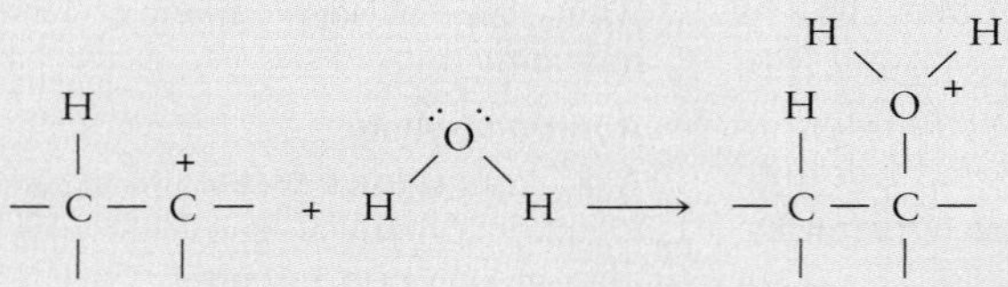

$$-\overset{\displaystyle H}{\underset{\displaystyle |}{C}}-\overset{+}{C}- \ + \ H\overset{\cdot\cdot}{\underset{}{O}}H \longrightarrow -\overset{\displaystyle H}{\underset{\displaystyle |}{C}}-\overset{\displaystyle H\ \diagdown \overset{+}{O}\ \diagup H}{\underset{\displaystyle |}{C}}-$$

In the third step, H^+ is lost to produce the alcohol

Step 3.

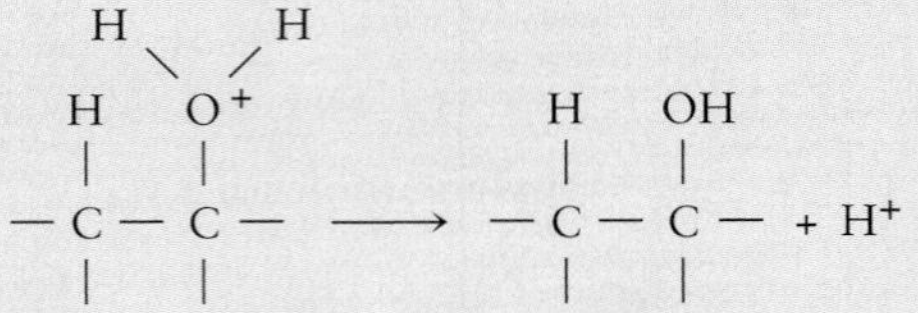

$$-\overset{\displaystyle H\ \diagdown \overset{+}{O}\ \diagup H}{\underset{\displaystyle |}{C}}-\overset{\displaystyle H}{\underset{\displaystyle |}{C}}- \longrightarrow -\overset{\displaystyle H}{\underset{\displaystyle |}{C}}-\overset{\displaystyle OH}{\underset{\displaystyle |}{C}}- \ + \ H^+$$

an alcohol

Notice that the acid catalyst, H^+, which initiated the reaction, is recovered unchanged in the final step of the mechanism.

By applying Step 2 of the mechanism for hydration, we can understand how HCl and HBr react with alkenes:

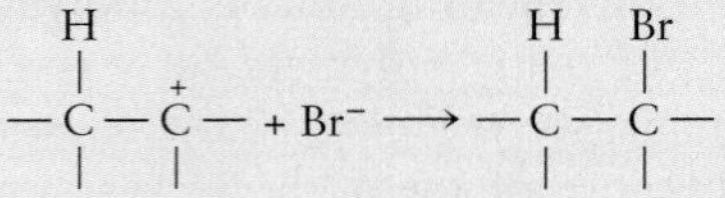

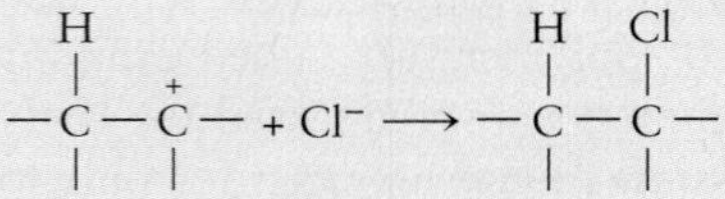

$$-\overset{\displaystyle H}{\underset{\displaystyle |}{C}}-\overset{+}{C}- \ + \ Br^- \longrightarrow -\overset{\displaystyle H}{\underset{\displaystyle |}{C}}-\overset{\displaystyle Br}{\underset{\displaystyle |}{C}}-$$

$$-\overset{\displaystyle H}{\underset{\displaystyle |}{C}}-\overset{+}{C}- \ + \ Cl^- \longrightarrow -\overset{\displaystyle H}{\underset{\displaystyle |}{C}}-\overset{\displaystyle Cl}{\underset{\displaystyle |}{C}}-$$

Carbocation
An ion of the form $-\overset{+}{\underset{\displaystyle |}{C}}-$.

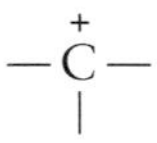

The lowercase n in Reaction 12.14 represents a large, unspecified number. From this reaction, we see that polyethylene is essentially a very long chain alkane. As a result, it has the chemical inertness of alkanes, a characteristic that makes polyethylene suitable for food storage containers, garbage bags, eating utensils, laboratory apparatus, and hospital equipment (see ■ Figure 12.7). Polymer characteristics are modified by using alkenes with different groups attached to either or both of the double-bonded carbons. For example, the polymerization of vinyl chloride gives the polymer poly (vinyl chloride), PVC:

$$n CH_2 = \overset{\displaystyle Cl}{\underset{\displaystyle |}{CH}} \ \xrightarrow[\text{catalysts}]{\text{Heat,}} \ +CH_2 - \overset{\displaystyle Cl}{\underset{\displaystyle |}{CH}}+_n \qquad (12.15)$$

vinyl chloride   poly (vinyl chloride)

Charles D. Winters

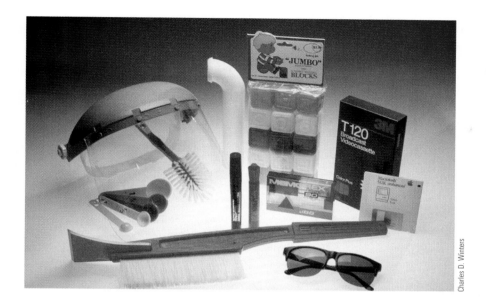

➤ **FIGURE 12.7** A large variety of items is made from polyethylene, including plastic containers, garbage bags, syringes, and laboratory equipment.

The commercial product Saran Wrap is an example of a **copolymer,** which is a polymer made up of two different monomers (Reaction 12.16):

Copolymer
An addition polymer formed by the reaction of two different monomers.

$$\underset{\text{vinyl chloride}}{n\mathrm{CH_2{=}\underset{\underset{\displaystyle Cl}{|}}{CH}}} + \underset{\text{vinylidene chloride}}{n\mathrm{CH_2{=}\overset{\overset{\displaystyle Cl}{|}}{\underset{\underset{\displaystyle Cl}{|}}{C}}}} \xrightarrow{\text{Catalyst}} \underset{\text{Saran Wrap}}{\left(\mathrm{CH_2 - \underset{\underset{\displaystyle Cl}{|}}{CH} - CH_2 - \overset{\overset{\displaystyle Cl}{|}}{\underset{\underset{\displaystyle Cl}{|}}{C}}} \right)_n} \qquad (12.16)$$

A number of the more important addition polymers are shown in ■ Table 12.3. As you can tell by looking at some of the typical uses, addition polymers have become nearly indispensable in modern life (see ■ Figure 12.8).

Learning Check 12.8

Draw the structural formula of a portion of polypropylene containing four repeating units of the monomer propylene,

$$\mathrm{CH_2{=}\overset{\overset{\displaystyle CH_3}{|}}{CH}}$$

© Frank Cezus/Tony Stone Worldwide

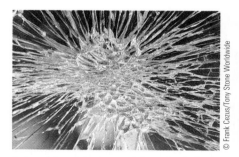

Automobile safety glass contains a sheet of poly (vinyl acetate) layered between two sheets of glass to prevent the formation of sharp fragments.

© Clark Taylor

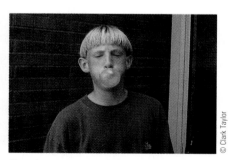

Chewing gum comes in many varieties.

➤ **FIGURE 12.8** Two uses of poly (vinyl acetate). What properties of poly (vinyl acetate) are exhibited in both of these products?

TABLE 12.3 Common addition polymers

Chemical name and trade name(s)	Monomer	Polymer	Typical uses
polyethylene	$CH_2=CH_2$	$-(CH_2-CH_2)_n$	Bottles, plastic bags, film
polypropylene	$CH_2=CH$ 丨 CH_3	$-(CH_2-CH)_n$ 丨 CH_3	Carpet fiber, pipes, bottles, artificial turf
poly (vinyl chloride) (PVC)	$CH_2=CH$ 丨 Cl	$-(CH_2-CH)_n$ 丨 Cl	Synthetic leather, floor tiles, garden hoses, water pipe
polytetraflouroethylene (Teflon®)	$CF_2=CF_2$	$-(CF_2-CF_2)_n$	Pan coatings, plumbers' tape, heart valves, fabrics
poly (methyl methacrylate) (Lucite®, Plexiglas®)	$CH_2=C$ 丨CH_3 $C-O-CH_3$ ‖O	$-(CH_2-C)_n$ 丨CH_3 $C-O-CH_3$ ‖O	Airplane windows, paint, contact lenses, fiber optics
poly (vinyl acetate)	$CH_2=CH$ 丨 $O-C-CH_3$ ‖O	$-(CH_2-CH)_n$ 丨 $O-C-CH_3$ ‖O	Adhesives, latex paint, chewing gum
polyacrylonitrile (Orlon®, Acrilan®)	$CH_2=CH$ 丨 CN	$-(CH_2-CH)_n$ 丨 CN	Carpets, fabrics
polystyrene (Styrofoam®)	$CH_2=CH$ 丨 (benzene ring)	$-(CH_2-CH)_n$ 丨 (benzene ring)	Food coolers, drinking cups, insulation

➤ 12.5 Alkynes

The characteristic feature of alkynes is the presence of a triple bond between carbon atoms. Thus, alkynes are also unsaturated hydrocarbons. Only a few compounds containing the carbon–carbon triple bond are found in nature. The simplest and most important compound of this series is ethyne, more commonly called acetylene (C_2H_2):

$$H-C\equiv C-H$$
acetylene

Acetylene is used in torches for welding steel and in making plastics and synthetic fibers.

Once again, orbital hybridization provides an explanation for the bonding of the carbon atoms. Structurally, the hydrogen and carbon atoms of acetylene molecules lie in a straight line. This same linearity of the triple bond and the two atoms attached to the triple-bonded carbons is found in all alkynes. These characteristics are explained by mixing a $2s$ and a single $2p$ orbital of each carbon to form a pair of sp hybrid orbitals. Two of the $2p$ orbitals of each carbon are unhybridized (see ■ Figures 12.9 and 12.10).

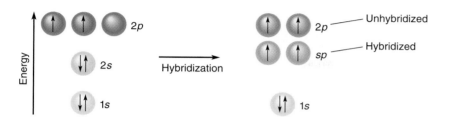

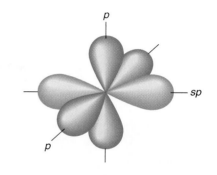

➤ **FIGURE 12.9** *sp* hybridization occurs when one of the 2*p* orbitals of carbon mixes with the 2*s* orbital. Two 2*p* orbitals remain unhybridized.

➤ **FIGURE 12.10** The unhybridized *p* orbitals are perpendicular to the two *sp* hybridized orbitals.

A carbon–carbon sigma bond in acetylene forms by the overlap of one *sp* hybrid orbital of each carbon. The other *sp* hybrid orbital of each carbon overlaps with a 1*s* orbital of a hydrogen to form a carbon–hydrogen sigma bond. The remaining pair of unhybridized *p* orbitals of each carbon overlap sideways to form a pair of pi bonds between the carbon atoms. Thus, each acetylene molecule contains three sigma bonds (two carbon–hydrogen and one carbon–carbon) and two pi bonds (both are carbon–carbon). This is shown in ■ Figure 12.11.

Alkynes are named in exactly the same ways as alkenes, except the ending *-yne* is used:

$$\overset{4}{C}H_3\overset{3}{C}H_2-\overset{2}{C}\equiv\overset{1}{C}-H \qquad H-\overset{1}{C}\equiv\overset{2}{C}-\overset{3}{C}H-\overset{4}{C}H_3$$
$$\qquad\qquad\qquad\qquad\qquad\qquad | \\ \qquad\qquad\qquad\qquad\qquad CH_3$$

1-butyne   3-methyl-1-butyne

Learning Check 12.9

Give the IUPAC name for each of the following:

a. $CH_3-CH_2-C\equiv C-CH_3$ **b.** $CH_3-\overset{CH_3}{\overset{|}{CH}}-CH_2-C\equiv C-CH_3$

The physical properties of the alkynes are nearly the same as those of the corresponding alkenes and alkanes: They are insoluble in water, less dense than water, and have relatively low melting and boiling points. Alkynes also resemble alkenes in their addition reactions. The same substances (Br_2, H_2, HCl, etc.) that add to double bonds also add to triple bonds. The one significant difference is that alkynes consume twice as many moles of addition reagent as alkenes in addition reactions that go on to completion.

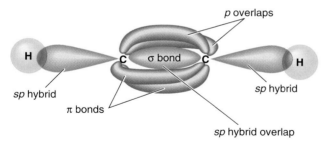

acetylene (ethyne)

➤ **FIGURE 12.11** The shape of acetylene is explained by sigma bonding between *sp* hybrid orbitals and pi bonding between unhybridized *p* orbitals.

➤ **FIGURE 12.12** Friedrich August Kekulé (1829–1896).

Aliphatic compound
Any organic compound that is not aromatic.

➤ 12.6 Aromatic Compounds and the Benzene Structure

Some of the early researchers in organic chemistry became intrigued by fragrant oils that could be extracted from certain plants. Oil of wintergreen and the flavor component of the vanilla bean are examples. The compounds responsible for the aromas had similar chemical properties. As a result, they were grouped together and called aromatic compounds.

As more and more aromatic compounds were isolated and studied, chemists gradually realized that aromatics contained at least six carbon atoms, had low hydrogen-to-carbon ratios (relative to other organic hydrocarbons), and were related to benzene (C_6H_6). For example, toluene, an aromatic compound from the bark of the South American tolu tree, has the formula C_7H_8.

Chemists also learned that the term *aromatic* was not always accurate. Many compounds that belong to the class because of chemical properties and structures are not at all fragrant. Conversely, there are many fragrant compounds that do not have aromatic compound properties or structures. Today, the old class name is used but with a different meaning. Aromatic compounds are those that contain the characteristic benzene ring or its structural relatives. Compounds that do not contain this structure (nonaromatic compounds) are referred to as **aliphatic compounds**. Alkanes, alkenes, and alkynes are, therefore, aliphatic compounds.

The molecular structure of benzene presented chemists with an intriguing puzzle after the compound was discovered in 1825 by Michael Faraday. The formula C_6H_6 indicated that the molecule was highly unsaturated. However, the compound did not show the typical reactivity of unsaturated hydrocarbons. Benzene underwent relatively few reactions, and these proceeded slowly and often required heat and catalysts. This was in marked contrast to alkenes, which reacted rapidly with many reagents, in some cases almost instantaneously. This apparent discrepancy between structure and reactivity plagued chemists until 1865, when Freidrich August Kekulé von Stradonitz (see ■ Figure 12.12), a German chemist, suggested that the benzene molecule might be represented by a ring arrangement of carbon atoms with alternating single and double bonds between the carbon atoms:

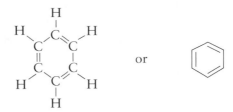

He later suggested that the double bonds alternate in their positions between carbon atoms to give two equivalent structures:

Kekulé structures

A modern interpretation of the benzene structure based on hybridization enables chemists to better understand and explain the chemical properties of benzene and other aromatic compounds. Each carbon atom in a benzene ring has three sp^2 hybrid orbitals and one unhybridized p orbital.

A single sigma bond between carbons of the benzene ring is formed by the overlap of two sp^2 orbitals, one from each of the double-bonded carbons. Because each carbon forms two single bonds, two of the sp^2 hybrid orbitals

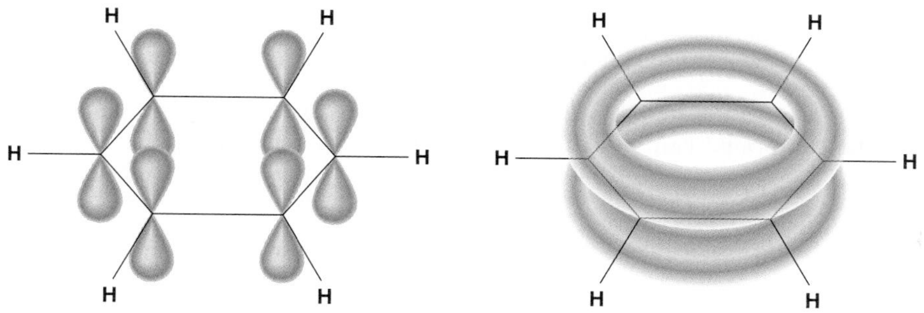

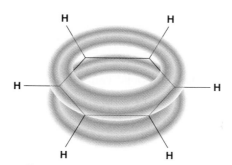

Each carbon atom forms three single bonds and has one electron in a nonhybridized *p* orbital.

Two lobes represent the delocalized pi bonding of the six *p* electrons.

of each carbon are involved. The third sp^2 hybrid orbital of each carbon forms a single sigma bond with a hydrogen by overlapping with a $1s$ orbital of hydrogen. The unhybridized p orbitals of each carbon overlap sideways above and below the plane of the carbon ring to form two delocalized pi lobes that run completely around the ring (see ■ Figure 12.13).

This interpretation leads to the conclusion that six bonding electrons move freely around the ring in the overlapping delocalized pi lobes. Because of this, the benzene structure is often represented by the symbol

with the circle representing the evenly distributed electrons in the pi lobes. All six carbon and six hydrogen atoms in benzene molecules lie in the same plane (see Figure 12.13). Therefore, substituted aromatic compounds do not exhibit cis-trans isomerism.

As you draw the structure of aromatic compounds, remember that only one hydrogen atom or group can be attached to a particular position on the benzene ring. For example, compounds (a) and (b) below exist, but (c) does not. Examination of the Kekulé structure of compound (c) shows that the carbon has five bonds attached to it in violation of the octet rule:

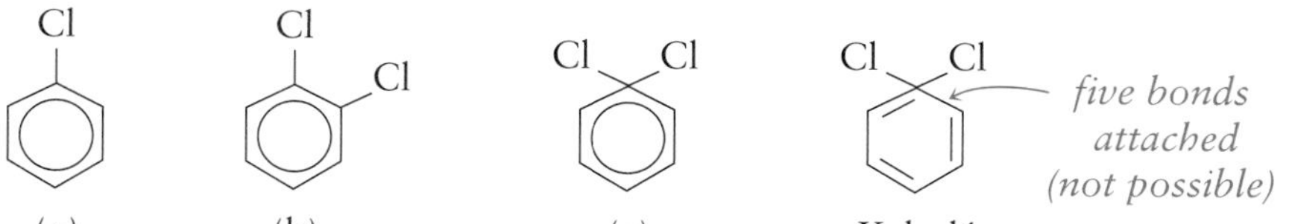

(a) exists   (b) exists   (c) does not exist   Kekulé structure   *five bonds attached (not possible)*

➤ 12.7 The Nomenclature of Benzene Derivatives

The chemistry of the aromatic compounds developed somewhat haphazardly for many years before systematic nomenclature schemes were developed. Some of the common names used earlier have acquired historical respectability and are used today; some have even been incorporated into the modern systematic nomenclature system.

The following guidelines are all based on the IUPAC aromatic nomenclature system. They are not complete, and you will not be able to name all aromatic compounds by using them. However, you will be able to name and recognize those used in this book.

Chemistry·⚛·Now™
Go to GOB Now and click to learn how to name aromatic compounds.

Guideline 1. When a single hydrogen of the benzene ring is replaced, the compound can be named as a derivative of benzene:

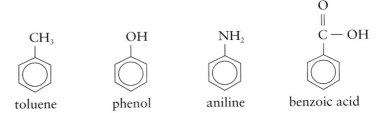

ethylbenzene   nitrobenzene   bromobenzene   chlorobenzene

Guideline 2. A number of benzene derivatives are known by common names that are also IUPAC-accepted and are used preferentially over other possibilities. Thus, toluene is favored over methylbenzene, and aniline is used rather than aminobenzene:

toluene   phenol   aniline   benzoic acid

Guideline 3. Compounds formed by replacing a hydrogen of benzene with a more complex hydrocarbon group can be named by designating the benzene ring as the group. When this is done, the benzene ring, C_6H_5—, or

Phenyl group
A benzene ring with one hydrogen absent, C_6H_5—.

is called a **phenyl group**:

$$CH_3 - CH_2 - CH_2 - CH - CH_2 - CH_2 - CH_3$$

4-phenylheptane   1,1-diphenylcyclobutane

It's easy to confuse the words *phenyl* and *phenol*. The key to keeping them straight is the ending: *-ol* means an alcohol (C_6H_5—OH), and *-yl* means a group (C_6H_5—).

Guideline 4. When two groups are attached to a benzene ring, three isomeric structures are possible. They can be designated by the prefixes *ortho (o)*, *meta (m)*, and *para (p)*:

o-dibromobenzene   *m*-dibromobenzene   *p*-dibromobenzene

Guideline 5. When two or more groups are attached to a benzene ring, their positions can be indicated by numbering the carbon atoms of the ring so as to obtain the lowest possible numbers for the attachment

CHEMISTRY AND YOUR HEALTH • 12.1

HOW TO LOWER YOUR CANCER RISK

The media constantly bombard us with reports that a certain food, chemical, or technology causes cancer. Most reports have been sensationalized and blown out of proportion, making it easy to lose sight of what has actually been learned about preventing and treating cancer. Cancer is second only to heart disease as the major fatal disease in the United States. Some cancers occur because the population is aging, and cancer is more common in the elderly. The incidence of some types such as breast cancer seems to be increasing, but part of this increase is thought to result from improved methods for diagnosis. The incidence of other types, including prostate, uterine, and stomach cancer, is declining.

No one really knows when or if a cure for cancer will be discovered, but a great deal of progress has been made toward understanding why and how cancer occurs. Many years of effort by many researchers has led to the accumulation of a great deal of data and knowledge concerning cancer. All researchers do not interpret the importance or meaning of research results the same way. However, the following suggestions about how to reduce your chances of contracting cancer are currently considered by most experts to be valid.

1. Do not smoke. Tobacco smoking causes more cancer worldwide than any other factor. The longer you smoke, the more dangerous the effects. Not only are smokers at risk to contract lung cancer, but smoking has also been shown to increase the risk of contracting mouth, throat, pancreatic, kidney, stomach, bladder, breast, and colon cancer. The incidence of cancer in the United States would drop by one-third if all tobacco use were to cease.

2. Increase your consumption of fruits, vegetables, and whole grains. Dietary factors are thought to cause almost as much cancer as tobacco smoking. Many studies have confirmed that people who have the highest rate of fruit and vegetable intake have the lowest rate of most cancers. The current recommendation of including at least five servings of fruits and vegetables in your daily diet is considered to be the most important dietary factor related to cancer prevention.

3. Reduce your intake of animal fat. Several studies have identified a diet high in animal fat, especially from red meat, as a risk factor for prostate and colon cancer and as a suspected risk factor for breast cancer. It is known that countries with high-fat diets do have the highest rates of breast and prostate cancer.

4. Don't grill meats at high temperatures. It has been shown that cooking meat over an open flame causes cancer-promoting compounds to form in the meat. Meat that is steamed, braised, baked, poached, stewed, or microwaved is considered safer than grilled meat.

5. Reduce your alcohol intake. Excessive alcohol use is known to cause cirrhosis and cancer of the liver. Studies show that women should consume no more than one glass of wine per day, and men should consume no more than two glasses. The combination of alcohol use with smoking has been shown to increase the risk of mouth, esophageal, and throat cancer.

6. Be active. Studies done during the last 10 years have shown that exercise, particularly when started by women early in life, protects against breast cancer. Recent studies have shown a similar effect for men and prostate cancer.

7. Maintain a proper weight. Obesity has been shown to increase the risk for heart disease, strokes, diabetes, and development of cancer. The correlation between body weight and cancer is not completely clear, but it is agreed that maintaining proper body weight reduces your risk of cancer.

8. Don't sunbathe. Limit your exposure to the sun. The incidence of skin cancer is known to increase with increased exposure to sunlight. Such cancers account for about 2% of the cancer deaths each year. When you must be in the sun, use a strong sunscreen.

9. Avoid exposure to known hazardous materials or conditions. When it is necessary to work with hazardous substances such as formaldehyde and benzene, or sources of radiation such as radioactive materials or X-ray devices, it is important to be aware of and use appropriate precautions and protective equipment. This is not generally a problem or concern for the majority of people.

---

positions. Groups are arranged in alphabetical order. If there is a choice of identical sets of numbers, the group that comes first in alphabetical order is given the lower number. IUPAC-acceptable common names may be used:

m-bromochlorobenzene
or 1-bromo-3-chlorobenzene

1,2,4-trichlorobenzene

3,5-dichlorobenzoic acid

Example 12.5

Name each of the following aromatic compounds:

a. 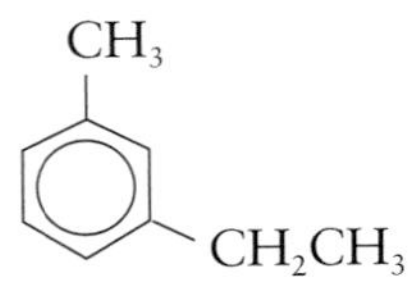
CH_3
CH_2CH_3

b. 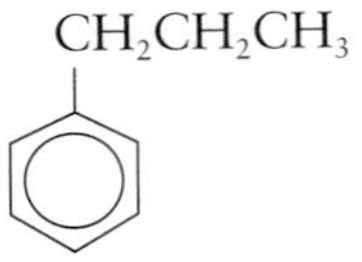
$CH_2CH_2CH_3$

c. 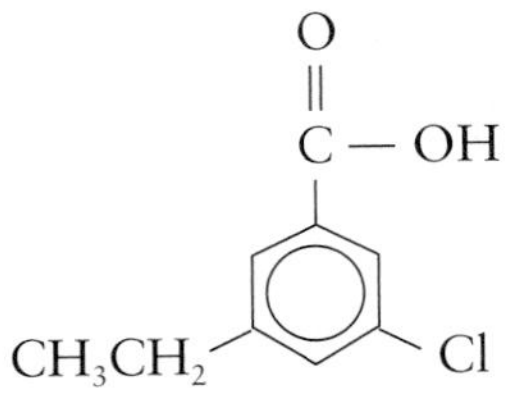
O
C — OH
CH_3CH_2
Cl

Solution

a. This compound is named as a substituted toluene. Both of the following are correct: 3-ethyltoluene and *m*-ethyltoluene. Note that in 3-ethyltoluene the methyl group, which is a part of the toluene structure, must be assigned to position number one.
b. This compound may be named as a substituted benzene or a substituted propane: propylbenzene, 1-phenylpropane.
c. When three groups are involved, the ring-numbering approach must be used: 3-chloro-5-ethylbenzoic acid.

Learning Check 12.10

Name the following aromatic compounds:

a. 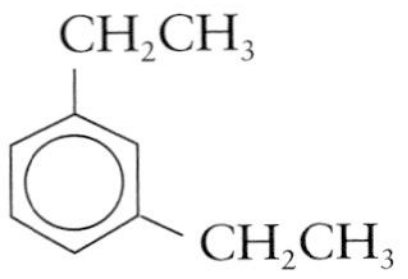
CH_2CH_3
CH_2CH_3

b. 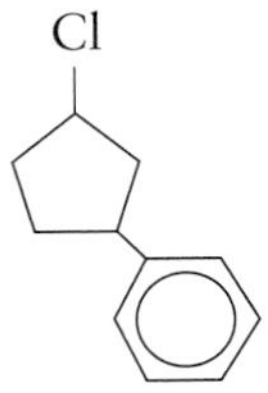
Cl

c. 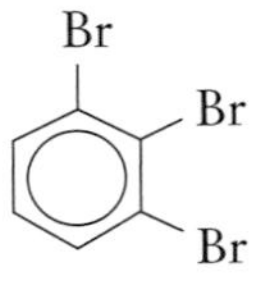
Br
Br
Br

d. 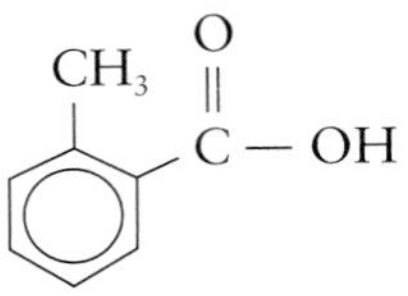
CH_3
O
C — OH

➤ 12.8 Properties and Uses of Aromatic Compounds

The physical properties of benzene and other aromatic hydrocarbons are similar to those of alkanes and alkenes. They are nonpolar and thus insoluble in water. This hydrophobic characteristic plays an important role in the chemistry of some proteins (Chapter 19).

Aromatic rings are relatively stable chemically. Because of this, benzene often reacts in such a way that the aromatic ring remains intact. Thus, benzene does not undergo the addition reactions that are characteristic of alkenes and alkynes. The predominant type of reaction of aromatic molecules is substitution, in which one of the ring hydrogens is replaced by some other group. Such aromatic reactions are of lesser importance for our purposes and are not shown here.

All aromatic compounds discussed to this point contain a single benzene ring. There are also substances called **polycyclic aromatic compounds,** which

Polycyclic aromatic compound
A derivative of benzene in which carbon atoms are shared between two or more benzene rings.

OVER THE COUNTER • 12.1

SMOKING: IT'S QUITTING TIME

Smoking is a difficult habit to break, especially if the attempt is made by stopping abruptly—going "cold turkey." Mark Twain described the more reasonable gradual approach when he said, "Habit is habit, and not to be flung out of the window by any man, but coaxed downstairs a step at a time." Smokers do not have to go "cold turkey" because there are several OTC aids available, as well as some new prescription products to help them gradually overcome the strong urge to smoke—and eventually quit.

The transdermal (absorbed through the skin) nicotine patch is available over the counter in doses of 7–22 mg. When used as directed, this method delivers a steady supply of nicotine to the bloodstream and helps minimize withdrawal symptoms. Nicotine gum helps reduce withdrawal symptoms when used correctly. The gum should be chewed briefly and then held next to the cheek, allowing the lining of the mouth to absorb the nicotine.

With a prescription, smokers can also obtain a nasal spray that provides a small dose of nicotine each time it is used. This product, called Nicotrol NS®, was approved by the FDA even though inhaling the nicotine poses a small risk that smokers will become as dependent on the mist as they are on cigarettes. The nicotine from the nasal spray gets into the bloodstream faster than nicotine from the gum or patch, providing immediate relief from cigarette craving. A squirt into each nostril gives a smoker 1 mg of nicotine, but it is not supposed to be used more than five times per hour.

The Nicotrol nicotine inhalation system has also received FDA approval. This inhaler, available only by prescription, consists of a plastic cylinder about the size of a cigarette that encloses a cartridge containing nicotine. When a smoker "puffs" on the device, nicotine vapors are absorbed through the lining of the mouth and throat. It takes about 80 puffs to deliver the amount of nicotine obtained from a single cigarette. An advantage of using the system is that a smoker still mimics the hand-to-mouth behavior of smoking, a part of the smoking habit that will be easier to break once nicotine withdrawal symptoms subside.

One of the newest prescription products designed to help break the smoking habit does not contain any nicotine. It is an antidepressant called bupropion that has been shown to be effective in the treatment of nicotine addiction. It is believed that bupropion mimics some of the action of nicotine by releasing the brain chemicals norepinephrine and dopamine, but it is not completely understood how it works. During treatment, a bupropion tablet (marketed as Zyban®) is taken once a day for 3 days and then twice daily during the week before smoking is stopped. Usually, the treatment is continued for the next 6 to 12 weeks to help curb the craving for cigarettes.

If a smoker truly wants to quit, these aids alone will not do it. A smoker must have some kind of support from a formal program, or at least informal support from family and friends. A smoker should also get rid of all tobacco products and avoid smoking triggers, such as other smokers, stress, and alcohol. Exercise can also be a distraction from smoking and can minimize the weight gain that sometimes accompanies giving up smoking.

---

contain two or more benzene rings sharing a common side, or "fused" together. The simplest of these compounds, naphthalene, is the original active ingredient in mothballs:

naphthalene

A number of more complex polycyclic aromatic compounds are known to be carcinogens—chemicals that cause cancer. Two of these compounds are

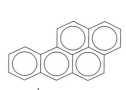   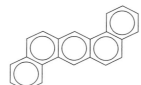

a benzopyrene   a dibenzanthracene

These cancer-producing compounds are often formed as a result of heating organic materials to high temperatures. They are present in tobacco smoke (see ■ Figure 12.14), automobile exhaust, and sometimes in burned or heavily browned food. Such compounds are believed to be at least partially respon-

Charles D. Winters

➤ **FIGURE 12.14** Cigarette smoke contains carcinogenic polycyclic aromatic compounds.

TABLE 12.4 Some important aromatic compounds

Name	Structural formula	Use
benzene		Industrial solvent and raw material
toluene	CH_3	Industrial solvent and raw material
phenol	OH	Manufacture of Bakelite® and Formica®
aniline	NH_2	Manufacture of drugs and dyes
styrene	$CH = CH_2$	Preparation of polystyrene products
phenylalanine	$CH_2CH - \overset{\overset{O}{\|}}{C} - OH$, NH_2	An essential amino acid
riboflavin		Vitamin B_2

sible for the high incidence of lung and lip cancer among cigarette smokers. Those who smoke heavily face an increased risk of getting cancer. Chemists have identified more than 4000 compounds in cigarette smoke, including 43 known carcinogens. The Environmental Protection Agency (EPA) considers tobacco smoke a Class A carcinogen.

The major sources of aromatic compounds are petroleum and coal tar, a sticky, dark-colored material derived from coal. As with many classes of organic compounds, the simplest structures are the most important commercial materials.

Benzene and toluene are excellent laboratory and industrial solvents. In addition, they are the starting materials for the synthesis of hundreds of other valuable aromatic compounds that are intermediates in the manufacture of a wide variety of commercial products, including the important polymers Bakelite® and polystyrene (see ■ Table 12.4).

A number of aromatic compounds are important in another respect: They must be present in our diet for proper nutrition. Unlike plants, which have the ability to synthesize the benzene ring from simpler materials, humans must obtain any necessary aromatic rings from their diet. This helps explain why certain amino acids, the building blocks of proteins, and some **vitamins** are dietary essentials (Table 12.4).

Vitamin
An organic nutrient that the body cannot produce in the small amounts needed for good health.

➤ **FOR FUTURE REFERENCE** DRUGS THAT MAY CAUSE PHOTOSENSITIVITY FROM SUN EXPOSURE

Normal skin will tan or burn when exposed to ultraviolet light from the sun (see Chemistry and Your Health 12.1). The use of some drugs can increase the sensitivity of the skin to the effects of UV light. The skin of sensitized individuals reacts by developing a rash or burning excessively. The following list contains the names of drugs that may cause such photosensitivity. Anyone taking one or more of these drugs should consult with a physician about the possibility of developing photosensitivity, and the precautions to take to prevent it.

acetazolamide
acetohexamide
alprazolam
amantadine
amiloride
aminobenzoic acid
amiodarone
amitriptyline
amoxapine
barbiturates
bendroflumethiazide
benzocaine
benzophenones
benzoyl peroxide
benzthiazide
captopril
carbamazepine
chlordiazepoxide
chloroquine
chlorothiazide
chlorpromazine
chlorpropamide
chlortetracycline
chlorthalidone
ciprofloxacin
clindamycin
clofazimine
clofibrate
clomipramine
cyproheptadine
dacarbazine
dapsone
demeclocycline
desipramine
desoximetasone
diethylstilbestrol
diflunisal
diltiazem
diphenhydramine
disopyramide

doxepin
doxycycline
enoxacin
estrogen
eretinate
flucytosine
fluorescein
fluorouracil
fluphenazine
flutamide
furosemide
glipizide
glyburide
gold preparations
griseofulvin
haloperidol
hexachlorophene
hydrochlorothiazide
hydroflumethiazide
ibuprofen
imipramine
indomethacin
isotretinoin
ketoprofen
lincomycin
lomefloxacin
maprotiline
mesoridazine
methacycline
methotrexate
methyclothiazide
methyldopa
metolazone
minocycline
minoxidil
nabumetone
nalidixic acid
naproxen
nifedipine
norfloxacin

nortriptyline
ofloxacin
oral contraceptives
oxyphenbutazone
oxytetracycline
para-aminobenzoic acid
perphenazine
phenelzine
phenobarbital
phenylbutazone
phenytoin
piroxicam
polythiazide
prochlorperazine
promazine
promethazine
protriptyline
pyrazinamide
quinidine
quinine
sulfonamides
sulindac
tetracycline
thiabendazole
thioridazine
thiothixene
tolazamide
tolbutamide
tranylcypromine
trazodone
tretinoin
triamterene
trichlormethiazide
trifluoperazine
triflupromazine
trimeprazine
trimethoprim
trimipramine
triprolidine
vinblastine

SOURCE: Adopted from Rybacki, J.J.; Long, J.W. *The Essential Guide to Prescription Drugs.* New York: Harper Perennial, 1999.

The Nomenclature of Alkenes. Compounds containing double or triple bonds between carbon atoms are said to be unsaturated. The alkenes contain double bonds, alkynes triple bonds. In the IUPAC nomenclature system, alkene names end in *-ene,* alkynes end in *-yne.*

The Geometry of Alkenes. In alkenes, the double-bonded carbons and the four groups attached to these carbons lie in the same plane. Because rotation about the double bond is restricted, alkenes may exist as geometric, or cis-trans, isomers. This type of stereoisomerism is possible when each double-bonded carbon is attached to two different groups.

Properties of Alkenes. The physical properties of alkenes are very similar to those of the alkanes. They are nonpolar, insoluble in water, less dense than water, and soluble in nonpolar solvents. Alkenes are quite reactive, and their characteristic reaction is addition to the double bond. Three important addition reactions are bromination (an example of halogenation) to give a dibrominated alkane, hydration to produce an alcohol, and the reaction with H—X to give an alkyl halide. The addition of H_2O and H—X are governed by Markovnikov's rule.

Addition Polymers. Addition polymers are formed from alkene monomers that undergo repeated addition reactions with each other. Many familiar and widely used materials, such as fibers and plastics, are addition polymers.

Alkynes. The alkynes contain triple bonds and possess a linear geometry of the two carbons and the two attached groups. Alkyne names end in *-yne.* The physical and chemical properties of alkynes are very similar to those of the alkenes.

Aromatic Compounds and the Benzene Structure. Benzene, the simplest aromatic compound, and other members of the aromatic class contain a six-membered ring with three double bonds. This aromatic ring is often drawn as a hexagon containing a circle, which represents the six electrons of the double bonds that move freely around the ring. All organic compounds that do not contain an aromatic ring are called aliphatic compounds.

The Nomenclature of Benzene Derivatives. Several acceptable IUPAC names are possible for many benzene compounds. Some IUPAC names are based on widely used common names such as toluene and aniline. Other compounds are named as derivatives of benzene or by designating the benzene ring as a phenyl group.

Properties and Uses of Aromatic Compounds. Aromatic hydrocarbons are nonpolar and have physical properties similar to those of the alkanes and alkenes. Benzene resists addition reactions typical of alkenes. Benzene and toluene are key industrial chemicals. Other important aromatics include phenol, aniline, and styrene.

You can get an approximate but quick idea of how well you have met the learning objectives given at the beginning of this chapter by working the selected end-of-chapter exercises given below. The answer to each exercise is given in Appendix B of the book.

Objective 1 (Section 12.1): Exercise 12.2

Objective 2 (Section 12.1): Exercise 12.4

Objective 3 (Section 12.2): Exercise 12.18

Objective 4 (Section 12.2): Exercise 12.20

Objective 5 (Section 12.3): Exercise 12.26

Objective 6 (Section 12.4): Exercise 12.36

Objective 7 (Section 12.5): Exercise 12.44

Objective 8 (Section 12.6): Exercise 12.48

Objective 9 (Section 12.7): Exercises 12.52 and 12.54

Objective 10 (Section 12.8): Exercise 12.66

Addition polymer (12.4)
Addition reaction (12.3)
Aliphatic compound (12.6)
Alkene (Introduction)
Alkyne (Introduction)
Aromatic hydrocarbon (Introduction)
Carbocation (12.4)

Copolymer (12.4)
Haloalkane or alkyl halide (12.3)
Hydration (12.3)
Hydrogenation (12.3)
Markovnikov's rule (12.3)
Monomer (12.4)
Phenyl group (12.7)

Polycyclic aromatic compound (12.8)
Polymer (12.4)
Polymerization (12.4)
Polyunsaturated (12.3)
Unsaturated hydrocarbon (Introduction)
Vitamin (12.8)

Key Reactions

1. Halogenation of an alkene (Section 12.3):

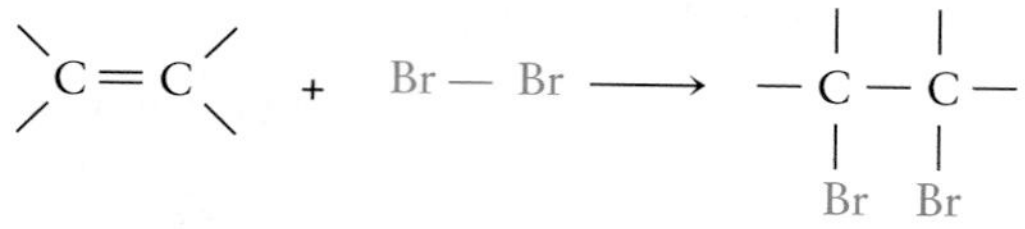

Reaction 12.2

2. Hydrogenation of an alkene (Section 12.3):

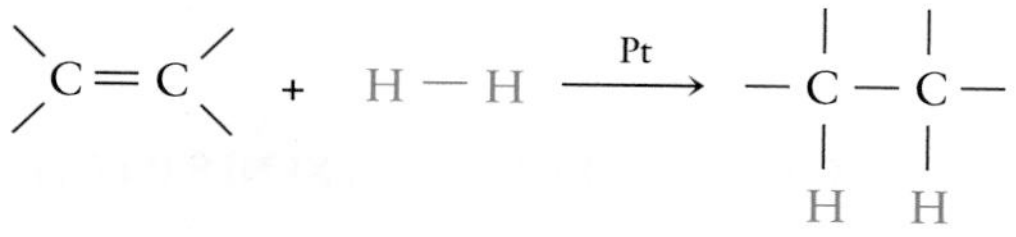

Reaction 12.4

3. Addition of H—X to an alkene (Section 12.3):

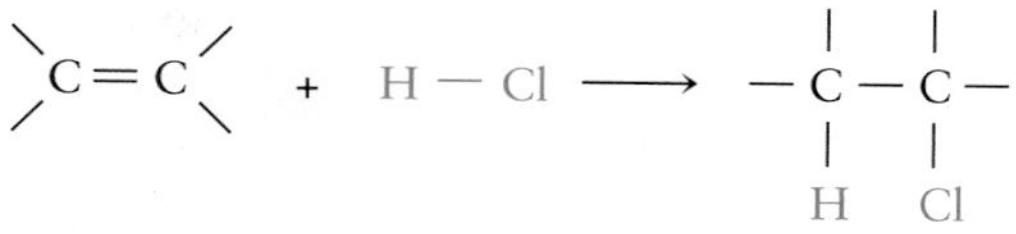

Reaction 12.6

4. Hydration of an alkene (Section 12.3):

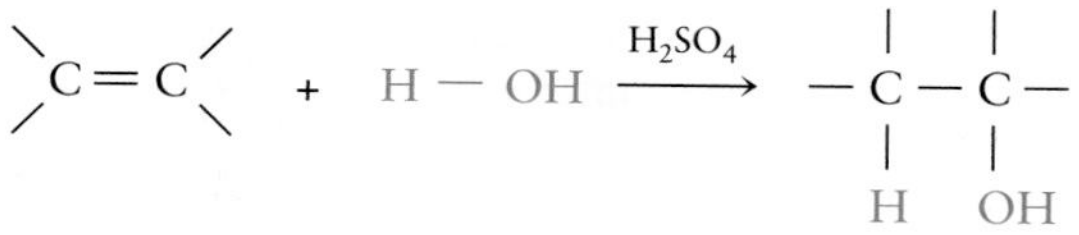

Reaction 12.9

5. Addition polymerization of an alkene (Section 12.4):

$$n\mathrm{CH_2{=}CH_2} \xrightarrow[\text{catalysts}]{\text{Heat, pressure}} \left(\mathrm{CH_2{-}CH_2}\right)_n$$

Reaction 12.14

Exercises

LEGEND: 1 = straightforward, 2 = intermediate, 3 = challenging. All even-numbered exercises are answered in Appendix B.

THE NOMENCLATURE OF ALKENES (SECTION 12.1) AND ALKYNES (SECTION 12.5)

12.1 What is the definition of an unsaturated hydrocarbon?

12.2 Define the terms alkene, alkyne, and aromatic hydrocarbon.

12.3 Select those compounds that can be correctly called *unsaturated* and classify each one as an *alkene* or an *alkyne:*

a. $CH_3—CH_2—CH_3$

b. $CH_3CH=CHCH_3$

c. 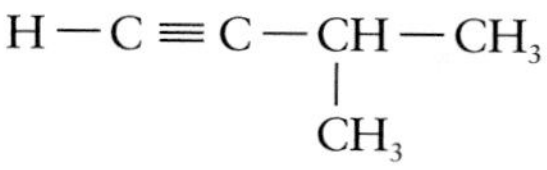

d. 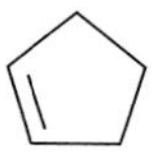

e. 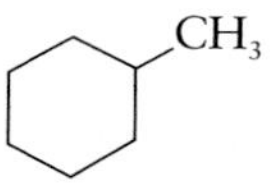

f. 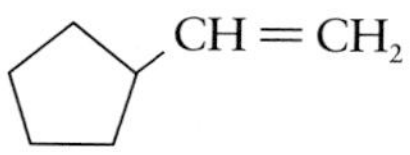

g. 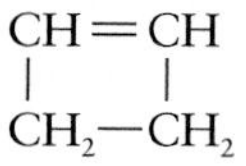

h. $CH_2=CHCH_2CH_3$

i. 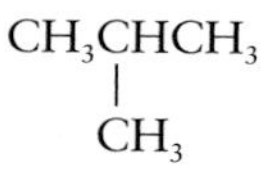

12.4 Give the IUPAC name for the following compounds:

a. $CH_3CH=CHCH_3$

b. 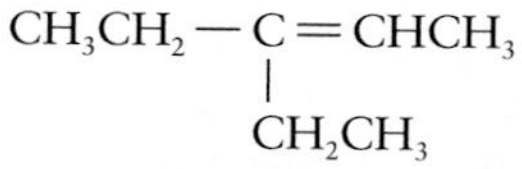

c. 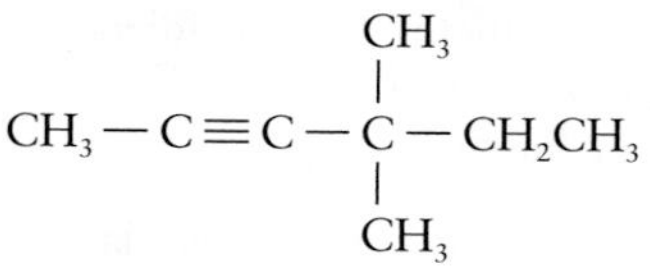

d. 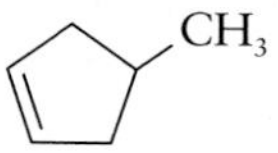

e. 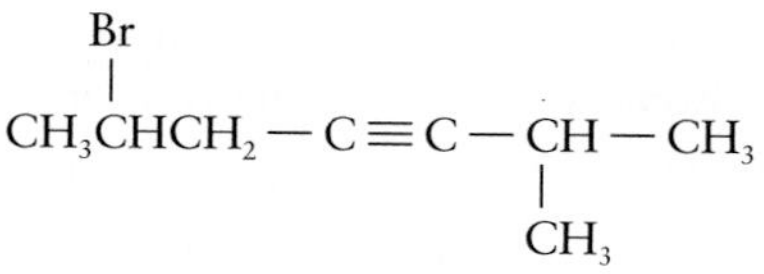

f. 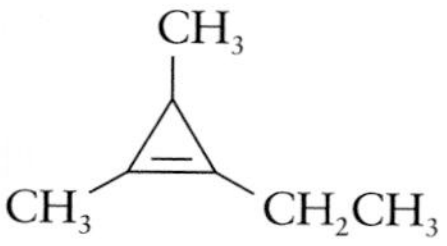

g. 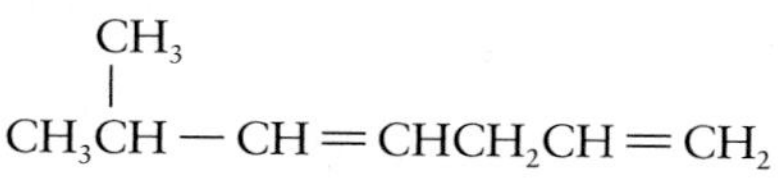

12.5 Give the IUPAC name for the following compounds:

a. CH₃CHCH=CHCH₂CH₃
      |
     CH₃

b. CH₃CH=CHCH=CHCHCH₃
                 |
                CH₃

c.

d. CH₃—C≡C—CH₂CH₃

e.
      CH₃
      |
  CH₂CHCH₃

CH₃

f. CH₂CH₂CH₃

g.
     CH₂CH₂CH₂CH₃
      |
CH₂=C—CH—CH=CHCH₂CH₃
        |
       CH₂CH₃

12.6 Draw structural formulas for the following compounds:
  a. 4-ethyl-2-heptene
  b. 4,4-diethyl-1-hexyne
  c. 3-ethyl-1,4-pentadiene
  d. 2,3-dimethylcyclohexene
  e. 1-methylcyclopropene

12.7 Draw structural formulas for the following compounds:
  a. 4,4,5-trimethyl-2-heptyne
  b. 1,3-cyclohexadiene
  c. 2-t-butyl-4,4-dimethyl-1-pentene
  d. 4-isopropyl-3,3-dimethyl-1,5-octadiene
  e. 2-methyl-1,3-cyclopentadiene
  f. 3-sec-butyl-3-t-butyl-1-heptyne

12.8 A compound has the molecular formula C₅H₈. Draw a structural formula for a compound with this formula that would be classified as (a) an alkyne, (b) a diene, and (c) a cyclic alkene. Give the IUPAC name for each compound.

12.9 Draw structural formulas and give IUPAC names for the 13 alkene isomers of C₆H₁₂. Ignore geometric isomers and cyclic structures.

12.10 α-farnesene is a constituent of the natural wax found on apples. Given that a 12-carbon chain is named as a dodecane, what is the IUPAC name of α-farnesene?

     CH₃   CH₃   CH₃
     | | |
CH₃C=CHCH₂CH₂C=CHCH₂CH=CCH=CH₂

α-farnesene

12.11 Each of the following names is wrong. Give the structure and correct name for each compound.
  a. 3-pentene
  b. 3-methyl-2-butene
  c. 2-ethyl-3-pentyne

12.12 Each of the following names is wrong. Give the structure and correct name for each compound.
  a. 2-ethyl-2-pentene
  b. 3,6-octadiene
  c. 5-ethylcyclopentene

THE GEOMETRY OF ALKENES (SECTION 12.2)

12.13 What type of hybridized orbital is present on carbon atoms bonded by a double bond? How many of these hybrid orbitals are on each carbon atom?

12.14 What type of orbital overlaps to form a pi bond in an alkene? What symbol is used to represent a pi bond? How many electrons are in a pi bond?

12.15 Describe the geometry of the carbon–carbon double bond and the two atoms attached to each of the double-bonded carbon atoms.

12.16 Explain the difference between geometric and structural isomers of alkenes.

12.17 Draw structural formulas and give IUPAC names for all the isomeric pentenes (C₅H₁₀) that are
  a. Alkenes that do not show geometric isomerism. There are four compounds.
  b. Alkenes that do show geometric isomerism. There is one cis and one trans compound.

12.18 Which of the following alkenes can exist as cis-trans isomers? Draw structural formulas and name the cis and trans isomers.

  a. CH₃CH₂CH₂CH₂CH=CH₂

  b. CH₃CH₂CH=CHCH₂CH₃

         CH₃
         |
  c. CH₃C=CHCH₂CH₃

12.19 Which of the following alkenes can exist as cis-trans isomers? Draw structural formulas and name the cis and trans isomers.

  a. H₂C=CH—CH₃

  b.   Br
        |
    CH₃C=CHCH₃

  c.   Cl
      |
   HC=CHCH₃

12.20 Draw structural formulas for the following:
  a. cis-2-pentene
  b. trans-2-hexene

12.21 Draw structural formulas for the following:
  a. trans-3,4-dibromo-3-heptene
  b. cis-1,4-dichloro-2-methyl-2-butene

PROPERTIES OF ALKENES (SECTION 12.3)

12.22 In what ways are the physical properties of alkenes similar to those of alkanes?

12.23 Which of the following reactions is an addition reaction?
a. $A_2 + C_3H_6 \rightarrow C_3H_6A_2$
b. $A_2 + C_6H_6 \rightarrow C_6H_5A + HA$
c. $HA + C_4H_8 \rightarrow C_4H_9A$
d. $3O_2 + C_2H_4 \rightarrow 2CO_2 + 2H_2O$
e. $C_7H_{16} \rightarrow C_7H_8 + 4H_2$

12.24 State Markovnikov's rule, and write a reaction that illustrates its application.

12.25 Complete the following reactions. Where more than one product is possible, show only the one expected according to Markovnikov's rule.

a.

$+ H_2 \xrightarrow{Pt}$

b. $CH_3CH_2CH{=}CH_2 + Cl_2 \rightarrow$

c.

$$\underset{\quad}{CH_3} \\ CH_3CH_2\overset{|}{C}{=}CH_2 + H_2O \xrightarrow{H_2SO_4}$$

d.

$$\underset{\quad}{CH_3} \\ CH_3\overset{|}{C}{=}CHCH_2CH_3 + HBr \rightarrow$$

12.26 Complete the following reactions. Where more than one product is possible, show only the one expected according to Markovnikov's rule.

a.

$+ Br_2 \longrightarrow$

b.

$+ H_2O \xrightarrow{H_2SO_4}$

c.

$$\underset{\qquad}{CH_3} \\ CH_2{=}\overset{|}{C}{-}CH{=}CH_2 + 2H_2 \xrightarrow{Pt}$$

d.

$+ HCl \longrightarrow$

12.27 Draw the structural formula for the alkenes with molecular formula C_5H_{10} that will react to give the following products. Show all correct structures if more than one starting material will react as shown.

a. $C_5H_{10} + Br_2 \longrightarrow CH_3CH_2\underset{\underset{Br}{|}}{C}H\underset{\underset{Br}{|}}{C}HCH_3$

b. $C_5H_{10} + H_2 \xrightarrow{Pt} CH_3CH_2CH_2CH_2CH_3$

c.

$$C_5H_{10} + H_2O \xrightarrow{H_2SO_4} CH_3\underset{\underset{OH}{|}}{\overset{\overset{CH_3}{|}}{C}}CH_2CH_3$$

d. $C_5H_{10} + HBr \longrightarrow CH_3\underset{\underset{Br}{|}}{C}HCH_2CH_2CH_3$

12.28 What reagents would you use to prepare each of the following from cyclohexene?

a.   b.   c.   d.

12.29 What is an important commercial application of hydrogenation?

12.30 Cyclohexane and 2-hexene both have the molecular formula C_6H_{12}. Describe a simple chemical test that would distinguish one from the other.

12.31 Terpin hydrate is used medicinally as an expectorant for coughs. It is prepared by the following addition reaction. What is the structure of terpin hydrate?

$+ 2H_2O \xrightarrow{H_2SO_4}$ terpin hydrate

ADDITION POLYMERS (SECTION 12.4)

12.32 Explain what is meant by each of the following terms: *monomer, polymer, addition polymer,* and *copolymer.*

12.33 A section of polypropylene containing three units of monomer can be shown as

$$\underset{\quad}{\overset{\overset{CH_3}{|}}{}}\quad \overset{\overset{CH_3}{|}}{}\quad \overset{\overset{CH_3}{|}}{} \\ {-}CH_2{-}CH{-}CH_2{-}CH{-}CH_2{-}CH{-}$$

Draw structural formulas for comparable three-unit sections of
a. Teflon
b. Orlon
c. Lucite

12.34 Identify a structural feature characteristic of all monomers listed in Table 12.3.

12.35 Rubber cement contains a polymer of 2-methylpropene (isobutylene) called polyisobutylene. Write an equation for the polymerization reaction.

12.36 Much of today's plumbing in newly built homes is made from a plastic called poly (vinyl chloride), or PVC. Using Table 12.3, write a reaction for the formation of poly (vinyl chloride).

12.37 Identify a major use for each of the following addition polymers:
a. Styrofoam
b. Acrilan
c. Plexiglas
d. PVC
e. polypropylene

ALKYNES (SECTION 12.5)

12.38 What type of hybridized orbital is present on carbon atoms bonded by a triple bond? How many of these hybrid orbitals are on each carbon atom?

12.39 How many sigma bonds and how many pi bonds make up the triple bond of an alkyne?

12.40 Describe the geometry in an alkyne of the carbon–carbon triple bond and the two attached atoms.

12.41 Explain why geometric isomerism is not possible in alkynes.

12.42 Give the common name and major uses of the simplest alkyne.

12.43 Describe the physical and chemical properties of alkynes.

12.44 Write the structural formulas and IUPAC names for all the isomeric alkynes with the formula C_5H_8.

AROMATIC COMPOUNDS AND THE BENZENE STRUCTURE (SECTION 12.6)

12.45 What type of hybridized orbital is present on the carbon atoms of a benzene ring? How many sigma bonds are formed by each carbon atom in a benzene ring?

12.46 What type of orbital overlaps to form the pi bonding in a benzene ring?

12.47 What does the circle within the hexagon represent in the structural formula for benzene?

12.48 Define the terms *aromatic* and *aliphatic.*

12.49 Limonene, which is present in citrus peelings, has a very pleasant lemonlike fragrance. However, it is not classified as an aromatic compound. Explain.

CH3
C
CH3
CH2

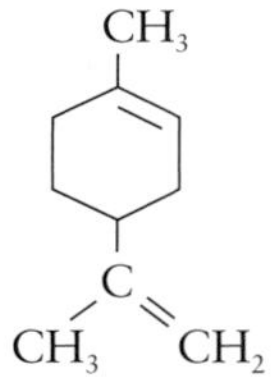

limonene

12.50 A disubstituted cycloalkane such as (a) exhibits cis-trans isomerism, whereas a disubstituted benzene (b) does not. Explain.

a.

Br
Br

b.

Br
Br

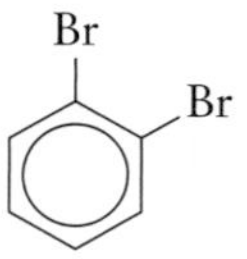

THE NOMENCLATURE OF BENZENE DERIVATIVES (SECTION 12.7)

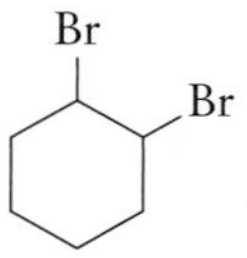

12.51 Give an IUPAC name for each of the following hydrocarbons as a derivative of benzene:

a.

CH2CH3

b.

Br
CH2CH3

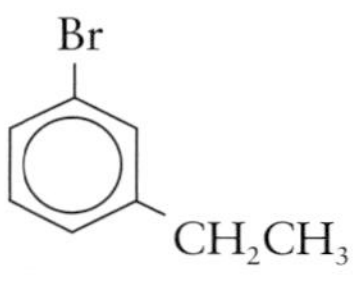

12.52 Give an IUPAC name for each of the following hydrocarbons as a derivative of benzene:

a.

CH3
CH3
CH3

b.

CH2CH3
CH2CH3

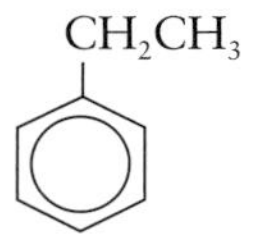

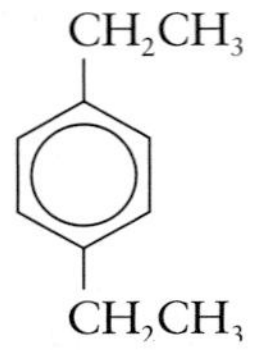

12.53 Give an IUPAC name for the following as hydrocarbons with the benzene ring as a group:

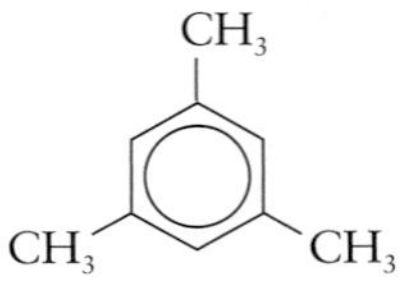

a.

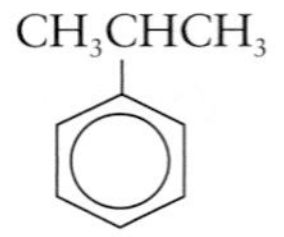

CH3CHCH3

b.

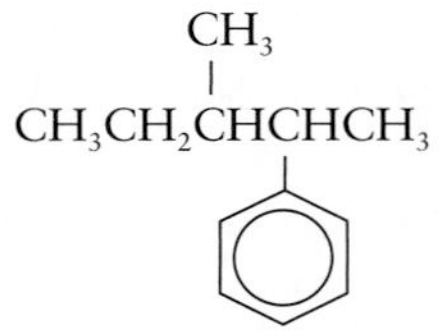

CH3
CH3CH2CHCHCH3

12.54 Give an IUPAC name for the following as hydrocarbons with the benzene ring as a substituent:

a.

CH3CH2CHCH=CH2

b.

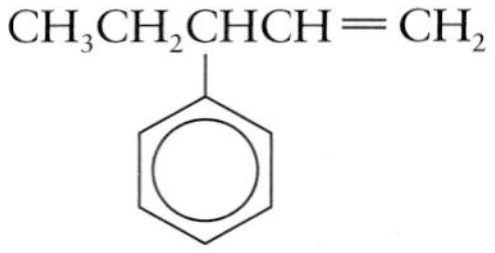

CH3CHCH2CH2CHCH3

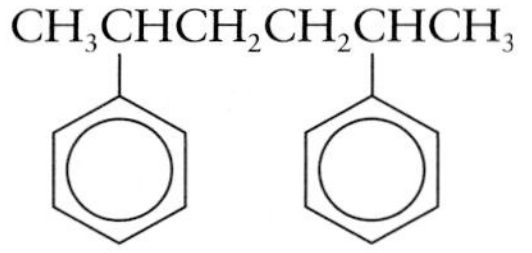

12.55 Name the following compounds, using the prefixed abbreviations for *ortho, meta,* and *para* and assigning IUPAC-acceptable common names:

a.

CH3
CH2CH3

b.

NH2
NO2

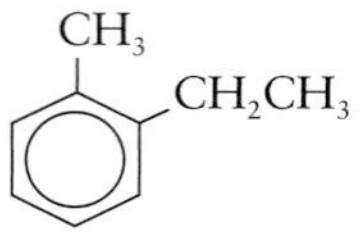

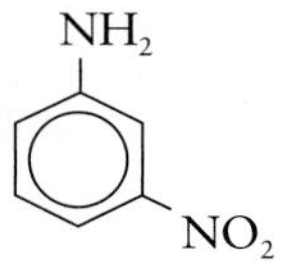

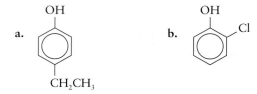

12.56 Name the following compounds, using the prefixed abbreviations for *ortho, meta,* and *para* and assigning IUPAC-acceptable common names:

a.

OH
CH2CH3

b.

OH
Cl

12.57 Name the following by numbering the benzene ring. IUPAC-acceptable common names may be used where appropriate:

a.

NO2
Cl
Cl

b.

OH
Cl
Cl

12.58 Name the following by numbering the benzene ring. IUPAC-acceptable common names may be used where appropriate:

a.

CH2CH3
CH3
CH2CH2CH3

b.

O
C—OH
Cl
Br

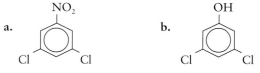

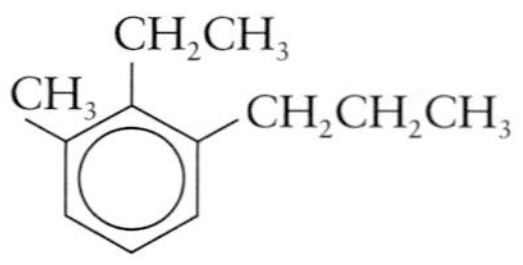

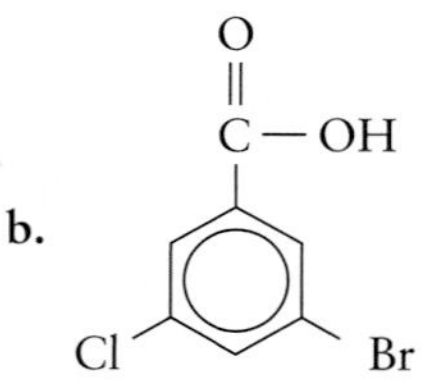

12.59 Draw structural formulas for the following:
   a. 2,4-diethylaniline
   b. 4-ethyltoluene
   c. *p*-ethyltoluene

12.60 Write structural formulas for the following:
   a. *o*-ethylphenol
   b. *m*-chlorobenzoic acid
   c. 3-methyl-3-phenylpentane

12.61 There are three bromonitrobenzene derivatives. Draw their structures and give an IUPAC name for each one.

PROPERTIES AND USES OF AROMATIC COMPOUNDS (SECTION 12.8)

12.62 Describe the chief physical properties of aromatic hydrocarbons.

12.63 Why does benzene not readily undergo addition reactions characteristic of other unsaturated compounds?

12.64 Compare the chemical behavior of benzene and cyclohexene.

12.65 For each of the following uses, list an appropriate aromatic compound:
   a. A solvent
   b. A vitamin
   c. An essential amino acid
   d. Starting material for dyes

12.66 For each of the following uses, list an appropriate aromatic compound:
   a. Used in the production of Formica®
   b. A starting material for polystyrene
   c. Used to manufacture drugs
   d. A starting material for Bakelite®

CHEMISTRY AROUND US, CHEMISTRY AND YOUR HEALTH, OVER THE COUNTER, AND HOW REACTIONS OCCUR

12.67 What role does geometric isomerism play in the chemistry of vision?

12.68 What are some good sources of lycopene in addition to watermelon?

12.69 What foods should be included in our diets to reduce the risk of cancer?

12.70 Why is baked meat considered safer than grilled meat?

12.71 What is a carbocation? How is it formed in the hydration of an alkene?

12.72 What is the active ingredient in patches, gum, and sprays designed to help break the smoking habit?

CHEMISTRY FOR THOUGHT

12.1 Napthalene is the simplest polycyclic aromatic compound:

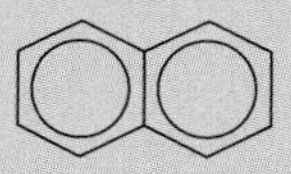

Draw a Kekulé structure for this compound like that shown for benzene in Section 12.6.

12.2 Why does propene not exhibit geometric isomerism?

12.3 Limonene is present in the rind of lemons and oranges. Based on its structure (see Exercise 12.49), would you consider it to be a solid, liquid, or gas at room temperature?

12.4 If the average molecular weight of polyethylene is 5.0×10^4 u, how many ethylene monomers ($CH_2=CH_2$) are contained in a molecule of the polymer?

12.5 Reactions to synthesize the benzene ring of aromatic compounds do not occur within the human body, and yet many essential body components involve the benzene structure. How does the human body get its supply of aromatic compounds?

12.6 Answer the question in the caption to Figure 12.8 pertaining to poly (vinyl acetate).

12.7 Why can't alkanes undergo addition polymerization?

12.8 Some polymers produce toxic fumes when they are burning. Which polymer in Table 12.3 produces hydrogen cyanide, HCN? Which produces hydrogen chloride, HCl?

12.9 "Super glue" is an addition polymer of the following monomer. Draw a structural formula for a three-unit section of super glue.

$$CH_2 = \overset{\overset{\displaystyle CN}{\displaystyle |}}{C} - O - \overset{\overset{\displaystyle O}{\displaystyle \|}}{C} - CH_3$$

12.10 One of the fragrant components in mint plants is menthene, a compound whose IUPAC name is 1-isopropyl-4-methylcyclohexene. Draw a structural formula for menthene.

InfoTrac College Edition Readings

"States need more lab capacity for pesticide-isomer oversight," *Pesticide & Toxic Chemical News,* Nov 11, 1999, 28(3). Record number A57633488.

"GP CLINICAL: Carbon monoxide poisoning guides," *GP,* Feb 3, 2003:78. Record number A97217655.

"Raising awareness about carbon monoxide: Is it time to require effective CO detectors in all buildings?" *Air Conditioning, Heating & Refrigeration News,* Jan 27, 2003, 218(4):24(3). Record number A97299219.

"Right isomer better than wrong isomer," *Drug Discovery & Technology News,* Feb 2002, 5(2). Record number A83437381.

GOB Chemistry • Now™

Assess your understanding of this chapter's topics with additional quizzing and conceptual-based problems at http://chemistry.brookscole.com/seager5e

APPENDIX A

The International System of Measurements

The International System of Units (SI units) was established in 1960 by the International Bureau of Weights and Measures. The system was established in an attempt to streamline the metric system, which included certain traditional units that had historical origins but that were not logically related to other metric units. The International System established fundamental units to represent seven basic physical quantities. These quantities and the fundamental units used to express them are given in Table A.1.

TABLE A.1 Fundamental SI units

Physical quantity	SI unit	Abbreviation
Length	meter	m
Mass	kilogram	kg
Temperature	kelvin	K
Amount of substance	mole	mol
Electrical current	ampere	A
Time	second	s
Luminous intensity	candela	cd

All other SI units are derived from the seven fundamental units. Prefixes are used to indicate multiples or fractions of the fundamental units (Table 1.2). In some cases, it has been found convenient to give derived units specific names. For example, the derived unit for force is $kg\ m/s^2$, which has been given the name newton (abbreviated N) in honor of Sir Isaac Newton (1642–1726). Some examples of derived units are given in Table A.2.

TABLE A.2 Examples of derived SI units

Physical quantity	Definition in fundamental units	Specific name	Abbreviation
Volume	m^3 (see NOTE below)	—	—
Force	$kg\ m/s^2$	newton	N
Energy	$kg\ m^2/s^2 = N\ m$	joule	J
Power	$kg\ m^2/s^3 = J/s$	watt	W
Pressure	$kg/m\ s^2 = N/m^2$	pascal	Pa
Electrical charge	$A\ s$	coulomb	C
Electrical potential	$kg\ m^2/A\ s^3 = W/A$	volt	V
Electrical resistance	$kg\ m^2/A^2\ s^3 = V/A$	ohm	Ω
Frequency	$1/s$	hertz	Hz

NOTE: The liter (L), a popular volume unit of the metric system, has been redefined in terms of SI units as $1\ dm^3$ or $1/1000\ m^3$. It and its fractions (mL, μL, etc.) are still widely used to express volume.

APPENDIX B

Answers to Even-Numbered End-of-Chapter Exercises

1.2 Mass is a measurement of the amount of matter in an object. Weight is a measurement of the gravitational force acting on an object.

1.4 a. The distance the ball could be thrown on the moon would be greater. Although the throwing force would be the same, the smaller gravitational force pulling the ball downward would allow the ball to travel farther before hitting the surface of the moon.

b. The rolling distance would be the same in both locations because gravity does not influence rolling motion on a smooth surface.

1.6 The gravitational force is greater the closer an object is to the center of Earth. Thus, a person would weigh more at the poles (closer to the center) and less at the equator.

1.8 a. Physical: Both pieces of the stick are still made of the same substance.

b. Chemical: The smoke is a new substance.

c. Physical: Each smaller piece is still the substance salt.

d. Chemical: The change in color indicates a change in the substances present.

1.10 a. Physical: The state of a substance (solid, liquid, or gas) can be changed without changing the identity of the substance.

b. Chemical: The word *reacts* indicates a chemical change.

c. Physical: Freezing is a physical change, so the temperature at which it occurs is a physical property.

d. Chemical: Because rusting is a chemical change (rust is a different substance from the metal), the ability to rust or not to rust is a chemical property.

e. Physical: Color is a physical property because it can be observed without trying to change the substance into another substance.

1.12 a. Yes: The change in melting point indicates a different substance has been formed. Also, the evolved gas is a different substance.

b. No: It must be different because its melting point is different from that of succinic acid.

c. The succinic acid molecules must be larger because the atoms of the succinic acid were divided between the molecules of the new solid and the molecules of the evolved gas.

d. Heteroatomic: Because the succinic acid changed into two different substances, molecules of succinic acid must contain at least two different kinds of atoms.

1.14 Heteroatomic: Because only a single substance is formed, its molecules must contain at least one atom from each of the two reacting substances.

1.16 Heteroatomic: The molecules of the products hydrogen and oxygen contain hydrogen atoms and oxygen atoms respectively. The two different kinds of atoms must have come from the water.

1.18 a. A is a compound because compounds are made up of heteroatomic molecules.

b. D is an element because elements are made up of homoatomic molecules.

c. E is a compound because it can be changed into the simpler substances G and J. The simpler substances G and J cannot be classified without conducting further tests on them to see if they can be changed into still other simpler substances.

1.20 a. R cannot be classified as an element or a compound. Even though none of the tests changed it into a simpler substance, all possible tests were not done on it, and some test might exist that would change it.

b. T can be classified as a compound because it was changed into simpler substances.

c. The remaining solid cannot be classified because no tests were done to see if it could be changed into simpler substances.

1.22 a. Homogeneous
b. Homogeneous
c. Heterogeneous
d. Heterogeneous
e. Homogeneous
f. Heterogeneous
g. Homogeneous

1.24 a. Pure substance
b. Solution
e. Solution
g. Solution

1.26 In a modern society, quantities must be expressed with precision. Goods, services, time, and so on must be clearly expressed as quantities in order for trade, purchases, salaries, and the like to be carried out or determined.

1.28 The stone would be useful for expressing relatively large quantities such as the weights of people, cows, or bags of grain. It is likely that originally "one stone" was equal to the weight of some field stone that had been accepted as a standard.

1.30 a. Metric
b. Metric
c. Nonmetric

d. Metric
e. Nonmetric
f. Metric

1.32 a. Liters or milliliters
c. Meters
e. Liters

1.34 a. 1 million microliters
b. 75 thousand watts
c. 15 million hertz
d. 200×10^{-12} meters

1.36 0.240 L, 240 cm^3

1.38 8.8 lbs

1.40 a. About 3 in.
b. 65 K
c. About 2 kg

1.42 a. 0.001 kg, or 1 g
b. About 68 fl oz
c. 333 mg, or about 300 mg

1.44 97.0° F in the morning and 99.0° F at bedtime

1.46 a. Incorrect: No zero should precede the first number in the nonexponential number. 2.7×10^{-3} is correct.
b. Correct
c. Incorrect: The decimal is not in the standard position in the nonexponential number. 7.19×10^{-5} is correct.
d. Incorrect: No nonexponential number is given. 1.0×10^3 is correct.
e. Incorrect: The decimal is not in the standard position in the nonexponential number. 4.05×10^{-4} is correct.
f. Incorrect: No exponent is used, and the decimal is not in the standard position in the nonexponential number. 1.19×10^{-1} is correct.

1.48 a. 1.4×10^4
b. 3.65×10^2
c. 2.04×10^{-3}
d. 4.618×10^2
e. 1.00×10^{-3}
f. 9.11×10^2

1.50 1.86×10^5 mi/s; 1.100×10^9 km/h

1.52 0.000 000 000 000 000 000 000 0299

1.54 a. 9.0×10^{-5}
b. 1.4×10^7
c. 7.6×10^{-2}
d. 1.9×10^1
e. 2.6×10^{12}

1.56 a. $(1.44 \times 10^2)(8.76 \times 10^{-2}) = 1.26 \times 10^1$
b. $(7.51 \times 10^2)(1.06 \times 10^2) = 7.96 \times 10^4$
c. $(4.22 \times 10^{-2})(1.19 \times 10^{-3}) = 5.02 \times 10^{-5}$
d. $(1.28000 \times 10^5)(3.16 \times 10^{-5}) = 4.04 \times 10^0$

1.58 a. 2.6×10^{-5}
b. 2.2×10^2
c. 6.4×10^{-4}
d. 7.25×10^{-4}
e. 3.1×10^{-2}

1.60 a. 1.7×10^{-1}

b. 1.0×10^{-9}
c. 2.6×10^6
d. 2.3×10^0
e. 1.5×10^7

1.62 a. 0.01 cm
b. 0.01 mm
c. 0.1°
d. 0.1 lb/in.^{2}

1.64 a. 6.00 mL
b. 37.0°C
c. 9.00 s
d. 15.5°

1.66 a. Measured: 5.06 lb; exact: 16 potatoes; 0.316 lb/potato
b. Measured: the individual percentages; exact: the 5 players; 71.34%/player. Note that the sum of the five percentages has four significant figures, so the calculated average must have four.

1.68 a. 3
b. 3
c. 4
d. 2
e. 4
f. 5

1.70 a. 5.2
b. 0.104
c. 0.518
d. 1.0×10^2
e. 2.52×10^{-18}

1.72 a. 6.2
b. 230
c. 0.589
d. 0.58
e. 27.75
f. 21.64

1.74 a. 4.48×10^{-3}
b. 2.208
c. 2.65
d. -13
e. 3
f. 0.81

1.76 a. Black rectangle: Area = 124.8 cm^2;
perimeter = 44.80 cm
Red rectangle: Area = 48.9 cm^2; perimeter = 45.24 cm
Green rectangle: Area = 8.11 cm^2; perimeter = 11.46 cm
Orange rectangle: Area = 9.0 cm^2; perimeter = 27.80 cm
b. Black rectangle: Area = 0.01248 m^2;
perimeter = 0.4480 m
Red rectangle: Area = 0.00489 m^2; perimeter = 0.4524 m
Green rectangle: Area = 0.000811 m^2;
perimeter = 0.1146 m
Orange rectangle: Area = 0.00090 m^2;
perimeter = 0.2780 m

c. No

1.78 a. $\dfrac{0.015 \text{ grain}}{1 \text{ mg}}$

b. $\dfrac{0.0338 \text{ fl oz}}{1 \text{ mL}}$

c. $\dfrac{1\text{ L}}{1.057\text{ qt}}$

d. $\dfrac{1\text{ m}}{1.094\text{ yd}}$

1.80 a. The factor is $\dfrac{1\text{ km}}{0.621\text{ mi}}$. The race is about 42 km long.

1.82 1.06 cups or 1 cup

1.84 39.6 lb; your baggage is not overweight.

1.86 1.31 g/L

1.88 5.500%

1.90 71%

1.92 IgG = 75.09%; IgA = 16.23%; IgM = 7.58%; IgD = 1.10%; IgE = 0.008%

1.94 a. 0.792 g/mL
b. 1.03 g/mL
c. 1.98 g/L
d. 8.90 g/cm^3

1.96 Volume = 63.0 cm^3; density = 11.4 g/cm^3

1.98 380 mL

·················· **CHAPTER 2** ··················

2.2 a.

b.

c.

d.

2.4 a. H_2O
b. H_2O_2
c. H_2SO_4
d. C_2H_6O

2.6 a. One carbon (C), four hydrogen (H)

b. One hydrogen (H), one chlorine (Cl), four oxygen (O)

c. One aluminum (Al), three chlorine (Cl)

d. Three carbon (C), eight hydrogen (H)

2.8 a. The two hydrogen atoms should be denoted by a subscript: H_2S
b. The letter L of chlorine should be lowercase: $HClO_2$
c. The coefficient 2 should not be used to indicate the number of atoms in the molecule. Subscripts should be used instead: H_2N_4
d. The numbers in the formula should be written as subscripts: C_2H_6

2.10 a. Charge = 4+; mass = 9 u
b. Charge = 9+; mass = 19 u

c. Charge = 20+; mass = 43 u
d. Charge = 47+; mass = 107 u

2.12 a. 4
b. 9
c. 20
d. 47

2.14 a. 19 electrons, 19 protons
b. 48 electrons, 48 protons
c. 51 electrons, 51 protons

2.16 a. 16 protons, 18 neutrons, 16 electrons
b. 40 protons, 51 neutrons, 40 electrons
c. 54 protons, 77 neutrons, 54 electrons

2.18 a. $^{28}_{14}\text{Si}$
b. $^{40}_{18}\text{Ar}$
c. $^{88}_{38}\text{Sr}$

2.20 a. Atomic number = 4; mass number = 9; $^{9}_{4}\text{Be}$
b. Atomic number = 9; mass number = 19; $^{19}_{9}\text{F}$
c. Atomic number = 20; mass number = 43; $^{43}_{20}\text{Ca}$
d. Atomic number = 47; mass number = 107; $^{107}_{47}\text{Ag}$

2.22 a. $^{37}_{17}\text{Cl}$
b. $^{65}_{29}\text{Cu}$
c. $^{66}_{30}\text{Zn}$

2.24 Three

2.26 Ca, calcium

2.28 N, nitrogen

2.30 a. 80.06 u
b. 92.09 u
c. 98.08 u
d. 28.02 u
e. 44.09 u

2.32 Ethylene, C_2H_4

2.34 $y = 3$

2.36 a. 14 neutrons
b. 27.0 u

2.38 Calculated value = 10.81 u
Periodic table value = 10.81 u

2.40 Calculated value = 63.55 u
periodic table value = 63.55 u

2.42 1.90 g F

2.44 a. 1 mol Si atoms = 6.02×10^{23} Si atoms
6.02×10^{23} Si atoms = 28.09 g Si
1 mol Si atoms = 28.09 g Si
b. 1 mol Ca atoms = 6.02×10^{23} Ca atoms
6.02×10^{23} Ca atoms = 40.08 g Ca
1 mol Ca atoms = 40.08 g Ca
c. 1 mol Ar atoms = 6.02×10^{23} Ar atoms
6.02×10^{23} Ar atoms = 39.95 g Ar
1 mol Ar atoms = 39.95 g Ar

2.46 a. 35.1 g Si
b. 6.66×10^{-23} g Ca
c. 3.09×10^{23} Ar atoms

2.48 Molecular weight $BF_3 = 67.8$ u
Molecular weight $H_2S = 34.1$ u
0.68 g BF_3

2.50 a. 1. 2 C_6H_6 molecules contain 12 C atoms and 12 H atoms.
2. 10 C_6H_6 molecules contain 60 C atoms and 60 H atoms.
3. 100 C_6H_6 molecules contain 600 C atoms and 600 H atoms.
4. 6.02×10^{23} C_6H_6 molecules contain 36.1×10^{23} C atoms and 36.1×10^{23} H atoms.
5. 1 mol of C_6H_6 molecules contains 6 mol of C atoms and 6 mol of H atoms.
6. 78.11 g of C_6H_6 contains 72.06 g of C and 6.05 g of H.
b. 1. 2 NO_2 molecules contain 2 N atoms and 4 O atoms.
2. 10 NO_2 molecules contain 10 N atoms and 20 O atoms.
3. 100 NO_2 molecules contain 100 N atoms and 200 O atoms.
4. 6.02×10^{23} NO_2 molecules contain 6.02×10^{23} N atoms and 12.04×10^{23} O atoms.
5. 1 mol of NO_2 molecules contains 1 mol of N atoms and 2 mol of O atoms.
6. 46.01 g of NO_2 contains 14.01 g of N and 32.00 g of O.
c. 1. 2 HCl molecules contain 2 H atoms and 2 Cl atoms.
2. 10 HCl molecules contain 10 H atoms and 10 Cl atoms.
3. 100 HCl molecules contain 100 H atoms and 100 Cl atoms.
4. 6.02×10^{23} HCl molecules contain 6.02×10^{23} H atoms and 6.02×10^{23} Cl atoms.
5. 1 mol of HCl molecules contains 1 mol of H atoms and 1 mol of Cl atoms.
6. 36.46 g of HCl contains 1.01 g of H and 35.45 g of Cl.

2.52 a. 4.5 mol H atoms
b. 6.02×10^{23} O atoms
c. 97.2% Cl

2.54 34.6 g C_2H_6O

2.56 In CH_4, mass % H = 25.14%.
In C_2H_6, mass % H = 20.11%.

2.58 Statements 4–6 are:
4. 6.02×10^{23} H_3PO_4 molecules contain 18.06×10^{23} H atoms, 6.02×10^{23} P atoms, and 24.08×10^{23} O atoms.
5. 1 mol of H_3PO_4 molecules contains 3 mol of H atoms, 1 mol of P atoms, and 4 mol of O atoms.
6. 97.99 g of H_3PO_4 contains 3.02 g of H, 30.97 g of P, and 64.00 g of O.
a. 1.44 g of H
b. 5.00 mol of O atoms
c. 8.42×10^{21} P atoms

2.60 Magnetite = 72.36% Fe; hematite = 69.94% Fe. Magnetite has the higher % Fe by mass.

CHAPTER 3

3.2 a. Group VA (15), period 5
b. Group IIB (12), period 5

c. Group IIIA (13), period 4
d. Group IVB (4), period 5

3.4 a. Kr, krypton
b. Sn, tin
c. Mo, molybdenum
d. W, tungsten

3.6 a. 4
b. 10
c. 18

3.8 a. Period
b. Period
c. Group
d. Group

3.10 Protons

3.12 a. 2
b. 6
c. 8

3.14 Four orbitals; one $2s$, and three $2p$

3.16 Seven orbitals in the subshell can contain a maximum of 14 electrons.

3.18 a. 8
b. 5
c. 4
d. 4

3.20 Cesium, Cs, has chemical properties most like sodium, Na. Both elements have one valence-shell electron.

3.22 An ore deposit that contained copper might also be expected to contain gold and silver because all three elements belong to group IB (11) of the periodic table and thus have similar chemical properties.

3.24 a. $1s^2 2s^2 2p^6 3s^2 3p^6 4s^2 3d^{10} 4p^6 5s^1$; one electron is unpaired.
b. $1s^2 2s^2 2p^6 3s^2 3p^2$; two electrons are unpaired.
c. $1s^2 2s^2 2p^6 3s^2 3p^6 4s^2 3d^2$; two electrons are unpaired.
d. $1s^2 2s^2 2p^6 3s^2 3p^6$; no electrons are unpaired.

3.26 a. $1s^2 2s^2 2p^6 3s^2 3p^6 4s^1$; seven s electrons.
b. $1s^2 2s^2 2p^6 3s^2 3p^1$; one unpaired electron.
c. $1s^2 2s^2 2p^6 3s^2$; four subshells are filled.

3.28 a. C, carbon
b. Na, sodium
c. Si, silicon, and S, sulfur
d. Nb, niobium
e. V, vanadium, and Co, cobalt

3.30 a. $[Ar]4s^2 3d^{10} 4p^3$
b. $[Ar]4s^2 3d^5$
c. $[Ne]3s^2 3p^2$
d. $[Kr]5s^2 4d^{10} 5p^5$

3.32 Na: $[Ne]3s^1$
Mg: $[Ne]3s^2$
Al: $[Ne]3s^2 3p^1$
Si: $[Ne]3s^2 3p^2$
P: $[Ne]3s^2 3p^3$
S: $[Ne]3s^2 3p^4$
Cl: $[Ne]3s^2 3p^5$
Ar: $[Ne]3s^2 3p^6$

3.34 a. p area
b. d area
c. f area
d. s area

3.36 a. Transition
b. Representative
c. Inner-transition
d. Noble gas
e. Representative

3.38 a. Nonmetal
b. Metal
c. Metalloid
d. Metalloid
e. Metal

3.40 a. K
b. Bi
c. Sr
d. Sn

3.42 a. Ga
b. Sb
c. C
d. Te

3.44 a. Si
b. Ba
c. Br
d. Ca

········· **CHAPTER 4** ·········

4.2 a. $:\ddot{\text{I}}\cdot$

b. $\text{Sr}:$

c. $\cdot\overset{\cdot}{\text{Sn}}\cdot$

d. $:\overset{\cdot\cdot}{\text{S}}\cdot$

4.4 a. $[\text{Kr}]5s^2 4d^{10}5p^3$
b. $[\text{Kr}]5s^1$
c. $[\text{Kr}]5s^2 4d^{10}5p^4$
d. $[\text{Xe}]6s^1$

4.6 a. $:\overset{\cdot\cdot}{\text{Sb}}\cdot$

b. $\text{Rb}\cdot$

c. $:\overset{\cdot\cdot}{\text{Te}}\cdot$

d. $\text{Cs}\cdot$

4.8 a. $\cdot\overset{\cdot}{\text{E}}\cdot$

b. $\cdot\overset{\cdot}{\text{E}}:$

4.10 a. Add 3 or lose 15
b. Add 17 or lose 1
c. Add 2 or lose 16
d. Add 31 or lose 1

4.12 a. Lose 1: $\text{Cs} \rightarrow \text{Cs}^+ + e^-$
b. Add 2: $\text{O} + 2e^- \rightarrow \text{O}^{2-}$
c. Add 3: $\text{N} + 3e^- \rightarrow \text{N}^{3-}$
d. Add 1: $\text{I} + e^- \rightarrow \text{I}^-$

4.14 a. Se^{2-}
b. Rb^+
c. Al^{3+}

4.16 a. sulfur, S
b. aluminum, Al
c. sodium, Na
d. chlorine, Cl

4.18 a. krypton, Kr
b. argon, Ar
c. krypton, Kr
d. neon, Ne

4.20 a. $\text{Ca} \rightarrow \text{Ca}^{2+} + 2e^-$; $\text{Cl} + e^- \rightarrow \text{Cl}^-$
Formula is CaCl_2.
b. $\text{Li} \rightarrow \text{Li}^+ + e^-$; $\text{Br} + e^- \rightarrow \text{Br}^-$
Formula is LiBr.
c. $\text{Mg} \rightarrow \text{Mg}^{2+} + 2e^-$; $\text{S} + 2e^- \rightarrow \text{S}^{2-}$
Formula is MgS.

4.22 a. BaTe
b. Ba_3N_2
c. BaF_2
d. Ba_3P_2

4.24 a. Binary
b. Binary
c. Not binary
d. Not binary
e. Binary

4.26 a. lithium ion
b. magnesium ion
c. barium ion
d. cesium ion

4.28 a. bromide ion
b. oxide ion
c. phosphide ion
d. telluride ion

4.30 a. calcium chloride
b. barium sulfide
c. zinc bromide
d. aluminum sulfide
e. strontium fluoride

4.32 a. tin(II) sulfide and tin(IV) sulfide
b. iron(II) chloride and iron(III) chloride
c. copper(I) oxide and copper(II) oxide
d. gold(I) chloride and gold(III) chloride

4.34 a. stannous sulfide and stannic sulfide
b. ferrous chloride and ferric chloride
c. cuprous oxide and cupric oxide
d. aurus chloride and auric chloride

4.36 a. PbO_2
b. CoCl_2
c. CuS
d. NiN
e. Pt_3P_2

4.38 a. 102.89 u
b. 78.08 u
c. 159.16 u
d. 34.83 u

4.40 a. Na^+ and Br^-
b. Ca^{2+} and F^-

   c. Cu^+ and S^{2-}
   d. Li^+ and N^{3-}

4.42 a. 22.99 g of Na^+ and 79.90 g of Br^-
   b. 40.08 g of Ca^{2+} and 38.00 g of F^-
   c. 127.10 g of Cu^+ and 32.06 g of S^{2-}
   d. 20.82 g of Li^+ and 14.01 g of N^{3-}

4.44 a. 6.02×10^{23} Na^+ ions and 6.02×10^{23} Br^- ions
   b. 6.02×10^{23} Ca^{2+} ions and 12.04×10^{23} F^- ions
   c. 12.04×10^{23} Cu^+ ions and 6.02×10^{23} S^{2-} ions
   d. 18.06×10^{23} Li^+ ions and 6.02×10^{23} N^{3-} ions

4.46

4.48 a. $H:\ddot{S}:H$ or $H-S-H$

   b. $:\ddot{\underset{..}{C}l}:\ddot{\underset{..}{F}}:$ or $Cl-F$

   c. $H:\ddot{\underset{..}{Br}}:$ or $H-Br$

   d. $:\ddot{\underset{..}{C}l}:\ddot{\underset{..}{O}}:H$ or $Cl-O-H$

4.50 a.

   b.

   c.

4.52 a. $H:\ddot{S}:H$ Bent

   b. $:\ddot{\underset{..}{C}l}:\underset{\underset{:\ddot{\underset{..}{C}l}:}{|}}{\ddot{P}}:\ddot{\underset{..}{C}l}:$ Triangular pyramid with P at the top

   c. $:\ddot{\underset{..}{F}}:\ddot{\underset{..}{O}}:\ddot{\underset{..}{F}}:$ Bent

   d.   Tetrahedral with Sn in center

4.54 a. $\left[:\ddot{\underset{..}{O}}:\ddot{N}::\ddot{\underset{..}{O}}:\right]^-$ Bent

   b. $\left[:\ddot{\underset{..}{O}}:\underset{\underset{:\ddot{\underset{..}{O}}:}{}}{\ddot{\underset{..}{C}l}}:\ddot{\underset{..}{O}}:\right]^-$ Triangular pyramid with Cl at the top

   c. $\left[:\ddot{\underset{..}{O}}:C::\ddot{\underset{..}{O}}:\right]^{2-}$ Flat triangle with C in the center

   d. $\left[H:\ddot{O}:H\right]^+$ Triangular pyramid with O at the top

4.56 a. $\overset{\delta^+}{H}-\overset{\delta^-}{I}$

   b. $\overset{\delta^{2+}}{S}=\overset{\delta^-}{O}$, $O^{\delta-}$

   c. Nonpolarized $O-O$, O

4.58 a. Polar covalent
   b. Polar covalent
   c. Nonpolar covalent
   d. Polar covalent
   e. Ionic

4.60 a. Polar
   b. Polar
   c. Nonpolar

4.62 a. Polar covalent
   b. Ionic
   c. Polar covalent

4.64 a. Polar $\overset{\delta^+}{C}\equiv\overset{\delta^-}{O}$

   b. Polar $\overset{\delta^+}{H}-\overset{\delta^{2-}}{Se}$, H^{δ^+}

   c. Nonpolar $I^{\delta-}$, $Al^{\delta^{3+}}$, $I^{\delta-}$, $I^{\delta-}$

4.66 a. phosphorus trichloride
   b. dinitrogen pentoxide
   c. carbon tetrachloride
   d. boron trifluoride
   e. carbon disulfide

4.68 a. N_2O_4
   b. SCl_6
   c. SiO_2
   d. OF_2

4.70 a. $Ca(NO_2)_2$, calcium nitrite
   b. $Mg(ClO)_2$, magnesium hypochlorite
   c. $Cs_2Cr_2O_7$, cesium dichromate
   d. K_2SO_3, potassium sulfite

4.72 a. $Ba(OH)_2$
   b. $MgSO_3$
   c. $CaCO_3$
   d. $(NH_4)_2SO_4$
   e. $LiHCO_3$

4.74 a. M_2SO_3
   b. $MC_2H_3O_2$
   c. MCr_2O_7
   d. MPO_4
   e. $M(NO_3)_3$

4.76 The stronger the attractive forces between molecules, the higher the melting and boiling points. Thus, we can conclude that the forces attracting dimethyl ether molecules to one another are weaker than the forces attracting ethyl alcohol molecules to one another. Because both compounds are covalently bonded, it is likely that the attractive forces are polar forces, and we might conclude that ethyl alcohol is more polar than dimethyl ether.

4.78 The predominant forces that exist between molecules of noble gases are dispersion forces because noble gas molecules are nonpolar. Dispersion forces increase with increasing mass of the particles involved, so the boiling points of the noble gases would increase in the order He, Ne, Ar, Kr, Xe, and Rn.

4.80 The relatively high melting point indicates that it is unlikely that weak dispersion forces are the predominant forces present.

Dipolar forces are also quite weak and therefore are also unlikely to be the forces present.

CHAPTER 5

5.2

	Reactants	Products
a.	H_2, O_2	H_2O
b.	$CaCO_3$	CaO, CO_2
c.	sodium, water	hydrogen, sodium hydroxide
d.	copper, silver nitrate	copper nitrate, silver

5.4 a. Not consistent: The number of O atoms is not the same on both sides.

b. Consistent

c. Not consistent: The masses of reactants and products are not equal.

d. Consistent

5.6

	Left side	Right side	Classification
a.	1Ag, 1Cu, 2N, 6O	1Ag, 1Cu, 1N, 3O	Unbalanced
b.	4N, 8O	4N, 8O	Balanced
c.	1Mg, 2O	2Mg, 2O	Unbalanced
d.	4H, 1S, 6O, 1Ca	4H, 1S, 6O, 1Ca	Balanced

5.8 a. $2KClO_3(s) \rightarrow 2KCl(s) + 3O_2(g)$

b. $2C_2H_6(g) + 7O_2(g) \rightarrow 4CO_2(g) + 6H_2O(\ell)$

c. $2N_2(g) + 5O_2(g) \rightarrow 2N_2O_5(g)$

d. $MgCl_2(s) + H_2O(g) \rightarrow MgO(s) + 2HCl(g)$

e. $CaH_2(s) + 2H_2O(\ell) \rightarrow Ca(OH)_2(s) + 2H_2(g)$

f. $2Al(s) + Fe_2O_3(s) \rightarrow Al_2O_3(s) + 2Fe(s)$

g. $2Al(s) + 3Br_2(g) \rightarrow 2AlBr_3(s)$

h. $2HgNO_3(aq) + 2NaCl(aq) \rightarrow Hg_2Cl_2(s) + 2NaNO_3(aq)$

5.10 a. +1

b. +3

c. +0

d. +1

e. +3

f. +5

5.12 a. N (+3)

b. B (+3)

c. C (+4)

d. S (+4)

e. Cl (+7)

f. I (+5)

5.14 a. Reduced

b. Neither

c. Oxidized

d. Reduced

e. Oxidized

5.16 a. $\underset{0}{4Al(s)} + \underset{0}{3\ O_2(g)} \rightarrow \underset{+3\ -2}{2Al_2O_3(s)}$

Oxidizing agent: O_2
Reducing agent: Al

b. $\underset{0}{2Na(s)} + \underset{+1\ -2}{2H_2O(\ell)} \rightarrow \underset{+1\ -2\ +1}{2NaOH(aq)} + \underset{0}{H_2(g)}$

Oxidizing agent: H_2O(H)
Reducing agent: Na

c. $\underset{+2\ -2}{FeO(s)} + \underset{+2\ -2}{CO(g)} \rightarrow \underset{0}{Fe(s)} + \underset{+4\ -2}{CO_2(g)}$

Oxidizing agent: FeO(Fe)
Reducing agent: CO(C)

d. $\underset{+7\ -2}{2MnO_4(aq)} + \underset{-1}{10Cl^-(aq)} + \underset{+1}{16H^+(aq)} \rightarrow$
$\underset{0}{5Cl_2(g)} + \underset{+2}{2Mn^{2+}(aq)} + \underset{+1\ -2}{8H_2O(\ell)}$

Oxidizing agent: MnO_4^- (Mn)
Reducing agent: Cl^-

e. $\underset{+1\ -2}{2N_2O(g)} + \underset{0}{3O_2(g)} \rightarrow \underset{+4\ -2}{4NO_2(g)}$

Oxidizing agent: O_2
Reducing agent: N_2O(N)

f. $\underset{+4\ -1}{2TiCl_4(s)} + \underset{0}{Zn(s)} \rightarrow \underset{+3\ -1}{2TiCl_3(s)} + \underset{+2\ -1}{ZnCl_2(s)}$

Oxidizing agent: $TiCl_4$(Ti)
Reducing agent: Zn

5.18 Oxidizing agent: NaOH (H)
Reducing agent: Al

5.20 a. Nonredox, decomposition

b. Redox, single replacement

c. Nonredox, double replacement

d. Nonredox, combination

e. Redox, combination

f. Redox, combination

5.22 Nonredox, decomposition

5.24 Redox

5.26 Redox

5.28 a. Ca^{2+}, Cl^-

b. Mg^{2+}, NO_3^-

c. NH_4^+, PO_4^{3-}

d. Li^+, OH^-

e. K^+, CrO_4^{2-}

f. Ca^{2+}, HCO_3^-

5.30 a. Total ionic: $SO_2(aq) + H_2O(\ell) \rightarrow 2H^+(aq) + SO_3^{2-}(aq)$
Spectator ions: none
Net ionic: $SO_2(aq) + H_2O(\ell) \rightarrow 2H^+(aq) + SO_3^{2-}(aq)$

b. Total ionic: $Cu^{2+}(aq) + SO_4^{2-}(aq) + Zn(s) \rightarrow$
$Cu(s) + Zn^{2+}(aq) + SO_4^{2-}(aq)$
Spectator ions: SO_4^{2-}
Net ionic: $Cu^{2+}(aq) + Zn(s) \rightarrow Cu(s) + Zn^{2+}(aq)$

c. Total ionic:
$2K^+(aq) + 2Br^-(aq) + 4H^+(aq) + 2SO_4^{2-}(aq) \rightarrow$
$Br_2(aq) + SO_2(aq) + 2K^+(aq) + SO_4^{2-}(aq) + 2H_2O(\ell)$
Spectator ions: K^+, one SO_4^{2-}
Net ionic: $2Br^-(aq) + 4H^+(aq) + SO_4^{2-}(aq) \rightarrow$
$Br_2(aq) + SO_2(aq) + 2H_2O(\ell)$

d. Total ionic:
$Ag^+(aq) + NO_3^-(aq) + Na^+(aq) + OH^-(aq) \rightarrow$
$AgOH(s) + Na^+(aq) + NO_3^-(aq)$
Spectator ions: Na^+, NO_3^-
Net ionic: $Ag^+(aq) + OH^-(aq) \rightarrow AgOH(s)$

e. Total ionic: $BaCO_3(s) + 2H^+(aq) + 2NO_3^-(aq) \rightarrow$
$Ba^{2+}(aq) + 2NO_3^-(aq) + CO_2(g) + H_2O(\ell)$
Spectator ions: NO_3^-
Net ionic:
$BaCO_3(s) + 2H^+(aq) \rightarrow Ba^{2+}(aq) + CO_2(g) + H_2O(\ell)$

f. Total ionic: $N_2O_5(aq) + H_2O(\ell) \rightarrow$
$2H^+(aq) + 2NO_3^-(aq)$

Spectator ions: none
Net ionic: $N_2O_5(aq) + H_2O(\ell) \rightarrow$
$$2H^+(aq) + 2NO_3^- (aq)$$

5.32 a. Total ionic: $H^+(aq) + I^-(aq) + K^+(aq) + OH^-(aq) \rightarrow$
$$K^+(aq) + I^-(aq) + H_2O(\ell)$$
Spectator ions: K^+, I^-
Net ionic: $H^+(aq) + OH^-(aq) \rightarrow H_2O(\ell)$
b. Total ionic:
$2H^+(aq) + SO_4^{2-}(aq) + 2NH_4^+(aq) + 2OH^-(aq) \rightarrow$
$$2NH_4^+(aq) + SO_4^{2-}(aq) + 2H_2O(\ell)$$
Spectator ions: SO_4^{2-}, NH_4^+
Net ionic: $2H^+(aq) + 2OH^-(aq) \rightarrow 2H_2O(\ell)$
c. Total ionic: $H^+(aq) + Cl^-(aq) + Li^+(aq) + OH^-(aq) \rightarrow$
$$Li^+(aq) + Cl^-(aq) + H_2O(\ell)$$
Spectator ions: Li^+, Cl^-
Net ionic: $H^+(aq) + OH^-(aq) \rightarrow H_2O(\ell)$

In every net ionic reaction, H^+ reacts with OH^- to form H_2O.

5.34 Exothermic

5.36 The insulation around the ice slows its rate of melting by preventing it from absorbing heat. However, this prevents the ice from performing its cooling function in the cooler. For the cooler to work properly, heat must move from the warmer food to the cooler ice, where it is absorbed as the ice melts.

5.38 a. $S(s) + O_2(g) \rightarrow SO_2(g)$
1. 1 S atom + 1 O_2 molecule → 1 SO_2 molecule
4. 6.02×10^{23} S atoms + 6.02×10^{23} O_2 molecules →
$$6.02 \times 10^{23} \ SO_2 \text{ molecules}$$
5. 1 mol S + 1 mol O_2 → 1 mol SO_2
6. 32.06 g S + 32.00 g O_2 → 64.06 g SO_2
b. $Sr(s) + 2H_2O(\ell) \rightarrow Sr(OH)_2(s) + H_2(g)$
1. 1 Sr atom + 2 H_2O molecules →
$$1 \ Sr(OH)_2 \text{ molecule} + 1 \ H_2 \text{ molecule}$$
4. 6.02×10^{23} Sr atoms + 12.04×10^{23} H_2O
molecules → 6.02×10^{23} $Sr(OH)_2$
molecules + 6.02×10^{23} H_2 molecules
5. 1 mol Sr + 2 mol H_2O → 1 mol $Sr(OH)_2$ + 1 mol H_2
6. 87.62 g Sr + 36.03 g H_2O → 121.64 g $Sr(OH)_2$ +
$$2.02 \text{ g } H_2$$
c. $2H_2S(g) + 3O_2(g) \rightarrow 2H_2O(g) + 2SO_2(g)$
1. 2 H_2S molecules + 3 O_2 molecules → 2 H_2O
molecules + 2 SO_2 molecules
4. 12.04×10^{23} H_2S molecules + 18.06×10^{23} O_2
molecules → 12.04×10^{23} H_2O molecules +
12.04×10^{23} SO_2 molecules
5. 2 mol H_2S + 3 mol O_2 → 2 mol H_2O + 2 mol SO_2
6. 68.15 g H_2S + 96.00 g O_2 → 36.03 g H_2O +
$$128.12 \text{ g } SO_2$$
d. $4NH_3(g) + 5O_2(g) \rightarrow 4NO(g) + 6H_2O(g)$
1. 4 NH_3 molecules + 5 O_2 molecules → 4 NO
molecules + 6 H_2O molecules
4. 24.08×10^{23} NH_3 molecules + 30.10×10^{23} O_2
molecules → 24.08×10^{23} NO molecules +
36.12×10^{23} H_2O molecules
5. 4 mol NH_3 + 5 mol O_2 → 4 mol NO + 6 mol H_2O
6. 68.14 g NH_3 + 160.00 g O_2 → 120.04 g NO +
$$108.10 \text{ g } H_2O$$
e. $CaO(s) + 3C(s) \rightarrow CaC_2(s) + CO(g)$
1. 1 CaO molecule + 3 C atoms → 1 CaC_2 molecule +
1 CO molecule
4. 6.02×10^{23} CaO molecules + 18.06×10^{23} C
atoms → 6.02×10^{23} CaC_2 molecules +
6.02×10^{23} CO molecules
5. 1 mol CaO + 3 mol C → 1 mol CaC_2 + 1 mol CO
6. 56.08 g CaO + 36.03 g C → 64.10 g CaC_2 +
$$28.01 \text{ g CO}$$

5.40 $2SO_2(g) + O_2(g) \rightarrow 2SO_3(g)$
1. 2 SO_2 molecules + 1 O_2 molecule → 2 SO_3 molecules
4. 12.0×10^{23} SO_2 molecules + 6.02×10^{23} O_2
molecules → 12.0×10^{23} SO_3 molecules
5. 2 mol SO_2 + 1 mol O_2 → 2 mol SO_3
6. 128 g SO_2 + 32.0 g O_2 → 160 g SO_3

Factors:
$$\frac{12.0 \times 10^{23} \ SO_2 \text{ molecules}}{6.02 \times 10^{23} \ O_2 \text{ molecules}}; \quad \frac{12.0 \times 10^{23} \ SO_3 \text{ molecules}}{12.0 \times 10^{23} \ SO_2 \text{ molecules}};$$

$$\frac{32.0 \text{ g } O_2}{128 \text{ g } SO_2}; \quad \frac{128 \text{ g } SO_2}{160 \text{ g } SO_3}; \quad \frac{32.0 \text{ g } O_2}{160 \text{ g } SO_3}; \quad \frac{128 \ SO_2}{32.0 \text{ g } O_2};$$

$$\frac{2 \text{ mol } SO_2}{1 \text{ mol } O_2}; \frac{2 \text{ mol } SO_3}{2 \text{ mol } SO_2}; \frac{1 \text{ mol } O_2}{2 \text{ mol } SO_3}; \frac{1 \text{ mol } O_2}{2 \text{ mol } SO_2}; \frac{2 \text{ mol } SO_3}{32.0 \text{ g } O_2};$$

$$\frac{128 \text{ g } SO_2}{2 \text{ mol } SO_2}; \frac{2 \text{ mol } SO_2}{32.0 \text{ g } O_2}; \frac{2 \text{ mol } SO_3}{1 \text{ mol } O_2}; \frac{160 \text{ g } SO_3}{32.0 \text{ g } O_2}; \frac{32.0 \text{ g } O_2}{1 \text{ mol } O_2}$$

This list does not include all possible factors.

5.42 892 g

5.44 445 g

5.46 495 g

5.48 1.01×10^3 g

5.50 256 g

5.52 a. O_2 is the limiting reactant.
b. 71.9 g NO_2

5.54 144 g

5.56 76.5%

5.58 72.5%

5.60 88.9%

CHAPTER 6

6.2 a. 82.5 mL
b. 107 mL
c. 46.2 mL

6.4 2 mL

6.6 a. The density will decrease because density is equal to mass/volume, and the volume of carbon dioxide will increase on heating while the mass remains unchanged.
b. 1.67 g/L

6.8 As the ball travels upward and slows, more and more of its kinetic energy is converted into potential energy. At the maximum height, when the ball stops before starting to fall, all the energy is in the form of potential energy. As the ball falls faster and faster, more and more of the potential energy is converted back into kinetic energy. If there were no energy losses from air resistance, the ball would gain back as much kinetic energy as it had when it was initially thrown.

6.10 Each type of molecule has the same kinetic energy of 3.18×10^{10} u cm²/s².

6.12 a. Particles (molecules) in a liquid move freely and therefore assume the container shape, but the cohesive forces between the particles prevent them from separating completely from each other and filling the container.

    b. There is very little space between the particles (molecules) of liquids and solids. As a result, increased pressure cannot push the particles closer together and will have little influence on their volumes.

    c. Gas molecules move around essentially unrestricted, and on average, equal numbers collide with each of the walls of a container in a given amount of time. The collisions cause the pressure on the walls, so the pressure on all the walls is the same.

6.14 Heat must be added to a substance to convert it from the solid to the liquid state, and then from the liquid to the gaseous state. If the added heat causes no temperature difference between the two states, as when a solid melts to a liquid at the melting point, then the added heat increases the potential energy of the particles of the substance. Thus, at the melting point the particles of both solid and liquid forms of a substance will have the same kinetic energy because they are at the same temperature. However, particles of the liquid will have more potential energy, corresponding to the energy that had to be added to change the solid to a liquid. Similar arguments apply for a comparison of the liquid and gaseous states of a substance.

6.16 a. Gaseous
    b. Solid
    c. Gaseous
    d. Gaseous

6.18 a. 0.926 atm
    b. 704 torr
    c. 13.6 psi
    d. 0.935 bar

6.20 a. 12.9 atm
    b. 13.0 bar
    c. 9.81×10^3 mmHg
    d. 386 in. Hg

6.22 a. 454 K
    b. $-272°C$
    c. 20 K

6.24 Column A: $P_f = 1.50$ atm; Column B: $V_f = 4.49$ L; Column C: $T_f = 221$ K

6.26 190 mL

6.28 4.09 atm

6.30 26.3 L

6.32 684 mL

6.34 5.0 L

6.36 2.7 L

6.38 80.0 atm

6.40 1.4×10^4 ft³

6.42 842 mL

6.44 5.00 g/L

6.46 a. 0.0179 mol
    b. 11.3 atm
    c. 13.9 L

6.48 3.99 atm

6.50 26.3 g

6.52 3.8 mol

6.54 46.0 g/mol

6.56 44.1 g/mol; the gas is most likely CO_2

6.58 370 torr

6.60 O_2 mass/H_2 mass = 16 according to Graham's law. From the periodic table, O_2 mass/H_2 mass = 32 u/2 u = 16. The agreement is very good.

6.62 The helium-filled balloon would appear to go flat first because helium molecules diffuse through the rubber balloon faster than do nitrogen molecules. The rates of diffusion according to Graham's law would be He rate/N_2 rate = $\sqrt{28.0 \text{ u}/4.00 \text{ u}} = 2.6$. Thus, He would be expected to escape through the balloon 2.6 times as fast as nitrogen.

6.64 a. Exothermic
    b. Exothermic
    c. Endothermic

6.66 A change in state refers to a process in which a substance changes from one of the three states of matter—solid, liquid, or gas—to another of the three states.

6.68 Liquid methylene chloride evaporates very rapidly, and when it evaporates, it absorbs heat. Thus, the liquid acted to cool and temporarily anesthetize any tissue on which it was sprayed.

6.70 The boiling points of the two liquids are different. Thus, a measurement of the temperature of each boiling liquid would differentiate between them. The ethylene glycol would have the higher boiling point of the two.

6.72 The temperature of boiling water is influenced by the atmospheric pressure on the water. On top of Mount Everest the atmospheric pressure is quite low, so water boils at a temperature significantly lower than the normal boiling point of 100°C. The temperature of the burning campfire would not be influenced by the altitude. Thus, the potato would cook most quickly if it were put into the campfire.

6.74 a. 1.1×10^3 cal
    b. 8.9×10^2 cal
    c. 2.03×10^6 cal

6.76 Even though the heat of fusion is high and a lot of heat could be stored by melting the salt, its high melting point of more than 1000°C makes it impractical.

6.78 3.67×10^5 cal

CHAPTER 7

7.2 a. Solvent: water; solute: sodium hypochlorite
    b. Solvent: isopropyl alcohol; solute: water
    c. Solvent: water; solute: hydrogen peroxide
    d. Solvent: SD alcohol; solutes: water, glycerin, fragrance, menthol, benzophenone, coloring

7.4 a. Not a solution because it is not homogeneous. It is cloudy rather than clear.

b. Solution; solvent: water; solutes: all dissolved components.

c. Not a solution because it is not homogeneous. Bits of pulp can be seen.

d. Solution if prepared carefully with a tea bag or good strainer; solvent: water; solutes: all dissolved components.

e. Not a solution because it is not homogeneous.

7.6 a. The oil and vinegar are immiscible.

b. The alcohol is soluble in the water.

c. The tar is soluble in the chloroform.

7.8 a. Unsaturated

b. Supersaturated

c. Saturated

7.10 The solution could be carefully cooled to a temperature lower than room temperature. Another method would be to allow some of the solvent to evaporate.

7.12 a. Soluble

b. Insoluble

c. Slightly soluble

d. Very soluble

e. Slightly soluble

7.14 Add pure water to the mixture. The $CaCl_2$ is soluble and will dissolve in the water. The liquid that contains the dissolved $CaCl_2$ can then be poured off carefully, leaving behind the insoluble solid $CaCO_3$.

7.16 a. Soluble in benzene

b. Soluble in benzene

c. Soluble in water

d. Soluble in benzene

7.18 The structure indicates that Freon-114 is a nonpolar material. Grease is also nonpolar, as indicated by its insolubility in water. Thus, grease would dissolve readily in Freon-114.

7.20 a. 0.364 M

b. 0.860 M

c. 1.75 M

7.22 a. 0.500 M

b. 0.200 M

c. 0.26 M

7.24 a. 0.376 mol

b. 0.0750 mol

c. 571 mL

7.26 a. 0.584 g

b. 0.157 L

c. 1.18×10^3 mL

7.28 a. 5.0%

b. 5.0%

c. 5.0%

7.30 a. 6.98%

b. 5.08%

c. 13.0%

d. 51.3%

7.32 a. 12.5%

b. 13.4%

c. 6.24%

7.34 a. 8.00%

b. 8.00%

c. 2.3%

d. 34.2%

7.36 0.025 L or 25 mL

7.38 a. 5.00%

b. 5.00%

c. 8.77%

7.40 About 40%

7.42 a. Put 0.0300 mol (4.26 g) of solid Na_2SO_4 into a 200 mL volumetric flask and add enough pure water to fill the flask to the mark after the solid dissolves.

b. Put 0.0625 mol (11.8 g) of solid $Zn(NO_3)_2$ into a 250 mL volumetric flask and add enough pure water to fill the flask to the mark after the solid dissolves.

c. Mix together 3.38 g of solid NaCl and 147 g (147 mL) of pure water. Allow the NaCl to dissolve.

d. Put 0.94 g of solid KCl into a 125 mL volumetric flask and add enough pure water to fill the flask to the mark after the solid dissolves.

7.44 %(w/w) = 25.0; %(w/v) = 23.8

7.46 a. 32.3 g

b. 0.700 mol

c. 31.3 mL

d. 2.10 g

7.48 a. Add 1.67 L of 18.0 M H_2SO_4 solution to enough water to give 5.00 L of solution. The 18.0 M acid should be added to water, and not the water to the acid. One way to make the desired solution is to add the 1.67 L of 18.0 M acid to about 3 L of pure water. Stir the resulting solution, then add enough additional water to bring the total volume to 5.00 L.

b. Put 41.7 mL of 3.00 M $CaCl_2$ solution into a 250-mL volumetric flask and add enough pure water to fill the flask to the mark when the contents are well mixed.

c. Put 30.0 mL of 10% solution into a calibrated volumetric container (such as a volumetric flask or large volumetric cylinder) and add enough pure water to bring the total volume of well-mixed solution to 200 mL.

d. Put 100 mL of 50.0% solution into a calibrated volumetric container and add enough pure water to bring the total volume of well-mixed solution to 500 mL.

7.50 a. 0.0113 M

b. 0.100 M

c. 0.0375 M

d. 0.0265 M

7.52 16.6 g

7.54 20.0 mL

7.56 50.7 mL

7.58 27.8 mL

7.60 1.3 g

7.62 Because ice cream is essentially a solution of substances dissolved in water, the freezing point of the solution is lower than that of pure water. The temperature of a mixture of ice and pure water is the same as the freezing point of pure water. Thus, a mixture of ice and water has a temperature higher than the freezing point of the ice cream solution. A mixture of ice, salt, and water has a temperature (freezing point) lower than that of pure water. If enough salt is added to the ice and water mixture, the temperature can be made lower than the freezing point of the ice cream solution, and it will freeze.

7.64 **a.** Boiling point = 101.56°C; freezing point = −5.58°C
  b. Boiling point = 100.78°C; freezing point = −2.79°C
  c. Boiling point = 102.34°C; freezing point = −8.37°C
  d. Boiling point = 103.12°C; freezing point = −11.2°C

7.66 **a.** Boiling point = 100.38°C; freezing point = −1.40°C
  b. Boiling point = 100.23°C; freezing point = −0.84°C
  c. Boiling point = 102.51°C; freezing point = −8.98°C

7.68 **a.** Osmolarity = 0.20
  b. Osmolarity = 0.60
  c. Osmolarity = 3.18

7.70 $\pi = 5.58 \times 10^3$ torr = 5.58×10^3 mmHg = 7.34 atm

7.72 $\pi = 1.12 \times 10^4$ torr = 1.12×10^4 mmHg = 14.7 atm

7.74 $\pi = 5.88 \times 10^4$ torr = 5.88×10^4 mmHg = 77.3 atm

7.76 $\pi = 3.5 \times 10^3$ torr = 3.5×10^3 mmHg = 4.6 atm

7.78 $\pi = 9.95 \times 10^3$ torr = 9.95×10^3 mmHg = 13.1 atm

7.80 When the bag functions as an osmotic membrane, only water molecules can pass through. As a result, there is a net flow of water into the bag from the surroundings. When the bag functions as a dialysis membrane, water, other small molecules, and hydrated ions can pass through. As a result, the large molecules that cannot pass through tend to increase in concentration inside the bag while the concentration of small molecules and hydrated ions decreases inside the bag.

7.82 The dispersing medium of a colloidal suspension is the material that is present in largest amount. Analogous to the solvent of a solution, it is the medium in which the colloidal particles are suspended. The dispersed phase consists of all the colloidal particles that are suspended in the dispersing medium. It is analogous to the solute of a solution. A colloidal emulsifying agent stabilizes the dispersed phase and prevents the particles of that phase from coagulating to form larger particles that would settle out of the suspension.

CHAPTER 8

8.2 **a.** Nonspontaneous: The rocket engines of the shuttle must operate continuously to push the shuttle into an orbit.
  b. Spontaneous: Once ignited, the fuel continues to burn without additional input of energy.
  c. Nonspontaneous: Heat must be supplied continuously to keep the water boiling.
  d. Nonspontaneous: The only way to raise the temperature of a substance is to supply heat.
  e. Nonspontaneous: To make your room orderly, someone must provide some work to move the clothing, etc. around. The work done represents an input of energy to the room.

8.4 **a.** Exergonic: The person doing the pushing is giving up energy to the automobile.
  b. Endergonic: The ice must gain energy in the form of heat in order to melt.
  c. Exergonic: The surroundings lose heat to the ice in order for the ice to melt.

d. Exergonic: In order to condense, the steam must give up some energy in the form of heat.
  e. Endergonic: The heat given up by the steam as it condenses is accepted by the surroundings.

8.6 **a.** Both the energy and the entropy of the water decrease when water freezes. The process is spontaneous if the temperature of the surroundings is lower than 0°C.
  b. The energy of the water decreases while the entropy stays essentially constant. The process is spontaneous.
  c. The energy of the perfume increases slightly as energy is absorbed during evaporation. The entropy of the perfume increases a great deal as the molecules evaporate from the liquid and disperse throughout the room. The process is spontaneous.

8.8 **a.** Entropy is highest at the end of the play because the players are in a more mixed-up state than they were in just before the play began.
  b. The answer is the 10% copper/gold alloy or the 2% alloy, depending on the atoms discussed. A lower concentration of copper atoms allows more disorder among the copper atoms but more order among the gold atoms. A higher concentration of copper allows less disorder for the copper atoms but increases the disorder of the gold atoms.
  c. Entropy is highest with the purse on the ground and the contents scattered.
  d. The mixture characteristic of coins in a piggy bank would have the highest entropy of the choices.
  e. The mixture of loose pearls in a box would have the highest entropy of the choices.

8.10 The answers are subjective and depend on the meaning given to the terms *very slow, slow,* and *fast.*
  a. Slow
  b. Very slow
  c. Slow
  d. Fast
  e. Very slow

8.12 **a.** Any of the following could be measured at different times: the weight of the block, the weight or volume of the water produced by the melting block, and the physical dimensions of the melting block.
  b. Either of the following could be measured at different times: the distance a weighted object penetrates into a sample, and the force required to break a specific-sized piece.
  c. Either of the following could be measured at different times: the length of the unburned candle and the weight of the candle.

8.14 **a.** Rate = 0.0330 M/min
  b. Rate = 0.0194 M/min

8.16 Rate = 3.96×10^{-3} M/min

8.18 Rate = 1.51×10^{-6} M/s

8.20 **a.** Assumption #1 is being violated. According to assumption #1, decreasing the concentrations of reactants would decrease the number of collisions and so decrease the rate at which the molecules would react.
  b. Assumption #2 is being violated. According to assumption #2, molecules must collide with a certain minimum energy in order to react. At a lower temperature fewer molecules would have the necessary minimum energy, so the reaction should go slower, not faster.

c. Assumption #1 is being violated. According to assumption #1, increasing the concentration of either of the reactants should increase the number of collisions and so increase the rate of the reaction.

8.22 An increase in temperature increases the speeds of the molecules involved in a reaction. The increased speed acts to increase the reaction rate in two ways. First, it causes more collisions to occur between molecules in a given amount of time. Second, it causes the energy of the colliding molecules to be greater so that more of the collisions will involve molecules that have at least the minimum energy needed for a reaction to take place.

8.24 a.

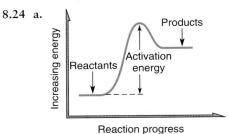

b.

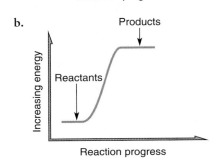

8.26

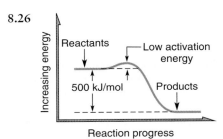

At room temp.

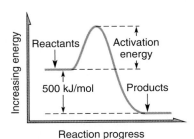

At 150° C

Both reactions are exothermic and the products have 500 kJ/mol less energy than the reactants, so both reactions give up 500 kJ/mol. The reaction that takes place at room temperature has a smaller activation energy than the reaction that must be heated to 150°C.

8.28 a. Rapid. The oppositely charged ions in solution will be attracted to each other and undergo many collisions per second.

b. Won't react. The reactant particles in solution both have a negative electrical charge and will repel each other. Collisions are highly unlikely.

c. Rapid. The reacting particles are in the gaseous state where they move about freely, making effective collisions likely.

d. Slow to rapid, depending on the state of division of the solid. If large lumps of CaO were exposed to gaseous CO_2, the reaction would be slow, since only the CaO on the surface of the lumps would be available. If the CaO were divided into dust-size particles and blown into gaseous CO_2, the reaction would go quite rapidly.

8.30 Heat the reaction mixture, add a catalyst, or increase the concentration of one or more of the reactants.

8.32 About 7.5 minutes

8.34 Catalysts provide reaction pathways with lower activation energy. They may do this by reacting to form an intermediate structure that yields products when it breaks up or by providing a surface on which the reactants react.

8.36 a. Equilibrium would be indicated when the intensity of the violet color of the reaction mixture remained constant with time.

b. Equilibrium would be indicated when the amount of undissolved sugar remained constant with time.

c. Equilibrium would be indicated when either the intensity of the red-brown color of the reaction mixture remained constant with time or the total pressure of the reaction mixture remained constant with time.

8.38 a. $K = \dfrac{[CO_2]}{[CO]^2[O_2]}$

b. $K = \dfrac{[NO_2]^2}{[N_2O_4]}$

c. $K = \dfrac{[CO_2]^4[H_2O]^6}{[C_2H_6]^2[O_2]^7}$

d. $K = \dfrac{[NO]^2\,[Cl_2]}{[NOCl]^2}$

e. $K = \dfrac{[O_2][ClO_2]^4}{[Cl_2O_5]^2}$

8.40 a. $K = \dfrac{[Fe(CN)_6{}^{3-}]}{[Fe^{3+}][CN^-]^6}$

b. $K = \dfrac{[Ag(NH_3)_2{}^+]}{[Ag^+][NH_3]^2}$

c. $K = \dfrac{[AuCl_4{}^-]}{[Au^{3+}][Cl^-]^4}$

8.42 a. $CH_4 + 2O_2 \rightleftarrows CO_2 + 2H_2O$
b. $3H_2 + CO \rightleftarrows CH_4 + H_2O$
c. $2O_3 \rightleftarrows 3O_2$
d. $4NO_2 + 6H_2O \rightleftarrows 4NH_3 + 7O_2$

8.44 $K = 0.47$

8.46 $K = 0.099$

8.48 a. The concentration of products will be larger than the concentration of reactants.

b. The concentration of products will be larger than the concentration of reactants.

c. The concentration of products will be smaller than the concentration of reactants.

d. The concentration of products will be smaller than the concentration of reactants.

8.50 a. Shift left
b. Shift right
c. Shift left

8.52 a. Shift right: The color will become less blue and more purple.
b. Shift right: The amount of solid $PbCl_2$ will increase.
c. Shift right: The intensity of the violet color of the mixture will decrease.
d. Shift right: The intensity of the violet color of the mixture will decrease.
e. Shift left: The intensity of the brown color of the mixture will increase.

8.54 a. The concentrations of NO_2 and O_2 will increase and the concentration of N_2O will decrease.
b. The concentrations of SO_2 and O_2 will increase and the concentration of SO_3 will decrease.
c. The concentration of CO_2 will increase and the concentrations of CO and O_2 will decrease.

8.56 a. Shift right
b. Shift left
c. Shift right
d. Shift left
e. No shift will occur
f. Shift left

········· **CHAPTER 9** ·········

9.2 a. $HBrO_2(aq) \rightarrow H^+(aq) + BrO_2^-(aq)$
b. $HS^-(aq) \rightarrow H^+(aq) + S^{2-}(aq)$
c. $HBr(aq) \rightarrow H^+(aq) + Br^-(aq)$
d. $HC_2H_3O_2(aq) \rightarrow H^+(aq) + C_2H_3O_2^-(aq)$

9.4 a. Arrhenius base: $LiOH(aq) \rightarrow Li^+(aq) + OH^-(aq)$
b. Not an Arrhenius base
c. Arrhenius base: $Sr(OH)_2(aq) \rightarrow Sr^{2+}(aq) + 2OH^-(aq)$
d. Not an Arrhenius base

9.6 a. Acids: $HC_2O_4^-$ and H_3O^+
Bases: H_2O and $C_2O_4^{2-}$
b. Acids: HNO_2 and H_3O^+
Bases: H_2O and NO_2^-
c. Acids: H_2O and HPO_4^{2-}
Bases: PO_4^{3-} and OH^-
d. Acids: H_2SO_3 and H_3O^+
Bases: H_2O and HSO_3^-
e. Acids: H_2O and HF
Bases: F^- and OH^-

9.8

	Acid	Conjugate base	Base	Conjugate acid
a.	$HC_2O_4^-$	$C_2O_4^{2-}$	H_2O	H_3O^+
b.	HNO_2	NO_2^-	H_2O	H_3O^+
c.	H_2O	OH^-	PO_4^{3-}	HPO_4^{2-}
d.	H_2SO_3	HSO_3^-	H_2O	H_3O^+
e.	H_2O	OH^-	F^-	HF

9.10 a. $HF + H_2O \rightleftarrows H_3O^+ + F^-$
b. $HClO_3 + H_2O \rightleftarrows H_3O^+ + ClO_3^-$
c. $HClO + H_2O \rightleftarrows H_3O^+ + ClO^-$
d. $HS^- + H_2O \rightleftarrows H_3O^+ + S^{2-}$

9.12 a. SO_4^{2-}
b. CH_3NH_2
c. ClO_4^-
d. NH_3
e. Cl^-

9.14 a. H_2CO_3
b. HS^-
c. H_2S
d. $H_2C_2O_4$
e. $H_2N_2O_2$

9.16 The substance added to complete each equation appears in blue.
a. $H_2AsO_4^-(aq) + NH_3(aq) \rightarrow NH_4^+(aq) + HAsO_4^{2-}(aq)$
b. $C_6H_5NH_2(aq) + H_2O(\ell) \rightarrow C_6H_5NH_3^+(aq) + OH^-(aq)$
c. $S^{2-}(aq) + H_2O(\ell) \rightarrow HS^-(aq) + OH^-(aq)$
d. $(CH_3)_2NH(aq) + HBr(aq) \rightarrow (CH_3)_2NH_2^+(aq) + Br^-(aq)$
e. $H_2PO_4^-(aq) + CH_3NH_2(aq) \rightarrow HPO_4^{2-}(aq) + CH_3NH_3^+(aq)$

9.18 a. $HS^-(aq) + NH_3(aq) \rightarrow S^{2-}(aq) + NH_4^+(aq)$
b. $H_2O(\ell) + ClO_3^-(aq) \rightarrow OH^-(aq) + HClO_3(aq)$
c. $H_2O(\ell) + NH_2^-(aq) \rightarrow OH^-(aq) + NH_3(aq)$
d. $HBO_3^{2-}(aq) + H_2O(\ell) \rightarrow BO_3^{3-}(aq) + H_3O^+(aq)$
e. $HNO_2(aq) + NH_3(aq) \rightarrow NO_2^-(aq) + NH_4^+(aq)$

9.20 hydrocyanic acid

9.22 a. hydrotelluric acid
b. hypochlorous acid
c. sulfurous acid
d. nitrous acid

9.24 succinic acid

9.26 H_2CO_3

9.28 a. 2.3×10^{-13} M
b. 7.7×10^{-11} M
c. 1.1×10^{-12} M
d. 1.3×10^{-5} M
e. 3.0×10^{-13} M

9.30 a. 1.4×10^{-10} M
b. 1.4×10^{-13} M
c. 2.0×10^{-15} M
d. 5.9×10^{-12} M
e. 1.1×10^{-6} M

9.32 In Exercise 9.28: (a) acidic, (b) acidic, (c) acidic, (d) basic, (e) acidic
In Exercise 9.30: (a) basic, (b) basic, (c) basic, (d) basic, (e) acidic

9.34 a. Acidic
b. Basic
c. Acidic
d. Basic
e. Acidic

9.36 a. 8.39, basic
b. 10.97, basic

c. 7.50, basic
d. 1.64, acidic
e. 4.71, acidic

9.38 a. 2.66, acidic
b. 11.41, basic
c. 5.12, acidic
d. 10.40, basic
e. 4.93, acidic

9.40 a. 5.4×10^{-10} M
b. 2.8×10^{-3} M
c. 3.8×10^{-6} M

9.42 a. $[H^+] = 1.1 \times 10^{-4}$ M; $[OH^-] = 8.9 \times 10^{-11}$ M
b. $[H^+] = 1.0 \times 10^{-4}$ M; $[OH^-] = 1.0 \times 10^{-10}$ M
c. $[H^+] = 1.4 \times 10^{-12}$ M; $[OH^-] = 7.2 \times 10^{-3}$ M

9.44 a. 8.9×10^{-9} M, basic
b. 1.2×10^{-4} M, acidic
c. 4.2×10^{-8} M, basic
d. 4.0×10^{-8} M, basic
e. 5.9×10^{-7} M, acidic

9.46 a. 1.2×10^{-3} M, acidic
b. 7.8×10^{-5} M, acidic
c. 4.8×10^{-3} M, acidic
d. 8.5×10^{-4} M, acidic

9.48 a. Add 1.0 L of dilute (6 M) HNO_3 to 1.0 L of pure water and mix well.
b. Add 50 mL of concentrated (15 M) NH_3 to 450 mL of pure water and mix well.
c. Add 83 mL of concentrated (12 M) HCl to 4.9 L of pure water and mix well.

9.50 a. $H_2SO_4(aq) + 2H_2O(\ell) \rightarrow 2\,H_3O^+(aq) + SO_4^{2-}(aq)$
b. $H_2SO_4(aq) + CaO(s) \rightarrow CaSO_4(aq) + H_2O(\ell)$
c. $H_2SO_4(aq) + Mg(OH)_2(s) \rightarrow MgSO_4(aq) + 2H_2O(\ell)$
d. $H_2SO_4(aq) + CuCO_3(s) \rightarrow CuSO_4(aq) + CO_2(g) + H_2O(\ell)$
e. $H_2SO_4(aq) + 2KHCO_3(s) \rightarrow K_2SO_4(aq) + 2CO_2(g) + 2H_2O(\ell)$
f. $H_2SO_4(aq) + Mg(s) \rightarrow MgSO_4(aq) + H_2(g)$

9.52 a. Total ionic: $2H^+(aq) + SO_4^{2-}(aq) + 2H_2O(\ell) \rightarrow 2H_3O^+(aq) + SO_4^{2-}(aq)$
Net ionic: $2H^+(aq) + 2H_2O(\ell) \rightarrow 2H_3O^+(aq)$
Simplified net ionic: $H^+(aq) + H_2O(\ell) \rightarrow H_3O^+(aq)$
b. Total ionic: $2H^+(aq) + SO_4^{2-}(aq) + CaO(s) \rightarrow Ca^{2+}(aq) + SO_4^{2-}(aq) + H_2O(\ell)$
Net ionic: $2H^+(aq) + CaO(s) \rightarrow Ca^{2+}(aq) + H_2O(\ell)$
c. Total ionic: $2H^+(aq) + SO_4^{2-}(aq) + Mg(OH)_2(s) \rightarrow Mg^{2+}(aq) + SO_4^{2-}(aq) + 2H_2O(\ell)$
Net ionic: $2H^+(aq) + Mg(OH)_2(s) \rightarrow Mg^{2+}(aq) + 2H_2O(\ell)$
d. Total ionic: $2H^+(aq) + SO_4^{2-}(aq) + CuCO_3(s) \rightarrow Cu^{2+}(aq) + SO_4^{2-}(aq) + CO_2(g) + H_2O(\ell)$
Net ionic: $2H^+(aq) + CuCO_3(s) \rightarrow Cu^{2+}(aq) + CO_2(g) + H_2O(\ell)$
e. Total ionic: $2H^+(aq) + SO_4^{2-}(aq) + 2KHCO_3(s) \rightarrow 2K^+(aq) + SO_4^{2-}(aq) + 2CO_2(g) + 2H_2O(\ell)$
Net ionic: $2H^+(aq) + 2KHCO_3(s) \rightarrow 2K^+(aq) + 2CO_2(g) + 2H_2O(\ell)$
Simplified net ionic: $H^+(aq) + KHCO_3(s) \rightarrow K^+(aq) + CO_2(g) + H_2O(\ell)$

f. Total ionic: $2H^+(aq) + SO_4^{2-}(aq) + Mg(s) \rightarrow Mg^{2+}(aq) + SO_4^{2-}(aq) + H_2(g)$
Net ionic: $2H^+(aq) + Mg(s) \rightarrow Mg^{2+}(aq) + H_2(g)$

9.54 $2HCl(aq) + MgO(s) \rightarrow MgCl_2(aq) + H_2O(\ell)$
$2HCl(aq) + Mg(OH)_2(s) \rightarrow MgCl_2(aq) + 2H_2O(\ell)$
$2HCl(aq) + MgCO_3(s) \rightarrow MgCl_2(aq) + H_2O(\ell) + CO_2(g)$
$2HCl(aq) + Mg(HCO_3)_2(aq) \rightarrow MgCl_2(aq) + 2H_2O(\ell) + 2CO_2(g)$

$2HCl(aq) + Mg(s) \rightarrow MgCl_2(aq) + H_2(g)$

9.56 a. Molecular: $Sn(s) + H_2SO_3(aq) \rightarrow SnSO_3(aq) + H_2(g)$
Total ionic: $Sn(s) + 2H^+(aq) + SO_3^{2-}(aq) \rightarrow Sn^{2+}(aq) + SO_3^{2-}(aq) + H_2(g)$
Net ionic: $Sn(s) + 2H^+(aq) \rightarrow Sn^{2+}(aq) + H_2(g)$
b. Molecular: $3Mg(s) + 2H_3PO_4(aq) \rightarrow Mg_3(PO_4)_2(s) + 3H_2(g)$
Total ionic: $3Mg(s) + 6H^+(aq) + 2PO_4^{3-}(aq) \rightarrow Mg_3(PO_4)_2(s) + 3H_2(g)$
Net ionic: $3Mg(s) + 6H^+(aq) + 2PO_4^{3-}(aq) \rightarrow Mg_3(PO_4)_2(s) + 3H_2(g)$
c. Molecular: $Ca(s) + HBr(aq) \rightarrow CaBr_2(aq) + H_2(g)$
Total ionic: $Ca(s) + 2H^+(aq) + 2Br^-(aq) \rightarrow Ca^{2+}(aq) + 2Br^-(aq) + H_2(g)$
Net ionic: $Ca(s) + 2H^+(aq) \rightarrow Ca^{2+}(aq) + H_2(g)$

9.58 a. Molecular: $RbOH(aq) + HBr(aq) \rightarrow RbBr(aq) + H_2O(\ell)$
Total ionic: $Rb^+(aq) + OH^-(aq) + H^+(aq) + Br^-(aq) \rightarrow Rb^+(aq) + Br^-(aq) + H_2O(\ell)$
Net ionic: $OH^-(aq) + H^+(aq) \rightarrow H_2O(\ell)$
b. Molecular: $2RbOH(aq) + H_2SO_3(aq) \rightarrow Rb_2SO_3(aq) + 2H_2O(\ell)$
Total ionic:
$2Rb^+(aq) + 2OH^-(aq) + 2H^+(aq) + SO_3^{2-}(aq) \rightarrow 2Rb^+(aq) + SO_3^{2-}(aq) + 2H_2O(\ell)$
Net ionic: $2OH^-(aq) + 2H^+(aq) \rightarrow 2H_2O(\ell)$
Simplified net ionic: $OH^-(aq) + H^+(aq) \rightarrow H_2O(\ell)$
c. Molecular:
$2RbOH(aq) + H_3PO_3(aq) \rightarrow Rb_2HPO_3(aq) + 2H_2O(\ell)$
Total ionic:
$2Rb^+(aq) + 2OH^-(aq) + 2H^+(aq) + HPO_3^{2-}(aq) \rightarrow 2Rb^+(aq) + HPO_3^{2-}(aq) + 2H_2O(\ell)$
Net ionic: $2OH^-(aq) + 2H^+(aq) \rightarrow 2H_2O(\ell)$
Simplified net ionic: $OH^-(aq) + H^+(aq) \rightarrow H_2O(\ell)$

9.60 a. Molecular: $2KOH(aq) + H_3PO_4(aq) \rightarrow K_2HPO_4(aq) + 2H_2O(\ell)$
Total ionic:
$2K^+(aq) + 2OH^-(aq) + 2H^+(aq) + HPO_4^{2-}(aq) \rightarrow 2K^+(aq) + HPO_4^{2-}(aq) + 2H_2O(\ell)$
Net ionic: $2OH^-(aq) + 2H^+(aq) \rightarrow 2H_2O(\ell)$
Simplified net ionic: $OH^-(aq) + H^+(aq) \rightarrow H_2O(\ell)$
b. Molecular: $3KOH(aq) + H_3PO_4(aq) \rightarrow K_3PO_4(aq) + 3H_2O(\ell)$
Total ionic:
$3K^+(aq) + 3OH^-(aq) + 3H^+(aq) + PO_4^{3-}(aq) \rightarrow 3K^+(aq) + PO_4^{3-}(aq) + 3H_2O(\ell)$
Net ionic: $3OH^- + 3H^+(aq) \rightarrow 3H_2O(\ell)$
Simplified net ionic: $OH^-(aq) + H^+(aq) \rightarrow H_2O(\ell)$
c. Molecular: $KOH(aq) + H_2C_2O_4(aq) \rightarrow KHC_2O_4(aq) + H_2O(\ell)$
Total ionic:

$K^+(aq) + OH^-(aq) + H^+(aq) + HC_2O_4^-(aq) \rightarrow$
$\qquad K^+(aq) + HC_2O_4^-(aq) + H_2O(\ell)$
Net ionic: $OH^-(aq) + H^+(aq) \rightarrow H_2O(\ell)$

9.62 a. Cation: Cu^{2+}; anion: Cl^-
  b. Cation: NH_4^+; anion: SO_4^{2-}
  c. Cation: Li^+; anion: PO_4^{3-}
  d. Cation: Mg^{2+}; anion: CO_3^{2-}
  e. Cation: Ca^{2+}; anion: $C_2H_3O_2^-$
  f. Cation: K^+; anion: NO_3^-

9.64 a. Acid: HCl; base: $Cu(OH)_2$
  b. Acid: H_2SO_4; base: NH_3 or NH_4OH
  c. Acid: H_3PO_4; base: LiOH
  d. Acid: H_2CO_3; base: $Mg(OH)_2$
  e. Acid: $HC_2H_3O_2$; base: $Ca(OH)_2$
  f. Acid: HNO_3; base: KOH

9.66 a. 126 g; the product would be $MgSO_4$
  b. 180 g; the product would be $Na_2B_4O_7$

9.68 a. Acid: HNO_3; solid: $MgCO_3$
  b. Acid: HCl; solid: CaO
  c. Acid: H_2SO_4; solid $RbHCO_3$

9.70 a. $2HNO_3(aq) + MgCO_3(s) \rightarrow$
  $\qquad Mg(NO_3)_2(aq) + H_2O(\ell) + CO_2(g)$
  b. $2HCl(aq) + CaO(s) \rightarrow CaCl_2(aq) + H_2O(\ell)$
  c. $H_2SO_4(aq) + 2RbHCO_3(s) \rightarrow Rb_2SO_4(aq) + 2H_2O(\ell)$
  $\qquad + 2CO_2(g)$

9.72 a. 1/2 mol
  b. 1/2 mol
  c. 1/3 mol

9.74 a. 0.44 eq; 4.4×10^2 meq
  b. 0.45 eq; 4.5×10^2 meq
  c. 6.24×10^{-2} eq; 62.4 meq

9.76 a. 3.50×10^{-2} eq; 35.0 meq
  b. 4.00×10^{-2} eq; 40.0 meq
  c. 1.35×10^{-1} eq; 135 meq
  d. 4.17×10^{-2} eq; 41.7 meq

9.78 9.98×10^{-3} mol, 1.74 g

9.80 B, A, C, D

9.82 a. Acid order is B, A, C, D.
  b. Conjugate base order is D, C, A, B.

9.84 a. $HNO_2(aq) \rightleftarrows H^+(aq) + NO_2^-(aq)$

$$K_a = \frac{[H^+][NO_2^-]}{[HNO_2]}$$

  b. $HCO_3^-(aq) \rightleftarrows H^+(aq) + CO_3^{2-}(aq)$

$$K_a = \frac{[H^+][CO_3^{2-}]}{[HCO_3^-]}$$

  c. $H_2PO_4^-(aq) \rightleftarrows H^+(aq) + HPO_4^{2-}(aq)$

$$K_a = \frac{[H^+][HPO_4^{2-}]}{[H_2PO_4^-]}$$

  d. $HS^-(aq) \rightleftarrows H^+(aq) + S^{2-}(aq)$

$$K_a = \frac{[H^+][S^{2-}]}{[HS^-]}$$

  e. $HBrO(aq) \rightleftarrows H^+(aq) + BrO^-(aq)$

$$K_a = \frac{[H^+][BrO^-]}{[HBrO]}$$

9.86 According to the definitions given in the chapter, the person should be given the 20% acetic acid solution. If the other solution were the one desired, the term *dilute* should have been used rather than *weak*.

9.88 Titrations are done in order to analyze acids or bases. The amount of acid or base contained in a titrated sample can be determined from the volume of titrant added and the concentration of the titrant.

9.90 a. The endpoint and equivalence point will be the same if the chosen indicator changes color at the same pH as the pH of the salt solution produced by the reaction of the acid and base involved in the titration.
  b. The endpoint and equivalence point will not be the same if the chosen indicator changes color at a pH that is different from the pH of the salt solution produced by the reaction of the acid and base involved in the titration.

9.92 a. 0.100 mol
  b. 0.225 mol

9.94 a. $2NaOH(aq) + H_2CrO_4(aq) \rightarrow Na_2CrO_4(aq) + 2H_2O(\ell)$
  b. $NaOH(aq) + HClO_3(aq) \rightarrow NaClO_3(aq) + H_2O(\ell)$
  c. $3NaOH(aq) + H_3AsO_3 \rightarrow Na_3AsO_3(aq) + 3H_2O(\ell)$

9.96 a. $2HCl(aq) + Sr(OH)_2(aq) \rightarrow SrCl_2(aq) + 2H_2O(\ell)$
  b. $3HCl(aq) + Ni(OH)_3(aq) \rightarrow NiCl_3(aq) + 3H_2O(\ell)$
  c. $2HCl(aq) + Zn(OH)_2(aq) \rightarrow ZnCl_2(aq) + 2H_2O(\ell)$

9.98 0.1660 M

9.100 a. 33.3 mL
  b. 41.7 mL
  c. 100 mL
  d. 120 mL
  e. 85.0 mL
  f. 83.3 mL

9.102 a. 3.50 M
  b. 0.891 M
  c. 12.4 M

9.104 122.0 g/mol

9.106 The ions produced when Na_2CO_3 dissolves in water are Na^+ and CO_3^{2-}. The Na^+ ion is the conjugate acid of the strong base NaOH and therefore is a very weak Brønsted acid. The CO_3^{2-} ion is the conjugate base of the weak acid HCO_3^- and therefore will act as a base in a hydrolysis reaction with water and will contribute to the pH of the solution. The reaction is

$$CO_3^{2-}(aq) + H_2O(\ell) \rightarrow HCO_3^-(aq) + OH^-(aq)$$

Thus, we see that the hydrolysis reaction contributes OH^- ions to the solution, causing it to have a pH greater than 7.

9.108 a. pH greater than 7. The ions produced when the salt dissolves in water are Na^+ and OCl^-. The Na^+ is the conjugate acid of a strong base (NaOH). It will be a weak acid and will not undergo a significant hydrolysis reaction. The OCl^- ion is the conjugate base of a weak acid (HOCl) and will therefore undergo a significant hydrolysis reaction that will contribute to the pH of the solution. The reaction is

$$OCl^-(aq) + H_2O(\ell) \rightarrow HOCl(aq) + OH^-$$

The OH^- ions produced by this reaction will cause the pH of the solution to be greater than 7.

b. pH greater than 7. The explanation is the same as in part a. The hydrolysis reaction of the CHO_2^- ion is

$$CHO_2^-(aq) + H_2O(\ell) \rightarrow HCHO_2(aq) + OH^-(aq)$$

c. pH equal to 7. The ions produced when the salt dissolves in water are K^+ and NO_3^-. The K^+ is the conjugate acid of the strong base KOH, and the NO_3^- is the conjugate base of the strong acid HNO_3. Since both ions come from strong sources, they are both weak in their respective behaviors as conjugate acid or base. As a result, neither of them will undergo a significant hydrolysis reaction that will influence the solution pH.

d. pH greater than 7. The explanation is the same as in part a. The hydrolysis reaction of the PO_4^{3-} ion is

$$PO_4^{3-}(aq) + H_2O(\ell) \rightarrow HPO_4^{2-}(aq) + OH^-(aq)$$

9.110 Different salts will hydrolyze to different extents in water solution and produce solutions of different pH. In titrations, the pH at the equivalence point is the pH characteristic of a solution of the salt produced by the reaction of the acid and base involved in the titration. Different acids and bases will produce different salts and possibly different solution pH values at the equivalence point. Thus, an indicator must be chosen that will change color at a pH as near as possible to the pH of the salt solution resulting from the titration.

9.112 The buffering anions present in the solution of the two salts are HPO_4^{2-} and $H_2PO_4^-$. When acid is added to the solution, the buffering reaction (like Equation 9.48) is

$$HPO_4^{2-}(aq) + H^+(aq) \rightleftharpoons H_2PO_4^-(aq)$$

When base is added, the buffering reaction (like Equation 9.49) is

$$H_2PO_4^-(aq) + OH^-(aq) \rightleftharpoons HPO_4^{2-}(aq) + H_2O(\ell)$$

9.114 $HCO_3^-(aq) + H^+(aq) \rightleftharpoons H_2CO_3(aq)$

9.116 a. pH = 3.85
   b. pH = 3.85
   c. The difference is that the buffer of part b would have a greater buffer capacity; that is, it could buffer against larger added amounts of acid or base.

9.118 a. pH = 4.54
   b. pH = 7.81
   c. pH = 6.12

9.120 $[HPO_4^{2-}]/[H_2PO_4^-] = 2.75$

.... **CHAPTER 10** ....

10.2 a. Beta, gamma, and positron
   b. Alpha and positron
   c. Gamma and neutron

10.4 a. A beta particle is identical with an electron.
   b. An alpha particle is made up of two protons and two neutrons.
   c. A positron is a positively charged electron.

10.6 a. The mass number is reduced by 4, and the atomic number is reduced by 2.
   b. The mass number is unchanged, and the atomic number is increased by 1.

c. The mass number is unchanged, and the atomic number is reduced by 1.
   d. The mass number and atomic number are both unchanged.
   e. The mass number is unchanged, and the atomic number is reduced by 1.

10.8 a. $^{54}_{26}Fe$
   b. $^{110}_{48}Cd$
   c. $^{48}_{22}Ti$

10.10 The quantity added to complete each equation appears in blue.
   a. $^{234}_{90}Th \rightarrow {}^{0}_{-1}\beta + {}^{234}_{91}Pa$
   b. $^{15}_{8}O \rightarrow {}^{0}_{1}\beta + {}^{15}_{7}N$
   c. $^{248}_{100}Fm \rightarrow {}^{4}_{2}\alpha + {}^{244}_{98}Cf$
   d. $^{226}_{88}Ra \rightarrow {}^{222}_{86}Rn + {}^{4}_{2}\alpha$
   e. $^{113}_{47}Ag \rightarrow {}^{113}_{48}Cd + {}^{00}_{-1}\beta$
   f. $^{41}_{18}Ar + {}^{0}_{-1}e \rightarrow {}^{41}_{17}Cl$

10.12 a. $^{121}_{50}Sn \rightarrow {}^{0}_{-1}\beta + {}^{121}_{51}Sb$
   b. $^{190}_{78}Pt \rightarrow {}^{4}_{2}\alpha + {}^{186}_{76}Os$
   c. $^{55}_{26}Fe + {}^{0}_{-1}e \rightarrow {}^{55}_{25}Mn$
   d. $^{72}_{31}Ga \rightarrow {}^{0}_{-1}\beta + {}^{72}_{32}Ge$
   e. $^{10}_{4}Be \rightarrow {}^{0}_{-1}\beta + {}^{10}_{5}B$
   f. $^{238}_{92}U \rightarrow {}^{4}_{2}\alpha + {}^{234}_{90}Th$

10.14 The half-life of a dozen cookies would be the time it takes for the number to be reduced to 6. The half-life of a $500 checking account would be the time it takes to reduce the amount in the account to $250.

10.16 0.56 ng remains

10.18 1.68×10^4 years old

10.20 2.24×10^4 years old

10.22 Long-term exposure to low-level radiation is more likely to cause changes in genetic material in cells. Such changes might result in genetic mutations, cancer, or other serious consequences. Short-term exposure to intense radiation destroys tissue rapidly and causes radiation sickness, which can lead to death, depending on the dose of radiation received.

10.24 A physical unit of radiation, such as a curie, indicates the activity of a source of radiation in terms such as the number of nuclear decays that occur per minute. A biological unit of radiation, such as a rem, indicates the damage caused by radiation in living tissue.

10.26 3.2 R of X rays

10.28 1.7×10^5 disintegrations/s

10.30 Hot spots are concentrations of radioisotope in tissue. Cold spots are areas of tissue that reject or keep out radioactive tracers.

10.32 $^{51}_{24}Cr + {}^{0}_{-1}e \rightarrow {}^{51}_{23}V$. The daughter is seen to be vanadium-51.

10.34 Use a radioactive isotope of oxygen to make water by burning hydrogen in it. Mix the radioactive water with hydrogen peroxide. Add the catalyst and collect the oxygen gas given off. If the gas is radioactive, it came from the water; if it is not radioactive, it came from the hydrogen peroxide.

10.36 Use a Geiger–Müller tube to measure the radioactivity of a gallon of water that contains the radioisotope. Pour that water into the pool and allow time for it to become completely and uniformly mixed with the water in the pool. Then, remove a gallon of the pool water and measure its radioactivity. For example, if the pool water has a radioactivity one ten-thousandth that of the original tracer-containing water, it means the original gallon of water has been diluted by a factor of 10,000, and therefore the pool contains 10,000 gal of water. Thus, the pool's volume is equal to the factor by which the radioactivity has been reduced.

10.38 $^{24}_{12}Mg + ^{4}_{2}\alpha \rightarrow ^{1}_{0}n + ^{27}_{14}Si$

10.40 $^{238}_{92}U + ^{1}_{0}n \rightarrow ^{239}_{94}Pu + 2\,^{0}_{-1}\beta$

10.42 A moderator slows neutrons so that their chances of being captured and interacting with a target nucleus are greater.

10.44 $^{66}_{30}Zn + ^{1}_{1}p \rightarrow ^{67}_{31}Ga$

10.46 $2\,^{1}_{1}H \rightarrow ^{2}_{1}H + ^{0}_{1}\beta$

10.48 The critical mass of fissionable material is the minimum amount needed to sustain a chain reaction. Without a critical mass, the fission chain reaction will not continue, but will slow down and stop.

10.50 $^{238}_{94}Pu \rightarrow ^{4}_{2}\alpha + ^{234}_{92}U$

············· **CHAPTER 11** ·············

11.2 Hair, skin, clothing, paper, plastic, carpet

11.4 They all contain carbon.

11.6 Covalent

11.8 **a.** Organic; most organic compounds are flammable.
b. Inorganic; a high-melting solid is typical of ionic compounds.
c. Organic; many organic compounds are insoluble in water.
d. Organic; most organic compounds can undergo combustion.
e. Organic; a low-melting solid is characteristic of covalent materials.

11.10 Organic compounds do not readily accept electrons to transport them, thus making the flow of electrical current impossible.

11.12 The ability of carbon to bond to itself repeatedly; isomerism.

11.14 The s orbital of the hydrogen overlaps with the sp^3 orbital of carbon.

11.16 A p orbital and an sp^3 orbital both have a two-lobed shape. A p orbital has equal-sized lobes, whereas an sp^3 orbital has different-sized lobes.

11.18

11.20 b, d, e

11.22 **a.** Incorrect; H cannot have two bonds.
b. Correct
c. Incorrect; the carbon with the C=O bond has five bonds.
d. Incorrect; both carbons of the C=C double bond have three bonds.
e. Correct

11.24 **a.** $CH_3CH_2CH_2-O-CH_2CH_2CH_3$
b.

11.26

11.28 **a.** Branched
b. Normal
c. Normal
d. Normal
e. Branched
f. Branched

11.30 **a.** Same
b. Same
c. Isomers
d. Same

11.32 **a.** Six
b. Six
c. Eight

11.34 **a.** 3-methylpentane
b. 2-methylpropane
c. 3,6-diethyl-4-methylnonane
d. 4,7-diethyl-5-methyldecane
e. 3-ethyl-5-methylnonane

11.36

a. 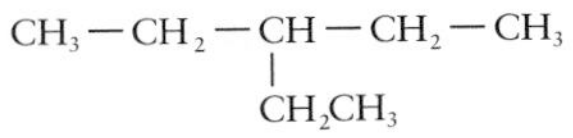 CH$_3$—CH$_2$—CH—CH$_2$—CH$_3$
　　　　　　　|
　　　　　　CH$_2$CH$_3$

b. 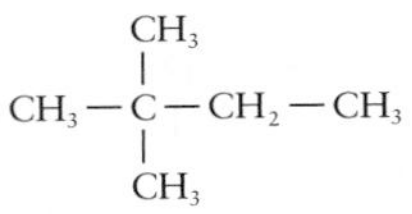 CH$_3$—C—CH$_2$—CH$_3$ (with CH$_3$ above central C and CH$_3$ below)

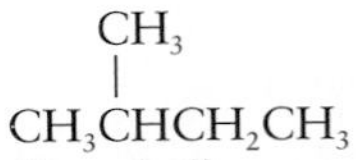

　　　CH$_3$
　　　|
b. CH$_3$—C—CH$_2$—CH$_3$
　　　|
　　　CH$_3$

c. CH$_3$—CH$_2$—C—CH—CH—CH$_2$—CH$_2$—CH$_2$—CH$_2$—CH$_3$
　　　　　　　　|　　|
　　　　　　CH$_3$　CH$_2$CH$_3$
(with CH$_3$ and CH$_2$CH$_2$CH$_3$ above)

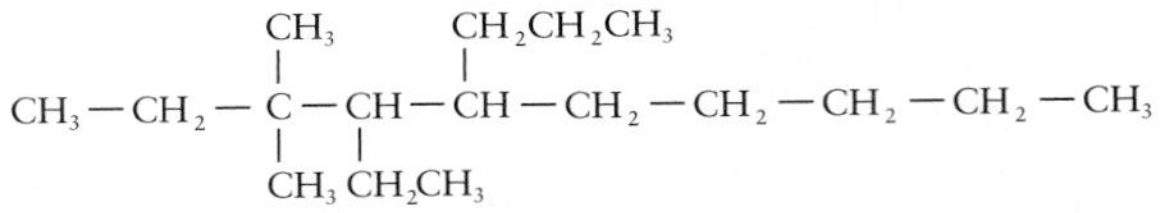

d. 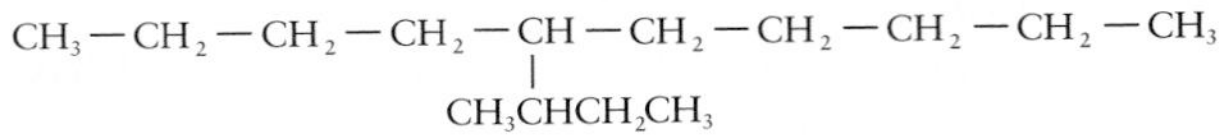 CH$_3$—CH$_2$—CH$_2$—CH$_2$—CH—CH$_2$—CH$_2$—CH$_2$—CH$_2$—CH$_3$
　　　　　　　　　　　　　|
　　　　　　　　　CH$_3$CHCH$_2$CH$_3$

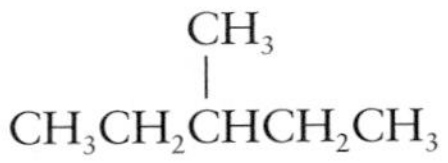

11.38

CH$_3$CH$_2$CH$_2$CH$_2$CH$_3$　　　CH$_3$CHCH$_2$CH$_3$ (with CH$_3$ above)
　　　pentane　　　　　　　　2-methylbutane

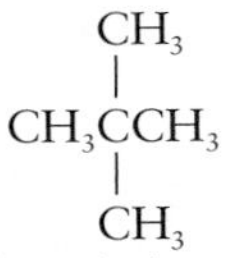

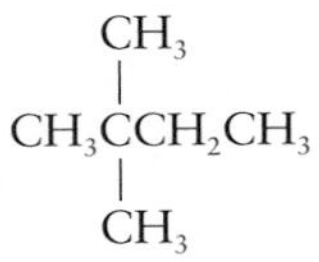

　　　　　　　　CH$_3$
　　　　　　　　|
　　　　　CH$_3$CCH$_3$
　　　　　　　　|
　　　　　　　　CH$_3$
　　2,2-dimethylpropane

11.40

CH$_3$CH$_2$CH$_2$CH$_2$CH$_2$CH$_3$　　　　　　　hexane

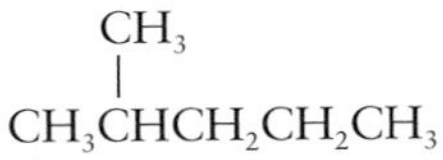

CH$_3$CHCH$_2$CH$_2$CH$_3$ (CH$_3$ above first CH)　　　2-methylpentane

CH$_3$CH$_2$CHCH$_2$CH$_3$ (CH$_3$ above middle CH)　　　3-methylpentane

CH$_3$CCH$_2$CH$_3$ (CH$_3$ above and CH$_3$ below central C)　　2,2-dimethylbutane

CH$_3$CHCHCH$_3$ (CH$_3$ above and CH$_3$ below)　　　2,3-dimethylbutane

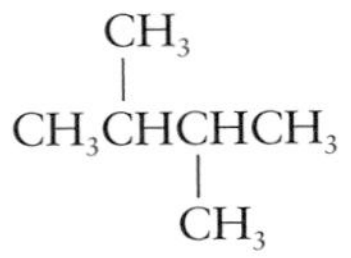

11.42 a. 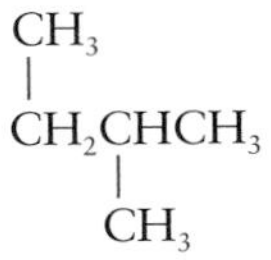 CH$_2$CHCH$_3$ (CH$_3$ above CH)

The longest carbon chain is four carbon atoms rather than three. The correct name is 2-methylbutane.

　　　　　　CH$_3$
　　　　　　|
b. 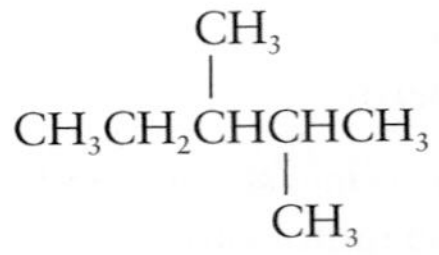 CH$_3$CH$_2$CHCHCH$_3$
　　　　　　　　|
　　　　　　　CH$_3$

The chain should be numbered from the right to give 2,3-dimethylpentane.

　　　　　　CH$_2$CH$_3$
　　　　　　|
c. 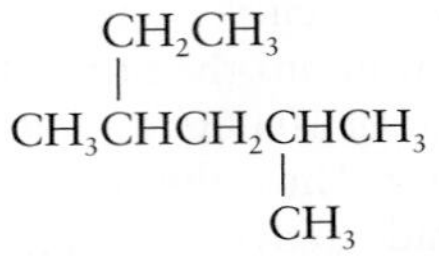 CH$_3$CHCH$_2$CHCH$_3$
　　　　　　　　|
　　　　　　　CH$_3$

The longest carbon chain is six carbon atoms and should be numbered from the right to give 2,4-dimethylhexane.

　　　　　Br
　　　　　|
d. 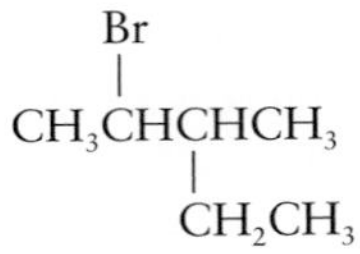 CH$_3$CHCHCH$_3$
　　　　　　|
　　　　　CH$_2$CH$_3$

The longest carbon chain is five carbon atoms rather than four. The correct name is 2-bromo-3-methylpentane.

11.44 a. cyclopentane
b. 1,2-dimethylcyclobutane
c. 1,1-dimethylcyclohexane
d. 1,2,3-trimethylcyclobutane

11.46

a.

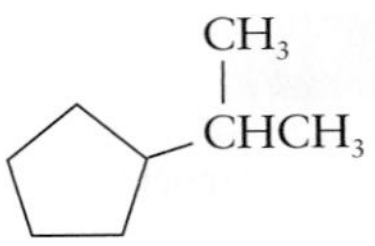

b.

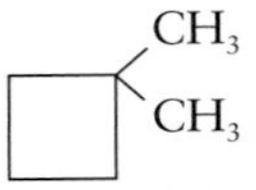

c.

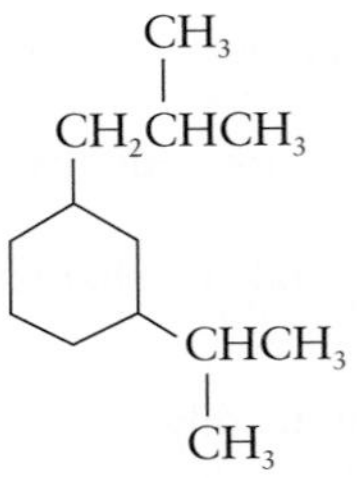

11.48 b, c, and d

11.50 So that the carbon atoms with attached hydrogens can assume a tetrahedral shape

11.52

a. none

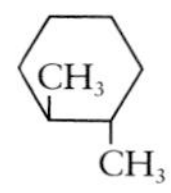

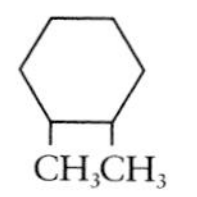

b. *trans*-1,2-dimethylcyclohexane   *cis*-1,2-dimethylcyclohexane

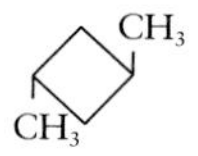

c. *trans*-1,3-dimethylcyclobutane   *cis*-1,3-dimethylcyclobutane

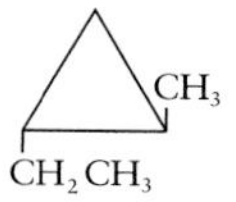

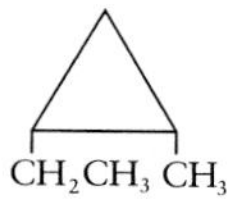

d. *trans*-1-ethyl-2-methylcyclopropane   *cis*-1-ethyl-2-methylcylopropane

11.54 a. *trans*-1-ethyl-2-methylcyclopropane
b. *cis*-1-bromo-2-chlorocyclopentane
c. *trans*-1-methyl-2-propylcyclobutane
d. *trans*-1,3-dimethylcyclohexane

11.56 a. Liquid
b. No
c. Yes
d. Less dense

11.58 2-methylheptane

11.60 a. $C_3H_8 + 5O_2 \longrightarrow 3CO_2 + 4H_2O$
b. $2C_4H_{10} + 13O_2 \longrightarrow 8CO_2 + 10H_2O$
c. $2C_5H_{10} + 15O_2 \longrightarrow 10CO_2 + 10H_2O$

11.62 $2C_6H_{14} + 13O_2 \longrightarrow 12CO + 14H_2O$

11.64 Organically grown fruits and vegetables are allowed to ripen longer on the tree or vine.

11.66 Mineral oil and petroleum jelly

11.68 Petroleum provides chemicals for synthetic products.

11.70 Oxygen

················· **CHAPTER 12** ·················

12.2 An alkene contains one or more C=C bonds.
An alkyne contains one or more C≡C bonds.
An aromatic hydrocarbon contains the characteristic benzene ring or similar feature.

12.4 a. 2-butene
b. 3-ethyl-2-pentene
c. 4,4-dimethyl-2-hexyne
d. 4-methylcyclopentene
e. 6-bromo-2-methyl-3-heptyne
f. 1-ethyl-2,3-dimethylcyclopropene
g. 6-methyl-1,4-heptadiene

12.6

a. 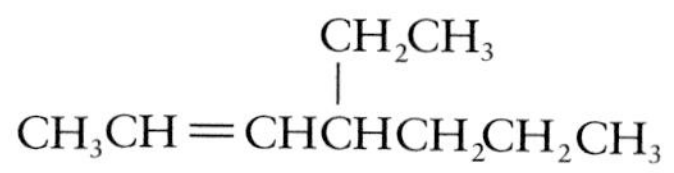 $CH_3CH{=}CHCHCH_2CH_2CH_3$ (with CH_2CH_3 substituent)

b. 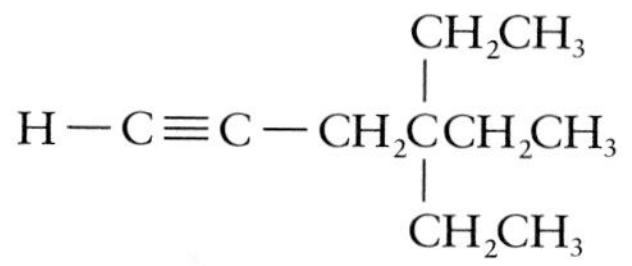 $H{-}C{\equiv}C{-}CH_2CCH_2CH_3$ (with CH_2CH_3 substituents above and below)

c. $H_2C{=}CH{-}CH{-}CH{=}CH_2$ (with CH_2CH_3 substituent)

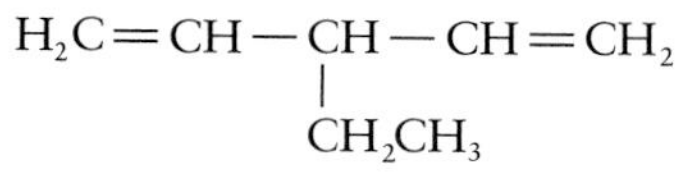

d.

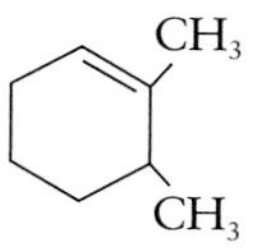

e.

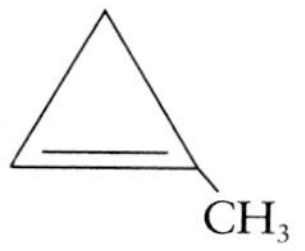

12.8 Several possibilities exist including these:

$CH{\equiv}CCH_2CH_2CH_3$ 1-pentyne

$CH{=}CHCH{=}CHCH_3$ 1,3-pentadiene

cyclopentene

12.10 3,7,11-trimethyl-1,3,6,10-dodecatetraene

12.12 a. 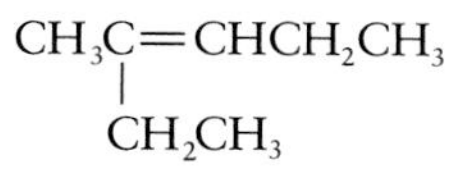 $CH_3C{=}CHCH_2CH_3$ (with CH_2CH_3 substituent)

3-methyl-3-hexene

The longest chain contains six carbon atoms.

b. $CH_3CH_2CH{=}CHCH_2CH{=}CHCH_3$

2,5-octadiene

The chain is numbered from the right.

c.

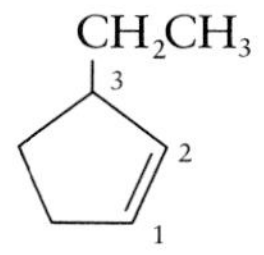

3-ethylcyclopentene

The ring should be numbered as shown.

12.14 The overlap of two *p* orbitals forms a π (pi) bond containing two electrons.

12.16 Structural isomers have a different order of linkage of atoms. Geometric isomers have the same order of linkage of atoms but different three-dimensional arrangements of their atoms in space.

12.18 a. a and c cannot have cis- and trans- isomers.

b.

$$H_2C=CH_2$$ cis-3-hexene

with CH_3CH_2 and CH_2CH_3 groups (cis-3-hexene)

trans-3-hexene

12.20 a. C=C with H, H, CH_3, CH_2CH_3

b. C=C with H, $CH_2CH_2CH_3$, CH_3, H

12.22 Both alkanes and alkenes are nonpolar compounds that are insoluble in water, less dense than water, and soluble in non-polar solvents.

12.24 When H—X adds to an alkene, the hydrogen becomes attached to the carbon atom that is already bonded to more hydrogens.

$$CH_2=CH-CH_3 + HCl \longrightarrow CH_3CHClCH_3$$

12.26

a. cyclohexyl—CH—CHCH$_3$ with Br, Br

b. cyclohexyl—OH

c. $CH_3CHClCH_2CH_3$ with CH$_3$

d. cyclopentane with CH$_3$ and Cl

12.28 a. Cl_2
b. H_2 with Pt catalyst
c. HCl
d. H_2O with H_2SO_4 catalyst

12.30 The addition of Br_2 to these samples will cause the cyclo-hexane solution to become light orange with unreacted Br_2, whereas the 2-hexene will react with the Br_2 and remain colorless.

12.32 A monomer is a starting material used in the preparation of polymers, long-chain molecules made up of many repeating units. Addition polymers are long-chain molecules prepared from alkene monomers through numerous addition reactions. A copolymer is prepared from two monomer starting materials.

12.34 A carbon–carbon double bond

12.36 $nCH_2=CH \xrightarrow{\text{catalyst}} (CH_2-CH)_n$ with Cl substituents

12.38 sp; two

12.40 Linear

12.42 Acetylene is used as a fuel for welding torches and in the synthesis of monomers for plastics and fibers.

12.44 $H-C\equiv C-CH_2CH_2CH_3$
1-pentyne

$CH_3-C\equiv C-CH_2CH_3$
2-pentyne

$CH_3CH-C\equiv C-H$ with CH_3
3-methyl-1-butyne

12.46 Unhydridized p orbitals

12.48 Aromatic refers to compounds containing the benzene ring. Aliphatic substances do not contain a benzene ring.

12.50 In cyclopentane, attached groups are located above and below the plane of the ring. In benzene attached groups are located in the same plane as the ring.

12.52 a. 1,3,5-trimethylbenzene
b. 1,4-diethylbenzene

12.54 a. 3-phenyl-1-pentene
b. 2,5-diphenylhexane

12.56 a. p-ethylphenol
b. o-chlorophenol

12.58 a. 2-ethyl-3-propyltoulene
b. 3-bromo-5-chlorobenzoic acid

12.60

a. benzene ring with OH and CH_2CH_3

b. benzene ring with $HO-C=O$ and Cl

c. benzene ring with $CH_3CH_2CCH_2CH_3$ and CH_3

12.62 Nonpolar and insoluble in water

12.64 Cyclohexene readily undergoes addition reactions. Benzene resists addition reactions and favors substitution reactions. Both undergo combustion.

12.66 a. phenol
b. styrene
c. aniline
d. phenol

12.68 Tomatoes, guava, pink grapefruit

12.70 Baked meat does not contain the cancer-promoting compounds that grilled meat may contain.

12.72 nicotine

APPENDIX C Solutions to Learning Checks

CHAPTER 1

1.1 A change is chemical if new substances are formed and physical if no new substances are formed.

a. Chemical: The changes in taste and odor indicate that new substances form.

b. Physical: The handkerchief doesn't change, and the evaporated water is still water.

c. Chemical: The changes in color and taste indicate that new substances form.

d. Chemical: The gases and smoke released indicate that new substances form.

e. Physical: The air is still air, as indicated by appearance, odor, and so on.

f. Physical: On being condensed, the steam forms water, which is the same substance present before the change.

1.2 a. Triatomic and heteroatomic: The three atoms of two kinds (oxygen and hydrogen) make it both triatomic (three atoms) and heteroatomic (two or more kinds of atoms).

b. Triatomic and homoatomic: The three atoms make it triatomic, but all three atoms are the same (oxygen).

c. Polyatomic and heteroatomic: The carbon and hydrogen atoms in the product had to come from the cellulose because it was reacted with pure oxygen. Thus, cellulose molecules must contain more than one atom (polyatomic), and they are different (heteroatomic).

1.3 Molecules of elements can be diatomic, triatomic, or polyatomic, but all atoms will be identical and will be represented by the same color. Molecules of compounds must be diatomic, triatomic, or polyatomic (more than one atom), and the atoms will not all be identical.

a. Compound: polyatomic and heteroatomic
b. Element: diatomic and homoatomic
c. Element: polyatomic and homoatomic
d. Compound: polyatomic and heteroatomic

1.4 The product would have to be a compound because it would be polyatomic (more than one atom because two substances were used) and heteroatomic (different atoms from the two different substances).

1.5 Substitute the given radius and π values into the equation

$$A = \pi r^2 = (3.14)(3.5 \text{ cm})^2 = (3.14)(12.25 \text{ cm}^2) = 38 \text{ cm}^2$$

1.6 The volume of a rectangular object is equal to the product of the three sides.

$$V = (30.0 \text{ cm})(20.0 \text{ cm})(15.0 \text{ cm})$$
$$= 9000 \text{ cm}^3, \text{ or } (9.00 \times 10^3 \text{ cm}^3)$$

$1 \text{ cm}^3 = 1 \text{ mL}$, so the volume is 9.00×10^3 mL. A liter is equal to 1000 mL, so 9000 mL = 9 L. The correct answers are 9.00×10^3 mL and 9.00 L.

1.7 Because 1 kg = 1000 g,

$$0.819 \text{ kg} \times \frac{1000 \text{ g}}{1 \text{ kg}} = 819 \text{ g}$$

1.8 Use Equation 1.4 and the fact that 77°F = 25°C, as shown in Example 1.7.

$$K = °C + 273 = 25°C + 273 = 298 \text{ K}$$

1.9 a. Scientific notation is used. In nonscientific notation, the number is written 588.

b. Scientific notation is not used. In scientific notation, the number is written 4.39×10^{-4}.

c. Scientific notation is used. In nonscientific notation, the number is written 0.0003915.

d. Scientific notation is not used. In scientific notation, the number is written 9.870×10^3.

e. Scientific notation is not used. In scientific notation, the number is written 3.677×10^1.

f. Scientific notation is not used. In scientific notation, the number is written 1.02×10^{-1}.

1.10 The two important things to note are the placement of the decimal in the standard position and the correct value of the exponent.

a. Written incorrectly: The decimal is not in standard position. Move the decimal 1 place to the left and increase the exponent by 1 to account for moving the decimal: 6.25×10^5.

b. Written incorrectly: The decimal is not in the standard position, and no exponent is used. Move the decimal 3 places to the right and use −3 for the exponent to account for moving the decimal: 9.8×10^{-3}.

c. Written incorrectly: The decimal is not in the standard position. Move the decimal 3 places to the right and reduce the exponent by 3 to account for moving the decimal: 4.1×10^{-6}.

d. Written correctly.

1.11 Perform multiplications and divisions separately for the nonexponential and exponential terms, then combine the results into the final answer.

a. $(2.4 \times 10^3)(1.5 \times 10^4) = (2.4 \times 1.5)(10^3 \times 10^4) = (3.6)(10^{3+4}) = 3.6 \times 10^7$

b. $(3.5 \times 10^2)(2.0 \times 10^{-3}) = (3.5 \times 2.0)(10^2 \times 10^{-3}) = (7.0)(10^{2+(-3)}) = 7.0 \times 10^{-1}$

c. $\dfrac{6.3 \times 10^5}{2.1 \times 10^3} = \left(\dfrac{6.3}{2.1}\right)\left(\dfrac{10^5}{10^3}\right) = (3.0)(10^{5-3})$

$$= 3.0 \times 10^2$$

d. $\dfrac{4.4 \times 10^{-2}}{8.8 \times 10^{-3}} = \left(\dfrac{4.4}{8.8}\right)\left(\dfrac{10^{-2}}{10^{-3}}\right) = (.50)(10^{-2-(-3)})$

$= .50 \times 10^1 = 5.0 \times 10^0$, or 5.0

1.12 The primary challenge is to interpret correctly the significance of zeros. Leading zeros to the left of nonzero numbers are not significant; all other zeros, including trailing zeros, are significant.

    a. 3, trailing zero is significant.
    b. 4
    c. 3, leading zeros are not significant.
    d. 2
    e. 1, leading zeros are not significant.
    f. 4, trailing zero is significant.

1.13 a. 3 significant figures; original decimal position is 2 places to the right of standard position, so exponent is $+2$. 1.01×10^2 m.
    b. 4 significant figures; original decimal position is 3 places to the right of standard position, so exponent is $+3$. 1.200×10^3 g.
    c. 3 significant figures; original decimal position is 3 places to the left of standard position, so exponent is -3. 2.30×10^{-3} kg.
    d. 4 significant figures; original decimal position is 3 places to the right of standard position, so exponent is $+3$. 1.296×10^{3}°C.
    e. 4 significant figures; original decimal position is 1 place to the right of standard position, so exponent is $+1$. 2.165×10^1 mL.
    f. 2 significant figures; original decimal position is 2 places to the left of standard position, so exponent is -2. 1.5×10^{-2} km.

1.14 a. Answer will be rounded to 2 significant figures to match 0.0019.

$$(0.0019)(21.39) = 0.04064 = 4.1 \times 10^{-2}$$

    b. Answer will be rounded to 2 significant figures to match 4.1.

$$\frac{8.321}{4.1} = 2.0295 = 2.0$$

    c. Answer will be rounded to 3 significant figures to match 0.0911 and 3.22.

$$\frac{(0.0911)(3.22)}{(1.379)} = 0.21272 = 0.213, \text{ or } 2.13 \times 10^{-1}$$

1.15 In each case, the sum or difference is rounded to have the same number of places to the right of the decimal as the least number of places in the terms added or subtracted.
    a. Answer will be rounded to have one place to the right of the decimal to match 3.2.

$$8.01 + 3.2 = 11.21 = 11.2$$

    b. Answer will be rounded to have no places to the right of the decimal to match 3000.

$$3000 + 20.3 + 0.009 = 3020.309 = 3020$$

    c. Answer will be rounded to have two places to the right of the decimal to match both 4.33 and 3.12.

$$4.33 - 3.12 = 1.21$$

    d. Answer will be rounded to have two places to the right of the decimal to match 2.42.

$$6.023 - 2.42 = 3.603 = 3.60$$

1.16 Two factors result from the relationship 1 g = 1000 mg. The factors are

$$\frac{1 \text{ g}}{1000 \text{ mg}} \quad \text{and} \quad \frac{1000 \text{ mg}}{1 \text{ g}}$$

The first factor is used because it cancels the mg unit and generates the g unit:

Step 1.   1.1 mg

Step 2.   1.1 mg   =   g

Step 3. $1.1 \; \cancel{\text{mg}} \times \dfrac{1 \text{ g}}{1000 \; \cancel{\text{mg}}}$ =   g

Step 4. $\dfrac{(1.1)(1 \text{ g})}{(1000)} = 0.0011 \text{ g} = 1.1 \times 10^{-3} \text{ g}$

The number of significant figures matches the number in 1.1 because the 1 and 1000 are exact numbers.

1.17 Two factors, 1 km/1000 m and 0.621 mi/1 km, are used to convert meters to miles, and two factors, 60 s/1 min and 60 min/1 h, are used to convert seconds in the denominator to hours.

$$10.0 \, \frac{\cancel{\text{m}}}{\cancel{\text{s}}} \times \frac{1 \, \cancel{\text{km}}}{1000 \, \cancel{\text{m}}} \times \frac{0.621 \text{ mi}}{1 \, \cancel{\text{km}}} \times \frac{60 \, \cancel{\text{s}}}{1 \, \cancel{\text{min}}}$$

$$\times \frac{60 \, \cancel{\text{min}}}{1 \text{ h}} = 22.356 \text{ mi/h}$$

Rounding to 3 significant figures gives 22.4 mi/h.

1.18 a. $\% = \dfrac{\text{part}}{\text{total}} \times 100; \; \% = \dfrac{\$988}{\$1200} \times 100 = 82.3\%$

    b. $\% = \dfrac{\text{part}}{\text{total}} \times 100$

$$90.4\% = \frac{\text{part}}{83} \times 100$$

$$\frac{(90.4\%)(83)}{100} = \text{part}$$

$$= \text{no. voting to not take final}$$

$$= 75.0$$

Number voting to take exam = 83 − 75 = 8.

1.19 The value of the density gives two factors that may be used to solve problems of this type:

$$\frac{2.7 \text{ g}}{1.0 \text{ cm}^3} \quad \text{and} \quad \frac{1.0 \text{ cm}^3}{2.7 \text{ g}}$$

    a. The sample volume is 60.0 cm³, and we wish to use a factor that will convert this to grams. The first factor given above will work.

$$60.0 \, \cancel{\text{cm}^3} \times \frac{2.7 \text{ g}}{1.0 \, \cancel{\text{cm}^3}} = 162 \text{ g (calculator answer)}$$

$$= 1.6 \times 10^2 \text{ g (properly rounded answer)}$$

    b. The sample mass is 98.5 g, and we wish to convert this to cm³. The second factor given above will work.

$$98.5 \cancel{g} \times \frac{1.0 \text{ cm}^3}{2.7 \cancel{g}} = 36.48 \text{ cm}^3 \text{ (calculator answer)}$$

$$= 3.6 \times 10^1 \text{ cm}^3 \text{ (properly rounded answer)}$$

1.20 a. The sample mass is equal to the difference between the mass of the container with the sample inside and the mass of the empty container:

$$m = 64.93 \text{ g} - 51.22 \text{ g} = 13.71 \text{ g}$$

The density of the sample is equal to the sample mass divided by the sample volume:

$$d = \frac{m}{V} = \frac{13.71 \text{ g}}{10.00 \text{ mL}}$$

$$= 1.371 \text{ g/mL (properly rounded answer)}$$

 b. The identity can be determined by calculating the density of the sample from the data and comparing the density with the known densities of the two possible metals. The volume of the sample is equal to the difference between the cylinder readings with the sample present and with the sample absent:

$$V = 25.2 \text{ mL} - 21.2 \text{ mL} = 4.0 \text{ mL}$$

According to Table 1.3, 1 mL = 1 cm^3, so the sample volume is equal to 4.0 cm^3. The density of the sample is equal to the sample mass divided by the sample volume:

$$d = \frac{m}{V} = \frac{35.66 \text{ g}}{4.0 \text{ cm}^3}$$

$$= 8.9 \text{ g/cm}^3 \text{ (properly rounded answer)}$$

A comparison with the known densities of nickel and chromium allows the metal to be identified as nickel.

·················· **CHAPTER 2** ··················

2.1 The number of atoms in each molecule is represented by a subscript in the formula.

 a. H$_3$PO$_4$ **b.** SO$_3$ **c.** C$_6$H$_{12}$O$_6$

2.2 a. The atomic number, Z, equals the number of protons: $Z = 4$. The mass number, A, equals the sum of the number of protons and the number of neutrons: $A = 4 + 5 = 9$. According to the periodic table, the element with an atomic number of 4 is beryllium, with the symbol Be. The isotope symbol is $_4^9$Be.

 b. According to the periodic table, chlorine has the symbol Cl and an atomic number, Z, of 17. The mass number, 37, is equal to the sum of the number of protons and neutrons: $A = \#p + \#n$. Therefore, the number of neutrons is equal to the mass number, A, minus the number of protons: $\#n = A - 17 = 37 - 17 = 20$.

 c. According to the periodic table, the element with the symbol Si is silicon, which has an atomic number of 14. Therefore, the atom contains 14 protons. Since $A = \#p + \#n$, we see that $\#n$, $= A - \#p = 28 - 14 = 14$.

2.3 a. Copper atoms have a mass of 63.55 u. Twice this value is 127.1 u. The element with atoms nearest this value is iodine, I, with atoms that have a mass of 126.9 u.

 b. Iron atoms have a mass of 55.85 u. One-fourth this value is 13.96 u, and the element with atoms closest to this value is nitrogen, N, with a value of 14.01 u.

 c. A carbon-12 atom has a mass defined as 12.00 u. Helium atoms have a mass of 4.003 u, so to the nearest whole helium atom, three helium atoms would equal the mass of a single carbon-12 atom.

2.4 Molecular weights are obtained by adding the atomic weights of the atoms in molecules. We have used four significant figures in atomic weights.

 a. H$_2$SO$_4$; molecular weight = 2(at. wt. H) + (at. wt. S) + 4(at. wt. O)

$$MW = 2(1.008 \text{ u}) + 32.06 \text{ u} + 4(16.00 \text{ u})$$
$$= 2.016 \text{ u} + 32.06 \text{ u} + 64.00 \text{ u}$$
$$= 98.08 \text{ u}$$

 b. C$_3$H$_8$O; molecular weight = 3(at. wt. C) + 8(at. wt. H) + (at. wt. O)

$$MW = 3(12.01 \text{ u}) + 8(1.008 \text{ u}) + 16.00 \text{ u}$$
$$= 36.03 \text{ u} + 8.064 \text{ u} + 16.00 \text{ u}$$
$$= 60.09 \text{ u}$$

2.5 a. Because naturally occuring fluorine consists of only a single isotope, the equation for calculating atomic weight becomes

$$\text{At. wt.} = \frac{(\% \text{ fluorine-19})(\text{mass fluorine-19})}{100}$$

$$= \frac{(100)(19.00 \text{ u})}{100} = 19.00 \text{ u}$$

The calculated value is the same as the periodic table value.

 b.

$$\text{At. wt.} = \frac{(\% \text{ Mg-24})(\text{mass Mg-24})}{100}$$

$$+ \frac{(\% \text{ Mg-25})(\text{mass Mg-25}) + (\% \text{ Mg-26})(\text{mass Mg-26})}{100}$$

$$= \frac{(78.70)(23.99 \text{ u}) + (10.13)(24.99 \text{ u})}{100}$$

$$+ \frac{(11.17)(25.98 \text{ u})}{100}$$

$$= 24.31 \text{ u}$$

Once again, the calculated value is the same as the periodic table value.

2.6 The mass ratio is obtained by dividing the mass of the magnesium sample by the mass of the carbon sample:

$$\frac{\text{Mg mass}}{\text{C mass}} = \frac{13.66 \text{ g}}{6.748 \text{ g}} = 2.024$$

The number of atoms in each sample is calculated using the factors obtained earlier from the mass of one atom of each element:

$$13.66 \text{ g } \cancel{Mg} \times \frac{1 \text{ Mg atom}}{4.037 \times 10^{-23} \text{ g } \cancel{Mg}}$$
$$= 3.384 \times 10^{23} \text{ Mg atoms}$$

$$6.748 \text{ g } \cancel{C} \times \frac{1 \text{ C atom}}{1.994 \times 10^{-23} \text{ g } \cancel{C}} = 3.384 \times 10^{23} \text{ C atoms}$$

2.7 The known quantity is one O atom, and the unit of the unknown is grams of O. The factor comes from the relationship

6.02×10^{23} O atoms = 16.00 g O. This relationship provides two factors:

$$\frac{6.02 \times 10^{23} \text{ O atoms}}{16.00 \text{ g O}} \quad \text{and} \quad \frac{16.00 \text{ g O}}{6.02 \times 10^{23} \text{ O atoms}}$$

The second factor is the one used:

$$(1 \text{ O atom})\left(\frac{16.00 \text{ g O}}{6.02 \times 10^{23} \text{ O atoms}}\right) = 2.66 \times 10^{-23} \text{ g O}$$

The ratio of this mass to the mass of a carbon atom is

$$\frac{2.66 \times 10^{-23} \text{ g O}}{1.994 \times 10^{-23} \text{ g C}} = 1.33 \text{ (properly rounded answer)}$$

The ratio of the atomic weights of oxygen and carbon is

$$\frac{16.00 \text{ u}}{12.01 \text{ u}} = 1.33 \text{ (rounded to 3 significant figures)}$$

The two ratios are the same to 3 significant figures.

2.8 The relationships between moles of atoms in 1 mol of molecules is given by the subscripts of the atoms. Thus, 1 mol of glucose molecules would contain 6 mol of C atoms; 12 mol of H atoms, and 6 mol of O atoms. One-half mol of glucose molecules would contain half as many moles of each atom, or 3 mol of C atoms, 6 mol of H atoms, and 3 mol of O atoms.

2.9 $\% = \dfrac{\text{part}}{\text{total}} \times 100$

In each compound, the part will be the mass of carbon associated with some mass of compound (the total). The mass relationships are easily obtained by assuming a sample size equal to 1.00 mol of each compound.

CO_2: 1.00 mol CO_2 molecules = 1.00 mol C atoms + 2.00 mol O atoms or, using atomic weights,

$$44.01 \text{ g CO}_2 = 12.01 \text{ g C} + 32.00 \text{ g O}$$

$$\% \text{ C} = \frac{(12.01 \text{ g C}) \times 100}{(44.01 \text{ g CO}_2)} = 27.29\% \text{ C}$$

CO: 1.00 mol CO molecules = 1.00 mol C atoms + 1.00 mol O atoms

$$28.01 \text{ g CO} = 12.01 \text{ g C} + 16.00 \text{ g O}$$

$$\% \text{ C} = \frac{(12.01 \text{ g C})}{(28.01 \text{ g CO})} \times 100 = 42.88\% \text{ C}$$

·········· **CHAPTER 3** ··········

3.1 Each element is found at the intersection of the given group (vertical columns) and period (horizontal rows).

    a. Ge   **b.** Re

3.2 a. Period 1 has 2 elements, H and He.

    b. Group IIB(12) has 4 elements, Zn, Cd, Hg, and element 112.

3.3 a. The maximum number of electrons that can be found in any orbital, including a $4p$ orbital, is 2.

    b. All d subshells, including the $5d$, contain 5 orbitals with a capacity of 2 electrons each. Thus, the $5d$ subshell can contain a maximum of 10 electrons.

    c. Shell number 1 contains only a single s orbital, the $1s$ orbital, and so can contain a maximum of 2 electrons.

3.4 The valence shell is the last shell with electrons. According to Table 3.2, group VA(15) elements all have 5 electrons in the valence shell. Group VIA(16) elements have 6, group VIIA(17) have 7, and the noble gases VIIIA(18) have 8.

3.5 a. Strontium (Sr) is in group IIA(2) and so contains 2 electrons in its valence shell.

    b. Once again, the roman numeral of the group heading gives the number of valence-shell electrons. The element is Ge, and because it is in group IVA(14), it has 4 electrons in the valence shell.

    c. The fifteenth element in period 4 is As, which is in group VA(15) and therefore has 5 electrons in the valence shell.

3.6 The subshell filling order is obtained from Figure 3.7 or Figure 3.8. All subshells except the last one are filled to their capacities as follows: s subshells—filled with 2, p subshells—filled with 6, d subshells—filled with 10, and f subshells—filled with 14. The last subshell has only enough electrons added to bring the total number of electrons to the desired value.

    a. Element 9 requires 9 electrons: $1s^2 2s^2 2p^5$; 1 unpaired

    b. Mg is element 12 and so needs 12 electrons:
$$1s^2 2s^2 2p^6 3s^2; \text{ 0 unpaired}$$

    c. The element in group VIA(16) and period 3 is S, which is element 16. Therefore, it needs 16 electrons:
$$1s^2 2s^2 2p^6 3s^2 3p^4; \text{ 2 unpaired}$$

    d. An atom with 23 protons needs 23 electrons:
$$1s^2 2s^2 2p^6 3s^2 3p^6 4s^2 3d^3; \text{ 3 unpaired}$$

3.7 a. From Learning Check 3.6, the electronic configuration of element number 9 is $1s^2 2s^2 2p^5$. The $1s^2$ electrons represent the configuration of helium. So, we can write the configuration as $[\text{He}]2s^2 2p^5$.

    b. The configuration of Mg is $1s^2 2s^2 2p^6 3s^2$. The first 10 electrons represent the configuration of neon: $[\text{Ne}]3s^2$.

    c. The configuration of S is $1s^2 2s^2 2p^6 3s^2 3p^4$. Once again, the first 10 electrons represent the configuration of neon: $[\text{Ne}]3s^2 3p^4$.

    d. The 23 electrons are represented by $1s^2 2s^2 2p^6 3s^2 3p^6 4s^2 3d^3$. The first 18 electrons represent the configuration of argon: $[\text{Ar}]4s^2 3d^3$.

3.8 a. On the basis of the location of each element in Figure 3.9, the distinguishing electrons are of the following types: element 38(Sr): s; element 47(Ag): d; element 50(Sn): p; element 86(Rn): p.

    b. Their classifications based on Figure 3.10 are element 38(Sr): representative element; element 47(Ag): transition element; element 50(Sn): representative element; element 86(Rn): noble gas.

3.9 The periodic table and Figure 3.12 are used.

    a. Xe: nonmetal
    b. As: metalloid
    c. Hg: metal
    d. Ba: metal
    e. Th: metal

3.10 a. The slowest reaction involves Li, and the fastest involves K.

    Li

    Na

    K

b. The rate of the reaction increases coming down the group, and according to Table 3.3, the ionization energy decreases coming down the group.

c. The decrease in ionization energy coming down the group means less energy is required to remove electrons from atoms of the elements located farther down the group. The reactions in Figure 3.16 involve the removal of an electron from atoms of the metals, so the easier it is to remove electrons (less energy required), the faster the reaction should go. This trend in speed is what is observed.

····················· **CHAPTER 4** ·····················

4.1 The subshell filling order is obtained from Figure 3.7 or Figure 3.8.

F; 9 electrons:	$1s^2 2s^2 2p^5$
K; 19 electrons:	$1s^2 2s^2 2p^6 3s^2 3p^6 4s^1$

The configuration of F can be changed to that of He ($1s^2$) by giving up 7 electrons or to that of Ne ($1s^2 2s^2 2p^6$) by accepting 1 electron. The configuration of K can be changed to that of Ar ($1s^2 2s^2 2p^6 3s^2 3p^6$) by giving up 1 electron or to that of Kr ($1s^2 2s^2 2p^6 3s^2 3p^6 4s^2 3d^{10} 4p^6$) by accepting 17 electrons.

4.2 a. Element number 9 is fluorine, which is in group VIIA(17) and so has 7 valence-shell electrons. The Lewis structure is $:\overset{..}{\underset{.}{F}}\cdot$.

b. Magnesium is in group IIA(2) and so has 2 valence-shell electrons. The Lewis structure is $\overset{.}{M}g\cdot$.

c. Sulfur is in group VIA(16) and so has 6 valence-shell electrons. The Lewis structure is $:\overset{..}{\underset{.}{S}}\cdot$.

d. Krypton is a noble gas and so has 8 valence-shell electrons. The Lewis structure is $:\overset{..}{\underset{..}{Kr}}:$.

4.3 a. Lithium contains 3 electrons, with one of them classified as a valence-shell electron: [He] $2s^1$ and Li· .

b. Bromine contains 35 electrons, with 7 of them classified as valence-shell electrons: [Ar]$4s^2 3d^{10} 4p^5$ and $:\overset{..}{\underset{..}{Br}}\cdot$.

c. Strontium contains 38 electrons, with 2 classified as valence-shell electrons: [Kr]$5s^2$ and $\overset{.}{S}r\cdot$.

d. Sulfur contains 16 electrons, with 6 classified as valence-shell electrons: [Ne]$3s^2 3p^4$ and $:\overset{..}{\underset{.}{S}}\cdot$.

4.4 The change that will actually take place is the one that involves the fewest electrons. Thus, F will gain 1 electron rather than give up 7, and K will give up 1 electron rather than gain 17:

$$F + 1e^- \rightarrow F^-$$
$$K \rightarrow K^+ + 1e^-$$

4.5 a. Element number 34 is a nonmetal. It is in group VIA(16) and will, when treated using the rules for nonmetals, accept 8 − 6 or 2 electrons during ionic bond formation:

$$Se + 2e^- \rightarrow Se^{2-}$$

b. Rb, a metal in group IA(1), will lose 1 electron during ionic bond formation:

$$Rb \rightarrow Rb^+ + 1e^-$$

c. Element 18 is a nonmetallic noble gas. It will not react to form ionic bonds.

d. In, a metal in group IIIA(13), will lose 3 electrons during ionic bond formation:

$$In \rightarrow In^{3+} + 3e^-$$

4.6 a. Mg, a metal in group IIA(2), will lose 2 electrons:

$$Mg \rightarrow Mg^{2+} + 2e^-$$

O, a nonmetal in group VIA(16), will gain 2 electrons:

$$O + 2e^- \rightarrow O^{2-}$$

The positive and negative ions are combined in the lowest numbers possible to give the compound formula. The combining requirement is that the total number of positive charges from the positive ion must equal the total number of negative charges from the negative ion. In the case of Mg^{2+} and O^{2-}, one of each ion satisfies the requirement, and the formula is MgO.

b. K, a group IA(1) metal, will lose 1 electron:

$$K \rightarrow K^+ + 1e^-$$

S, a group VIA(16) nonmetal, will gain 2 electrons:

$$S + 2e^- \rightarrow S^{2-}$$

In the formula, $2K^+$ will be needed for each S^{2-}. Thus, the compound formula is K_2S.

c. Ca, a group IIA(2) metal, will lose 2 electrons:

$$Ca \rightarrow Ca^{2+} + 2e^-$$

Br, a group VIIA(17) nonmetal, will gain 1 electron:

$$Br + 1e^- \rightarrow Br^-$$

In the formula, $2Br^-$ will combine with each Ca^{2+}: $CaBr_2$.

4.7 a. magnesium oxide
 b. potassium sulfide
 c. calcium bromide

4.8 a. Charge balance requires that $2Br^-$ combine with each Co^{2+}: $CoBr_2$. Names are cobalt(II) bromide and cobaltous bromide.

b. Charge balance requires that $3Br^-$ combine with each Co^{3+}: $CoBr_3$. Names are cobalt(III) bromide and cobaltic bromide.

4.9 a. For H_2S, the molecular weight is the sum of the atomic weights of the atoms in the formula:

$$\begin{aligned} MW &= (2)(\text{at. wt. H}) + (1)(\text{at. wt. S}) \\ &= (2)(1.01 \text{ u}) + (1)(32.1 \text{ u}) \\ MW &= 34.1 \text{ u} \end{aligned}$$

For CaO, the formula weight is also equal to the sum of the atomic weights of the atoms in the formula:

$$\begin{aligned} FW &= (1)(\text{at. wt. Ca}) + (1)(\text{at. wt. O}) \\ &= (1)(40.1 \text{ u}) + (1)(16.0 \text{ u}) \\ FW &= 56.1 \text{ u} \end{aligned}$$

b. According to Section 2.6, 1.00 mol of a molecular compound has a mass in grams equal to the molecular weight of the compound. Thus, 1.00 mol H_2S = 34.1 g H_2S. For ionic compounds, 1.00 mol of compound has a mass in grams equal to the formula weight of the compound. Thus, 1.00 mol CaO = 56.1 g CaO.

c. In Section 2.6, we learned that 1.00 mol of a molecular compound contains Avogadro's number of molecules. Thus, 1.00 mol H_2S = 6.02×10^{23} molecules of H_2S. In the case of ionic compounds, 1.00 mol of compound contains Avogadro's number of formula units. That is, 1.00 mol of calcium oxide contains Avogadro's number of CaO units, where each unit represents one Ca^{2+} ion and one O^{2-} ion. Thus:

$$1.00 \text{ mol CaO} = 6.02 \times 10^{23} \text{ CaO units} \qquad \text{or}$$
$$1.00 \text{ mol CaO} = 6.02 \times 10^{23} \text{ Ca}^{2+} \text{ ions}$$
$$+ \ 6.02 \times 10^{23} \text{ O}^{2-} \text{ ions}$$

4.10 a. N is in group VA(15) and therefore has 5 valence-shell electrons. Two N atoms will provide 10 total electrons to satisfy the octets. At least 2 electrons are needed to bond the atoms. This leaves only 8 electrons to satisfy the remaining octet requirements of both atoms. This deficiency of electrons means some multiple bonds will be needed between the atoms:

$$:\!\overset{\cdot\cdot}{N}\!\cdot \ + \ \cdot\overset{\cdot\cdot}{N}\!: \longrightarrow \ :N:::N: \qquad \text{or} \qquad N\equiv N$$

b. Br is in group VIIA(17) and therefore has 7 valence-shell electrons. Two atoms will provide 14 electrons, and the bonding is readily shown:

$$:\!\overset{\cdot\cdot}{\underset{\cdot\cdot}{Br}}\!\cdot \ + \ \cdot\overset{\cdot\cdot}{\underset{\cdot\cdot}{Br}}\!: \longrightarrow \ :\!\overset{\cdot\cdot}{\underset{\cdot\cdot}{Br}}\!:\!\overset{\cdot\cdot}{\underset{\cdot\cdot}{Br}}\!: \qquad \text{or} \qquad Br\!-\!Br$$

4.11 a. Each C contributes 4 valence-shell electrons, and each H contributes 1. The total available is 8:

$$H\!:\!\overset{\displaystyle H}{\underset{\displaystyle H}{\overset{\cdot\cdot}{C}}}\!:\!H \qquad \text{or} \qquad H\!-\!\overset{\displaystyle H}{\underset{\displaystyle H}{\overset{|}{C}}}\!-\!H$$

b. Each H contributes 1 electron, the C contributes 4, and the O contributes 6. The total of 12 electrons cannot satisfy all octets unless two pair are shared between the C and O atoms:

$$H\!:\!\overset{\displaystyle \ }{\underset{\displaystyle H}{\overset{\cdot\cdot}{C}}}\!:\!:\!\overset{\cdot\cdot}{O}\!: \qquad \text{or} \qquad H\!-\!\overset{\displaystyle \ }{\underset{\displaystyle H}{\overset{|}{C}}}\!=\!O$$

c. Each H contributes 1 electron, each O contributes 6, and the N contributes 5. The 24 total electrons can satisfy all octets only if two pair are shared between the N and one of the O atoms:

$$H\!:\!\overset{\cdot\cdot}{\underset{\cdot\cdot}{O}}\!:\!\underset{\displaystyle :\overset{\cdot\cdot}{\underset{\cdot\cdot}{O}}:}{N}\!:\!:\!\overset{\cdot\cdot}{O}\!: \qquad \text{or} \qquad H\!-\!O\!-\!\underset{\displaystyle O}{\overset{\displaystyle \ }{N}}\!=\!O$$

4.12 a. The P provides 5 electrons, each O provides 6, and the 3− charge indicates an additional 3 electrons. The total number of electrons is therefore 32, which is enough to satisfy all octets without any multiple bonds:

$$\left[\ :\!\overset{\cdot\cdot}{\underset{\cdot\cdot}{O}}\!:\!\underset{\displaystyle :\overset{\cdot\cdot}{\underset{\cdot\cdot}{O}}:}{\overset{\displaystyle :\overset{\cdot\cdot}{O}:}{P}}\!:\!\overset{\cdot\cdot}{\underset{\cdot\cdot}{O}}\!:\ \right]^{3-}$$

b. The S provides 6 electrons, each O provides 6, and the 2− charge indicates an additional 2 electrons. The total number of electrons is therefore 26, which is enough to satisfy all octets without any multiple bonds:

$$\left[\ :\!\overset{\cdot\cdot}{\underset{\cdot\cdot}{O}}\!:\!\underset{\displaystyle :\overset{\cdot\cdot}{\underset{\cdot\cdot}{O}}:}{\overset{\displaystyle :\overset{\cdot\cdot}{O}:}{S}}\!:\!\overset{\cdot\cdot}{\underset{\cdot\cdot}{O}}\!:\ \right]^{2-}$$

c. The N provides 5 electrons, each H provides 1, and the + charge indicates that 1 must be subtracted. The total number of electrons is therefore 8, which is enough to satisfy the octet of N, and each H requirement of 2 electrons without any multiple bonds:

$$\left[\ H\!:\!\overset{\displaystyle H}{\underset{\displaystyle H}{\overset{\cdot\cdot}{N}}}\!:\!H\ \right]^{+}$$

4.13 a. The 3 electron pairs around the B atom will arrange themselves in a flat triangle around the B. Therefore, the molecule will be flat (planar) with the F atoms forming a triangle around the B.

b. The Lewis structure is

$$:\!\overset{\cdot\cdot}{\underset{\cdot\cdot}{Cl}}\!: \\ :\!\overset{\cdot\cdot}{\underset{\cdot\cdot}{Cl}}\!:\!\overset{\cdot\cdot}{\underset{\cdot\cdot}{C}}\!:\!\overset{\cdot\cdot}{\underset{\cdot\cdot}{Cl}}\!: \\ :\!\overset{\cdot\cdot}{\underset{\cdot\cdot}{Cl}}\!:$$

The 4 electron pairs around the C atom will arrange themselves in a tetrahedral orientation. Thus, the molecule will be tetrahedral with C in the center and a Cl atom at each of the corners of the tetrahedron around the C.

c. The Lewis structure is

$$:\!\overset{\cdot\cdot}{\underset{\cdot\cdot}{O}}\!:\!\overset{\cdot\cdot}{\underset{\cdot\cdot}{S}}\!:\!:\!\overset{\cdot\cdot}{\underset{\cdot\cdot}{O}}\!:$$

The 3 electron pairs around the S will occupy the corners of a triangle around the S. This will give rise to a bent molecule

$$O\!-\!\overset{\cdot\cdot}{S}\!\underset{\displaystyle O}{\searrow}$$

d. The Lewis structure is

$$:S\!:\!:C\!:\!:\overset{\cdot\cdot}{S}\!:$$

Each double bond on the central C atom is treated like a single pair of electrons when VSEPR theory is used. Thus, the double bonds are treated like 2 pairs and will arrange themselves to be on opposite sides of the C atom. A linear molecule results: S=C=S.

4.14 a. The Lewis structure from Learning Check 4.12 is

$$\left[\ :\!\overset{\cdot\cdot}{\underset{\cdot\cdot}{O}}\!:\!\underset{\displaystyle :\overset{\cdot\cdot}{\underset{\cdot\cdot}{O}}:}{\overset{\displaystyle :\overset{\cdot\cdot}{O}:}{P}}\!:\!\overset{\cdot\cdot}{\underset{\cdot\cdot}{O}}\!:\ \right]^{3-}$$

We see that phosphorus is the central atom, and it has four electron pairs around it. The four pairs will be located at the corners of a tetrahedron with the P in the middle. The four O atoms will occupy the corners of the tetrahedron, giving the same shape as that of the SO_4^{2-} ion of Example 4.14a.

b. The Lewis structure from Learning Check 4.12 is

$$\left[\ :\!\overset{\cdot\cdot}{\underset{\cdot\cdot}{O}}\!:\!\underset{\displaystyle :\overset{\cdot\cdot}{\underset{\cdot\cdot}{O}}:}{\overset{\displaystyle :\overset{\cdot\cdot}{O}:}{S}}\!:\!\overset{\cdot\cdot}{\underset{\cdot\cdot}{O}}\!:\ \right]^{2-}$$

We see that sulfur is the central atom, and it has four electron pairs around it. The four pairs will be located at the corners of a tetrahedron with the S in the middle. The shape of the ion is determined by the location of the oxygen atoms and the S atom. The O atoms will occupy three of the corners of the tetrahedron, and one pair of electrons will occupy the fourth. Thus, the shape of the ion will be a triangular-based pyramid with the O atoms at the three corners of the base and the P atom at the top.

c. The Lewis structure from Learning Check 4.12 is

$$\left[\begin{array}{c} \text{H} \\ \text{H} \text{:} \ddot{\text{N}} \text{:} \text{H} \\ \ddot{\text{H}} \end{array} \right]^{+}$$

We see that nitrogen is the central atom, and it has four electron pairs around it. The four pairs will be located at the corners of a tetrahedron with the N in the middle. The four H atoms will thus be located at the corners of the tetrahedron with the N in the middle, giving a tetrahedral-shaped ion like the SO_4^{2-} ion of Example 4.14a.

4.15 a. No polarization because bound atoms are identical.

b. Br is located higher in the group and so is more electronegative than I:

$$\overset{\delta^+ \quad \delta^-}{\text{I} - \text{Br}}$$
$$\longrightarrow$$

c. Br is farther toward the right in the periodic table and so is more electronegative than H:

$$\overset{\delta^+ \quad \delta^-}{\text{H} - \text{Br}}$$
$$\longrightarrow$$

4.16 The values of ΔEN are used.

a. $\Delta EN = 4.0 - 0.8 = 3.2$. Bond is classified as ionic.

b. $\Delta EN = 3.5 - 3.0 = 0.5$. Bond is classified as polar covalent.

c. $\Delta EN = 3.0 - 1.5 = 1.5$. Bond is classified as polar covalent.

d. $\Delta EN = 3.5 - 0.8 = 2.7$. Bond is classified as ionic.

4.17 a. sulfur trioxide
b. boron trifluoride
c. disulfur heptoxide
d. carbon tetrachloride

4.18 a. Ca is a group IIA(2) metal and so forms Ca^{2+} ions. Formula: $CaHPO_4$; name: calcium hydrogen phosphate

b. Mg is a group IIA(2) metal and so forms Mg^{2+} ions. Formula: $Mg_3(PO_4)_2$; name: magnesium phosphate

c. K is a group IA(1) metal and so forms K^+ ions. Formula: $KMnO_4$; name: potassium permanganate

d. The ammonium ion, NH_4^+, acts as the metal ion in this compound. Formula: $(NH_4)_2Cr_2O_7$; name: ammonium dichromate

4.19 Assuming that dispersion forces are significant in establishing the melting and boiling points of these elements leads to the conclusion that, in each pair, the element with higher atomic weight would have the higher melting and boiling point.

a. Se   **b.** Sb   **c.** Ne

················· **CHAPTER 5** ·····················

5.1 Both subscripts and coefficients are used.
N_2 : 2 atoms—on the basis of subscript
$3H_2$: 6 atoms—on the basis of subscript and coefficient
$2NH_3$: 2 N atoms (on the basis of coefficient) and 6 H atoms (on the basis of subscript and coefficient)

5.2 $2SO_2 + O_2 \rightarrow 2SO_3$

5.3 a. Oxygen is -2 (Rule 5).
Sulfur is $+6$ (Rule 6).
b. Calcium is $+2$ (Rule 3).
Oxygen is -2 (Rule 5).
Chlorine is $+5$ (Rule 6).
c. Oxygen is -2 (Rule 5).
Chlorine is $+7$ (Rule 7).

5.4 a. $Zn = 0$ (Rule 1), $H^+ = +1$ (Rule 2), $Zn^{2+} = +2$ (Rule 2), $H_2 = 0$ (Rule 1).
Zn is oxidized, so it is the reducing agent.
H^+ is reduced, so it is the oxidizing agent.
b. In KI: $K = +1$ (Rule 3), $I = -1$ (Rule 6).
$Cl_2 = 0$ (Rule 1).
In KCl: $K = +1$ (Rule 3), $Cl = -1$ (Rule 6).
$I_2 = 0$ (Rule 1).
I in KI is oxidized, so KI is the reducing agent.
Cl_2 is reduced, so Cl_2 is the oxidizing agent.
c. In IO_3^-: $O = -2$ (Rule 5), $I = +5$ (Rule 7).
In HSO_3^-: $H = +1$ (Rule 4), $O = -2$ (Rule 5), $S = +4$ (Rule 7).
$I^- = -1$ (Rule 2).
In HSO_4^-: $H = +1$ (Rule 4), $O = -2$ (Rule 5), $S = +6$ (Rule 7).
S in HSO_3^- is oxidized, so HSO_3^- is the reducing agent.
I in IO_3^- is reduced, so IO_3^- is the oxidizing agent.

5.5 a. The O.N. of H changes from $+1$ to 0, and the O.N. of I changes from -1 to 0. Reaction is redox. Because one substance changes into two substances, the reaction is a decomposition.

b. The O.N. of H does not change. The O.N. of O changes from -1 (in a peroxide) to -2 (in water) and 0 (in O_2). This is an example in which the same element is both oxidized and reduced. The reaction is redox. Because one substance changes into two substances, the reaction is a decomposition.

c. No oxidation numbers change. The reaction is nonredox. This is a double-replacement (metathesis) reaction.

d. The O.N. of P changes from 0 to $+5$, and the O.N. of O changes from 0 to -2. The reaction is redox. Because two substances combine to form a single substance, the reaction is a combination.

e. The O.N. of Na does not change. The O.N. of I changes from -1 to 0, and the O.N. of Cl changes from 0 to -1. The reaction is redox. Because the Cl simply replaces the I in the compound, this is a single-replacement reaction.

5.6 a. Total ionic: $2Na^+(aq) + 2I^-(aq) + Cl_2(aq) \rightarrow 2Na^+(aq) + 2Cl^-(aq) + I_2(aq)$
Na^+ is a spectator ion.
Net ionic: $2I^-(aq) + Cl_2(aq) \rightarrow 2Cl^-(aq) + I_2(aq)$
b. Total ionic: $Ca^{2+}(aq) + 2Cl^-(aq) + 2Na^+(aq) + CO_3^{2-}(aq) \rightarrow 2Na^+(aq) + 2Cl^-(aq) + CaCO_3(s)$
Na^+ and Cl^- are spectator ions.
Net ionic: $Ca^{2+}(aq) + CO_3^{2-}(aq) \rightarrow CaCO_3(s)$
c. Total ionic: $Ba^{2+}(aq) + 2OH^-(aq) + 2H^+(aq) + SO_4^{2-}(aq) \rightarrow 2H_2O(\ell) + BaSO_4(s)$
There are no spectator ions, so the net ionic reaction is the same as the total ionic reaction.

5.7 a. $\dfrac{3 \text{ mol } H_2}{2 \text{ mol } NH_3}$ and $\dfrac{2 \text{ mol } NH_3}{3 \text{ mol } H_2}$

b. The pattern is mol $A \rightarrow$ mol B, and the pathway is mol $H_2 \rightarrow$ mol NH_3.

Step 1. 2.11 mol H_2

Step 2. 2.11 mol H_2 $\qquad\qquad$ = mol NH_3

Step 3. 2.11 mol H_2 $\times \dfrac{2 \text{ mol } NH_3}{3 \text{ mol } H_2}$ = mol NH_3

Step 4. $(2.11)\left(\dfrac{2 \text{ mol } NH_3}{3}\right) = 1.41$ mol NH_3

c. The pattern is g $A \to$ g B, and the pathway is g $H_2 \to$ mol $H_2 \to$ mol $N_2 \to$ g N_2. The combined method gives:

Step 1. 9.47 g H_2

Step 2. 9.47 g H_2 $\qquad\qquad\qquad$ = g N_2

Step 3. 9.47 g H_2 $\times \dfrac{1 \text{ mol } H_2}{2.016 \text{ g } H_2} \times \dfrac{1 \text{ mol } N_2}{3 \text{ mol } H_2}$

$\qquad\qquad \times \dfrac{28.02 \text{ g } N_2}{1 \text{ mol } N_2}$ $\qquad$ = g N_2

Step 4. $(9.47)\left(\dfrac{1}{2.016}\right)\left(\dfrac{1}{3}\right)\left(\dfrac{28.02 \text{ g } N_2}{1}\right) = 43.9$ g N_2

5.8 The mass of NH_3 possible from each reactant is calculated. The reactant giving the smallest mass of NH_3 is the limiting reactant, and the mass of NH_3 calculated for the limiting reactant is the amount that will be produced:

$(2.00 \text{ mol } H_2)\left(\dfrac{2 \text{ mol } NH_3}{3 \text{ mol } H_2}\right)\left(\dfrac{17.0 \text{ g } NH_3}{1 \text{ mol } NH_3}\right) = 22.7$ g NH_3

$(15.5 \text{ g } N_2)\left(\dfrac{1 \text{ mol } N_2}{28.02 \text{ g } N_2}\right)\left(\dfrac{2 \text{ mol } NH_3}{1 \text{ mol } N_2}\right)\left(\dfrac{17.0 \text{ g } NH_3}{1 \text{ mol } NH_3}\right)$
$\qquad\qquad\qquad\qquad\qquad\qquad\qquad = 18.8$ g NH_3

The results show the limiting reactant to be the 15.5 g of N_2, and the amount of NH_3 produced will be 18.8 g.

5.9 a. % yield $= \dfrac{\text{actual yield}}{\text{theoretical yield}} \times 100$

$\qquad = \dfrac{17.43 \text{ g}}{21.34 \text{ g}} \times 100 = 81.68\%$

b. The theoretical yield must be calculated. The pattern is g $CaCO_3 \to$ g CaO, and the pathway is g $CaCO_3 \to$ mol $CaCO_3 \to$ mol CaO $\to$ g CaO. In combined form, the solution is

$(510 \text{ g } CaCO_3)\left(\dfrac{1 \text{ mol } CaCO_3}{100.1 \text{ g } CaCO_3}\right)\left(\dfrac{1 \text{ mol } CaO}{1 \text{ mol } CaCO_3}\right)$

$\qquad\qquad\qquad \times \left(\dfrac{56.08 \text{ g } CaO}{1 \text{ mol } CaO}\right) = 286$ g CaO

The % yield is then calculated using this theoretical yield:

% yield $= \dfrac{\text{actual yield}}{\text{theoretical yield}} \times 100$

$\qquad = \dfrac{235 \text{ g}}{286 \text{ g}} \times 100 = 82.2\%$

························ **CHAPTER 6** ························

6.1 a.
$$d = \dfrac{\text{mass}}{\text{volume}}$$

copper: $d = \dfrac{114.2 \text{ g}}{12.8 \text{ mL}} = 8.92$ g/mL

glycerin: $d = \dfrac{63.0 \text{ g}}{50.0 \text{ mL}} = 1.26$ g/mL

helium: $d = \dfrac{0.286 \text{ g}}{1500 \text{ mL}} = 1.91 \times 10^{-4}$ g/mL

b. Copper: Doubling the pressure will not influence the volume, mass, or density of a solid.
Glycerin: Doubling the pressure will not significantly influence the volume and will not influence the mass or density of a liquid.
Helium: The increased pressure will cause the volume to decrease. The mass will not be changed, so the density will increase.

6.2 $d = \dfrac{m}{V}$, so $m = Vd$

$\qquad m = (1200 \text{ mL})(1.18 \times 10^{-3} \text{ g/mL})$

$\qquad\quad = 1.42$ g

6.3 KE $= \frac{1}{2}mv^2 = \frac{1}{2}(3.00 \text{ g})(10.0 \text{ cm/s})^2 =$

$\qquad\qquad\qquad\qquad\qquad\qquad 1.50 \times 10^2$ g cm^2/s^2

KE $= \frac{1}{2}mv^2 = \frac{1}{2}(3.00 \text{ g})(20.0 \text{ cm/s})^2 =$

$\qquad\qquad\qquad\qquad\qquad\qquad 6.00 \times 10^2$ g cm^2/s^2

6.4 The factors used come from the information in Table 6.3.

a. 670 torr $\times \dfrac{1 \text{ atm}}{760 \text{ torr}} = 0.882$ atm

b. 670 torr $\times \dfrac{14.7 \text{ psi}}{760 \text{ torr}} = 13.0$ psi

6.5 a. K = °C + 273 = 27 + 273 = 300 K
b. K = °C + 273 = 0 + 273 = 273 K
c. °C = K − 273 = 0 − 273 = − 273°C
d. °C = K − 273 = 100 − 273 = − 173°C

6.6 a. Equation 6.8 is used. Temperatures must be converted to kelvins:

$\begin{array}{ll} P_i = 1.90 \text{ atm} & P_f = 1.00 \text{ atm} \\ V_i = 10.0 \text{ L} & V_f = ? \\ T_i = 30°C = 303 \text{ K} & T_f = 10.2°C = 262.8 \text{ K} \end{array}$

$$\dfrac{P_i V_i}{T_i} = \dfrac{P_f V_f}{T_f}$$

$$\dfrac{(1.90 \text{ atm})(10.0 \text{ L})}{303 \text{ K}} = \dfrac{(1.00 \text{ atm})(V_f)}{262.8 \text{ K}}$$

Solve for V_f:

$$V_f = \dfrac{(1.90 \text{ atm})(10.0 \text{ L})(262.8 \text{ K})}{(1.00 \text{ atm})(303 \text{ K})} = 16.5 \text{ L}$$

b. $\begin{array}{ll} P_i = 800 \text{ torr} & P_f = 900 \text{ torr} \\ V_i = 500 \text{ mL} & V_f = 250 \text{ mL} \\ T_i = 300 \text{ K} & T_f = ? \end{array}$

$$\dfrac{P_i V_i}{T_i} = \dfrac{P_f V_f}{T_f}$$

$$\dfrac{(800 \text{ torr})(500 \text{ mL})}{300 \text{ K}} = \dfrac{(900 \text{ torr})(250 \text{ mL})}{T_f}$$

Invert both sides and solve for T_f:

$$T_f = \dfrac{(300 \text{ K})(900 \text{ torr})(250 \text{ mL})}{(800 \text{ torr})(500 \text{ mL})} = 169 \text{ K}$$

°C = K − 273 = 169 K − 273 = − 104°C

6.7 To use the ideal gas law, all quantities must have units to match those of R. The only unit that needs to be changed is the temperature:

$$K = °C + 273 = 30°C + 273 = 303 \text{ K}$$

$$PV = nRT$$

Solve for P:

$$P = \frac{nRT}{V}$$

$$= \frac{(2.15 \text{ mol})\left(0.0821 \dfrac{L \text{ atm}}{\text{mol K}}\right)(303 \text{ K})}{(12.6 \text{ L})}$$

$$= 4.24 \text{ atm}$$

6.8 The ideal gas law in the form $PV = mRT/MW$ is solved for MW:

$$MW = \frac{mRT}{PV}$$

The temperature must be converted to kelvins:

$$K = 27°C + 273 = 300 \text{ K}$$

$$MW = \frac{(3.35 \text{ g})\left(0.0821 \dfrac{L \text{ atm}}{\text{mol K}}\right)(300 \text{ K})}{(1.21 \text{ atm})(2.00 \text{ L})}$$

$$= 34.1 \text{ g/mol}$$

This matches the molecular weight of H_2S.

6.9 $P_{total} = P_{He} + P_{N_2} + P_{O_2}$
The partial pressures of N_2 and O_2 are changed to torr before adding. The factors used come from information in Table 6.3.

$$P_{N_2}: \quad 0.200 \text{ atm} \times \frac{760 \text{ torr}}{1 \text{ atm}} = 152 \text{ torr}$$

$$P_{O_2}: \quad 7.35 \text{ psi} \times \frac{760 \text{ torr}}{14.7 \text{ psi}} = 380 \text{ torr}$$

$P_{total} = P_{He} + P_{N_2} + P_{O_2} = 310 \text{ torr} + 152 \text{ torr} + 380 \text{ torr} = 842 \text{ torr}$

6.10 The lighter (least massive) helium molecules will diffuse faster than the neon molecules. The relationship between the diffusion rates is given by Graham's law.

$$\frac{\text{Rate of He}}{\text{Rate of Ne}} = \sqrt{\frac{\text{mass of Ne}}{\text{mass of He}}} = \sqrt{\frac{20.18}{4.003}}$$

$$= \sqrt{5.041} = 2.25$$

Thus, rate of He = (2.25) rate of Ne. So, helium is seen to diffuse at a rate more than twice that of neon.

6.11 a. Evaporation is endothermic.
 b. Freezing is exothermic.
 c. Melting is endothermic.

6.12 The lower the forces between particles, the higher will be the vapor pressure.
 a. Both substances form hydrogen bonds, but dispersion forces are stronger in propyl alcohol. Thus, methyl alcohol has the higher vapor pressure.
 b. The only forces acting between the molecules in each case are dispersion forces. These are stronger between the heavier nitrogen molecules, so helium has the higher vapor pressure.

 c. Molecules of HF form strong hydrogen bonds with one another, whereas the only forces between neon molecules are weak dispersion forces. Thus, neon has the higher vapor pressure.

6.13 Heat absorbed = (mass)(specific heat)(temp. change). The specific heat of helium is obtained from Table 6.8.

Heat absorbed = (1000 g)(1.25 cal/g °C)(700°C − 25°C)

$$= 8.44 \times 10^5 \text{ cal}$$

6.14 Heat absorbed = (mass)(heat of vaporization). The heat of vaporization for water was given in the text.

Heat absorbed = (5000 g)(540 cal/g) = 2.70×10^6 cal

CHAPTER 7

7.1 a. Solvent is gold; solutes are copper, zinc, and nickel.
 b. Solvent is nitrogen; solutes are oxygen, argon, and the other gases.

7.2 At 80°C, a saturated solution of KNO_3 contains about 181 g of solute/100 g of H_2O, and a saturated solution of NaBr contains about 118 g of solute per 100 g H_2O. At 50°C both solutions are still saturated, but the KNO_3 solution contains only about 102 g of solute per 100 g H_2O, while the NaBr solution contains about 117 g of solute per 100g H_2O. Thus, the solubility of KNO_3 is seen to be much more temperature dependent than the solubility of NaBr.

7.3 a. The solubility of nitrates and lack of solubility of sulfates and carbonates of barium makes barium nitrate, $Ba(NO_3)_2$, the salt to use. More of it would dissolve and would release more Ba^{2+} ions.
 b. Like dissolves like, so oil will dissolve best in oily solvents such as light mineral oil or gasoline. Light mineral oil is the solvent used because it leaves a residue that is similar to the natural oily secretions found on the bird feathers. Gasoline would be hazardous to use (flammable) and leaves no such useful residue.

7.4 a. The data may be substituted directly into Equation 7.5:

$$M = \frac{1.25 \text{ mol solute}}{2.50 \text{ L solution}} = 0.500 \frac{\text{mol solute}}{\text{L solution}}$$

The solution is 0.500 molar, or 0.500 M.
 b. In this problem the volume of solution must be converted to liters before the data are substituted into Equation 7.5:

$$(225 \text{ mL solution})\left(\frac{1 \text{ L}}{1000 \text{ mL}}\right) = 0.225 \text{ L solution}$$

$$M = \frac{0.486 \text{ mol solute}}{0.225 \text{ L solution}} = 2.16 \frac{\text{mol solute}}{\text{L solution}}$$

The solution is 2.16 molar, or 2.16 M.
 c. In this problem, the volume of solution must be converted to liters and the mass of solute must be converted to moles before the data can be substituted into Equation 7.5:

$$(100 \text{ mL solution})\left(\frac{1 \text{ L}}{1000 \text{ mL}}\right) = 0.100 \text{ L solution}$$

$$(2.60 \text{ g NaCl})\left(\frac{1 \text{ mol NaCl}}{58.44 \text{ g NaCl}}\right) = 0.0445 \text{ mol NaCl}$$

In the last calculation, the factor $\dfrac{1 \text{ mol NaCl}}{58.44 \text{ g NaCl}}$ comes

from the calculated formula weight of 58.44 u for NaCl. The data are now substituted into Equation 7.5:

$$M = \frac{0.0445 \text{ mol NaCl}}{0.100 \text{ L solution}} = 0.445 \frac{\text{mol NaCl}}{\text{L solution}}$$

The solution is 0.445 molar, or 0.445 M.

7.5 a. To calculate %(w/w), the mass of solute contained in a specific mass of solution is needed. The mass of solute is 0.900 g, and the mass of solution is equal to the solvent mass (100 g) plus the solute mass (0.900 g):

$$\%(\text{w/w}) = \frac{\text{solute mass}}{\text{solution mass}} \times 100 = \frac{0.900 \text{ g}}{100.9 \text{ g}} \times 100$$
$$= 0.892 \ \%(\text{w/w})$$

To calculate %(w/v), the number of grams of solute must be known along with the number of milliliters of solution. The mass of solute is 0.900 g, and the solution volume is 100 mL:

$$\%(\text{w/v}) = \frac{\text{g of solute}}{\text{mL of solution}} \times 100 = \frac{0.900 \text{ g}}{100 \text{ mL}} \times 100$$
$$= 0.900 \ \%(\text{w/v})$$

b. The given quantity is 30 mL of beverage, and the desired quantity is milliliters of pure alcohol. The %(v/v) provides the necessary factor:

$$30 \text{ mL beverage} \times \frac{45 \text{ mL alcohol}}{100 \text{ mL beverage}} = 14 \text{ mL alcohol}$$

7.6 a. $M = \dfrac{\text{mol of solute}}{\text{liters of solution}}$

$$\begin{aligned}\text{mol of solute} &= (M)(\text{liters of solution})\\ &= (1.00 \text{ M})(0.500 \text{ L})\\ &= 0.500 \text{ mol of solute}\end{aligned}$$

The solute is $MgCl_2$, which has a formula weight of 95.2 u. The formula weight provides the factor in the following calculation:

$$0.500 \text{ mol MgCl}_2 \times \frac{95.2 \text{ g MgCl}_2}{1 \text{ mol MgCl}_2} = 47.6 \text{ g MgCl}_2$$

The solution is prepared by putting 47.6 g of $MgCl_2$ into a 500-mL flask and adding pure water to the mark, making certain all the solid solute dissolves and the resulting solution is well mixed.

b.
$$\%(\text{w/v}) = \frac{\text{g of solute}}{\text{mL of solution}} \times 100$$

$$\text{g of solute} = \frac{\%(\text{w/v})(\text{mL of solution})}{100}$$

$$= \frac{(12.0\%)(100 \text{ mL})}{100} = 12.0 \text{ g}$$

The solution is prepared by putting 12.0 g of $MgCl_2$ into a 100-mL flask and adding pure water to the mark, making certain all the solid solute dissolves and the resulting solution is well mixed.

c.
$$\%(\text{v/v}) = \frac{\text{solute volume}}{\text{solution volume}} \times 100$$

$$\text{solute volume} = \frac{\%(\text{v/v})(\text{solution volume})}{100}$$

$$= \frac{(20.0\%)(1.00 \text{ L})}{100} = 0.200 \text{ L}$$
$$= 200 \text{ mL}$$

The solution is prepared by putting 200 mL of ethylene glycol into a 1.00-L flask and adding pure water to the mark, making certain the two liquids are completely mixed to form the final solution.

7.7 Equation 7.9 is used to calculate the volume of 6.00 M NaOH solution needed.

$$(C_c)(V_c) = (C_d)(V_d)$$
$$(6.00\text{M})(V_c) = (0.250 \text{ M})(500 \text{ mL})$$
$$V_c = \frac{(0.250 \text{ M})(500 \text{ mL})}{(6.00 \text{ M})} = 20.8 \text{ mL}$$

The solution is prepared by putting 20.8 mL of 6.00M NaOH solution into a 500-mL flask, adding pure water to the mark, and making certain the resulting solution is well mixed.

7.8 a. The pattern is mol $A \to$ liters A solution, and the pathway is mol $H_2SO_4 \to$ liters H_2SO_4 solution:

Step 1. 0.150 mol H_2SO_4

Step 2. 0.150 mol H_2SO_4 = L H_2SO_4 solution

Step 3. $0.150 \text{ mol H}_2\text{SO}_4 \times \dfrac{1 \text{ L H}_2\text{SO}_4 \text{ solution}}{0.200 \text{ mol H}_2\text{SO}_4}$
$$= \text{L H}_2\text{SO}_4 \text{ solution}$$

Step 4. $(0.150)\left(\dfrac{1 \text{ L H}_2\text{SO}_4 \text{ solution}}{0.200}\right)$
$$= 0.750 \text{ L H}_2\text{SO}_4 \text{ solution}$$
$$= 750 \text{ mL H}_2\text{SO}_4 \text{ solution}$$

b. The pattern is mol $A \to$ liters B solution, and the pathway is mol $H_2SO_4 \to$ mol NaOH $\to$ liters NaOH solution. In combined form, the solution is

$$(0.125 \text{ mol H}_2\text{SO}_4)\left(\frac{2 \text{ mol NaOH}}{1 \text{ mol H}_2\text{SO}_4}\right)\left(\frac{1 \text{ L NaOH solution}}{0.255 \text{ mol NaOH}}\right)$$
$$= 0.980 \text{ L NaOH solution} = 980 \text{ mL NaOH solution}$$

c. The pattern is liters A solution $\to$ liters B solution, and the pathway is liters H_2SO_4 solution $\to$ mol $H_2SO_4 \to$ mol NaOH $\to$ liters NaOH solution. The volumes are converted to liters, and the solution in combined form is

$$(0.025 \text{ L H}_2\text{SO}_4 \text{ solution})\left(\frac{0.120 \text{ mol H}_2\text{SO}_4}{1 \text{ L H}_2\text{SO}_4 \text{ solution}}\right)$$
$$\times \left(\frac{2 \text{ mol NaOH}}{1 \text{ mol H}_2\text{SO}_4}\right)\left(\frac{1 \text{ L NaOH solution}}{0.250 \text{ mol NaOH}}\right)$$
$$= 0.0240 \text{ L NaOH solution} = 24.0 \text{ mL NaOH solution}$$

7.9 a. Because $CaCl_2$ is a strong electrolyte, it dissociates completely:

$$CaCl_2 \to Ca^{2+} + 2Cl^-. \text{ Thus, } n = 3.$$

$$\Delta t_b = nK_bM = (3)(0.52°\text{C/M})(0.100 \text{ M}) = 0.16°\text{C}$$

Because the boiling point of a solution is higher than the boiling point of the pure solvent, Δt_b is added to the boiling point of the pure solvent.

Solution B.P. $= 100.00°C + 0.16°C = 100.16°C$

$\Delta t_f = nK_f M = (3)(1.86°C/M)(0.100 \, M) = 0.558°C$

Because the freezing point of a solution is lower than the freezing point of the pure solvent, Δt_f is subtracted from the freezing point of the solvent.

Solution F.P. $= 0.00°C - 0.558°C = -0.558°C = -0.56°C$

$\pi = nMRT$
$= (3)(0.100 \, mol/L)(62.4 \, L \, torr/K \, mol)(300 \, K)$
$= 5.62 \times 10^3 \, torr = 7.39 \, atm$

b. Because ethylene glycol does not dissociate in solution, $n = 1$.

$\Delta t_b = nK_b M = (1)(0.52°C/M)(0.100 \, M) = 0.052°C$

Solution B.P. $= 100.00°C + 0.052°C = 100.05°C$

$\Delta t_f = nK_f M = (1)(1.86°C/M)(0.100 \, M) = 0.186°C$

Solution F.P. $= 0.00°C - 0.186°C = -0.186°C$
$= -0.19°C$

$\pi = nMRT$
$= (1)(0.100 \, mol/L)(62.4 \, L \, torr/K \, mol)(300 \, K)$
$= 1.87 \times 10^3 \, torr = 2.46 \, atm$

CHAPTER 8

8.1 Because only Ce^{4+} and Fe^{2+} are mixed, the initial concentration of Ce^{3+} is 0.

$$Rate = \frac{C_t - C_0}{\Delta t} = \frac{1.5 \times 10^{-5} \, mol/L - 0 \, mol/L}{75 \, s}$$

$$= 2.0 \times 10^{-7} \frac{mol/L}{s}$$

8.2 a. $K = \dfrac{[NO_2]^2}{[N_2O_4]}$

b. $K = \dfrac{[N_2O][NO_2]}{[NO]^3}$

8.3 a. $K = \dfrac{[IBr]^2}{[Br_2][I_2]}$

$$= \frac{(1.96 \times 10^{-2})^2}{(1.50 \times 10^{-1})(5.00 \times 10^{-2})}$$

$$= 5.12 \times 10^{-2}$$

The small value of K indicates the equilibrium position is toward the left.

b. $K = \dfrac{[NH_3]^2}{[N_2][H_2]^3}$

$$= \frac{(9.00)^2}{(9.23 \times 10^{-3})(2.77 \times 10^{-2})^3}$$

$$= 4.13 \times 10^8$$

The large K value indicates the equilibrium position is toward the right.

8.4 a. Heat $+ NH_4NO_3(s) \rightleftarrows NH_4^+(aq) + NO_3^-(aq)$

The equilibrium will shift to the right when heat is added in an attempt to use up the added heat. This shift corresponds to dissolving more of the solid at the higher temperature. The solubility is higher at a higher temperature.

b. $2SO_2(g) + O_2(g) \rightleftarrows 2SO_3(g)$

If O_2 were removed from an equilibrium mixture, the equilibrium would shift left in an attempt to replenish the removed O_2. This shift would reduce the amount of SO_3 present at equilibrium, which means the concentration of SO_3 would be lower than it was in the initial equilibrium.

CHAPTER 9

9.1 a.

b.

c.

9.2 a. The anhydrous compound is called hydrogen iodide. The name of the water solution is obtained by dropping *hydrogen* from the anhydrous compound name and adding the prefix *hydro-* to the stem *iod*. The *-ide* suffix on the *iod* stem is replaced by the suffix *-ic* to give the name *hydroiodic*. The word *acid* is added, giving the final name *hydroiodic acid* for the water solution.

b. The anhydrous compound is called hydrogen bromide. The name of the water solution is obtained the same way it was for HI(aq) in part a, using the stem *brom* in place of *iod*. The resulting name for the water solution is *hydrobromic acid*.

9.3 The removal of the two H^+ ions leaves behind the CO_3^{2-} polyatomic ion. From Table 4.6 this ion is the carbonate ion. According to rules 1–4, the *-ate* suffix is replaced by the *-ic* suffix to give the name *carbonic acid*.

9.4 a. Solution is basic because $[OH^-]$ is greater than $1.0 \times 10^{-7} \, mol/L$:

$$[H_3O^+] = \frac{1.0 \times 10^{-14}(mol/L)^2}{[OH^-]} = \frac{1.0 \times 10^{-14}(mol/L)^2}{1.0 \times 10^{-5} \, mol/L}$$

$$= 1.0 \times 10^{-9} \, mol/L$$

b. Solution is basic because $[H_3O^+]$ is less than $1.0 \times 10^{-7} \, mol/L$:

$$[OH^-] = \frac{1.0 \times 10^{-14}(mol/L)^2}{[H_3O^+]} = \frac{1.0 \times 10^{-14}(mol/L)^2}{1.0 \times 10^{-9} \, mol/L}$$

$$= 1.0 \times 10^{-5} \, mol/L$$

This solution has the same $[H_3O^+]$ and $[OH^-]$ as the solution in part a.

c. Solution is acidic because $[H_3O^+]$ is greater than 1.0×10^{-7} mol/L:

$$[OH^-] = \frac{1.0 \times 10^{-14}(\text{mol/L})^2}{[H_3O^+]} = \frac{1.0 \times 10^{-14}(\text{mol/L})^2}{1.0 \times 10^{-2} \text{ mol/L}}$$

$$= 1.0 \times 10^{-12} \text{ mol/L}$$

9.5 a. $[H^+] = 1 \times 10^{-14}$ mol/L. pH is the negative of the exponent used to express $[H^+]$, so pH $= -(-14) = 14.0$

b. $[OH^-] = 1.0$ mol/L. The value of $[H^+]$ is calculated:

$$[H^+] = \frac{1 \times 10^{-14}(\text{mol/L})^2}{[OH^-]} = \frac{1 \times 10^{-14}(\text{mol/L})^2}{1.0 \text{ mol/L}}$$

$$= 1 \times 10^{-14} \text{ mol/L}$$

pH is the negative of the exponent used to express $[H^+]$, so pH $= -(-14) = 14.0$.

c. $[OH^-] = 1 \times 10^{-8}$ mol/L. The value of $[H^+]$ is calculated:

$$[H^+] = \frac{1 \times 10^{-14}(\text{mol/L})^2}{[OH^-]} = \frac{1 \times 10^{-14}(\text{mol/L})^2}{1 \times 10^{-8} \text{ mol/L}}$$

$$= 1 \times 10^{-6} \text{ mol/L}$$

pH is the negative of the exponent used to express $[H^+]$, so pH $= -(-6) = 6.0$.

9.6 a. pH $= 10.0$, and

$$[H^+] = 1 \times 10^{-\text{pH}} = 1 \times 10^{-10} \text{ mol/L}.$$

$$[OH^-] = \frac{K_w}{[H^+]} = \frac{1 \times 10^{-14} (\text{mol/L})^2}{1 \times 10^{-10} \text{ mol/L}} = 1 \times 10^{-4} \text{ mol/L}$$

b. pH $= 4.0$, and $[H^+] = 1 \times 10^{-\text{pH}} = 1 \times 10^{-4}$.

$$[OH^-] = \frac{K_w}{[H^+]} = \frac{1 \times 10^{-14} (\text{mol/L})^2}{1 \times 10^{-4} \text{ mol/L}} = 1 \times 10^{-10} \text{ mol/L}$$

c. pH $= 5.0$, and $[H^+] = 1 \times 10^{-\text{pH}} = 1 \times 10^{-5}$.

$$[OH^-] = \frac{K_w}{[H^+]} = \frac{1 \times 10^{-14} (\text{mol/L})^2}{1 \times 10^{-5} \text{ mol/L}} = 1 \times 10^{-9} \text{ mol/L}$$

9.7 a. The calculator answer is 4.3768. The correctly rounded value is 4.38.

b. The calculator answer is 8.0915. The correctly rounded value is 8.09.

9.8 a. The calculator answer is 0.0017783, or 1.7783×10^{-3}. The correctly rounded value should have two significant figures to match the two figures to the right of the decimal in the pH value. The answer is 1.8×10^{-3} mol/L.

b. The calculator answer is 4.6774×10^{-9}. The correctly rounded answer is 4.7×10^{-9} mol/L.

9.9 According to Table 9.4, concentrated NH_3 stock solution is 15 M. Equation 7.9 is used to determine the amount of this solution needed.

$$C_c V_c = C_d V_d$$

$$(15 \text{ M})(V_c) = (3.0 \text{ M})(500 \text{ mL})$$

$$V_c = \frac{(3.0 \text{ M})(500 \text{ mL})}{(15 \text{ M})}$$

$$= 100 \text{ mL}$$

Because the resulting solution molarity is given using only two significant figures, the volumes can be measured using graduated cylinders. Measure 100 mL of stock (15 M) aqueous ammonia,

put it into a container and add 400 mL of distilled water. Mix the resulting solution well.

9.10 $2HCl(aq) + Sr(HCO_3)_2(s) \rightarrow$
$$SrCl_2(aq) + 2CO_2(g) + 2H_2O(\ell)$$

Total Ionic: $2H^+(aq) + 2Cl^-(aq) + Sr(HCO_3)_2(s) \rightarrow$
$$Sr^{2+}(aq) + 2Cl^-(aq) + 2CO_2(g) + 2H_2O(\ell)$$
Cl^- is a spectator ion.

Net Ionic: $2H^+(aq) + Sr(HCO_3)_2(s) \rightarrow$
$$Sr^{2+}(aq) + 2CO_2(g) + 2H_2O(\ell)$$

9.11 a. Molecular: $Ca(s) + 2H_2O(\ell) \rightarrow$
$$H_2(g) + Ca(OH)_2(s)$$

Total Ionic: $Ca(s) + 2H_2O(\ell) \rightarrow H_2(g) + Ca(OH)_2(s)$
Net Ionic: $Ca(s) + 2H_2O(\ell) \rightarrow H_2(g) + Ca(OH)_2(s)$
All three equations are the same because none of the reactants or products form ions in solution.

b. Molecular: $Mg(s) + H_2SO_4(aq) \rightarrow$
$$H_2(g) + MgSO_4(aq)$$

Total Ionic: $Mg(s) + 2H^+(aq) + SO_4^{2-}(aq) \rightarrow$
$$H_2(g) + Mg^{2+}(aq) + SO_4^{2-}(aq)$$
The SO_4^{2-} is a spectator ion.

Net Ionic: $Mg(s) + 2H^+(aq) \rightarrow H_2(g) + Mg^{2+}(aq)$

9.12 a. Molecular: $HNO_3(aq) + NaOH(aq) \rightarrow$
$$H_2O(\ell) + NaNO_3(aq)$$

Total Ionic: $H^+(aq) + NO_3^-(aq) + Na^+(aq) + OH^-(aq) \rightarrow$
$$H_2O(\ell) + Na^+(aq) + NO_3^-(aq)$$
Na^+ and NO_3^- are spectator ions.

Net Ionic: $H^+(aq) + OH^-(aq) \rightarrow H_2O(\ell)$

b. Molecular: $H_2SO_4(aq) + 2KOH(aq) \rightarrow$
$$2H_2O(\ell) + K_2SO_4(aq)$$

Total Ionic: $2H^+(aq) + SO_4^{2-}(aq) + 2K^+(aq) + 2OH^-(aq) \rightarrow$
$$2H_2O(\ell) + 2K^+(aq) + SO_4^{2-}(aq)$$
K^+ and SO_4^{2-} are spectator ions.

Net Ionic: $2H^+(aq) + 2OH^-(aq) \rightarrow 2H_2O(\ell)$

9.13 $6HCl(aq) + 2Al(s) \rightarrow 2AlCl_3(aq) + 3H_2(g)$
$6HCl(aq) + Al_2O_3(s) \rightarrow 2AlCl_3(aq) + 3H_2O(\ell)$
$3HCl(aq) + Al(OH)_3(s) \rightarrow AlCl_3(aq) + 3H_2O(\ell)$
$6HCl(aq) + Al_2(CO_3)_3(s) \rightarrow 2AlCl_3(aq) + 3H_2O(\ell) + 3CO_2(g)$
$3HCl(aq) + Al(HCO_3)_3(s) \rightarrow AlCl_3(aq) + 3H_2O(\ell) + 3CO_2(g)$

9.14 a. The dissociation reaction is $NaCl(aq) \rightarrow Na^+(aq) + Cl^-(aq)$. Thus, 1 mol of NaCl produces 1 mol of Na^+ or 1 mol of positive charges. Thus, 1 mol NaCl = 1 eq NaCl. Therefore,

$$0.10 \text{ mol NaCl} \times \frac{1 \text{ eq NaCl}}{1 \text{ mol NaCl}} = 0.10 \text{ eq NaCl}$$

Also, because 1 eq = 1000 meq,

$$0.10 \text{ eq} \times \frac{1000 \text{ meq}}{1 \text{ eq}} = 1.0 \times 10^2 \text{ meq}$$

b. The dissociation reaction is $Mg(NO_3)_2(aq) \rightarrow Mg^{2+}(aq) + 2NO_3^-(aq)$. Thus, 1 mol of $Mg(NO_3)_2$ produces 1 mol of Mg^{2+} or 2 mol of positive charges. Thus, 1 mol $Mg(NO_3)_2$ = 2 eq $Mg(NO_3)_2$. Therefore,

$$0.10 \text{ mol Mg(NO}_3)_2 \times \frac{2 \text{ eq Mg(NO}_3)_2}{1 \text{ mol Mg(NO}_3)_2}$$
$$= 0.20 \text{ eq Mg(NO}_3)_2$$

Also, because 1 eq = 1000 meq,

$$0.20 \text{ eq} \times \frac{1000 \text{ meq}}{1 \text{ eq}} = 2.0 \times 10^2 \text{ meq}$$

9.15 The Cl^- ion has a single charge, so we may write

1.00 mol NaCl = 1.00 mol Cl^- = 1.00 eq Cl^- = 1.00 eq NaCl

As in Example 9.11, the pattern for the calculation is liters solution $A \rightarrow$ eq A, and the pathway is liters Cl^- solution $\rightarrow$ eq Cl^-. The conversion factor comes from the concentration of 0.103 eq/L given in the problem:

$$(0.250 \ \cancel{L} \ Cl^- \ \cancel{solution}) \left(\frac{0.103 \ eq \ Cl^-}{1.00 \ \cancel{L} \ \cancel{solution}} \right) = 0.0258 \ eq \ Cl^-$$

The equivalents of Cl^- are converted to the quantities asked for by using factors from the relationships given above and the formula weight for NaCl, 58.44 u:

$$(0.0258 \ \cancel{eq \ Cl^-}) \left(\frac{1.00 \ eq \ NaCl}{1.00 \ \cancel{eq \ Cl^-}} \right) = 0.0258 \ eq \ NaCl$$

$$(0.0258 \ \cancel{eq \ NaCl}) \left(\frac{1.00 \ mol \ NaCl}{1.00 \ \cancel{eq \ NaCl}} \right) = 0.0258 \ mol \ NaCl$$

$$(0.0258 \ \cancel{mol \ NaCl}) \left(\frac{58.44 \ g \ NaCl}{1.00 \ \cancel{mol \ NaCl}} \right) = 1.51 \ g \ NaCl$$

9.16 a. $HPO_4^{2-} \rightleftharpoons H^+ + PO_4^{3-}$

$$K_a = \frac{[H^+][PO_4^{3-}]}{[HPO_4^{2-}]}$$

b. $HNO_2 \rightleftharpoons H^+ + NO_2^-$

$$K_a = \frac{[H^+][NO_2^-]}{[HNO_2]}$$

c. $HF \rightleftharpoons H^+ + F^-$

$$K_a = \frac{[H^+][F^-]}{[HF]}$$

9.17 a. In a series of similar oxyacids, acid strength increases with increases in number of oxygens: $HClO_3$, $HClO_2$, $HClO$.

b. The anion from the weakest acid will be the strongest base. HNO_3 is a stronger acid than HNO_2, so NO_2^- is a stronger base than NO_3^-.

c. The anion of an acid behaves as a weaker acid than the acid from which it came. The anion actually behaves as a base in this case, so $HC_2H_3O_2$ is a stronger acid than $C_2H_3O_2^-$.

9.18 The pattern is liters solution $A \rightarrow$ mol B, and the pathway is liters NaOH solution $\rightarrow$ mol NaOH $\rightarrow$ mol H_3PO_4. In combined form, the steps in the factor-unit method are

Step 1. 0.0141 L NaOH solution

Step 2. 0.0141 L NaOH solution = mol H_3PO_4

Step 3. $0.0141 \ \cancel{L \ NaOH \ solution} \times \dfrac{0.250 \ \cancel{mol \ NaOH}}{1 \ \cancel{L \ NaOH \ solution}}$

$$\times \frac{1 \ mol \ H_3PO_4}{3 \ \cancel{mol \ NaOH}} = mol \ H_3PO_4$$

Step 4. $(0.0141) \left(\dfrac{0.250}{1} \right) \left(\dfrac{1 \ mol \ H_3PO_4}{3} \right)$

$$= 0.00118 \ mol \ H_3PO_4$$

The molarity of the H_3PO_4 solution is calculated:

$$M = \frac{mol \ H_3PO_4}{L \ H_3PO_4 \ solution}$$

$$= \frac{0.00118 \ mol \ H_3PO_4}{0.0250 \ L \ H_3PO_4 \ solution}$$

$$= \frac{0.0472 \ mol \ H_3PO_4}{L \ H_3PO_4 \ solution} = 0.0472 \ M$$

9.19 a. The ions resulting from dissociation are Na^+ and NO_3^-. Na^+ is the cation of a strong base, and NO_3^- is the anion of a strong acid. Neither ion will hydrolyze significantly in water, so the solution pH will be about 7.

b. The ions resulting from dissociation are Na^+ and NO_2^-. Na^+ is the cation of a strong base and will not hydrolyze significantly. NO_2^- is the anion of a weak acid and will behave as a base in the hydrolysis reaction. The solution will therefore be basic, and the pH will be higher than 7.

c. The ions resulting from dissociation are K^+ and BO_3^{3-}. K^+ is the cation of a strong base and will not hydrolyze significantly. BO_3^{3-} is the anion of a weak acid and will behave as a base in the hydrolysis reaction. The solution will therefore be basic, and the pH will be higher than 7.

9.20 a. Because the acid (HCOOH) and conjugate base ($HCOO^-$) concentrations are the same, the pH will equal pK_a for the acid. This is shown by the following calculation:

$$pH = pK_a + \log \frac{[B^-]}{[HB]}$$

$$= 3.74 + \log \frac{[HCOO^-]}{[HCOOH]}$$

$$= 3.74 + \log \frac{0.22 \ mol/L}{0.22 \ mol/L}$$

$$= 3.74 + \log 1$$

$$= 3.74 + 0 = 3.74$$

b. The acid is H_2SO_3, and the conjugate base is HSO_3^- (produced when the $NaHSO_3$ salt dissociates).

$$pH = pK_a + \log \frac{[B^-]}{[HB]}$$

$$= 1.82 + \log \frac{(0.10 \ mol/L)}{(0.25 \ mol/L)}$$

$$= 1.82 + \log 0.40$$

$$= 1.82 - 0.40$$

$$= 1.42$$

c. The acid is H_2CO_3 and the conjugate base is HCO_3^-. The desired pH is 7.40. Substitution into Equation 9.54 gives

$$pH = pK_a + \log \frac{[HCO_3^-]}{[H_2CO_3]}$$

$$7.40 = 6.37 + \log \frac{[HCO_3^-]}{[H_2CO_3]}$$

$$7.40 - 6.37 = \log \frac{[HCO_3^-]}{[H_2CO_3]}$$

$$1.03 = \log \frac{[HCO_3^-]}{[H_2CO_3]}$$

We now know the log of the desired ratio. Use Step 3 in Table 9.3 to get the antilog of 1.03. The correctly rounded value is 11. Thus, the concentration of HCO_3^- must be 11 times the concentration of H_2CO_3 in the blood to maintain the desired pH.

CHAPTER 10

10.1 **a.** $^{131}_{53}I$ **b.** $^{0}_{-1}\beta$ **c.** $^{4}_{4}X$ **d.** $^{0}_{-4}X$

10.2 $^{234}_{90}Th \rightarrow {}^{0}_{-1}\beta + {}^{234}_{91}Pa$

In general: $^{A}_{Z}X \rightarrow {}^{0}_{-1}\beta + {}^{A}_{Z+1}Y$

When a β particle is emitted, the daughter has the same mass number and an atomic number higher by 1 than that of the decaying nucleus.

10.3 $^{A}_{Z}X \rightarrow {}^{0}_{1}\beta + {}^{A}_{Z-1}Y$

The atomic number of the daughter is 1 less than that of the decaying nucleus.

10.4 $^{A}_{Z}X + {}^{0}_{-1}e \rightarrow {}^{A}_{Z-1}Y$

The atomic number of the daughter is 1 less than that of the decaying nucleus.

10.5 **a.** $^{50}_{25}Mn \rightarrow {}^{0}_{1}\beta + {}^{50}_{24}Cr$
     b. $^{54}_{25}Mn + {}^{0}_{-1}e \rightarrow {}^{54}_{24}Cr$
     c. $^{56}_{25}Mn \rightarrow {}^{0}_{-1}\beta + {}^{56}_{26}Fe$
     d. $^{224}_{88}Ra \rightarrow {}^{4}_{2}\alpha + {}^{220}_{86}Rn$

10.6 During the first half-life, the original 200.00 mg will be reduced to 100.00 mg. This will be reduced to 50.00 mg during the second half-life and to 25.00 mg during the third. Thus, 3 half-lives or 3(7.7 min) = 23.1 min will be required.

10.7 The elapsed time of 79.8 hours is 6 half-lives (79.8/13.3 = 6). The fraction remaining is $1/2 \times 1/2 \times 1/2 \times 1/2 \times 1/2 \times 1/2$ = 1/64 of the original diagnostic dose.

10.8 Equation 10.2 is used: $\dfrac{I_x}{I_y} = \dfrac{d_y^2}{d_x^2}$

$$\frac{10.0 \text{ units}}{I_5} = \frac{(5 \text{ ft})^2}{(25 \text{ ft})^2}$$

$$I_5 = \frac{(10.0 \text{ units})(25 \text{ ft})^2}{(5 \text{ ft})^2} = 250 \text{ units}$$

10.9 $^{55}_{25}Mn + {}^{1}_{1}p \rightarrow {}^{1}_{0}n + {}^{55}_{26}Fe$

The product is iron-55.

CHAPTER 11

11.1 Look for the presence of carbon atoms.

    a. Inorganic   **d.** Inorganic
    b. Organic   **e.** Organic
    c. Organic   **f.** Inorganic

11.2 Refer to Table 11.1 for help.
    a. Organic
    b. Organic
    c. Inorganic

11.3 Compound (a) is a structural isomer because it has the same molecular formula but a different structural formula.

11.4 **a.** $CH_3 - \overset{\displaystyle OH}{\overset{\displaystyle |}{CH}} - CH_2 - CH_2 - CH_3$

    b. $CH_3 - \underset{\underset{\displaystyle CH_3}{|}}{CH} - CH_2 - \overset{\overset{\displaystyle O}{\|}}{C} - OH$

11.5 **a.** The number of hydrogen atoms is twice the carbon atoms plus two: C_8H_{18}.
    b. $CH_3{-}CH_2{-}CH_2{-}CH_2{-}CH_2{-}CH_2{-}CH_2{-}CH_3$

11.6 **a.** Same molecule: In both molecules, the five carbons are bonded in a continuous chain.
    b. Same molecule: In both molecules, there is a continuous chain of four carbons with a branch at position 2.

$$\underset{1}{CH_3} - \underset{2}{\underset{\underset{\displaystyle CH_3}{|}}{CH}} - \underset{3}{CH_2} - \underset{4}{CH_3} \qquad \underset{1}{CH_3} - \underset{2}{\overset{\overset{\displaystyle \underset{3}{CH_2} - \underset{4}{CH_3}}{|}}{CH}} - CH_3$$

    c. Structural isomers: Both molecules have a continuous chain of five carbons, but the branch is located at different positions.

11.7 **a.** $\underset{2}{CH_2} - \underset{3}{CH_2} - \underset{4}{CH_3}$ a butane
    with $\overset{\underset{1}{CH_3}}{|}$ above

    b. $\underset{2}{\underset{\underset{1}{CH_3}}{\overset{|}{CH_2}}} - \underset{3}{\underset{\underset{\displaystyle CH_3}{|}}{CH}} - \underset{4}{\underset{\underset{5}{CH_3}}{\overset{|}{CH_2}}}$ a pentane

    c. $\underset{3}{CH_3} - \underset{2}{\underset{\underset{4}{CH}}{|}}{CH} - \underset{}{CH_3}$ ... or

    $CH_3 - \underset{2}{\underset{\underset{1}{CH_3}}{|}}{CH} - \underset{}{CH_3}$ a butane

11.8 **a.** $CH_3 - \underset{2}{\underset{\underset{1}{CH_3}}{|}}{}$ ...

$$\underset{2}{CH_3} - \underset{3}{\overset{\overset{\displaystyle \underset{2}{CH_2}}{|}}{CH}} - \underset{4}{CH_2} - \underset{5}{CH_2} - \underset{6}{CH_3}$$
with $\underset{1}{CH_3}$ below

A CH_3 group is located at position 3. If the chain had been numbered beginning at the right, the CH_3 group would have been at position 4.
    b. Numbering from the left, groups are located at positions 5, 5, 7. From the right, the groups are at positions 2, 4, 4. The first difference occurs with the first number, so numbering from the right (2, 4, 4) is correct.

11.9 Proceeding from the left, the groups are a methyl, propyl, isopropyl, sec-butyl, and ethyl.

11.10 **a.** The chain is numbered from the right to give 2-methylhexane.

b. 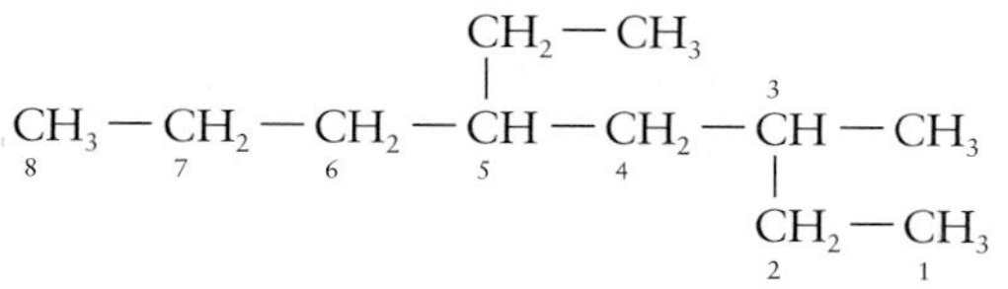
$$CH_3-CH_2-CH_2-\underset{\underset{5}{|}}{\overset{\overset{CH_2-CH_3}{|}}{\underset{4}{CH}}}-CH_2-\underset{\underset{CH_2-CH_3}{|}}{\overset{\overset{3}{}}{\underset{}{CH}}}-CH_3$$

(positions labeled 8 7 6 5 4 3 with CH₂—CH₃ as 2 1)

The chain is numbered from the right to give 5-ethyl-3-methyloctane.

c. The chain is numbered from the right to give 4-isopropyl-2,3-dimethylheptane.

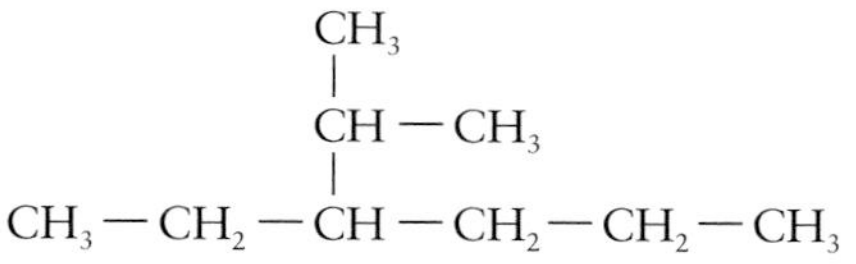

11.11 a. 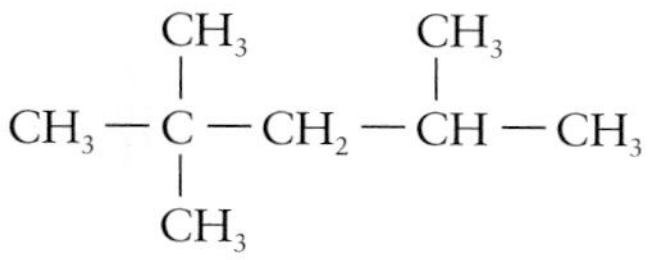
$$CH_3-\underset{\underset{CH_3}{|}}{\overset{\overset{CH_3}{|}}{C}}-CH_2-\underset{}{CH}-CH_3$$
(CH at position 4 bears CH₃)

b.
$$CH_3-CH_2-\underset{\underset{|}{}}{\overset{\overset{CH_3}{|}}{\underset{CH-CH_3}{}}}CH-CH_2-CH_2-CH_3$$

c.
$$CH_3-CH-\underset{\underset{CH_3}{|}}{CH}-\underset{\underset{CH_3}{|}}{\overset{\overset{CH_2-CH_3}{|}}{CH}}-CH-CH_2-CH_2-CH_3$$

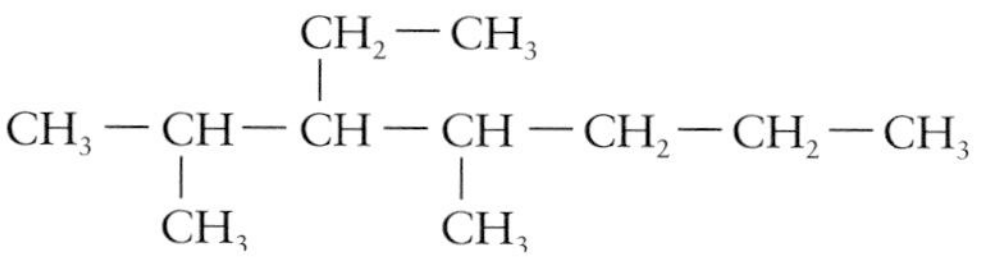

11.12 a. 1,4-dimethylcyclohexane

b. The correct name is ethylcyclopropane. When only one group is attached to a ring, the position is not designated.

c. The name 1-ethyl-2-methylcyclopentane is correct, whereas 2-ethyl-1-methylcyclopentane is incorrect because the ring numbering begins with the carbon attached to the first group alphabetically.

11.13 a. (1) *Trans* because the two Br's are on opposite sides of the ring.
(2) *Cis*; both Cl's are on the same side.
(3) *Cis*; the two groups are on the same side.

b. In showing geometric isomers of ring compounds, it helps to draw the ring in perspective:

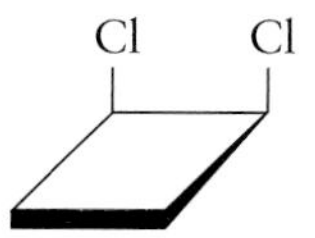

Cl   Cl

················· **CHAPTER 12** ·····················

12.1 In each of these alkenes, the double-bonded carbons occur at positions 1 and 2.

a. 3-bromo-1-propene
b. 2-ethyl-1-pentene
c. 3,4-dimethylcyclohexene

12.2 a. The chain is correctly numbered from the right to give 2-methyl-1,3-butadiene.

b. The chain is correctly numbered from the left to give 2-methyl-1,3,6-octatriene.

c. 7-bromo-1,3,5-cycloheptatriene

12.3 a. This structure does not exhibit geometric isomerism because there are two H's attached to the carbon at position 1.

b. 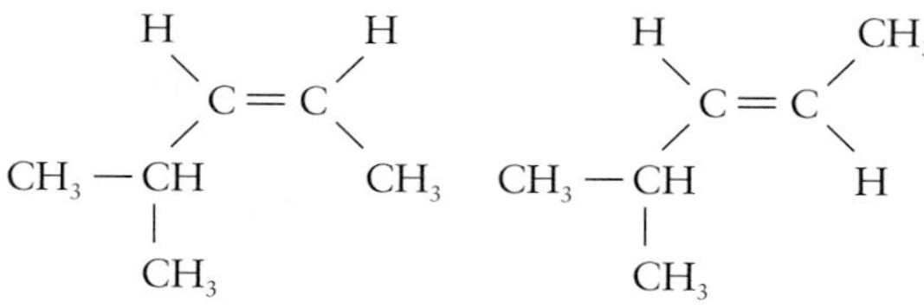
$$\underset{\underset{CH_3}{|}}{\overset{\overset{H}{\diagdown}}{CH_3-CH}}C=C\overset{\overset{H}{\diagup}}{\underset{}{CH_3}} \qquad \underset{\underset{CH_3}{|}}{\overset{\overset{H}{\diagdown}}{CH_3-CH}}C=C\overset{\overset{CH_3}{\diagup}}{\underset{}{H}}$$
cis   *trans*

c. This structure does not exhibit geometric isomerism because there are two methyl groups attached to the left double-bonded carbon:

12.4

a. 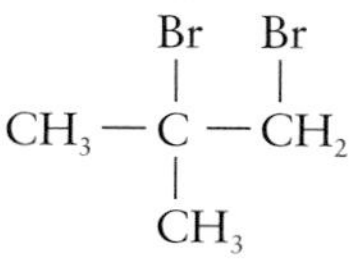
$$CH_3-\underset{\underset{CH_3}{|}}{\overset{\overset{Br\ \ Br}{|\ \ |}}{C}}-CH_2$$

b. 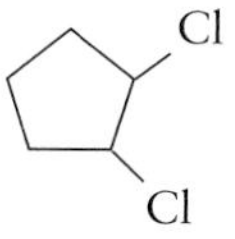
(cyclopentane ring with Cl and Cl substituents)

12.5 a. 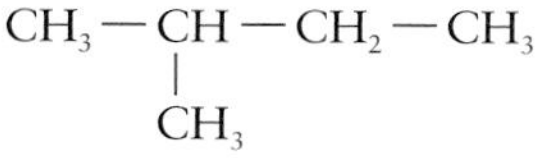
$$CH_3-\underset{\underset{CH_3}{|}}{CH}-CH_2-CH_3$$

b. 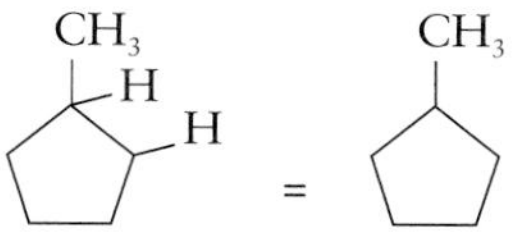
(cyclopentane with CH₃ and H) = (methylcyclopentane)

12.6 a. The major product will be the one where H attaches to the CH carbon:

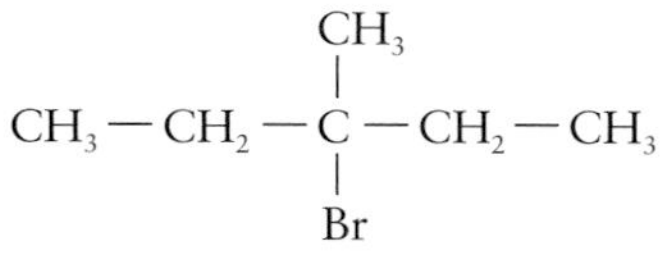

$$CH_3-CH_2-\underset{\underset{Br}{|}}{\overset{\overset{CH_3}{|}}{C}}-CH_2-CH_3$$

b. Position 1 on the ring has an attached hydrogen, whereas position 2 does not have any attached hydrogens:

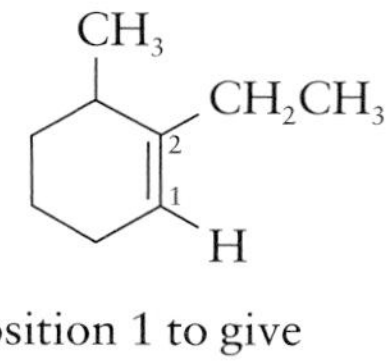

(cyclohexene ring with CH₃ and CH₂CH₃ groups, positions 1 and 2, H shown)

Thus, H attaches at position 1 to give

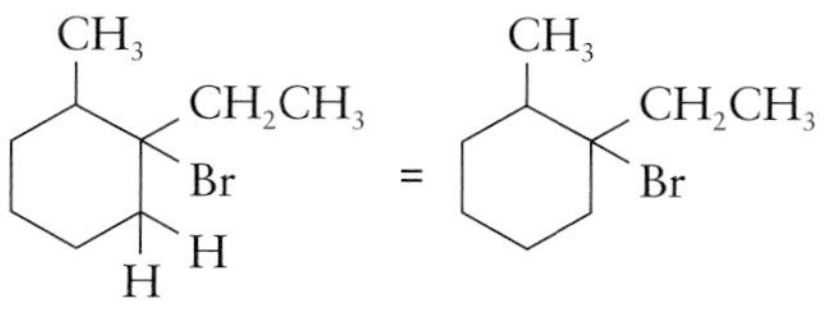

(cyclohexane ring with CH₃, CH₂CH₃, Br, H) = (cyclohexane ring with CH₃, CH₂CH₃, Br)

12.7 a. Markovnikov's rule predicts that H will attach at position 1 to give:

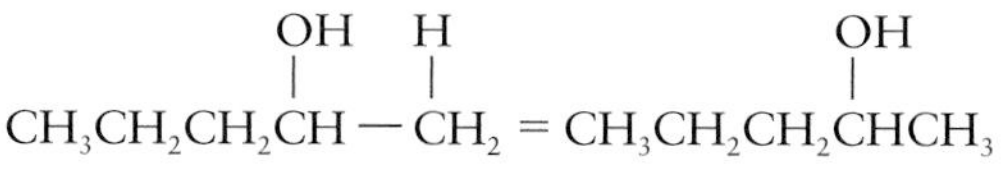

$$CH_3CH_2CH_2\underset{}{CH}\underset{\overset{|}{OH}}{}-\underset{\overset{|}{H}}{CH_2}=CH_3CH_2CH_2\underset{\overset{|}{OH}}{CH}CH_3$$

b. 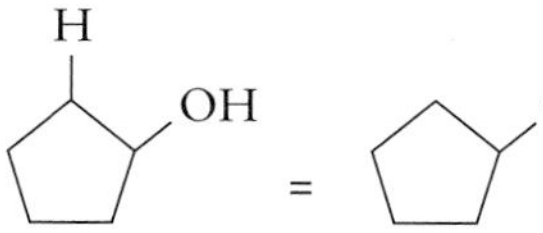
(cyclopentane with H and OH) = (cyclopentane with OH)

12.8 The double bond becomes a single bond:

$$\underset{1}{\underbrace{-CH_2-\overset{\overset{\displaystyle CH_3}{\displaystyle |}}{CH}}}-\underset{2}{\underbrace{CH_2-\overset{\overset{\displaystyle CH_3}{\displaystyle |}}{CH}}}-\underset{3}{\underbrace{CH_2-\overset{\overset{\displaystyle CH_3}{\displaystyle |}}{CH}}}-\underset{4}{\underbrace{CH_2-\overset{\overset{\displaystyle CH_3}{\displaystyle |}}{CH}}}-$$

12.9 Each chain is correctly numbered from the right.
   a. 2-pentyne   **b.** 5-methyl-2-hexyne

12.10 **a.** Numbers or the term *meta* may be used:
   1,3-diethylbenzene   or   *m*-diethylbenzene

b. The compound must be named as a derivative of cyclopentane. The correct name is 1-chloro-3-phenylcyclopentane.

c. Numbers must be used when there are three groups: 1,2,3-tribromobenzene.

d. If the compound is named as a derivative of benzoic acid, then the methyl group is at position 2. The name is 2-methylbenzoic acid.

GLOSSARY

Absolute zero The temperature at which all motion stops; a value of 0 on the Kelvin scale.

Acid dissociation constant The ratio of product molarities (H^+ and acid anion) multiplied together and divided by the molarity of undissociated acid. All molarities are measured at equilibrium.

Acidic solution A solution in which the concentration of H_3O^+ is greater than the concentration of OH^-; also, a solution in which pH is less than 7.

Activation energy Energy needed to start some spontaneous processes. Once started, the processes continue without further stimulus or energy from an outside source.

Activity series A tabular representation of the tendencies of metals to react with H^+.

Addition polymer A polymer formed by the linking together of many alkene molecules through addition reactions.

Addition reaction A reaction in which a compound adds to a multiple bond.

Aliphatic compound Any organic compound that is not aromatic.

Alkaline solution See **Basic solution**.

Alkane A hydrocarbon that contains only single bonds.

Alkene A hydrocarbon that contains one or more double bonds.

Alkyl group A group differing by one hydrogen from an alkane.

Alkyl halide See **Haloalkane**.

Alkyne A hydrocarbon that contains one or more triple bonds.

Alpha particle The particle that makes up alpha rays; identical to the helium nucleus and composed of two protons and two neutrons.

Anion A negatively charged ion.

Aromatic hydrocarbon Any organic compound that contains the characteristic benzene ring or similar feature.

Arrhenius acid Any substance that provides H^+ ions when dissolved in water.

Arrhenius base Any substance that provides OH^- ions when dissolved in water.

Atom The limit of chemical subdivision for matter.

Atomic mass unit (u) A unit used to express the relative masses of atoms. One u is equal to $\frac{1}{12}$ the mass of an atom of carbon-12.

Atomic number of an atom A number equal to the number of protons in the nucleus of an atom. Symbolically it is represented by Z.

Atomic orbital A volume of space around atomic nuclei in which electrons of the same energy move. Groups of orbitals with the same n values form subshells.

Atomic weight The mass of an average atom of an element expressed in atomic mass units.

Avogadro's law Equal volumes of gases measured at the same temperature and pressure contain equal numbers of molecules.

Balanced equation An equation in which the number of atoms of each element in the reactants is the same as the number of atoms of that same element in the products.

Basic solution A solution in which the concentration of OH^- is greater than the concentration of H_3O^+; also, a solution in which pH is greater than 7; also called alkaline solution.

Basic unit of measurement A specific unit from which other units for the same quantity are obtained by multiplication or division.

Becquerel A physical unit of radiation measurement corresponding to one nuclear disintegration per second.

Beta particle The particle that makes up beta rays; identical to an electron but produced in the nucleus when a neutron is changed into a proton and an electron.

Binary compound A compound made up of two different elements.

Biological unit of radiation A radiation measurement unit that indicates the damage caused by radiation in living tissue.

Boiling point The temperature at which the vapor pressure of a liquid is equal to the prevailing atmospheric pressure.

Bond polarization   A result of the attraction of shared electrons to the more electronegative atom of a bonded pair of atoms.

Boyle's law   A gas law that describes the pressure and volume behavior of a gas sample kept at constant temperature. Mathematically, it is $PV = k$.

Branched alkane   An alkane in which at least one carbon atom is not part of a continuous chain.

Branching chain reaction   See **Expanding chain reaction**.

Breeder reaction   A nuclear reaction in which isotopes that will not undergo spontaneous fission are changed into isotopes that will.

Brønsted acid   Any hydrogen-containing substance that is capable of donating a proton (H^+) to another substance.

Brønsted base   Any substance capable of accepting a proton from another substance.

Buffer   A solution with the ability to resist changing pH when acids (H^+) or bases (OH^-) are added.

Buffer capacity   The amount of acid (H^+) or base (OH^-) that can be absorbed by a buffer without causing a significant change in pH.

Carbocation   An ion of the form $-\overset{+}{\underset{|}{C}}-$.

Catalyst   A substance that changes (usually increases) reaction rates without being used up in the reaction.

Cation   A positively charged ion.

Chain reaction   A nuclear reaction in which the products of one reaction cause a repeat of the reaction to take place.

Charles's law   A gas law that describes the temperature and volume behavior of a gas sample kept at constant pressure. Mathematically, it is $V/T = k'$.

Chemical change   A change matter undergoes that involves changes in composition.

Chemical property   A property matter demonstrates when attempts are made to change it into new substances.

Cis-   On the same side (as applied to geometric isomers).

Cohesive force   The attractive force between particles; it is associated with potential energy.

Cold spot   Tissue from which a radioactive tracer is excluded or rejected.

Colligative property   A solution property that depends only on the concentration of solute particles in solution.

Colloid   A homogeneous mixture of two or more substances in which the dispersed substances are present as larger particles than are found in solutions.

Combination reaction   A chemical reaction in which two or more substances react to form a single substance.

Combined gas law   A gas law that describes the pressure, volume, and temperature behavior of a gas sample. Mathematically, it is $PV/T = k''$.

Compound   A pure substance consisting of two or more kinds of atoms in the form of heteroatomic molecules or individual atoms.

Compound formula   A symbol for the molecule of a compound, consisting of the symbols of the atoms found in the molecule. Atoms present in numbers higher than 1 have the number indicated by a subscript.

Compressibility   The change in volume of a sample resulting from a pressure change acting on the sample.

Concentration   The amount of solute contained in a specific amount of solution.

Condensation   An exothermic process in which a gas or vapor is changed to a liquid or solid.

Condensed structural formula   A structural molecular formula showing the general arrangement of atoms but without showing all the covalent bonds.

Conformations   The different arrangements of atoms in space achieved by rotation about single bonds.

Conjugate acid–base pair   A Brønsted acid and its conjugate base.

Conjugate base   The species remaining when a Brønsted acid donates a proton.

Copolymer   An addition polymer formed by the reaction of two different monomers.

Covalent bond   The attractive force that results between two atoms that are both attracted to a shared pair of electrons.

Critical mass   The minimum amount of fissionable material needed to sustain a critical chain reaction at a constant rate.

Critical reaction   A constant-rate chain reaction.

Crystal lattice   A rigid three-dimensional arrangement of particles.

Curie   A physical unit of radiation measurement corresponding to 3.7×10^{10} nuclear disintegrations per second.

Cycloalkane   An alkane in which carbon atoms form a ring.

Cyclotron   A cyclic particle accelerator that works by changing electrical polarities as charged particles cross a gap. The particles are kept in a spiral path by a strong magnetic field.

Dalton's law of partial pressures   The total pressure exerted by a mixture of gases is equal to the sum of the partial pressures of the gases in the mixture.

Daughter nuclei   The new nuclei produced when unstable nuclei undergo radioactive decay.

Decomposition   A change in chemical composition that can result from heating.

Decomposition reaction   A chemical reaction in which a single substance reacts to form two or more simpler substances.

Density   The number given when the mass of a sample of a substance is divided by the volume of the same sample.

Derived unit of measurement   A unit obtained by multiplication or division of one or more basic units.

Dialysis   A process in which solvent molecules, other small molecules, and hydrated ions pass from a solution through a membrane.

Dialyzing membrane   A semipermeable membrane with pores large enough to allow solvent molecules, other small molecules, and hydrated ions to pass through.

Diatomic molecule   A molecule that contains two atoms.

Diffusion   A process that causes gases to spontaneously intermingle when they are brought together.

Dipolar force   The attractive force that exists between the positive end of one polar molecule and the negative end of another.

Diprotic acid   An acid that gives up two protons (H^+) per molecule when dissolved.

Dispersed phase   The substance present in a colloidal dispersion in amounts less than the amount of dispersing medium.

Dispersing medium   The substance present in a collodial dispersion in the largest amount.

Dispersion force   A very weak attraction force acting between the particles of all matter; results from momentary nonsymmetric electron distributions in molecules or atoms.

Disruptive force   The force resulting from particle motion; associated with kinetic energy.

Dissolving   A term used to describe the process of solution formation when the solvent and solutes form a homogeneous mixture.

Distinguishing electron   The last or highest-energy electron found in an element.

Double bond   The bond resulting from the sharing of two pairs of electrons.

Double-replacement reaction   A chemical reaction in which two compounds react and exchange partners to form two new compounds.

Effective collision   A collision that causes a reaction to occur between colliding molecules.

Effusion   A process in which a gas escapes from a container through a small hole.

Electrolyte   A solute that when dissolved in water forms a solution that conducts electricity.

Electron capture   A mode of radioactive decay for some unstable nuclei in which an electron from outside the nucleus is drawn into the nucleus, where it combines with a proton to form a neutron.

Electronegativity   The tendency of an atom to attract shared electrons of a covalent bond.

Electronic configurations   The detailed arrangements of electrons indicated by a specific notation, $1s^2 2s^2 2p^4$, and so on.

Element   A pure substance consisting of only one kind of atom in the form of homoatomic molecules or individual atoms.

Elemental symbol   A symbol assigned to an element that is based on the name of the element and consists of one capital letter or a capital letter followed by a lowercase letter.

Emulsifying agent (stabilizing agent)   A substance that when added to colloids prevents them from coalescing and settling.

Endergonic process   A process that gains or accepts energy as it takes place.

Endothermic reaction   A reaction that absorbs heat.

Endpoint of a titration   The point at which the titration is stopped on the basis of an indicator color change or pH meter reading.

Entropy   A measurement or indication of the disorder or randomness of a system; the more disorderly a system, the higher its entropy.

Equilibrium concentration   The unchanging concentration of reactants and products in a reaction system that is in a state of equilibrium.

Equilibrium constant   A constant that relates the equilibrium concentrations of products to those of reactants, each being raised to exponents obtained from stoichiometric coefficients.

Equilibrium expression   An equation relating the equilibrium constant and reactant and product concentrations.

Equivalence point of a titration   The point at which the unknown solution has exactly reacted with the known solution. Neither is in excess.

Equivalent of salt   The amount that will produce 1 mol of positive charges on dissolving and dissociating into ions.

Evaporation   An endothermic process in which a liquid is changed to a gas; also called vaporization.

Exact numbers   Numbers that have no uncertainty; numbers from defined relationships, counting numbers, and numbers that are part of reduced simple fractions.

Exergonic process   A process that gives up energy as it takes place.

Exothermic reaction   A reaction that liberates heat.

Expanded structural formula   A structural molecular formula showing all the covalent bonds.

Expanding chain reaction   A reaction in which the products of one reaction cause more than one more reaction to occur; also called branching chain reaction.

Factors used in the factor-unit method   Fractions obtained from fixed relationships between quantities.

Family of the periodic table   See **Group of the periodic table.**

First ionization energy   The energy required to remove the first electron from a neutral atom.

Formula weight   The sum of the atomic weights of the atoms shown in the formula of an ionic compound.

Functional group   A unique reactive combination of atoms that differentiates molecules of organic compounds of one class from those of another.

Gamma ray   A high-energy ray that is like an X ray but with a higher energy.

Gas law   A mathematical relationship that describes the behavior of gases as they are mixed, subjected to pressure or temperature changes, or allowed to diffuse.

Geiger–Müller tube   A radiation-detection device operating on the principle that ions form when radiation passes through a tube filled with low-pressure gas.

Geometric isomers   Molecules that differ in the three-dimensional arrangements of their atoms in space and not in the order of linkage of atoms.

Graham's law   A mathematical expression that relates rates of effusion or diffusion of two gases to the masses of the molecules of the two gases.

Gray   A biological unit of radiation measurement that corresponds to the transfer of 1 J of energy to 1 kg of tissue.

Group of the periodic table   A vertical column of elements that have similar chemical properties; also called family of the periodic table.

Half-life   The time required for one-half the unstable nuclei in a sample to undergo radioactive decay.

Haloalkane   A derivative of an alkane in which one or more hydrogens are replaced by halogens; also called alkyl halide.

Heat of fusion   The amount of heat energy required to melt exactly 1 g of a solid substance at constant temperature.

Heat of vaporization   The amount of heat energy required to vaporize exactly 1 g of a liquid substance at constant temperature.

Henderson–Hasselbalch equation   A relationship between the pH of a buffer, pK_a, and the concentrations of acid and salt in the buffer.

Heteroatomic molecule   A molecule that contains two or more kinds of atoms.

Heterogeneous matter   Matter with properties that are not the same throughout the sample.

Heterogeneous catalyst   A catalytic substance normally used in the form of a solid with a large surface area on which reactions take place; also called surface catalyst.

Homoatomic molecule   A molecule that contains only one kind of atom.

Homogeneous catalyst   A catalytic substance that is distributed uniformly throughout the reaction mixture.

Homogeneous matter   Matter that has the same properties throughout the sample.

Homologous series   Compounds of the same functional class that differ by a —CH_2— group.

Hot spot   Tissue in which a radioactive tracer concentrates.

Hund's rule   A statement of the behavior of electrons when they occupy orbitals: Electrons will not join other electrons in an orbital if an empty orbital of the same energy is available for occupancy.

Hybrid orbital   An orbital produced from the combination of two or more nonequivalent orbitals of an atom.

Hydrate   A salt that contains specific numbers of water molecules as part of the solid crystalline structure.

Hydrated ion   An ion in solution that is surrounded by water molecules.

Hydration   The addition of water to a multiple bond.

Hydrocarbon   An organic compound that contains only carbon and hydrogen.

Hydrogen bonding   The result of attractive dipolar forces between molecules in which hydrogen atoms are covalently bonded to very electronegative elements (O, N, or F).

Hydrogenation   A reaction in which the addition of hydrogen takes place.

Hydrolysis reaction   Any reaction with water; for salts, a reaction of the acidic cation and/or basic anion of the salt with water.

Hydrophobic   Characterizing molecules or parts of molecules that repel (are insoluble in) water.

Ideal gas law   A gas law that relates the pressure, volume, temperature, and number of moles in a gas sample. The equation is $PV = nRT$.

Immiscible   A term used to describe liquids that are insoluble in each other.

Inhibitor   A substance that decreases reaction rates.

Inner-transition element   An element in which the distinguishing electron is found in an f subshell.

Inorganic chemistry   The study of the elements and all noncarbon compounds.

Insoluble substance   A substance that does not dissolve to a significant extent in the solvent.

Internal energy   The energy associated with vibrations within molecules.

Inverse square law of radiation   A mathematical way of saying that the intensity of radiation is inversely proportional to the square of the distance from the source of the radiation.

Ionic bond   The attractive force that holds together ions of opposite charge.

Ion product of water   The equilibrium constant for the dissociation of pure water into H_3O^+ and OH^-.

Isoelectronic   A term meaning "same electronic," used to describe atoms or ions that have identical electronic configurations.

Isomerism   A property in which two or more compounds have the same molecular formula but different arrangements of atoms.

Isotopes   Atoms having the same atomic number but different mass numbers; that is, they are atoms of the same element that contain different numbers of neutrons in their nuclei.

Kinetic energy   The energy a particle has as a result of its motion. Mathematically, it is $KE = \frac{1}{2}mv^2$.

Lattice site   The individual location occupied by a particle in a crystal lattice.

Law of conservation of matter   Atoms are neither created nor destroyed in chemical reactions.

Le Châtelier's principle   The position of equilibrium will shift to compensate for changes made that upset a system at equilibrium.

Lewis structure   A representation of an atom or ion in which the elemental symbol represents the atomic nucleus and all but the valence-shell electrons. The valence-shell electrons are represented by dots arranged around the elemental symbol.

Limiting reactant   The reactant present in a reaction in the least amount based on its reaction coefficients and molecular weight.

The limiting reactant determines the maximum amount of product that can be formed.

Limiting-reactant principle The maximum amount of product possible from a reaction is determined by the amount of reactant present in the least amount, based on its reaction coefficient and molecular weight.

Linear accelerator A particle accelerator that works by changing electrical polarities as charged particles cross gaps between segments of a long tube.

Markovnikov's rule In the addition of H—X to an alkene, the hydrogen becomes attached to the carbon atom that is already bonded to more hydrogens.

Mass A measurement of the amount of matter in an object.

Mass number of an atom A number equal to the sum of the number of protons and neutrons in the nucleus of an atom. Symbolically, it is represented by A.

Matter Anything that has mass and occupies space.

Melting point The temperature at which a solid changes to a liquid; the solid and liquid have the same vapor pressure.

Metal An element found in the left two-thirds of the periodic table. Most have the following properties: high thermal and electrical conductivities, high malleability and ductility, and a metallic luster.

Metallic bond An attractive force responsible for holding solid metals together, originating from the attraction between positively charged atomic kernels that occupy the lattice sites and mobile electrons that move freely through the lattice.

Metalloids Elements that form a narrow diagonal band in the periodic table between the metals and nonmetals; they have properties somewhat between those of metals and nonmetals.

Mixture A physical blend of matter that can be physically separated into two or more components.

Moderator A material capable of slowing down neutrons that pass through it.

Molarity A solution concentration that is expressed in terms of the number of moles of solute contained in 1 L of solution.

Mole The number of particles (atoms or molecules) contained in a sample of element or compound with a mass in grams equal to the atomic or molecular weight, respectively.

Molecular equation An equation written with each compound represented by its un-ionized formula.

Molecular weight The relative mass of a molecule expressed in atomic mass units and calculated by adding the atomic weights of the atoms in the molecule.

Molecule The smallest particle of a pure substance that has the properties of that substance.

Monomer The starting material that becomes the repeating unit of polymers.

Monoprotic acid An acid that gives up only one proton (H^+) per molecule when dissolved.

Net ionic equation An equation that contains only un-ionized or insoluble materials and ions that undergo changes as the reaction proceeds. All spectator ions are eliminated.

Network solid A solid in which the lattice sites are occupied by atoms that are covalently bonded to each other.

Neutral A term used to describe any water solution in which the concentrations of H_3O^+ and OH^- are equal; also, a water solution with pH = 7.

Neutralization reaction A reaction in which an acid and base react completely, leaving a solution that contains only a salt and water.

Noble gas configuration An electronic configuration in which the last eight electrons occupy and fill the s and p subshells of the highest-occupied shell.

Nonelectrolyte A solute that when dissolved in water forms a solution that does not conduct electricity.

Nonmetal An element found in the right one-third of the periodic table. Most occur as brittle, powdery solids or gases and have properties generally opposite those of metals.

Nonpolar covalent bond A covalent bond in which the bonding pair of electrons is shared equally by the bonded atoms.

Nonpolar molecule A molecule that contains no polarized bonds, or a molecule containing polarized bonds in which the resulting charges are distributed symmetrically throughout the molecule.

Normal alkane Any alkane in which all the carbon atoms are aligned in a continuous chain.

Normal boiling point The temperature at which the vapor pressure of a liquid is equal to one standard atmosphere (760 torr); also called standard boiling point.

Nuclear fission A process in which large nuclei split into smaller, approximately equal-sized nuclei when bombarded by neutrons.

Nuclear fusion A process in which small nuclei combine or fuse to form larger nuclei.

Nucleus The central core of atoms that contains protons, neutrons, and most of the mass of atoms.

Octet rule A rule for predicting electron behavior in reacting atoms: Atoms will gain or lose sufficient electrons to achieve an outer electron arrangement identical to that of a noble gas. This arrangement usually consists of eight electrons in the valence shell.

Organic chemistry The study of carbon-containing compounds.

Organic compound A compound that contains the element carbon.

Osmolarity The product of n and M in the equation $\pi = nMRT$.

Osmosis The process in which solvent flows through a semipermeable membrane into a solution.

Osmotic pressure The hydrostatic pressure required to prevent the net flow of solvent through a semipermeable membrane into a solution.

Oxidation Originally, a process involving a reaction with oxygen. Today it means a number of things, including a process in which electrons are given up, hydrogen is lost, or an oxidation number increases.

Oxidation number   A positive or negative number assigned to the elements in chemical formulas according to a specific set of rules; also called oxidation state.

Oxidation state   See **Oxidation number.**

Oxidizing agent   The substance that contains an element that is reduced during a chemical reaction.

Partial pressure   The pressure an individual gas of a mixture would exert if it were in the container alone at the same temperature as the mixture.

Pauli exclusion principle   A statement of the behavior of electrons when they occupy orbitals: Only electrons spinning in opposite directions can simultaneously occupy the same orbital.

Percent   A solution concentration that expresses the amount of solute in 100 parts of solution.

Percentage yield   The percentage of the theoretical amount of product actually produced by a reaction.

Periodic law   A statement about the behavior of the elements when they are arranged in a specific order. In its present form, it is stated as follows: Elements with similar chemical properties occur at regular (periodic) intervals when the elements are arranged in order of increasing atomic numbers.

Period of the periodic table   A horizontal row of elements.

pH   The negative logarithm of the molar concentration of H^+ (H_3O^+) in a solution.

Phenyl group   A benzene ring with one hydrogen absent, C_6H_5—.

Physical change   A change matter undergoes without changing composition.

Physical property   A property of matter that can be observed or measured without trying to change the composition of the matter being studied.

Physical unit of radiation   A radiation measurement unit that indicates the activity of the source of the radiation; for example, the number of nuclear decays per minute.

pK_a   The negative logarithm of K_a.

Polar covalent bond   A covalent bond that shows bond polarization; that is, the bonding electrons are shared unequally.

Polar molecule   A molecule that contains polarized bonds and in which the resulting charges are distributed asymmetrically throughout the molecule.

Polyatomic ion   A covalently bonded group of atoms that carries a net electrical charge.

Polyatomic molecule   A molecule that contains three or more atoms.

Polycyclic aromatic compound   A derivative of benzene in which carbon atoms are shared between two or more benzene rings.

Polymer   A very large molecule made up of repeating units.

Polymerization   A reaction that produces a polymer.

Polyunsaturated   A term usually applied to molecules with several double bonds.

Position of equilibrium   An indication of the relative amounts of reactants and products present at equilibrium.

Positron   A positively charged electron.

Potential energy   The energy a particle has as a result of attractive or repulsive forces acting on it.

Pressure   A force per unit area of surface on which the force acts. In measurements and calculations involving gases, it is often expressed in units related to measurements of atmospheric pressure.

Product of a reaction   A substance produced as a result of a chemical reaction; written on the right side of the equation representing the reaction.

Pure substance   Matter that has a constant composition and fixed properties.

Rad   A biological unit of radiation measurement that corresponds to the transfer of 1×10^{-2} J or 2.4×10^{-3} cal of energy to 1 kg of tissue.

Radiation sickness   A condition following short-term exposure to intense radiation.

Radical   An electron-deficient particle that is very reactive; also called free radical.

Radioactive dating   A process for determining the age of artifacts and rocks, based on the amount and half-life of radioisotopes contained in the object.

Radioactive decay   A process in which an unstable nucleus changes energy states and in the process emits radiation.

Radioactive nuclei   Nuclei that undergo spontaneous changes and emit energy in the form of radiation.

Reactant of a reaction   A substance that undergoes chemical change during a reaction; written on the left side of the equation representing the reaction.

Reaction mechanism   A detailed explanation of how a reaction actually takes place.

Reaction rate   The speed at which a chemical reaction takes place.

Reducing agent   The substance that contains an element that is oxidized during a chemical reaction.

Reduction   Originally, a process in which oxygen was lost. Today it means a number of things, including a process in which electrons are gained, hydrogen is accepted, or an oxidation number decreases.

Rem   A biological unit of radiation measurement that corresponds to the health effect produced by one roentgen of gamma rays or X rays regardless of the type of radiation involved.

Representative element   An element in which the distinguishing electron is found in an s or a p subshell.

Roentgen   A biological unit of radiation measurement used with X rays and gamma rays; the quantity of radiation that generates 2.1×10^9 ion pairs per cubic centimeter of dry air or 1.8×10^{12} ion pairs per gram of tissue.

Salt   A solid crystalline ionic compound at room temperature that contains the cation of a base and the anion of an acid.

Saturated hydrocarbon   Another name for an alkane.

Saturated solution   A solution that contains the maximum amount possible of dissolved solute in a stable situation under the prevailing conditions of temperature and pressure.

Scientific models   Explanations for observed behavior in nature.

Scientific notation   A way of representing numbers consisting of a product between a nonexponential number and 10 raised to a whole-number exponent that may be positive or negative.

Scintillation counter   A radiation-detection device that works on the principle that phosphors give off light when struck by radiation.

Shell   A location and energy of electrons around a nucleus that is designated by a value for n, where $n = 1, 2, 3$, etc.

Side reaction   A reaction that does not give the desired product of a reaction.

Significant figures   The numbers in a measurement that represent the certainty of the measurement, plus one number representing an estimate.

Simple ion   An atom that has acquired a net positive or negative charge by losing or gaining electrons.

Single-replacement reaction   A chemical reaction in which an element reacts with a compound and displaces another element from the compound.

Solubility   The maximum amount of solute that can be dissolved in a specific amount of solvent under specific conditions of temperature and pressure.

Soluble substance   A substance that dissolves to a significant extent in the solvent.

Solute   One or more substances present in a solution in amounts less than that of the solvent.

Solution   A homogeneous mixture of two or more pure substances.

Solvent   The substance present in a solution in the largest amount.

Specific heat   The amount of heat energy required to raise the temperature of exactly 1 g of a substance by exactly 1°C.

Spectator ion   An ion in a total ionic reaction that is not changed as the reaction proceeds. It appears in identical forms on the left and right sides of the equation.

Spontaneous process   A process that takes place naturally with no apparent cause or stimulus.

Stable substance   A substance that does not undergo spontaneous changes under the surrounding conditions.

Standard atmosphere   The pressure needed to support a 760-mm column of mercury in a barometer tube.

Standard boiling point   See **Normal boiling point**.

Standard conditions   A set of specific temperature and pressure values used for gas measurements.

Standard position for a decimal   In scientific notation, the position to the right of the first nonzero digit in the nonexponential number.

State of equilibrium   A condition in a reaction system in which the rates of the forward and reverse reactions are equal.

Stereoisomers   Compounds with the same structural formula but different spatial arrangements of atoms.

Stoichiometry   The study of mass relationships in chemical reactions.

Strong acid or base   An acid or base that dissociates (ionizes) essentially completely when dissolved to form a solution.

Structural isomers   Compounds that have the same molecular formula but in which the atoms bond in different patterns.

Sublimation   The endothermic process in which a solid is changed directly to a gas without first becoming a liquid.

Subshell   A component of a shell that is designated by a letter from the group s, p, d, and f.

Supercritical mass   The minimum amount of fissionable material that must be present to cause a branching chain reaction to occur.

Supercritical reaction   A branching chain reaction.

Supersaturated solution   An unstable solution that contains an amount of solute greater than the solute solubility under the prevailing conditions of temperature and pressure.

Surface catalyst   See **Heterogeneous catalyst**.

Thermal expansion   A change in volume of a sample resulting from a change in temperature of the sample.

Thermonuclear reaction   A nuclear fusion reaction that requires a very high temperature to start.

Titration   An analytical procedure in which one solution (often a base) of known concentration is slowly added to a measured volume of an unknown solution (often an acid). The volume of the added solution is measured with a buret.

Torr   The pressure needed to support a 1-mm column of mercury in a barometer tube.

Total ionic equation   An equation written with all soluble ionic substances represented by the ions they form in solution.

Tracer   A radioisotope used medically because its progress through the body or localization in specific organs can be followed.

Trans-   On opposite sides (as applied to geometric isomers).

Transition element   An element in which the distinguishing electron is found in a d subshell.

Transuranium element   A synthetic element with atomic number greater than that of uranium.

Triatomic molecule   A molecule that contains three atoms.

Triple bond   The bond resulting from the sharing of three pairs of electrons.

Triprotic acid   An acid that gives up three protons (H^+) per molecule when dissolved.

Tyndall effect   A property of colloids in which the path of a beam of light through the colloid is visible because the light is scattered.

Universal gas constant   The constant that relates pressure, volume, temperature, and the number of moles of gas in the ideal gas law.

Unsaturated hydrocarbon   A hydrocarbon containing one or more multiple bonds.

Valence shell   The outermost (highest-energy) shell that contains electrons.

Vaporization   See **Evaporation**.

Vapor pressure   The pressure exerted by vapor that is in equilibrium with its liquid.

Vitamin   An organic nutrient that the body cannot produce in the small amounts needed for good health.

Volume/volume percent   A concentration that expresses the volume of liquid solute contained in 100 volumes of solution.

VSEPR theory   A theory based on the mutual repulsion of electron pairs; used to predict molecular shapes.

Water of hydration   Water retained as a part of the solid crystalline structure of some salts.

Weak (or moderately weak) acid or base   An acid or base that dissociates (ionizes) less than completely when dissolved to form a solution.

Weight   A measurement of the gravitational force acting on an object.

Weight/volume percent   A concentration that expresses the grams of solute contained in 100 mL of solution.

Weight/weight percent   A concentration that expresses the mass of solute contained in 100 mass units of solution.

INDEX

Features Within the Text